A BIRD'S-EYE VIEW OF VETERINARY MEDICINE

Edited by **Carlos C. Perez-Marin**

A Bird's-Eye View of Veterinary Medicine
Edited by Carlos C. Perez-Marin

Published by InTech
Janeza Trdine 9, 51000 Rijeka, Croatia

Publishing Process Manager Oliver Kurelic

Technical Editor Teodora Smiljanic

Cover Designer InTech Design Team

Additional hard copies can be obtained from orders@intechopen.com

A Bird's-Eye View of Veterinary Medicine, Edited by Carlos C. Perez-Marin
p. cm.
ISBN: 978-953-51-0031-7

For EU product safety concerns:
IN TECH d.o.o., Prolaz Marije Krucifikse Kozulić 3, 51000 Rijeka, Croatia, info@intechopen.com or visit our website at intechopen.com.

Meet the editor

Dr Carlos C. Perez-Marin was born in Cordoba (Spain), where he obtained his veterinary degree. In 1996 he earned his Masters in Equine Science and obtained the Doctoral thesis in January 2001. He has worked as a teacher in the Department of Animal Medicine and Surgery at the University of Cordoba (Spain) since 1997. His teaching and research interests are in animal reproduction, with special reference to equine and bovine embryo transfer and preservation, cryopreservation of feline semen, and estrous control in small ruminants. He is involved in providing a clinical service in the Veterinary Teaching Hospital of the University of Cordoba and collaborates with some external veterinarian teams. Drawing on his clinical background, Dr Perez-Marin also co-ordinates the veterinary students' clinical extramural placements, where students gain experience with veterinary practitioners.

Contents

Preface

Veterinary medicine is advancing at a very rapid pace, particularly given the breadth of the discipline. This book examines new developments covering a wide range of issues from health and welfare in livestock, pets, and wild animals to public health supervision and biomedical research.

As well as containing reviews offering fresh insight into specific issues, this book includes a selection of scientific articles which help to chart the advance of this science. The book is divided into several sections. The opening chapters cover the veterinary profession and veterinary science in general, while later chapters look at specific aspects of applied veterinary medicine in pets and in livestock. Finally, research papers are grouped by specialisms with a view to exploring progress in areas such as organ transplantation, therapeutic use of natural substances, and the use of new diagnostic techniques for disease control.

This book was produced during World Veterinary Year 2011, which marked the 250th anniversary of the veterinary profession. It provides a fittingly concise and enjoyable overview of the whole science of veterinary medicine.

Carlos C. Perez-Marin
Department of Animal Medicine and Surgery,
Faculty of Veterinary Medicine,
University of Cordoba,
Spain

Part 1

Veterinary Medicine: General Aspects

1

The Veterinary Business Landscape: Contemporary Issues and Emerging Trends

Colette Henry and Lorna Treanor
Royal Veterinary College,
University of London,
Hatfield (Herts),
UK

1. Introduction

The veterinary sector has witnessed considerable change in recent years, and this has had a significant impact on the wider veterinary business landscape. Most noteworthy, perhaps, has been the change at the global level in the focus of animal welfare, with a growth in small/companion animal care and a reduction in large animal work (Lowe, 2009). This has resulted in increased competition amongst veterinary practices and the recognition that veterinary services need to be appropriately marketed. Furthermore, it has been noted that the increase in and improved access to on-line veterinary pharmacies is beginning to 'squeeze' practices' profit margins on the sale of veterinary medicines, prompting the realisation that pricing structures for core veterinary services may need to be significantly revised in order to compensate.

With specific regard to large animal work, equine services now tend to be handled by specialists, and farm animal work is considered to be in decline, with the latter prompting particular concern. Indeed, significant changes to the farming sector (i.e. declining livestock populations, a reduction in the number/size of farm holdings, new legislation and a radical change in the relationship between government and veterinarians) mean that farm animal veterinary practices now face significant competitive and sustainability issues.

There have also been significant changes to existing veterinary business models, with a marked decrease in the number of small private, independent practices and a drive toward partnerships, groups and large corporate structures with shared resources and strengthened buying powers. As a result of the latter, new management career opportunities have opened up within the corporate sector for appropriately trained and entrepreneurially motivated veterinarians to run individual clinics or sites as semi-independent business units.

Collectively, the above changes have created an extremely dynamic and highly competitive veterinary business landscape. Indeed, it is now widely recognised that managers within the broader veterinary industry need to be much more commercially aware than in the past, prompting the recognition of a significant skills gap in business and management (Lowe, 2009). This skills gap is recognised beyond the traditional practice management role, with some acknowledgement that managers in the veterinary science, diagnostics and genetics areas of the pharmaceutical, bio-science and pet food industries are experiencing similar difficulties. Recent literatures further platform the need for other veterinary professionals, such as advisors in government departments and those responsible for intensive livestock production units, to acquire business and management skills. Such trends, are not confined to the UK, rather, they would appear to be global phenomena.

Against this backdrop, and by way of providing valuable context for some of the more clinically oriented chapters in this book, our chapter considers some of the recent changes and emerging trends within the broader veterinary sector and the actual and potential impact of these on the veterinary business landscape. In particular, we focus our discussion on changes and trends within three key areas related to veterinary medicine: firstly, the recent changes within the agricultural sector; secondly, the global gender shift toward a predominately female profession and, finally, the evolving veterinary educational curriculum with an emerging trend toward the incorporation of business and management topics. We argue that, collectively, changes in these areas will have a considerable impact on the veterinary sector from a business perspective, and that such impact will be both positive and negative in nature. Given that the extent of the changes and emerging trends in these particular areas is only beginning to gain recognition, little by way of robust academic attention has been paid to their potential impact from a business perspective. As a consequence, there is a dearth of research at both the conceptual and empirical level in these areas, with a clear need for extant literature to be developed and expanded.

In this conceptual chapter, we explore the above issues by reviewing relevant literatures, identifying current trends and discussing the potential impact of these on the veterinary profession. With few studies in this particular area, the authors adopt an inductive approach, designed to generate new understanding and propositions for further research (Rosa & Dawson, 2006). Following this introductory section, we examine some of the recent changes within the agricultural sector and the impact of these on the veterinary profession. This is followed by a discussion of the gender shift currently being experienced within the profession and, subsequently, the emerging trend within veterinary curricula to include business and entrepreneurship modules. With regard to the latter, the potential impact of such curricula developments on employable skills is explored. The chapter concludes with a discussion on the implications of all these changes for the veterinary sector and offers some suggestions for future research. Thus, the authors lay the theoretical foundation for future empirical work in this field by debating some key concepts and generating hypotheses for future study. Specifically, the authors posit that veterinary curricula will need to continue to adapt to ensure graduates are equipped with the relevant skills and abilities to enable them to keep pace with the dynamic and highly competitive industry in which they will eventually work. In this regard, the incorporation of business and entrepreneurship education will be critical.

2. The agricultural sector

According to Lincoln (2004), the veterinary profession of old was primarily concerned with horses and other large working animals; this was not surprising, given the strong agricultural basis of earlier local economies. However, as national economies moved away from dominant agricultural bases, the mass manufacture of farm machinery made life somewhat less laboursome for farming communities, reducing the dependence on large working animals and manual labour. Eventually, modernization processes led to the industrialization of both arable and livestock farming (Ilbery & Maye, 2010), with assembly-line production systems incorporated into farming practice to help boost production and lower costs. Through a combination of mechanisation, genetics and artificial inputs, farmers were able to improve yields (Boulton, Rushton, Wathes & Wathes, 2011). As a result, in the dairy sector, milk quotas had to be introduced in the '80s to address overproduction, and this was followed by the deregulation of milk markets in the '90s. Dairy producer numbers declined, lactation yields per cow increased and there was a significant reduction in the number of dairy cattle (Boulton et al; 2011). The impact of such a radical change on the veterinary profession was becoming clear - fewer dairy cattle would eventually mean fewer 'patients' for farm animal vets. Indeed, the gradual reduction in veterinary 'patient' numbers has not been confined to the diary sector. The pig population has also halved over the past fifteen years resulting from a combination of factors including the Euro-pound exchange rate, changes to animal welfare regulations and disease problems (Henry, Baillie & Rushton, 2011). The 'Foot and Mouth' (FMD) epidemic of 2001, in conjunction with changes to EU subsidy payments, has also resulted in a reduced sheep population. According to the Department for Environment, Food and Rural Affairs (Defra), the gradual decline in livestock numbers witnessed by the agricultural sector has been developing over recent decades (Defra, 2010). The net effect of this raft of changes and events has been a reduction in the number of family-run, working farms in the UK, whilst a growth in larger corporate holdings and 'hobby-style' or supplementary-income farms has been noted (Defra, 2009; Rushton & McLeod, 2006; Henry et al., 2011).

EU enlargement, the CAP reform, changing consumer demands and globalization have also impacted negatively on the agricultural sector (Rudmann, 2008), with the scale of support provided to farmers drastically reduced (Alsos, Carter, Ljunggren & Welter, 2011). A gradual decline in income from traditional farming activities has forced many farming businesses to diversify in order to remain viable. Diversification requires farmers to combine other, typically non-agricultural activities with their core farm business. In this regard, conversion of farm buildings for alternative, tourism-related product offerings became a popular choice. Indeed, about 50% of farms in the UK have supplemented traditional incomes through some type of farm diversification (Defra, 2008, as cited by McElwee & Bosworth, 2010: 820). Again, all of this has had a negative impact on the veterinary community, as the trend toward smaller farms with fewer livestock and diversified activities has resulted in less demand for veterinary services overall.

3. The gender shift

The veterinary profession has traditionally been a male domain, however, in recent years, there has been an unprecedented increase in the number of women entering veterinary medicine (BVA/AVS, 2008; Lowe, 2009). By way of example, in 1998, one in three vets in the UK were female, with 57% of those under thirty being women; by 2002, 37% of vets registered with RCVS were female (RCVS, 2002), whereas in 2010, seventy one percent of UK vets under 40 were female (RCVS, 2010: viii). The recent '2010 Survey of the UK Veterinary and Veterinary Nursing Professions' (RCVS, 2010) highlighted that fifty-four percent of respondent veterinarians working in practice were women; furthermore, eighty-four percent of respondents in veterinary practice were employed within clinical veterinary practice and fifty-seven percent of those were women. Indeed, in 2011, women outnumber men as working veterinarians in the UK; moreover, eighty percent of veterinary undergraduates are female (RCVS, 2010). This significant shift in the gender make-up of both the student and practicing professional veterinary populations raises a number of important issues relating to educational strategies, career orientation, pay, working practices, and the general future direction of the profession. All of these, as will be discussed in section five of this chapter, will have a direct impact on the future veterinary business landscape.

Not surprisingly, perhaps, the trend toward a predominantly female profession is not restricted to the UK. According to statistics from the American Veterinary Medical Association (AVMA), the number of female veterinarians in the US more than doubled between 1991 and 2002 (Zhao, 2002). Whereas only 5% of applications to US vet schools were from females in the late 1960s, by the end of the 1990s, over seventy percent of applications were from young women.

The increasing numbers and dominance of women within the veterinary profession have also been discussed in several other countries (see, for example, Heath (2007) in relation to Australia; Lofstedt (2003) in relation to Canada, and Basagac Gul et al. (2008) with regard to Turkey). Collectively, such data suggest that the 'feminization' of the veterinary profession is not only a global phenomenon but also looks set to continue for some time.

With regard to the 'cause' of this phenomenon, several explanations have been offered, including improvement in chemical restraints available for large animals, the elimination of gender-based admission discrimination and the generally caring portrayal of veterinarians in both the literature and on television (Lofstedt, 2003:534). Controversially, perhaps, the marked increase in women entering the veterinary profession and, in parallel, the decline in men's participation, may also be attributed to the relatively lower salary levels of veterinary professionals (Lofstedt, 2003). This hypothesis builds on Tuchman's (1989) argument that professions deemed to be lacking in appropriate economic incentives and, by association, cultural legitimacy, may quickly lose their attraction for men. It must be noted, however, that despite the increasing number of women entering veterinary medicine and the fact that women now make up more than half the veterinary workforce, most veterinary businesses, whether based on traditional practice or the commercialisation of scientific research, are still led by men. Thus, to date, women have not featured prominently as business leaders within the veterinary profession, nor have they been perceived as potential entrepreneurs or innovators. This point is particularly important, because based on what we know about women and business ownership/entrepreneurship generally (see, for example, Brush, 1997; Brush, Carter, Gatewood, Green & Hart, 2001; Henry & Treanor, 2010), it is unlikely that current levels of veterinary entrepreneurship will be maintained or, indeed, increased in a female dominated veterinary sector.

As with the majority of professional disciplines, veterinary colleges are the 'gatekeepers to the profession' (Andrews, 2009). However, the merit-based admission system now in place, whilst undoubtedly facilitating women's entry into the profession, means that it is veterinary 'applicants', rather than veterinary 'students' who should be profiled to gauge the attractiveness of (or barriers to) veterinary medicine (Andrews 2009; Lincoln, 2004). For example, in the US, Zhao (2002) reports that both vet students and veterinarians consider the salary to be off-putting to male students. The salary of veterinary practitioners is significantly less than that of their human medicine counterparts and, therefore, men may be more inclined to train in human medicine. Zhao (2002) indicates that males and females choose their careers for different reasons - women will select something they are passionate about and think they will enjoy, whereas financial reward tends to shape the options considered by men. Zhao (2002) points to anecdotal evidence that men consider their longer-term futures and ability to provide for their families; similarly, he posits that an attraction for women to veterinary medicine is the ability to work part-time when raising young children, an option that has not traditionally been open to medical professionals to the same extent.

Interestingly, Lincoln (2010) found that young males were more likely to be deterred from applying to veterinary schools where faculty is predominately female. Research also suggests that men are dissuaded from entering female dominated professions as 'women's work' tends to attract lower salaries (Reskin & Roos, 1990; Lincoln 2004). It certainly seems that the veterinary profession is likely to become increasingly feminised in the medium to long term, and that the implications of such feminisation for future veterinary businesses now need to be given serious consideration.

4. Veterinary education

Admission to veterinary programmes around the globe is a highly competitive process, with candidates typically required to attain top grades in science-based subjects to qualify for entry. Owing to the large number of applicants to veterinary schools, interviews often form part of the selection process. Although curricula will vary from provider to provider, veterinary programmes are extremely intensive courses of study and typically include pre-clinical, para-clinical and clinical components. In addition, as part of their extramural study (EMS), veterinary students are required to spend a considerable amount of time over the course of their degree in a range of veterinary practices and hospitals to gain practical clinical experience. In the UK, course duration is typically five years, although it can be longer elsewhere.

In addition to an in-depth knowledge and understanding of the scientific and clinical aspects, veterinary students also need to have a grasp of the legal, ethical and social elements of veterinary practice. Thus, communication and interpersonal skills; responsible and professional behaviour, and an understanding of the business context of veterinary practice are now recognized as important areas that should be incorporated into veterinary curricula. With particular regard to the latter, it was a group of women veterinary surgeons in the UK who, in 1941, first called for business training to be included within veterinary programmes (Aitken, 1994). While such calls are gaining momentum, the Lowe Report (2009) highlights that UK veterinary graduates remain insufficiently prepared for effective business management or leadership, this despite the fact that most will probably work within or lead a small veterinary business at some stage in their careers. More specifically, Lowe (2009) called for the inclusion of business planning, marketing, human resource management and an awareness of the veterinary business environment to be provided to veterinary students as part of their undergraduate training so that, upon entering the workplace, they would be equipped to offer clients treatments and prevention strategies that are both cost-effective and aligned with the needs of the individual veterinary business within which they work (Lowe, 2009: 55). Similar calls have been made for the inclusion of such business and management training in North America (Ilgen, 2002; Kogen et al., 2005).

Currently, where business and management topics are included in the veterinary curriculum, they tend to be incorporated under the banner of 'Professional Studies', a strand of education running though most veterinary degrees, which, as its label suggests, covers the professional behaviour aspects of veterinary medicine. This strand typically includes the non-clinical skills and abilities associated with working as a veterinary professional, thus, communications, legal and ethical considerations, insurance, veterinary certification and compliance issues, all form part of this element of the programme. With particular regard to business and management, a range of topics are taught, including enterprising skills and abilities (i.e. creativity and idea generation), practice management (i.e. marketing, finance and human resource management for existing veterinary practices) and entrepreneurship (i.e. new venture creation or the development of a new service/clinic within the veterinary practice sector). Indeed, there appears to be growing recognition that any type of entrepreneurship education can be extremely beneficial for students regardless of their discipline area. Such education not only prepares individuals for starting new businesses but, in increasingly competitive environments, equips students with the knowledge, skills and competences to engage in a more enterprising, innovative and flexible manner (Hynes & Richardson, 2007: 733), preparing them for a world where they will increasingly need to manage their own careers in an entrepreneurial way (Hytti & O'Gorman, 2004: 11-12) and add value to their employers' businesses (Henry & Treanor, 2010).

Notwithstanding the above, the incorporation of business education, including entrepreneurship education, within veterinary curricula is still relatively new, despite the long-standing recognition that the provision of business skills will benefit both veterinary graduates and employers. However, the typical positioning of business topics within the professional studies strand of the veterinary curriculum has been somewhat problematic, with anecdotal evidence suggesting that such non-clinical topics tend to be viewed by students as peripheral rather than core. While this is, to some degree, understandable, given the already overcrowded veterinary curriculum, it does pose a number of problems for veterinary business educators tasked with preparing students for the competitive business environments in which they will eventually work.

While pedagogical approaches to veterinary education will vary from provider to provider, most veterinary schools tend to employ a combination of formal lectures, directed learning sessions, tutorials, practicals and case-based learning (QAA, 2002), with assessment methodologies ranging from multiple-choice questions (MCQs) through to portfolios and objective structured clinical examinations (OSCEs) (Baillie and Rhind, 2008). Veterinary business topics also tend to be examined in the traditional MCQ or EMQ (extended matching questions) format, however, the more practical and reflective assessment methods normally associated with mainstream business education are beginning to be introduced. While the current trend within veterinary curricula is clearly toward the inclusion rather than the exclusion of business education, it is widely acknowledged that the business curriculum is still evolving and is undergoing continuous refinement as a result of student feedback and industry consultation.

5. Discussion and implications

The changes discussed above under the three headings of the agricultural sector, the gender shift and veterinary education have significant implications for the veterinary business landscape across the broader veterinary sector, impacting beyond the practice wing of the profession. As mentioned at the outset of this chapter, to date, robust academic attention has not been paid to the implications of changes in these areas. Such implications will now be discussed under the relevant headings below.

5.1 Implications of changes within the agricultural sector

A review of the literature highlights a number of changes within the agricultural sector, all of which have a direct impact on the provision of farm animal veterinary services. In the first instance, and at the fundamental level, reduction in livestock numbers, farm holdings and farm sizes mean that there are going to be significantly fewer farm 'patients' requiring veterinary services. Although it is now recognised that farm animal veterinary practice is in decline (Lowe, 2009), there is still a substantial farming community around the globe, and thus this valuable specialist service still needs to be provided to those who require it. As a result, farm animal veterinary practices are coming under increasing pressure to review their operational business model to ensure their practice is both viable and sustainable.

In parallel, with reduced government support, a decline in income from traditional farming activities, and forced to diversify to sustain their own businesses, farmers now have less disposable income to spend on veterinary services. As a result, with regard to veterinary work, there is a growing trend toward farmers doing more for themselves rather than immediately referring work to the vet. Therefore, the average spend on veterinary services per farm holding is going to reduce. The business implications of this for rural farm animal veterinary practice are quite serious: turnover and profit margins will be affected. As a result, farm animal veterinary practices will need to revisit their operational costs to ensure the service they are offering is both viable and aligned with their clients' needs. In this regard, future veterinary graduates interested in farm animal work will need to have a thorough understanding of their clients' needs; appreciate the changes taking place within the farming sector and be able to advise farmers on the more strategic issues related to farm animal health and welfare.

It would appear too that marketing will be crucial to the survival of farm animal veterinary practice. With reduced demand for services but increased competition from existing providers, farm animal vets will need to engage more closely with their client base to determine farmers' needs, adapt their service and market it appropriately. It seems inevitable that, in the future, farmers will look for a more holistic veterinary service offering. It would appear, therefore, that in parallel with farming enterprises themselves, farm animal veterinary practices will need to explore diversification strategies if they are to remain viable from a business perspective.

5.2 Implications of the gender shift

Interestingly, the global trend towards a predominately female veterinary profession, as discussed above, has raised a number of serious concerns that will directly impact on the future veterinary business landscape. Such concerns range from a potential reduction in mean salaries for veterinarians overall (as evidence from other sectors suggests that women tend to accept lower salaries), through to potential ramifications for large animal practice (as women tend to favour small, companion animal work (Lowe, 2009)). There are also fears that women will remain less likely to own practices than their male counterparts (RCVS, 2006, 2010), which may result in larger corporate chains becoming increasingly commonplace.

Rubin (2001: 1753) cites a succinct list of 'critical issues or concerns' set out by the Gender Issues Task Force (GITF), established by the National Commission on Veterinary Economic Issues (NCVEI) in 2000, as:

- Women's income lags that of men
- Women work fewer hours
- Women may be less likely to own practices; may achieve practice ownership later
- Women may define job satisfaction more subjectively, with less objective rewards
- Women may price services lower than men
- Women's business acumen may lag that of men
- Women may be more highly satisfied with lower income levels
- Women's income may be negatively influencing income levels for all veterinarians
- Women tend to bear more of the burden of balancing home, childcare, and eldercare responsibilities with professional responsibilities
- Women may be more reluctant to take on job or management responsibilities in the face of home management responsibilities.

By way of example of the potential negative impact of the 'feminization' of the veterinary profession, Smith (2002) highlights the gender shift that has occurred in the United States since the 1980s and outlines the parallel decline in veterinarians' economic standing over that period. In a similar vein, other commentators have reported that male veterinary graduates in the USA have received more offers of employment, larger salaries and, typically, greater benefit packages than their female counterparts (Gehrke, 1997; Slater & Slater, 2000; Lincoln, 2004). Similarly, the Brakke Management and Behaviour Study conducted in the USA in 1998 showed that women earned significantly less than their male peers even when factors such as ownership status, years experience and hours worked were the same.

Similar concerns relating to reduced salaries and pay gaps between genders have been voiced in the UK. For example, the Inter-Professional Group study of 'Women Professionals in the EC' (1993) highlighted unfavourable salary trends for UK women vets (Aitken, 1994). This issue was further highlighted in the RCVS Survey of the Profession (2010), which highlighted that: "females account for 59% of those who are leaving [the profession] on grounds of pay" (RCVS, 2010: 40). Thus, the veterinary sector could face significant staff shortages in the future, with veterinary businesses unable to deliver the range of services required by their clients.

As women come to dominate the profession numerically, the lower pay they receive ultimately means that the average salary for veterinarians may fall in real terms. For women in particular, the impact is greater, with existing pay gaps and potentially lower salary levels restricting women from amassing the requisite capital to buy into a partnership or establish their own practice. This will have a direct impact on the veterinary business landscape, with potentially fewer independent practices and a continued growth in large corporate structures.

5.3 Implications of changes within veterinary education

In contrast with the two areas discussed above, the trend in veterinary education toward the incorporation of business and management topics within the curriculum should have a significant positive impact on the veterinary business landscape. Regardless of their discipline area, graduates in the 21st century face an increasingly competitive and volatile employment market and it is critical that they are equipped with skills that enable them to maximise their full potential (HEA, 2005). Future graduates will need to have greater ownership of their employable skills and be able to cope with economic upheavals to capitalise on career opportunities (McNair, 2003, as cited in HEA, 2005:12). Universities are under pressure to demonstrate how they are enhancing graduate employability (Dearing, 1997; BIS, 2009) and, in this regard, veterinary schools are no exception. According to BIS (2009), employable skills are modern workplace skills such as team working, business awareness and communication skills; essentially, they are skills that prepare students for the world of work. Thus, interpersonal, leadership and team skills; confidence, motivation and networking skills; problem-solving and business acumen are all deemed critical to enhancing graduate employability (Hawkins, 1999). In this regard, business and management education, more specifically, entrepreneurship education will have a particularly valuable role to play in veterinary curricula in the future. Indeed, there is already evidence in the literature to show that employers recruiting postgraduates place considerable value on such skills in the context of leadership, analytical thinking, problem-solving and idea generation in addition, of course, to actual work experience (CIHE, 2010).

Due to the highly competitive nature of the current veterinary business landscape, veterinary practices need to operate as viable business entities more so than ever before. Thus, veterinary graduates will need to have at least some foundation in marketing and finance. Furthermore, the new career opportunities created by the corporate sector will require entrepreneurially minded individuals who have an understanding of the broader veterinary business sector in addition to their veterinary degree. Through joint venture partnership arrangements, appropriately oriented veterinary graduates will have the opportunity to run their own business without all of the costs and risks associated with setting up a new independent practice or raising the finances required to buy into an existing partnership. If current trends in veterinary (business) education continue, then future veterinary graduates should be well equipped to not only exploit such opportunities as they emerge, but also to add value to the business dimension of all types of existing veterinary practice by helping to develop effective marketing strategies, contributing toward financial sustainability and enhancing the overall veterinary service offering. Furthermore, given that the majority of veterinary graduates take up roles within traditional practice settings, there would appear to be considerable scope for the commercialisation of research and new venture creation outside of this realm. In this regard, increasing the entrepreneurship dimension of the veterinary business curriculum will have obvious benefits.

6. Conclusions

This chapter has considered some of the recent changes and emerging trends within the broader veterinary sector and the actual and potential impact of these on the veterinary business landscape. Focusing our discussion on three key areas: the agricultural sector, the gender shift toward a predominately female profession and the evolving veterinary educational curriculum we reviewed the relevant literatures and explored the implications involved for the veterinary business landscape. Essentially, given the dearth of research in these particular areas, our primary contribution in this chapter has been to raise awareness of the potential impact of recent changes on the broader veterinary business landscape. In so doing, we enhance understanding of the business dimension, platform concerns, and encourage debate around future potential strategies that could help alleviate the negative impact of some of the changes we have discussed. This is important, given that little by way of concerted academic attention has been paid to the business dimension of the broader veterinary sector.

We demonstrated that the above changes have created an extremely dynamic and highly competitive veterinary business landscape. In the first instance, changes to the agricultural sector have had a significant negative impact on farm animal veterinary businesses, with fewer 'patients' for farm animal vets. Rural veterinary service providers will have to rethink their service offering and, possibly, diversify to ensure business viability in what is essentially a declining but still competitive marketplace. To some degree, this parallels the measures that farming clients have had to implement in order to sustain their own farm business. Undoubtedly, in the future, we will see significantly fewer farming enterprises and fewer farm animal veterinary practices in the veterinary business landscape. However, those farm animal veterinary service providers remaining in the sector will be offering a much more holistic service much more aligned to the changing needs of their farming clients.

The gender shift toward a predominately female veterinary profession will, as highlighted above, have a considerable impact on the veterinary business landscape. This area, however, has been particularly slow to gain attention in the literature and now requires urgent action if some of the potential negative impacts outlined in the earlier part of our chapter are to be avoided. The impact of career breaks, maternity leave, lower pay and, potentially, a continued trend toward companion animal care only, all suggest a future veterinary business landscape that fails to appropriately cater for farm animal, exotic species or specialist services and procedures. Furthermore, if women make up the majority of the veterinary workforce in the future, then their needs, working behaviours and career aspirations will need to be appropriately catered for and managed. This will have to lead to the introduction of new, more family-friendly working practices. In addition, if, as our discussion suggests, women remain less likely to pursue business leadership roles, then the veterinary business landscape will be 'stocked' with corporate enterprises managed, very possibly, by non-veterinary business managers. While further exploration is clearly needed, this could potentially be detrimental for the veterinary profession as a whole.

With regard to the veterinary curriculum, while the case for business and entrepreneurship education within veterinary medicine has never been stronger, the challenge now for educationalists is to figure out *what* exactly should be taught and *how* best to teach it. In what is already a crowded curriculum with large student groups to cater for, prioritizing the business and enterprise agenda over the more clinical topics or 'ologies' (i.e. immunology, pathology, epidemiology, etc) is not an easy task. Given all the changes that are taking place in the broader veterinary sector, we argue the case for including the full spectrum of business and entrepreneurship education - 'enterprise skills for life', core elements of 'practice management' and opportunities for 'new venture creation' within the five-year veterinary medicine degree programme. Given that relatively few graduates work outside the pure practice wing of the profession, the opportunities to encourage entrepreneurship are significant. Indeed, while a full discussion on this particular topic is outside the scope of this chapter, the potential for innovation and new venture creation in the context of commercialisation of veterinary clinical research, development of new drugs and treatments, and the creation of new products and services for supply to veterinary practices to date has not been fully explored. Developments in these areas could only serve to enhance the veterinary business landscape in the future.

6.1 Future research

Looking to the future, collaboration and partnership arrangements amongst veterinary service and education providers, both nationally and internationally, is likely to offer considerable mutual benefit for all those involved in the veterinary sector. In this regard, future research should consider how such partnerships might be encouraged and where they might be most beneficial. In addition, and with specific regard to the gender shift, future research should focus on increasing the business leadership and entrepreneurial intentions of female veterinary students. This is particularly important if the veterinary business landscape is to continue to include practices led by veterinary professionals. Finally, from our discussion in this chapter, we posit that there is considerable potential within veterinary educational curricula to enhance students' understanding of the business dimension of the veterinary sector, to ensure they are fully aware of the changes taking place within the sector and how these will impact on the veterinary businesses in which they will work in the future. In this regard, the evolving veterinary educational curriculum needs to continue its trend toward a business orientation.

7. References

Aitken, M.M. (1994) "Women in the Veterinary profession" The Veterinary Record, 134, 546-551.

Alsos, G. A., Carter, S., Ljunggren, E. and Welter, F. (2011), "Introduction: researching entrepreneurship in agriculture and rural development", in G.A Alsos., S. Carter., E. Ljunggren and F. Welter (eds), *The handbook of research in entrepreneurship in agriculture and rural development*, Cheltenham: Edward Elgar, pp.1-18.

Andrews, F.M. (2009) 'Veterinary School admissions in the United Kingdom: attracting students to veterinary careers to meet the expanding needs of the profession and of global society', Rev. Sci. Tech., 28 (2): 699-707.

Baillie, S. and Rhind, S. (2008), "A guide to assessment methods in veterinary medicine," A 'Blue Sky' project funded by the RCVS Trust, available from http://www.live.ac.uk/documents/assessment_guide.pdf, last accessed 20th May 2010.

Basagac Gul, R.T., Ozkul, T., Ackay, A. & Ozen, A. (2008) 'Historical Profile of Gender in Turkish Veterinary Education' JVME, 35 (2), pp. 305-309.

BIS - Department for Business Innovation and Skills. (2009). "Higher ambitions: The future of universities in a knowledge economy," available from www.bis.gov.uk

Boulton, A., Rushton, J., Wathes, C. M. & Wathes, D. C. (2011). "Three decades of change: a review of the UK dairy industry", Royal Agricultural Society Journal (forthcoming).

Brush, C. (1997). Women owned businesses: Obstacles and opportunities. *Journal of Developmental Entrepreneurship*, 2(1), 1–25.

Brush, C., Carter, N., Gatewood, E., Green, P., & Hart, M. (2001). "The Diana Project, women business owners and equity capital: the Myths dispelled." Insight Report, Kansas City, MO: Kauffman Center for Entrepreneurial Leadership.

BVA - British Veterinary Association / AVS - Association of Veterinary Students. (2008). "Survey 2008," available from: http:www.bva.co.uk/student_centre/Student_survey.aspx, last accessed 9th April 2010.

CIHE - Council for Industry and Higher Education. (2010), "Talent fishing: What businesses want from postgraduates," A CIHE report for the Department of Business Innovation and Skills, CIHE: London (www.cihe.co.uk)

Dearing, R. (1997), "The Dearing Report," available from: https://bei.leeds.ac.uk/Partners/NCIHE

Defra. (2009), "Agriculture in the United Kingdom", available from:
http://www.defra.gov.uk/evidence/statistics/foodfarm/general/auk/latest/index.htm (last accessed 17 May 2011).

Defra. (2010), "June Census of Agriculture and Horticulture - UK", 25 May, available from: http://archive.defra.gov.uk/evidence/statistics/foodfarm/landuselivestock/junesurvey/documents/June-labour-2010.pdf (last accessed 30 May 2011).

Gehrke, BC. (1997) 'Employment of Male and Female Graduates of US veterinary medical colleges" Jamva, 212: 208-209

Hawkins, P. (1999), "The art of building windmills: Career tactics for the 21st century

HEA - Higher Education Academy (2005), "Embedding employability in the curriculum: enhancing students' career planning skills," report available from http://www.heacademy.ac.uk, last accessed 10th May 2010.

Heath TJ., (2007) "Longitudinal study of Veterinary Students and veterinarians: The First 20 Years", Australian Veterinary Journal, Vol 85 No. 7, pp. 281-289

Henry, C. & Treanor, L. (2010). Entrepreneurship education and Veterinary Medicine: Enhancing Employable Skills. Education & Training, Vol. No. (December, 2010), pp. 607-623.

Henry, C., Baillie, S. & Rushton, J. (2011). Exploring the future sustainability of farm animal veterinary practice, paper presented at the Rural Enterprise Conference, Nottingham Trent University, June.

Hynes, B. & Richardson, T. (2007), "Creating an entrepreneurial mindset: getting the process right for information and communication technology students", in Lowry, G. (Ed.), Information Systems and Technology Education: From the University to the Workplace, IGI Global, Hershey, PA.

Hytti, U. & O'Gorman, C. (2004), "What is 'enterprise education'? An analysis of the objectives and methods of enterprise education programmes in four European countries", Education & Training, Vol. 46 No. 1, pp. 11-23.

Ilbery, B. & Maye, D. (2010), "Agricultural Restructuring and Changing Food Networks in the UK," in N. Coe and A. Jones, (eds), The Economic Geography of the UK, London: Sage, pp. 166-180.

Ilgen, D.R. (2002), "Skills, knowledge, aptitudes and interests for veterinary practice management: fitting personal characteristics to situational demands", JVME, Vol. 29 No. 3, pp. 153-6.

Kogan, L.R., McConnell, S.L. & Schoenfeld-Tacher, R. (2005), "Perspectives in professional education: response of a veterinary college to career development needs identified in the KPMG LLP study", Journal of the American Veterinary Medical Association, Vol. 226 No. 7.

Lincoln, A., (2004) "Vetting Feminization Beyond Motives: Structural and Economic Influences on Applications to American Veterinary Medical Colleges, 1976-1988", Paper presented at the annual meeting of the American Sociological Association, Hilton San Francisco & Renaissance Parc 55 Hotel, San Francisco, CA, Online Last accessed: 19th January 2011 via: http://www.allacademic.com/meta/p110046_index.html

Lincoln A. E. (2010) The shifting supply of men and women to occupations: feminization in veterinary education. Social Forces 88, 1969–1998. doi: 10.1353/sof.2010.0043

Lofstedt, J. (2003), "Gender and veterinary medicine", Canadian Veterinary Journal, Vol. 44 No. 7, pp. 533-5.

Lowe, P. (2009), Unlocking Potential – A Report on Veterinary Expertise in Food Animal Production, Department for Environment and Food Rural Affairs (DEFRA), London.

McElwee, G. and Bosworth, G. (2010). "Exploring the strategic skills of farmers across a typology of farm diversification approaches," Journal of Farm Management, 13 (12), pp. 819-838.

McNair, S. (2003), "Employability in Higher Education," LTSN Generic Centre, University of Surrey.

QAA – Quality Assurance Agency. (2002), "Veterinary Science – Subject Benchmark Statements," Quality Assurance for Higher Education, Gloucester.

RCVS – Royal College of Veterinary Surgeons. (2006). "The UK Veterinary Profession in 2006: The findings of a survey of the profession, conducted by the Royal College of Veterinary Surgeons", London: Royal College of Veterinary Surgeons. Available at: http://www.rcvs.org.uk/shared_asp_files/uploadedfiles/97d3daf0-9567-4b2f-9972-b0a86cfb13c7_surveyprofession2006.pdf (last accessed: 8th January, 2010).

RCVS. (2010), "The 2010 RCVS Survey of the UK Veterinary and Veterinary Nursing Profession", London: Royal College of Veterinary Surgeons. Available at: http://www.rcvs.org.uk/Templates/SinglePublication.asp?NodeID=7828174 (last accessed: 4 November 2010).

Reskin, B. & Roos, P. Job queues, gender queues. Explaining women's inroads into male occupations. Philadelphia: Temple University Press, 1990

Rosa, P. and Dawson, A. (2006), "Gender and the commercialization of university science: Academic founders of spinout companies," Entrepreneurship and Regional Development, July, pp.341-366.

Rubin, H. E. (2001). "Generation and gender - a critical mix," JAVMA, 218(11), pp.1753-1754.

Rudmann, C. (Ed) (2008), "Entrepreneurial skills and their role in enhancing the relative independence of farmers", Research Institute of Organic Agriculture FiBL, Ackerstrasse, available from: www.fibl.org

Rushton, J. & McLeod, A. (2006). "Animal health economics - UK veterinary services and practices", presentation to Defra.

Slater, R.S. & Slater, M.S. (2000). "Women in veterinary medicine" Journal of the American Veterinary Medical Association, 217(4), pp. 472-476.

Smith, C.A. (2002) "Gender and Work: What veterinarians can learn from research about women, men and work", JAVMA, Vol 220, No. 9, pp. 1304 - 1311.

Tuchman, G. (1989). "Edging women out: Victorian novelists, publishers and social change", New Haven, CT: Yale University Press.

Zhao, Yilu. "Women Soon to be Majority of Veterinarians." New York Times 9 June 2002. Melissa Kaplan's Herp Care Collection. 2 Mar. 2009 <http://www.anapsid.org/vets/vetdermos.html

2

Artificial Insemination in Veterinary Science

Annett Heise
Faculty of Veterinary Science
University of Pretoria
South Africa

1. Introduction

Artificial insemination (AI) is widely used by veterinarians and veterinary specialists in most domestic as well as wild animal species. Reasons for use of AI instead of natural matings are diverse and different for individual species. As in humans, AI can be used as a tool to increase conception rate in animals that have fertility problems. This is usually performed in companion animals, like horses and dogs, where individuals are important as breeding animals. In these cases, breeding animals are mostly chosen for performance and pedigree instead of breeding soundness. Breeders go through a lot of effort to produce offspring of a specific female or male animal irrespective of potential reproductive problems.

In production animals, AI is a way to increase reproductive efficiency and production. AI has proven to be a very effective reproductive technology that selectively increases genetic gain through increased selection pressure on males. In Holstein cattle, for example, AI supported selection for the milk production trait and within 40 years milk production has nearly doubled. Farm animals, males as well as females, are usually chosen for breeding programs based on breeding soundness examinations (BSEs). These BSEs determine suitability and likelihood of females or males to participate successfully in breeding programs. Animals that do not fulfill certain criteria are identified. These "problem" animals are excluded from insemination programs. Estrus cycles of females can be manipulated to institute efficient insemination programs. With the use of these estrus synchronization programs, large groups of females can be inseminated at the same time. This does not only have the advantage of concentrating work on specific days during breeding, but will ultimately also simplify the herd management before and after the offspring are born. Group feeding of pregnant animals, partus observation, vaccination programs for calves and tail docking of lambs are just a few examples of improved herd management areas through the group effect achieved through AI.

Another reason for AI is to ensure effective use of semen. An increased number of offspring from a superior sire can be produced when AI is employed. For example, a stallion's ejaculate can be sufficient to inseminate 5-10 mares at the same time when split into doses instead of one live cover on a mare. Ram ejaculates can also be split into up to 15 fresh AI doses. Freezing bull semen can provide up to 200 straws of frozen semen from one ejaculate, equaling 200 AI doses. Overuse of males is prevented and commercial distribution facilitated. Other important aspects are the prevention of venereal disease transmission that plays a major role in the economic system of offspring production, and increased safety for valuable breeding animals as mating related injuries are avoided.

Venereal diseases that play a major economic role in cattle production are for example Trichomonosis and Campylobacteriosis both of which decrease reproductive efficiency through decreased pregnancy rates, high return rates to estrus and increased pregnancy losses. In general, shipping fresh and frozen semen nationally and internationally involves fewer health risks and welfare implications than transporting live animals.

Import and export of frozen semen is a huge economic market. Semen of a specific bloodline, a specific individual male or breed is imported. This is especially important in countries where breeds were introduced and are somewhat isolated with a small genetic pool. To eventually prevent breeding setbacks due to inbreeding and to expand the genetic pool, imported semen is used. Furthermore, AI can be used for frozen semen from males that have died or are not physically available for matings due to distance or physical inability. A great advantage of frozen semen in general is that it can be stored indefinitely and has the potential to outlive the male donor animal by years.

2. First developments and milestones in the field of veterinary Artificial Insemination

Artificial insemination (AI) was the first assisted reproductive technique applied to control and improve reproduction as well as genetics. The first successful insemination was performed by the Italian physiologist and priest Abbe Lazzaro Spallanzani (1784) in a dog which whelped three pups 62 days later (Foote, 2002). The establishment of AI as a practical procedure was initiated in Russia in 1899 by Ivanov who studied AI in domestic farm animals, dogs, foxes, rabbits and poultry. He also developed semen extenders. Milanov, another Russian scientist and successor of Ivanov, started large scale breeding programs for cattle and sheep, and designed and made artificial vaginas. Horse breeding programs and research was initiated at the same time in Japan even though translations of the original research only became available to the western world after 1958. Some AI work, especially in horses and cattle, had been done in Denmark in the early 1900s. It was Danish veterinarians who established the method of rectovaginal fixation of the cervix for insemination in cattle which enabled semen deposition deep into the cervix or into the uterine body (Foote, 2002). This technique is still used today. Another Danish invention was the straw for packaging semen. These straws have been further developed and modified by the French and are now used worldwide for processing and storage of frozen semen. Research on artificial insemination in Italy led to the development of an artificial vagina for dogs in 1914 and to the establishment of the "International Congress on AI and Animal Reproduction" in 1948. This congress is held every four years since (Foote, 2002). Rapid development of AI in dairy cattle occurred in the USA in the 1940s. One of the important milestones was the establishment of the "Dairy breeding research centre" on the campus of the Pennsylvania State University in 1949 to assist in the development of artificial insemination in dairy cattle. Interest in and development of frozen semen started with successful cryopreservation of gametes from a variety of animal species after discovery of the protective action of glycerol by scientists in Cambridge, England, in 1949 (Amann & Pickett, 1987). Early research focused on bull spermatozoa; methods successful to cryopreserve bull spermatozoa have been less successful for other species. Barker and Gandier reported the first pregnancy from frozen stallion semen in 1957 (Barker & GaPndier, 1957). Poor pregnancy results with the use of frozen-thawed stallion semen limited research funding for breeding trials in this species (Amann & Pickett, 1987). Only later the Chinese inseminated more than 110'000 mares with frozen-thawed stallion semen (between 1980 and 1985). The United States also did not have interest in developing improved procedures for freezing stallion spermatozoa in the seventies. Only in 1980 the Animal Reproduction Laboratory of Colorado State University began a long-range research program aimed at developing satisfactory procedures to freeze stallion semen. The acceptance of frozen semen as a method to produce registered foals by two of the largest breed associations, the American Quarter Horse and the American Paint Horse association in 2001 has furthermore stimulated new interest in the frozen semen technology (Loomis, 2001).

3. AI techniques in different species

Insemination techniques used depend on species, type of semen, breeding system, availability of equipment and expertise. Intravaginal (dogs), intracervical (sheep), transcervical intrauterine (cattle, horses, dogs, sheep), transcervical deep horn intrauterine (horses, cattle, pigs), laparoscopic (sheep), surgical intrauterine (dogs) insemination as well as endoscopic semen deposition at the uterotubal junction (horses) is available.

3.1 AI techniques in dogs

In dogs, fresh, chilled extended and frozen semen can be used. AI using fresh semen is usually performed if the animals will not or cannot copulate naturally. There are certain breeds, like English bulldogs and other brachycephalic breeds that almost always require AI due to their anatomical incompatibility to mate naturally. AI is also used if time constraints are an issue as fresh semen AI in the bitch is much quicker than a natural mating where the coital tie between male and female can last up to 40 minutes or more. AI can be used in bitches with congenital vaginal abnormalities like vaginal strictures or septae which might cause copulation failure. These cases often have ethical implications as well. Owners need to understand that even though AI can be performed easily these bitches might require Cesarian sections as the vaginal abnormalities might also impair natural delivery or that certain vaginal abnormalities are heritable and offspring with the same problem might be produced. AI is also used if the semen quality of the specific male is poor and addition of semen extender or pooling of more than one ejaculate is required for one insemination dose. Fresh semen is deposited into the cranial vagina of the bitch using an insemination pipette that is inserted through the vulva and directed into the vagina. Once the semen has been deposited, the hindquarters of the bitch are usually elevated for a few minutes to facilitate movement of the semen from the vagina through the cervix into the uterus. Vaginal contractions can be elicited for the same reason at the same time by either tickling the vaginal wall or massaging the clitoris. An insemination dose of at least 150 x 10^6 sperm is recommended and most commonly the whole ejaculate is used which may contain 250-2500 x 10^6 sperm (Johnston et al., 2001).

Pregnancy rates with the use of fresh semen for insemination is reported to vary between 65-84% on average depending on semen quality, timing of insemination and correct site of semen deposition (Johnston et al., 2001; Linde-Forsberg & Forsberg, 1989).

Chilled extended semen is usually used if the male and female are geographically separated. An extender is added to the collected semen before shipment in order to prolong the lifespan of sperm. For the same reason, the extended semen is slowly cooled down to 4 degrees Celsius. The entire ejaculate is usually used for the AI but an insemination dose of at least 150 x 10^6 sperm is recommended. Average pregnancy rate for chilled extended semen is 50% but can vary between 33-89% (Johnston et al., 2001). Extended semen can be used for up to 4 days after collection but should be inseminated as soon as possible. The insemination technique is the same as described above for fresh semen or transcervical intrauterine insemination can be used. Transcervical AI is performed using special catheters (e.g. Norwegian catheters) or endoscopes. The anatomy and location of the cervix in the bitch is such that it is difficult to penetrate the cervix and perform transcervical intrauterine inseminations due to a vaginal fold that is obscurring direct access to the cervix. Therefore, special equipment is needed for transcervical intrauterine AI. Norwegian catheters have been developed in the 1970s (Andersen, 1972, 1975) and are available in three different sizes to accommodate different sized bitches. For this technique, the cervix must be fixed through abdominal palpation and the catheter is moved slowly to find the cervical canal while the cervix is pulled downward. This method requires a lot of practice and there is a potential danger that inexperienced veterinarians might traumatize the cervix. Used by experienced clinicians this technique offers an inexpensive, fast way of inseminating non anesthetized bitches. Using a rigid endoscope for transcervical AI enables the operator to visualize the cervix, straighten the cervical canal and insert a catheter into the uterus to deposit the semen. The advantage of this technique is that it is not blind and the danger of trauma to the reproductive tract is minimal. Bitches do not usually have to be sedated for the procedure and AI can be repeated a few times if necessary.

For frozen semen, deposition of the semen into the uterine lumen is recommended in order to achieve good pregnancy rates (Thomasse et al., 2001; Thomassen et al., 2006). Even though the first report of successful AI with frozen semen in dogs was a result of intravaginal deposition of the semen (Seager, 1969) it has been shown that success rates with intrauterine inseminations are far superior (Thomassen et al., 2001; Thomassen et al., 2006). Intrauterine insemination can be performed by transcervical intrauterine AI as described above or surgical AI via laparotomy. In cases where transcervical intrauterine AI is unsuccessful or the equipment is not readily available, surgical intrauterine AI is performed. It is usually a quick procedure that requires general anesthesia for about 20 minutes. A small abdominal incision is made, the uterus exteriorized and the semen injected directly into the uterine lumen through the uterine wall. Disadvantages of this procedure are the relatively high costs as well as the anesthetic and surgical risk. Surgical insemination is considered an unethical procedure in some countries (Linde-Forsberg, 1991). Insemination doses are usually composed of 100-150 x 10^6 motile spermatozoa. Whelping rates for frozen semen AI are reported to be as high as 70-75% if the timing of insemination is accurate, good quality frozen-thawed semen is used and the semen deposition is correct (Thomassen et al., 2001; Thomassen et al., 2006).

3.2 AI techniques in horses

AI is a widely practiced breeding method in most sport horse breeds worldwide. Fresh, chilled extended and frozen-thawed semen can be used in horses. Common reasons for the use of insemination for fresh semen instead of natural mating are the reduced risk of injury to the often very valuable stallion by avoiding live coverings, insemination of mares that do not want to allow a mating or splitting of an ejaculate into several insemination doses if more than one mare is to be inseminated at the same time. Generally accepted insemination doses for fresh semen as supported by the World Breeding Federation of Sport Horses (WBFSH) are 300 x 10^6 straight forward swimming (progressively motile) spermatozoa (Katila, 2005) which is slightly lower than the originally recommended fresh semen AI dose of 500 x 10^6 progressively motile spermatozoa (PMS) (Pickett & Voss, 1975). First cycle pregnancy rate for fresh semen AI is approximately 60%. Chilled extended semen is usually used if the mare and the stallion are in distant places. Semen is collected, extended with a suitable semen extender to provide buffers, nutrients, antibiotics etc. in order to prolong lifespan of the spermatozoa and slowly cooled down to 4 degrees Celsius. Usually, special transport containers that form cooling units (e.g. Equitainer®, Hamilton Thorne) are used and the semen of most stallions is viable for at least 48 hours. Most insemination centres try to inseminate mares with chilled extended semen within 24 hours of collection. One billion spermatozoa or 600 x 10^6 progressively motile spermatozoa (according to WBFSH (Katila, 2005)) form one insemination dose and pregnancy rates of about 60 % (Sieme et al., 2003) can be expected depending on timing of insemination relative to ovulation, AI dose and semen quality after chilling. Some stallions' semen is not suitable for chilling and the semen quality deteriorates very rapidly. It is therefore recommended to perform a trial cooling of equine semen before a shipping is done and to evaluate semen quality parameters over time to assess longevity.

The insemination technique for fresh and chilled extended semen is similar in the horse. A transcervical intrauterine insemination is performed where the semen is deposited into the uterine body. A commercially available insemination pipette is inserted manually into the vagina, the external opening of the cervix is located using the index finger and the pipette is guided through the cervical canal into the uterus. The penetration of the cervix is generally very easy as the cervical tissue is smooth muscle that relaxes under estrogen influence. That allows easy access to the equine uterus at the time of insemination. Frozen-thawed semen has first been used to inseminate mares in 1957 (Barker & Gandier, 1957) even though it only gained increasing popularity over the last 15-20 years. Using frozen semen has a lot of benefits: accessibility to semen from stallions in competition or stallions that become ill, injured or overbooked during the breeding season. Using frozen semen eliminates the need to organize stallion availability at the optimum time for breeding of the mare and disease transmission. Chances for injury are decreased as direct mare-stallion contact is avoided. Pregnancy rate per cycle for frozen semen varies between 30-50% on average (Leipold et al., 1998; Metcalf, 2007; Sieme et al., 2003; Vidament, 2005; Vidament et al., 1997). There are, however, some disadvantages regarding frozen semen. The management of mares during estrus is more intense which increases the costs involved for frozen semen inseminations. This is necessary as inseminations with frozen semen should be done as close to ovulation as possible since frozen-thawed spermatozoa have a shortened lifespan. Frozen-thawed spermatozoa survive for about 12 hours in the reproductive tract of mares whereas fresh semen can survive for 48-72 hours. Frozen-thawed equine semen can be inseminated into the uterine body as described for fresh and chilled distended semen if an insemination dose of 250 x 10^6 progressively motile spermatozoa is available (Katila, 2005). If the insemination dose is lower or a very small volume is to be inseminated (0.25-0.5ml), the semen can be deposited deeply into the uterine horn ipsilateral to the ovary where ovulation will take place or has taken place (Katila, 2005). Other possibilities for low dose (5-25 x 10^6

progressively motile sperm) or very low volume (0.02-0.2 ml) inseminations (Katila, 2005) are endoscopic deposition of semen at the utero-tubal junction right in the tip of the uterine horn ipsilateral to the ovary where ovulation will take place (Lindsey et al. , 2001; Lindsey et al. , 2002; L. H. A. Morris, 2004), and surgical deposition of spermatozoa directly into the oviduct (McCue et al. , 2000).

3.3 AI techniques in dairy cattle

In Britain, AI in dairy cattle began to be available in 1942, and by 1950 20% of dairy cattle were being inseminated. By 1960, more than 2 million cows were inseminated yearly, which was about 80% of the maximum level that AI would reach (Brassley, 2007). The established procedure for AI in cattle since the 1960s is transcervical deposition of semen into the uterine body. This technique replaced the original vaginal or shallow cervical insemination performed in the 1940s as the intrauterine method proved to be more efficient and resulted in higher fertility (Lopez-Gatius, 2000). Transcervical intrauterine AI involves the technique of cervical fixation per rectum to facilitate easier penetration of the cervical rings with a stainless steel Cassou device (AI pistolette) (Noakes et al. , 2001). Other techniques like unicornual or bicornual insemination where semen is deposited into one or both uterine horns, and intraperitoneal insemination have been investigated (Lopez-Gatius, 2000) but could not replace the transcervical intrauterine AI with semen deposition into the uterine body. For commercial AI in cattle, frozen-thawed semen is routinely used and a generally accepted insemination dose contains 10-20 x 10^6 spermatozoa. Deep horn AI close to the uterotubal junction has been investigated and facilitates AI with a conventional number of spermatozoa reduced x 100 or if very small volumes of semen (0.1-0.25 ml) are to be used (Hunter, 2003). Potential advantages of deep horn AI include: raising fertility of genetically valuable bulls whose non-return rates to estrus are sub-optimal, reducing the number of sperm per AI dose, facilitating the use of limited numbers of sex selected sperm cells available from flow cytometry, and breeding from valuable but oligospermic (too few sperm in ejaculate) bulls (Hunter, 2003).

3.4 AI techniques in pigs

Even though AI in pigs has been used since the 1930s, wider commercial application only started in the 1980s, with a dramatic increase in the use of AI in pork production in the last 20 years. International comparison shows that in 1990, 80% of pigs in former Eastern Germany were bred by AI, compared to only 23% in West Germany, while the figures for Norway were 71% and for the Netherlands 51%. France and the UK were about the same with 10% and the USA was only 7% (Brassley, 2007). By the end of the 1990s, close to 50% of the worldwide gilts and sows were inseminated (Roca et al., 2006). Nowadays, AI is used routinely in pig companies. In some European countries, like Belgium, Italy, the Netherlands, Norway and Spain, more than 80% of females are bred by AI and in North America and Brazil the percentage has reached 75% (Roca et al., 2006). One reason for the slow take-off for AI in pigs were initial conception rates which were as low as 18.1% in 1955 rising to 62% in 1961 while natural conception rates were expected to be close to 90% (Brassley, 2007). Conception rates could be increased to about 80% by the mid 1970s. Consistent 80-90% fertility rates with the use of liquid extended semen are common on many farms by now with an overall fertility success (measured as farrowing rates and litter sizes) similar to or better than those resulting from natural matings (Roca et al., 2006). Apart from good conception rates, there are other reasons for the dramatic increase in AI utilization for pig breeding systems. On one hand there was the change of breeding and farrowing units that became larger and more specialized, and that the application of AI mating technologies became more feasible and cost effective on the other hand (Singleton, 2001). At the same time, the pork industry initiated payment programs based on actual carcass value instead of live weight basis. Genetic evaluation programs were used to implement genetic improvement programs through the use of AI to produce higher quality pork carcasses. For most commercial breeders it was the prospect of genetic improvement that was the major incentive to engage in AI (Brassley, 2007). Semen processing centres (boar studs) or on-farm collection facilities provide boar semen used for AI programs. More than 99% of AIs are performed with semen extended in a liquid state that can be stored at 15-20°C for up to 3 days (cooled semen)(Roca et al., 2006). The number of viable sperm per dose ranges between 2.5- 4 billion motile spermatozoa. Dose volumes range from 80-100ml for fresh liquid semen. The insemination procedure, called intra-cervical insemination (intra-CAI), involves the deposition of spermatozoa into the posterior part of the cervix using a catheter that engages with the folds of the cervix, stimulating the corkscrew tie of the boar's penis (Roca et al., 2006). This easy and quick procedure has been developed in the mid 1950s, standardized in the 1970s and is still used worldwide today (Roca et al., 2006). The remaining 1% of AIs utilizes frozen-thawed semen at doses of 5-6 x 10^9 spermatozoa. Even though large sperm numbers are used, fertility is substantially lower than that obtained with cooled semen. Due to the lowered reproductive performance, frozen-thawed boar semen is mostly limited to specialized breeding programs, research and for export puposes (Singleton, 2001; Wongtawan et al. , 2006). There are two insemination procedures available that allow insemination with low numbers of spermatozoa: post-cervical insemination (post-CAI) and deep uterine insemination (DUI). Both techniques facilitate deposition of semen into the uterus. Using post-CAI, semen is placed into the uterine body while DUI facilitates placement of semen in the proximal 1/3 of one uterine horn (Martinez et al., 2005). Similar fertility results can be expected for DUI using 600 x 10^6 spermatozoa and 1-1.5 x 10^9 spermatozoa for post-CAI as compared to intra-CAI using 3000 x 10^9 sperm per dose (Roca et al., 2006). Or in other words, a threefold reduction of fresh sperm numbers using post cervical AI, and for DUI a 20-fold reduction for fresh and sixfold reduction for frozen semen can achieve acceptable pregnancy rates (J. M. Vazquez et al., 2008). Other possibilities for inseminations with low sperm numbers are surgical intrauterine inseminations where high pregnancy rates (89%) can be obtained with as few as 10 x 10^6 cooled stored spermatozoa that are placed on the uterotubal junction (Krueger & Rath, 2000; Krueger et al. , 1999). A further reduction in sperm numbers to 5 x 10^6 can be achieved if spermatozoa are placed close to the uterotubal junction by laparoscopy (Fantinati et al., 2005). A new procedure where semen is placed into the oviduct via laparoscopy also displays an opportunity for the use of diluted and sex sorted spermatozoa (Vazquez et al., 2008). Other insemination techniques like intraperitoneal insemination, where semen was deposited directly into the abdominal cavity, have been used to investigate insemination possibilities other than intra-CAI (Hunter, 1978). This technique, however, has never come to use in the AI industry due to ineffective sperm transport to the oviducts, reduced fertility and difficulty of the procedure when compared to intra-CAI. It has also been shown that specific boar stimuli (such as olfactory and tactile stimuli) at or around the time of AI positively influence reproductive performance through an effect on reproductive processes like sperm transport and ovulation (Soede, 1993).

3.5 AI in sheep (Youngquist & Threlfall, 2007)

Fresh, chilled and frozen semen is utilized for AI in sheep. Methods used are vaginal, cervical, laparoscopic intrauterine and transcervical intrauterine insemination. Vaginal insemination can be used for fresh and chilled semen and is also referred to as the "shot in the dark", or SID, method as the semen is blindly deposited into the cranial vagina. Pregnancy rates performing vaginal AI for fresh and chilled semen are acceptable while they are not for frozen semen. The technique is very fast but the use of semen is inefficient as AI doses have to contain large numbers of sperm. For cervical insemination, the hindquarters are elevated while the ewe is restrained "over the rail". The cervix is visualized using a speculum and a light source. Semen is deposited into the cervix with an angled tip insemination gun.

In contrast to horses and cattle, a successful method for transcervical intrauterine insemination in sheep has not been well established due to the specific anatomy of the ovine cervix (Halbert et al. , 1990). The cervical canal is approximately 7 cm long with a series of 6-8 rearward facing, offset rings that make transcervical insemination difficult to impossible. The Guelph system for transcervical AI (GST-AI) has been developed which requires special positioning of the ewe, cervical retraction and stabilization, and the use of specially designed instruments. Trained, experienced inseminators may penetrate the cervix in as much as 75-85% of ewes. Cervical injury, abscesses, infections and poor pregnancy rates are associated with this technique. Surgical AIs (i.e. via midventral laparotomy) are effective, but they are costly, time consuming, require technical proficiency, limit the number of times ewes can be used and require anesthesia (Evans & Maxwell, 1987). Laparoscopic intrauterine AI is used for frozen semen, and requires laparoscopic equipment as well as technical expertise. Animals are usually sedated and restrained in a laparoscopic cradle in dorsal recumbency. The laparoscope is used to identify the uterus and a loaded insemination pipette directed to the uterus to facilitate intrauterine deposition of the semen. Three hundred or more ewes can be inseminated per day using this technique if an experienced team works in a well-equipped and organized operation. Insemination doses for fresh as well as frozen-thawed semen are 400×10^6, 200×10^6, 20×10^6, 100×10^6 progressively motile spermatozoa for vaginal, cervical, laparoscopic and transcervical AI, respectively. Expected lambing rates for fresh semen are 20-60%, 40-80%, 70-100%, 40-80%, and 5-20%, 25-60%, 40-80%, 30-70% for frozen semen for vaginal, cervical, laparoscopic and transcervical AI, respectively.

3.6 AI in other species

Since the 1920s, AI has been a reproductive tool in commercial rabbitries that permits more controlled management and better planning (e.g. in batch parturition and weaning) than natural mating (Morrell, 1995). Conception rates after AI can be equivalent to or better than that achieved with natural breeding (Morrell, 1995). Natural matings in healthy individuals can result in conception rates of 85%. A single ejaculate can be split into 20-50 insemination doses and used for AI (Lavara et al. , 2005). For the AI procedure, the female is placed into a restraining box, the tail is lifted and an insemination pipette with a bend approximately 8cm from the end is inserted into the vagina at an angle of 45° in order progress beyond the pelvic rim (Morrell, 1995). Semen is deposited intravaginally. Due to the fact that rabbits have two separate uterine horns and cervices, intracervical AI is not performed as it would be required to release semen into both cervices. Other peculiarities regarding AI in rabbits are that females should be kept separate from males for about 19 days before breeding to exclude pseudopregnancy, that timing of AI is based on general behaviour and vulva colour of the doe, and that rabbits are induced ovulators that always require induction of ovulation when AI is used. Three methods for induction of ovulation are available: 1) mating with a vasectomized buck, 2) administration of human chorionic gonadotropin (hCG) or 3) administration of gonadotropin releasing hormone (GnRH) analogues (Morrell, 1995).

AI in non-human primates is based on captive colony management and propagation of endangered or valuable founder animals. Its application, (although limited due to high costs, substantial technical requirements and limited available captive populations), has found use in Great Apes as well as Old World and New World Macaques (Wolf, 2009). Initially, intravaginal AI was performed while intrauterine insemination is the technique of choice today.

Application of AI in South American camelids has been challenging due to the inconsistent success in collecting semen from males. Llamas as well as alpacas copulate for extended periods of time (10-60 min), display a recumbent mating posture, deposit semen into the uterus, and the semen is very viscous (Adams et al., 2009). Camelids are not as easily trained to mount an AV as rams, bulls or stallions. It is necessary to maintain a stable AV temperature during prolonged copulation. Other semen collection attempts using condoms or intravaginal sacs, vaginal sponges, electro-ejaculation, post-coital vaginal aspiration, and fistulation of the penile urethra were associated with recovery of poor semen samples and contamination of semen samples with blood (Adams et al., 2009). For AI, semen is deposited into the uterus transcervically or via laparoscopy. Reported pregnancy rates after AI vary widely between 2-68% (Adams et al., 2009).

4. Use of epididymal spermatozoa for AI

Another interesting field for application of AI is the use of epididymal spermatozoa. Harvesting of epididymal sperm enables storage and usage of valuable genetic material of males after death or shortly before death if unexpected accidents or health problems occur. The epididymis is part of the male reproductive tract, is connected to the testis and forms a site for sperm maturation and storage. The epididymis can be dissected free from the testis after castration and epididymal spermatozoa can be harvested by flushing or slice-and-dice techniques (Bruemmer, 2006). Aspiration of spermatozoa from the epididymis has also been performed. Interestingly, the first reported pregnancy in a mare after AI was achieved with frozen-thawed epididymal stallion spermatozoa in 1957 (Barker & Gandier, 1957).

Epididymal spermatozoa have been harvested from a variety of species like cats (Filliers et al., 2008; Hermansson & AxnÈr, 2007), dogs (Garcia-Macias et al., 2006; Hewitt et al., 2001; Nothling et al., 2007; Ponglowhapan et al., 2006), rats (Yamashiro et al., 2007), horses (Braun et al., 1994; Bruemmer, 2006; Heise et al., 2011; Johnson & Coutinho da Silva, 2008; Melo et al., 2008), cattle (Goovaerts et al., 2006; Martins, Rumpf, Pereira, & Dode, 2007), pigs (Ikeda et al., 2002), sheep (Garcia-Macias et al., 2006), goats (Blash, Melican, & Gavin, 2000), red deer (Fernandez-Santos et al., 2006; Garcia-Macias et al., 2006; MartÌnez-Pastor et al.,

2006), Spanish Ibex (Santiago-Moreno et al., 2006), African buffalo (Herold et al., 2006), North American buffalo (Lessard at al., 2009) and monkeys (Goff et al., 2009; Ng et al., 2002). Pregnancies and offspring after AI with epididymal spermatozoa have been produced amongst others in horses (Barker & Gandier, 1957; Heise et al., 2010; Morris et al., 2002; Papa et al., 2008), dogs (Hori, Hagiuda, Kawakami, & Tsutsui, 2005) and Spanish Ibex (Santiago-Moreno et al., 2006). Application of AI for epididymal spermatozoa holds tremendous potential for future use of valuable genetics not only in domestic but also especially in wild animal species.

5. Use of sexed semen for AI

Another aspect of artificial insemination in animals is the use of sex sorted spermatozoa. Separation of the X and Y bearing sperm is desirable in animals as one sex has significantly more value than the other in certain species. For dairy cattle for example, cows as the "milk producers" are the main source of income for the industry while bull calves are of less value as lactation is limited to females. For other food producing animals a higher percentage of male offspring might be beneficial as they grow faster and produce more meat. In many large, long-lived species like elephants, rhinoceroses, dolphins that are kept in captivity, sex selection could ease and avoid housing problems with males of these species (Durrant, 2009).

It has first been established for human spermatids in 1979 that there is a difference in DNA content between the mammalian X-chromosome-bearing spermatozoa and the Y-chromosome (Otto et al., 1979). Since then, DNA content measurements have been used to identify the sex-chromosome bearing sperm populations with good accuracy in semen from at least 23 mammalian species (Garner, 2006; Garner et al., 1983; Lu et al., 2010; Pinkel et al., 1982), and offspring have been produced from sexed sperm of at least seven species, including rabbits (Johnson et al., 1989), humans (Levinson et al., 1995), cattle (Cran et al., 1993), horses (Buchanan et al., 2000), sheep (Catt et al., 1996), dogs (Meyers et al., 2008), cats (Pope et al., 2008), elk (Schenk & DeGrofft, 2003), buffalo (Presicce et al., 2005) and dolphins (O'Brien & Robeck, 2006). The first offspring born with flow cytometrically sex sorted spermatozoa was in rabbits after surgical AI into the oviduct (Johnson et al., 1989).

The Beltsville Sperm Sexing technology (Garner, 2006) uses the difference in DNA content of the X- and Y-chromosome to sort the sex-determining gametes. The procedure is called fluorescence-activated cell separation (FACS) where ejaculated spermatozoa are treated with a DNA stain (called a flourochrome) and due to the fact that X-chromosomes contain more DNA, the stain take-up will be higher for the X-chromosome bearing spermatozoa than the Y-chromosome bearing spermatozoa. This difference in stain absorption is used in a flow cytometer chamber where the fluorescent stain in the spermatozoa is excited by a laser. Each live sperm produces an emission with an intensity that is directly related to the quantity of DNA within the sperm head. The X bearing spermatozoa emit more intense light than the Y bearing spermatozoa. A high speed computer is used to analyze the relative fluorescence of the X- and Y-sperm populations as they flow through a cytometer chamber. Spermatozoa are then assigned either a negative or positive charge depending on the DNA content while passing by charged plates. An electromagnetic field separates the X- and Y-chromosome bearing spermatozoa (Senger, 2003). This separation technology has a 85-95% success rate (Garner, 2006).

Not all mammalian sperm are equally suitable for sex sorting of spermatozoa. Apart from the DNA content difference of the X- and Y-bearing spermatozoa, the head shape of the spermatozoa plays a role as well. Flattened, oval shaped sperm heads (e.g. bull, boar, ram spermatozoa) are more readily oriented in a sperm sorter using hydrodynamics than those gametes with more round or angular head shapes (rodent spermatozoa) (Garner, 2006). The area of the flat profile of the sperm head can be multiplied times the difference in DNA content of the X-and Y-chromosome bearing sperm to give the sorting index. This index suggests that bull and boar sperm are well suited for separation in a flow sorter.

Initially, the use of sex sorted spermatozoa was limited due to the slow separation process where only a few hundred thousand sperm per hour could be sorted. Newer sperm sorter systems are able to sort 20,000 sperm/s resulting in up to 6000 or more sperm/s each of X- and Y-sperm at 90% accuracy (Garner & Seidel Jr, 2008). Due to low sperm numbers acquired with sex sorting, initial efforts to predetermine the sex required surgical insemination. Later, with improvement of the equipment, quantities were sufficient for in vitro fertilization (IVF). Today, sexed sperm are commercially available for cattle where the standard insemination doses of 2×10^6 sexed sperm achieve 70-80% of the pregnancy rates achieved with non-sorted sperm in doses of $10\text{-}20 \times 10^6$ (Bodmer et al., 2005; Garner, 2006). In pigs, low dose (70×10^6 spermatozoa) AI with flow cytometrically sorted sperm deep into the uterine horn resulted in pregnancy rates of 35-45.6% (Vazquez et al., 2003). In horses, hysteroscopic insemination into the uterine horn (Lindsey et al., 2002; Morris & Allen, 2002) and ultrasound guided deep uterine AI were performed using sex sorted spermatozoa in low concentrations (5×10^6 sperm cells/ dose).

6. Conclusion

AI has been and still is the most used reproductive technique in animals. A lot of research has been done over the last few decades, constantly improving techniques, methods and applications of AI. Routine AI procedures as well as specialized techniques like low-dose inseminations or use of sexed semen offer a wide variety for application of AI in domestic as well as wild and endangered animal species.

7. References

Adams, G. P., Ratto, M. H., Collins, C. W., & Bergfelt, D. R. (2009). Artificial insemination in South American camelids and wild equids. *Theriogenology, 71*(1), 166-175.

Amann, R. P., & Pickett, B. W. (1987). Principles of cryopreservation and a review of cryopreservation of stallion spermatozoa. *Journal of Equine Veterinary Science, 7*(3), 145-173.

Andersen, K. (1972). Fertility of frozen dog semen. *Acta Veterinaria Scandinavia, 13,* 128-130.

Andersen, K. (1975). Insemination with frozen dog semen based on a new insemination technique. *Zuchthygiene, 10,* 1-4.

Barker, C. A. V., & Gandier, J. C. C. (1957). Pregnancy in a mare resulted from frozen epididymal spermatozoa. *Canadian Journal of Comparative Veterinary Medical Science, 21,* 47-51.

Blash, S., Melican, D., & Gavin, W. (2000). Cryopreservation of epididymal sperm obtained at necropsy from goats. *Theriogenology, 54*(6), 899-905.

Bodmer, M., Janett, F., Hoessig, M., Daas, N. d., Reichert, P., & Thun, R. (2005). Fertility in heifers and cows after low dose insemination with sex-sorted and non-sorted sperm under field conditions. *Theriogenology, 64*(7), 1647-1655.

Brassley, P. (2007). Cutting across nature? The history of artificial insemination in pigs in the United Kingdom. *Studies in History and Philosophy of Science Part C: Studies in History and Philosophy of Biological and Biomedical Sciences, 38*(2), 442-461.

Braun, J., Sakai, M., Hochi, S., & Oguri, N. (1994). Preservation of ejaculated and epididymal stallion spermatozoa by cooling and freezing. *Theriogenology, 41*(4), 809-818.

Bruemmer, J. E. (2006). Collection and Freezing of Epididymal Stallion Sperm. *Veterinary Clinics of North America: Equine Practice, 22*(3), 677-682.

Buchanan, B. R., Seidel, G. E., McCue, P. M., Schenk, J. L., Herickhoff, L. A., & Squires, E. L. (2000). Inseminations of mares with low numbers of either unsexed or sexed spermatozoa. *Theriogenology, 53*, 1333-1344.

Catt, S. L., Catt, J. W., Gomez, M. C., Maxwell, W. M. C., & Evans, G. (1996). Birth of a male lamb derived from an in vitro matured oocyte fertilised by intracytoplasmic injection of a single presumptive male sperm. *Veterinary Record, 139*, 494-495.

Cran, D. G., Johnson , L. A., Miller, N. G. A., Cochrane, D., & Polge, C. (1993). Production of bovine calves following separation of X-chromosome and Y-chromosome bearing sperm and *in vitro* fertilization. *Veterinary Record, 132*, 40-41.

Durrant, B. S. (2009). The importance and potential of artificial insemination in CANDES (companion animals, non-domestic, endangered species). *Theriogenology, 71*(1), 113-122.

Evans, G., & Maxwell, W. M. C. (1987). *Salamon's artificial insemination of sheep and goats.* Sydney: Butterworths.

Fantinati, P., Zannoni, A., Bernardini, C., Webster, N., Lavitrano, M., Forni, M., et al. (2005). Laparoscopic insemination technique with low numbers of spermatozoa in superovulated prepuberal gilts for biotechnological application. *Theriogenology, 63*(3), 806-817.

Fernandez-Santos, M. R., Esteso, M. C., Montoro, V., Soler, A. J., & Garde, J. J. (2006). Cryopreservation of Iberian red deer (Cervus elaphus hispanicus) epididymal spermatozoa: Effects of egg yolk, glycerol and cooling rate. *Theriogenology, 66*(8), 1931-1942.

Filliers, M., Rijsselaere, T., Bossaert, P., De Causmaecker, V., Dewulf, J., Pope, C. E., et al. (2008). Computer-assisted sperm analysis of fresh epididymal cat spermatozoa and the impact of cool storage (4°C) on sperm quality. *Theriogenology, 70*(9), 1550-1559.

Foote, R. H. (2002). The history of artificial insemination: Selected notes and notables. *Journal of Animal Science, 80*, 1-10.

Garcia-Macias, V., Martinez-Pastor, F., Alvarez, M., Garde, J. J., Anel, E., Anel, L., et al. (2006). Assessment of chromatin status (SCSAÆ) in epididymal and ejaculated sperm in Iberian red deer, ram and domestic dog. *Theriogenology, 66*(8), 1921-1930.

Garner, D. L. (2006). Flow cytometric sexing of mammalian sperm. *Theriogenology, 65*(5), 943-957.

Garner, D. L., Gledhill, B. L., Pinkel, D., Lake, S., Stephenson, D., & Van Dilla, M. A. (1983). Quantification of the X- and Y-chromosome bearing spermatozoa of domestic animals by flow cytometry. *Biology of Reproduction, 28*, 312-321.

Garner, D. L., & Seidel Jr, G. E. (2008). History of commercializing sexed semen for cattle. *Theriogenology, 69*(7), 886-895.

Goff, K., Liukkonen, J., & Kubisch, H. M. (2009). Postmortem recovery and cryopreservation of spermatozoa from the vas deferens of rhesus macaques (Macaca mulatta). *Theriogenology, 72*(6), 834-840.

Goovaerts, I. G. F., Hoflack, G. G., Van Soom, A., Dewulf, J., Nichi, M., de Kruif, A., et al. (2006). Evaluation of epididymal semen quality using the Hamilton-Thorne analyser indicates variation between the two caudae epididymides of the same bull. *Theriogenology, 66*(2), 323-330.

Halbert, G. W., Dobson, H., Walton, J. S., & Buckrell, B. C. (1990). The structure of the cervical canal of the ewe. *Theriogenology, 33*, 977-992.

Heise, A., K‰hn, W., Volkmann, D. H., Thompson, P. N., & Gerber, D. (2010). Influence of seminal plasma on fertility of fresh and frozen-thawed stallion epididymal spermatozoa. *Animal Reproduction Science, 118*(1), 48-53.

Heise, A., Thompson, P. N., & Gerber, D. (2011). Influence of seminal plasma on fresh and post-thaw parameters of stallion epididymal spermatozoa. *Animal Reproduction Science, 123*(3-4), 192-201.

Hermansson, U., & AxnÈr, E. (2007). Epididymal and ejaculated cat spermatozoa are resistant to cold shock but egg yolk promotes sperm longevity during cold storage at 4°C. *Theriogenology, 67*(7), 1239-1248.

Herold, F. C., de Haas, K., Colenbrander, B., & Gerber, D. (2006). Comparison of equilibration times when freezing epididymal sperm from African buffalo (Syncerus caffer) using Triladyl(TM) or AndroMed®. *Theriogenology, 66*(5), 1123-1130.

Hewitt, D. A., Leahy, R., Sheldon, I. M., & England, G. C. W. (2001). Cryopreservation of epididymal dog sperm. *Animal Reproduction Science, 67*(1-2), 101-111.

Hori, T., Hagiuda, K., Kawakami, E., & Tsutsui, T. (2005). Unilateral intrauterine insemination with prostatic fluid-sensitized frozen caudal epididymal sperm in beagle dogs. *Theriogenology, 63*(6), 1573-1583.

Hunter, R. H. F. (1978). Intraperitoneal insemination, sperm transport and capacitation in the pig. *Animal Reproduction Science, 1*(2), 167-179.

Hunter, R. H. F. (2003). Advances in deep uterine insemination: a fruitful way forward to exploit new sperm technologies in cattle. *Animal Reproduction Science, 79*(3-4), 157-170.

Ikeda, H., Kikuchi, K., Noguchi, J., Takeda, H., Shimada, A., Mizokami, T., et al. (2002). Effect of preincubation of cryopreserved porcine epididymal sperm. *Theriogenology, 57*(4), 1309-1318.

Johnson, A. E. M., & Coutinho da Silva, M. A. (2008). Effects of recovery technique, freezing extender and antioxidants on motility parameters of cryopreserved stallion epididymal sperm. *Theriogenology, 70*(3), 579-580.

Johnson, L. A., Flook, J. P., & Hawk, H. W. (1989). Sex pre-selection in rabbits: live birth from X and Y sperm separated by DNA and cell sorting. *Biology of Reproduction, 41*, 199-203.

Johnston, S. D., Root Kustritz, M. V., & Olson, P. N. S. (2001). *Canine and Feline Theriogenology*: W. B. Saunders Company.

Katila, T. (2005). Effect of the inseminate and the site of insemination on the uterus and pregnancy rates of mares. *Animal Reproduction Science, 89*(1-4), 31-38.

Krueger, C., & Rath, D. (2000). Intrauterine insemination in sows with reduced sperm number. *Reproduction, Fertility and Development, 12*, 113-117.

Krueger, C., Rath, D., & Johnson, L. A. (1999). Low dose insemination in synchronized gilts. *Theriogenology, 52*, 1363-1373.

Lavara, R., Mocé, E., Lavara, F., de Castro, M. P. V., & Vicente, J. S. (2005). Do parameters of seminal quality correlate with the results of on-farm inseminations in rabbits? *Theriogenology, 64,* 1130-1141.

Leipold, S. D., Graham, J. K., Squires, E. L., McCue, P. M., Brinsko, S. P., & Vanderwall, D. K. (1998). Effect of spermatozoal concentration and number on fertility of frozen equine semen. *Theriogenology, 49*(8), 1537-1543.

Lessard, C., Danielson, J., Rajapaksha, K., Adams, G. P., & McCorkell, R. (2009). Banking North American buffalo semen. *Theriogenology, 71*(7), 1112-1119.

Levinson, G., Keyvanfar, K., Wu, J. C., Fugger, E. F., Fields, R. A., & Harton, G. L. (1995). DNA-based X-enriched sperm separation as an adjunct to preimplantation genetic testing for the prevention of X-linked disease. *Human Reproduction, 10,* 797-782.

Linde-Forsberg, C. (1991). Achieving canine pregnancy by using frozen or chilled extended semen. *Veterinary Clinics of North America, 21,* 467-485.

Linde-Forsberg, C., & Forsberg, M. (1989). Fertility in dogs in relation to semen quality and the time and site of insemination with fresh and frozen semen. *Journal of Reproduction and Fertility Supplement, 39,* 299-310.

Lindsey, A. C., Bruemmer, J. E., & Squires, E. L. (2001). Low dose insemination of mares using non-sorted and sex-sorted sperm. *Animal Reproduction Science, 68*(3-4), 279-289.

Lindsey, A. C., Schenk, J. L., Graham, J. K., Bruemmer, J. E., & Squires, E. L. (2002). Hysteroscopic insemination of low numbers of non sorted or flow-sorted spermatozoa. *Equine Veterinary Journal, 34,* 128-132.

Loomis, P. R. (2001). The equine frozen semen industry. *Animal Reproduction Science, 68*(3/4), 191-200.

Lopez-Gatius, F. (2000). Site of semen deposition in cattle: A review. *Theriogenology, 53*(7), 1407-1414.

Lu, Y., Zhang, M., Lu, S., Xu, D., Huang, W., Meng, B., et al. (2010). Sex-preselected buffalo (Bubalus bubalis) calves derived from artificial insemination with sexed sperm. *Animal Reproduction Science, 119*(3-4), 169-171.

Martinez, E. A., Vazquez, J. M., Roca, J., Cuello, C., Gil, M. A., Parrilla, I., et al. (2005). An update on reproductive technologies with potential short-term application in pig production. *Reproduction in Domestic Animals, 40,* 300-309.

MartÌnez-Pastor, F., Anel, L., Guerra, C., ¡lvarez, M., Soler, A. J., Garde, J. J. n., et al. (2006). Seminal plasma improves cryopreservation of Iberian red deer epididymal sperm. *Theriogenology, 66*(8), 1847-1856.

Martins, C. F., Rumpf, R., Pereira, D. C., & Dode, M. N. (2007). Cryopreservation of epididymal bovine spermatozoa from dead animals and its uses in vitro embryo production. *Animal Reproduction Science, 101*(3-4), 326-331.

McCue, P. M., Fleury, J. J., & Denniston, D. J. (2000). Oviductal insemination of mares. *Journal of Reproduction and Fertility Supplement, 56,* 499-502.

Melo, C. M., Papa, F. O., Fioratti, E. G., Villaverde, A. I. S. B., Avanzi, B. R., Monteiro, G., et al. (2008). Comparison of three different extenders for freezing epididymal stallion sperm. *Animal Reproduction Science, 107*(3-4), 331-331.

Metcalf, E. S. (2007). The efficient use of equine cryopreserved semen. *Theriogenology, 68*(3), 423-428.

Meyers, M. A., Burns, G., Arn, D., & Schenk, J. L. (2008). Birth of canine offspring following insemination of a bitch with flow-sorted spermatozoa. [Abstract]. *Reproduction, Fertility and Development, 20*, 213.

Morrell, J. M. (1995). Artificial insemination in rabbits. *British Veterinary Journal, 151*(5), 477-488.

Morris, L. H., & Allen, W. R. (2002). An overview of low dose insemination in the mare. *Reproduction in Domestic Animals, 37*, 206-210.

Morris, L. H. A. (2004). Low dose insemination in the mare: an update. *Animal Reproduction Science, 82-83*, 625-632.

Morris, L. H. A., Tiplady, C., & Allen, W. R. (2002). The in vivo fertility of cauda epididymal spermatozoa in the horse. *Theriogenology, 58*, 643-646.

Ng, S. C., Martelli, P., Liow, S. L., Herbert, S., & Oh, S. H. (2002). Intracytoplasmic injection of frozen-thawed epididymal spermatozoa in a nonhuman primate model, the cynomolgus monkey (Macaca fascicularis). *Theriogenology, 58*(7), 1385-1397.

Noakes, D. E., Parkinson, T. J., & England, G. C. W. (2001). *Veterinary reproduction and obstetrics* (8th ed.): W. B. Saunders.

Nothling, J. O., Gerber, D., Colenbrander, B., Dijkstra, M., Bakker, T., & De Cramer, K. (2007). The effect of homologous prostatic fluid on motility and morphology of dog epididymal spermatozoa extended and frozen in Biladyl with Equex STM paste or Andromed. *Theriogenology, 67*(2), 264-275.

O'Brien, J. K., & Robeck, T. R. (2006). Development of sperm sexing and associated reproductive technology for sex preselection of captive bottlenose dolphins (Tursiops truncatus). *Reproduction, Fertility and Development, 18*, 319-329.

Otto, F. J., Hacker, U., Zante, J., Schumann, J., Göhde, W., & Meistrich, M. L. (1979). Flow cytometry of human sperm. *Histochemistry, 62*, 249-254.

Papa, F. O., Melo, C. M., Fioratti, E. G., Dell'Aqua Jr, J. A., Zahn, F. S., & Alvarenga, M. A. (2008). Freezing of stallion epididymal sperm. *Animal Reproduction Science, 107*(3-4), 293-301.

Pickett, B. W., & Voss, J. L. (1975). The effect of semen extenders and sperm numbers on mare fertility. *Journal of Reproduction and Fertility Supplement, 23*, 95-98.

Pinkel, D., Lake, S., Gledhill, B. L., Van Dilla, M. A., Stephenson, D., & Watchmaker, G. (1982). High resolution DNA content measurements of mammalian sperm. *Cytometry, 3*, 1-9.

Ponglowhapan, S., Chatdarong, K., Sirivaidyapong, S., & Lohachit, C. (2006). Freezing of epididymal spermatozoa from dogs after cool storage for 2 or 4 days. *Theriogenology, 66*(6-7), 1633-1636.

Pope, C. E., Crichton, E. B., Gomez, M. C., Dumas, C., & Dresser, B. (2008). Birth of domestic cat kittens of predetermined sex after transfer of embryos produced by in vitro fertilization of oocytes with flow-sorted spermatozoa. [Abstract]. *Reproduction, Fertility and Development, 20*, 213-214.

Presicce, G. A., Rath, D., Klinc, P., Senatore, E. M., & Pascale, M. (2005). Buffalo calves born following AI with sexed semen. [Abstract]. *Reproduction in Domestic Animals, 40*, 349.

Roca, J., Vaszques, J. M., Gil, M. A., Cuello, C., Parrilla, I., & Martinez, E. A. (2006). Challenges in Pig Artificial Insemination. *Reproduction in Domestic Animals, 41*(2), 43-53.

Santiago-Moreno, J. n., Toledano-DÌaz, A., Pulido-Pastor, A., Dorado, J. s., GÛmez-Brunet, A., & LÛpez-Sebasti n, A. (2006). Effect of egg yolk concentration on cryopreserving

Spanish ibex (Capra pyrenaica) epididymal spermatozoa. *Theriogenology, 66*(5), 1219-1226.

Santiago-Moreno, J. n., Toledano-DÌaz, A., Pulido-Pastor, A., GÛmez-Brunet, A., & LÛpez-Sebasti n, A. (2006). Birth of live Spanish ibex (Capra pyrenaica hispanica) derived from artificial insemination with epididymal spermatozoa retrieved after death. *Theriogenology, 66*(2), 283-291.

Schenk, J. L., & DeGrofft, D. L. (2003). Insemination of cow elk with sexed frozen semen. [Abstract]. *Theriogenology, 59*, 514.

Seager, S. W. J. (1969). Successful pregnancies utilizing frozen dog semen. *AI Digest, 17*, 6-16.

Senger, P. L. (2003). *Pathways to Pregnancy and Parturition* (Second ed.): Current Conceptions,Inc.

Sieme, H., Sch‰fer, T., Stout, T. A. E., Klug, E., & Waberski, D. (2003). The effects of different insemination regimes on fertility in mares. *Theriogenology, 60*(6), 1153-1164.

Singleton, W. L. (2001). State of the art in artificial insemination of pigs in the United States. *Theriogenology, 56*(8), 1305-1310.

Soede, N. M. (1993). Boar stimuli around insemination affect reproductive processes in pigs: A review. *Animal Reproduction Science, 32*(1-2), 107-125.

Thomassen, R., Farstad, W., Krogenaes, A., Fougner, J. A., & Andersen Berg, K. (2001). Artificial insemination with frozen semen in dogs: a retrospective study. *Journal of Reproduction and Fertility Supplement, 57*, 341-346.

Thomassen, R., Sanson, G., Krogenes, A., Fougner, J. A., Berg, K. A., & Farstad, W. (2006). Artificial insemination with frozen semen in dogs: A retrospective study of 10 years using a non-surgical approach. *Theriogenology, 66*(6-7), 1645-1650.

Vazquez, J. M., Martinez, E. A., Parrilla, I., Roca, J., Gil, M. A., & Vazquez, J. L. (2003). Birth of piglets after deep intrauterine insemination with flow cytometrically sorted boar spermatozoa. *Theriogenology, 59*(7), 1605-1614.

Vazquez, J. M., Roca, J., Gil, M. A., Cuello, C., Parrilla, I., Vazquez, J. L., et al. (2008). New developments in low-dose insemination technology. *Theriogenology, 70*(8), 1216-1224.

Vidament, M. (2005). French field results (1985-2005) on factors affecting fertility of frozen stallion semen. *Animal Reproduction Science, 89*(1-4), 115-136.

Vidament, M., Dupere, A. M., Julienne, P., Evain, A., Noue, P., & Palmer, E. (1997). Equine frozen semen: Freezability and fertility field results. *Theriogenology, 48*(6), 907-917.

Wolf, D. P. (2009). Artificial insemination and the assisted reproductive technologies in non-human primates. *Theriogenology, 71*, 123-129.

Wongtawan, T., Saravia, F., Wallgren, M., Caballero, I., & RodrÌguez-MartÌnez, H. (2006). Fertility after deep intra-uterine artificial insemination of concentrated low-volume boar semen doses. *Theriogenology, 65*(4), 773-787.

Yamashiro, H., Han, Y.-J., Sugawara, A., Tomioka, I., Hoshino, Y., & Sato, E. (2007). Freezability of rat epididymal sperm induced by raffinose in modified Krebs-Ringer bicarbonate (mKRB) based extender solution. *Cryobiology, 55*(3), 285-294.

Youngquist, R. S., & Threlfall, W. R. (2007). *Current Therapy in Large Animal Theriogenology* (2nd ed.): Saunders Elsevier.

3

Steroid Hormones in Food Producing Animals: Regulatory Situation in Europe

Annamaria Passantino
Department of Veterinary Public Health,
Faculty of Veterinary Medicine,
University of Messina,
Italy

1. Introduction

Hormones are chemicals produced by animals to co-ordinate their physiological activities. They act as messengers, produced in and released from one kind of tissue to gradually stimulate or inhibit some process in a different tissue over a long period.

Steroid hormones fulfill an important role at different stages of mammalian development comprising prenatal development, growth, reproduction and sexual and social behavior.

The importance of individual hormones varies between sexes and age and a disruption of the endocrine equilibrium may result in multiple biological effects.

One hormone can have multiple actions, e.g. the male hormone testosterone controls many processes from the development of the foetus, to libido in the adult. Alternatively, one function may be controlled by multiple hormones, e.g. the menstrual cycle involves oestradiol, progesterone, follicle-stimulating hormone and luteinising hormone.

Hormones produced by the bodies of humans and animals are called endogenous or natural hormones. Compounds chemically synthesised to mimic the effect of natural hormones are called synthetic or xenobiotic hormones.

Hormones are vital in normal development, maturation and physiological functioning of many vital organs and processes in the body. However, like any other chemicals of natural or synthetic origin, hormones may be toxic to living organisms under certain circumstances. The toxicity may be due to an excess of its normal ('physiological') action. This may be the result of excessive exposure to the substance, for example following absorption of a large dose, or because the physicochemical nature of the substance gives it greater or more prolonged activity of the same type, or because the hormonal action occurs at an abnormal time during development or adult life, or is an action on an organism of the inappropriate sex. Hormones, like other chemicals, may also exert direct toxic actions not related to their endocrine ('physiological') effects.

Due to the obvious ability to improve weight gain and feed efficiency in meat producing animals, natural hormones and/or the synthetic surrogates have been used in agricultural practice for several decades (table 1).

Substances	Form	Main use - Animals
Oestrogens alone:		
DES	Feed additive	Steers, heifers
DES	Implant	Steers
DES	Oil solution	Veal calves
Hexoestrol	Implant	Steers, sheep, calves, poultry
Zeranol	Implant	Steers, sheep
Gestagens alone:		
Melengestrol acetate		Heifers
Androgens alone:		
TBA	Implant	Heifers, culled cows
Combined preparations:		
DES and Testosterone	Implant	Calves
DES and Methyl-testosterone	Feed additive	Swine
Hexoestrol and TBA	Implant	Steers
Zeranol and TBA	Implant	Steers
Oestradiol-17β and TBA	Implant	Bulls, steers, calves, sheep
Oestradiol-17β benzoate and testosterone propionate	Implant	Heifers, calves
Oestradiol-17β benzoate and progesterone	Implant	Steers

Table 1. Hormonally-active substance used in animal production (by http://www.fao.org/DOCREP/004/X6533E/X6533E01.htm, modified)

Implanting hormonal growth promoters is currently widespread in the beef cattle industry of many non-EU countries for the better performance in growth and improvement of feed efficiency. These hormonal implants may enhance growth during suckling, growing and finishing stages of production (Mader, 1997; Platter *et al.,* 2003).

Growth hormones are implanted under the skin (usually behind the ear) of the animal in the form of depot capsules, where they release a specific dose of hormones over a fixed period of time.

The five hormone types most widely used in meat production include three natural hormones, oestradiol 17-β, testosterone, and progesterone, and two synthetic substances, trenbolone and zeranol.

Oestradiol 17-β has oestrogenic action (i.e. responsible for female characteristics); testosterone has androgenic action (i.e. responsible for male characteristics); and progesterone has gestagenic action (i.e. responsible for maintaining pregnancy). The other two hormones, as aforesaid, mimic the biological activity of the natural hormones: trenbolone mimics the action of testosterone, and zeranol mimics oestradiol 17-β.

1.1 Hormonally active substances

1.1.1 Oestradiol 17-β

Oestradiol 17-β is the most active of the female sex hormones synthesized and secreted mainly by the ovary, the adrenals and the testis.

Oestradiol is synthesized and secreted in early stages of embryogenesis and has an active role in the normal development of the female sex accessories during the lifetime of females.

It has been used to induce parturition (birth) especially in sheep, a species in which an associated oestradiol-induced increase in mothering ability has also been recorded (Poindron, 2005).

In non-pregnant animals, oestradiol has been used clinically to increase uterine contractions and cervical softening for the expulsion of unwanted uterine contents in the absence of a corpus luteum (i.e. to remove a dead fetus or infected material especially in cattle) (Elmore, 1992; Pepper &, Dobson, 1987; Sheldon & Noakes, 1998).

Oestradiol has been used in the past in turkeys and other poultry to castrate young birds. Implants would be placed subcutaneously at 5-6 weeks of age, or in slightly older birds, but certainly 4 weeks before killing. Alternatively, preparations were available as feed-additives. This approach is not now used in Europe although it was used in slower growing Spanish breeds. There are very few reproductive problems in rabbits.

Fetal mummifications and macerations occur, as do endometritis and pyometra but treatment with oestradiol has not been reported (Flecknell, 2000).

In fish, administration of oestradiol at first-fry feeding (approximately 40-70 days of age depending on species) will induce ovarian development and female characteristics in salmonids, flat-fish and eels (Shepherd & Bromage, 1988). Sex-control in this way depresses or inhibits maturation to ensure that metabolism is channelled into body growth (i.e., more saleable flesh). However, this approach is no longer commercially adopted.

Another use of very low doses of oestradiol is as a growth promoter via appetite stimulating and increased food-conversion properties. Occasionally in the past, this approach has been taken to advance the onset of puberty and thus alleviate potential gynaecological problems in slower maturing species. However, as use of hormonal growth promoters is prohibited within the European Union[1].

1.1.2 Testosterone

Testosterone and its more active metabolite, 5α-dihydrotestosterone (DHT), are the main sex hormones secreted by males. Testosterone is responsible for the early development, and the appearance and maintenance of male secondary sex accessory organs (prostate, secretory glands, penis size, etc.) during adulthood. Testosterone secretion is also affected by the complex interaction among all endocrine glands, especially with those in the brain.

Testosterone is metabolized and as a result, metabolites of different activity are generated. Some of these metabolites play a more active role in certain organs than in others.

The actions of both testosterone and DHT are mediated through their high affinity and high specificity binding and activation of an intracellular protein, the androgen receptor (AR). This AR protein is a member of the steroid hormone superfamily. The ligand-activated androgen receptor mediates its effects on cell growth and differentiation through the

[1] See the following paragraph: Background of the European Union legislation

activation and/or suppression of specific gene transcription in target organs. Androgen receptors are detected in tissues of females, as well as males. The presence of this receptor in organs such as the ovary indicates significant activity of androgens in both sexes. Furthermore, the androgen receptor is thought to be involved in ovarian tumorigenesis, as it has been detected in 67 percent of ovarian tumors. Although current information indicates the presence of only a single androgen receptor, it is known that different subsets of genes may be activated by either testosterone or DHT (Chang *et al.*, 1995).

In animals, testosterone or testosterone propionate, alone or in combination with other hormonally active substances, is used primarily to improve the rate of weight gain and feed efficiency. This effect is most likely a consequence of the anabolic action of androgens.

1.1.3 Progesterone

Progesterone is synthesized and secreted mainly by the corpus luteum in the ovary of cycling females, and, during pregnancy, by the placenta.

As all hormones, progesterone synthesis and secretion is regulated by a series of positive and negative feedback mechanisms in which polypeptidic hormones secreted by the brain (hypothalamus, pituitary) affect circulating progesterone levels.

Progesterone and synthetic progestins are used pharmacologically in women in conjunction with ovulation stimulation drugs as well as during early pregnancy in cases of luteal phase dysfunction. Although results have been conflicting, some studies find an association between pregnancy-related intake of progestins and increased risk of hypospadias (congenital malformation of the urethral opening on the penis) in the male offspring (Carmichael *et al.*, 2005). It should however be noted that this was observed in relation to pharmacological doses of progestins, and as progesterone levels are normally high during pregnancy, minor additional exogenous progestagenic activity would presumably be without significant effects in the presence of a high endogenous activity, unless the synthetic progestins act at different sites and by different mechanisms. In contrast, serum levels of progestins in children and postmenopausal women are very low. Data on effects of progesterone in the prepubertal child are scarce and no new data have been identified. Likewise, no animal studies on the effects of progesterone during the postnatal development have been published recently.

It is well established that progesterone not only serves as the precursor of all the major steroid hormones (androgens, oestrogens, corticosteroids) in the gonads and adrenals, but also is converted into one or more metabolites by most tissues in the body (Wiebe, 2006).

1.1.4 Trenbolone acetate

Trenbolone acetate (TBA) is a synthetic steroid with an anabolic potency that may exceed that of testosterone. It is a prodrug that converts into its active form 17β-trenbolone, which isomerises into 17α-trenbolone.

17β-trenbolone is the major form occurring in muscle tissue, whereas the 17α-epimer is the major metabolite occurring in liver and in the excreta including bile. It is assumed to exert its anabolic action via interaction with androgen and glucocorticoid receptors (Danhaive and Rousseau, 1986, 1988). Experiments with cattle tissues have shown that 17β-trenbolone binds to the androgen receptor with similar affinity as dihydrotestosterone. It also binds to the progesterone receptor with an affinity that exceeds that of progesterone. The other metabolites of TBA, including 17α-trenbolone (17α-hydroxy-estra-4,9,11-trien-3-one) and TBO (estra-4,9,11-triene-3,17-dione) show a significantly lower binding affinity to both types of receptors (Bauer *et al.*, 2000).

Reports regarding the (mis)use of TBA as an anabolic agent in sports people describe several adverse effects, including liver cell injury with an increase in liver-specific enzymes in serum, cholestatic jaundice, peliosis hepatitis and various neoplastic lesions. Moreover, decreased endogenous testosterone production and spermatogenesis, oligospermia and testicular atrophy may be associated with the repeated use of TBA as anabolic (Bahrke and Yesalis, 2004; Maravelias *et al.*, 2005).

1.1.5 Zeranol

Zeranol is derived from the naturally occurring mycoestrogen zearalenone, and is a potent oestrogen receptor agonist *in vivo* and *in vitro* (Leffers *et al.*, 2001; Le Guevel and Pakdel, 2001; Takemura *et al.*, 2007; Yuri *et al.*, 2006). Its actions resemble those of oestradiol. (Leffers *et al.*, 2001).

Zeranol stimulates the proliferation of ER-dependent cell proliferation in MCF-7 human breast cancer cells (which are widely used in the assessment of estrogenic activity) and in transfected cells (Leffers *et al.*, 2001; Le Guevel and Pakdel, 2001; Liu and Ling, 2004).

It is used alone or in combination with TBA as a hormonal growth promoter in various products.

2. Background of the European Union legislation

The use of hormonal growth promoters in food-producing animals has been a sensitive issue of debate in the EU and elsewhere for several decades.

Prior to 1981, the EC had no universal policy on the use of growth promoting hormones in meat animals.

The use of hormones had been banned in Italy since 1961, in Denmark since 1963, and in Germany since 1977. Belgium and Greece had never permitted the use of hormones for fattening purposes. However, Spain, the United Kingdom, France and Netherlands permitted the use of most hormones for speeding growth in beef cattle[2].

The move to impose a Europe wide ban was spurred by the worrying discovery in 1977 of breast enlargement in girls and boys attending a school in Milan (Italy) (Scaglioni *et al.*, 1978). Although oestrogen contamination was not detected when samples of school meals were tested, an uncontrolled supply of poultry and beef was hypothesized as being the cause of this outbreak (Fara *et al.*, 1979).

---

[2] See http://www.fao.org/DOCREP/004/X6533E/X6533E03.htm#refeecstr Accessed August 19, 2011.

In 1980 the discovery of 30,000 jars of baby food containing dethylstilboestrol, commonly known as DES, contaminated French veal was reported. European consumer organizations called for a boycott of veal, and the market for veal was severely affected. On September 20, 1980, the EC Council of Agriculture Ministers adopted a declaration in favor of a ban on the use of oestrogen and endorsed the principle of greater harmonization of legislation on veterinary medicines and of greater control on animal rearing, both at the production and slaughtering stages.

On October 31, 1980, the EC Commission proposed even more rigorous legislation that would ban the use of all hormone products in meat production, except for therapeutic purposes[3]. This proposal was expanded later by documents COM(80)920 and COM(80)922, presented on 6 January 1981. These allowed for the controlled use for therapeutic and zootechnical purposes of three natural hormone products, and introduced a number of control measures on the production and handling of such products, together with proposals on the testing of animals. Discussions in the European Parliament revealed that three EC member States (Belgium, Ireland and the United Kingdom) favored the use of some hormones to promote growth in meat animals, and Ireland and the United Kingdom also argued for the retention of the synthetic hormones, trenbolone and zeranol. Third countries, including The United States, Argentina, Australia, Canada, New Zealand and South Africa raised concerns concerning the potential impact of a ban on their exports to European Communities[4].

Subsequently, the European Council adopted its first directive on the hormones issues in July 1981 (Directive 81/602/EEC)[5].

Directive 81/602/EEC prohibits the administering to farm animals of substances having a *thyrostatic action* or substances having an *oestrogenic, androgenic or gestagenic* action; the placing on the market or slaughtering of farm animals to which these substances have been administered; the placing on the market of meat from such animals; the processing of meat from such animals and the placing on the market of meat products prepared from or with such meat. The Directive provides two exceptions to the prohibition: one exception is provided for substances with an oestrogenic, androgenic or gestagenic action when they are used for therapeutic or zootechnical purposes and administered by a veterinarian or under a veterinarian's responsibility. The other exception was for oestradiol-17β, progesterone, testosterone, TBA and zeranol - when they were used for growth promotion purposes and their use was governed according to the individual regulatory schemes maintained by EC member States. This exception was made pending an examination of the effects of these hormones on the health of consumers and the adoption of an EC rule. EC member States are obliged to apply their regulatory schemes to imports from third countries in a manner not more favorable than that applied to intra-EC trade.

[3] Commission of the European Communities 1980. Proposal for a Council Regulation (EEC) Concerning the Uses of Substances with a Hormonal Action and those having a Thyrostatic Action in Domestic Animals. COM (80) 614, 31 October, Brussels: Commission of the European Communities.

[4] WTO, 1997. EC measures concerning meat and meat products (hormones), complaint by the United States. Report of the WTO Panel, WT/DS26/R/USA, August 18, 1997, 2: 26-28. Available at http://www.sice.oas.org/dispute/wto/horm-us.asp, Accessed August 20, 2011.

[5] Council Directive 81/602/EEC of 31 July 1981 concerning the prohibition of certain substances having a hormonal action and of any substances having a thyrostatic action. Official Journal of the European Communities, L Series, No. 222, pp. 32-33.

The relevant international organizations - FAO, WHO, OIE and Codex - started to seriously examine the safety of these hormones in meat production only during the 1980s.

The first substantive scientific report had been published by OIE in 1983. The Joint FAO/WHO Expert Committee on Food Additives (JECFA)[6] had discussed and issued a scientific report on these hormones only in 1988[7].

There were two other international reports comprising collective scientific work: the 1984 Scientific Report published by the European Commission (based on the Lamming Report[8]) and the Proceedings of the 1995 EC Scientific Conference[9].

---

[6] It is an independent expert group which deals with specific commodity issues or general health and safety matters related to food. The JECFA focuses on the scientific evaluation of a veterinary drug and does not consider government policies and politics.

[7] The JECFA Report, on which the Codex standard for zeranol is based, noted that zeranol was a weak oestrogen which mimicked the action of oestradiol-17β. The report concluded that the toxic (*in casu* tumorigenic) effect of zeranol is associated with its hormonal (i.e. oestrogenic) properties and that an ADI could thus be established on the basis of a no-hormonal-effect level. Adopting what it considered to be a conservative approach by using as a basis studies on ovariectomized female cynomolgus monkeys (highly sensitive to oestrogenic substances) and using a safety factor of 100, JECFA set an ADI for human beings of 0-0.5 µg/kg of body weight. For a 70 kg person consuming 500 g of meat daily over an entire lifetime, the maximum permissible or safe level of zeranol residues in meat would then, according to JECFA, be 70 µg/kg of edible tissue. However, the report noted that when zeranol is administered to cattle according to good animal husbandry practice, the maximum mean residue levels did not exceed 0.2 µg/kg in muscle, 10 µg/kg in liver, 2 µg/kg in kindney, and 0.3 µg/kg in fat at any time after implantation. These residue levels obtained on the basis of good animal husbandry practice are thus below the maximum permissible level of 70 µg/kg. However, in order to set a level which is detectable by routine residue analysis methods, the Codex MRL was increased to 2 µg/kg in muscle and set at 10 µg/kg in liver.
With respect to trenbolone acetate (TBA), the Report concluded that its potential toxic effects only arise as a consequence of its hormonal activity. The report further concluded that, therefore, an ADI could be established on the basis of a no-hormonal-effect level. Adopting what it considered to be a conservative approach by using as a basis studies on castrated male rhesus macaque monkeys (which are highly sensitive to compounds with antigonadotropic activity) and pigs (which are a sensitive model for assessing hormonal effects of TBA) and using a safety factor of 100, JECFA later set an ADI for human beings of 0-0.02 µg/kg of body weight (34th JECFA Report of 1989). The maximum ADI for a 60 kg person would thus be 1.2 µg of TBA residues. JECFA then set MRL's for -trenbolone in muscle and -trenbolone in liver of 2 µg/kg and 10 µg/kg, respectively, based on average residue levels in heifers at 15-30 days after implantation of 300 mg TBA, noting that concentrations would be even lower at proposed GPVD. According to JECFA, the MRL's thus obtained on the basis of conservative estimates should not exceed the Codex ADI or safe level at any time after implantation of the drug, that is, irrespective of the withdrawal period used.

[8] The Lamming Group's interim report, issued in September 1982, found that the three natural hormones (oestradiol-17β, testosterone and progesterone) "would not present any harmful effects to the health of the consumer when used under the appropriate conditions as growth promoters in farm animals".
For the findings of the scientific working group, see Lamming, G.E., Ballarini, G., Baulieu, E.E. *et al.*, (1987). Scientific Report on Anabolic Agents in Animal Production. *The Veterinary Record*, 121, at 389-392.

[9] The 1995 EC Scientific Conference on Growth Promotion in Meat Production concluded that: "At present, there is no evidence for possible health risks to the consumer due to the use of natural sex hormones for growth promotion, since: Residue levels of these substances measured in meat of treated animals fall within the physiological range observed in meat of comparable untreated animals. The daily production of sex hormones by humans is much higher than the amounts possibly consumed from meat, even in the most sensitive humans (prepubertal children and menopausal women).

The European Communities stressed, however, that these reports did not constitute the entire body of scientific knowledge on the issue of safe use of these hormones for growth promotion. There were also important studies made by individual scientists, and other specialized institutions like the International Agency for Research on Cancer (IARC).

The EC Scientific Veterinary Committee gave its reaction to the Lamming Report on November 9, 1982, followed by the EC Scientific Committee for Animal Nutrition on November 17, 1982 and by the EC Scientific Committee for Food on February 4, 1983. These Committees supported the conclusions and recommendations of the Lamming Report, but stressed the need to lay down provisions regarding the establishment of proper programmes to control and monitor the use of anabolic agents with regard, in particular, to instructions for use, surveillance programmes and analysis methods. In January 1984, the Commission asked a group of experts within the EC Scientific Committee on Anabolic Agents to review the information on trenbolone and zeranol. On June 12, 1984, the Commission published a proposal (COM(84)295 final) for a Council Directive amending Directive 81/602/EEC, which envisaged the controlled use of the three natural hormones for growth promotion purposes and proposed re-examining the ban on the two synthetic hormones after their scientific evaluation had been completed. However, the European Parliament, the EC Economic and Social Committee and the EC Council of Ministers rejected the Commission's proposal.

The EC Commission amended its proposal accordingly and on December 31, 1985 the EC Council adopted Directive 85/649/EEC[10]. This Directive banned the use of all the substances concerned for growth promotion purposes and established more detailed provisions concerning authorized therapeutic uses. Its preamble began by emphasizing that differing rules on hormone use in different member countries had distorted trade in the European market and that "[…] these distortions of competition and barriers to trade must therefore be removed […]".

The Directive was challenged in the European Court of Justice, which annulled it on procedural grounds. The proposals were re-introduced by the EC Commission and re-adopted by the EC Council as Council Directive 88/146/EEC on March 7, 1988[11]. This Directive extends the prohibition imposed by Directive 81/602/EEC to the administration to farm animals of trenbolone acetate and zeranol for any purpose, and oestradiol-17β,

Due to an extensive first-pass metabolism, the bioavailability of ingested hormones is low, thus providing a further safety margin".

With regard to the synthetic hormones, zeranol and trenbolone, the 1995 EC Scientific Conference concluded that: "At the doses needed for growth promotion, residue levels [of trenbolone and zeranol] are well below the levels regarded as safe (the MRLs). There are, at present, no indications of a possible human health risk from the low levels of covalently-bound residues of trenbolone".

See Assessment of Health Risk - Working Group II", in 1995 EC Scientific Conference Proceedings, pp. 20-21.

[10] Council Directive 85/649/EEC of 31 December 1985 prohibiting the use in livestock faring of certain substances having hormonal action. *Official Journal of the European Communities*, L Series, No. 382, pp. 228-231.

[11] Council Directive 88/146/EEC of 7 March 1988 prohibiting the use in livestock farming of certain substances having a hormonal action. *Official Journal of the European Union*, L Series, No. 70, pp. 16-18.

testosterone and progesterone for fattening purposes. However, the Directive maintains the permission to administer these three natural hormones to animals for therapeutic and zootechnical purposes under prescribed conditions; in particular, therapeutic treatment is defined to mean the administering to an individual animal of any of the substances which are authorized to treat a fertility problem diagnosed on examination by a veterinarian. The products which are used for therapeutic treatment may be administered only by a veterinarian, in the form of an injection (to the exclusion of implantation) to farm animals which have been clearly identified. Such treatment must be registered by the veterinarian and these animals may not be slaughtered before expiry of the period fixed. In the case of animals at the end of their reproductive career, the treatments are prohibited from being administered during the fattening period following the end of their breeding life. Article 4 of Directive 88/146/EEC explicitly requires that undertakings in the EC member States producing the prohibited hormones, those companies authorized to market these hormones for whatever purposes and undertakings producing pharmaceutical and veterinary products based on those substances, must keep a detailed register recording (in chronological order) the quantities produced or acquired and those sold or used for the production of pharmaceutical and veterinary products. The importation from third countries of animals and meat from animals to which have been administered substances with thyrostatic, oestrogenic, androgenic or gestagenic action is prohibited. However, under certain conditions, Article 7 of Directive 88/146/EEC allows trade in those animals and meat from those animals treated for therapeutic or zootechnical purposes, including imports from third countries.

Directive 88/299/EEC[12] lays down the conditions for applying the derogations, provided for in Article 7 of Directive 88/146/EEC, from the prohibition on trade in certain categories of animals and their meat. The first derogation of the Directive requires EC member States to authorize trade in animals intended for reproduction and reproductive animals at the end of their career (and of meat of such animals) which, during their reproductive career, have undergone one of two categories of treatments. The first category is therapeutic treatment with one of the following substances: oestradiol-17β, testosterone and progesterone; and those derivatives which readily yield the parent compound on hydrolysis after absorption at the site of application which appear in a list of approved products. The second category is the administration of substances having an oestrogenic, androgenic or gestagenic action for synchronization of oestrus, termination of unwanted gestation, the improvement of fertility and the preparation of donors and recipients for the implantation of embryos, provided that the products in which they are contained appear on a list of approved products and with the respect of strict conditions of use concerning, in particular, the respect of the withdrawal period, the monitoring of those conditions of use and of the means of identification of the animals. In addition, Articles 3 and 4 of this Directive provide that trade between the EC member States of the European Communities in animals intended for reproduction and reproductive animals and meat from such animals is allowed only if all the conditions laid down in the Directive are respected, in particular as regards the waiting period and the requirement that animals have not received any of the above treatments with any of the

[12] Council Directive 88/299/EEC of 17 May 1988 on trade in animals treated with certain substances having a hormonal action and their meat, as referred to in Article 7 of Directive 88/146/EEC. *Official Journal of the European Union*, Series L, No. 87, pp. 36-38.

above substances during the fattening period following the end of their breeding life. The EC stamp may be affixed to the meat only if the waiting time ended before the animals are slaughtered. The second derogation in Directive 88/299/EEC allows imports from third countries of treated animals and meat of such animals under guarantees equivalent to those for domestic animals and meat.

Following reports of significant use of illegal growth-promoting hormonal substances in a number of EC member States, on September 26, 1988 the European Parliament established a Committee of Enquiry into the Problem of Quality in the Meat Sector. The conclusions of this Committee were published in a document known as the "Pimenta Report", which recognized the ban on the use of hormones[13]. On March 29, 1989, the European Parliament adopted its recommendations to maintain and expand the ban.

The European Parliament adopted another report on the issue of use of hormones for animal growth promotion, the "Collins Report" of February 7, 1989[14]. This report argued that: "Current licensing systems for the regulation of veterinary medicines (including at present, growth promoting products) require that a new product satisfy three criteria: safety, quality and efficacy. These criteria may well be satisfactory for therapeutic drugs. They are by no means sufficient for growth promoting products. For the latter it is proposed here that the Community's veterinary medicine licensing system be adapted to include a "fourth hurdle", entailing an objective socio-economic and environmental impact assessment". In the Commission's July 1988 draft proposals for the reform of veterinary medicine licensing in the Community this idea was accepted in principle. The final version of the proposals (December 1988) does not include this concept. It is clear, however, that the social, agricultural and environmental implications of the use of growth and yield promoting pharmaceuticals require a licensing system somewhat different from that which exists for these products when used for therapeutic purposes.

Directive 96/22/EC[15] replaces Directives 81/602/EEC, 88/146/EEC and 88/299/EEC. It maintains the prohibition on the use of these hormones for growth promotion purposes; extends the prohibition on the use of beta-agonists; restricts the use of the hormones at issue for therapeutic or zootechnical purposes, reinforcing in particular the role of the

---

[13] The scientific conclusions regarding the use of natural hormones rested upon strict conditions of use which it believed could not in reality be attained. The Committee was of the opinion that use of the natural/nature-identical hormones carries the risk of inexperienced application, incorrect dosage and unsupervised injection which could pose a risk to the animal and the consumer, and also noted doubts with regard to long-term cumulative and interactive potential carcinogenicity. In addition, the Committee believed that proven necessity and socio-economic desirability should be criteria of acceptability for the use of (bio)chemical growth promoters in animal-rearing. In brief, the essential findings of the Pimenta Report were that the prohibition of hormonal substances for non-therapeutic (i.e. growth-promoting) purposes must be maintained and expanded.

[14] European Parliament, Committee on the Environment, Public Health and Consumer Protection, Report on "The USA's Refusal to comply with Community legislation on slaughterhouses and hormones and the consequences of this refusal", EP 128 381/B, 7 February 1989, named after its reporter Mr. Collins, MEP.

15 Council Directive 96/22/EC of 29 April 1996 concerning the prohibition on the use in stockfarming of certain substances having a hormonal or thyrostatic action and of beta-agonists, and repealing Directives 81/609/EEC, 88/146/ECC and 88/299/EEC. Official Journal of the European Union, L Series, No. 125, pp. 5-9.

veterinarian; and reinforces the provisions on control and testing. Penalties and sanctions in case of violations are to be increased where checks detect the presence of prohibited substances or products or residues of substances administered illegally.

In Directive 96/23/EC measures are specified to control the ban[16] (Passantino *et al.*, 2001). The control should be performed by special, dedicated institutes. Analysis of the samples taken is performed by routine or field laboratories (RFLs). In each member state the RFLs are coordinated and controlled by at least one national reference laboratory (NRL) designated by the national government.

Finally, the NRLs are supported, advised and controlled by four community reference laboratories (CRLs), which were designated in 1991 by the EU and implemented in 1993 (Stephany *et al.*, 1994). Annually a residue monitoring programme must be made in which the results of the controls of the previous year and the targets for control in the new year are given. People working on the farms and veterinarians are made co-responsible for the control of the ban. Samples can be taken at production plants for banned substances and animal feeds, and at farms, slaughterhouses and butchers. There should be at least one national reference laboratory for every banned substance. Indications should be available for sampling and analysis. The main sanction on the use of banned substances is the destruction of the positive animals. The farmer has to pay for the additional controls that are performed. When meat is imported from third countries, it must also be controlled. When the products give a positive result the European Commission should be informed and additional controls are indicated. Eventually this could lead to a ban on the import from a certain country.

It is important underlines that the Opinion of the Scientific Committee on Veterinary Measures relating to Public Health (SCVPH)[17] of 30 April 1999 on potential adverse effects to human

[16] A principal objective of Council Directive 96/23/EC is to detect illegal use of substances in animal production as well as detecting the misuse of authorized veterinary medicinal products. The Directive lays down measures requiring European Member States to monitor the substances and their residues in both live animals and animal products, listed in Annex I to the Directive.
In particular, this directive describes guidelines for residue control and divides all pharmacologically active substances into two groups:
· Group A compounds, which comprise prohibited substances (listed in the Directive 96/22/EC and in Annex IV of the Regulation 2377/90/EC);
· Group B compounds, which comprise substances with final and provisional MRLs (listed in Annexes I and III of the Regulation 2377/90/EC). See, Council Directive 96/23/EC of 29 April 1996 on measures to monitor certain substances and residues thereof in live animals and animal products and repealing Directives 85/358/EEC and 86/469/EEC and Decisions 89/187/EEC and 91/664/EEC. Official Journal of the European Union, L Series, No. 125, pp. 10-32.
[17] The United States and Canada contested the prohibition imposed by the European Communities on the import from third countries of treated animals with the hormones and, in 1997, a panel of the World Trade Organisation (WTO) ruled that the measure was not in line with the Agreement on the Application of Sanitary and Phytosanitary Measures (SPS). The EU appealed against this ruling and, in 1998, the WTO Appellate Body reversed most of the findings of the panel. The WTO Appellate Body only upheld the finding that prohibition of imports of meat from hormone-treated animals to the EU did not comply with the requirement that such a measure should be based on a relevant assessment of the risks to human health. In reaction to these findings, the EU carried out a complementary risk assessment and mandated the Scientific Committee on Veterinary measures relating to Public Health (SCVPH) to evaluate the risks to human health from hormone residues in bovine meat and meat products treated with six hormones for growth promotion.

health from hormone residues in bovine meat and meat products (which was reviewed on 3 May 2000 and confirmed on 10 April 2002[18]) established that there is a substantial body of recent evidence suggesting that oestradiol 17ß has to be considered as a complete carcinogen, as it exerts both tumour-initiating and tumour-promoting effects, and that the data currently available do not make it possible to give a quantitative estimate of the risk to human health[19].

In the light of the conclusions of the 1999, 2000, and 2002 Opinions, the European Communities adopted Directive 2003/74/EC[20], which amends Directive 96/22/EC in relation to the prohibition permanently of the use of hormones in stockfarming.

Directive 2003/74/EC maintains the permanent prohibition of the placing on the market of meat and meat products from animals treated with oestradiol-17β for growth-promotion purposes originally contained in Directive 96/22/EC[21].

---

[18] The EU Scientific Committee confirmed that the use of hormones as growth promoters for cattle poses a potential health risk to consumers, following a review of 17 studies and other recent scientific data.

Subsequent to the adoption of the 1999 Opinion, additional scientific information was made available to the European Commission in the form of scientific studies conducted by: (i) the United Kingdom's Veterinary Products Committee sub-group on the 1999 Opinion (October 1999); (ii) the Committee for Veterinary Medicinal Products ("CVMP") of the European Union (a subcommittee of the European Medicines Agency (EMEA)) (December 1999); and (iii) the Joint FAO/WHO Expert Committee on Food Additives ("JECFA") (February 2000). At the request of the European Commission, the SCVPH examined this scientific information and, on 3 May 2000, issued a review of its 1999 Opinion in which it declined to alter the conclusions contained therein (the "2000 Opinion").

On 10 April 2002, a second review of the 1999 Opinion was issued by the SCVPH (the "2002 Opinion") on the basis of more recent scientific data collected since the previous review. The scientific data reviewed by the SCVPH included the final results of all 17 studies that had been commissioned by the European Commission. Publishing its third opinion on the risks to human health from hormone residues in beef products, the SCVPH found no reason to change its previous opinions of 1999 and 2000.

See, Opinion of the Scientific Committee on Veterinary Measures relating to Public Health on "Assessment of potential risks to human health from hormone residues in bovine meat and meat products, adopted on 30 April 1999. Available at

http://ec.europa.eu/food/fs/sc/scv/out21_en.pdf Accessed August 19, 2011.

See also Opinion of the Scientific Committee on Veterinary Measures relating to Public Health on "Review of specific documents relating to the SCVPH opinion of 30 April 1999 on the potential risks to human health from hormone residues in bovine meat and meat products", adopted on 3 May 2000, and on "Review of specific documents relating to the SCVPH opinion of 30 April 1999 and 3 May 2000 on the potential risks to human health from hormone residues in bovine meat and meat products", adopted on 10 April 2002. Available at http://ec.europa.eu/food/fs/sc/scv/out50_en.pdf Accessed August 19, 2011.

[19] In this context, it is important underlines that Article 168 of the Treaty establishing the European Union (EU) states that a high level of human health protection shall be ensured in the definition and implementation of all EU policies and activities. A comprehensive body of EU legislation has been put in place to achieve this objective. All of this legislation is publicly available at http://eur-lex.europa.eu/en/index.htm Accessed August 23, 2011.

[20] Directive 2003/74/EC of the European Parliament and of the Council of 22 September 2003 amending Council Directive 96/22/EC concerning the prohibition on the use in stockfarming of certain substances having a hormonal or thyrostatic action and of beta-agonists. Official Journal of the European Union, L Series, No. 262, pp. 17-21.

In relation to the five other hormones (testosterone, progesterone, TBA, zeranol, and MGA) Directive 2003/74/EC continues to apply the prohibition contained in Directive 96/22/EC, but on a provisional basis[22].

This Directive specifies that, even though the scientific information available showed the existence of risks associated with these substances, "the current state of knowledge does not make it possible to give a quantitative estimate of the risk to consumers"[23]. Accordingly, the prohibition of these five hormones should apply "while the Community seeks more complete scientific information from any source, which could shed light and clarify the gaps in the present state of knowledge of these substances"[24].

On 27 October 2003, the European Communities notified the Dispute Settlement Body (DSB) of the adoption, publication, and entry into force of Directive 2003/74/EC, as well as the 1999, 2000, and 2002 Opinions, which it considered to be risk assessments that sufficiently justified the permanent and provisional import prohibitions under the *SPS Agreement* (Agreement on the Application of Sanitary and Phytosanitary Measures)[25].

The latter use has to be phased out by until 14 October 2006 and for the rest of the uses the Commission was to present a report in October 2005.

The report was presented on 11 October 2005 to Council and Parliament. It comes to the conclusion that the use of the alternative substances such as prostaglandins is already common.

Veterinarians predict an insignificant impact of future unavailability of oestradiol 17β and its ester like derivates on farmers and on animal welfare. It was moreover observed that the unavailability of oestradiol and its ester like derivates would have minimal economic effect. This is because the incidence of fetal mummification and fetal maceration is low, and although the incidence of pyometra is higher, methods of prevention not involving use of oestradiol do exist and would be preferable.

Article 11a of Council Directive 96/22/EC as amended by Council Directive 2003/74/EC requires the presentation of a report on the necessity of the use of the hormone oestradiol 17β in food animal production. A report concerning the availability of alternative veterinary medicinal products to those containing oestradiol 17β or its ester-like derivatives for the treatment of fetal maceration or mummification in cattle, and for the treatment of pyometra prepared (later addressed as the Report) by an independent scientist has been presented by the Commission in October 2006[26]. The Report concludes that oestradiol 17β is not essential in food animal production.

---

[21] Directive 2003/74/EC, supra, footnote 20, Recital 10 and Article 1 (amending Articles 2 and 3 of Directive 96/22/EC).

[22] Ibid.

[23] Directive 2003/74/EC, supra, footnote 20, Recital 7.

[24] Ibid., Recital 10.

[25] European Communities – Measures concerning meat and meat products (hormones), Communication from the European Communities, WT/DS26/22, WT/DS48/20, 28 October 2003. Available at http://trade.ec.europa.eu/doclib/docs/2003/november/tradoc_114641.pdf Accessed August 23, 2011.

[26] See, Report concerning the availability of alternative veterinary medicinal products to those containing oestradiol 17β or its ester-like derivatives for the treatment of fetal maceration or

3. Situation in other countries

Whereas the EU has banned the use of all hormones, other countries do allow the use of steroid hormones and hormone-like substances in various combinations with the aim to improve weight gain and feed efficiency in livestock farming. Recommended application occurs in the form of small implants or devices, containing the active hormones, into the subcutaneous tissue of the ears. Both ears are completely discharged at slaughter[27] (Galbraith, 2002). Pharmaceutically, these implants represent slow-release devices, containing relatively large quantities of hormones, which are fractionally released over a period of several months.

In the United States[28], Canada, Australia, New Zealand and in some countries in South America, Asia and Africa the natural hormones - testosterone, 17ß-oestradiol and progesterone - and the (semi-) synthetic hormones trenbolone, zeranol and melengestrol acetate can be used to promote growth.

In particular, in the USA[29] five hormones are authorized as the active component of solid ear implants (17β-estradiol - as such or as benzoate, testosterone - as such or as propionate,

---

mummification in cattle, and for the treatment of pyometra. Commission of the European Communities - SEC(2005) 1303 of 11.10.2005.

[27] See European Commission, 1999. Scientific Committee on Veterinary Measures relating to Public Health (SCVPH). Assessment of potential risks to human health from hormone residues in bovine meat and meat products. Available at http://europa.eu/comm/food/fs/sc/scv/out21_en.html Accessed August 23, 2011.

[28] The U.S. Food and Drug Administration (FDA) and the U.S. Department of Agriculture (USDA) cooperate in regulating growth promotants for livestock. Both of these agencies maintain that hormones in beef from an implanted animal have no physiological significance for humans. All animal drug products are approved for safety and effectiveness under the Federal Food, Drug, and Cosmetic Act (21 U.S.C. 301 et seq.). Information on approved hormone products are at 21 CFR Parts 522, 556, and 558. FDA requirements for the review and approval of new animal drug applications is at http://www.fda.gov/AnimalVeterinary/GuidanceCompliance Enforcement/GuidanceforIndustry/ucm123821.htm Accessed August 26, 2011.

[29] The United States continues to maintain that U.S. beef from cattle treated with certain approved growth hormones pose no public health risk. Overall, the official U.S. position is that "there is a clear world-wide scientific consensus supporting the safety of these approved and licensed hormones when used according to good veterinary practice".

The United States claims that this position is supported by "scientific reviews of the six hormones, international standards pertaining to their use, and a longstanding history of administering the six hormones to cattle for growth promotion purposes". Accordingly, the United States claims that the use of these hormones as growth promoters in beef production is safe, when applied in accordance with good veterinary practices.

The United States has criticized the EU's scientific opinions for focusing on only one growth promotant, estradiol-17β, and on its potential genotoxicity, while directing relatively little attention toward the other natural and synthetic hormones. The United States also claims that the "EU failed to use solid evaluative methods in their studies and completely disregarded the large body of evidence from epidemiological studies that indicate that estradiol does not contribute to any increased cancer risk and that meat from animals tested with estradiol is safe for consumers".

Regarding the EU's more recent reviews, the United States claims they fail to provide any new evidence that would call into question the findings and conclusions of other authoritative reviews.

More broadly, the United States also disputes whether the EU's scientific reviews serve as a risk assessment. The United States claims: "There has been no new risk assessment based on scientific information and reasoning presented by the EU," further claiming that the "17 studies" funded by the Commission beginning in 1998 were "not intended as a to serve as a risk assessment, but instead were progesterone, trenbolone acetate and zeranol) and two hormones as feed additives: melengestrol acetate (MGA) for feedlot heifers[30] (Berende & Ruitenberg, 1983; Meissonnier & Mitchell-Vigneron, 1983) and ractopamine for swine (Marchant Forde *et al.*, 2003).

As aforementioned, the hormone ban of the EU is currently the cause of a dispute between the EU and these third countries, led by the United States and Canada. The reason for this dispute is that those countries want to export meat to EU nations from animals treated with in their view acceptable hormones. In their opinion the EU blocks international trade on improper grounds and against international law.

The National Cattlemen's Beef Association (NCBA), the largest national group of cattle producers, has long opposed the EU's ban on imports of U.S. hormone-treated beef, claiming that the ban is scientifically unjustified and fails to satisfy the EU's WTO requirements under the SPS. Similar concerns have been expressed by other U.S. farm groups, including American Farm Bureau Federation (AFBF), the Animal Health Institute (AHI), and the American Meat Institute (AMI)[31]. Many trade analysts believe that the United States has a strong case against the hormone ban under WTO rules that require SPS restrictions to be based on risk assessment and to have a scientific justification. These various interest groups have continued to exert pressure on U.S. trade policy officials to hold to their position regarding the EU's meat hormone ban.

Recently, On May 13, 2009, following a series of negotiations, the United States and the EU signed a memorandum of understanding (MOU) implementing an agreement that could resolve this long standing dispute[32]. On July 13, 2009, the EC adopted regulations opening a tariff quota for imports of High Quality Beef, effective August 1, 2009[33]

4. Existing community legislation

Currently Council Directive 2008/97/EC[34] has been published amending Council Directive 96/22/EC concerning the prohibition on the use in stock farming of certain substances

---

to fill in the gaps". Accordingly, the United States claims, the EU's 2003 update to its hormone ban is not in compliance with its WTO obligations and should be discontinued.

For more detailed discussion see USDA, Foreign Agriculture Service, Historic Overview and Chronology of EU's Hormone Ban, GAIN Report E23206, Nov. 7, 2003, available at http://www.fas.usda.gov/gainfiles/200311/145986773.pdf Accessed August 26, 2011; USDA, FAS, EU Presentation on Hormone Ban Directive (2003/74/EC), GAIN Report E23217, Nov. 13, 2003, available at http://www.fas.usda.gov/gainfiles/200311/145986807.pdf Accessed August 26, 2011.

[30] http://www.fda.gov/AnimalVeterinary/SafetyHealth/ProductSafetyInformation/ucm055436.htm Accessed August 23, 2011.

[31] See, "Coalition Statement on EU's Latest Pronouncement on Hormones," May 14, 2002 by AFBF, AHI, AMI, and NCBA, Available at http://www.meatami.com/ht/d/sp/i/1482/pid/1482 Accessed August 23, 2011.

[32] For other information, see WTO, European Communities - Measures Concerning Meat And Meat Products (Hormones), Joint Communication from the European Communities and the United States, WT/DS26/28, September 30, 2009; 74 Federal Register 40864, August 13, 2009; 74 Federal Register 48808, September 24, 2009; and USTR press releases.

[33] Council Regulations (EC) No 617/2009 of 13 July 2009 opening an autonomous tariff quota for imports of high-quality beef. Official Journal of the European Union, L series, No. 182, 1..

[34] Directive 2008/97/EC of the European Parliament and of the Council of 19 November 2008 amending Council Directive 96/22/EC concerning the prohibition on the use in stockfarming of certain substances

having a hormonal or thyreostatic action and of β-agonists, to exclude companion animals from the prohibition.

In fact, experience gained in particular with national residue plans submitted under Directive 96/23/EC has shown that the misuse of product presentations intended for pet animals does not play a role as a source of abuse or misuse. That is partly because it is economically unattractive to use presentations intended for pet animals for growth promotion in food-producing animals.

It was considered therefore appropriate to limit the scope of Directive 96/22/EC only to food-producing animals and withdraw the prohibition for pet animals, as well as to adjust the definition of therapeutic treatment.

5. Conclusions

The EU continues to maintain that "there is a lack of data on the type and amount of [growthpromoting hormone] residues in meat on which to make a quantitative exposure assessment" that would change the EU's understanding of the "possible risks to human health" associated with hormone-treated meat and meat products. It claims that this position is supported by a series of commissioned research studies and scientific reviews conducted by the EU, although there has been no conclusive testing on the issue.

The most recent review, conducted in 2007 by the European Food Safety Authority (EFSA), cites evidence supporting that estradiol-17β be considered as a carcinogen, and states that all six hormones may pose endocrine, developmental, immunological, neurobiological, immunotoxic, genotoxic, and carcinogenic effects, particularly for susceptible risk groups (such as prepubertal children). The toxicological and epidemiological data reviewed by the Commission panels do not allow a quantitative estimate of the risk, leading to the panel's conclusions that no threshold levels can be defined for any of the six hormones

Based on this series of reviews, the Commission maintains that these reviews "reaffirmed public health concerns about the large scale use of hormones administered to cattle for growth promoting purposes," and therefore "provided the scientific basis for community legislation not allowing the use of hormones for growth promoting purposes in the EU".[35]

6. References

Bahrke, M.S., & Yesalis C.E. (2004). Abuse of anabolic androgenic steroids and related substances in sport and exercise. *Current Opinion in Pharmacology*, 4, 614-620, ISSN: 1471-4892.

Bauer, E.R., Daxenberger, A., Petri, T., Sauerwein, H., &Meyer, H.H. (2000). Characterization of the affinity of different anabolic and synthetic hormones to the human androgen receptor, human sex hormone binding globulin and to the bovine progestin

having a hormonal or thyrostatic action and of beta-agonists. Official Journal of the European Union, L series, No. 318, 9-11.

[35] EFSA, "EFSA Concludes Review of New Scientific Data on Potential Risks to Human Health from Certain Hormone Residues in Beef," Press Release dated July 18, 2007, Available at http://www.efsa.europa.eu/EFSA/efsa_locale-1178620753812_1178622723847.htm. Accessed August 26, 2011.

receptor. *Acta Pathologica Microbiologica et Immunologica Scandinavica*, 108, 838-846, ISSN: 0903-4641.

Berende, P.L.M., & Ruitenberg, E.J. (1983). Domestication, conservation and use of animal resources. Elsevier, Amsterdam, pp 191-233, ISBN: 0-444.42068-1.

Carmichael, S.L., Shaw, G.M., Laurent, C., Croughan, M.S., Olney, R.S., & Lammer, E.J. (2005). Maternal progestin intake and risk of hypospadias. *Archives of Pediatrics & Adolescent Medicine*, 159(10), 957-62, ISSN: 1072-4710.

Chang, C., Saltzman, A., Yeh, S., Young, W., Keller, E., Lee, H-J, Wang, C., & Mizokami, A. (1995). Androgen Receptor: an overview. *Critical Reviews in Eukaryotic Gene Expression*, 5, 97-125, ISSN: 1045-4403.

Danhaive, P.A., & Rousseau, G.G. (1986). Binding of glucocorticoid antagonists to androgen and glucocorticoid hormone receptors in rat skeletal muscle. *The Journal of Steroid Biochemistry and Molecular Biology*, 24, 481-487, ISSN: 0960-0760.

Danhaive, P.A., & Rousseau, G.G. (1988). Evidence for sex-dependent anabolic response to androgenic steroids mediated by muscle glucocorticoid receptors in the rat. *The Journal of Steroid Biochemistry and Molecular Biology*, 29, 575-581, ISSN: 0960-0760.

Elmore, R.G. (1992). Focus on bovine reproductive disorders: managing cases of fetal mummification. *Veterinary Medicine*, 87, 155-159, ISSN: 8750-7943.

Fara, G.M., Del Corvo, G., Bernuzzi, S., Bigatello, A., Di Pietro, C., Scaglioni, S., & Chiumello, G. (1979). Epidemic of breast enlargement in an Italian school. *Lancet*, 314, 295-297, ISSN: 0140-6736.

Flecknell, P. (2000). Manual of rabbit medicine and surgery. Publishers: British Small Animal Veterinary Association, ISBN-10: 0905214463, Quedgeley, Gloucs (UK).

Galbraith, H. (2002). Hormones in international meat production: biological, sociological and consumer issues. *Nutrition Research Reviews*, 15, 243-314, ISSN: 0954-4224.

Le Guevel, R., & Pakdel, F. (2001). Assessment of oestrogenic potency of chemicals used as growth promoter by in-vitro methods. *Human Reproduction*, 16, 1030-6, ISSN: 0268-1161.

Leffers, H., Naesby, M., Vendelbo, B., Skakkebaek, N.E., & Jorgensen, M. (2001).Oestrogenic potencies of zeranol, oestradiol, diethylstilboestrol, bisphenol-A and genistein: implications for exposure assessment of potential endocrine disrupters. *Human Reproduction*, 16, 1037-45, ISSN: 0268-1161.

Liu, S., & Lin, Y.C. (2004). Transformation of MCF-10A human breast epithelial cells by zeranol and estradiol-17beta. *The Breast Journal*, 10(6), 514-21, ISSN: 1075-122X.

Mader, T.L. (1997). Carryover and lifetime effects of growth promoting implants. Proc. OSU Symposium: Impact of implants on performance and carcass value of beef cattle, pp 88-94, Oklahoma Agric. Exp. Stn., Stillwater, OK, 21-23 November 1996.

Maravelias, C., Dona, A., Stefanidou, M., & Spiliopoulou, C. (2005). Adverse effects of anabolic steroids in athletes. A constant threat. *Toxicology Letters*, 158, 167-175, ISSN: 0378-4274.

Marchant Forde, J.N., Lay, D.C. Jr, Pajor, E.A., Richert, B.T., & Schinckel, A.P. (2003). The effects of ractopamine on behavior and physiology of finishing pigs. *Journal of Animal Science*, 81, 416-422, ISSN:0021-8812.

Meissonnier, E., & Mitchell-Vigneron, J. (1983) Anabolics in animal production. Office International des Epizooties, ISBN 92-9044-118-6, Paris.

Passantino, A., Fenga, C., Venza, M., Maressa, P., & Passantino M. (2001). The future of hormonal substances in zootechnical breeding: legislative aspects. Preprints of III

Congress of the European Society for Agricultural and Food Ethics, Florence (Italy), 3-5 October 2001, 483-484.

Pepper, R., & Dobson, H. (1987). Preliminary results of treatment and endocrinology of chronic endometritis in the dairy cow. *The Veterinary Record*, 120, 53-56, ISSN: 0042-4900.

Platter, W.J., Tatum, J.D., Belk, K.E., Scanga, J.A., & Smith, G.C. (2003). Effects of repetitive use of hormonal implants on beef carcass quality, tenderness, and consumer ratings of beef palatability. *Journal of Animal Science*, 81, 984-996, ISSN:0021-8812.

Poindron, P. (2005). Mechanisms of activation of maternal behaviour in mammals. *Reproduction Nutrition Development*, 45, 341-51, ISSN: 0926-5287.

Scaglioni, S., Di Pietro, C., Bigatello, A., & Chiumello, G. (1978). Breast enlargement in an Italian school. *Lancet*, 311, 551-552, ISSN: 0140-6736.

Sheldon, I.M., & Noakes, D.E. (1998). Comparison of three treatments for bovine endometritis. *The Veterinary Record*, 142, 575-579, ISSN: 0042-4900.

Shepherd, C.J., & Bromage, N.R. (1988). Intensive fish farming. Pub: BSP Professional Books, pp 103-105, ISBN: 0632019042, Oxford, Boston.

Stephany, R.W., van Ginkel, L.A., & Schothorst, R.C. (1994). Perspective. European Union reference laboratories for residue analysis: a quality challenge. *Analyst*, 119, 2707-2711, ISSN: 0003-2654.

Takemura, H., Shim, J.Y., Sayama, K., Tsubura, A., Zhu, B.T., & Shimoi, K. (2007). Characterization of the estrogenic activities of zearalenone and zeranol in vivo and in vitro. *The Journal of Steroid Biochemistry and Molecular Biology*, 103(2), 170-7, ISSN: 0960-0760.

Wiebe, J.P. (2006). Progesterone metabolites in breast cancer. *Endocrine-Related Cancer*, 13, 717-738, ISSN: 1351-0088.

Yuri, T., Tsukamoto, R., Miki, K., Uehara, N., Matsuoka, Y., & Tsubura, A. (2006). Biphasic effects of zeranol on the growth of estrogen receptor-positive human breast carcinoma cells. *Oncology Reports*, 16(6), 1307-12, ISSN: 1021335X-.

4

How Experience Can Be Useful in Veterinary Pathological Anatomy

Paulo Tomé[1] and Helena Vala[2]
[1]*Algoritmi Research Center/Polytechnic Institute of Viseu*
[2]*Center for Studies in Education, and Health Technologies, CI&DETS, Polytechnic Institute of Viseu*
Portugal

1. Introduction

Veterinary anatomical pathology is a medical specialty which is very similar to human anatomical pathology. This specialty consists in diagnosing diseases based mainly on gross and microscopic examination of organs and tissues obtained by surgical procedure or necropsy. It consists in applying criteria based on knowledge of gross and histological lesions in order to obtain a final pathological diagnosis. The aim of this specific diagnosis is, generally speaking, to find which disease is affecting the animal, if it is alive, or what caused its death.

The main difficulties in veterinary pathological diagnosis are related to the various animal species a pathologist has to deal with, the use of exhaustive classifications of different types, costs, subjectivity, including disagreement between the clinical and pathological diagnoses, and the urgency in issuing the report. So, the urgent need to compile all available information to obtain a diagnosis in a short time is the constant challenge professional experts face.

Pathology consists in using scientific methods to study structural and functional changes in cells, tissues and organs that underlie disease (Cotran et al., 1999). It is divided into two branches: anatomical pathology, dedicated to examining organs, tissues and cadavers and clinical pathology, dedicated to laboratory analysis of body fluids and/or tissues. The pathologist has the professional expertise devoted to the practice of both, anatomical and clinical pathology (Langone Medical Center, Department of Pathology, 2011).

Past knowledge is an important resource in veterinary pathological anatomy. Professionals often use their previous experience (Dungworth et al., 1999; Goldschmidt et al., 1998; Head et al., 2003; Hendrick et al., 1998; Kennedy et al., 1998; Kiupel et al., 2008; Koestner et al., 1999; Maxie, 2007; Meuten et al., 2004; Misdorp et al., 1999; Scott et al., 2001; Slayter et al., 1994a; Valli et al., 2002; Wilcock et al., 2002)[1].

This chapter will be devoted to veterinary anatomical pathology, not in terms of further development of its study methods, extensively detailed in the specialty bibliography, but from a different approach, with the aim of better understanding routine work in the pathology lab.

[1] The twelve books: (Dungworth et al., 1999; Goldschmidt et al., 1998; Head et al., 2003; Hendrick et al., 1998; Kennedy et al., 1998; Kiupel et al., 2008; Koestner et al., 1999; Meuten et al., 2004; Misdorp et al., 1999; Slayter et al., 1994a; Valli et al., 2002; Wilcock et al., 2002) were published by World Health Organization (WHO).

Emphasis will be given to the difficulties faced by the veterinary pathologist every day, in order to justify the development of new solutions to support the issuance of more accurate diagnoses.

The authors describe a framework and a software system that can be useful for diagnosing diseases. The developed framework is based on analysis of the bibliography in this field. Both proposals can be useful to yield better diagnoses. Pathological diagnosis is a supplementary diagnostic test usually ordered by the veterinarian clinician, in order to direct treatment to administer to the sick animal, or living members of a group of animals (e.g. cattle), in which a death occurred. Thus, although the methods used are similar to those used in human medicine, the ends are quite different.

As far as is known, no artificial intelligence system has ever been applied or implemented to areas of diagnosis in veterinary medicine. The system proposed in this chapter is applied to generating solutions for real clinical cases, submitted to complementary pathological diagnosis exams.

In the following sections it will be described pathological diagnosis, as well as the main re-using knowledge mechanisms created in the field of Artificial Intelligence, framework system and the software application. Finally, in section 5 are reported final conclusions and beliefs about the framework and software application.

2. Pathological diagnosis

Veterinary anatomical pathology is quite similar to human anatomical pathology, usually referred as anatomical pathology. It is a medical specialty which diagnoses diseases, based on gross, microscopic, chemical, immunologic and molecular examination of organs and tissues, obtained by surgical procedure and by necropsy (in animal species; autopsy refers to Human species) (Miller et al., 2009).

Anatomical pathology is itself divided in subspecialties, the main ones being:

- surgical pathology, which is the subspecialty pathologists devote more time to. It consists in gross and microscopic examination of surgical specimens, as well as biopsies, submitted mostly by clinicians, dermatologists and surgeons (Grzybicki et al., 2004);
- cytopathology, the branch of anatomical pathology, dedicated to the microscopic examination of cytological specimens, obtained from smears or fine needle aspirates from organs, masses or cysts, used mostly as a complementary diagnosis of the histological exam, since it has more interpretative limitations (Meyer, 2001);
- forensic pathology, whose main aim is to determine cause of death of the animal and is applied mostly in cases of death without previous disease, including criminal causes. This subspecialty is applied to legal purposes. In veterinary medicine, the forensic pathologists are required mainly due to suspicious death by criminal causes (almost always attributed to acts of revenge by neighbors) but in reality, the outcome of most cases culminates in the diagnosis of natural causes, mainly diseases of dietary, infectious or parasitic aetiology, which is very common in domestic animals (Williams et al., 1998);
- necropsy pathology (equivalent to autopsy pathology in human medicine) consists in performing necropsies, i.e. complete and careful examination of the animal cadaver or post-mortem examination, assessing body cavities, liquid presence,

position of organs, gross appearance of organs, sampling injured organs, focus on the transition between normal and injured aspect, in each organ, in order to determine disease factors which caused animal's death and to contribute to pre-diagnosis of disease.

This subspecialty has a greater importance in veterinary medicine, since it allows other cohabitants of a group (herd or flock) to be saved or an infected animal to be eliminated, acting preventatively to avoid transmission to other animals or humans, contributing favourably to public health. Also, euthanasia is acceptable and legally provided in veterinary medicine, which reinforces the appeal and importance of this branch of anatomical pathology.

In all subspecialties, additional tests from other medical specialities could be required to determine the cause of death or to obtain a definite diagnosis. The most commonly required tests are linked to toxicology, virology, bacteriology and genetics.

Pathological diagnosis is the medical specialty that deals with the examination of gross and microscopic lesions (Miller et al., 2009). It also consists in the further microscopic study of tissues and cells, in order to provide a complementary means for diagnosis. The solid background of expertise is constructed based on medical literature, educational programs, training activities, meetings, technical rules and the cognitive skills of the pathologist which allows him to describe lesions, interpret histological slides and make decisions. In this long and complex process, the pathologist takes into account the animal data, clinical history (including physical exam), results of other complementary exams (including biochemical analysis and x-rays), using a strategy to arrive at a diagnosis which is not qualitatively different from those used by clinicians (Pena & Andrade-Filho, 2009).

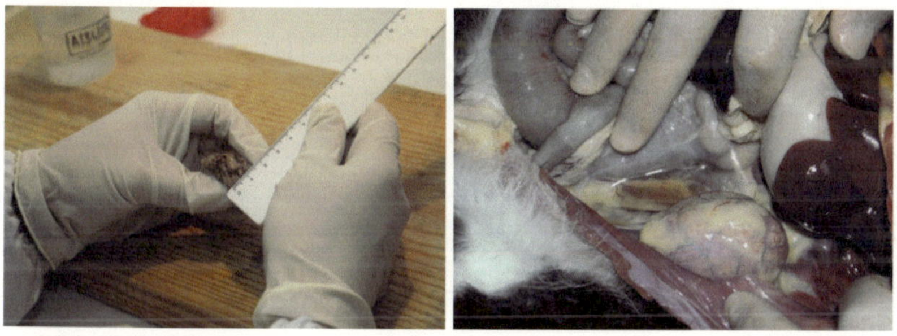

(a) Description and measurement of a surgical specimen

(b) Post-mortem exam of dead animal cadaver

Fig. 1. Gross examination

The most commonly used procedures in the pathological diagnosis processes include gross examination, histopathology or microscopic examination, evaluating histological aspects, including cellular characteristics and organizational patterns, immunohistochemistry and cytopathology, among other more specific tests. Gross examination consists in describing the specimen with the naked eye. It is based on observation, description and measurement of gross lesions, found in a surgical specimen, submitted to the pathology lab (figure 1 a)), or found in animal cadavers, if such is the case, during the meticulous post-mortem exam (figure 1 b)), especially if the animal is injured. It is also during this step, that the pathologist decides which areas and specimens need to be processed for histopathology and microscopic evaluation (University of Utah of Spencer S. Eccles, 2011; Zarbo & Nakhleh, 2009).

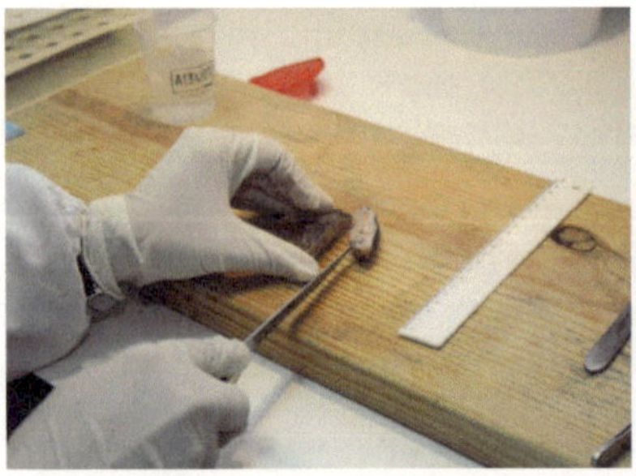

(a) Cut of formalin fixed specimens

(b) Selection of areas to be processed

(c) Placing the specimen in standard plastic histological cassettes

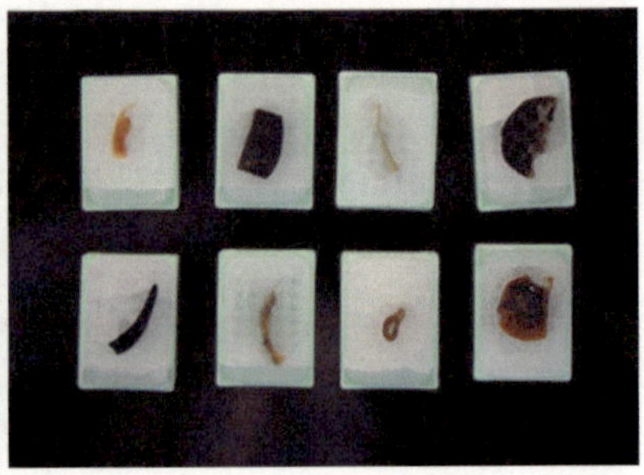

(d) Paraffin blocks

(e) Execution of histological sections with a rotary microtome

(f) Staining by standard haematoxylin and eosin

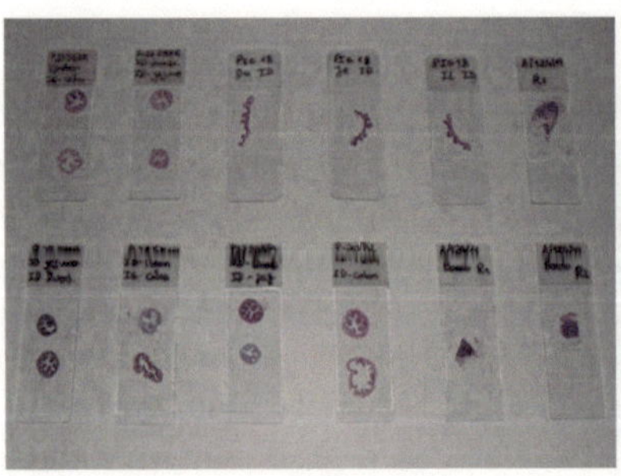

(g) Final preparations

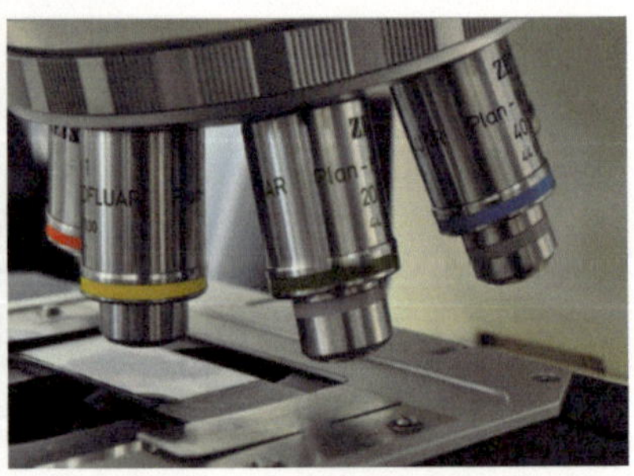

(h) Microscopic observation

Fig. 2. Routine histological technique

To perform tissue observations under the microscope, a complex and relatively time consuming procedure is performed, in order to obtain very thin sections that can be traversed by the microscope condenser's light. The microscope is an extension of the pathologist's vision. It is a fundamental instrument throughout the process of diagnosis and with which the pathologist must feel very comfortable, spending many hours of his life at the microscope. Without a microscope the pathologist is blind and incomplete.

The basis of routine histological technique (figure 2) culminates obtaining histological sections (with 3 μm) that will subsequently be stained by standard haematoxylin and eosin or other more specific methods, to obtain a more accurate histopathological diagnosis, namely histochemistry. Histochemistry consists in using chemical reactions to localize chemical compounds of cells and tissues (figure 3 a) and b)) (Pellicciari, 2009) and immunohistochemistry, a technique for identifying cellular or tissue constituents (antigens) by means of antigen-antibody interactions (figure 3 c) and d)). These methods are performed in the pathological diagnostic field, mainly in oncology diagnosis (Miller, 2002).

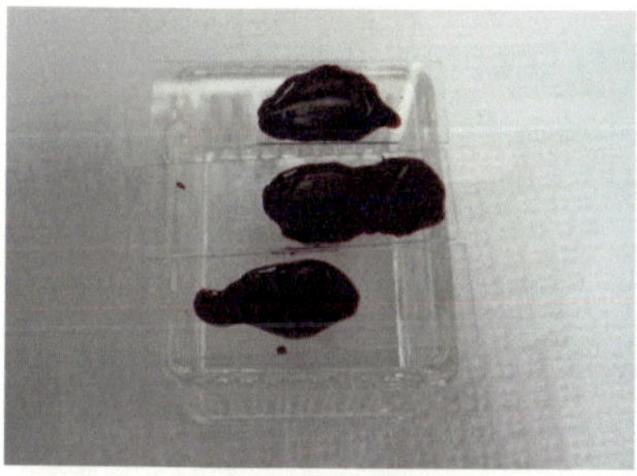

(a) Histochemistry

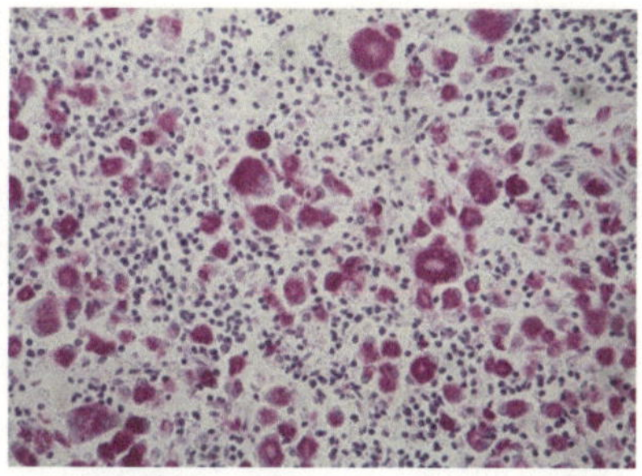

(b) Histochemistry. Ziehl-Neelsen Staining. Bar = 25 μm

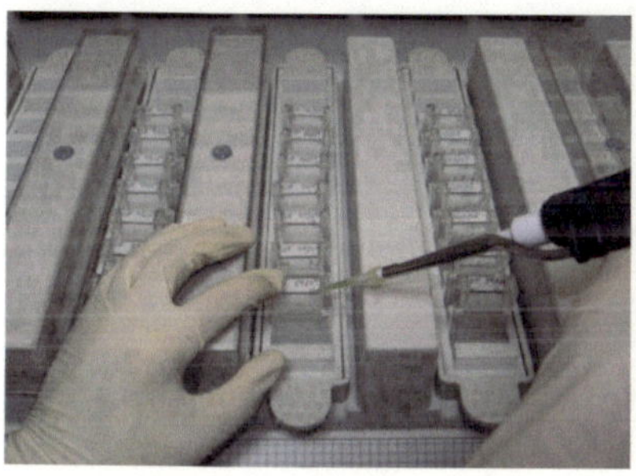

(c) Immunohistochemistry

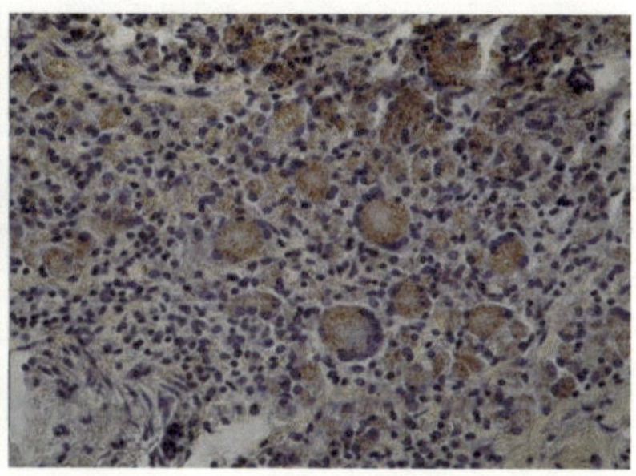

(d) Immunohistochemistry. Staining with a peroxidase-labeled-(strept)avidin-biotin (LAB-SA) method. Bar = 25 μm

Fig. 3. Specific histological techniques

Finally, the professional veterinary pathologist produces an understandable and grammatically correct report (Pena & Andrade-Filho, 2009) which includes a complete description of macroscopic and histological observations and a definite diagnosis issued based on the specific knowledge of these observations and statements.

The pathological report must be understood directly by the clinician which operates under the same system of rules (Pena & Andrade-Filho, 2009), without triggering an array of interpretive questions which will increase the time spent on a particular case. Also, misinterpretation of a pathological diagnosis may lead to an incorrect or surgical therapeutic approach to the lesion concerned (Pena & Andrade-Filho, 2009). So, the development of a system of rules will also help to solve this problem.

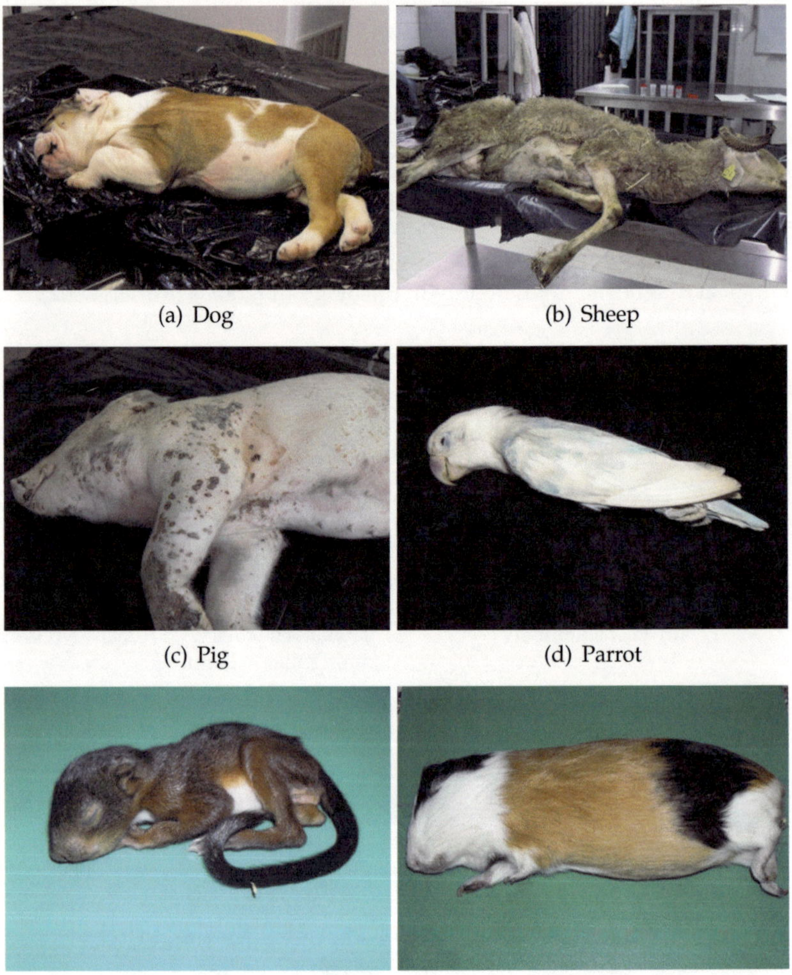

(a) Dog (b) Sheep

(c) Pig (d) Parrot

(e) Squirrel (f) Guinea pig

Fig. 4. Different species of animals to necropsy

One of the major difficulties associated with veterinary medicine is, that unlike human medicine, which is dedicated to a single species, its target is a number of different species of animals to study/diagnose. To overcome this difficulty, the animal species are usually grouped into specialties, based in their affinity (e.g. pets - dogs and cats, ruminants - cows, sheep and goats, swine, equine species, exotic species, etc.) (figure 4). In parallel, veterinary anatomical pathology deals with different animal species, which requires knowledge of the anatomy of each species, species-specific diseases, their lesions and lesion patterns. To achieve a correct and definitive diagnosis per species, the ideal situation would be the constitution of multidisciplinary teams, in which each expert would focus on a single species or group of species. This may, however, be impossible for smaller institutions wishing to offer a diagnosis service in the field of veterinary pathology.

Also, pathological anatomy uses different types of exhaustive classifications (e.g. infectious diseases, metabolic diseases, endocrine diseases, neoplasm, etc.). It can be highlighted, as reference books to the veterinary pathologist's diagnostic decisions: Dungworth et al. (1999); Goldschmidt et al. (1998); Head et al. (2003); Hendrick et al. (1998); Kennedy et al. (1998); Kiupel et al. (2008); Koestner et al. (1999); Maxie (2007); Meuten (2002); Meuten et al. (2004); Misdorp et al. (1999); Scott et al. (2001); Slayter et al. (1994a); Valli et al. (2002); Wilcock et al. (2002), among several others dedicated to the pathology of different animal species.

If, on the one hand, not following a clinical case directly or contacting the owner or veterinary clinician personally may be beneficial, in terms of exemption in reaching conclusions, on the other hand, the summary of the clinical information in the requested analysis, may omit crucial information to the final conclusion/diagnosis.

Furthermore, reading that the animal is two years old and has intense itching, is not the same as visiting a young animal that that is constantly scratching itself. So, during the long process of making a diagnosis, this type of information may easily lose its proper influence, which would not happen if it had been obtained by direct observation, especially if the pathologist is seeing abnormal cells in the microscope, i.e. with features of high malignancy, which would tend to overlap, in terms of influencing the summary of the clinical history reported indirectly.

Often, the pathologist has no notice knowledge of the outcome, performed treatments or results of outside consultations, which prevents him from verifying the authenticity of his diagnosis and evaluate eventual failures, when he fails, making it difficult to improve and learn from their mistakes (Pena & Andrade-Filho, 2009).

For these reasons, pathologists may enhance their diagnostic abilities, acquiring specific abilities related to other fields by means of expressing relevant warrants and backings to their conclusions, deducing and expressing intrinsic rules that guide the different actions related to the diagnostic task (Pena & Andrade-Filho, 2009).

In complex cases, if the pathologist needs to use additional analytical methods to reach a definitive diagnosis, such as histochemistry, immunohistochemistry or other medical speciality tests, the authors propose a prior consultation with the client, as these additional studies increase diagnostic costs, and thus, should be subjected to client discretion. This information should be included in the first report, with the explanation for the necessity of more diagnostic procedures, along with the budget of all the expected costs. Procedures and respective costs must, preferably, be presented in a phased and sequential manner. As an example, a report for an undifferentiated round cell tumour should describe macroscopic and microscopic features, the preliminary diagnosis as "consistent with round cell tumour", and a detailed account for the need of additional diagnostic workup: "to determine definitive histogenesis, further protocols are recommend to rule out possible differential diagnosis: toluidine blue for mast cell (cost value), CD1a for histiocytoma (cost value) and CD3 and CD79α for lymphoma (cost value). These protocols will be realized in this order and executed only to the point needed to reach a definitive diagnosis.

Although the pathologist is protected in his lab from contact with the owner and patient (animal), he should be sensitive to their economic difficulties, especially in the absence of health insurance for animals. In addition, he must be aware that the diagnosis is only a precursor to the stage of healing, i.e. the treatment stage remains indispensable and is also expensive. For these reasons, and according to proper medical conduct, the proposed plan

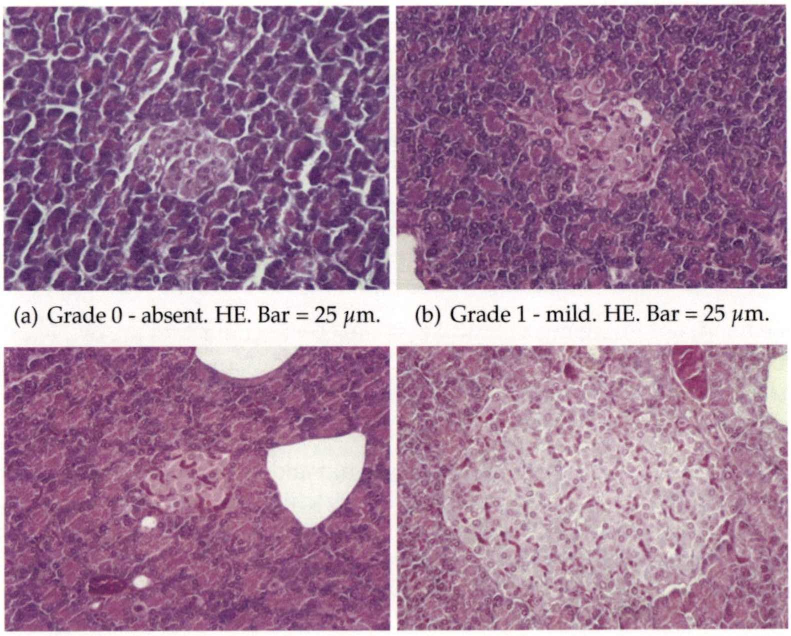

(a) Grade 0 - absent. HE. Bar = 25 μm. (b) Grade 1 - mild. HE. Bar = 25 μm.

(c) Grade 2 - moderate. HE. Bar = 25 μm (d) Grade 3 - severe. HE. Bar = 25 μm

Fig. 5. Semi-quantitative access to congestion

must hereinafter follow one of the following two rules: from the lowest to highest cost or from the most likely diagnosis to the less likely. The goal is that customers do not have to go through the whole plan outlined by the pathologist, and achieve a definitive diagnosis as early as possible, thereby providing savings to owners, who still have to invest heavily in the therapeutic approach/process that follows.

In recent literature (Pena & Andrade-Filho, 2009), the pathologic diagnosis was envisioned as a human task, involving people - the pathologist which can make a diagnosis, based on the strategy called pattern recognition. In other words, pattern recognition is the realization that the histological picture conforms to a previously learned picture of a certain disease, which produces a diagnosis interpretation. Also, other kinds of diagnosis strategies, usually adopted by pathologists, were described in this paper, some leading to pathological difficulties or errors, which need to be compensated with new action plans.

Faced with clinical information that a mass located in the mammary gland was punched, the pathologist will inevitably direct his diagnosis to one of the multiple hypotheses included in the classification of mammary tumours available. He is being driven by location and can hardly be impartial and independent from the information provided. If instead, the clinician punched a herniated organ placed at this location, it is not impossible to get a diagnosis of a mammary adenoma, as the pathologist's brain was already formatted to diagnose of one of the several types of mammary tumours.

Recent reports showed disagreement between clinical and pathologic diagnoses, except in necropsy, which is the best method to assess overall diagnostic accuracy (Kent et al., 2004; Vos et al., 2005). Also, autopsy is still the most accurate method of determining the cause of death and auditing accuracy of clinical diagnosis, diagnostic tests and death certification

(Roulson et al., 2005). Increased availability of teaching funds may promote efforts to have necropsies performed in veterinary teaching hospitals (Kent et al., 2004). However, in this aspect the problem is getting worse in human medicine, in which the number of autopsies has shown a tendency to decline (Roulson et al., 2005).

The pressure of having to issue a diagnosis that will help save the animal's life combined with deterioration of the clinical situation increases the responsibility of having to make decisions quickly and can shorten the long and complex process that the pathologist needs to go through to be in possession of the necessary information to make decisions. So, the pathologist is already under sufficient pressure.

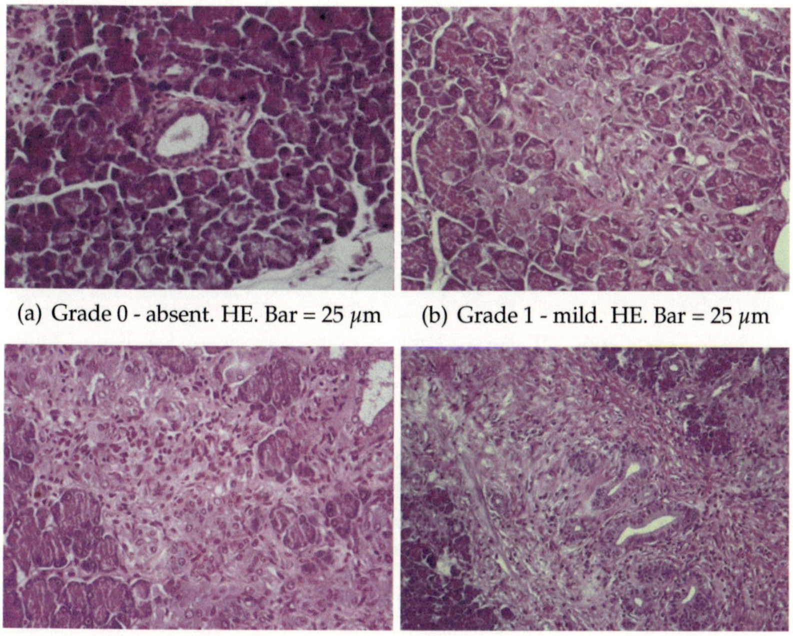

(a) Grade 0 - absent. HE. Bar = 25 μm (b) Grade 1 - mild. HE. Bar = 25 μm

(c) Grade 2 - moderate. HE. Bar = 25 μm (d) Grade 3 - severe. HE. Bar = 50 μm

Fig. 6. Semi-quantitative access to fibrosis

More pressure can be caused by the phone ringing continuously in the lab, which will not help to reach an accurate pathological diagnosis. The pathology lab must be a sacred place, where silence and peace must prevail. It is the only way to bring information together, to interpret the observed images correctly. Only then will the images seen on macroscopic exam interact with those observed under the microscope and in the pathologist brain come together to form a conclusion. We must be aware that the pressure clinicians or owners place on the pathologist may precipitate a insufficiently accurate diagnosis.

Nowadays several semi-quantitative and quantitative methods to access macroscopic and microscopic lesions, namely congestion (figure 5), fibrosis (figure 6), inflammatory infiltrate (figure 7), cellular degeneration, mitosis rate, malignancy degrees, among several others, have been developed and published to promote international uniformity, reducing the subjectivity of descriptions depending on the observer, and also to allow statistical studies. Just as a reference, it can be mentioned two: classification and grading of canine mammary tumours (Goldschmidt et al., 2011) and classification of diabetic nephropathy (Tervaert et al., 2010).

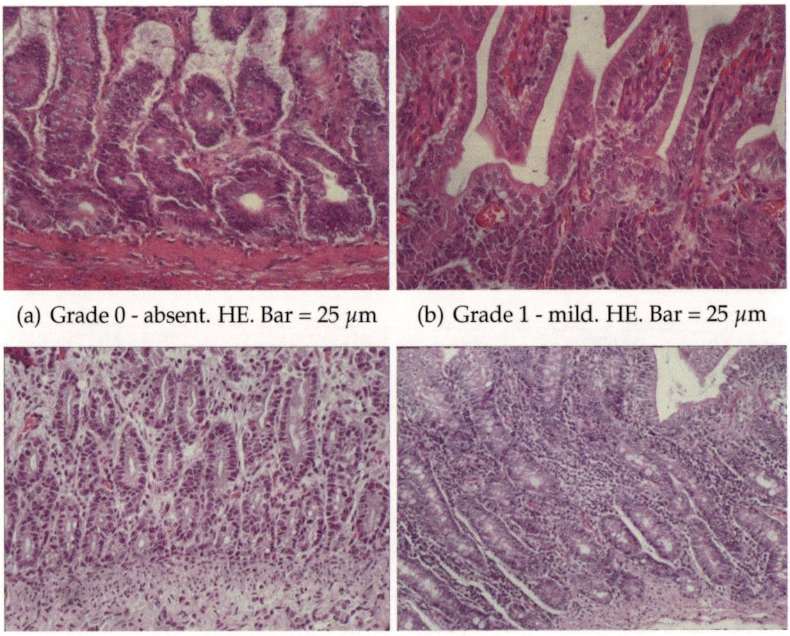

(a) Grade 0 - absent. HE. Bar = 25 μm (b) Grade 1 - mild. HE. Bar = 25 μm

(c) Grade 2 - moderate. HE. Bar = 25 μm (d) Grade 3 - severe. HE. Bar = 50 μm

Fig. 7. Semi-quantitative access to inflammatory infiltrate

In research projects two or more observers are required to score lesions independently, in order to enhance the credibility of the results, which is time consuming, expensive and fastidious. Also, some authors defend a standardization of the pathological diagnosis, as a measure to minimize diagnostic variability and error (Foucar, 2000).

The pathologist cannot be reduced to a cognitive instrument for assessing a diagnosis (Pena & Andrade-Filho, 2009) but needs useful instruments to improve his performance.

So the urgent need to compile all available information to obtain a diagnosis in a short period of timeframe, which will help to save the animal's life, is the constant challenge set before the professionals.

A system of rules that allows pathologists to standardize descriptions, easily interpreted by clinicians, will permit an association of descriptions to a conclusive final diagnosis, which would simplify stages in the complex system of diagnosis, thus increasing accuracy. So, the authors propose a model that provides less variability, discrepancies and uncertainties.

3. The re-using knowledge mechanisms

The Artificial Intelligence research community has already proposed some techniques that can be used in several domains to re-use experience. These techniques are described in several bibliographic sources, such as Bratko (2000); Cawsey (1998); Nilsson (1998); Russel & Norvig (2003); Shapiro (1987). In this section, only will be described techniques that have been used in related areas, namely Rule-Based, Pattern and Case-Based Reasoning (CBR).

Rule-Based is the oldest technique. This mechanism emerged from early work on practical and intuitive systems for logical inference (Russel & Norvig, 2003). The rule-based systems have three properties: locality, detachment and truth-functionality. Proposals by Bachmann et al. (2004); Lloyd-Williams (1994); Tauzovich (1990) are examples of Rule-based systems.

Pattern techniques (Shapiro, 1987), not pattern recognition which is described for example in Duda et al. (2001), are used in many areas. They generally involve the definition of patterns and the use of pattern-matching algorithms. A pattern is a structural sketch of an item with some pieces left undefined. Pattern matching is the process of comparing patterns to see if they are similar. In the Information Systems domain there are several works, such as Batra (2005); Conte et al. (2004); Rieu et al. (2002); Seruca & Loucopoulos (2003), which use pattern techniques.

Finally, the CBR method uses past resolved situations to solve new problems (Aamodt & Plaza, 1994; Kolodner, 1993; Mantaras et al., 2006; Riesbeck & Schank, 1989; Watson, 1999). According to Althoff et al. (1995), CBR is a good method to apply in ill situations, such as modelling tasks. CBR comprises two main elements: Case and Process model.

The *case* is formed by the problem and its solution (Kolodner, 1993). The objective and the characteristics of the situation are described by the problem. The solution consists of the solution itself, the solution evaluation and reasoning. Identification of case types constitutes the major step forward in the development of the CBR system. The set of cases of the CBR system is called case memory. An important issue related to cases is indexing. The index is a label associated with the case that will allow us to remember it.

The solving process (or process model), usually called CBR Cycle, begins with the problem description and ends with the solution. The CBR cycle has two main models: 4Rs proposed by Aamodt & Plaza (1994) and the one proposed by Kolodner (1993). The CBR cycle, as illustrated in figure 8, generally involves the following activities:

- case search to find similar cases;
- similarity evaluation to measure the level of similarity between the problem that needs solving and the stored ones;
- adaptation to adjust one or several solutions to the current problem;
- case retain to store the newly resolved problem.

Case search is based on problem description. Similarity evaluation is based on similarity functions (Althoff et al., 1995). Consequently, the new solution is built by adapting old solutions to the needs of the current problem. The last task of the CBR cycle is the inclusion of the case in the case memory. Given the fact that a new case is added to the system, the CBR systems could be said to have the ability to learn. It is worth mentioning, that there are a lot of domains where the CBR methodology has been used (Kolodner, 1993; Mántaras & Plaza, 1997; Watson, 1996). CBR application areas consist of software development, health, architectural design (Pearce et al., 1992), meal planning and legal reasoning systems, for example.

4. The framework and the system

It is important to say that this work was developed after a first approach (published in Tomé & Vala (2010)) of applying CBR to a set of 641 with good results. As shown in table 1, 47.1% of the solutions have a similarity greater than 40% of the solution proposed by the veterinary pathologists. Only 16.8% of cases have solutions in the range 0 to 20%. Although

[2] Adapted from (Leake, 1996).

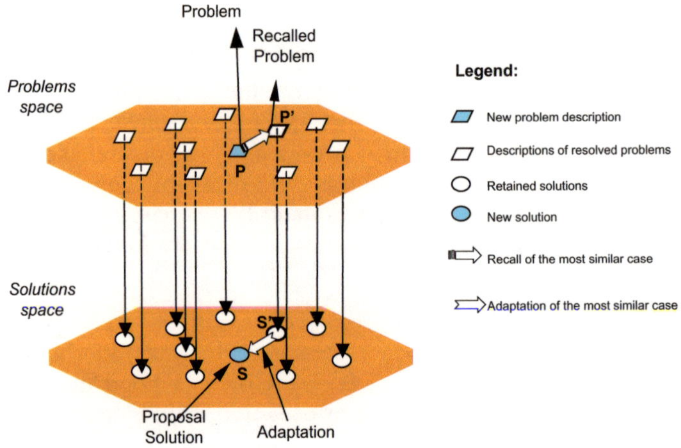

Fig. 8. The CBR method[2]

these results are positive, it is important to mention that the cases are described by regular text which constrains the searching process.

Level of solutions' similarities	[0 , 0.2[	[0.2 , 0,4[	[0.4 , 0.6[	[0.6 , 0.8 [	[0.8 , 1]
	108	231	166	80	56

Table 1. Analysis of level of similarity between solutions

As previously mentioned, several authors proposed a diagnosis system (also called a classification schema) based on some animals aspects. These diagnosis systems are used in disease diagnosis by veterinarian pathologists. The WHO proposed a classification schema for animal tumours in twelve, published three books with a classification schema for animal disease based on most affected organs.

The veterinary pathologists, faced with a disease in a particular species, with a certain clinical information, assess which of the body system, and/or specific organ, is most affected. The specialists process a set of likely differential diagnoses, directing the gross and microscopic diagnosis in order to exclude the less likely, until the possession of a set of information consistent with a pathologic diagnosis of a specific aetiology, mostly grouped in bibliography in: congenital and hereditary; abnormalities of development and growth; degenerative; circulatory; traumatic; foreign bodies; metabolic; nutritional; endocrine; toxic; immune-mediated; inflammatory (viral; bacterial; fungal; protozoal; parasitic; miscellaneous); unknown cause (Maxie, 2007).

The veterinary pathologists, faced with a neoplastic disease in a particular species (mostly pets whose lifetime is longer, having a higher predisposition to this pathologic entity) with a certain clinical information, assess in which body system, and/or specific organ, the tumour is primary located, directing the gross and microscopic diagnosis to exclude the less likely diagnosis, using international histopathological classifications. These provided a widely accepted standard nomenclature of domestic animals tumours, promoting advancements in veterinary pathology and facilitating communication between veterinary and medical pathologists, clinicians and researchers. These classifications also pretended to cross geographic and academic boundaries and establish an international consensus pertaining to the nomenclature of domestic animal tumours (Kiupel et al., 2008).

These classifications include a preliminary assessment of the mass, equivalent to gross examination (measurements; superficial and cut section aspects - colour, consistency of the tissues; evaluation of local invasion; attachment to surrounding tissues; evaluation of possible regional lymph node involvement) and then the evaluation of histological criteria which includes cellular phenotype which includes, in turn, nuclear characteristics - shape, colour and staining characteristics, type of chromatin, nucleoli and cytoplasmic characteristics - shape, colour and staining characteristics, intercellular unions); presence or absence of extracellular matrix and its type, when present; cellular pattern organization and cellular differentiation - anaplasia, anysocariosis, anysocytosis, pleomorphism; mitotic rate.

Based on these cellular features and differentiation degree, the tumours are subdivided primarily into benign and malignant groups and then subdivided, based on phenotype and amount of cellular matrix, in mesenchymal or epithelial neoplasias. The third major sub-division consists in the determination of the specific subtype from the classification for the specific target organ. In general, to determine the specific subtype of a given classification of mesenchymal neoplasias, the specialists must take into account the predominant cellular type recognized and the type of cellular matrix produced by the cells (Slayter et al., 1994b). In a given classification of epithelial neoplasias, the veterinary pathologists must take into account the predominant cell type recognized and the cellular organizational pattern.

Based on the previous descriptions, it can be concluded that each diagnosis:

- is applied for a specific type of disease;
- is based on a set of characteristics;
- is related to a specific type of animal/part..

Based in the previous aspects, it can be concluded that the framework for classification of animal diseases can have the structure illustrated in figure 9.

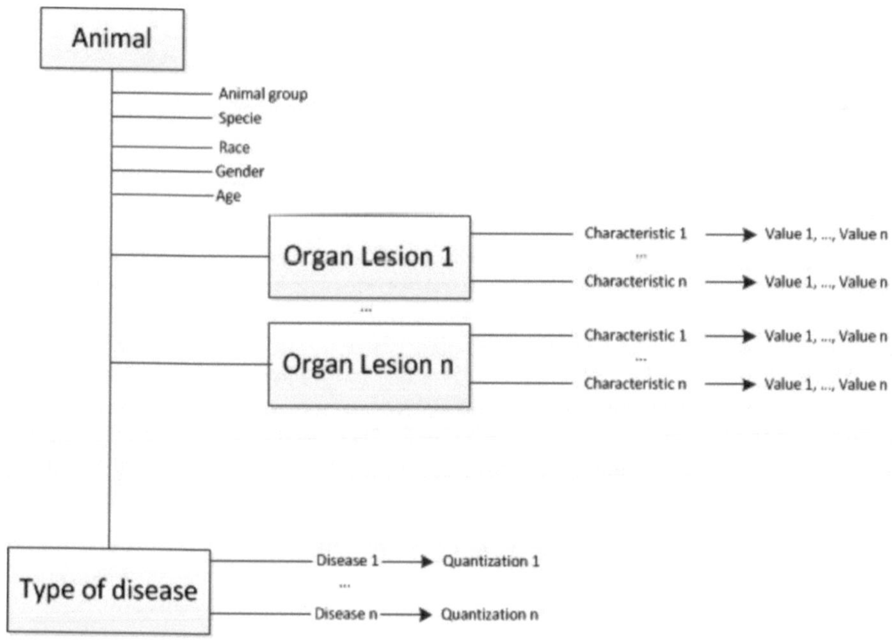

Fig. 9. Classification Schema

For instance, the Adenoma disease (defined in (Meuten et al., 2004)) is characterized by:

> "... rare tumor in domestic animals. ... . Adenonomas are discrete, solitary, tan-white masses confined to the renal cortex. They are small (usually < 2cm) in dogs and cats."

Based on the previous description it can be concluded that Adenoma is characterized by the following attributes: *Size, Color, Number, Place, Aspect, Shape, Differentiation, Pattern, Cellular Phenotype, Cytoplasm, Nuclei, Nucleoli, Mitotic Rate* and *Stroma*. The value of each of each attribute is *< 2 cm, Tan-White, Solitary, Renal cortex, Discrete and Masses, Well-differentiated, Tubular, Cuboidal epithelial, Ample, Single, Single, Rare* and *Poor*. How these attributes and values are stored in a software application is shown in figure 11.

The way the attributes and values of each type of disease are structured improves the search of previous cases of the animal's disease. As mentioned at the beginning of this section, searching previous situations with diagnostics described in regular text it is not an easy task because similar situations can be described with the slightest differences.

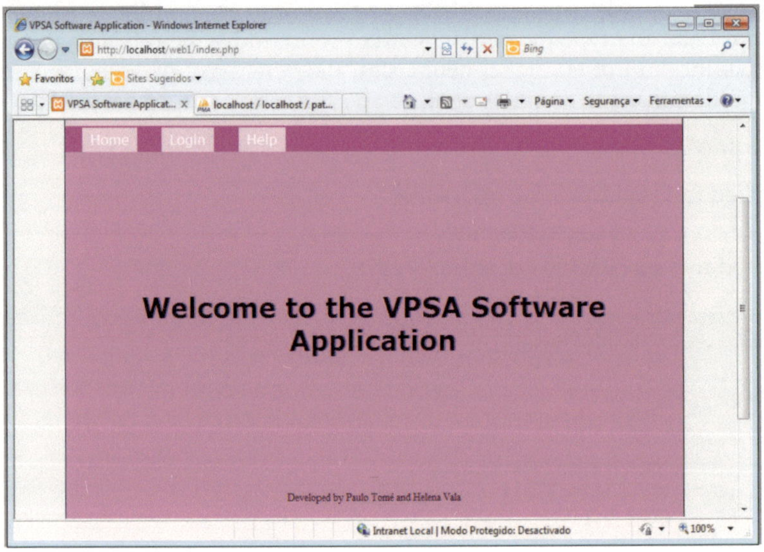

Fig. 10. Software Application

As previously mentioned, it was developed a software application, of which the main window is shown in figure 10 that can be used for storing knowledge about diseases and cases of animal diseases. The first part of the application - knowledge about diseases - is a library of descriptions of animal diseases described in the bibliography of the domain (such as (Dungworth et al., 1999; Goldschmidt et al., 1998; Head et al., 2003; Hendrick et al., 1998; Kennedy et al., 1998; Kiupel et al., 2008; Koestner et al., 1999; Meuten et al., 2004; Misdorp et al., 1999; Slayter et al., 1994a; Valli et al., 2002; Wilcock et al., 2002)). The second part of the application stores data about cases of animal diseases. In this part it is applied the CBR method described in section 3.

In the CBR part (the structure is shown in figure 12) the four phase cycle of Aamodt and Plaza was implemented (Aamodt & Plaza, 1994). The system implements the retrieve, reuse, revise and retain phases and has a repository with cases and knowledge domain.

As shown in figure 12, the system has components that store data and components that do some of the systems tasks. The components: case memory, vocabulary of domain, adaptation

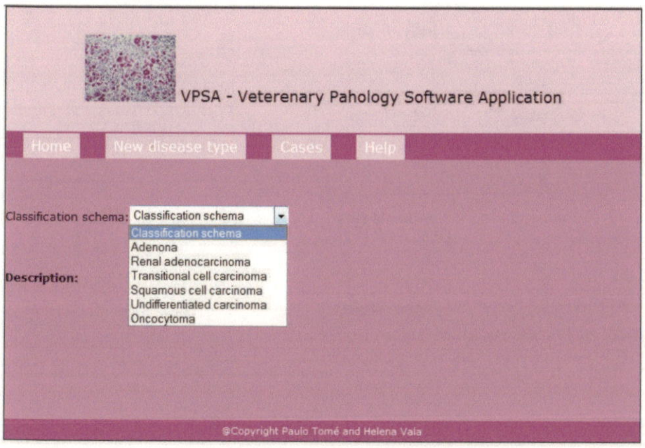

(a) Definition of Disease types

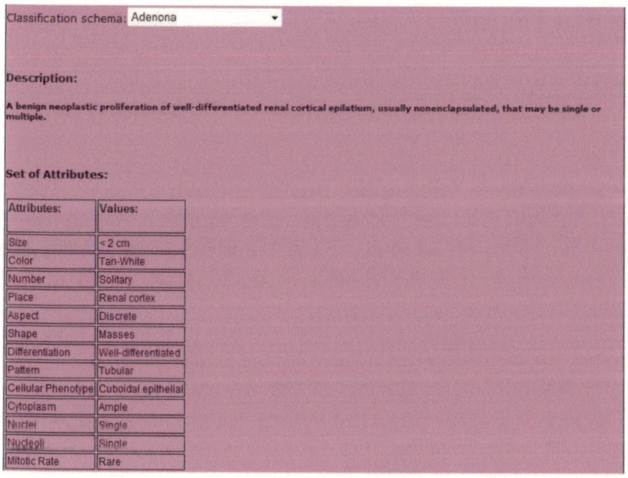

(b) Adenoma Disease

Fig. 11. Definition of Diseases types

rules, case description, inference rules and metric system, store data; while the components: retrieve, reusing, revision and retaining, implement the system tasks.

Each case is stored as a frame (Minsky, 1974). Each case is divided according the Kolodner (1993) proposal into: objective, characteristics and solution. The characteristics and the solutions are pairs of name x value.

In the domain vocabulary module the most frequent expressions used to describe cases of the domain are stored.

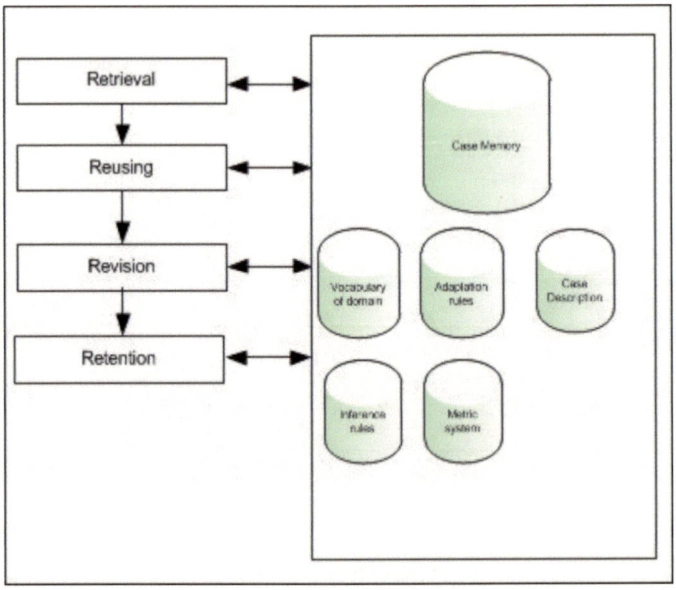

Fig. 12. Structure of the CBR part

The adaptation rules module stores information about how diagnosis can be adapted. In this module if-then rules [3] can be stored which define how diagnosis can be adapted based on the case's characteristics.

The Inference rules module is also built out of if-then rules. These rules are based on the knowledge defined in the books of the domain (such as (Dungworth et al., 1999; Goldschmidt et al., 1998; Head et al., 2003; Hendrick et al., 1998; Kennedy et al., 1998; Kiupel et al., 2008; Koestner et al., 1999; Meuten et al., 2004; Misdorp et al., 1999; Slayter et al., 1994a; Valli et al., 2002; Wilcock et al., 2002)).

In figure 13 the process of definition of a new case disease is shown. As can be seen in the figure, the first three elements (dropdowns of the figure) are elements of the Classification Schema illustrated in figure 9. The elements *Values usually used* and *Attributes also used* were designed to help the user in the process of defining a new case.

The element *Values usually used* can be used for retrieving values that are frequently used for a specific attribute. For instance, in the Adenoma situation the attribute size is usually characterized by the value < *2cm*. The use of this facility reduces the chance of defining equal values by different ways.

The element *Attributes also used* aims to help the veterinary pathologist in the definition of the characteristics of the case. Every time the veterinary pathologist defines an attribute, the set of *Attributes also used* shows the attributes which were not defined but may defined regarding the diseases where they appear. For instance, consider the disease and attributes shown in table 2. If the veterinary pathologist already defined the attributes x and y the attributes shown in *Attributes also used* are z,w,a and b. The attributes c,d,e,f,g and h do not appear because they are not in diseases where the attributes x and y appear.

[3] Stored according Cawsey (1998).

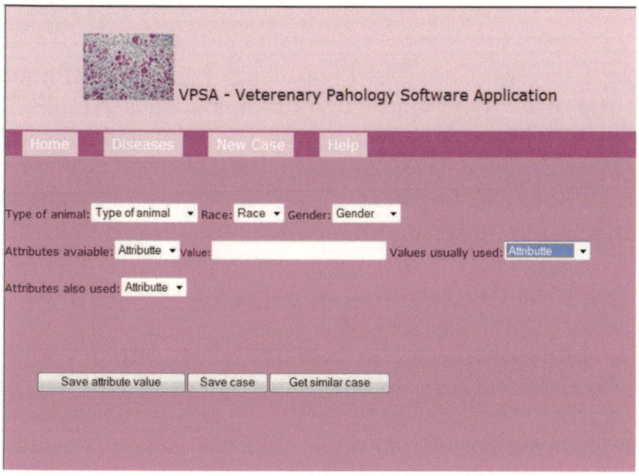

Fig. 13. Definition of Disease types

Diseases	Attributes
A	*x,y,z,w*
B	*x,y,a,b*
C	*a,b,c,d*
D	*e,f,g,h*

Table 2. Example of Diseases and attributes

Finally, it will be explained the functionality of solution finding based on previously resolved/solved cases. This functionality applies CBR to get a case which is the most similar to the current one. The system searches in the Case memory for a case with similarities to the current one. For all the cases in the Case memory the similarity to the current one based on the values of the attributes and on the knowledge of the domain is determined. The case with highest similarity value is selected.

5. Conclusion and future remarks

In this chapter, it is proposed a framework (Conceptual schema) that can be used to structure knowledge about causes of animal death. The framework was developed based on the analysis of the bibliography of the field.

The framework can be used to develop new databases to store data about animal's death diagnosis and also to improve applications that use previous knowledge of causes of animal death in order to make new diagnoses.

The use of previous knowledge about animal death can improve the task of making diagnoses. First the veterinarian is able to diagnose quickly because he can easily retrieve information about how a similar past situation was diagnosed. Besides that, past knowledge can be considered a library of information where the veterinarian can get useful help or information.

Finally, it is important to highlight that the framework is supported (and was tested) by a software application. Through the developing and testing of the software application, it can be guaranteed that the framework fits into aim for which it was developed.

6. References

Aamodt, A. & Plaza, E. (1994). Case-based reasoning: Foundational issues, methodological variations and systems approaches, *AI-Communications* 7(1): 39–52.

Althoff, K. D., Auriol, E., Barletta, R. & Manago, M. (1995). A review of industrial case-based reasoning tools, *Technical report*, AI Intelligence.

Bachmann, F., Bass, L., Klein, M. & Shelton, C. (2004). Experience using an expert system to assist an architect in designing for modifiability, *Fouth Working IEEE/IFIP Conference on Software Architecture*.

Batra, D. (2005). Conceptual data modeling patterns: Representation and validation, *Journal of Database Management* 16(2): 84–106.

Bratko, I. (2000). *Prolog Programming for Artificial Intelligence*, 3th edn, Addison-Wesley.

Cawsey, A. (1998). *The Essence of Artificial Intelligence*, Prentice-Hall.

Conte, A., Hassine, I., Rieu, D. & Tasted, L. (2004). An information system development tool based on pattern reuse, *ICEIS*, Porto.

Cotran, R., Kumar, V. & Collins, T. (1999). *Robbins Pathologic Basis of Disease (6th ed.)*, W.B. Saunders, Philadelphia.

Duda, R. O., Hart, P. E. & Stork, D. G. (2001). *Pattern Classification*, second edn, John Wiley & Sons, Inc.

Dungworth, D. L., Hauser, B., Hahn, F. F., D.W., W. & Haenichen, T. andHarkema, J. (1999). *International classification of tumours of domestic animals histological classification of tumors of the respiratory system of domestic animals*, Vol. VI, Armed Forces Institute of Pathology.

Foucar, E. (2000). Individuality in the specialty of surgical pathology: self-expression or just another source of diagnostic error?, *Am J Surg Pathol* 24: 1573–1576.

Goldschmidt, M., Dunstan, R., Stannard, A., von Tscharner, C., Walder, E. & Yager, J. (1998). *International classification of tumours of domestic animals histological classification of epithelial and melanocytic tumors of skin of domestic animals*, Vol. III, Armed Forces Institute of Pathology.

Goldschmidt, M., Peña, L., Rasotto, R. & Zappulli, V. (2011). Classification and grading of canine mammary tumors, *Vet Pathol* 48: 117–131.

Grzybicki, D., Vrbin, C., Reilly, T., Zarbo, R. & Raab, S. (2004). Use of physician extenders in surgical pathology practice, *Archives of Pathology & Laboratory Medicine* 128(2).

Head, K., Cullen, J. M., Dubielzig, R., Else, R.W. andMisdorp, W., Patnaik, A., Tateyama, S. & Van der Gaag, I. (2003). *International classification of tumours of domestic animals histological classification of tumors of the alimentary system of domestic animals*, Vol. X, Armed Forces Institute of Pathology.

Hendrick, M., Mahaffey, E., Moore, F. & Vos, J.H. andWalder, E. (1998). *International classification of tumours of domestic animals. Histological classification of mesenchymal tumors of skin ande soft tissues of domestic animals*, Vol. II, Armed Forces Institute of Pathology.

Kennedy, P., Cullen, J., Edwards, J., Goldschmidt, M. H. andLarsen, S., Munson, L. & Nielsen, S. (1998). *International classification of tumours of domestic animals histological classification of tumors of the genital system of domestic animals*, Vol. IV, Armed Forces Institute of Pathology.

Kent, M., Lucroy, M., Dank, G., Lehenbauer, T. & Madewell, B. (2004). Concurrence between clinical and pathologic diagnoses in a veterinary medical teaching hospital: 623 cases (1989 and 1999), *Journal of the American Veterinary Medical Association* 224(3): 403–406.

Kiupel, M., Capen, C., Miller, M. & Smedley, R. (2008). *International classification of tumours of domestic animals histological classification of tumors of the Endocrine System of domestic animals*, Vol. Second Series; XII, Armed Forces Institute of Pathology.

Koestner, A., Bilzer, T., Fatzer, R., Schulman, F., Summers, B. A. & Van Winkle, T. (1999). *International classification of tumours of domestic animals histological classification of tumors of the nervous system of domestic animals*, Vol. V, Armed Forces Institute of Pathology.

Kolodner, J. (1993). *Case-Based Reasoning*, Morgan Kaufmann Publishers.

Langone Medical Center, Department of Pathology (2011). For patients: Practice of pathology. Consulted in July of 2011.
URL: *http://pathology.med.nyu.edu/patient-care/for-patients/practice-of-pathology*

Leake, D. B. (1996). Cbr in context: The present and future, *in* D. B. Leake (ed.), *Case-Based Reasoning - Experiences, Lessons, & Future Directions*, AAAI Press/The MIT Press, pp. 3 – 30. ISBN - 0-262-62110-X.

Lloyd-Williams, M. (1994). Expert system support for object-oriented database design, *International Journal of Applied Expert Systems* 1(3).

Mantaras, R. L., Mcsherry, D., Bridge, D., Leake, D., Smyth, B., Craw, S., Faltings, B., Maher, M. L., Cox, M. T., Forbus, K., Keane, M., Aamodt, A. & Watson, I. (2006). Retrieval, reuse, revison and retention in case-based reasoning, *The Knowledge Engineering Review* 20(3): 215–240.

Maxie, M. G. (2007). *Jubb, Kennedy & Palmer's Pathology of Domestic Animals*, Vol. I, II, III, fifth edn, Elsevier Saunders, Edinburgh.

Meuten, D. (2002). *Tumors in domestic animals*, 4th edn, Blackwell Publishing Company.

Meuten, D. J., Everitt, J., Inskeep, W., Jacobs, R. M., Peleteiro, M. & Thompson, K. (2004). *International classification of tumours of domestic animals histological classification of tumors of the urinary system of domestic animals*, Vol. XI, Armed Forces Institute of Pathology.

Meyer, D. (2001). The acquisition and management of cytology specimens, *in* R. Raskin, D. Meyer, R. Raskin & D. Meyer (eds), *Atlas of canine and feline cytology*, W.B. Saunders Company, pp. 1 – 5.

Miller, F. P., Vandome, A. F. & McBrewster, J. (2009). *Anatomical Pathology*, VDM Publishing House Ltd.

Miller, K. (2002). Immunocytochemical techniques, *in* J. Bancroft, M. Gamble, J. Bancrofth & M. Gamble (eds), *Theory and pratice of histological techniques*, 5th edn, Churchill Livingstone, pp. 421–424.

Minsky, M. (1974). A framework for representing knowledge, *Technical report*, Massachusetts Institute of Technology.

Misdorp, W., Else, R., Hellmén, E. & Lipscomb, T. (1999). *International classification of tumours of domestic animals histological classification of mammary tumors of the dog and cat*, Vol. VII, Armed Forces Institute of Pathology.

Mántaras, R. L. & Plaza, E. (1997). Case-based reasoning: An overview, *AI Comunication* 10(1): 21–29.

Nilsson, N. J. (1998). *Artificial Intelligence: a New Synthesis*, Morgan Kaufmann Publishers.

Pearce, M., Goel, A. G., Kolodner, J. L., Zimring, C., Sentosa, L. & Billington, R. (1992). Case-based design support: A case study in architectural design, *IEEE Expert* October: 14–20.

Pellicciari, C. (2009). Histochemistry through the years, browsing a long-established journal: novelties in traditional subjects, *Eur J Histochem* 54(4).

Pena, G. & Andrade-Filho, S. (2009). How does a pathologist make a diagnosis?, *Arch Pathol Lab Med* 133: 124–132.

Riesbeck, C. K. & Schank, R. C. (1989). *Inside Case-Based Reasoning*, Lawrence Erlbaum Associates, Publishers.

Rieu, D., Conte, A. & Giraudin, J. P. (2002). Pattern-based environments for information systems development, *The Sciences of Design*, Lyon, France.

Roulson, J., Benbow, E. & Hasleton, P. (2005). Discrepancies between clinical and autopsy diagnosis and the value of post mortem histology; a meta-analysis and review, *Histopathology* 47: 551–559.

Russel, S. & Norvig, R. (2003). *Artificial Intelligence, a Modern Approach*, Prentice Hall.

Scott, D., Miller, W. & Griffin, C. (2001). *Muller and Kirk's Small Animal Dermatology*, 6th edn, Saunders, Philadelphia.

Seruca, I. & Loucopoulos, P. (2003). Towards a systematic approach to the capture of patterns within a business domain, *The Journal of Systems and Software* 67: 1–18.

Shapiro, S. C. (1987). *Encyclopedia of Artificial Intelligence*, Vol. 1 e 2, John Wiley & Sons.

Slayter, M. V., Boosinger, T. R., Pool, R. R., Dämmrich, K., Misdorp, W. & Larsen, S. (1994a). *International classification of tumours of domestic animals histological classification of bone and joint tumors of domestic animals*, Vol. I, Armed Forces Institute of Pathology.

Slayter, M. V., Boosinger, T. R., Pool, R. R., Dämmrich, K., Misdorp, W. & Larsen, S. (1994b). *International classification of tumours of domestic animals histological classification of bone and joint tumors of domestic animals animals*, Vol. Second Series , Vol. I, Armed Forces Institute of Pathology.

Tauzovich, B. (1990). An expert system for conceptual data modelling, *8th International Conference on the Entity-Relationship Approach*, Toronto, Canada.

Tervaert, T., Mooyaart, A., Amann, K., Cohen, A., Cook, H. & Drachenberg, C. (2010). Pathologic classification of diabetic nephropathy, *J Am Soc Nephrol* 21: 556–63.

Tomé, P. & Vala, H. (2010). Cbr technique in veterinary pathological anatomy, *in* J. E. Q. Varajão, M. M. Cruz-Cunha & G. D. Putnik (eds), *Enterprise Information Systems*, Vol. 2, Springer, Viana do Castelo - Portugal, pp. 401–405.

University of Utah of Spencer S. Eccles (2011). For patients: Practice of pathology. Consulted in March of 2011.
URL: *http://library.med.utah.edu/WebPath/HISTHTML/HISTOTCH/HISTOTCH.html*

Valli, V., Jacobs, R.M. andParodi, A. a. W. & Moore, P. (2002). *International classification of tumours of domestic animals histological classification of hematopoietic tumors of domestic animals*, Vol. VIII, Armed Forces Institute of Pathology.

Vos, J., Borst, G., Visser, I., Soethout, K., de Haan, L. & Haffmans, F. (2005). Comparison of clinical and pathological diagnoses in dogs, *Vet Q* 27(1): 2–10.

Watson, I. (1996). Case-based reasoning tools: an review, *2nd UK Workshop on Case-Based Reasoning*, AI-CBR/SGES Publications, University of Salford, pp. 71–78.

Watson, I. (1999). Cbr is a methodology not a technology, *Knowledge Based Systems Journal* 12(5-6).

Wilcock, B., Dubielzig, R. R. & Render, J. A. (2002) *Histological classification of ocular and otic tumors ofdomestic animals*, Vol. IX, Armed Forces Institute of Pathology.

Williams, D. J., Ansford, A. J. S. D. & Forrest, A. S. (1998). *Colour Guide Forensic Pathology*, Churchill Livingstone.

Zarbo, R. & Nakhleh, R. (2009). Surgical pathology specimens for gross examination only and exempt from submission, *Archives of Pathology & Laboratory Medicine* 123(2): 133–139.

5

Veterinary Dysmorphology

Enio Moura[1] and Cláudia T. Pimpão[2]
[1]Service of Medical Genetics,
[2]Laboratory of Pharmacology and Toxicology,
Faculty of Veterinary Medicine, Campus São José,
Agricultural and Environmental Sciences Center,
Pontifícia Universidade Católica do Paraná, Curitiba
Brazil

1. Introduction

Dysmorphology is a branch of clinical genetics that deals with birth defects on a multidisciplinary basis. It involves knowledge of genetics, embryology, pathology, pediatrics and clinical medicine in order to diagnose birth defects and treat patients. Literally, dysmorphology means "the study of abnormal form". It is a word of Greek origin: *dys*, defect, disorder; *morphé*, form; and *lógos*, word, study. The term was coined by David Smith[1] in the 1960s (Aase, 1990), emphasizing its multidisciplinary aspects to differentiate it from traditional teratology. Due to its multidisciplinary nature, a specialist in dysmorphology works alongside cardiologists, ophthalmologists, neurologists, urologists, orthopedists, imaginologists, pathologists, biochemists and other specialists. It is a field that requires not only medical and biological knowledge but also high level of humanitarian feeling and respect for life, with dedication and perception to reduce the suffering of others. This chapter introduces the fundamentals of dysmorphology and offers an overview of this science in the context of veterinary medicine, with significant examples and emphasis on nosological, pathological, embryological, genetic and clinical aspects. The purpose of this is to provide veterinarians with an approach to congenital defects rather than the usual approach, emphasizing that much can be done instead of simply submitting affected animals to euthanasia. Therapeutic methods and surgical procedures will not be discussed, as each condition requires its own procedures, and these can easily be found in textbooks on medical therapy or surgery and also in scientific journals on veterinarian or human medicine.

---

[1] David Weyhe Smith (Oakland, 1926 – Seattle, 1981) was an American pediatrician and professor who dedicated his life to children affected by congenital defects, also providing guidance and support to their parents. He was a pioneer of dysmorphology, publishing a large number of articles on this science, of which he was the grand master. One of his books, *Recognizable Patterns of Human Malformation,* became a worldwide classic in the field of human dysmorphology (Wiedemann, 1991), and is now in its sixth edition, updated by Jones (2006). Aase (1990), in a posthumous homage, referred to David W. Smith as "the best and finest man I ever knew".

2. Dysmorphology in veterinary medicine

In veterinary medicine, dysmorphology is still a neglected field of knowledge, but has begun to take shape in line with the advances in veterinary medical genetics. Its basis is derived from human dysmorphology due to current knowledge of the genomic similarities between man and other vertebrates, especially mammals, showing that morphogenesis is evolutionarily conserved throughout the zoological scale. The inductive molecular mechanisms that form the embryonic pattern are identical in all vertebrates (Opitz et al., 2002). Furthermore, veterinary medicine and human medicine share most of the same methods and techniques, both in terms of diagnosis and therapy. However, in the current stage of the development of veterinary dysmorphology, some minute criteria used in human dysmorphology, especially those concerning the extension of concepts, could not always be adopted here.

2.1 Importance

The meaning of dysmorphology in veterinary medicine becomes clear when one considers that congenital defects in animals cause different types of impact: 1) the obvious suffering of the affected individual; 2) the psychological stress for the owners, who are more affectionate towards animals; 3) the abandonment of affected animals by many owners; and 4) the economic loss suffered by breeders, both of companion and production animals. All of these situations are directly linked to the professional work of the veterinarian, meaning different goals in terms of intervention.

2.2 Objectives

The main objectives of veterinary dysmorphology are humanitarian, professional, scientific, preventive and educational. The goal in humanitarian terms is to strive to minimize the suffering of affected individuals and have to do with the perception that animals are sentient beings, i.e., they can feel pain, discomfort, a number of difficulties and other feelings when they are affected by a defect or illness and, consequently, they suffer. The professional goals are to offer qualified assistance to clients who seek guidance and a solution to the congenital health problems of their animals. With these aims in mind more and more people are seeking veterinary clinics and hospitals, especially those dedicated to small animals. The scientific goals have to do with producing knowledge that will result in benefits for the animals themselves and also human beings. The study of spontaneous animal models of human diseases generates knowledge that could not be obtained from human patients for legal and ethical reasons. There are a great number of these animal models recognized today. New medical and molecular biology technology greatly facilitate these studies. The preventive goals are to attempt to advise owners and breeders of animals to help them make decisions and prevent the birth of new individuals with congenital defects. Scientifically based advice is an important step to avoiding the perpetuation of abnormalities, be they of a genetic nature or caused by environmental factors. The educational objectives include raising awareness that veterinarians, like any other professional, should contribute to the development of a better society in every way possible, encouraging ethical values and respect for human beings, animals, nature and life in general.

3. Dysmorphisms: Concept and classification

3.1 Concept

Dysmorphism[2] is the generic name for abnormalities of the form and structure of an animal's body, being used especially for congenital abnormalities, thus constituting an *inborn error of development*. It is important to remember that the adjective "congenital" means born with the individual, independent of the cause. However, it is not synonymous with either "genetic" or "hereditary". Therefore, there are congenital defects that have a genetic cause (the majority) while others have an environmental cause. The word dysmorphism has a broad meaning and can also refer to alterations that are expressed in deviations of size, position, number and even the coloring of one or more parts of the body. Thus, all of the following examples are dysmorphisms: cleft lip and palate (*cheilopalatoschisis*); absence of the cranial vault and cerebral hemispheres (*anencephaly*); very small jaw (*micrognathia*); heart with the apex to the right-hand side (*dextrocardia*); an out of place kidney (*renal ectopia*); one or two missing eyes (*anophthalmia*); extra toes (*polydactyly*); two heads (*dicephaly*); a simple and hyperplasic congenital stain on the skin (*congenital nevus*). Some are extremely serious and lead to death in early life, such as anencephaly; others cause little or no harm to the animal, such as polydactyly (Fig.1).

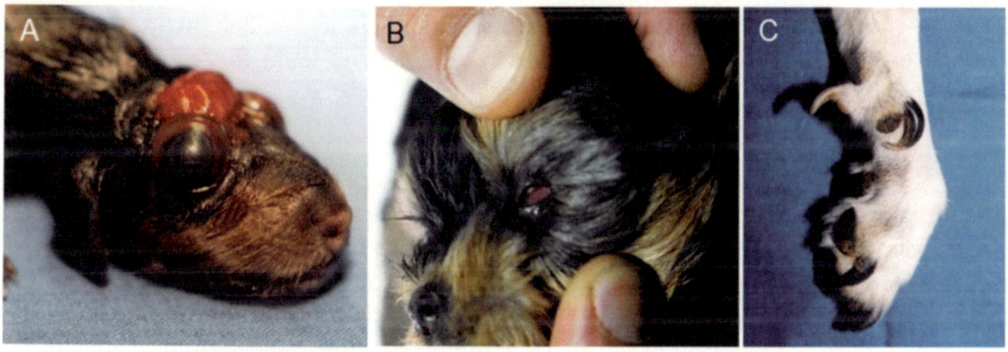

Fig. 1. Examples of dysmorphisms in dogs: A) Anencephaly; B) Anophthalmia; C) Polydactyly.

3.2 Classification

According to the nature of the alteration that causes the dysmorphism and the stage at which it manifests, the phenomenon can be divided into three categories: *malformation, disruption and deformation*. Malformations begin earlier, during the embryonic period, while disruptions appear later. Most deformations begin during the fetal stage, which is when the conceptus grows more rapidly (Kumar & Burton, 2008). If subjacent histological alterations are also considered, a fourth category can be added: *dysplasia*.

[2] Dysmorphism should not be confused with dimorphism, which means the existence of "two different and normal forms" of a given characteristic (from the Greek *di*, two; *morphé*, form), commonly used in the expression "sexual dimorphism" to indicate the natural differences between male and female in a given species.

3.2.1 Malformation

This is a morphological defect of an organ, part of an organ, or a larger region of the body, resulting from an intrinsically abnormal developmental process (Spranger, 1982)[3]. This definition highlights that malformations are disorders in the formation of organs and manifest early, i.e., they are *abnormal processes since the time of their origin*. The affected organ can show varying degrees of imperfection or simply not exist. Malformations arise during blastogenesis or organogenesis and are causally heterogeneous and those that appear earlier (blastogenesis) tend to affect more than one region of the body (Opitz et al., 2002). Although they can be induced by environmental factors, they are commonly caused by gene mutations, chromosomal aberrations or a combination of genetic and environmental factors (Kumar & Burton, 2008). Examples: 1) A nonsyndromic cleft lip and palate can be caused by a recessive autosomal mutation in Brittany spaniels (Richtsmeier et al., 1994), Pyrenees shepherds (Kemp et al., 2009) and boxers (Moura et al., 2011); 2) In Rocky Mountain horses ciliary body cysts, iridal hypoplasia, iridocorneal adhesions and megalocornea are caused by codominant autosomal mutation, with cysts expressed in the heterozygotes and multiple ocular anomalies expressed in the homozygotes (Ewart et al., 2000), whose locus was mapped to the long arm of chromosome 6 (Andersson et al. 2008); 3) The trisomy 18 in cattle causes brachygnathia and is lethal (Herzog, 1974); 4) Cardiac malformations usually have a multifactorial etiology, such as a patent ductus arteriosus in poodles (Buchanan & Patterson, 2003); 5) Neural tube closure defects, such as spina bifida and anencephaly, also have a multifactorial etiology, including mutations in genes involved in folate metabolism (De Marco et al., 2011).

3.2.2 Disruption

A disruption is a morphological defect of an organ, part of an organ or a larger region of the body, resulting from a disturbance in an originally normal developmental process (Spranger, 1982)[4]. This definition highlights that the *formation process of the organs was initially normal, but suffered negative interference during its development*, having affected the organ during the embryonic or fetal stage. In a disruption there is damage to structures of the conceptus because of interference in the blood supply, anoxia, infection or mechanical force (Epstein, 2004), affecting several different tissues and the defect does not respect the boundaries imposed by embryonic developmental process (Aase, 1990). In general, disruptions are caused by environmental factors, but genetic factors can predispose to the development of a disruption (Kumar & Burton, 2008). Examples: 1) A number of therapeutic drugs can cause damage to the embryo when used during pregnancy, especially drugs for continuous use, such as the anticonvulsant valproic acid (Jentink et al., 2010); 2) The feline panleukopenia virus has long been known to cause cerebellar hypoplasia and eventually

[3] In the early 1980s, due to divergences and mistaken nomenclature at that time, specialists in congenital defects formed an International Work Group (IWG) to classify and define terms and expressions for morphogenesis disorders. Their recommendations were published in 1982 (Spranger et al., 1982) and adopted by most dysmorphologists and are still in use today, with very few modifications.

[4] The definitions of malformation and disruption used here correspond, respectively, to the *primary malformation* and *secondary malformation* that were previously in use in dysmorphology. In 1982, the IWG restricted the use of malformation to intrinsically abnormal or primary defects (Opitz, 1984).

hydranencephaly in cats (Résibois et al., 2007); 3) The intrauterine infection caused by the blue tongue virus (BTV serotype 8) causes hydranencephaly in cattle (Wouda et al., 2009); 4) A well known cause of disruption in human beings and which has also been found in the rhesus monkey (Tarantal & Hendrickx, 1987), are amniotic bands that can cause constriction of body structures, leading to facial clefts, amputation of fingers and toes or limbs and other abnormalities.

3.2.3 Deformation

This is a defect of the shape or position of part of the body caused by mechanical forces (Spranger, 1982). These forces may be of internal or external origin and are most evident in the bones, cartilage and joints, as soft tissues tend to return to their original form as soon as the force ceases. Deformations set in during the fetal stage and are normally caused by forces that act late in pregnancy (Epstein, 2004). A large number of deformations have a spontaneous resolution after birth, although this normally takes place slowly. Sometimes, however, treatment is necessary to correct a deformation (Aase, 1990). Like disruptions, although later on, deformations also cause damage to previously intact developed structures with no intrinsic tissue abnormality (Kumar & Burton, 2008). Examples: 1) Club foot as a result of too little space in the uterus and muscular hypotonia or joint laxity, impeding movement of the extremities. Among the factors that reduce uterine space are a very large fetus and a reduction in amniotic fluid, a condition known as oligohydramnios; 2) Increased cephalic perimeter in the hydrocephaly[5] due to pressure from the accumulation of cerebrospinal fluid. The first example is caused by an external force (pressure from the uterus) whereas the second one is caused by an internal force (pressure from the cerebrospinal fluid). It is worth bearing in mind, however, that not all deformations are congenital. They can also develop in postnatal life. For example, bones can be bent because of a lack of calcium or the joints can lose alignment because of the laxity of the joints or lack of calcium. These structures become deformed because of the weight of the body.

3.2.4 Dysplasia

Dysplasia is a morphological defect that results from a disorganization of cells (Spranger, 1982) and other components of a tissue, which, in turn, does not have normal architecture. Therefore, dysplasias are disorders of histogenesis and the alterations occur on a microscopic level and are reflected in the imperfect appearance of the organ or other affected regions of the body. Examples: 1) Congenital alopecia and dental abnormalities in dogs with X-linked hypohidrotic ectodermal dysplasia. There is also absence of piloglandular units in the areas of alopecia (Moura & Cirio, 2004). 2) Thoracic and pelvic limbs are short and bent in dogs affected by chondrodysplasia as can be seen in Great Pyrenees dogs (Bingel & Sande, 1994) or in Labrador retrievers (Smit et al., 2011). Figure 2 shows a visual comparison between normal development and the four types of inborn errors of development (malformation, disruption, deformation and dysplasia).

---

[5] The cause of hydrocephaly is a malformation or disruption that blocks the drainage of cerebrospinal fluid which, in turn, dilates the cerebral ventricles and separates the cranial bones, *deforming* the head.

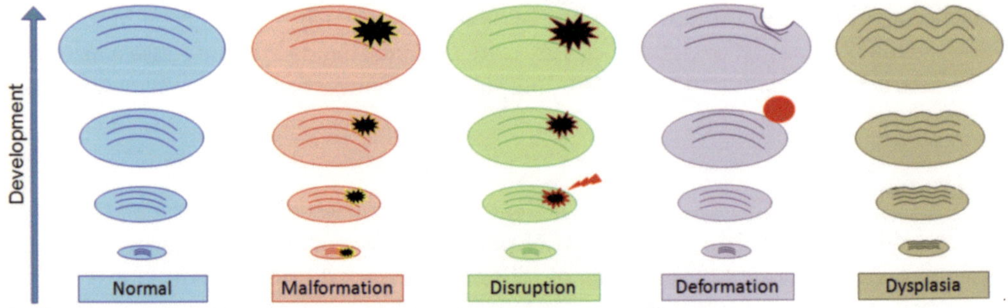

Fig. 2. Types of dysmorphisms compared with normal development.

3.3 Further considerations

Dysmorphisms can also be classified based on criteria other than those presented above. When the medical consequences are considered, they are divided into major and minor dysmorphisms. The former have a significant effect on the quality of life of the affected individual, such as oral clefts, or even impede survival (lethal dysmorphisms), such as anencephaly. The latter have little or no effect on the quality of life of the affected individual and do not require treatment or can be easily treated, such as polydactyly. Dysmorphisms can occur due to increased size or number of body parts and are known as excess dysmorphisms, e.g., extra fingers and toes (polydactyly), duplication of the face (diprosopia) and two heads (dicephaly); or missing body parts, known as reduction dysmorphisms, such as unilateral or bilateral absence of the radius (radial agenesis), missing pinna (anotia) and missing tail (anury). Figure 3 shows some of these defects.

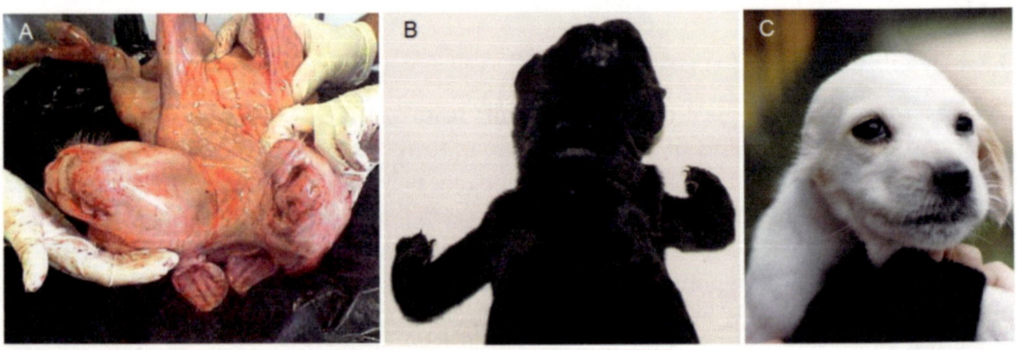

Fig. 3. A) Dicephaly (excess dysmorphism) in extreme premature calf); B) Bilateral radial agenesis in a dog (reduction dysmorphism); C) Anotia in a dog (reduction dysmorphism). Photograph **A** courtesy of Dr. Antonia M. R. B. do Prado and Dr. Joséli Maria Büchele, Laboratory of Anatomy, Faculty of Veterinary Medicine, Pontíficia Universidade Católica do Paraná

4. Epidemiological aspects of dysmorphisms

On its own, a certain type of congenital defect may be rare, but congenital defects as a whole are relatively common in all domestic species and their different breeds and are a significant cause of neonatal and infant mortality and morbidity. Preliminary data suggest that the frequency of dysmorphisms is greater in pigs, followed in descending order by dogs, horses, cattle and cats[6] (Hámori, 1983).

4.1 Frequency, mortality and morbidity

The *frequency* of major dysmorphisms in humans is estimated at 3% of live births (Marden et al, 1964). This number practically doubles by the end of the first year of life, with the identification of defects that manifest later (Kumar & Burton, 2008). In domestic animals, these numbers must be similar or even higher because inbreeding is common in selective breeding both for companion animals and for farm animals, increasing the coefficient of consanguinity. One study (Priester et al, 1970), which included all domestic species, analyzed almost 138,000 individuals and found a rate of 4.68%. However, minor dysmorphisms are not always identified and are seldom included in statistics. The same may occur in the case of major dysmorphisms when they affect internal structures and cause intrauterine or neonatal death and are not always reported. Rates of occurrence vary according to each abnormality and can also vary according to species, breed and geographical location. *Mortality* varies a great deal from one dysmorphism to another. Some are so serious that they cause intrauterine or perinatal death, such as combined heart-lung defects; others are incompatible with life and the affected individual dies soon after birth, as is the case with anencephaly. However, a large number of major dysmorphisms can be corrected surgically or be given some form of treatment, making life viable. Furthermore, minor dysmorphisms have no meaning in terms of mortality. Like mortality, *morbidity* also varies according to the dysmorphism. In many cases, treatment definitively cures the defect and the individual goes on to live a normal life; in other cases, treatment can improve the quality of life, but the individual will have chronic difficulties until the end of his days.

4.2 Risk factors

The artificial selection in breeding to obtain animals with a new or better appearance or for financial gain often means consanguineous unions, increasing the risk of the recurrence of a given defect in future generations. The environment also poses many risk factors during pregnancy, such as: 1) residue from pesticides in water, pasture and other foods; 2) certain toxic plants[7] in the pasture; 3) inadequate conservation of animal feed and other foodstuff

[6] Cats are generally under-represented in samples involving several species. However, veterinarians of small animals often come across congenital defects in purebred cats, which are suggestive of a relatively high rate of dysmorphisms in this species, which is not always reported in journals. They have seen conjoined twins, poliotia, radial agenesis, frontonasal dysplasia, cyclopia, diprosopia, meningocele, anencephaly, spina bifida, hydrocephaly, polydactyly, syndactyly, anal atresia, hypospadias, true hermaphroditism, pseudohermaphroditism and others.

[7] Examples include certain species of the genus *Lupinus* (lupines) that contain teratogenic alkaloids (anagyrine, ammodendrine) that cause a form of congenital arthrogryposis in calves, known as "crooked calf disease". This disease occurs when pregnant cows ingest the plants between the fortieth

which results in mycotoxins; 4) excessive levels of nitrogen-containing food preservatives[8] in low quality animal feed; 5) medicine[9] in the early stages of pregnancy; 6) medicines for continuous use during pregnancy, such as some anticonvulsants; 7) occurrence of certain infectious or metabolic diseases during pregnancy.

5. Etiopathogenesis of dysmorphisms

5.1 Etiology

The *etiological agents* of dysmorphisms can be divided into three large groups: *genetic, environmental and multifactorial*. Although, presumably, any cause of congenital defect is included in one of these groups, a large number of dysmorphsisms have no identified cause and are provisionally separated as a group of *unknown etiology*.

5.1.1 Genetic etiology

In humans, genetic factors are responsible for most dysmorphisms with a known cause (Kumar & Burton, 2008). Considering that mammals in general share most of their genes with humans and that the genes responsible for morphogenesis are highly conserved, the genetic factors must be equally responsible for most of the congenital defects of domestic mammals. These factors are: 1) *dominant autosomal mutations*, such as the fibrillin-1 gene (*FBN 1*), which causes Marfan syndrome in cattle (Singleton et al., 2005). The affected animals have long, thin limbs, laxity of the joints, lens abnormalities, aortic dilation, etc. (Besser et al., 1990); 2) *recessive autosomal mutations*, such as the one that causes anotia, cleft palate and bifid tongue in St. Bernard dogs (Villagómez & Alonso, 1998); 3) *recessive X-linked mutations*, such as the one that occurs in the ectodysplasin A1 and A2 gene isoforms (*EDA-A1* and *EDA-A2*) and causes X-linked hypohidrotic ectodermal dysplasia in dogs (Casal et al. 2005). Affected individuals have imperfect teeth, oligodontia and cutaneous areas with no piloglandular units, as shown in Figure 4 (Moura & Cirio, 2004); 4) *deletions of Y chromosome genes*, such as the one of the sex-determining region Y gene (*SRY*), which determines sex reversal in horses, i.e., it leaves individuals with an male XY karyotype with a female phenotype and ovaries (Raudsepp et al., 2010); 5) *chromosomal aberrations*, such as trysomy of chromosome 30 in horses, which makes them smaller than normal, with a

---

and hundredth day of pregnancy. The affected calves characteristically have bent extremities and can also have scoliosis, torticollis and cleft palate. Ingestion of toxic plants by farm animals can have a serious economic impact on livestock producers in given geographical areas through death losses, abortions, and birth defects (Lee et al., 2008).

[8] It has not been completely established that *N*-nitroso compounds, such as nitrosamine and its precursors (nitrites and nitrates) naturally cause congenital diseases. Studies in animals indicate that they are highly carcinogenic and at higher levels can be teratogenic, but there is a need for further studies (Griesenbeck et al. 2010). Likewise, mycotoxins should also be considered. Fumonisins (frequent contaminants of corn) and ochratoxin-A (often found in stored grains) experimentally cause craniofacial and neural tube defects in mice (Marasas et al., 2004; Ueta et al. 2010).

[9] An example of routine use in veterinary medicine is griseofulvin, an antifungal drug for animals with dermatomycosis. Scott et al. (1975) observed multiple congenital defects in newborn cats whose mothers orally received high doses of griseofulvin (500-1000 mg, at weekly intervals). There were cerebral defects, skeletal defects, cleft palate, anophthalmia with absence of optic nerves, anal atresia, lack of atrioventricular valves and other defects.

serious angular deviation of the thoracic limbs and polydactyly (Bowling & Millon, 1990); 6) *structural chromosomal aberrations*, such as the translocation between the X chromosome and an autosome described by Schelling et al., (2001) in an intersex Yorkshire terrier.

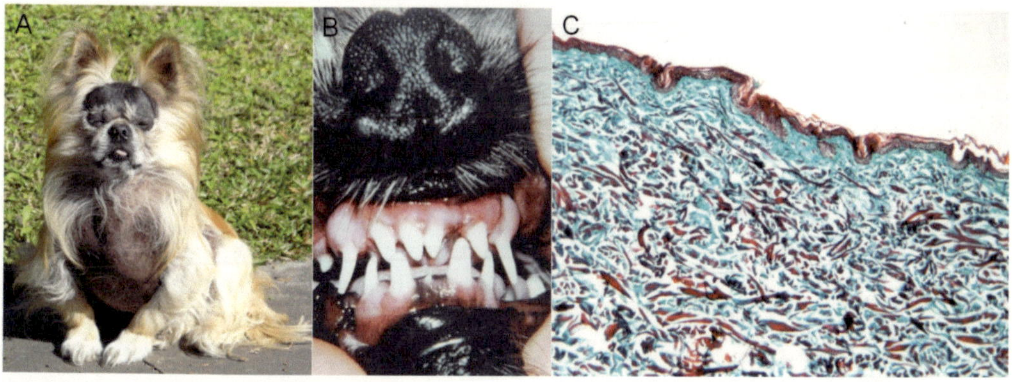

Fig. 4. X-linked hypohidrotic ectodermal dysplasia: A) Congenital absence of hair (congenital alopecia); B) Dental defects (conical teeth and oligodontia); C) Microphotograph of the skin (from an alopecic area) showing absence of hair follicles, sebaceous glands and sweat glands. Shorr stain, X10 objective. Photograph **C** courtesy of Dr. Silvana M. Cirio, Laboratory of Pathology, Faculty of Veterinary Medicine, FEPAR.

5.1.2 Environmental etiology

The environment can be a source of numerous potential teratogens, an important cause of congenital defects in both humans and animals because their presence is not always obvious or the harmful effect of certain products is unknown to the public at large, which increases the risk of maternal exposure to them. *Environmental teratogens* are agents in the *environment* that can negatively impact embryonic development, causing congenital defects. Many chemical products and medicines have been considered potentially teratogenic for decades. Some have been confirmed as such and others have been refused, while the potential of others is doubtful or has been mistakenly confirmed (Koren & Nickel, 2010; 2011). In principle, the risk of a defect is related to the frequency and degree of maternal exposure (Holmes, 2011). Agents that are commonly considered as potentially teratogenic can be separated into four groups: 1) *environmental contaminants,* such as mercury[10], lead, polychlorinated biphenyls, organochlorines and dioxins. There is also considerable doubt

[10] Mercury does not cause gross birth defects, but rather lesions in the central nervous system which cause psychomotor retard, convulsions and other neurological signs (Lancaster, 2011). The sad history of *Minamata disease* was first reported in 1953, but its cause was identified only three years later: a chemical plant discharged water containing inorganic mercury into Minamata Bay in Japan, and this was transformed into organic mercury (methylmercury) by marine microorganisms. Contaminated fish and clams with mythylmercury were consumed as food by the inhabitants of the town of Minamata and this was the source of poisoning which, in the case of pregnant women, affected the fetus, causing *congenital Mimata disease* (Moura, 1993). The number of people affected by the disease was officially established as 2,252, of whom 1,043 had already died 36 years later (Harada, 1995).

concerning their real effect on teratogenic processes. For example, dioxins have been held responsible for a variety of defects. However, that scientific data have so far only confirmed a link to spina bifida (Ngo, 2010); 2) *medicines*[11] such as phenytoin, valproic acid, coumarins and antibiotics; 3) *physical agents* such as heat and radiation[12]; 4) *infectious agents* such as viruses, bacteria and protozoa. Thus, teratogens are of a chemical (dioxins, drugs), physical (heat, radiation) and biological nature (microbes). Non-infectious maternal diseases can also cause dysmorphisms: diabetes mellitus, iodine deficiency, uterine myomas, autoimmune diseases, etc. Uterine dysmorphisms and abnormalities in the extra-embryonic membranes can also cause fetus defects.

5.1.3 Multifactorial etiology

The joint action of several genes (polygenic inheritance) can also play a significant role in the occurrence of dysmorphisms. Deleterious alleles of the different genes play an additive effect until they reach a threshold at which the defect occurs. Environmental factors usually add to this mechanism. Thus, an individual with a given genotype becomes more susceptible to developing a defect, which can be unleashed when environmental forces are at play (multifactorial inheritance). Neural tube closure defects (anencephaly, spina bifida) and conotruncal defects (truncus arteriosus, tetralogy of Fallot) generally have a multifactorial etiology.

5.2 Pathogenesis

The process that gives rise to an abnormal form is a *morphogenetic disturbance* and is known as dysmorphogenesis. The mechanisms involved in dysmorphogenesis are complex, compromising a number of the embryonic developmental phenomena (division of cells, cellular migration, cell adhesion, differentiation, etc.), including cell signaling pathways, e.g., the Sonic hedgehog pathway. They can also include hypoxia resulting from blocked

[11] The tragic history of thalidomide is a warning that the use of medicine during pregnancy should be strictly limited to what is absolutely necessary and should be taken only with medical advice. Thalidomide was launched in 1956, initially as a sedative. By the early 1960s it had become a popular medicine to control morning sickness during pregnancy and could be purchased without a prescription. The following year, Australian obstetrician *William McBride* and German pediatrician *Widukind Lenz* independently saw the link between thalidomide in the early stages of pregnancy and *phocomelia*. The children born with this defect had short limbs close to their bodies and malformed fingers and toes. The abnormality could affect one limb, both arms or all four limbs and often affected internal organs such as the heart and duodenum, constituting a pattern of abnormalities that became known as *thalidomide embryopathy*. It is estimated that there were 8,000 to 10,000 cases worldwide (von Moos et al. 2003). However, thalidomide has been used successfully to treat diseases such as cancer, leprosy and auto-immune diseases.

[12] The *Chernobyl nuclear accident* (Ukraine), in 1986, is a blatant and relatively recent example of the catastrophic effect of *ionizing radiation* on living beings. A number of disorders, including a variety of *malformations*, were registered in animals, from fish to mammals, and even in invertebrates. Congenital defects of the mouth, anus, legs and head were reported in domestic animals. Until today, in animal populations of the contaminated region, there is *transgenerational genomic instability* and *gene mutation rates* are significantly high. In some areas of Europe, the radionuclide levels remain high in mammals, birds, amphibians and fish (Yablokov, 2009).

circulation or non-formation of blood vessels like that caused by antiangiogenic drugs, such as thalidomide and retinoids (Holaday & Berkowitz, 2009). They manifest at the histological level in different ways, such as aplasia, hypoplasia, dysplasia, atrophy, hypertrophy, etc. To follow, we present the fundamental concepts of the developmental field theory, the knowledge of which facilitates the understanding of dysmorphogenesis.

5.2.1 Developmental fields

Developmental field is, initially, the entire embryo in the early stages of its development and, later, it is a region or part of the body of the embryo which responds as a unit to embryonic induction and gives rise to multiple or complex anatomic structures (Spranger et al., 1982; Opitz et al., 2002). Developmental fields are systems that control the progressive differentiation of the structure and size, and also the temporal and spatial distribution of the complex organ components. To understand better the developmental field concept, it is important to remember the meaning of the terms *blastogenesis, organogenesis, morphogenesis, histogenesis* and *phenogenesis* in the context of dysmorphology. *Blastogenesis* is the set of events of embryonic development from fertilization to the end of gastrulation, i.e., it includes phenomena such as the formation of the morula, blastocyst, ectoderm, endoderm, mesoderm, neural tube and midline, in addition to cardioangiogenesis, mesonephrogenesis and curving of the embryo, which then takes the shape of a C (Opitz et al., 2002). At the end of the gastrulation, *organogenesis* begins. This is the set of events that lead to the formation of the organs and other parts of the body (*morphogenesis*) and includes the differentiation of the cells and tissues (*histogenesis*). On average, the duration of blastogenesis is similar in the majority of mammals, but the duration of the phenomena that follow varies from one species to another, especially the fetal period, which is reflected in the duration of pregnancy. The development that ranges from the fetal period and postnatal period to puberty is called *phenogenesis* (Opitz et al., 2002). During the fetal period, phenogenesis is characterized by growth and maturation, preparing the individual for birth. The *primary developmental field* is the field represented by the entire embryo in the early stages of blastogenesis. The initial phenomena of this period include pattern formation, generating components known as **progenitor fields**, which are the primordia of all final structures (Davidson, 1993; Martínez-Frías et al., 1998), as they give rise to the heart, central nervous system and limb buds (Opitz et al. 2002). When the components of a field remain contiguous, they constitute a *monotopic field*, in other words, a developmental field related to the formation of a single area of the body. However, there are fields in which the components separate from one another to give rise to distant final structures among themselves, which are known as *polytopic fields*. Thus, a polytopic field has to do with the formation of structures situated in different areas of the body (Opitz, 1982). *Secondary developmental fields* are the fields formed by the subdivision of progenitor fields, and each of them originates a determined final structure during organogenesis (Martínez-Frías et al., 1998).

5.2.2 Congenital defects and developmental fields

Malformations, disruptions and dysplasias are the result of disorders that occur in one of more developmental field. If they result from alterations that occur during blastogenesis, when the progenitor fields are formed, they are *primary field defects;* if they are the result of alterations that occur during organogenesis, after the formation of the progenitor fields, they are *secondary field defects*. Those that are localized, i.e., the affect only one structure or part of the body, which was differentiated during organogenesis, are *monotopic field defects*; those affecting structures situated in different parts of the body are *polytopic field defects* and originate in the early stages of blastogenesis (Martínez-Frías, 1998).

5.2.3 Heterogeneity, pleiotropy and sequence

Developmental fields are embryonic morphogenetic units and, thus, respond to an inductive stimulus as an epimorphically hierarchical self-organizing unit that is temporally synchronized and spatially coordinated (Opitz, 1982). Consequently, a given abnormal phenotype can originate as a response of a field to different causes. This becomes clearer if we remember that different fields can share parts of the same cell signaling pathway involved in embryonic development. For example, the *Sonic hedgehog cell signaling pathway* is composed of genes that codify proteins that act as intercellular signals in many development processes. Mutations in these genes can result in malformations such as holoprosencephaly, syndactyly, polydactyly, heart defects, kidney defects and anal atresia, or even in neoplasias such as meningioma and squamous cell carcinoma (Cohen Jr, 2004). A congenital defect can be the result of mutations in different genes, chromosomal aberrations or the action of mutagenic agents or teratogens, i.e., it is *causally heterogeneous*. Evidently, the meaning of this is that defects that are clinically the same can have a genetic or environmental cause (Fig. 5). A defect caused by an environmental factor, but which is the same as another caused genetically, is known as a *phenocopy*.

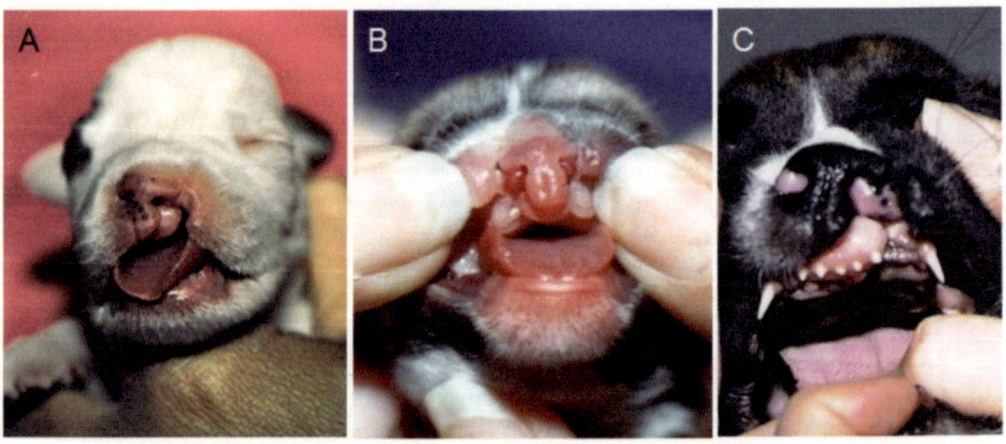

Fig. 5. Cleft lip and palate in dogs are phenotypically and etiologically heterogeneous. They can be unilateral (A) or bilateral (B). They can affect the entire extension of the palate or only the primary palate (C). They can have a multifactorial origin or a monogenic autosomal recessive origin, or be caused by environmental factors (phenocopy).

On the other hand, the same gene can give rise to a series of morphogenetic events that are necessary for the formation of different final structures. This genetic phenomenon in which a gene is responsible for several different phenotypic characteristics is known as *pleiotropy* and is also one of the phenomena that can occur in dysmorphogenesis. For example, a mutation in the gene that codifies ectodysplasin, a protein that regulates the initial epithelial-mesenchymal interaction necessary for the formation of ectodermal derivatives, causes X-linked hypohidrotic ectodermal dysplasia, which occurs in humans, cattle, mice and dogs, and is characterized by defects in the teeth and in the skin (absence of piloglandular units) and other ectodermal derivatives (Moura & Cirio, 2004). In addition to heterogeneity and pleiotropy, another phenomenon may be behind multiple abnormalities: they may be the consequence of only one initial defect that provokes the occurrence of new events in a chain reaction, characterizing a *dysmorphogenetic sequence*. For instance, micrognathia can unleash retroglossoptosis which, in turn, causes cleft palate and respiratory distress, a condition known as Pierre Robin sequence (Mackay, 2011).

5.2.4 More frequently affected systems and selected examples of abnormalities

Cardiovascular system: communication between the pulmonary trunk and the aorta artery (*patent ductus arteriosus*, Fig. 6-A); communication between the atria (*atrial septal defect*) or between the ventricles (*ventricular septal defect*); combination of pulmonary stenosis with a ventricular septal defect, hypertrophy of the right ventricle and an overriding aorta, i.e., the aorta is positioned directly over a ventricular septal defect (*tetratology of Fallot*); imperfect mitral valve and support structures (*mitral dysplasia*); imperfect tricuspid valve and support structures (*tricuspid dysplasia)*; right or left atrium divided by a membrane, leaving it with two chambers (*cor triatriatum*); absence of communication between the right atrium and ventricle (*tricuspid atresia*); lack of differentiation of the origins of the pulmonary trunk, aorta and coronary arteries, forming a common arterial trunk (*truncus arteriosus*).

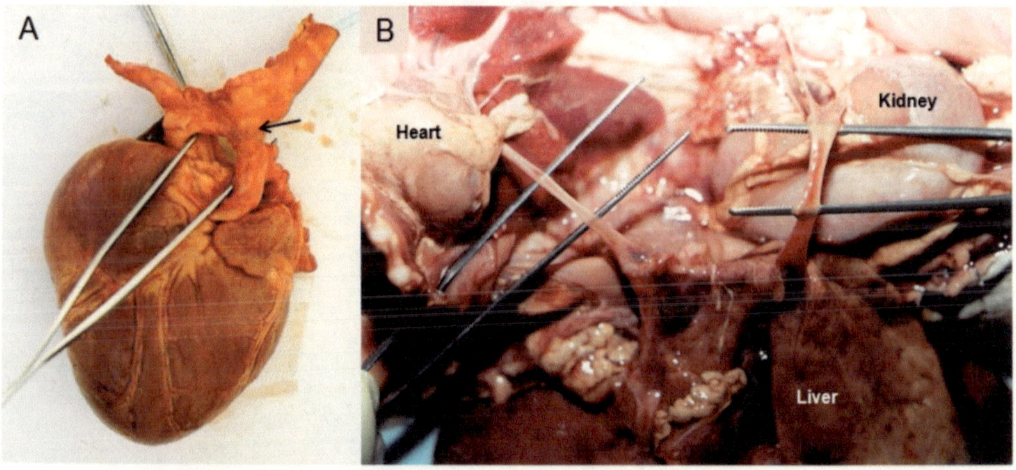

Fig. 6. Examples of abnormalities of the cardiovascular system: A) Patent ductus arteriosus (arrow); B) Abnormality of the caudal vena cava in a dachshund dog with Caroli disease. Note that this vein (raised by forceps) does not have normal morphology. Photograph **B** courtesy of Dr. Marconi R. de Farias, Companion Animal Veterinary Hospital, Faculty of Veterinary Medicine, Pontifícia Universidade Católica do Paraná.

Musculoskeletal system: missing jaw (*agnathia*); shorter-than-normal jaw (*brachygnathia*); abnormally small jaw (*micrognathia*); incompletely formed vertebra (*hemivertebra*, Fig. 7-A and B); missing vertebrae in the lumbar and sacral region (*lumbosacral agenesis*); absence of the radius (*radial agenesis, radial hemimelia*) which may be unilateral or bilateral; absence of one or more limbs (*amelia*); fused digits (*syndactyly*); missing digits and respective metacarpal bones (*ectrodactyly*, Fig. 7-C); existence of extra digits (*polydactyly*); presence of one or more supernumerary limbs (*polymelia*); shorter-than-normal bones due to deficient endochondral ossification (*chondrodysplasia*); incongruence of the elbow joint (*elbow dysplasia*); incongruence of the coxofemoral joint (*hip dysplasia*); congenital joint immobility (*arthrogryposis*); paraumbilical defect in the abdomen with evisceration (*gastroschisis*, Fig. 7-D); congenital umbilical hernia containing viscera (*omphalocele*); congenital cleft of the thoracic wall (*thoracoschisis*), with exteriorization of the heart (*ectopia cordis*).

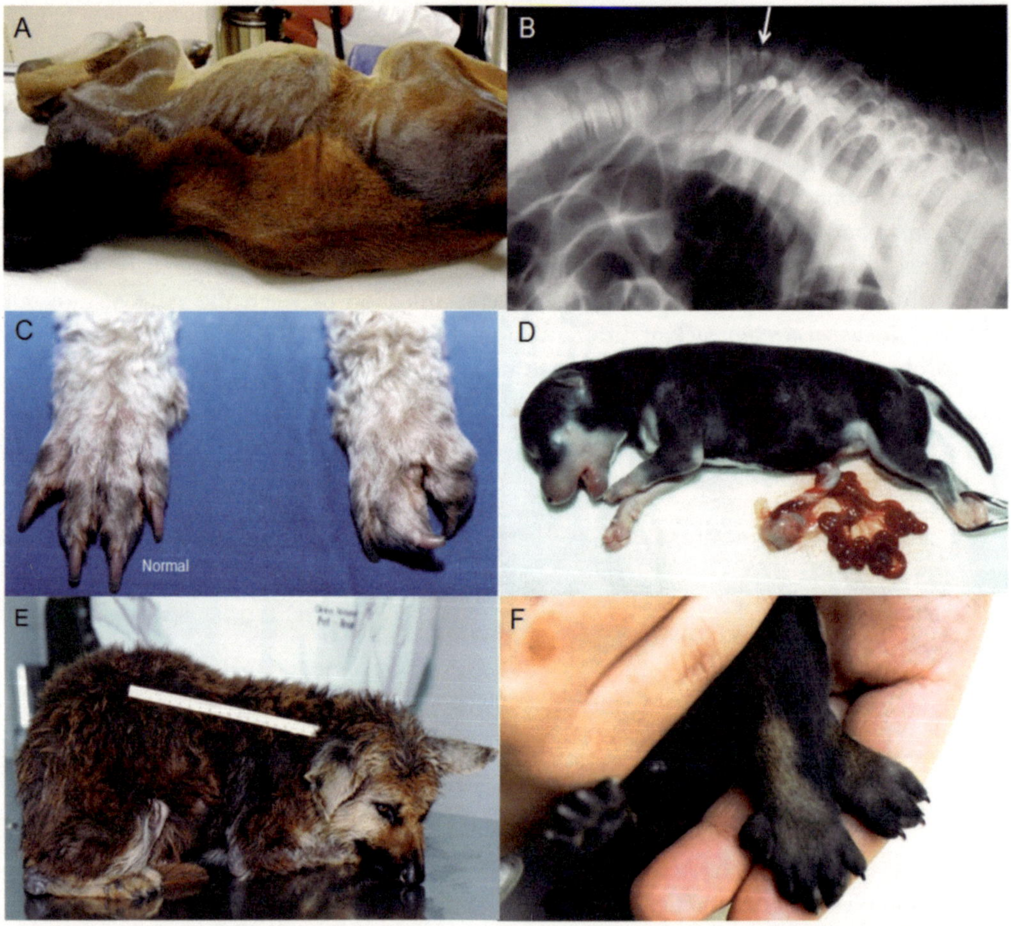

Fig. 7. Examples of abnormalities of the musculoskeletal system: A) Newborn foal with serious scoliosis as a result of spinal defects (hemivertebrae); B) Radiograph of the foal shown in A, revealing the hemivertebrae (arrow); C) Ectrodactyly in a poodle dog (left thoracic limb); D) Gastroschisis in a newborn Siberian husky; E) Pituitary dwarfism in a German shepherd at the age of 4 years; F) Duplication of the left hind paw of a newborn Rottweiler (polypodia). Photographs **A** and **B** courtesy of Dr. Pedro V. Michelloto Junior, Equine Veterinary Hospital, Faculty of Veterinary Medicine, Pontifícia Universidade Católica do Paraná.

Nervous system: absence of the upper parts of the brain and skull (*anencephaly,* Fig.1-A); retroflection of the head and partial or total absence of the cervical vertebrae associated with serious defects of the central nervous system (*iniencephaly*); stenosis of the aqueduct of Sylvius or another part of the brain's drainage system, leading to accumulated cerebrospinal fluid in the cerebral ventricles (*hydrocephaly,* Fig. 8-A); meningeal herniation, i.e., a protrusion of the meninges at some point of the spine (*meningocele,* Fig. 8-B); meningeal herniation containing a segment of the spinal cord (*meningomyelocele*); meningeal herniation containing part of the brain (*meningoencephalocele*); absence of the major connection between the two cerebral hemispheres (*agenesis of the corpus callosum*); failed development of the prosencephalon (forebrain), compromising the separation of the cerebral hemispheres and causing defects in the midline of the face (*holoprosencephaly*); absence of brain gyri (*lissencephaly*); presence of small and multiple brain gyri (*polymicrogiria*); enlargement of the brain gyri (*pachygiria*); lower than normal development of the cerebellum (*cerebellar hypoplasia*).

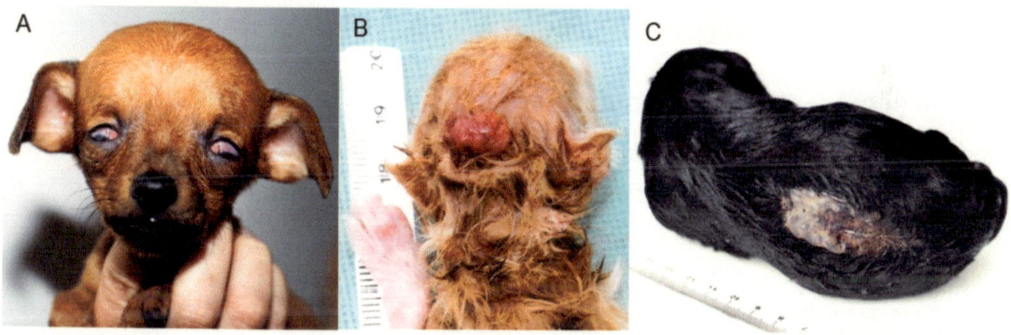

Fig. 8. Examples of abnormalities of the central nervous system: A) Hydrocephaly in a dachshund puppy ; B) Meningocele in a newborn Persian kitty; C) Spina bifida occulta with neural involvement in a newborn Rottweiler puppy.

Genitourinary System: missing kidney (*renal agenesis*) that can be unilateral or bilateral; disorganized renal tissue (*renal dysplasia*); cysts in the kidneys (*polycystic renal disease*); fusion of the kidneys by one of the poles, generally the lower, with a silhouette forming in the shape of a horseshoe (*horseshoe kidney*); proximal bifurcation of the renal pelvis (*bifid renal pelvis*); ureter flowing into an organ other than the bladder, commonly in the urethra or the vagina and even the uterus (*ureteral ectopia*); failure to close the urachus (*patent urachus*); urethral opening located on a point of the ventral side of the penis, scrotum or perineum (*hypospadias,* Fig. 9-A); urethra with a connection to the rectum (*urethrorectal fistula*); cystic tumors formed of tissue from different embryonic germ layers that are strangers to the affected organ (*teratoma*). They are more frequent in ovaries and testicles, but can occur in the prostate or in organs from other systems; absence of vagina (vaginal agenesis); larger-than-normal clitoris (*clitoromegaly,* Fig. 9-B); much smaller-than-normal penis (*micropenis*); testicle retained in the abdomen or inguinal canal (*cryptorchidism*), which can be unilateral or bilateral; presence of a membrane that totally or partially closes the vaginal opening (*persistent hymen*); presence of a fibrous ligament between the gland and the foreskin (*persistent penile frenulum*). In domestic mammals, unlike human beings where the hymen and the penile frenulum are characteristics found in adults, there is a regression of these structures during development.

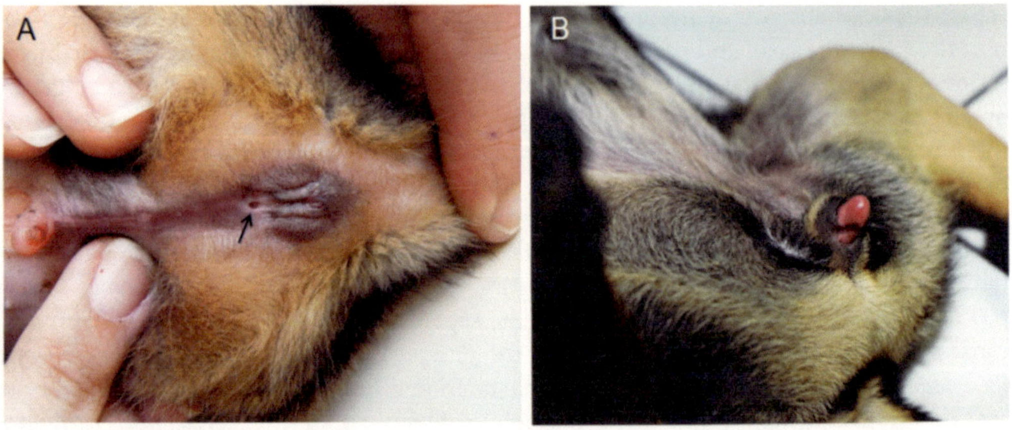

Fig. 9. Examples of abnormalities of the genitourinary system: A) Perineal hypospadias in a dog. Observe the urethral opening (arrow) located in the perineum, near the anus; B) Ambiguous genitalia in a dog with XX true hermaphroditism.

6. Clinical presentation of dysmorphisms

Dysmorphisms can affect an individual as a *single defect* (Fig. 10-A and B) or a set of defects (*multiple congenital anomalies*). The latter include syndromes, developmental field defects, associations and sequences. All arise from disorders in one of more developmental field.

6.1 Single defect

A *single defect* (*isolated defect*) affects only one specific part of an organ or a single organ or a local region of the body. Single defects include an eyelid coloboma, a cleft palate, a patent ductus arteriosus and a hypospadias. Most congenital defects are single and their etiology is often multifactorial. A given type of single defect may be viewed as part of certain syndromes, pointing out that different causes can affect the same developmental field, leading to the same result (Aase, 1990). Single defects are monotopic defects, which are the result of a disturbance in a single monotopic primary field or in a secondary field.

6.2 Multiple congenital anomalies

The occurrence of two or more defects in one or more parts of the body of the same individual is characterized as a *multiple congenital anomaly* (Fig. 10-C). However, the concomitant occurrence of anomalies does not define the etiology. A given pattern of defects may indeed have a specific etiology or may be caused by a variety of possible agents that affect the same developmental fields or a same monotopic or polytopic field.

6.2.1 Syndrome

A syndrome is a *set of defects* that occur simultaneously in the same individual and result from the *same cause*. Therefore, the pattern of defects must be repeated in all affected individuals. Syndromes are defined after several cases have been described, enabling one to determine which defects actually compose the syndrome and which are accidental. Strictly speaking, however, the cause is not always known, but it is presumed. Generally, syndromes are considered as having resulted from a disorder in more than one developmental field (Spranger, 1982). Examples: 1) Hurler Syndrome (mucopolysaccharidosis I), a recessive autosomal syndrome in which there is a deficiency of alpha-L-iduronidase. Affected dogs have corneal opacity, facial dysmorphism, joint abnormalities, aortic dilation and thickened heart valves (Traas et al., 2007); 2) Hunter Syndrome (mucopolysaccharidosis II), an X-linked recessive syndrome in which there is iduronate-sulfatase deficiency. Affected dogs have a course facial appearance, macrodactyly, corneal dystrophy and neurological alterations (Wilkerson, 1998).

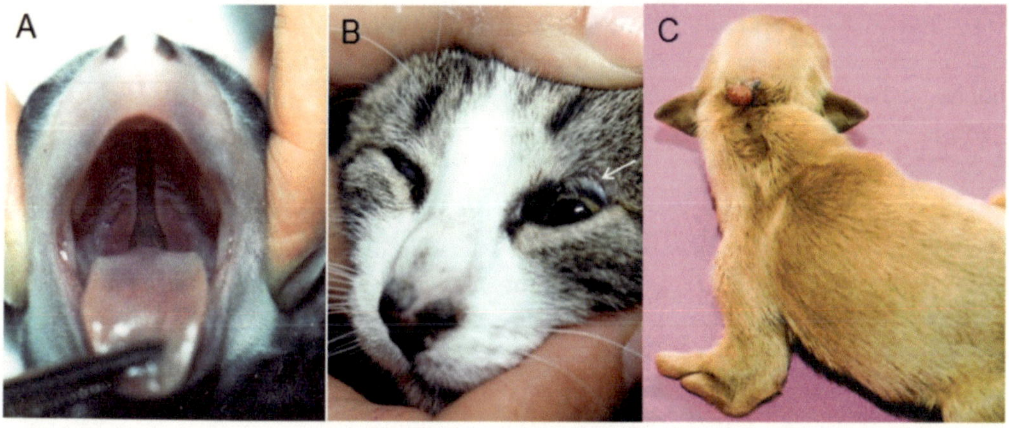

Fig. 10. Examples of single defect: A) Dog with cleft palate; B) Eyelid coloboma (arrow) in a cat; C) Example of a multiple congenital anomaly in a dog (meningocele and peromelia).

6.2.2 Developmental field defects

Developmental field defects are resulting from a *dysmorphogenetic response* of a developmental field (Hersch et al., 2002) and are causally heterogeneous. This means that the *same field reacts in a similar way to different factors*. They may be multiple anomalies that affect an area of anatomically related structures, being monotopic field defects; or they may be anomalies in different areas of the body, being polytopic field defects. Holoprosencephaly in its different degrees is a monotopic field, such as semilobar holoprosencephaly in Morgan horses (Kock et al., 2005), and may be associated with otocephaly, as reported by Martínez et al., (2006) in a Rottweiler dog. Lumbosacral agenesis in its different degrees is a polytopic field defect and has been described in a number of domestic species, such as sheep (Denis, 1975), cattle (Jones, 1999; Son et al., 2008), pigs (Avedillo & Camón, 2007) and dogs (Araújo et al., 2008).

6.2.3 Association

Association is a *combination of two or more defects* in the same individual and occurs in the population more frequently than could be expected merely at random, but it does *not have a specific cause*. Like syndromes, associations must be defects involving more than one developmental field (or only one polytopic field), but are etiologically heterogeneous.

Example: VACTERL (or VATER) is an acronym for the non-random co-occurrence of vertebral anomalies (V), anal atresia (A), cardiac malformations (C), tracheoesophageal fistula (TE), renal anomalies (R) and limb defects (L). The presence of two of these anomalies in the same patient is considered sufficient for diagnosis (Källén et al., 2001), although other researchers are of the opinion that at least three are required (Martínez-Frías, 1994; Faivre et al., 2005). The defects that compose VACTERL association originate during blastogenesis and the existence of a characteristic pattern of abnormalities caused by different factors suggests a dysmorphogenetic response from a primary developmental field (Hersch et al., 2002). For this reason, Martínez-Frías & Frías (1999) consider it a primary, polytopic developmental field defects. In animals, the first record of a spontaneous occurrence of VACTERL association was made by Moura et al. (2010) in a female cat with costovertebral defects, anal atresia, heart and kidney defects, bilateral radial agenesis and persistent cloaca, but no tracheoesophageal fistula.

6.2.4 Sequence

This is a set of defects in an individual and results from an *initial defect* (malformation, disruption, dysplasia or deformation) that interferes in the development process, *leading to additional defects*. For example, as the result of a persistent right aortic arch, a ring is formed around the thoracic esophagus, causing an increased diameter of the esophagus in the cervical segment (a pouch is formed cranially to the point of constriction). A bilateral elbow dysplasia can cause a luxation of the elbows, which leads to medial rotation of the forearms, deforming the carpus. Micrognatia causes retroglossoptosis, which in turn can cause a cleft palate and respiratory distress (Pierre Robin sequence).

6.3 Further considerations

The same dysmorphism can appear in isolation or be part of syndromes or associations. Single defects or multiple defects of the same area (monotopic defects) without anomalies in other areas of the body are often called *nonsyndromic defects* to distinguish them from the same defects in an individual affected by a syndrome. A given defect can be a manifestation of the same anomaly expressed in different degrees, as is common in the developmental field defects. Any degree of expression of these cases is part of the total spectrum of a multiple anomaly. For instance, a single upper central incisor or an iris coloboma may be a minimum clinical sign of the total spectrum of holopresencephaly (Fig. 11-A), which at the most serious degree includes anomalies such as cyclopia, proboscis, absence of optic nerves and cleft lip and palate (Spranger, 1982; Jones, 2006). Dysmorphisms that affect the right and left side of the body (*symmetrical dysmorphisms*) are more likely to have a genetic cause than those that affect only one side (*asymmetric dysmorphisms*). However, this is not a rule (Turnpenny & Ellard, 2007). When a dysmorphism is extenal it is easily observed, especially if it is a major dysmorphism. Internal dysmorphisms, in turn, can only be confirmed by imaging tests, endoscopic viewing or a postmortem examination. Nevertheless, affected individuals will show clinical signs (Fig.11-B and C) at some time during their development, enabling a diagnosis using the aforementioned examinations. Internal congenital defects must be on the list of possible diagnoses in stillborn and critically ill newborn animals, although they may look normal externally. Internal congenital defects too must be considered a possible cause of clinical conditions identified in the first months of life. Although most of these defects are identified in the first six months, there are cases that manifest clinically later on and are diagnosed in adult life. For example, many cases of dogs with pulmonic stenosis show no clinical signs for years (Ware, 2011).

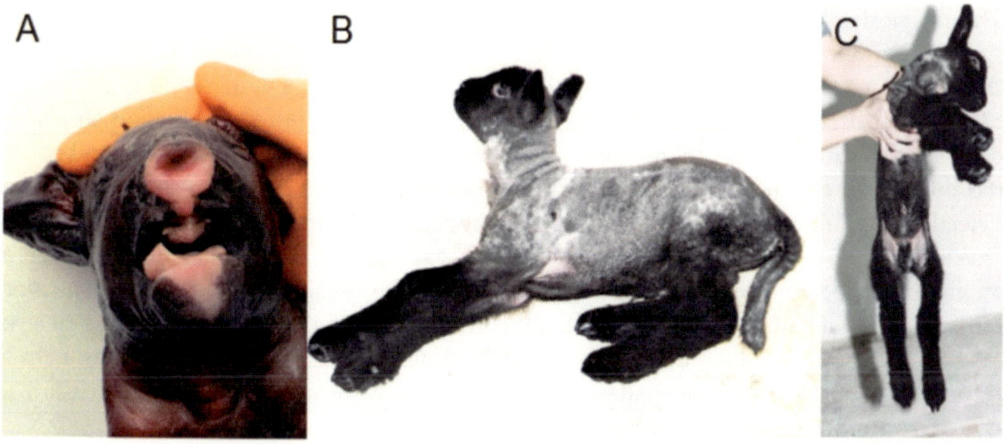

Fig. 11. A) Holoprosencephaly in a pig; B and C) Clinical signs of cerebellar disorder in a lamb (backward contraction of the neck, thoracic limbs in hyperextension and flexed pelvic limbs) as a result of cerebellar hypoplasia.

7. Medical history, physical examination and management of dysmorphic patients

In dysmorphology, the clinical activities that lead to the diagnosis (medical history, physical examination, complementary examinations, etc.) and the therapeutic procedures are that same as those of clinical medicine and/or surgery, with the inherent aspects of each specialty and any adaptations that each dysmorphic condition may require. In any situations, a correct diagnosis is very important for making the best decision. It has implications in terms of prognosis and possible treatment. All procedures should be noted according to well defined criteria in order to obtain a complete and consistent record, whose importance in terms of diagnosis and therapy transcends each case, enabling comparisons with new cases and generating innovations. The following topics introduce some general observations concerning these aspects.

7.1 Medical history

A detailed and solid medical history is one of the pillars of the diagnosis. It should include information about the environment in which the mother lives or lived, maternal diseases (metabolic disorders, infectious diseases, etc.), the mother's food, any medication given to the mother during pregnancy, and particularities of pregnancy and labor. Whenever possible, the veterinarian should seek not only similar cases but also different defects between sibs of the affected animal and other relatives. A pedigree should be drawn that is as complete as possible to facilitate the genetic analysis and enable the identification of an inheritance pattern when there is one.

7.2 Physical examination

The physical examination of a dysmorphic patient follows clinical principles, as in any specialty, and should be detailed. In addition to obvious defects, other abnormalities that the animal may have should be sought, even if they are only minimal. All anatomical regions and organic systems should be examined. All inspection and palpation should be done carefully in order not to miss any minor anomalies. One should always remember that the animal hair coat may make it difficult to see an anomaly but palpation contact facilitates detection. Minor anomalies, which may at first appear irrelevant, may help in the diagnosis. One should also remember that external defects, even if there is only one, may be a sign of an internal defect, meaning that specific examinations are required to locate them. The more minor defects at patient has, the greater the chance that it will also have major defects. Measurements of various segments of the body should be made routinely as they help to recognize patterns or define patterns of congenital malformations. An examination of normal relatives could also be useful.

7.3 Diagnostic tests

In many cases the diagnosis can be obtained with a minimum number of diagnostic tests. However, in other cases, a large number of tests is required, which significantly increases the cost, which often makes a more in-depth evaluation impossible in veterinary medicine. Some tests can only be conducted at universities, research centers or large private clinics. Imaging tests (simple X-ray, contrast X-ray, tomography, magnetic resonance imaging, ultrasound, endoscopy) are especially useful, as are karyotype analyses, biochemical examinations and, when available, DNA tests. The affected animals that do not survive should be submitted to a detailed postmortem examination. This examination provides diagnostic information and can help to develop treatment strategies for future cases.

7.4 Care and treatment

Therapy varies according to the type of defect. Very serious dysmorphisms often make life impossible, while less serious types often require no treatment. When treatment is possible, it can vary from methods that improve the quality of life of the patient to full correction of the defect. For instance, in the first case, a system of wheels for a dog that was born with no pelvic limbs (nowadays, there are "wheelchairs" for dogs and cats that are sold commercially); in the second case, a surgery to correct a cleft lip and palate or to correct a patent ductus arteriosus. There are cases where simple care is all that is required, such as giving a dog with oligodontia soft and previously cut food.

8. Prevention of dysmorphisms

Preventing dysmorphisms involves three main points: *genetic counseling, screening* and *monitoring during pregnancy*. The former two apply when the defect is hereditary or is influenced by genetic factors. The latter is useful in any situation.

8.1 Genetic counseling

The basic recommendation in the case of genetic diseases in animals is that affected individuals should not breed and that normal couples with affected descendents should not breed again. However, owners of normal animals that have had affected descendents are not usually willing to follow this advice. With this in mind, when there is a correct diagnosis and the genetic nature of the defect is known, other decisions are possible, but the risk of recurrence should be taken into consideration. To avoid autosomal recessive phenotypes, an important strategy is never to mate individuals who are known to be heterozygotes one with another, such as those who have already had an affected descendent. For recessive X-linked phenotypes, daughters of affected individuals are all carriers (heterozygotes) and should not be mated, even when the males are normal. When there is a family history of a recessive defect and the zygosity of an individual is unknown, consanguineous unions should be avoided.

8.2 Screening

When breeding small or large animals, identifying heterozygotes (recessive phenotypes and dominant phenotypes with reduced penetrance) can be an economically advantageous procedure and allows people to make appropriate preventive decisions concerning the mating of such individuals. For several recessive heredopathies commercial DNA tests are now available, significantly reducing costs. For instance, there is a test for arthrogryposis multiplex in cattle; for pulmonary hypoplasia with anasarca (PHA) in Dexter cattle; for polycystic kidney disease in cats, for Sly disease (mucopolysaccharidosis VII) in dogs; for hereditary microphthalmia in sheep. In abnormalities in which there is incomplete dominance between the mutant and normal allele, the heterozygotes can be identified visually because they have an abnormal phenotype, although it is less serious than that of homozygotes for the mutant allele. For example, there is a type of chondrodysplasia in cattle, common in the Dexter breed, in which the heterozygotes have shorter than normal limbs. If two chondrodysplasic individuals breed, there is a 50% risk of recurrence of the parents' phenotype (heterozygotes) and a 25% risk of a very serious and lethal chondrodysplasia, popularly known as a "bulldog calf", when the affected animals have very short limbs and spine, and their heads somewhat resemble that of a bulldog.

8.3 Monitoring during pregnancy

Several aspects deserve attention during pregnancy: 1) Care should be taken not to expose the pregnant animal to teratogens in the environment (mercury, herbicides, toxic plants, organochlorides, etc.); 2) The animals should not be given medicine with teratogenic potential; 3) They should not be submitted to radiographic examinations. If this cannot be avoided, the abdomen of the pregnant animal should be protected with a lead apron or the examinations should be conducted in the later stages of the pregnancy; 4) Examinations that use radioactive contrast (scintigraphy) should not be conducted; 5) Care should be taken concerning the quality and conservations of animal food. Feed that is exposed to dampness could lead to contamination by mycotoxins; 6) Food should be supplemented with folic acid[13]; 7) Pregnant animals that suffer from diabetes should be monitored. In humans, diabetes mellitus is a significant risk factor when it comes to congenital defects. This may also be the case in animals.

9. Ethical aspects in veterinary dysmorphology

When a dysmorphic animal is born, its owners can have many different reactions: pity, fear or repulsion, to name a few. They also make different decisions: some will care for the animal, while others may abandon it or kill it. Those who seek a veterinarian wish to give the animal to a new owner, treat it or sacrifice it. What will be done depends on the moral and ethical principles of each owner and each veterinarian. The most common argument is that, to avoid a life of suffering, the affected animal should be euthanatized. In this case, some points have to be considered: 1) In the case of serious malformations that are incompatible with life or when there are no technical resources to treat the animal properly, the ethical choice is undoubtedly to shorten the suffering; 2) Many cases that initially seem to be incompatible with life can be fully resolved with adequate treatment, while in other cases the animal may not even require surgery to live well, as long as it received proper care and affection. This includes some cases of conjoined twins. Unlike humans, animals have no awareness of their appearance and do not suffer because of it and can live happy lives if their owner gives them care and affection; 3) If the defects are not extensive, there are owners who decide not to submit their animal to surgery when they realize that it is living well and is not suffering and they develop an emotional tie and grow used to its different appearance; 4) There are cases when medical treatment is enough to make the animal comfortable, even though it may not completely cure the deficiency; 5) There are people who cannot bear the idea of looking after dysmorphic animals, while others do not mind and feel gratified when they see that they are keeping the animal alive. These latter are often willing to adopt animals with deficiencies; 6) The treatment and monitoring of dysmorphic animals generates technical and scientific knowledge that results in improved efficiency and quality of medical and nursing care for new cases in both animals and humans. However that may be, the decisions reflect the socioeconomic and cultural status of each society.

10. Acknowledgements

The authors would like to thank all their veterinarian colleagues who kindly provided photographs to illustrate this chapter and also those who over the years have referred patients with dysmorphisms. Dr. Moura would especially like to thank Dr. Ana Maria Della Torre (Clínica Veterinária Vida Animal), Dr. Antonia Maria do Rocio Binder do Prado (Pontíficia Universidade Católica do Paraná), Dr. Enio Celso Heller (Clínica Veterinária

[13] Supplementing food with folic acid has been recommended to prevent neural tube defects and cleft lip and palate in humans. This food supplement for pregnant Boston terriers, a breed of dog that is particularly susceptible to cleft lip and palate, significantly reduced the occurrence of these abnormalities in controlled studies (Elwood & Colquhoum, 1997; Guilloteau et al., 2006).

Guaíra), Dr. Jaime Trevisan Ribeiro (Clínica Veterinária Pedigree), Dr. José Leônidas Wagner (Clínica Veterinária Pet House), Dr. Joséli Maria Büchele (Pontíficia Universidade Católica do Paraná), Dr. Luiz Carlos Leite (Universidade Estadual do Centro-Oeste, PR), Dr. Marconi Rodrigues de Farias (Pontíficia Universidade Católica do Paraná), Dr. Pedro Vicente Michelloto Junior (Pontíficia Universidade Católica do Paraná), Dr. Robson Gomes da Costa Gouveia (Clínica Veterinária Guaíra), Dr. Silvana Maris Cirio (FEPAR) and Dr. Valter da Silva Queiroz (Pontíficia Universidade Católica do Paraná).

11. References

Aase, J. M. (1990). *Diagnostic dysmorphology*, Plenum Medical Book Company, ISBN 0-306-43444-X, New York, USA

Andersson, L.S.; Juras, R.; Ramsey, D.T.; Eason-Butler, J.; Ewart, S.; Cothran, G. & Lindgren, G. (2008). Equine Multiple Congenital Ocular Anomalies maps to a 4.9 megabase interval on horse chromosome 6. *BMC Genetics*, Vol.9, No.88, pp. 1-10, ISSN 1471-2156

Araújo, B. M.; Kemper, B.; Figueiredo, M. L.; Chioratto, R. & Tudury, E. A. (2008). Perosomus elumbus em cão beagle. *Ciência Veterinária nos Trópicos*, Vol.11, No.1, pp. 36-39, ISSN 1415-6326

Avedillo, L. J. & Camón, J. (2007). Perosomus elumbis in a pig. *Veterinary Record*, Vol.160, No.4, pp. 127-129, ISSN 0042-4900

Besser, T. E.; Potter, K. A.; Bryan, G.M. & Knowlen, G. G. (1990). An animal model of the Marfan syndrome. *American Journal of Medical Genetics*, Vol. 7, No.1, pp. 159-165, ISSN 0148-7299

Bingel, S.A. & Sande, R. D. (1994). Chondrodysplasia in five Great Pyrenees. *Journal of the American Veterinary Medical Association*, Vol.205, No.6, pp. 845-848, ISSN 0003-1488

Bowling, A.T. & Millon, L.V. (1990). Two autosomal trisomies in the horse: 64,XX,-26,+t(26q26q) and 65,XX,+30. *Genome / National Research Council Canada*, vol.33, No.5, pp.679-682.

Buchanan, J. W. & Patterson, D. F. (2003) Etiology of patent ductus arteriosus in dogs. *Journal of Veterinary Internal Medicine*, Vol.17, No.2, pp. 167-171, ISSN 0891-6640

Casal, M. L.; Scheidt, J. L.; Rhodes, J. L.; Henthorn, P. S., & Werner, P. (2005) Mutation identification in a canine model of X-linked ectodermal dysplasia. *Mammalian Genome*, Vol.16, No.7, pp. 524-531, ISSN 0938-8990

Cohen Jr, M. M. (2004). An introduction to sonic hedgehog signaling, In: Epstein, C. J.; Erickson, R.P. & Wynshaw-Boris, A. *Inborn errors of development*, pp. 210-228, Oxford, ISBN 0-19-514502-X, New York, USA

Davidson, E.H. (1993) Later embryogenesis: regulatory circuitry in morphogenetic fields. *Development*, Vol.118, No.3, pp. 665-90, ISSN 0950-1991

De Marco, P.; Merello, E.; Cama, A.; Kibar, Z. & Capra V. (2011) Human neural tube defects: Genetic causes and prevention. *Biofactors*, article first published online: 14 jun 2011, doi: 10.1002/biof.170, ISSN 1872-8081

Dennis, S. M. (1975) Perosomus elumbis in sheep. *Australian Veterinary Journal*, 1975 Vol.51, No.3, pp. 135-136, ISSN 0005-0423

Elwood, J.M. & Colquhoun, T.A. (1997). Observations on the prevention of cleft palate in dogs by folic acid and potential relevance to humans. *New Zealand Veterinary Journal*, Vol.45, No.6, pp. 254-6, ISSN 0048-0169

Epstein, C. J. (2004). Human Malformations and their genetic basis, In: Epstein, C. J., Erickson, R.P., Wynshaw-Boris, A. *Inborn errors of development*, pp. 3-9, Oxford, ISBN 0-19-514502-X, New York, USA

Ewart, S.L.; Ramsey, D.T.; Xu, J. & Meyers, D. (2000). The horse homolog of congenital aniridia conforms to codominant inheritance. *Journal of Heredity*, Vol.91, No.2, pp.93-98, ISSN 0022-1503

Faivre, L.; Portnoï, M.F.; Pals, G.; Stoppa-Lyonnet, D.; Le Merrer, M; Thauvin-Robinet, C.; Huet, F.; Mathew, C.G.; Joenje, H.; Verloes, A. & Baumann, C. (2005). Should chromosome breakage studies be performed in patients with VACTERL association? *American Journal of Medical Genetics A*, Vol.137, No.1, pp. 55-58, ISSN 1552-4825

Griesenbeck, J.S.; Brender, J.D.; Sharkey, J.R.; Steck, M.D.; Huber Jr, J.C.; Rene, A.A.; McDonald, T.J.; Romitti, P.A.; Canfield, M.A.; Langlois, P.H.; Suarez, L. & National Birth Defects Prevention Study. (2010). Maternal characteristics associated with the dietary intake of nitrates, nitrites, and nitrosamines in women of child-bearing age: a cross-sectional study. *Environmental Health*, Vol.9, No.10, pp.1-17, ISSN 1476-069X

Guilloteau, A.; Servet, E.; Biourge, V. & Ecochard, C. (2006). Folic acid and cleft palate in brachycephalic dogs. *WALTHAM Focus*, Vol.16, No.2, pp. 30-33

Hámori, D. (1983). *Constitutional disorders and hereditary diseases in domestic animals*, Elsevier, ISBN 0-444-99683-4, Amsterdam, Netherlands

Harada, M. (1995) Minamata disease: methylmercury poisoning in Japan caused by environmental pollution. *Critical Reviews in Toxicology*,Vol.25, No.1, pp. 1-24.

Hersh, J.H.; Angle, B.; Fox, T.L.; Barth, R.F.; Bendon, R.W. & Gowans, G. (2002) Developmental field defects: coming together of associations and sequences during blastogenesis. *American Journal of Medical Genetics*, Vol.110, No.4, pp. 320-323, ISSN 0148-7299

Herzog, A. (1974). Autosomal trisomy in lethal brachygnathia of cattle (bovine trisomia-brachygnathia-syndrome). [Abstract in English]. *Deutsche Tierärztliche Wochenschrift*, Vol.81, No.4, pp.78-80, ISSN 0012-0847

Holaday, J. W. & Berkowitz, B. A. (2009) Antiangiogenic drugs: insights into drug development from endostatin, avastin and thalidomide. *Molecular Interventions*, Vol.9, No.4, pp. 157-166, ISSN:1534-0384

Holmes, L.B. (2011) Human teratogens: update 2010. *Birth Defects Research. Part A, Clinical and Molecular Teratology*, Vol.91, No.1, pp. 1-7, ISSN 1542-0752

Jentink, J.; Loane, M.A.; Dolk, H.; Barisic, I.; Garne, E.; Morris, J.K. & de Jong-van den Berg, L.T. (2010). Valproic acid monotherapy in pregnancy and major congenital malformations. EUROCAT Antiepileptic Study Working Group. *The New England Journal of Medicine*, Vol.362, No.23, pp. 2185-2193, ISSN 0028-4793

Jones, C.J. (1999). Perosomus elumbis (vertebral agenesis and arthrogryposis) in a stillborn Holstein calf. *Veterinary Pathology*, Vol.36, No.1, pp. 64-70, ISSN 0300-9858

Jones, K. L. (2006) *Smith's Recognizable Patterns of Human Malformation*, Elsevier Saunders, ISBN 0-7216-0615-6, Philadelphia, USA

Källén, K.; Mastroiacovo, P.; Castilla, E.E.; Robert, E. & Källén, B. (2001). VATER non-random association of congenital malformations: study based on data from four malformation registers. *American Journal of Medical Genetics*, Vol.101, No.1, pp. 26-32, ISSN 0148-7299

Kemp, C.; Thiele, H.; Dankof, A.; Schmidt, G.; Lauster, C.; Fernahl, G. & Lauster R. (2009). Cleft lip and/or palate with monogenic autosomal recessive transmission in Pyrenees shepherd dogs. *Cleft Palate-Craniofacial Journal,* Vol.46, No.1, pp. 81-88, ISSN 1055-6656

Koch, T.G.; Loretti, A.P.; de Lahunta, A.; Kendall, A.; Russell, D. & Bienzle, D. (2005). Semilobar holoprosencephaly in a Morgan horse. *Journal of Veterinary Internal Medicine*, Vol.19, No.3, pp. 367-372, ISSN 0891-6640

Koren, G. & Nickel, S. (2010). Sources of bias in signals of pharmaceutical safety in pregnancy. *Clinical and investigative medicine*, Vol.33, No.6, pp. e349-355, ISSN 1488-2353

Koren, G.; & Nickel, C. (2011). Perpetuating fears: bias against the null hypothesis in fetal safety of drugs as expressed in scientific citations. *Journal of Population Therapeutics and Clinical Pharmacology*, Vol.18, No.1, pp. e28-32, ISSN 1710-6222

Kumar, P. & Burton, B. K. (2008). *Congenital malformations – Evidence-based evaluation and management*, McGraw Hill, ISBN 978-0-07-147189-3, New York, USA.

Lancaster, P. A. (2011). Causes of birth defects: lessons from history. *Congenital Anomalies*, Vol.51, No.1, pp. 2-5, ISSN 0914-3505

Lee, S.T.; Panter, K.E.; Pfister, J.A.; Gardner, D.R. & Welch, K. D. (2008). The effect of body condition on serum concentrations of two teratogenic alkaloids (anagyrine and ammodendrine) from lupines (Lupinus species) that cause crooked calf disease. *Journal of Animal Science*, Vol.86, No.10, pp. 2771-2778, ISSN 0021-8812

Mackay, D.R. (2011). Controversies in the diagnosis and management of the Robin sequence. *Journal of Craniofacial Surgery*, Vol.22, No.2, pp. 415-420, ISSN 1049-2275

Marasas, W.F.; Riley, R.T.; Hendricks, K.A.; Stevens, V.L.; Sadler, T.W.; Gelineau-van Waes, J.; Missmer, S.A.; Cabrera, J.; Torres, O.; Gelderblom, W.C.; Allegood, J.; Martínez, C.; Maddox, J.; Miller, J.D.; Starr, L.; Sullards, M.C.; Roman, A.V.; Voss, K.A.; Wang, E. & Merrill Jr, A.H. (2004). Fumonisins disrupt sphingolipid metabolism, folate transport, and neural tube development in embryo culture and in vivo: a potential risk factor for human neural tube defects among populations consuming fumonisin-contaminated maize. *Journal of Nutrition*, Vol.134, No.4, pp. 711-716, ISSN 0022-3166

Marden, P.M.; Smith, D.W. & McDonald, M. J. (1964). Congenital anomalies in the newborn infant, including minor variations. A study of 4,412 babies by surface examination for anomalies and buccal smear for sex chromatin. *Jornal de Pediatria*, Vol.64, pp. 357-371.

Martínez-Frías, M.L. (1994). Developmental field defects and associations: epidemiological evidence of their relationship. *American Journal of Medical Genetics,* Vol.49, No.1, pp. 45-51, ISSN 0148-7299

Martínez-Frías, M.L.; Frías, J.L. & Opitz, J.M. (1998). Errors of morphogenesis and developmental field theory. *American Journal of Medical Genetics*, Vol.76, No.4, pp. 291-296, ISSN 0148-7299

Martínez-Frías, M.L. & Frías, J.L. (1999). VACTERL as primary, polytopic developmental field defects. *American Journal of Medical Genetics*, Vol.83, No.1, pp. 13-16, ISSN 0148-7299

Martínez, J.S.; Velázquez, I.R.; Reyes, H. & Fajardo, R. (2006). Congenital holoprosencephaly with severe otocephaly in a rottweiler puppy. *Veterinary Record*, Vol.158, No. 15, pp. 518-519.

Moura, E. (1993). *Biologia educacional*, Editora Moderna, ISBN 85-16-00924-6, São Paulo, Brazil

Moura, E. & Cirio, S.M. (2004). Clinical and genetic aspects of X-linked ectodermal dysplasia in the dog -- a review including three new spontaneous cases. *Veterinary Dermatology*, Vol.15, No.5, pp. 269-277, ISSN 0959-4493

Moura, E.; Cirio, S. M. & Villanova Jr, J. A. (2010). VACTERL association in a cat. *American Journal of Medical Genetics A*, Vol.152, No.3, pp. 777-780, ISSN 1552-4825

Moura, E., Cirio, S. M. & Pimpão, C. T. (2011). Nonsyndromic cleft lip and palate in boxer dogs: evidence of monogenic autosomal recessive inheritance. *Cleft Palate-Craniofacial Journal*, Epub ahead of print: 2011-08-01, *ISSN* 1545-1569

Ngo, A.D.; Taylor, R. & Roberts, C.L. (2010). Paternal exposure to agent orange and spina bifida: a meta-analysis. *European Journal of Epidemiology*, Vol.25, No.1, pp. 37-44, ISSN 0393-2990

Opitz J. M. (1982). The developmental field concept in clinical genetics. *Journal of Pediatrics*, Vol.101, No.5, pp. 805-809, ISSN 0022-3476

Opitz J. M. (1984). *Tópicos recentes de genética clínica*, Sociedade Brasileira de Genética, Ribeirão Preto, Brazil.

Opitz, J. M.; Zanni, G.; Reynolds Jr, J. F. & Gilbert-Barness, E. (2002). Defects of Blastogenesis. *American Journal of Medical Genetics* (*Seminars in Medical Genetics*), Vol.115, No.4, pp. 269–286, ISSN 0148-7299

Priester, W.A.; Glass, A.G. & Waggoner, N.S. (1970). Congenital defects in domesticated animals: general considerations. *American Journal of Veterinary Research*, Vol.31, No.10, pp. 1871-1879, ISSN 0002-9645

Raudsepp, T.; Durkin, K.; Lear, T.L.; Das, P.J.; Avila, F.; Kachroo, P. & Chowdhary, B.P. (2010). Molecular heterogeneity of XY sex reversal in horses. *Animal Genetics*, Vol.41, No.2 (Suppl.), pp. 41-52, ISSN 0268-9146

Résibois, A.; Coppens, A. & Poncelet, L. (2007). Naturally occurring parvovirus-associated feline hypogranular cerebellar hypoplasia—a comparison to experimentally-induced lesions using immunohistology. *Veterinary Pathology*, Vol.44, No.6, pp. 831–841, ISSN 0300-9858

Richtsmeier, J.T.; Sack Jr, G.H.; Grausz, H.M. & Cork, L.C. (1994). Cleft palate with autosomal recessive transmission in Brittany spaniels. *Cleft Palate-Craniofacial Journal*, Vol.31, No.5, pp. 364-371, ISSN 1055-6656

Schelling, C.; Pieńkowska, A.; Arnold, S.; Hauser, B. & Switoński, M. (2001). A male to female sex-reversed dog with a reciprocal translocation. *Journal of Reproduction and Fertility. Supplement,* Vol.57, pp. 435-438, ISSN 0449-3087

Scott, F.W.; Lahunta, A.; Schultz, R.D.; Bistner, S.I. & Riis, R.C. (1975). Teratogenesis in cats associated with griseofulvin therapy. *Teratology,* Vol.11, No.1, pp. 79-86, ISSN 0040-3709

Singleton, A.C.; Mitchell, A.L.; Byers, P.H.; Potter, K.A. & Pace, J. M. (2005). Bovine model of Marfan syndrome results from an amino acid change (c.3598G > A, p.E1200K) in a calcium-binding epidermal growth factor-like domain of fibrillin-1. *Human Mutation,* Vol.25, No.4, pp. 348-352, ISSN 1059-7794

Smit, J.J.; Temwitchitr, J.; Brocks, B.A.; Nikkels, P.G.; Hazewinkel, H.A. & Leegwater, P.A. (2011). Evaluation of candidate genes as a cause of chondrodysplasia in Labrador retrievers. *Veterinary Journal,* Vol.187, No.2, pp. 269-271, ISSN 1090-0233

Son, J.M.; Yong, H.Y.; Lee, D.S.; Choi, H.J.; Jeong, S.M.; Lee, Y.W.; Cho, S.W.; Shin, S.T. & Cho, J.K. (2008). A case of perosomus elumbis in a Holstein calf. *Journal of Veterinary Medical Science,* Vol.70, No.5, pp. 521-523, ISSN 0916-7250

Spranger, J.; Benirschke, K.; Hall, J. G.; Lenz, W.; Lowry, R. B.; Opitz, J. M.; Pinsky, L.; Schwarzacher, H. G. & Smith, D. W. (1982). Errors of morphogenesis: Concepts and terms. Recommendations of an International Work Group. *Journal of Pediatrics,* Vol. 100, No.1, pp. 160-165, ISSN 0022-3476

Tarantal, A.F. & Hendrickx, A.G. (1987). Amniotic band syndrome in a rhesus monkey: a case report. *Journal of Medical Primatology,* Vol.16, No.5, pp. 291-9, ISSN 0047-2565

Traas, A.M.; Wang, P.; Ma, X.; Tittiger, M.; Schaller, L.; O'donnell, P.; Sleeper, M.M.; Vite, C.; Herati, R.; Aguirre, G.D.; Haskins, M. & Ponder K. P. (2007). Correction of clinical manifestations of canine mucopolysaccharidosis I with neonatal retroviral vector gene therapy. *Molecular Therapy: The Journal of the American Society of Gene Therapy,* Vol.15, No.8, pp. 1423-1431, ISSN 1525-0016

Turnpenny, P. & Ellard, S. (2007). *Emery's Elements of medical genetics.* Churchill Livingstone, ISBN 978-0-7020-2917-2, Edinburgh, England

Ueta, E.; Kodama, M.; Sumino, Y.; Kurome, M.; Ohta, K.; Katagiri, R. & Naruse, I. (2010). Gender-dependent differences in the incidence of ochratoxin A-induced neural tube defects in the Pdn/Pdn mouse. *Congenital Anomalies,* Vol.50, No.1, pp. 29-39, ISSN 0914-3505

Villagómez, D.A. & Alonso, R.A. (1998). A distinct Mendelian autosomal recessive syndrome involving the association of anotia, palate agenesis, bifid tongue, and polydactyly in the dog. *Canadian Veterinary Journal,* Vol.39, No.10, pp. 642-643, ISSN 0008-5286

von Moos, R.; Stolz, R.; Cerny, T. & Gillessen, S. (2003). Thalidomide: from tragedy to promise. *Swiss Medical Weekly,* Vol.133, No.5-6, pp. 77-87, ISSN 1424-7860

Ware, W. A. (2011). *Cardiovascular disease in small animal medicine,* pp. 234-238, Manson Publishing, ISBN 978-1-84076-153-5, London, England

Wiedemann, H. R. (1991). The pioneer of pediatric medicine. *European Journal of Pediatrics,* Vol.150, No.8, pp. 533, ISSN 0340-6199

Wilkerson, M.J.; Lewis, D.C.; Marks, S.L. & Prieur, D.J. (1998). Clinical and morphologic features of mucopolysaccharidosis type II in a dog: naturally occurring model of Hunter syndrome. *Veterinary Pathology*, Vol.35, No.3, pp. 230-233, ISSN 0300-9858

Wouda, W.; Peperkamp, N.H.; Roumen, M.P.; Muskens, J.; van Rijn, A. & Vellema, P. (2009). Epizootic congenital hydranencephaly and abortion in cattle due to bluetongue virus serotype 8 in the Netherlands. *Tijdschrift voor Diergeneeskunde*, Vol.134, No.10, pp. 422-427, ISSN 0040-7453

Yablokov, A.V. (2009). Chernobyl's radioactive impact on fauna. *Annals of the New York Academy of Sciences*, Vol.1181, pp. 255-280, ISSN 0077-8923

6

Use of Dual-Energy X-Ray Absorptiometry (DXA) with Non-Human Vertebrates: Application, Challenges, and Practical Considerations for Research and Clinical Practice

Matthew D. Stone and Alec J. Turner
Kutztown University,
USA

1. Introduction

The applications of Dual-energy X-ray Absorptiometry (DXA) to vertebrate research and veterinary practice are many. DXA has been used successfully to rapidly and non-invasively quantify bone density and body composition in a variety of animals. The use of DXA has been limited, primarily, to basic and applied research, but DXA technology has great promise for clinical practice involving animals. Despite this potential, a number of limitations hinder its use in veterinary practice. These issues must be resolved before DXA can be widely used in traditional veterinary practice and a goal of this chapter is to discuss these limitations. This chapter reviews the past and current uses of DXA in basic and applied research involving non-human vertebrates.

2. DXA theory

DXA is a non-invasive technique for the determination of body composition. Users of DXA are able to rapidly quantify lean tissue mass, fat mass, total body mass, bone mineral mass, and bone mineral density. A comprehensive review of how DXA quantifies body composition can be found in (Adams, 1997; Jebb, 1997; Peppler & Mazess, 1981; Pietrobelli et al., 1996). The method by which DXA estimates body composition is based on the principle that the intensity of X-rays as they pass through tissues is attenuated in proportion to tissue mass (Figure 1).

The attenuation of a single intensity X-ray beam, as it passes through a single-tissue model (e.g. bone) of unknown mass, can be calculated based on the equation below (modified from Jebb, 1997):

$$R = M_B(R_B) \tag{1}$$

where R is the degree of attenuation, M_B is the mass of the tissue (e.g. bone), and R_B is the tissue-specific attenuation coefficient. In this scenario, bone mass is the unknown of interest

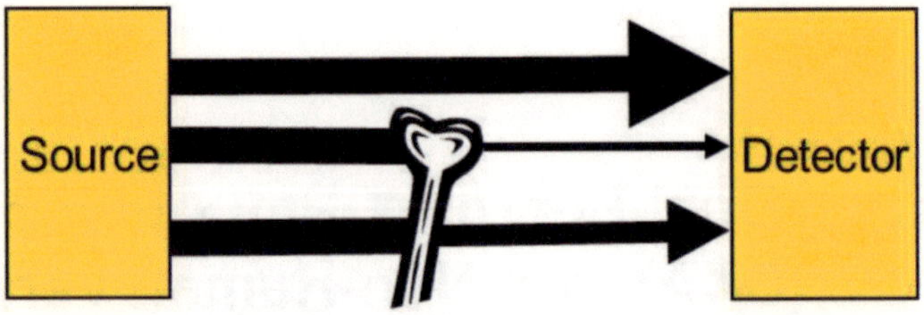

Fig. 1. Depiction of X-ray intensity attenuation as it passes through a tissue. X-ray attenuation is proportional to tissue mass.

while the degree of attenuation is determined by measuring the difference in intensity of X-rays between the source and detector. The attenuation coefficient is a known constant for a particular tissue that has been derived theoretically and empirically (Jebb, 1997). If more than one tissue type is present, the attenuation of the X-ray is a function of each of the individual tissue components contribution to the total beam attenuation (Figure 2).

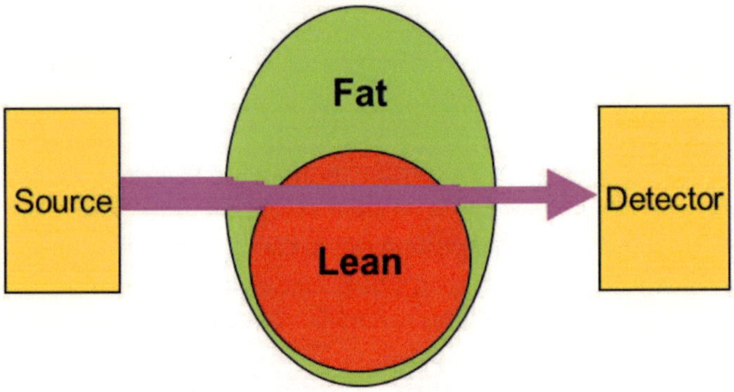

Fig. 2. Depiction of X-ray intensity as it passes through two tissue types. The total beam attenuation is a combination of the individual contributions of the two tissues, each at a different rate.

The relationship between the individual tissue contributions to total beam attenuation in a two tissue model (e.g. fat and lean tissue) can be summarized in the following equation, which is an extension of equation 1 (modified from Jebb, 1997).

$$R = M_F(R_F) + M_L(R_L) \tag{2}$$

Subscripts F and L denote fat and lean tissue contributions to total beam attenuation. In this equation there are two tissues of unknown mass contributing to beam attenuation. Directly solving for both of these unknowns is impossible, so the use of single energy X-ray absorptiometry is limited by its ability to distinguish various tissue components. DXA technology was developed to overcome this limitation. DXA utilizes X-rays of two different peak energies (high and low) and the attenuation of these beams can be used to calculate both unknowns. Each of these beams is attenuated differently when they pass through specific tissues, as indicated in equation 3.

$$R_{High} = M_F(R_F) + M_L(R_L) \tag{3}$$

$$R_{Low} = M_F(R_F) + M_L(R_L)$$

When more than two tissue components exist (e.g. bone, fat, and lean tissue) DXA cannot directly estimate the relative proportion of all three components. DXA indirectly estimates these three components by first distinguishing areas of the scan that contain soft tissue (lean & fat mass combined) from areas containing bone and soft tissue (Figure 3).

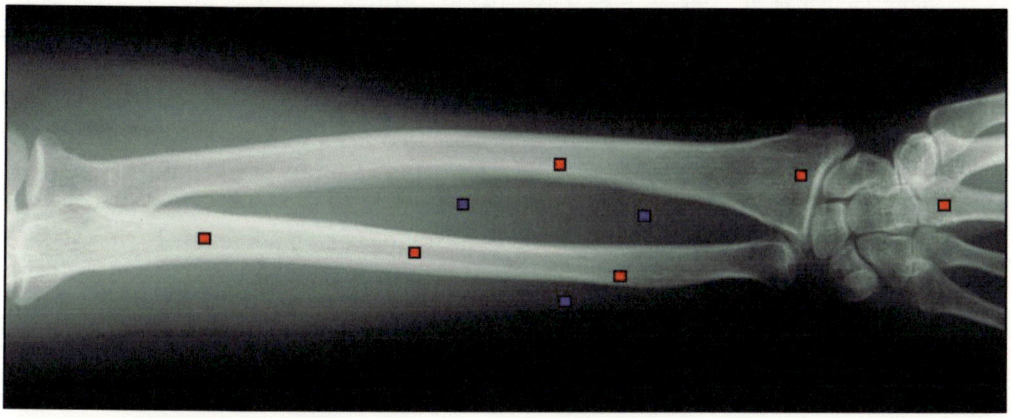

Fig. 3. X-ray image depicting the method by which DXA estimates body composition from a scan. DXA distinguishes regions (pixels) containing bone (red squares) from regions without bone (blue pixels).

In areas lacking bone, DXA can directly estimate the proportion of fat and lean tissue. For areas that contain bone, DXA determines the proportion of bone and soft tissue. Since DXA cannot distinguish between lean and fat tissue in these areas, DXA applies the proportion of fat and lean tissue, determined at neighboring non-bone areas, to the soft-tissue component of bone containing regions (Pietrobelli et al., 1996).

3. Basic applications of DXA

DXA is a non-invasive technique that was originally designed for the purpose of predicting current and future risk of bone fracture in humans by measuring bone mineral density (Grier et al., 1996). DXA can rapidly quantify total body mass, fat mass, lean tissue mass, and bone content and density, so it has potential applications in a variety of research and clinical fields. DXA has been used in and/or has practical applications to the fields of nutrition, sports and exercise science, physical therapy, animal science and nutrition, food science, pharmacology, pathology, metabolism, endocrinology, dentistry, and veterinary medicine. Because DXA is a non-invasive technique it is well suited for longitudinal studies. This advantage allows for reduced sampling error and potentially reduces required sample sizes and reduces those associated costs.

4. DXA use with animals

To date DXA has been applied to a diversity of small and large animal species, representing most major taxonomic groups. Of the major taxonomic groups, mammals have received the most attention, in part, due to their role as human models of osteoporosis or their role in the food industry. Even though mammals have received the most attention, other taxa are becoming increasingly well represented.

In mammals, DXA has been used most extensively with rodents. Due to the extent of published studies covering the use of DXA with rodents we elected to avoid their inclusion in this chapter except in a few instances for comparison. DXA has been used with species of agricultural concern including sheep (Mercier et al., 2006; Ponnampalam et al., 2007; Pouilles et al., 2000; Turner et al., 1995a), pigs (Clarys et al., 2010; Elowsson et al., 1998; Lee et al., 2011; Losel et al., 2010; Lukaski et al., 1999; Mitchell et al., 1998; Nielson et al., 2004), horses (McClure et al., 2001; Secombe et al., 2002), goats (Corten et al., 1997), and cattle (Zotti et al., 2010). Additionally, DXA has been applied to wildlife such as the grizzly bear (Felicetti, 2003 as cited in Stevenson & van Tets, 2008). DXA has been used to analyze excised bones of several species of marine mammals including Blainville's beaked whale (Zotti et al., 2009), Mediterranean monk seals (Mo et al., 2009), and bottlenose dolphins (Lucic et al., 2010). Finally, DXA has been applied to domestic pets including cats (Buelund et al., 2011; Turner et al., 1995b), rabbits (Castaneda et al., 2006; Hanafusa et al., 1995), guinea pigs (Fink et al., 2002), and dogs (Lorinson, 2009; Markel & Bogdanske, 1994a, 1994b; Mawby et al., 2004; Schneider et al., 2004; Toll et al., 1994; Zotti et al., 2004a). The use of DXA with ferrets has been noted (Grier et al., 1996); however few published studies have been cited.

In birds, DXA has been used primarily with species in the food science industry such as domestic poultry, including the red junglefowl (Jensen et al., 2005), white leghorns (Kim et al., 2006; Jensen et al., 2005), and turkey (Zotti et al., 2003). It has also been applied to wildlife including wild turkey, ruffed grouse, bobwhite quail (Dirrigl et al., 2004) and small passerines (Korine et al., 2004).

In reptiles, fewer species have been used with DXA, but most major taxonomic groups are represented, including snakes (Secor & Nagy, 2003); turtles (Fledelius et al., 2005; Stone et al., 2010), and lizards (Zotti et al., 2004b). The focus of these studies was validating the use of DXA with these species; however, Zotti et al., (2004b) used DXA to study metabolic bone disease in the green iguana (*Iguana iguana*) and Fledelius et al, (2005) used DXA to investigate how supplementing calcium in the diet of tortoises impacts bone density.

DXA has been used successfully and/or validated in a diverse number of animal species; however, there are a number of common household pets and research model species to which DXA has never been applied. This is especially true for exotic animals. Further research is needed to validate the use of DXA with these species before DXA can be put into practical use in animal research or medicine. To our knowledge, DXA has not been used or validated with amphibians or fish, despite the importance of these groups to animal research. In addition to filling in these "species gaps," further research is needed with species to which DXA has already been applied. Establishing a series of reference intervals for body composition in these species will provide important baseline data for future studies and will provide a frame of reference from which clinicians can use to diagnose pathology (Jebb, 1997).

5. Applications to veterinary medicine

5.1 Patient monitoring of bone density and body composition

As a non-invasive methodology, DXA is effective for longitudinal quantification of bone health and nutritional status of patients. Thus, DXA has potential to be a powerful tool for preventative and post-operative health care because it allows veterinary practitioners to quantify and systematically monitor patients' body condition and bone health over time.

DXA has been previously used to monitor bone density changes during fracture healing in rodents (Millett et al., 1998). Significant increases in femur bone density were measured during fracture healing of Sprague-Dawley rats. The region of the bone closest to the callus showed the largest difference in bone mineral density between the experimental and control groups (non-fractured). A similar study was conducted in dogs, where DXA was used to quantify changes in bone mineral density at the site of fracture following ostectomies of various widths (Markel & Bogdanske, 1994b). The results of these and other studies suggest that DXA can be used to effectively monitor changes in bone density after fracture. The ability to conduct analyses at particular regions of interest (ROIs), in addition to whole body analyses, makes this an extremely effective tool for post-operative monitoring.

In addition to post-operative monitoring of bone healing, DXA shows great promise in preventative medicine, such as early identification/diagnosis of metabolic bone diseases. This would serve an important service to exotic animal practice, because metabolic bone disease is the most common disease of some captive reptiles (Raiti & Haramati, 1997). Furthermore, DXA has the potential for diagnosing increases in bone density found in third carpal bone disease of horses; however, its utility was not deemed practical due to the logistical issues of scanning time with large animals (Secombe et al., 2002).

Monitoring of changes in fat mass in patients is another effective use of DXA. In the domestic dog and cat, obesity is the most common nutritional disorder (Mawby et al., 2004; German, 2006). Therefore, DXA is potentially a useful tool to monitor the efficacy of therapies, nutrition, and weight reduction regimes for obese or overweight pets. Effective monitoring programs have been shown to be critical for successful weight loss (Yaissle et al., 2004 as cited in German, 2006) and long-term maintenance of a healthy body mass (Laflamme & Kuhlman, 1995 as cited in German, 2006).

5.2 Visual diagnosis of gravidity & foreign objects

Because DXA scanners provide a digital image of the patient, DXA can be used for basic radiographic applications. The resolution of images produced by DXA is relatively poor compared to traditional X-ray radiography; therefore, DXA scanning is unlikely to replace traditional radiography for most applications requiring high resolution imaging (e.g. diagnosing of minor stress fractures). The poor resolution of images produced by DXA is a software issue, rather than a technological limitation of the technique per se. For instance, certain DXA imaging procedures provide sufficient resolution for visual diagnosis for minute quantities of calcification such as with aortic atherosclerosis (Wilson, 2006). Despite the limitations in resolution of routinely produced DXA images, image quality is sufficient to diagnose a variety of conditions. For example, in our research we have used DXA to determine if female turtles are gravid (Figure 4).

Additionally, we have used DXA to visually detect the presence of foreign objects in wildlife. Specifically, we have identified the ingestion of fish hooks in turtles (Figure 5). Fishhook injuries are common in aquatic wildlife such as turtles. Swallowing of fishhooks in household pets is also relatively common (Michels et al., 1995).

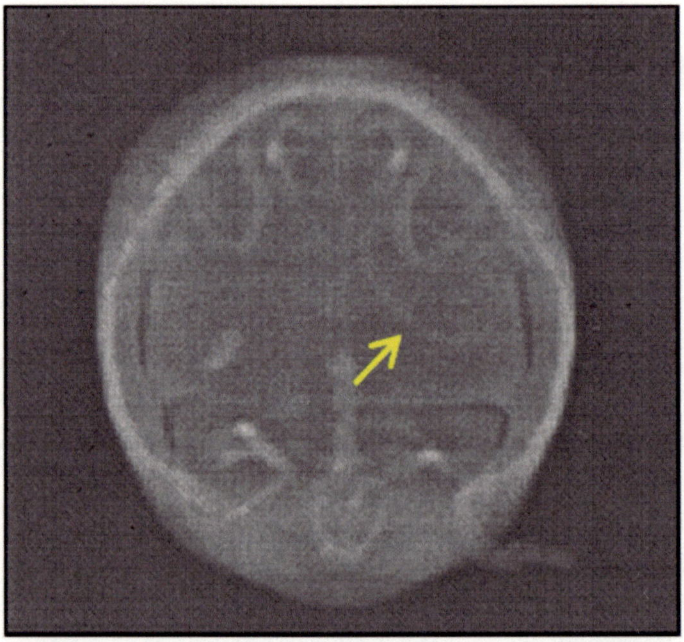

Fig. 4. X-ray image of a gravid female Eastern Box Turtle (*Terrapene carolina*). Arrow identifies a single egg.

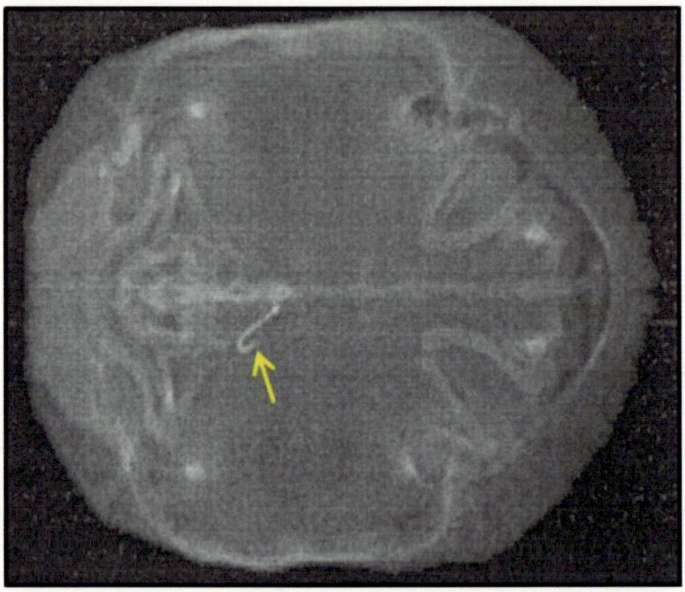

Fig. 5. X-ray image produced from a DXA scan of a wild-caught red-eared slider turtle (*Trachemys scripta*). Arrow indicates the presence of a fish hook lodged in the gastrointestinal tract.

Finally, we have used DXA to diagnose a gunshot wound in an adult wild-caught red-eared slider turtle (*Trachemys scripta;* Figure 6). In this case the subject showed no obvious signs of trauma and the external wound had completely healed. Upon our initial external inspection of the subject we failed to recognize the injury. It wasn't until we performed a DXA scan that we discovered a bullet located in the forelimb. After the diagnosis from the X-ray image, we visually identified the object as a hard palpable mass. The 0.22 caliber bullet was encysted in connective tissue surrounding the humerus, which was fractured and had not healed from the gunshot wound.

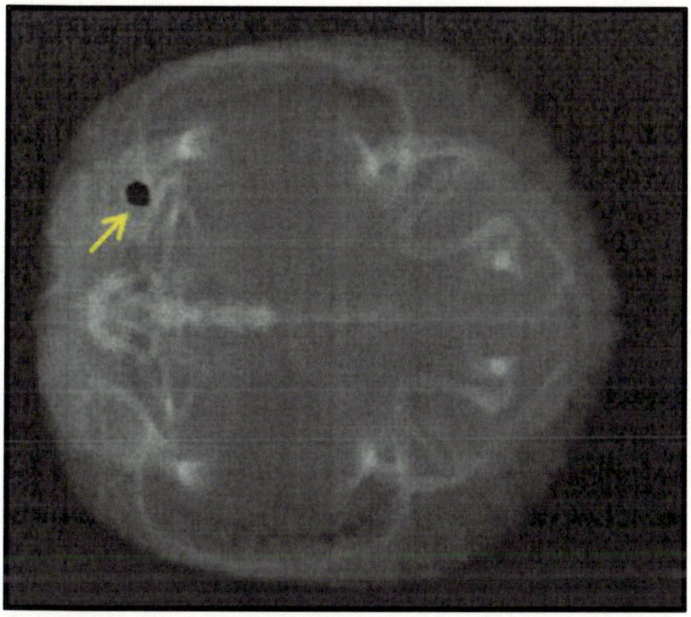

Fig. 6. X-ray image of a red-eared slider turtle (*Trachemys scripta*) that had been shot by a 0.22 caliber bullet (yellow arrow).

Even though DXA scans typically result in lower image quality, there are benefits to using it over traditional X-ray radiography. For instance, DXA offers important health benefits over traditional radiography for both the radiographer and the patient. X-ray exposure using DXA is a fraction of that produced by other means. Patient X-ray exposure during a typical whole-body DXA scan is a fraction of that during a typical chest X-ray. Furthermore, the technician's exposure to X-rays is generally negligible if the unit is operated several meters away from the source, where scatter radiation is minimal.

6. DXA precision in non-human vertebrates

Critical to the successful use of DXA in clinical and research settings is its ability to make accurate and, more importantly, precise estimates of body composition. Accuracy is of lesser importance because estimates can be corrected when systematic biases exist with a particular technique (Stone et al, 2010). On the other hand, precise estimates of body composition are critical to be able to detect longitudinal changes. We reviewed the precision of DXA in estimating body composition among a variety of taxa (Table 1). The precision among taxa was relatively low in most cases. Fat mass tended to be the least precisely estimated parameter of body composition; however, in most cases was within ranges seen in humans. A notable exception is with turtles. DXA's poor ability to estimate fat content is a result of the relatively high proportion of bone in turtles (Stone et al., 2010, discussed below).

Source	Study Animal	Location	Subject Treatment	Subject moved between scans? (#scans)	CV%			
					Lean Tissue Mass	Fat Mass	Bone Mineral Content	Bone Mineral Density
Secor & Nagy, 2003	Snake	Whole body	Euthanasia	No (2)	0.6	9.2	1.0	NR
Zotti et al., 2004b	Iguana	Femur[1]	Anesthesia	Yes (5)	NR	NR	NR	1.7
Zotti et al., 2004b	Iguana	Lumbar Spine	Anesthesia	Yes (5)	NR	NR	NR	1.6
Zotti et al., 2004b	Iguana	Head	Anesthesia	Yes (5)	NR	NR	NR	1.3
Stone et al., 2010	Turtle	Whole body	Anesthesia	No (2)	1.05	28.54	1.00	0.97
Nagy & Clair, 2000	Mouse	Whole body excluding head	Anesthesia	Yes (3)	0.86	2.20	1.60	0.84
Rose et al., 1998	Zucker Rat	Whole body	Anesthesia	Yes (3)	2.88	12.16	6.34	NR
Kastl et al., 2002	Lewis Rat	Humerus	Excised bone	No (6)	NR	NR	0.90	0.76
Kastl et al., 2002	Lewis Rat	Humerus	Excised bone	Yes (6)	NR	NR	1.32	0.86
Stevenson & van Tets, 2008	Vole	Whole body	Euthanized	Yes (>5)	1.6	6.8	2.3	3.6
Elowsson et al., 1998	Pig	Processed carcass	Euthanized	No (3)	0.94	13.51	1.91	NR
Lukaski et al.,1999	Pig	Whole body	Anesthesia	(3)	0.72	2.37	1.12	NR
Korine et al., 2004	Bird	Whole body	Physical restraint	? (2-3)	1.28	4.92	NR	NR
Korine et al., 2004	Bird	Whole body	Euthanized	? (2-3)	0.47	1.71	NR	NR
Korine et al., 2004	Bird	Whole body (feathers removed)	Euthanized	? (2-3)	0.16	2.06	NR	NR
Castaneda et al., 2006	Rabbit	Lumbar spine	Anesthesia	Yes (3)	NR	NR	NR	7.8
Black et al., 2001	Rhesus Monkey	Whole body	Anesthesia	Yes (5)	2.3	10.3	1.2	NR
Toll et al., 1994	Dog	Whole body	Anesthesia	No (6)	0.51	1.55	1.40	0.79

Table 1. Literature review of the precision, as determined by mean intraindividual coefficients of variation, of DXA estimates of lean tissue mass, fat mass, bone mineral content, and bone mineral density in various non-human vertebrates. NR = not reported. Table adapted from Stone (2009).

7. Limitations/impediments for use of DXA in clinical practice

Even though there are a variety of important applications of DXA to veterinary research and clinical practice, there are a number of logistical issues that preclude its widespread use. These limitations include the expense to purchase, operate, and maintain DXA equipment, the space required to house a unit, the time to scan a subject, the need to restrain the test subjects during scanning, and the potential for certain confounding variables to influence accurate/precise estimates of body composition as a result of technological limitations of DXA. In this section we discuss these limitations and offer some potential solutions.

7.1 Expense

In most veterinary practices the purchase of DXA equipment is likely to be cost prohibitive. The cost of a new unit averages $35,000 USD (Walpert, 2000). Additionally, there are a number of hidden costs such as software upgrades, equipment repair and maintenance, technician training, and remodeling costs associated with installation (e.g. electrical, space, etc.) Also, considering that the majority of diagnosis in a veterinary clinic would utilize traditional X-ray units, it is unlikely that clinics will purchase both traditional X-ray equipment and a dual-energy X-ray absorptiometer. Even though the costs of purchasing a DXA unit might be prohibitive, alternatives might exist. Potential users might be able to contract DXA services from local clinical or academic institutions.

7.2 Size

Space limitations are potentially major impediments for use of DXA in small animal practice. Since most DXA scanners are designed to be large enough to perform a whole body scan of an adult human, the housing of DXA equipment necessitates a dedicated room, which will likely deter or preclude its use in most small animal veterinary businesses. Some manufacturers, after recognizing this limitation, have designed DXA models that double as an exam table when not in use. Smaller models exist (e.g. PIXImus, GE Medical Systems), but are limited to small rodent-sized species, and not likely useful for most veterinary applications. Despite the space demands that a full-size DXA scanner necessitates, the size offers potential for its use with large-animal practices and has already proven useful for food-industry research. Even though there are benefits of the large scanner size for these applications, there are upper limits in body size that DXA can handle. Although DXA has been used previously with horses and cattle, its uses have been limited to analyses of bone density on excised bones (Secombe et al., 2002; Zotti et al., 2010). Currently, scanners are not large enough to allow for full-body scanning of larger animals, with the exception of carcass analysis.

7.3 Time

The use of DXA incurs a variable, and in some cases a significant, time component; however, the total time involved might not vary much more than it takes to produce a traditional X-ray radiograph. The total time it takes to scan an individual will vary depending on the type of equipment, its application, and the species under investigation. The time involved in producing a completed DXA scan of a patient involves four separate phases, with the vast majority of time allocated to machine and subject preparation. Phase 1 includes unit powering, calibration, and quality control. Phase 2 includes subject preparation, handling, orientation, and restraint. Phase 3 includes subject scanning time. Phase 4 is computational analyses. Subject scanning time can vary depending on a variety of factors including scanner brand, beam type (pencil vs. fan-beam), scanner resolution, and subject size/region of interest. Phase 1 is typically performed once each day that scanning takes place. The total time to power the unit and perform calibrations and quality control will likely vary depending on the unit manufacturer. In our experience this takes approximately 20-30 minutes. Phase 2 is likely to vary drastically with the patient under investigation due to the nuances on each organism's physiology and their body size. For instance, larger organisms pose more of a challenge to manipulate and orient on the scanner bed and may require more advanced planning. Phase 3 will vary primarily with patient size. For small patients such as rodents, where high-resolution small-animal software is necessary, scan time will take up to a few minutes. Scan time will increase with body size of the subject. The completion of DXA computational analyses (Phase 4) exemplifies the power of this tool. DXA nearly instantaneously calculates body composition at the completion of scanning. The relatively rapid scanning time combined with quick analyses makes DXA a potentially powerful tool in a clinical setting.

7.4 Need for chemical/physical restraint

Despite the relatively rapid scanning time of DXA, the accuracy and precision of DXA estimates of body composition are sensitive to the movements of subjects on the scanner bed (Cawkwell, 1998; Engelke et al., 1995). Due to this caveat, chemical or physical restraint of subjects is typically required with the use of DXA. The choice and efficacy of a particular method of restraint will vary among taxonomic groups. Use of restraint for full - body scanning of mammals and birds will invariably require general anesthesia to prevent movement artifacts. In birds, other methods of restraint have been used. Korine et al. (2004) covered the heads of small birds to immobilize them during scanning, but they found this method decreased precision in estimating lean tissue and fat mass compared to estimates determined after euthanasia (Table 1). In other exotic animals, reptiles in particular, additional methods are available. Cooling of body temperature is an effective method to immobilize subjects during scanning (Stone et al., 2010). This method requires planning on the part of the practitioner because safe cooling of core body temperature can take several hours for large reptiles. Despite its potential as a viable source of restraint, cooling of body temperature in reptiles tended to result in less precise estimates of body composition compared to anesthesia and euthanasia. (Stone et al., 2010)

7.5 Intestinal artifacts: A case study on the ingestion of foreign particles and their associated impacts on DXA estimates of body composition

DXA estimates of body composition from whole body scans are likely to be influenced by superimposing the contents of the gastrointestinal tract with the composition of the subject's tissues. Even though the effects of intestinal contents are thought to have a negligible impact on DXA estimates of body composition in humans, the consumption of certain items in animals might impact DXA estimates. For instance, the consumption of calcified particles such as sand, stones, or bone might influence estimates of bone mineral content because they will result in similar beam attenuation as bone. In reptiles and birds the consumption of calcified particles is common as these items might aid in digestion or provide particular nutrients (Walde et al., 2007). In this study we investigated whether ingestion of intestinal artifacts (sand in this case) impacts DXA estimates of body composition in turtles.

To determine the effects of intestinal artifacts on DXA estimates of body composition, we performed DXA analysis on 55 captive box turtles, *Terrapene carolina*. This captive population of turtles was housed outdoors in a pen with a sand substrate. Prior to scanning, straight carapace length (SCL) and mass were recorded for each individual. All specimens, who were not fasted prior to analysis, were then cooled for a minimum of eight hours at 4°C to prevent movement artifacts while scanning. Following the same procedure for turtles as Stone et al. (2010), scanning was performed using a Hologic QDR-4500A fan-beam scanner equipped with a small-animal software program on three occasions: 13 May 2004, 15 October 2005, and 10 May 2005. During each scanning day, duplicate scans, without movement between scans, were performed on each individual. If the same individual was scanned on multiple dates, only scans from the first day were used in analyses. Prior to statistical analyses, duplicate DXA estimates of body composition (bone mineral content (BMC), fat mass (FM), lean tissue mass (LTM), and body mass (BM)) were averaged for each individual. All turtles were qualitatively scored based on the amount of sand in the gastrointestinal tract. Turtles were scored as having zero (n=27), moderate (n=20), or heavy levels (n=8) of intestinal artifacts (Figure 7). Following DXA analyses, fecal samples were collected to confirm the presence of sand in the intestine.

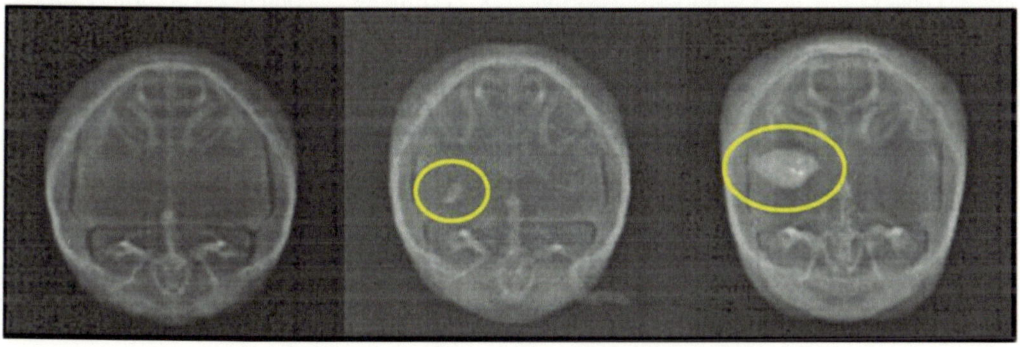

Fig. 7. DXA scans of box turtles (*Terrapene carolina*) showing, from left to right, (a.) zero, (b.) moderate, and (c.) heavy levels of intestinal artifacts (yellow circles).

To examine the effects of intestinal artifacts on DXA estimates of bone mineral content and lean tissue mass, we performed analysis of covariance (ANCOVA). Prior to analyses, we $\log_{10}$ transformed bone mineral content and lean tissue mass (dependent variables) to linearize the relationship between covariate (straight carapace length) and the response variable. Prior to analysis, all assumptions of ANCOVA were tested and not violated, including the assumption of parallelism (BMC, $F_{2,48}=0.24$, $p>0.05$; LTM, $F_{2,48}=0.11$, $p=0.89$). To compare fat mass among turtles with different artifact scores we used ANOVA, because body size did not significantly influence fat mass in this study. Data were left untransformed. For all analyses, when significant differences among intestinal artifact levels were found, Tukey multiple-comparisons were used to determine where differences occurred. Statistical analyses were performed using the general linear model procedure on Minitab version 13.2.

DXA estimates for bone mineral content were significantly different among turtles with intestinal artifact scores (ANCOVA; $F_{2,50}$=3.12, p=0.05). Bone mineral content was not different between turtles with zero and moderate levels of intestinal sand (T=1.12, p=0.52), as well as between turtles with moderate and heavy levels (T=1.62, p=0.25). A significant difference was found in mean BMC between turtles with zero and heavy levels of intestinal artifacts (T=2.48, p=0.04; Figure 8).

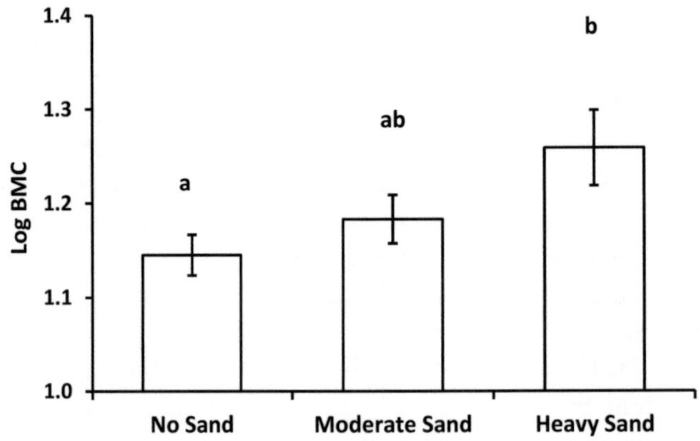

Fig. 8. Least square means of log-transformed DXA estimates of bone mineral content for turtles with no sand, moderate sand, and heavy sand in the gastrointestinal tract. Significant differences (P<0.05) are indicated by different letters. Data are presented as mean ± 1SE.

DXA estimates for lean tissue mass were not significantly different among turtles with different intestinal artifact scores ($F_{2,50}$=1.73, p=0.19; Figure 9).

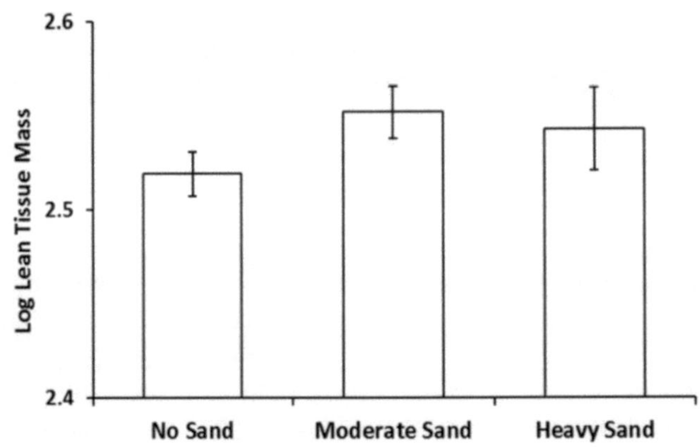

Fig. 9. Least square means of log-transformed DXA estimates of lean tissue mass for turtles with no sand, moderate sand, and heavy sand in the gastrointestinal tract. Data are presented as mean ± 1SE.

The effects of intestinal artifacts on estimates of fat mass followed similar trends to lean tissue mass. We observed a no significant effect of intestinal contents on DXA estimates of fat mass (F = 1.76, P=0.18; Figure 10).

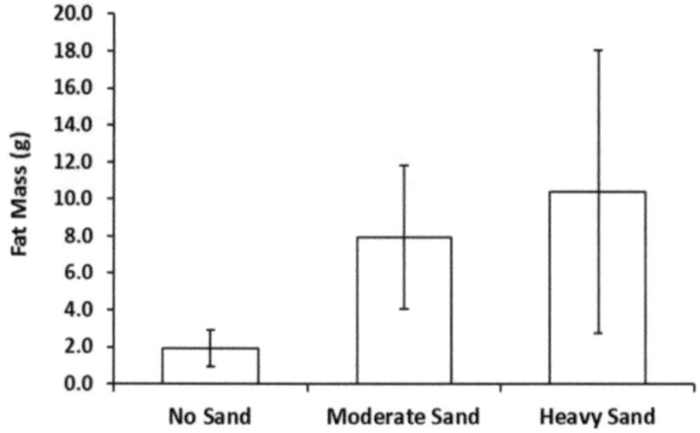

Fig. 10. Mean fat mass estimates for turtles with no sand, moderate sand, and heavy sand in the gastrointestinal tract. Data are presented as mean ± 1SE.

The results of this study suggest that DXA estimates of bone mineral content are influenced by ingestion of mineralized foreign particles. DXA tended to overestimate bone mineral content in individuals that ingested sand. Therefore, it is important to consider whether individuals have access to foreign particles when utilizing DXA to quantify bone mineral content in exotic animal practice or wildlife research involving birds or reptiles. In cases where ingestion of foreign particles is possible, fasting may be required to ensure evacuation of the gastrointestinal tract prior to DXA scanning.

7.6 Taxon-specific considerations

Technological limitations exist with the use of DXA with certain taxa. As we discussed previously, DXA is unable to directly estimate the proportion of lean, fat, and bone when they occur concurrently at a single pixel on the scan. This is a limitation with the use of two peak energies. To distinguish all three tissue components a third peak energy would be required (Swanpalmer et al., 1998). DXA overcomes this limitation by using neighboring pixels without bone to estimate the proportion of fat and lean tissue and then applies them to pixels containing bone. This limitation presents a serious impediment for the use of this technique with species that contain a high proportion of bone. Stone et al. (2010) found that DXA was unable to accurately or precisely predict fat mass in turtles as a result of the high proportion of pixels containing bone. Despite the limitation, DXA was able to effectively estimate bone and lean tissue mass, and so this methodology could be useful in a clinical setting for predicting and monitoring metabolic bone diseases in turtles (Stone et al, 2010).

8. Conclusions

Dual-energy X-ray absorptiometry can effectively quantify lean tissue, fat, and bone mineral mass in humans and most animals. As a result of its high degree of precision and accuracy, it has proven useful for research in fracture risk and healing, obesity, metabolism, pathology, and nutrition. Despite the plethora of potential and realized uses for DXA in veterinary research, the use of DXA remains restricted, primarily, to the human health industry or animal research. Currently, a number of limitations exist that prevent the widespread use of DXA in veterinary medicine. Most notably, the high purchase and maintenance costs and relatively short life-span of DXA units, combined with their size, are major limitations for DXA and hinder its widespread use in traditional veterinary practice. Until these limitations are resolved, DXA will continue to be restricted to human-based clinical settings or with animal research institutions.

9. Acknowledgment

We thank the Oklahoma State University Department of Nutritional Sciences for providing logistical support and access to their DXA equipment.

10. References

Adams, J.E. (1997). Single and dual energy X-ray absorptiometry. *European Radiology*, Vol. 7, Suppl. 2, pp. (S20-S31), ISSN: 0938-7994

Buelund, L., Nielsen, D., Mcevoy, F., Svalastoga, E., & Bjornvad, C. (2011). Measurement of body composition in cats using computed tomography and dual energy X-ray absorptiometry. *Veterinary Radiology and Ultrasound*, Vol. 52, No. 2, (March-April 2011), pp. (179-184), ISSN: 1058-8183

Black, A., Tilmont, E.M., Baer, D.J., Rumpler, W.V., Ingram, D.K. Roth, G.S., & Lane, M.A. (2001). Accuracy and precision of dual-energy X-ray absorptiometry for body composition measurements in Rhesus monkeys. *Journal of Medical Primatology*, Vol. 30, pp. (94-99), ISSN:0047-2565

Castaneda, S., Largo, R., Calvo, E., Rodriguez-Salvanes, F., Marcos, M.E., az-Curiel, M., & Herrero-Beaumont, G. (2006). Bone mineral measurements of subchondral and trabecular bone in healthy and osteoporotic rabbits. *Skeletal Radiology*, Vol. 35, No. 1, (January 2006), pp. (34-41), ISSN: 0364-2348

Cawkwell, G.D. (1998). Movement artifact and dual X-ray absorptiometry. *Journal of Clinical Densitometry*, Vol. 1, No. 2, pp. 141-147, ISSN 1094-6950

Clarys, J.P., Scafoglieri, A., Provyn, S., Louis, O., Wallace, J.A., & De Mey, J. (2010). A macro-quality evaluation of DXA variables using whole dissection, ashing, and computer tomography in pigs. *Obesity*, Vol. 18, No. 8, (August 2010), pp. (1477-1485), ISSN: 1930-7381

Corten, F.G.A., Caulier, H., van der Waerden, J.P.C.M., Kalk, W., Corstens, F.H.M., & Jansen, J.A. (1997). Assessment of bone surrounding implants in goats: Ex vivo measurements by dual X-ray absorptiometry. *Biomaterials*, Vol. 18, No. 6, (March 1997), pp. (495-501), ISSN: 0142-9612

Dirrigl, F.J., Dalsky, G.P., & Warner, S.E. (2004). Dual-energy X-ray absorptiometry of birds: An examination of excised skeletal specimens. *Journal of Veterinary Medicine: Series A*, Vol. 51, No. 6, (August 2004), pp. (313-319), ISSN: 0931-184X

Elowsson, P., Forslund, A.H., Mallmin, H., Feuk, U., Hansson, I., & Carlsten, J. (1998). An evaluation of dual-energy X-ray absorptiometry and underwater weighing to estimate body composition by means of carcass analysis in piglets. *The Journal of Nutrition*, Vol. 128, No. 9, (September 1998), pp. (1543-1549), ISSN: 0022-3166

Engelke, K., Gluer, C.C., & Genant, H.K. (1995). Factors influencing short-term precision of dual X-ray bone absorptiometry (DXA) of spine and femur. *Calcified Tissue International*, Vol. 56, No. 1, (January 1995), pp. (19-25), ISSN: 0171-967X

Felicetti, L.A., Robbins, C.T., & Shipley, L.A. (2003). Dietary protein content alters energy expenditure and composition of the mass gain in grizzly bears (*Ursus arctos horribilis*). *Physiological and Biochemical Zoology*, Vol. 76, No. 2, (March-April 2003), pp. (256-261), ISSN: 1522-2152

Fink, C., Cooper, H.J., Huebner, J.L., Guilak, F., & Kraus, V.B. (2002). Precision and accuracy of a transportable dual-energy X-ray absorptiometry unit for bone mineral measurements in guinea pigs. *Calcified Tissue International*, Vol. 70, No. 3, (March 2002), pp. (164-169), ISSN: 0171-967X

Fledelius, B., Jorgensen, G.W., Jensen, H.E., & Brimer, L. (2005). Influence of the calcium content of the diet offered to leopard tortoises (*Geochelone pardalis*). *Veterinary Record*, Vol. 56, No. 26, (June 2005), pp. (831-835), ISSN 0042-4900

German, A.J. (2006). The growing problem of obesity in dogs and cats. *The Journal of Nutrition*, Vol. 136, No. 7, (July 2006), pp. (1940S-1946S), ISSN: 0022-3166

Grier, S.J., Turner, A.S., & Alvis, M.R. (1996). The use of dual-energy x-ray absorptiometry in animals. *Investigative Radiology*, Vol. 31, No. 1, (January 1996), pp. (50-62), ISSN: 0020-9996

Hanafusa, S., Matsusue, Y., Yasunaga, T., Yamamuro, T., Oka, M., Shikinami, Y., & Ikada, Y. (1995). Biodegradable plate fixation of rabbit femoral-shaft osteotomies: A comparative study. *Clinical Orthopaedics and Related Research*, Vol. 315, (June 1995), pp. (262-271), ISSN: 0009-921X

Jebb, S.A. (1997). Measurement of soft tissue composition by dual energy X-ray absorptiometry. *British Journal of Nutrition*, Vol. 77, No. 2, (February 1997), pp. (151-163), ISSN: 0007-1145

Jensen, P., Keeling, L., Schutz, K., Andersson, L., Mormede, P., Brandstrom, H., Forkman, B., Kerje, S., Fredriksson, R., Ohlsson, C., Larsson, S., Mallmin, H., & Kindmark, A. (2005). Feather pecking in chickens is genetically related to behavioural and developmental traits. *Physiology and Behavior*, Vol. 86, (September 2005), pp. (52-60), ISSN: 0031-9384

Kastl, S., Sommer, T., Klein, P., Hohenberger, W., & Engelke, K. (2002). Accuracy and precision of bone mineral density and bone mineral content in excised rat humeri using fan beam dual-energy X-ray absorptiometry. *Bone*, Vol. 30, No. 1, (January 2002), p. (243-246), ISSN 8756-3282

Kim, W.K., Donalson, L.M., Mitchell, A.D., Kubena, L.F., Nisbet, D.J., & Ricke, S.C. (2006). Effects of alfalfa and fructooligosaccharide on molting parameters and bone qualities using dual energy X-ray absorptiometry and conventional bone assays. *Poultry Science*, Vol. 85, No. 1, (January 2006), pp. (15-20), ISSN: 0032-5791

Korine, C., Daniel, S., van Tets, I.G., Yosef, R., & Pinshow, B. (2004). Measuring fat mass in small birds by dual-energy X-ray absorptiometry. *Physiological and Biochemical Zoology*, Vol. 77, No. 3, (May-June 2004), pp. (522-529), ISSN: 1522-2152

Laflamme, D.P., & Kuhlman, G. (1995). The effect of weight loss regimen on subsequent weight maintenance in dogs. *Nutrition Research*, Vol. 15, No. 7, (July 1995), pp. (1019–1028), ISSN: 0271-5317

Lee, C.Y., Chan, S.H., Lai, H.Y., & Lee, S.T. (2011). A method to develop an in vitro osteoporosis model of porcine vertebrae: Histological and biomechanical study Laboratory investigation. *Journal of Neurosurgery: Spine*, Vol. 14, No. 6, (June 2011), pp. (789-798), ISSN: 1547-5654

Lorinson, K., Lobcke, S., Skalicky, M., & Lorinson, D. (2009). Densitometric properties of canine appendicular bones based on dual-energy X-ray absorptiometry measurements. *Wiener Tierarztliche Monatsschrift*, Vol. 96, pp. (72-77), ISSN 0043-535X

Losel, D., Kremer, P., Albrecht, E., & Scholz, A.M. (2010). Comparison of a GE Lunar DPX-IQ and a Norland XR-26 dual energy X-ray absorptiometry scanner for body composition measurements in pigs - in vivo. *Archiv Tierzucht*, Vol. 53, No. 2, pp. (162-175), ISSN 0003-9438

Lucic, H., Vukovic, S., Posavac, V., Gomercic, M.D., Gomercic, T., Galov, A., Skrtic, D., Curkovic, S., & Gomercic, H. (2010). Application of dual energy X-ray absorptiometry method for small animals in measuring bone mineral density of the humerus of bottlenose dolphins (*Tursiops truncatus*) from the Adriatic Sea. *Veterinarski Arhiv*, Vol. 80, No. 2, pp. (299-310), ISSN 0372-5480

Lukaski, H.C., Marchello, M.J., Hall, C.B., Schafer, D.M., & Siders, W.A. (1999). Soft tissue composition of pigs measured with dual X-ray absorptiometry. *Nutrition*, Vol. 15, No. 9, (September 1999), pp. (697-703), ISSN 0899-9007

Markel, M.D., & Bogdanske, J.J. (1994a). Dual-energy X-ray absorptiometry of canine femurs with and without fracture fixation devices. *American Journal of Veterinary Research*, Vol. 55, No. 6, (June 1994), pp. (862-866), ISSN 0002-9645

Markel, M.D., & Bogdanske, J.J. (1994b). The effect of increasing gap width on localized densitometric changes within tibial ostectomies in a canine model. *Calcified Tissue International*, Vol. 54, No. 2, (February 1994), pp. (155-159), ISSN 0171-967X

Mawby, D.I., Bartges, J.W., d'Avignon, A., Laflamme, D.P., Moyers, T.D., & Cottrell, T. (2004). Comparison of various methods for estimating body fat in dogs. *Journal of the American Animal Hospital Association*, Vol. 40, No. 2, (March-April 2004), pp. (109-114), ISSN 0587-2871

McClure, S.R., Glickman, L.T., Glickman, N.W., & Weaver, C.M. (2001). Evaluation of dual energy X-ray absorptiometry for in situ measurement of bone mineral density of equine metacarpi. *American Journal of Veterinary Research*, Vol. 62, No. 5, (May 2001), pp. (752-756), ISSN 0002-9645

Mercier, J., Pomar, C., Marcoux, M., Goulet, F., Theriault, M., & Castonguay, F.W. (2006). The use of dual-energy X-ray absorptiometry to estimate the dissected composition of lamb carcasses. *Meat Science*, Vol. 73, No. 2, (June 2006), pp. (249-257), ISSN 0309-1740

Michels, G.M., Jones, B.D., Huss, B.T., & Wagner-Mann, C. (1995). Endoscopic and surgical retrieval of fishhooks from the stomach and esophagus in dogs and cats: 75 cases (1977-1993). *Journal of the American Veterinary Medical Association*, Vol. 207, No. 9, (November 1995), pp. (1194-1197), ISSN 0003-1488

Millett, P.J., Cohen, B., Allen, M.J., & Rushton, N. (1998). Bone mineral density changes during fracture healing: a densitometric study in rats, In: *Internet Journal of Orthopedic Surgery and related Subjects*, 18.8.2011, Available from: http://www.rz.uni-duesseldorf.de/WWW/MedFak/Orthopaedie/journal/, ISSN 0949-2607

Mitchell, A.D., Scholz, A.M., & Conway, J.M. (1998). Body composition analysis of small pigs by dual-energy X-ray absorptiometry. *Journal of Animal Science*, Vol. 76, No. 9, (September 1998), pp. (2392-2398), ISSN 0021-8812

Mo, G., Zotti, A., Agnesi, S., Finola, M.G., Bernardini, D., & Cozzi, B. (2009). Age classes and sex differences in the skull of the Mediterranean monk seal, *Monachus monachus* (Hermann, 1779): A Study Based on Bone Shape and Density. *Anatomical Record: Advances in Integrative Anatomy and Evolutionary Biology*, Vol. 292, No. 4, (April 2009), pp. (544-556), ISSN 1932-8486

Nagy, T.R., & Clair, A.L. (2000). Precision and accuracy of dual-energy X-ray absorptiometry for determining in vivo body composition of mice. *Obesity Research*, Vol. 8, No. 5, (August 2000), p. (392-398), ISSN 1071-7323

Nielson, D.H., McEvoy, F.J., Poulsen, H.L., Madsen, M.T., Buelund, L.E., & Svalastoga, E. (2004). Dual-energy X-ray absorptiometry of the pig: Protocol development and evaluation. *Meat Science*, Vol. 68, No. 2, (October 2004), pp. (235-241), ISSN 0309-1740

Peppler, W.W., & Mazess, R.B. (1981). Total body mineral and lean body mass by dual-photon absorptiometry: theory and measurement procedure. *Calcified Tissue International*, Vol. 33, No. 1, (December 1981), pp. (353-359), ISSN 0171-967X

Pietrobelli, A., Formica, C., Wang, Z.M., & Heymsfield, S.B. (1996). Dual-energy X-ray absorptiometry body composition model: Review of physical concepts. *American Journal of Physiology: Endocrinology and Metabolism*, Vol. 271, No. 6, (December 1996), pp. (E941-E951), ISSN 0193-1849

Ponnampalam, E.N., Hopkins, D.L., Dunshea, F.R., Pethick, D.W., Butler, K.L., & Warner, R.D. (2007). Genotype and age effects on sheep meat production: Carcass composition predicted by dual energy X-ray absorptiometry. *Australian Journal of Experimental Agriculture*, Vol. 47, No. 10, pp. (1172-1179), ISSN 0816-1089

Pouilles, J.M., Collard, P., Tremollieres, F., Frayssinet, P., Railhac, J.J., Cahuzac, J.P., Autefage, A., & Ribot, C. (2000). Accuracy and precision of in vivo bone mineral measurements in sheep using dual-energy X-ray absorptiometry. *Calcified Tissue International*, Vol. 66, No. 1, (January 2000), pp. (70-73), ISSN 0171-967X

Raiti, P., & Haramati, N. (1997). Magnetic resonance imaging and computerized tomography of a gravid leopard tortoise (*Geochelone pardalis pardalis*) with metabolic bone disease. *Journal of Zoo and Wildlife Medicine*, Vol. 28, No. 2, (June 1997), p. (189-197), ISSN 1042-7260

Rose, B.S., Flatt, W.P., Martin, R.J., & Lewis, R.D. (1998). Whole body composition of rats determined by dual energy X-ray absorptiometry is correlated with chemical analysis. *Journal of Nutrition*, Vol. 128, No. 2, (February 1998), pp. (246-250), ISSN 0022-3166

Schneider, S., Breit, S.M., Grampp, S., Kunzel, W.W.F., Liesegang, A., Mayrhofer, E., & Zentek, F. (2004). Comparative assessment of bone mineral measurements obtained by use of dual-energy X-ray absorptiometry, peripheral quantitative computed tomography, and chemical-physical analyses in femurs of juvenile and adult dogs. *American Journal of Veterinary Research*, Vol. 65, No. 7, (July 2004), pp. (891-900), ISSN 0002-9645

Secombe, C.J., Firth, E.C., Perkins, N.R., & Anderson, B.H. (2002). Pathophysiology and diagnosis of third carpal bone disease in horses: A review. *New Zealand Veterinary Journal*, Vol. 50, No. 1, (February 2002), pp. (2-8), ISSN 0048-0169

Secor, S.M., & Nagy, T.R. (2003). Non-invasive measure of body composition of snakes using dual-energy X-ray absorptiometry. *Comparative Biochemistry and Physiology A: Molecular and Integrative Physiology*, Vol. 136, No. 2, (October 2003), pp. (379-389), ISSN 1095-6433

Stevenson, K.T., & van Tets, I.G. (2008). Dual-energy X-ray absorptiometry (DXA) can accurately and nondestructively measure the body composition of small, free-living rodents. *Physiological and Biochemical Zoology*, Vol. 81, No. 3, (May-June 2008), pp. (373-382), ISSN 1522-2152

Stone, M.D. (2009). Effects of season, sex, and age on the calcium physiology and bone dynamics of turtles. Oklahoma State University, Ph.D. dissertation, Stillwater, OK

Stone, M.D., Arjmandi, B.H., & Lovern, M.B. (2010). Dual-energy X-ray absorptiometry (DXA) as a non-invasive tool for the prediction of bone density and body composition of turtles. *Herpetological Review*, Vol. 41, No. 1, (March 2010), pp. (36-42), ISSN 0018-084X

Swanpalmer, J., Kullenberg, R., & Hansson, T. (1998). Measurement of bone mineral using multiple-energy X-ray absorptiometry. *Physics in Medicine and Biology*, Vol. 43, pp. (379-387), ISSN 0031-9155

Toll, P.W., Gross, K.L., Berryhill, S.A., & Jewell, D.E. (1994). Usefulness of dual energy X-ray absorptiometry for body composition measurement in adult dogs. *The Journal of Nutrition*, Vol. 124, Suppl. 12, (December 1994), pp. (2601S-2603S), ISSN 0022-3166

Turner, A.S., Mallinckrodt, C.H., Alvis, M.R., & Bryant, H.U. (1995a). Dual-energy X-ray absorptiometry in sheep: Experiences with in-vivo and ex-vivo studies. *Bone*, Vol. 17, No. 4, (October 1995), pp. (S381-S387), ISSN 8756-3282

Turner, A.S., Norrdin, R.W., Gaarde, S., Connally, H.E., & Thrall, M.A. (1995b). Bone mineral density in feline mucopolysaccharidosis VI measured using dual-energy X-ray absorptiometry. *Calcified Tissue International*, Vol. 57, No. 3, (September 1995), pp. (191-195), ISSN 0171-967X

Walde, A.D., Delaney, D.K., Harless, M.L., & Pater, L.L. (2007). Osteophagy by the desert tortoise (*Gopherus agassizii*). *Southwestern Naturalist*, Vol. 52, No. 1, (March 2007), pp. (147-149), ISSN 0038-4909

Walpert, B. (2000). With ancillary services, you can do well by doing good. In: *American College of Physicians Observer*, October, 8.2.2011, Available from: http://www.acpinternist.org

Wilson, K.E. (2006). Detection of abdominal aortic calcification with IVA. 8.1.2011, Available from: http://www.hologic.com/data/W_158_AAC_05-06.pdf

Yaissle, J.E., Holloway, C., & Buffington, C.A.T. (2004). Evaluation of owner education as a component of obesity treatment programs for dogs. *Journal of the American Veterinary Medical Association*, Vol. 224, No. 12, (June 2004), pp. (1932–1935), ISSN 0003-1488

Zotti, A., Gianesella, M., Ceccato, C., & Morgante, M. (2010). Physiological values and factors affecting the metacarpal bone density of healthy feedlot beef cattle as measured by dual-energy X-ray absorptiometry. *Journal of Animal Physiology and Animal Nutrition*, Vol. 94, No. 5, (October 2010), pp. (615-622), ISSN 0931-2439

Zotti, A., Isola, M., Sturaro, E., Menegazzo, L., Piccinini, P., & Bernardini, D. (2004a). Vertebral mineral density measured by dual-energy X-ray absorptiometry (DEXA) in a group of healthy Italian boxer dogs. *Journal of Veterinary Medicine: Series A*, Vol. 51, No. 5, (June 2004), pp. (254-258), ISSN 0931-184X

Zotti, A., Poggi, R., & Cozzi, B. (2009). Exceptional bone density DXA values of the rostrum of a deep-diving marine mammal: A new technical insight in the adaptation of bone to aquatic life. *Skeletal Radiology*, Vol. 38, No. 12, (December 2009), pp. (1123-1125), ISSN 0364-2348

Zotti, A., Rizzi, C., Chiericato, G., & Bernardini, D. (2003). Accuracy and precision of dual-energy X-ray absorptiometry for ex vivo determination of mineral content in turkey poult bones. *Veterinary Radiology and Ultrasound*, Vol. 44, No. 1, (January-February 2003), pp. (49-52), ISSN 1058-8183

Zotti, A., Selleri, P., Carnier, P., Morgante, M., & Bernardini, D. (2004b). Relationship between metabolic bone disease and bone mineral density measured by dual-energy X-ray absorptiometry in the green iguana (*Iguana iguana*). *Veterinary Radiology and Ultrasound*, Vol. 45, No. 1, (January 2004), pp. (10-16), ISSN 1058-8183

Part 2

Clinical Attention in Pets

7

Periodontal Disease in Dogs

Fábio Alessandro Pieri, Ana Paula Falci Daibert,
Elisa Bourguignon and Maria Aparecida Scatamburlo Moreira
Federal University of Viçosa,
Brazil

1. Introduction

Periodontics is a science that aims to study the periodontium and the diagnosis, prevention and treatment of periodontal diseases, in order to promote and restore the periodontal health (Harvey & Emily, 1993; Roman et al., 1995).

The periodontium is the set of adjacent structures to the teeth that provides them with support and protection. These structures are: gingiva, cementum, alveolar bone and periodontal ligament (Harvey & Emily, 1993; Roman et al., 1995; De Marco & Gioso, 1997; Clarke, 2001).

Periodontal disease is the most common oral disease in dogs with up of 80% of animals affected (Riggio et al., 2011). This disease is progressive and involves two stages: gingivitis (reversible) and periodontitis (irreversible, but often controllable). It is caused by plaque buildup on teeth. The plaque is a smooth membrane, adhesive, contaminated with saliva bacteria and debris. Bacteria and bacterial products cause inflammation of soft tissue. The plaque becomes mineralised to form calculus, which migrates into the gingival sulcus, causing additional inflammation, loss of periodontal ligament, bone loss and ultimately tooth loss (Ford & Mazzaferro, 2007).

Medical problems that affect the oral cavity should be identified in its early stages, so that the animals can be treated before showing serious secondary systemic disorders related to malnutrition and/or infections (Pachaly, 2006). One should also be aware of ways to prevent the disease, as animal tooth brushing and the use of antimicrobials as an adjunct in periodontal therapy (De Marco & Gioso, 1997).

2. Dental anatomy of dogs

As in most domestic mammals and in humans, dogs have diphyodont dentition, featuring two sets of teeth, a deciduous or primary and a permanent, although edentulous at birth (Harvey, 1992).

The oral anatomy of dogs has subdivisions and similar structures to those of humans (Figure 1), differing in the shape of the cavity, which also varies between breeds (Roza, 2004), anatomy and quantity of teeth and in the teeth apex (Harvey & Emily, 1993). Dogs have, like humans, incisors, canines, premolars and molars, differing among themselves in functions and numbers of roots (Roza, 2004; Mitchell, 2005).

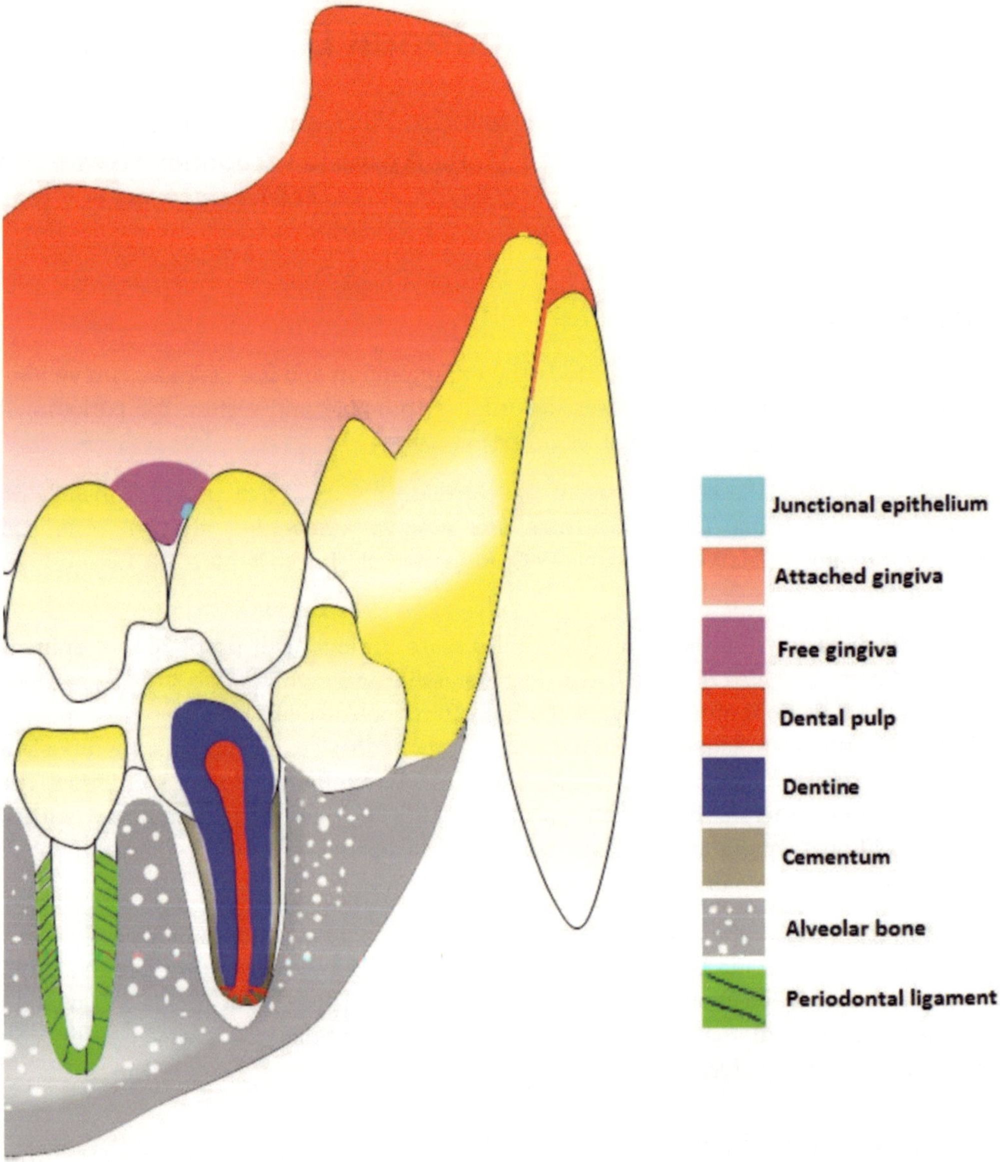

Fig. 1. Structures of dental organ in dogs

Domestic dogs have in their primary teeth, 28 teeth (12 incisors, four canines, 8 premolars and 4 molars), and in the permanent, 42 teeth (12 incisors, four canines, 16 premolars and 10 molars) (Roza, 2004). Regardless of the number of roots, function, size and shape, the teeth have subdivisions that are common to all types (Roza, 2004), and form the dental organ together with some adjacent structures (Gioso, 2007).

For a better understanding of periodontal disease it is important to have further information about a set of structures that constitute the alveolar-dental joint, the periodontium (Picosse, 1987).

The periodontium (peri, around; dental, tooth) (Lindhe & Karring, 1997) is the set of hard and soft tissues (Mitchell, 2005) that support (Harvey & Emily, 1993, Domingues et al., 1999), by fixing, adhering (Lindhe & Karring, 1997; Roza, 2004) and protecting the tooth in the alveolar bone (Roza, 2004). The structures that comprise the periodontium are the periodontal ligament, cementum, gingiva and alveolar bone (Carranza, 1983; Lindhe & Karring, 1997; Wiggs & Lobprise, 1997). There is a division of these structures according to their functions, so there is a periodontal support formed by the cementum, the periodontal ligament, alveolar bone, and gingiva that besides participating in the support also comprises the protection periodontium (Roza, 2004).

The gingiva (Figure 1) is the part of the masticatory mucosa that surrounds the cervical portion of the tooth and covers the alveolar process (Lindhe & Karring, 1997). Its main function is to protect structures adjacent to the tooth, being the first line of defence against periodontal disease (Harvey & Emily, 1993). Two parts can be distinguished: the free and attached gingiva (Harvey & Emily, 1993; Lindhe & Karring, 1997).

The free gingiva can be pink or pigmented in some breeds, with firm consistency and an opaque surface (Lindhe & Karring, 1997). The margin of the free gingiva is the edge of it. Between the free gingiva and the tooth, a groove is formed (Mitchell, 2005) known as the gingival sulcus, which, in normal conditions in the dogs, varies in depth from one to three millimeters (Harvey & Emily, 1993; Roza, 2004). The sulcus is surrounded by an adhered epithelium that secretes a fluid with inflammation mediatory cells, immunoglobulins and antibacterial substances important in the physical and immunological protection of the junctional epithelium and deeper tissues (Pope, 1993).

The junctional epithelium (Figure 1) is located at the bottom of the sulcus, with flat and elongated cells (Hennet, 1995; Wiggs & Lobprise, 1997) adhering to the enamel through hemidesmosomes, promoting the junction between the gingiva and the tooth (Harvey & Emily, 1993). The junctional epithelium ends in the cementum-enamel junction (Roza, 2004).

In processes such as inflammation, hyperplasia or in both, the junctional epithelium can recede apically or the gingiva can increase, making deeper the gingival sulcus (Harvey & Emily, 1993). In gingival hyperplasia the deepening of the sulcus occurs without loss of periodontal tissue, named false pocket, although when there is a loss of the support tissue and protection of the tooth, the sulcus is called periodontal pockets, which can be of two types: suprabone, when the bottom of the sulcus is coronal to the support alveolar bone, and intrabone, when the bottom is located apically in relation to the adjacent alveolar bone. The pocket depth can vary between regions of the mouth and even between neighbouring teeth (Newman et al., 2004).

The attached gingiva (Figure 1) is the continuation of the free gingiva, that, is firmly attached to the underlying bone, and extends to the mucogingival junction (Carranza, 1983).

Cementum (Figure 1) is a hard tissue with no vascularity (Harvey et al. 1994; Hennet, 1995) that covers the tooth root (Mitchell, 2005), is composed of collagen fibres embedded in an organic matrix, and has in its mineral portion, which is responsible for about 65% of its weight, hydroxyapatite crystals (Lindhe & Karring, 1997). It has as its main functions the insertion of periodontal ligament fibres into the root of the tooth (Harvey & Emily, 1993; Lindhe & Karring, 1997; Roza, 2004), the contribution to the process of repair of the root surface and the maintenance of the periodontal ligament fibres (Picasso, 1987; Lindhe & Karring, 1997).

There are two types of cementum: the first, called primary or acellular cementum, is formed in association with root formation and teeth eruption (Lindhe & Karring, 1997). It occupies the coronal and middle thirds of the tooth root and is constituted mostly of Sharpey's fibres, which are the periodontal ligament collagen fibres that attach at one end to the cementum and at the other to the alveolar bone (Roza, 2004). The other type of cementum is called secondary or cellular cementum, formed after the teeth (Lindhe & Karring, 1997) and usually located in the periapical region. It is secreted by cementocytes or cementoblasts, which are cells that are trapped into the organic matrix of cementum. It does not have vascularization, therefore is nourished from the periodontal ligament. The cells secrete cellular cementum in response to functional demands (Wiggs & Lobprise, 1997). Despite the description of the location of each type of cementum, in some cases they can occur alternately in some areas of the root surface (Lindhe & Karring, 1997).

The alveolar bone (Figure 1) involves the maxilla, incisor bone and jawbone that support the teeth in cavities where they are inserted. These cavities are called alveolus (Picasso, 1987; Harvey & Emily, 1993; Roza, 2004).

Composed of 65% minerals (Wiggs & Lobprise, 1997), this bone has a hard consistency and is very dense and compact, but differs from the root cementum because it has innervation, blood and lymphatic vasculature (Lindhe & Karring, 1997).

The interior of the alveolus is where the cribriform plate is located, which radiographically is known as lamina dura, characterised as a radiopaque line around the alveolus. The fibres of the periodontal ligament that attach to the tooth are connected to this plaque and it is where the vessels pass for ligament irrigation and for the nutrition of the cementum organic matrix (Harvey & Emily, 1993; Roza, 2004)

The alveolar bone can be resorbed or remodelled, according to the stimuli that it may suffer (Harvey & Emily, 1993; Roza, 2004).

The periodontal ligament (Figure 1) is a connective tissue structure that binds the tooth to its alveolus, fixing it (Lindhe & Karring, 1997; Figueiredo & Parra, 2002). It originates from mesenchymal cells of the dental sac (Picasso, 1987; Wiggs & Lobprise, 1997).

The periodontal ligament contains nerves and great vascularity, with vessels emanating from the maxillary artery in the case of the maxilla and from the inferior alveolar artery in the case of the jaw, and other cells. It is located between the root cementum and the cribriform plate. Its height, width, quality and condition are crucial to give the tooth its characteristic mobility (Harvey & Emily, 1993; Lindhe & Karring, 1997; Roza, 2004).

There are three different categories of fibres in the periodontal ligament: the fibres of the gingival grouping, composed by dental gingival fibres (connect the cementum to the gingiva), the alveologingival fibres (connects the cribriform plate to the gingiva), the circular (surrounding the tooth at the free gingiva), the transseptal fibres (connects the supra cementum alveolus of neighbour teeth), the fibres of the dental alveolar group, which are the fibres of the alveolar ridge (connecting the alveolar ridge to the cementum, obliquely), the horizontal fires (connect the cementum to the ridge, horizontally), the oblique fibres (connect the cementum to the alveolar bone, has a higher number of ligaments), apical fibres (connecting the bone to the cementum around the apex) and the inter-root fibres (that are between roots of multirooted teeth) (Wiggs & Lobprise 1997; Roza, 2004).

The periodontal ligament plays several roles in the tooth, such as:

- physical features, support, shock absorption caused by the chewing strength and transmission of occlusal forces to the bone (Figueiredo & Parra, 2002);
- formation, by osteoblasts, cementoblasts and fibroblasts (Clarke, 2001);
- reabsorption by osteoclasts, cementoclast and fibroclasts (Clarke, 2001);
- sensory, because it is abundantly innervated by sensory nerve fibres that are able to transmit tactile sensations of pressure and pain by the trigeminal pathways (Carranza, 1983; Figueiredo & Parra, 2002);
- nutritive, since it has blood vessels that provide nutrients and other substances required by the ligament tissues by the cementocytes and the more superficial osteocytes of alveolar bone (Carranza, 1983; Figueiredo & Parra, 2002);
- and homeostasis, because of its ability to absorb and synthesise the intercellular substance of the ligament connective tissue, alveolar bone and cementum (Carranza, 1983; Figueiredo & Parra, 2002).

In cases where the complete avulsion of tooth occurs, there is a possibility of reintegration of the tooth to the body (Pieri, 2004) if it returns to the alveolus quickly, since the periodontal ligament has ability to rejoin the cementum (Harvey & Orr, 1990). In these cases the endodontic treatment should be performed since the apical vascularisation of the tooth was ruptured (Pieri, 2004).

3. Periodontal disease

Periodontal disease is a condition that affects the periodontium, therefore the structures that surrounds the teeth, whose role is to protect and provide support to it (De Marco & Gioso, 1997). It is an infectious disease (Mitchell, 2005) that affects more than 80% of dogs (Riggio, 2011), and climbing to about 85% of dogs over four years old (Roman et al., 1995; De Marco & Gioso, 1997). This fact makes it the most prevalent disease in dogs (Harvey & Emily, 1993; Gioso & Carvalho, 2004; Mitchell, 2005). It has been described as a multifactorial infection with aetiological factors such as bacterial plaque, microflora, immune status, the amount of saliva, breed, age, routine of prophylactic cleaning and type of food. However, plaque is the primary aetiological agent, which consists predominantly of gram-positive, aerobic, non-motile bacteria early in the infection, and anaerobic, gram-negative and motile bacteria in the later stages of infection (Harvey & Emily, 1993; Gioso, 2007).

The disease is caused by the accumulation of bacterial plaque on teeth and gingiva (Harvey & Emily, 1993; Gioso & Carvalho, 2004), by toxic metabolism products of these microorganisms and the host immune response against infection (Mitchell, 2005) that triggers the inflammatory process. Initially this process affects only the gingiva tissue, which characterises the gingivitis that later may worsen and develop into a process of periodontitis which involves changes in other periodontium tissues and can cause bone, periodontal ligament, and in some cases, cementum or even tooth loss (Harvey & Emily, 1993).

Bacterial plaque is a sticky, yellowish material that colonises the entire mouth (Gioso, 2007), the faces of the teeth in their enamel structure (Slee & O'Connor, 1983; Katsura et al., 2001) and gingival sulcus (Domingues et al., 1999). This plaque is a biofilm (Dupont, 1997; Roza, 2004) or an undefined microbial community associated with the tooth surface (Wilderer; Charaklis, 1989, Lang et al., 1997), and is considered the leading cause of pathological process (aetiologic agent) (Tanzer et al. 1977; McPhee & Cowley, 1981). The plaque has as its main constituents: salivary glycoproteins, minerals, oral bacteria, extracellular polysaccharides that adhere to the tooth surface, desquamated epithelial cells, leukocytes, macrophages and lipids (Harvey & Emily, 1993; Roza, 2004)

Initially there is a pellicle formation upon the tooth surfaces and other areas of the mouth, called attached pellicle, which is an organic film derived from the saliva that, at first, has no micro-organisms (Sans & Newman, 1997). In the acquired pellicle begins the formation of a biofilm through the adhesion of the first microorganisms that are mostly gram-positive aerobic bacteria (Figure 2) (Lang et al. 1997; Sans & Newman, 1997; Gioso, 2007), mainly of the *Streptococcus* genus, which produce an exopolysaccharide, a substance that acts like a "glue", facilitating the attachment of these bacteria to the surfaces in question (Wiggs & Lobprise, 1997; Gioso, 2007; Roza, 2004) especially in places where there are small irregularities, cracks or roughness (Sans & Newman, 1997).

The first layer is usually single-celled and appears irregularly distributed on the tooth surface. With the microbial growth, the layer starts to come out of these areas of irregularities onto the enamel surface and increases in volume. The isolated plaques start to coalesce, forming a single plate. Over time in the process of plaque formation, a new phase is present, where new microbial strains with less ability to adhere to tooth structure, adhere to the already formed plaque, featuring new microcolonies and increasing the biodiversity of the plaque (Figure 2) (Sans & Newman, 1997). This process is called the organisation of the bacterial plaque, in which the bacteria need approximately 24 to 48 hours to get organised enough to cause the disease (Gioso, 2007). At this stage, the fight against bacterial plaque, with the simple disruption of it, is able to stall and reverse the process (Slee et al., 1983).

According to the location, the plaque can be classified as supra or sub-gingival. The supragingival plaque corresponds mainly to microbial aggregates found on tooth surfaces (mostly in the gingival third of the crown), however may extend into the gingival sulcus, where they are in immediate contact with the marginal gingiva. The subgingival plaque corresponds to bacterial aggregates found entirely within the gingival sulcus or periodontal pockets (Harvey & Emily 1993).

The bacterial constituents present in dental plaque are modified according to the disease evolution. In healthy gingiva, the cocci represent nearly two-thirds of the bacteria, followed by non-motile small rods. The bacteria present are mostly gram-positive and there is no significant representation of more virulent bacterial types. A new work from Federal University of Viçosa, Brazil, identified bacteria that are present in initial supra-gingival plaque in ten young dogs and found the genera *Streptococcus*, *Staphylococcus* and *Enterococcus* as the main components (personal data). At the stage of gingivitis, the gram-positive rods (non-motile) gradually increase, surpassing the cocci, and the number of gram-negative bacteria also grows. This change continues to go on until the periodontium involvement phase (periodontitis), when the more pathogenic gram-negative microorganisms become the majority, so that the spirochetes represent almost half of the bacteria, while gram-positives are underrepresented (Niezengard et al., 1997; Riggio et al., 2011).

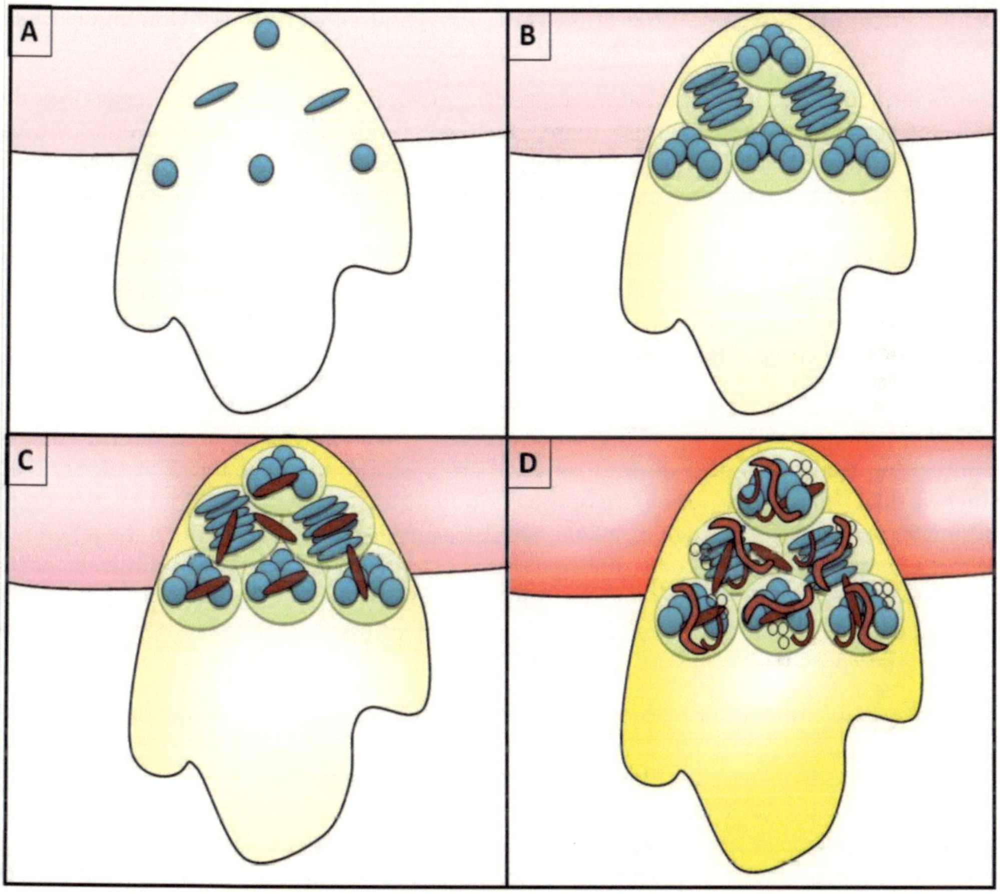

Fig. 2. Formation phases of dental plaque: A) Adhesion of Gram-positive cocci and rods; B) cellular proliferation and exopolysaccharide production; C) adhesion of Gram-negative bacteria; D) maturation of plaque with increase of bacterial biodiversity.

Without interference in the plaque formation an inflammatory process can occur, which marks the beginning of a periodontal disease and that provides a favourable environment for change in the microbial composition of the plaque, that become a biofilm with more pathogenic characteristics, continuing with later stages of the disease (Sans & Newman, 1997; Loesche & Grossman, 2001; Gioso, 2007).

The main bacteria involved in the formation of dental plaque are *Streptococcus* sp. (Gibbons, 1972; Tanzer et al. 1974; Duchin & Houte, 1978; Slee & O'Connor, 1983; Slee et al. 1983; Corner et al., 1988; Murray et al., 1992; Harvey & Emily, 1993; Lang et al., 1997; Loesche &

Grossman, 2001; Katsura et al., 2001; Drummond et al., 2004; Swerts et al., 2005), *Actinomyces* sp. (Slee & O'Connor, 1983; Katsura et al., 2001), and *Lactobacillus* sp. (Drummond et al. 2004; Roza, 2004). These colonise initially the adhered film of the enamel and then start to multiply and aggregate. Thereafter, surface receptors on the cocci and gram-positive rods allow adherence of gram-negative bacteria and, over time, they present the greatest biodiversity and pathogenic potential (Lang et al., 1997). In dogs, the most important bacteria of this group are *Veillonella, Bacterioides, Prevotella* (Domingues et al., 1999), *Fusobacterium* (Murray et al., 1992; Braga et al., 2005) and *Porphyromonas* (Domingues et al., 1999; Katsura et al., 2001; Braga et al., 2005; Senhorinho et al., 2011).

The bacteria involved in the periodontal disease process may migrate to other regions of the body by bacteraemia and colonise there, causing various diseases such as endocarditis, nephritis (Harvey & Emily, 1993; Debowes, 1996; Gioso, 2007), hepatitis (Debowes, 1996; Gioso, 2007) and myocarditis (Harvey & Emily, 1993). Another disease that can be caused by the microorganisms involved in periodontal disease, is a pathological process known as periodontic-endodontic lesion. There is an infection and inflammation of the dental pulp caused by the migration of periodontal bacteria at the apex of the tooth and root canal penetration through the foramen, which are small holes through which the vascular supply, lymphatics and nerves of the tooth passes (Pieri, 2004).

In dogs, the progression of periodontal disease can be divided into stages based on the clinical appearance of the gingiva (classified according to an index ranging from zero degree representing healthy gingiva, to four, when there is severe involvement of it (Roza, 2004; Mitchell, 2005) and loss of adherence of the periodontium, which is measured by inserting a millimeter probe into the gingival sulcus. This measure represents how much the junctional epithelium has migrated toward the apex of the tooth. The result in millimeters is the distance between the bottom of the gingival sulcus or periodontal pocket to the cementoenamel junction. The rate of adhesion loss is expressed by the ratio between the loss and the total normal adhesion (Mitchell, 2005).

In the process of disease formation, the healthy periodontium is considered stage 0 (Figure 3) (Harvey & Emily, 1993) when the gingiva presents a uniform colour, the bacterial plaque is imperceptible and there are little pathogenic characteristics, and there is no halitosis (De Marco & Gioso, 1997). With the increase in bacterial plaque and its change regarding the amount and specificity of the present microorganisms, such as its pathogenicity, begins to emerge an inflammation that marks the beginning of periodontal disease itself, which can be divided into stages of gingivitis (stage 1), initial periodontitis (stage 2), moderate periodontitis (stage 3) and severe periodontitis (stage 4) (Harvey & Emily, 1993).

The first stage of periodontal disease begins with the emergence of bacteria which, through its metabolites, cause inflammation of the gingival tissue, called gingivitis (Figure 3 and Figure 4). This is the first defence of the tooth, without much pathogenicity and without damaging the structures of the support periodontium. Such inflammation is similar to that seen in other connective tissues. Vasodilation, leukocyte marginalisation, cell migration, production of prostaglandins and destructive enzymes also occurs (Gioso, 2007), making the gingiva red, swollen and painful, and may cause halitosis (De Marco & Gioso, 1997). At this stage the possibility of regression of the disease remains through the proper management of the oral health of the animal with measures such as brushing, promoting removal of the aetiological agent (bacterial plaque) (Gorrel & Rawlings, 1996; Gioso, 2007). At this moment it is also possible to use mouthwash in conjunction with brushing to obtain increased efficiency of bacterial removal (Clarke, 2001). Periodontal disease can remain in this state or progress to the loss of insertion of the gingiva to the tooth, thus, the loss of the junctional epithelium adhesion and consequently the formation of periodontal pockets characterising the beginning of the pathologic stage called initial periodontitis (Hennet, 2002).

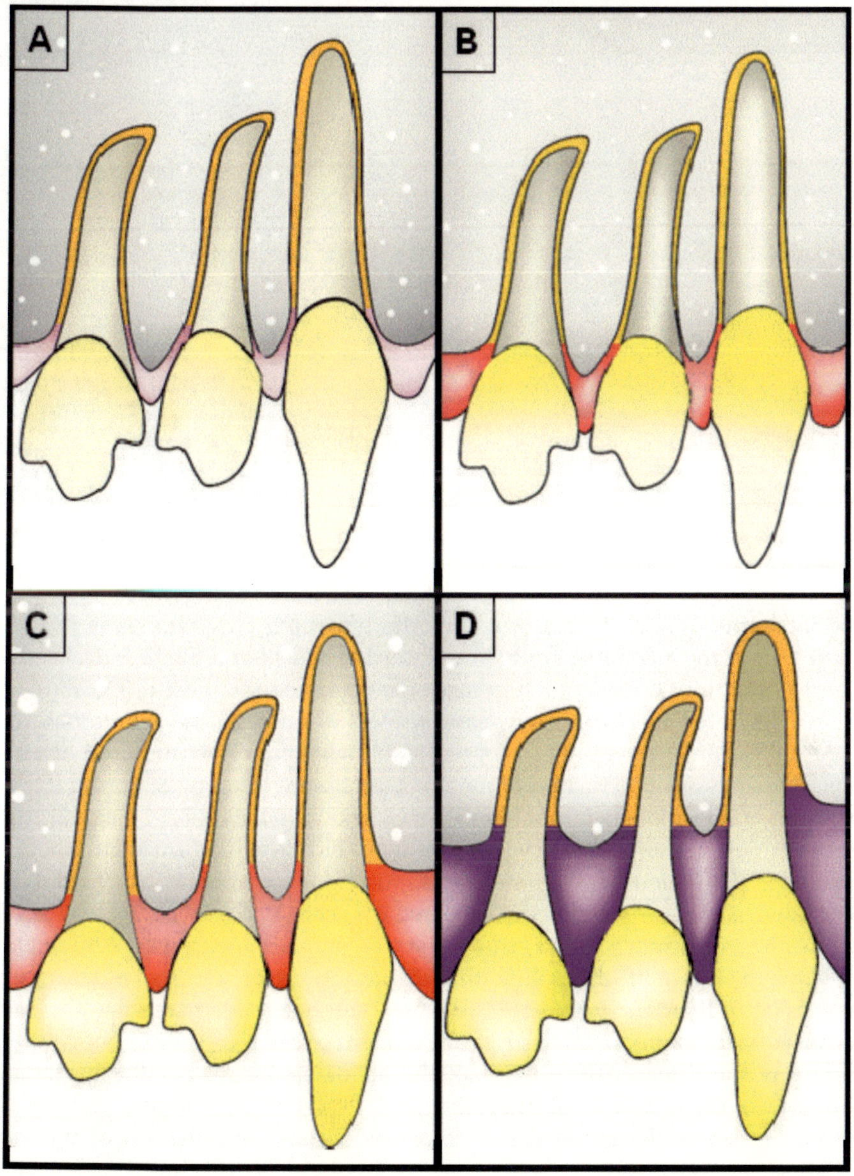

Fig. 3. Periodontal disease progression: A) healthy periodontium; B) gingivitis; C) Initial periodontitis – begins with the loss of periodontal tissue; D) moderate and severe periodontitis - The loss of periodontal tissue is more severe (25 to 50% loss) and the gum can appear cyanotic..

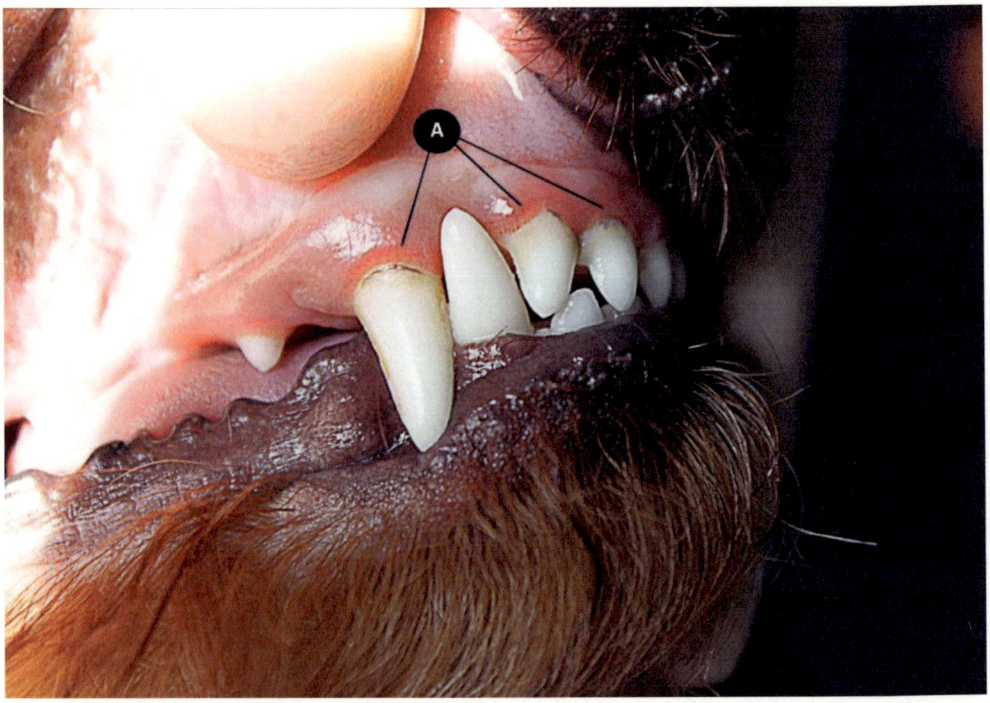

Fig. 4. Dog with gingivitis: A) red line on the cervical region of the tooth indicating gingival inflammation in a two years old poodle

Initial periodontitis (Figure 3) is a sequel of the untreated gingivitis. It is the phase that begins with loss of the junctional epithelium insertion considered the first irreversible phase of the disease, and from this point it becomes only possible to stabilise it (De Marco & Gioso, 1997; Gioso, 2007; Roza, 2004). This phase is the extension of the inflammatory process toward the supporting periodontium (periodontal ligament, cementum and alveolar bone) and leads to progressive destruction of the reported tissues (Harvey & Emily, 1993; Hennet, 2002). The gingiva is still in its normal topography and may have slight gingival recession in some breeds, but when it is very sore it may bleed when touched. There is mild inflammation in the periodontal ligament and little evidence of bone loss (Harvey & Emily, 1993). Intensive halitosis is presented (De Marco & Gioso, 1997). There is also the formation of dental calculus, commonly called tartar, which is nothing more than the mineralisation of dental plaque by salts of the saliva that makes it easier for the adherence of new microorganisms with more pathogenic features, subsequently increasing the severity of injuries caused to the periodontal tissues. The amount of calculus should not be used for the classification of the disease, since it is not the aetiologic agent, but the bacterial plaque. Large amounts of calculus are not indicative of the stage of the disease, and the diagnosis must be established in the gingival poll to assess the loss of adhesion of the epithelium (Gioso, 2007; Roza, 2004).

Moderate (Figure 5) and severe (Figure 6) periodontitis are the most advanced stages of periodontal disease and differ only in the degree of injury (Harvey & Emily, 1993). Some authors do not report the presence of moderate periodontitis, considering only the severe periodontitis after the initial process of aggression to the support tissues of the teeth (De

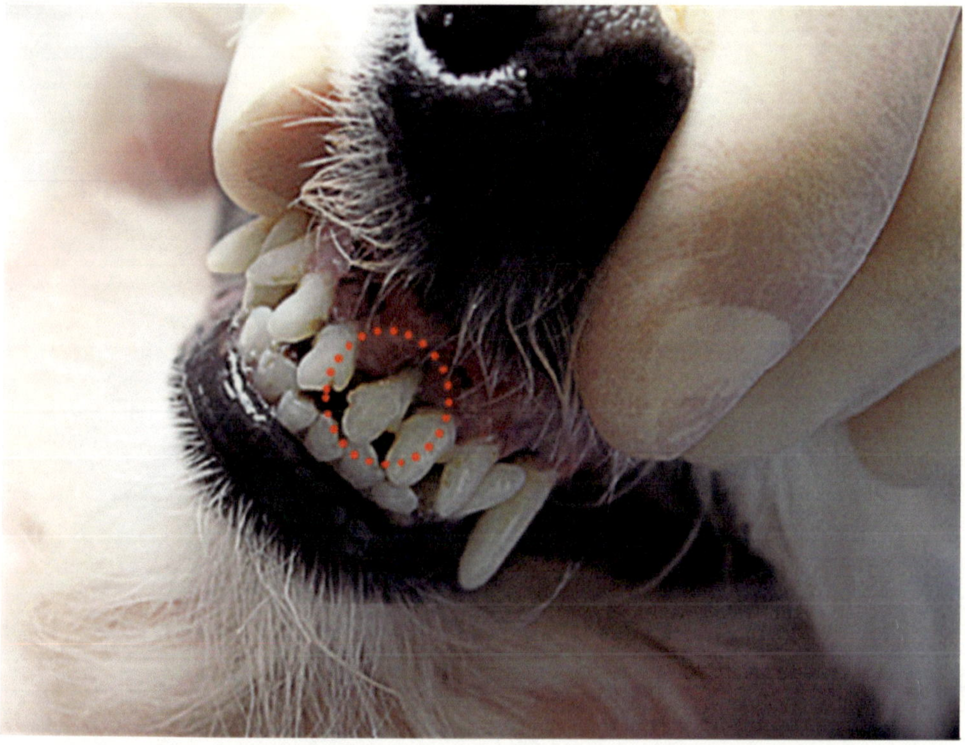

Fig. 5. Into de red circle: root exposure in the upper left central incisor, indicating moderate periodontitis in a poodle with five years.

Marco & Gioso, 1997). In these stages the microbiota of dental plaque is completely different from the initial one, having only a small amount of non-motile gram-positive cocci and large percentages of motile gram-negative spirochetes, virtually absent in healthy individuals (Niezengard et al., 1997). This phase shows severe inflammation, eventually cyanotic (Figure 3), with bleeding in response to the minimum stimulus, greater accumulation of bacterial plaque, intense halitosis (De Marco & Gioso, 1997), mild (in moderate) or large (in severe) tooth mobility (Harvey & Emily, 1993), and in most cases, there is also receding gingiva and a large accumulation of dental calculus. The bacteria enter the bloodstream and cause damage to the heart, liver, kidneys, joints and other organs (De Marco & Gioso, 1997; Gioso, 2007). In moderate periodontitis, there is between a 25 and 50% loss of supporting periodontium (which can be masked by large gingival hyperplasia in some animals), creating the need for treatment and care for the maintenance of the teeth in the socket. In severe periodontitis there is a loss of more than 50% of the supporting periodontal tissue and furcation may be exposed in multirooted teeth and in many cases leading to exfoliation of teeth (Harvey & Emily, 1993).

4. Prevention of periodontal disease

Prevention s considered essential for the maintenance of the animals' teeth throughout their lives, making it impossible for the formation of the periodontal disease process (Lyon, 1991). Brushing, chewable products, promoters of friction and the use of antimicrobial substances

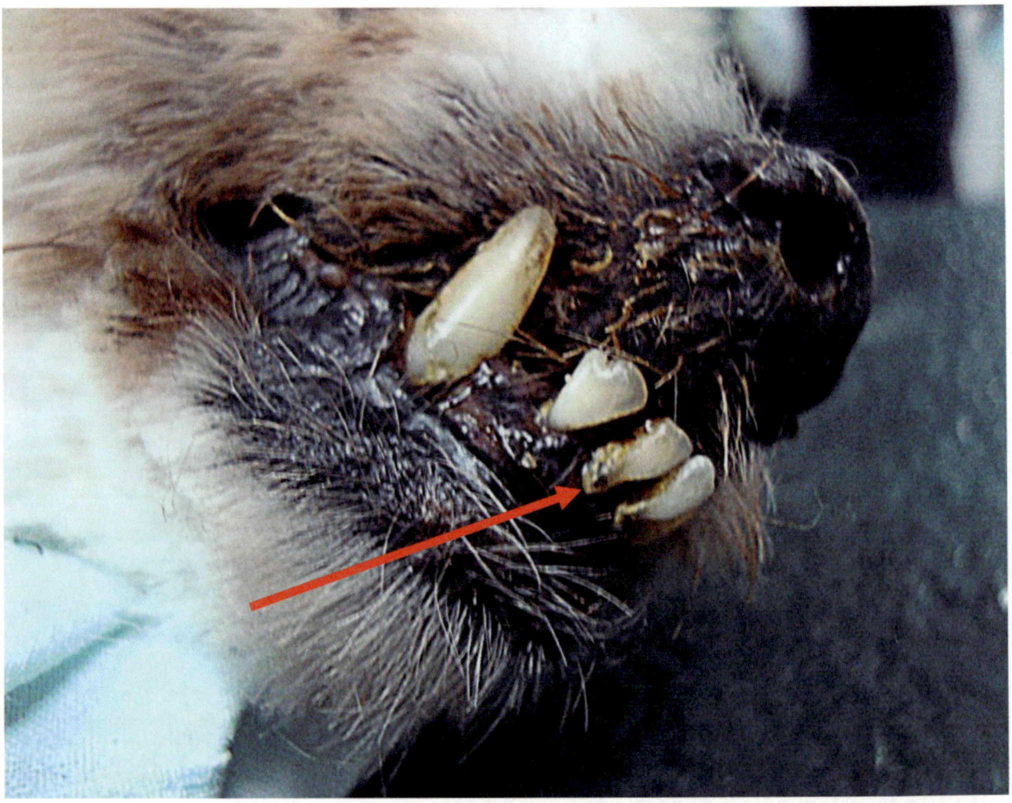

Fig. 6. At the end of the red arrow: extensive root exposure with dental calculus in the lower right medial incisor, indicating severe periodontitis in an old mongrel dog.

are considered preventive techniques that remove supra and sub-gingival plaque. The effectiveness of the preventive technique should be monitored by a veterinarian and in most cases will require their intervention, performing dental prophylaxis in order to eliminate residual plaque and calculus in places of difficult access in the teeth (Lima et al., 2004).

The tooth brushing (Figure 7), which acts by removing the biofilm through friction (Dupont, 1998), is considered a technique with greater effect to reduce the buildup of plaque on the tooth (Hennet, 2002). The frequency of tooth brushing in the animals should be daily, to constantly avoid the formation of dental plaque (Niemiec, 2008) and to establish a routine between the owner and the animal. However, because less than 10% of owners agree with these recommendations for the dental care of their dogs (Lima et al., 2004) and because of the time required for the organisation of the plaque, dog tooth brushing procedure has been recommended three times a week with satisfactory results (Dupont, 1998; Niemiec, 2008).

There are numerous veterinary toothbrushes on the market today, but a child brush with soft bristles is considered effective for removal of the biofilm in dogs. Many veterinary toothpastes are also available, with flavours to facilitate animal acceptance of the brushing, besides the fact that it increases friction promoted with brushing (Niemiec, 2008). It is clear, however, that the use of correct brushing techniques minimises the need for this additive effect of toothpaste on the attrition of the teeth (Lima et al., 2004). Antimicrobial substances

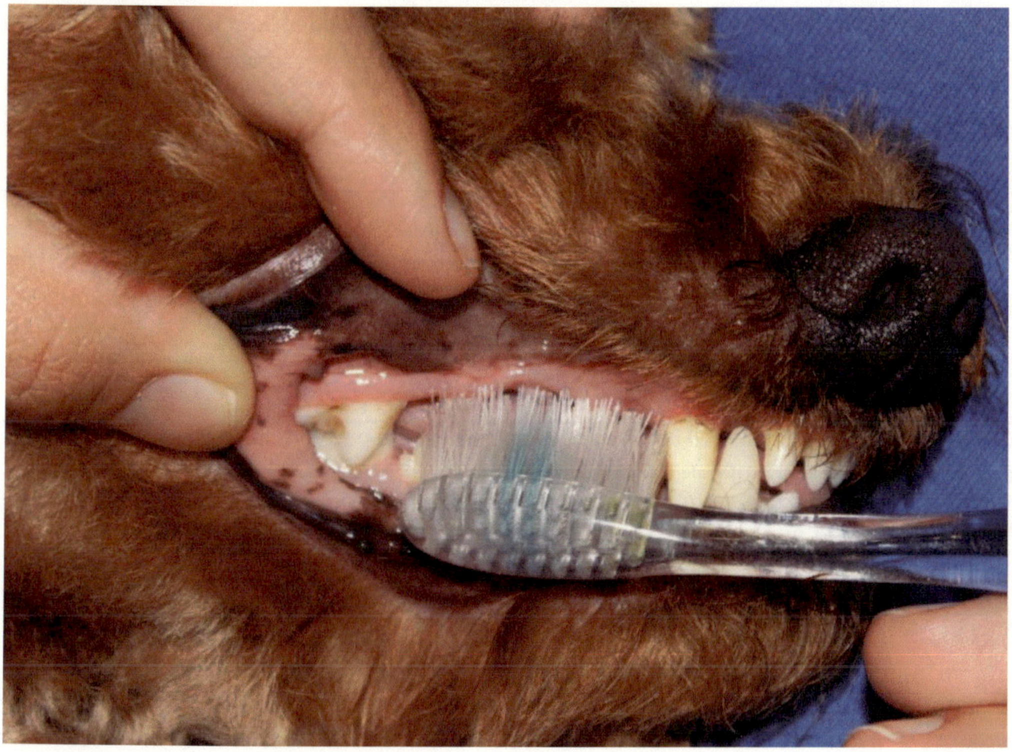

Fig. 7. Dog's Tooth brushing by an owner

may also help in effective biofilm removal (Jensen et al., 1995; Lima et al. 2004; Niemiec, 2008), used in conjunction with a toothbrush, added to toothpaste or employed as solutions, minimising the ineffective brushing technique in some inaccessible places in the animal's mouth.

The brushing technique should be used properly because inappropriate use can increase the prevalence of periodontal disease, as occurs in humans (Jongenelis & Wiedemann, 1997). One of the indicated techniques uses circular movements on the dental surfaces, with the brush tilted at an angle of 45 ° from the gingival margin (Niemiec, 2008).

Some studies have been conducted with the intention of establishing new therapies to combat dental plaque accumulation on the face of the tooth, both for the aid in brushing as for the use as a preventive agent of choice against periodontal disease when the practice of brushing is inefficient, as in wayward animals that do not allow the mechanical handling of oral practices (Gioso & Carvalho, 2004; Pieri, 2010).

Various forms of introduction of these therapies in the animal management have been tested, including the use of cookies and chewing objects for oral hygiene (Gioso & Carvalho, 2004; Niemiec, 2008), additives to drinking water with an inhibitory effect on the growth of bacteria (Clarke, 2006) and oral rinse solutions as well as leather and biscuits with the addition of antimicrobial agents (Addy, 1997; Niemiec, 2008).

Topical application of a drug to control the disease is considered desirable, in view of the lower incidence of side effects when compared to other routes of application. For the prevention of periodontal disease, taking into account the least harmful nature of bacteria in the onset of the disease and higher prevalence of supragingival plaque, it is recommended that the use of topical oral solutions such as mouthwashes is sufficient to combat bacteria in question with great advantage because of its easy application in most patients (Ciancio & Niezengard, 1997).

Among the chemicals, that can be used this way to reduce the accumulation of plaque on dental surfaces, the bisguanids, quaternary ammonia and phenols have been widely evaluated. Chlorhexidine appears as a substance that has the greatest efficacy in the inhibition of oral plaque (Hennet, 2002) and has good antiseptic activity against all oral pathogens, more directly on the bacterial plaque organisms (Harvey & Emily, 1993). Its main concentration is the commercial use of alcoholic solution at 0.12% and it is also found in alcohol-free solutions and in gel form (Robinson, 1995).

Despite the above indications for the use of chlorhexidine in the fight against dental plaque, it presents a series of unpleasant effects when used for prolonged therapy, such as loss of taste by the patient, pigmentation of the enamel, burning and even ulceration of the buccal mucosa (Zanini et al., 1995). These effects justify the use of this material only for few days (Gioso, 2007), which makes its application not recommended in the prevention of periodontal disease, which requires a prolonged use of the antimicrobial agent chosen for this purpose (Lascala & Moussalli, 1995).

The natural sweetener xylitol has been used in human patients in chewing gum, mouthwashes and toothpastes in order to reduce plaque. There is also one xylitol based product in the veterinary market today, which when added to the animal drinking water has a lowering effect on oral bacterial plaque formation (Dunayer & Gwaltney-Brant, 2006).

Recently, the ozonised sunflower oil was tested, with positive results on microbial reduction in human patients with periodontal disease (Fiorini et al., 2006) and copaiba oil was applied topically on dogs and the results were equal to those obtained with chlorhexidine on the oral microbial population (Pieri, 2010). Additionally, some in vitro tests were performed to analyse the antimicrobial activity of Copaiba oil on plaque-forming bacteria (Simões, 2004; Pieri, 2010; Valdevite et al., 2007) and the evaluation of the inhibition of *Streptococcus* sp. adherence in glass capillaries caused by the same phytotherapic (Pieri, 2010), obtaining in both cases positive results. A actual work has been conduced to evaluate the copaiba oil as antimicrobial against bacterial isolates from initial dental plaque of dogs, aiming identify a potential drug to prevent the plaque formation and consequently the periodontal disease (personal data).

Many researchers continue their analysis looking for natural drugs as propolis (Swerts et al., 2005), *Camellia sinensis* (Chang et al., 2009), *Mimosa tenuiflora* (Macedo-Costa, 2009), *Vitis amurensis* (Yim et al., 2010), *Rhinacanthus nasutus* (Puttarak et al., 2010), *Murraya koenigii, Allium sativum* and *Melaleuca alternifolia* (Prabhakar *et al., 2009)* to its use in the prevention of periodontal disease by inhibiting plaque formation.It important that this drug combine properties such as antimicrobial activity that does not induce bacterial resistance, and inhibition of microbial adherence on tooth surfaces that suggest a great potential for use in therapies in the oral cavity and as an aid in oral hygiene (Sudo et al., 1986; Corner et al., 1988; Pieri, 2010). For the use in the treatment of domestic animals it is suggested the inclusion of this antimicrobial and non-adherent agent in formulations containing the base flavours of chicken, beef, fish, etc. (De Marco & Gioso, 1997).

5. Diagnosis

The diagnosis of periodontal disease is based on history, clinical examination and radiological evaluation. Any changes in apprehension and chewing of food, as well as in general conditions and in the behaviour of animals, can be associated with oral disorders. Certain physical and behavioral changes are highly suggestive of dental disorders, including abnormal ways of eating and drinking, acute reactions to the ingestion of cold water, selective appetite (preference for soft foods), anorexia and weight loss, salivation, bleeding, epitaxy, digging of the ground, behavior of rubbing their feet on the face, shaking of the head, oronasal fistulas, abnormal aggressive behavior (because of pain) and distress and anguish (Emily & Penman, 1994; Pachaly, 2006; Gorrel, 2004). When it comes to periodontal disease, the main complaint of the owner will always be halitosis (Emily & Penman, 1994; Gorrel, 2004; Gioso, 2007) due to tissue decay and bacterial fermentation in the sulcus or periodontal pocket (Gioso, 2007).

Like any other clinical examination, the examination in dentistry should be preceded by thorough history and general physical examination. At the end, the oral cavity should be examined. It is necessary to do a complete oral examination to assess the presence of periodontal disease and other diseases, such as fractures or dental malocclusions. The intra- and extraoral structures should be assessed, including bone surfaces, the jaw muscles, salivary glands and regional cervical lymph nodes (Gorrel., 2004). Ideally, the complete periodontal examination should be performed in anaesthetised dogs (Harvey, 1992; Gorrel, 2004; Gioso, 2007). The evaluation of the tooth must be made with an explorer and periodontal probe (Gioso, 2007). The examination must be careful; incorrect handling of the probe may damage the soft tissues and lead to misdiagnosis of periodontal lesions (Gorrel, 2004). The changes observed should be recorded in an appropriate medical record and serve as the basis for the therapeutic treatment (Pachaly, 2006).

The periodontal examination includes the evaluation of teeth mobility, of injuries or furcation exposure, gingival retracting or hyperplasia, the evaluation of the depth, the presence of dental plaque, of gingivitis and dental calculus The furcation is the area between the roots of teeth that have more than one root. This area is usually filled with alveolar bone. During exploration, a depression can be felt while passing the extremity of a probe perpendicular to the tooth crown and below the gingival margin. In the presence of periodontitis, the furcation bone can be resorbed and probe inserted between the roots. Changes in the furcation are classified on a scale ranging from 0 to 3, where in grade 3 lesions the probe passes freely through the furcation, from the vestibular part to the lingual/palatal tooth (Gorrel, 2004).

The gingival sulcus is the space between the free gingiva and the tooth crown. In dogs, the depth of the gingival sulcus should be less than 3mm, and in giant breed dogs less than 4 mm (Gioso, 2007). When periodontitis is established, the junctional epithelium, the region of the gingival tissue inserted to the tooth surface, migrates apically along the root. If the apical migration is not accompanied by a receding gingiva then the periodontal pocket is formed, which has a depth greater than 3mm (Gorrel, 2004). Values above 3 mm mean loss of clinical attachment of the junctional epithelium with bone destruction (periodontitis) and periodontal pocket formation (Gioso, 2007)

The periodontal probe is essential in the examination and diagnosis of periodontal disease. This thin probe has a tip calibrated in millimetres, measuring the depth of the gingival sulcus when it is inserted between the gingiva and the tooth (Grove, 1998). The probe depth is defined as the distance between the coronal margin of the free gingiva and apical junctional epithelium (Gorrel, 2004). It is measured by positioning the tip of the periodontal probe parallel to the long axis of the tooth (or following the contour of the crown), and gently inserting between the teeth and free gingiva until the bottom of the sulcus is felt. In cases of gingival recession, periodontal destruction usually does not cause the formation of periodontal pockets. Gingival recession is measured in millimetres from the cementoenamel junction, where the gingival attachment should be normally at the gingival margin. The most profound measure for each tooth must be registered in the dental chart. Normally the junctional epithelium is located near the cementoenamel junction. In cases of gingival hyperplasia, in other words, in the presence of excessive amounts of soft tissues, the pseudopocket formed is measured with the probe, defined as the distance between the junctional epithelium and gingival margin. Since the areas deeper than 5 mm are difficult to clean mechanically, surgery might be needed to remove the deep pockets and pseudopockets, depending on the care undertaken by the owner (Gorrel, 2004).

The accumulation of dental deposits (plaque and calculus) and the severity of gingivitis can be quantified by standardised indices that correspond to the numerical expression of the presence or absence of disease severity. These indexes are extremely useful when there is a need for assessment of periodontal disease. The accumulation of plaque and calculus can be quantified in terms of coverage or thickness for all teeth (Gorrel, 2004).

The plaque is not always visible to the dental inspection, therefore solutions that highlight the plaque may be used (Gorrel, 2004; Gioso, 2007). The calculus is evident, presenting as a hard mass on the tooth surface, intra-or extra-sulcular, yellowish, brownish, sometimes greenish, which is not removed by scraping or brushing with gauze. The calculus most frequently occurs in fourth premolar and first superior molar teeth, as close to them are the openings of the parotid ducts and zygomatic glands, however, over time, almost all teeth can be affected (Gioso, 2007).

Bleeding during the survey, which indicates an inflammatory process in the connective tissues within the junctional epithelium, is a particularly useful method for evaluating an active gingivitis (Grove, 1998). Dogs rarely show signs of pain due to periodontal disease, even when there is loss of many teeth or exposed root dentine, which can cause sensitivity. There may be ulcers on the buccal mucosa (cheek) or on the tongue, because of the direct contact with areas of severe periodontal disease (Gioso, 2007). However, to form a definitive diagnosis, loss of tooth support must be present (Grove, 1998). The full-mouth radiographic examination is mandatory for patients with periodontal disease to get information from bone and periodontal structures (Gorrel, 2004). The results obtained by clinical and radiographic examinations are complementary and the diagnosis requires the completion of both (Harvey & Emily, 1993). Although radiographs provide essential data for determining the state of periodontal disease, this diagnostic test has low sensitivity to assess the progression of periodontitis. This is due to the inability to accurately repeat the positions, exposure and development time. Thus, any comparison between two different radiographs of the same animal becomes limited (Gorrel, 2004).

Radiographs are evaluated for changes in alveolar bone, interdental bone height, presence of lamina dura, trabecular pattern, periodontal ligament and severity of bone loss. X-rays show two-dimensional representation of three-dimensional structures. Sometimes, the radiographs do not show adequately the severity of the disease. Early lesions of bone destruction are sometimes not observed radiographically. Buccal and lingual alveolar bones are particularly difficult to assess because of the overlap. In addition to the radiological findings, the clinician must rely on clinical examination, including sulcular depths, tooth mobility, and gingival appearance, in order to decide on the diagnosis and treatment plan (Bellows, 2001).

The earliest radiographic sign of periodontitis is loss of definition of the bone ridge. In healthy animals, the bony ridge appears as a radiopaque line, which follows one or two millimetres in the apical direction, in parallel to an imaginary line drawn between the cementoenamel junction of two adjacent teeth (Harvey, 1992). This loss of definition of the bone ridge is always accompanied by progressive demineralisation of the lamina dura (Harvey & Emily, 1993). Other radiographic signs of periodontal disease include rounding of the alveolar margin, the discontinuity of the lamina dura, widening of the periodontal space and the gradual disappearance of the alveolar bone (Gorrel, 2004).

In some cases, pathological fractures of the jaw are seen as a consequence of severe bone loss. This situation occurs especially in small breed dogs, typically in the inferior first molar, whose roots reach the ventral cortex of the lower jaw (Gorrel, 2004).

Once the diagnosis is established, treatment plan should be developed that will range from just dental curettage and polishing to extraction. In some situations the extraction of the involved tooth is the best treatment option, especially if the loss of adhesion and mobility is very pronounced. However, for moderate disease cases there are a variety of other treatment strategies (Wiggs & Lobprise, 1997).

6. Treatment

Harvey and Emily (1993) described that the goal of periodontal treatment is to control microorganisms, restore normal anatomy and physiology and avoid new adhesion of bacterial plaque on tooth surfaces. Furthermore, periodontal pockets should be eliminated and re-adhesion of tissue to the tooth should be promoted, aiming, wherever possible, to do this by destroying the minimum of healthy tissue and keeping the gingiva.

The periodontal soft tissues quickly re-adhere to the cementum after the debridement, which removes the dead space of the pocket. However, this union could be weaker than the original depending on the type of tissue that repopulates the root surface; gingival epithelium, gingival connective tissue, alveolar bone and periodontal ligament; the latter being more desirable. When replacement by the alveolar bone occurs, the result is root resorption or ankylosis. Epithelium and gingival connective tissue are not desirable because they are extremely weak. Grafts or barrier materials can be used to delay or exclude the gingival tissue growth, favouring the growth of periodontal ligament growth (Wiggs & Lobprise, 1997).

According to Gioso (2007), treatment is based on the elimination of plaque or calculus, normal gingival depth restoration and monitoring through a preventive program. General anaesthesia is essential to perform the scraping and it is a procedure that can last around 2 to 3 hours in more advanced cases. The main treatment options for periodontal curettage are supragingival curettage, subgingival curettage, root planing, gingivectomy, gingivoplasty and gingival grafts.

Manfra-Marretta et al. (1992) highlighted the importance of subgingival curettage, where the plaques accumulated in the marginal gingiva that cause inflammation and affect the supporting structures of the tooth are removed. A curette to subgingival scrapping may be used. In some cases it is necessary to do a gingivectomy of periodontal pockets.

After removing all of the dental calculus, teeth must undergo a polishing with a rubber cup. It is not necessary to put a lot of pressure on the tooth and polishing should not exceed 15 seconds per tooth (Manfra-Marretta et al., 1992).

If there is severe bone loss, with roots exposure, an elevation of a gingival flap, complete curettage and displacement of the gingival margin closer to the apex of the tooth may be necessary, followed by fixating it with sutures. In the case of failure of this treatment, the extraction is the next step (Gioso, 2007).

Some surgical techniques are also indicated in the treatment of periodontal disease. According to Gioso (2007), when the periodontal pocket has more than 2 mm depth, partial gingivectomy is indicated, eliminating the pocket and re-forming the normal depth of the gingival sulcus (gingivoplasty). Another indication of the gingivectomy is gingival hyperplasia. In this surgery, excess gingiva is excised with a scalpel blade or electrocautery, which controls bleeding and is therefore preferred.

Another surgical technique described is the simple gingiva flap (or retail), which is indicated to obtain access to deeper periodontal structures through the creation of mucogingival flap. For its realisation an incision along the longitudinal axis of the root should be made, preserving the interdental papillae. Subsequently, the gingival flap should be completely translocated with the aid of a periosteum elevator and then root planing and repair of bone defects should be proceeded with. At the end of the procedure the flap should be sutured to its source with separate sutures (Wiggs & Lobprise, 1997).

The sliding flap can also be done in order to cover the root in cases of secondary exposure to gingival defect or periodontal disease. In order to do this an adjacent donor site must be identified and a flap at least 2.5 times wider than the defect to be covered must be created. The flap should contain full thickness of the epithelium and connective tissues. The periosteum should be maintained at the donor site, which will be left exposed. The proceeding finishes with the lateral slip of the flap and its suturing with simple interrupted technique, with stitches 1.5 mm apart, in its receptor place (Wiggs & Lobprise, 1997). This surgical technique is also indicated for closure of oronasal fistulas, which are abnormal communications between the oral and nasal cavities. In these cases it is very important to provide hermetic sealing of the suture without tension (Bolson & Pachaly, 2004).

7. Acknowledgments

The authors would like to thank Pró-Reitoria de Extensão e Cultura from Federal University of Viçosa (UFV), FAPEMIG (Fundação de Amparo à Pesquisa do Estado de Minas Gerais), CAPES (Coordenação de Aperfeiçoamento de Pessoal de Nível Superior) and CNPq (Conselho Nacional de Desenvolvimento Científico e Tecnológico) for financial support8. References

Addy, M. (1997). Anti-sépticos na terapia periodontal. In: Lindhe J. *Tratado de periodontia clínica e implantologia oral.* 3ed. p.332-349, Guanabara Koogan, Rio de Janeiro..

Bellows, J. (2001) *All pets dental - The why, when, and how of small animal dental radiology.* Avaiable from: http://www.dentalvet.com/vets/basicdentistry/whywhenhow_radiology.htm

Bolson, J.; Pachaly, J.R. (2004) Fístula oronasal em cães (*Canis familiaris* Linnaeus, 1758) - Revisão de literatura. *Arquivos de Ciências Veterinárias e Zoologia,* Vol.7 No.1 pp. 53-56.

Braga, C.; Rezende, C.; Pestana, A.; Carmo, L.; Costa, J.; Silva, L.; Assis, L.; Lima, L.; Farias, L.; Carvalho, M. (2005) Isolamento e Identificação da microbiota periodontal de cães da raça pastor alemão. *Ciência Rural,* Vol.35, pp.385-390.

Carranza, F.A. (1983) *Periodontia Clínica de Glickman.* Editora interamericana, Rio de Janeiro, Brazil.

Chang, H.; Hwang, H.; Kang, E.; Lee, J.; Chung, D.; Yang, W.; Chung, W.; Kim, H. (2009) The effect of Green tea bag in dogs with periodontal disease. *Journal of Veterinary Clinics,* Vol.26, pp.41-47.

Ciancio, S.; Niezengard, R. (1997) Controle e prevenção da doença periodontal In: Niezengard, R.; Newman, M. *Microbiologia oral e imunologia.* 2ed. Guanabara Koogan, pp.309-330, Rio de Janeiro.

Clarke, D.E. (2006) Drinking water additive decreases plaque and calculus accumulation in cats. *Journal of Veterinary Dentistry,* Vol.23, pp.79-82.

Clarke, D.E. (2001) Clinical and microbiological effects of oral zinc ascorbate gel in dogs. *Journal of Veterinary Dentistry,* Vol.18, pp.177-183.

Corner, A.; Dolan, M.M.; Yankell, S.L.; Malamud, D. (1988) C31G, a new agent for oral use with potent antimicrobial and antiadherence properties. *Antimicrobial Agents and Chemotherapy,* Vol.32, pp.350-353.

Debowes, L.; Mosier, D.; Logan, E.; Harvey, C.E.; Lowry, S.; Richardson, D.C. (1996) Association of periodontal disease and histologic lesions in multiple organs from 45 dogs. *Journal of Veterinary Dentistry,* Vol.13, pp.57-60.

De Marco, V.; Gioso, M.A. (1997) Doença periodontal em cães e gatos: profilaxia e manejo dietético. *Clinica Veterinária* (São Paulo), Vol.2, pp.24-28.

Domingues, L.M.; Alessi, A.C.; Schoken-Iturrino, R.P.; Dutra LS (1999) Microbiota saprófita associada à doença periodontal em cães. *Arquivo Brasileiro de Medicina Veterinária e Zootecnia,* Vol.51, pp.329-332.

Drumond, M.R.; Castro, R.D.; Almeida, R.V.; Pereira, M.S.; Padilha, W.W. (2004) Comparative study in vitro of the antibacterial activity from phytotherapeutic products against cariogenical bacteria. *Pesquisa Brasileira em Odontopediatria e Clínica Integrada,* Vol.4, pp.33-38.

Duchin S; Houte V (1978) Colonization of teeth in humans by Streptococcus mutans as related to its concentration in saliva and host age. *Infection and Immunity,*Vol.20, pp.120-125.

Dunayer, E.K.; Gwaltney-Brant, S.M. (2006) Acute hepatic failure and coagulopathy associated with xylitol ingestion in eight dogs. *Journal of the American Veterinary Medical Association,* Vol.229, pp.1113-1117.

Dupont, G.A. (1997) Understanding dental plaque: biofilm dynamics. *Journal of Veterinary Dentistry,* Vol.14, pp.91-94.

Dupont, G.A. (1998) Prevention of periodontal disease. *Veterinary Clinics of North American: small animal practice*, Vol.28, pp.1129-1145.

Emily PP; Penman S (1994) *Handbook of small animal dentistry*. Pergamon, p.35-53, Oxford.

Figueiredo, M.C.; Parra, S.L. (2002) *Aspectos normais da membrana periodontal e osso alveolar*. Avaiable from: http://www.odontologia.com.br.

Fiorini, J.M.; Cardoso, C.; Macedo, S.; Schneedorf, J.M.; Fiorini, J.E. (2006) Ação do óleo ozonizado como coadjuvante na terapia da doença periodontal. *Revista Internacional de Periodontia Clínica*, Vol.3, pp.55-99.

Ford, R.B.; Mazzaferro, E.M. (2007) *Manual de procedimentos veterinários e tratamento emergencial segundo Kirk e Bistner*. Editora Roca, pp.279-365, São Paulo.

Gibbons, R.J. (1972) Ecology and cariogenic potential of oral streptococci. In: Wannamaker, L.W.; Matsen, J. *Streptococci and streptococcal diseases*. Academic press Inc, pp.371-385, New York.

Gioso, M.A. (2007) *Odontologia para o clínico de pequenos animais*. Ed.Manole, São Paulo.

Gioso, M.A.; Carvalho, V.G. (2004) Métodos Preventivos para a manutenção da boa saúde bucal em cães e gatos. *Clínica Veterinária (São Paulo)*, Vol.9, pp.68-76.

Gorrel, C.; Rawlings, J.M. (1996) The role of dental hygiene chew in maintaining periodontal health in dogs. *Journal of Veterinary Dentistry*, Vol.13, pp.31-34.

Gorrel, C. (2004) *Veterinary dentistry for the general practitioner*. W.B. Saunders, pp.87- 110, Philadelphia.

Grove, T.K. (1998) Treatment of periodontal disease. *Veterinary Clinics North American*: small anim pract, Vol.28, No.5, pp.1147-1164,

Harvey, C.E. (1992) Distúrbios orais, faringianos e das glândulas salivares. In: Ettinger, S.J. *Tratado de medicina interna veterinária*. Ed. Manole, pp.1265-1290, São Paulo.

Harvey, C.; Emily, P. (1993) *Small Animal Dentistry*. Mosby - year book inc, St. Louis.

Harvey, C.E.; Orr, S. (1990) *Manual of small animal dentistry*. British small animal veterinary association, Cheltenham;

Harvey, C.; Shofer, F.S.; Laster, L. (1994) Association of age and body weight with periodontal disease in north American dogs. *Journal of Veterinary Dentistry*, Vol.11, pp.94-105.

Hennet, P. (1995) Dental anatomy and physiology of small carnivores. In: Crossley, D.A.; Penmann, S. *Manual of Small Animal Dentistry*. British small animal veterinary association pp. 93-104, Cheltenham.

Hennet, P. (2002) Effectiveness of a dental gel to reduce plaque in beagle dogs. *Journal of Veterinary Dentistry*, Vol.19, pp.11-14.

Jensen, L.; Logan, E.; Finney, O.; Lowry, S.; Smith, M.; Hefferren, J.; Simone, A.; Richardson, D. (1995) Resuction in accumulation of plaque, stain, and calculus in dogs by dietary means. *Journal of Veterinary Dentistry*, Vol.12, pp.161-163.

Jongenelis, A.P.; Wiedemann, W (1997) A comparison of plaque removal effectiveness of electric versus a manual toothbrush in children. *Journal of Dentistry for Children*, Vol.64, pp.176-182.

Katsura, H.; Tsukiyama, R.; Suzuky, A.; Kobayashi, M.; (2001) In vitro antimicrobial activities of bakuchiol against oral microorganisms. *Antimicrobial agents and chemotherapy*, Vol.45, pp.3009-3013.

Lang, N.; Mombelli, A.; Attström, R. (1997) Placa e cálculo dentais. In: Lindhe, Jan. *Tratado de periodontia clínica e implantologia oral.* 3.ed. Guanabara Koogan. pp.66-91, Rio de Janeiro.

Lascala, N. T.; Moussalli, N.H. (1995) *Compêndio terapêutico periodontal.* 2.ed. Editora Artes Médicas, São Paulo.

Lima, T.B.; Eurides, D.; Rezende, R. Milken, V; Silva, L.; Fioravanti, M. (2004) Escova dental e dedeira na remoção da placa bacteriana dental em cães. *Ciência Rural,* Vol.34, pp.155-158.

Lindhe, J; Karring, T. (1997) *Tratado de periodontia clínica e implantologia oral.* 3ed. Guanabara Koogan, Rio de Janeiro.

Loesche, W.; Grossman, N. (2001) Periodontal disease as a specific, albeit chronic, infection: diagnosis and treatment. *Clinical Microbiologicals Reviews,* Vol.14, pp.727-752.

Lyon, K.F. (1991) Dental home care. *Journal of Veterinary Dentistry,*Vol.8, pp.26-30.

Macêdo-Costa, M.; Pereira, M.; Pereira, L.; Pereira, A.; Rodriques, O. (2009) Atividade Antimicrobiana do Extrato da mimosa tenuiflora. *Pesquisa Brasileira de Odontopediatria Clínica Integrada,* Vol.9, pp.161-165.

Manfra-Marretta, S.; Cchloss, A.J.; Klippert, L.S. (1992) Classification and prognostic factors of endodontic-periodontic lesions in the dog. *Journal of Veterinary Dentistry,* Vol.9, No.2, pp.27-30.

McPhee, T.; Cowley, G. (1981) *Essentials of periodontology and periodontics.* 3ed. Blackwell scientific, Oxford.

Mitchell, P. (2005) *Odontologia de Pequenos Animais.* Editora Roca, São Paulo.

Murray, P.; Prakobphol, A.; Lee, T.; Hoover, C.; Fisher, S. (1992) Adherence of oral streptococci to salivary glycoproteins. *Infection and Immunity,* Vol.60, pp.31-38.

Newman, M.G.; Carranza, F.A.; Takei, F.A.; Henry, H. (2004) *Periodontia Clínica.* 9.ed. Guanabara Koogan, Rio de Janeiro..

Niemiec, B.A. (2008) Periodontal Therapy. *Topics in Companion Animal Medicine,* Vol.23, pp.81-90.

Niezengard, R.; Newman, M.G.; Zambom, J.J. (1997) Doença Periodontal In: Niezengard, R.; Newman, M.G. *Microbiologia Oral e Imunologia.* 2.ed. Guanabara Koogan. pp.309-330, Rio de Janeiro.

Pachaly, J.R. (2006) Odontoestomatologia. In: Cubas, Z.S.; Silva, J.C.R.; Catão-dias, Z.S. *Tratado de animais selvagens.* Editora Roca, pp. 1068-1091, São Paulo.

Picosse, M. (1987) *Anatomia dentária.* 4.ed. Sarvier, São Paulo.

Pieri, F.A. (2004) *Tratamento endodôntico em pequenos animais.* Faculdade de Medicina Veterinária da Universidade Federal Rural de Pernambuco, Recife. 33p.

Pieri, F.A. (2010) Clinical and microbiological effects of copaiba oil (Copaifera officinalis) on dental plaque forming bacteria in dogs. *Arquivo Brasileiro de Medicina Veterinária e Zootecnia,* Vol 62, No. 3, pp.578-585.

Pope, E.R. (1993) Periodontal and endodontic disease. In: Bojrab, M.J. *Disease Mechanisms in Small Animal Surgery.* 3ed. Lea Febiger, p.187-190, Philadelphia.

Prabhakar, A.R.; Vipin, A.; Basappa, N. (2009) Effect of curry leaves, garlic and tea tree oil on Streptococcus mutans and Lactobacilli in children. *Pesquisa Brasileira de Odontopediatria Clínica Integrada,* Vol.9, pp.259-263.

Puttarak, P.; Charoonratana, T. ; Panichayupakaranant, P. (2010) Antimicrobial activity and stability of rhinacanthins-rich Rhinacanthus nasutus extract. *Phytomedicine,* Vol.17, pp.323-327.

Riggio, M.P.; Lennon, A.; Taylor, D.J.; Bennett, D. (2011) Molecular identification of bacteria associated with canine periodontal disease. *Veterinary Microbiology, Vol.150, No.3-4*, pp. 394-400.

Robinson, J. (1995) Chlorhexidine Gluconate - the solution for dental problems. *Journal of Veterinary Dentistry*, Vol.12, pp.29-31.

Roman, F.S.; Cancio, S.; Cediel, R.; Garcia, P.; Sanches, M. (1995) Periodoncia. *Canis et felis*, Vol.16, pp.37-38.

Roza, M.R. (2004) *Odontologia em pequenos animais*. L.F. Livros de Veterinária, Rio de Janeiro.

Sans, M.; Newman, M.G. (1997) Placa dental e cálculo. In: Niezengard, R.J.; Newman, M.G. (1997) *Microbiologia Oral e Imunologia*. 2.ed. Guanabara Koogan, pp.275-292, Rio de Janeiro.

Senhorinho, G.N.A., Nakano, V., Liu, C., Song, Y., Finegold, S.M., Avila-Campos, M.J. (2011) Detection of Porphyromonas gulae from subgingival biofilms of dogs with and without periodontitis. *Anaerobe*, in press.

Simões, C.A. (2004) Brazilian Patent n°PI0404266-2. Instituto Nacional de Propriedade Industrial, Rio de Janeiro.

Slee, A.; O'Connor, J. (1983) In vitro activity of octenidine dihydrochloride (WIN 41464-2) against preformed plaques of selected oral plaque forming microorganisms. *Antimicrobial Agents and Chemotherapy*, Vol.23, pp.379-384.

Slee, A.; O'Connor, J.; Bailey, D. (1983) Relationship between structure and antiplaque and antimicrobial activities for a series of bispyridines. *Antimicrobial Agents and Chemotherapy*, Vol.23, pp.531-535.

Sudo, S.; Schotzko, N.K.; Floke, L.E. (1976) Use of hydroxyapatite-coated glass beads for preclinical testing of potential antiplaque agents. *Applied and Environmental Microbiology*, Vol.32, pp.428-432.

Swerts, M.S.; Costa, A.M.; Fiorini, J.E. (2005) Efeito da associação de clorexidina e própolis na inibição da aderência de *Streptococcus spp. Revista Internacional de Periodontia Clínica*, Vol.2, pp.10-16.

Tanzer, J.M.; Freedman M.L.; Fitzgerald, R.J.; Larson, R.H. (1974) Diminished virulence of glucan synthesis-defective of *Streptococcus mutans. Infection and Immunity*. Vol.16, pp.197-203.

Tanzer JM, Slee A, Kamay B ; Scheer E (1977) In vitro evaluation of three iodine-containing compounds as antiplaque agents. Antimicrobial Agents and Chemotherapy, 12:107-113.

Valdevite, L.M.; Leitão, D.P.; Leite, M.F.; Polizello, A.C.; Freitas, O.; Spadaro, A. (2007) Study of the in vitro effect of copaíba oil upon virulence factors of the cariogenic bacterium Streptococcus mutans. In: Anals of 10th IUBMB Conference e 36ª Reunião Anual da SBBq, Salvador, Brazil.

Wiggs, R.B.; Lobprise, H. (1997) *Veterinary dentistry principles and practice*. Lippincott-Raven, Philadelphia.

Wilderer, P.A.; Charaklis, W.G. (1989) *Structure and function of biofilms*. John Wiley, Chichester.

Yim, N.; Ha, D.; Trung, T.; Kim, J.; Lee, S.; Na, M.; Jung, H.; Kim, H.; Kim, Y.; Bae, K. (2010) The antimicrobial activity of compounds from the leaf and stem of Vitis amurensis against two oral pathogens. *Bioorganic & Medicinal Chemistry Letters*, Vol.20, pp.1165-1168

Zanini, A.C.; Basile, A.C.; Martin, M.I.; Oga, S. (1995) *Guia de Medicamentos*. Atheneu, São Paulo:

8

Continuous Electrocardiography in Dogs and Cats

Guilherme Albuquerque de Oliveira Cavalcanti
Federal University of Minas Gerais,
Brazil

1. Introduction

A new electrocardiogram (ECG) was started in 1961 when J. Norman Holter introduced a system to record the ECG of a patient during their activities for long periods of time. This method was identified as a method of continuous electrocardiography (CE), long-term electrocardiogram, dynamic electrocardiography, ambulatory electrocardiographic, or simply Holter. In subsequent years, the development of technology has introduced a great number of advances and improvements in the fidelity of the recording, size and weight of equipment, systems acquisition and data analysis. The intermittent recording capacity triggered by the owner or by the occurrence of arrhythmia, has become available. Systems to record and reproduce the CE continue to evolve and improvements in the capacity of analysis and editing of commercial systems have frequently been introduced.

The entry of the CE in veterinary medicine contributes to a better understanding of cardiac arrhythmias, to study the cardiac response to drugs and to diagnose heart disease. But before you do the examination, the classic electrocardiographic (ECG) should be performed, because this modality is able to observe the cardiac depolarization for more angles (leads) that the CE.

2. The continuous electrocardiographic in diagnosis support

The most appropriate method to evaluate the heart rate in dogs is the CE system. This test allows assessment of rhythm for 24 hours, while conventional electrocardiography can only do about 0.2% of this period. Thus, this method allows the recognition of rhythm with the animal in different types of activity (sleep, exercise, feeding) and, consequently, under different physiological states.

Currently this method is most widely using the following conditions: a) confirmation of an arrhythmia as the cause of symptoms that occur during daily activities, b) prediction of future cardiac events, c) evidence of therapeutic efficacy of antiarrhythmic agents, d) detection of myocardial ischemia (MacKie et al., 2010).

There are some arrhythmias that occur at intervals greater than 24 hours, in such cases it is recommended either to repeat the exam, or use of cardiac event recorders (CER). When the owner presses a button on the device of CER, for example in the moment when the clinical sign occurs, the device records the moments before, during and after the clinical signs which should be evaluated. The CER device is smaller and can be used for a week or more, but requires constant observation of the animal, although there are some devices that can function as EC conventionally or as CER at the same time.

In humans, medicine is defined in three classes of indications to perform CE: Class I - there is a consensus that the CE is a useful tool and reliable examination, class II - CE is useful and reliable, but there was no consensus, class III – there is consensus that the CE is unnecessary. In all situations in which the role of CE is discussed, it should be taken in account the history and the physical examination of the patient, and the review of the usefulness of the test. The decision to conduct a CE examination and interpretation of their results cannot be performed independently.

The major use of this examination in dogs and cats is to identify the type and degree of cardiac arrhythmias.

Being the main indications:

- Animals that have unexplained episodes of syncope, collapse, transient weakness and ataxia, even without previous history of heart disease.
- Prediction of dilated cardiomyopathy (DCM) in healthy adult dogs of the predisposed breeds (mostly Doberman and Boxer).
- Animals with relatives who have sudden death with no identified cause or relatives who had DCM.
- Identification of arrhythmias with sporadic or intermittent clinical signs.
- The monitoring of the antiarrhythmic treatment.

In moments of bradycardia, usually myocytes specialized in electrical conduction, trigger stimulus that can result in repeated arrhythmias. Therefore, due to the fact that some arrhythmias begin in a moment of great parasympathetic stimulation, it is logical to think that in an examination that is performed for 24 hours, with the animal performing its daily activities, it would be superior to the classic ECG in which the animal is contained on a clinic table. However, the ECG provides a panoramic view of the heart in ten different views (leads), not being replaced by the CE examination.

3. Attaching the holter monitor

The Holter apparatus is a handheld device that runs tests and consists of an ECG recording unit that has conductors of electromagnetic waves coupled to the animal by adhesives or suction cups. The recording of the examination is decoded by a computer program that specifically allows the evaluation of cardiac rhythm throughout the recording period.

In order for the recorders to meet the standards required by the *American Heart Association*, they should be able to record undistorted signals of high and low frequency, between 0.05 and 100Hz. These systems are able to record the signal ST. The recording can be common on cassette or memory card (more usual today) that stores the fully digitized signals captured by the electrodes. The advantage of scanning equipment is to reduce the weight and size of the recorders and the elimination of mechanical parts such as motor and reduction system. The recording should be in two leads. Devices that record in three channels can replace the two channels monitors.

The systems analysis should always allow classification of the forms of waves and broad interaction with the device analyst for corrections and to eliminate possible artifacts.

The two channels apparatus have five electrodes and the three channels have seven. When using two channels device, the electrodes are fixed to the skin on the chest of the animal, according to the Ware scheme (1998):

Channel 1 - White electrode (negative): the sixth right intercostal space, two fingers above the costochondral articulation;

- Red lead (positive): the sixth right intercostal space, in the location of costochondral joint;
- Green lead (ground): the sixth right intercostal space near the sternum.

Channel 2 - Brown electrode (negative): the sixth left intercostal space, two fingers above the costochondral articulation;

- Black electrode (positive): the sixth left intercostal space toward intermediate between red and green electrodes positioned on the opposite side.

In equipment with four electrodes (Cardioflash® or Cardiolight®: www.cardios.com.br), the local standard ECG precordial examination can been adapted for CE as follows: position of red electrode in V2 place (sixth left intercostal space adjacent to the sternum), black in V4 (sixth left intercostal space in the costochondral junction location), orange electrode between black and red, and white electrode in the V1 position of conventional ECG (fifth right intercostal space near the sternum), as shown in figure 1.

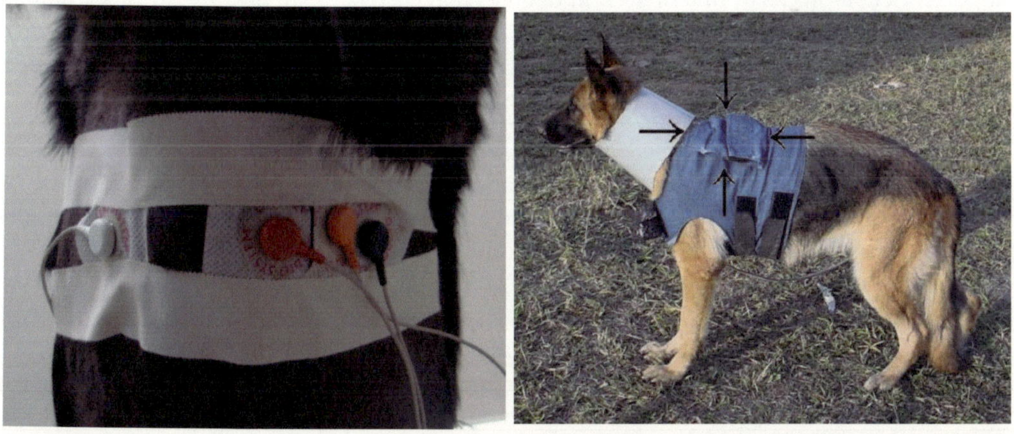

Fig. 1. Left side of the figure - Animal with electrodes attached to the chest for the conduct of examination by equipment with four electrodes (See the text for more details). Note the shaved area with the setting of adhesives and the coupling of electrodes. Right side of the figure – Dog using cervical collar and being subjected to examination. Note that the device is packed in the side pocket of its waistcoat (arrows).

In the thoracic region, where the adhesive electrodes are positioned, there must be ample shaving, cleaning of the area with alcohol to remove the hair and skin oils. After the gel electrodes are fixed with adhesive tape, dry the skin (Figure 1) and it should be held with a protective bandage to prevent the electrodes to come off.

Later, you put on a denim jacket or something similar, which will remain on the animal during the exam. The CE device will be packaged in the side pocket of the jacket and the cables will remain protected (Figure 1).

There is the possibility of performing the examination without the use of waistcoat only conditioning the device with bandage. After placing the recording monitor, cervical or Elizabethan collar should be used to prevent dog bites and damages to the unit. In the literature, the use of cervical or Elizabethan collars did not influence the heart rate in healthy dogs (Cavalcanti et al., 2007). However, it is clear that some dogs tolerate the use of cervical collar better than the Elizabethan one.

The weight of the device and the size of the vest should be compatible with the size of the animal; cats should use smaller devices

4. The result of the examination

Every animal should have a sheet of notes taken during the activities along the course of the recording. It should be noted that the starting time of the examination correlates with the observed daily heart rate (HR) and, possibly, arrhythmias. The report should contain the results of the exam starting time, duration of the examination, the level of recording quality, the maximum, average and minimum HR in each hour and the same HR data on the total examination time. Some supraventricular and ventricular premature complexes may occur in healthy dogs and cats, and in humans these premature complexes increased with age. Atrioventricular blocks of first or second degrees may also occur in healthy dogs. Variations and changes unique to the T wave are not considered diagnostic of CE in small animals.

4.1 Heart rate

Athlete human beings and athlete dogs have mean HR and maximum HR lower than sedentary individuals, the physical activity modulates the sinus node causing decrease of the average of these HR (Cavalcanti et al., 2009, Martinelli et al., 2005), and increasing HR variability and the average RR interval. Endurance exercise training has been established to alter autonomic nervous system activity, resulting in an apparent increase in cardiac parasympathetic tone coupled with decreases in sympathetic activity. For example, in both humans and animals, the heart rate at submaximal workloads was reduced in trained individuals compared with sedentary controls (Smith et al., 1989). A resting bradycardia is a well-established consequence of exercise training and is, in fact, used as a marker that the exercise-trained state has been achieved. Both acetylcholine content and cholineacetyl transferase were increased in the hearts of trained rats compared with control rats (DeSchryver and Mertens-Strythaggen, 1975).

The circadian variation in HR is similar to that illustrated in Figure 2. Obviously, the limits of the upper and lower HR should be calibrated in the software, according to the animal being studied.

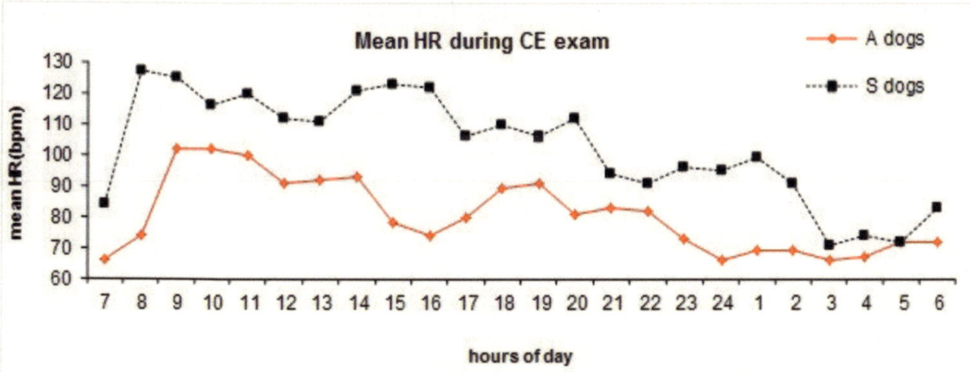

Fig. 2. Graphics show the average of the mean HR at different times of day, observed in sedentary (S) and athlete (A) dogs of the German shepherd breed. Note that the two groups of dog exhibit the same trend to increase and decrease in mean HR over 24 hours, and lower heart rates were observed from three to six hours of the morning.

4.2 Indication of the CE to evaluate clinical signs related to arrhythmias

Monitoring with the CE for 24 hours is especially useful because it provides details of the patient's heart rhythm, the total number and type of abnormal complexes, as well as the specific time of the day when they occurred. The clinician can assess the animal's activities in the period of recording and correlate exam results with clinical signs manifested at every moment.

The method of CE does the ECG recording for long periods and during the patient's daily activities. This allows us to observe spontaneous changes and those caused by activities or situations experienced by the patient's daily routine and, above all, make sure that the reported symptom related to whether or not an electrocardiographic changes. Symptoms that may be caused by changes in heart rhythm should occur often enough to be surprised while doing the recording of the ECG; however, the direct relationship is not always present and can have situations like:

a. a) Without clinical signs and without arrhythmia
b. b) Without clinical sign and with arrhythmia
c. c) With clinical sign and without arrhythmia
d. d) With clinical sign and with arrhythmia, but no correlation.

In the situation "a" it is impossible to establish any relationship and the examination should be repeated depending on the severity of the clinical suspicion, in "b", though symptoms do not occur, the type of arrhythmia observed may suggest a possible correlation with the sign mentioned above; in "c", discards the possibility of a relationship of symptoms with arrhythmia and another cause should be investigated, in "d", each case must be individualized, because we have patients that fall in both "b" and "c".

4.3 Supraventricular arrhythmias

The CE detects the same supraventricular arrhythmias than the ECG, however there are some considerations to be taken.

The presence of supraventricular arrhythmias is typically observed in less than 25% of healthy small animals within 24 hours. The evaluation of this type of arrhythmia should be cautious; the software which reads the tapes are calibrated to values of human waves and sinus arrhythmia of normal dogs may be misinterpreted as supraventricular premature complex, and even if there are trembling in the baseline sinus, arrhythmia associated with the software can be interpreted incorrectly, such as atrial fibrillation. Solid memory devices (e.g. Memory card) work similarly to cassette and offer the advantage to be programmed to work with range of values of dogs and cats ranges, minimizing the number of premature supraventricular complexes false positive in the exam. So when the reading is done, someone familiar with sinus arrhythmia of dogs must be present to review this arrhythmia.

4.4 Sinus Pauses (SP)

The long SP (more than 2 seconds) or in high quantity may be a factor in the initiation of arrhythmias. The SP generally occurs when the animal's FC is the lowest. Among the breeds studied: 40% of healthy German shepherd dogs had SP. All healthy Doberman and Boxer dogs and 70% of healthy dogs of the English cocker spaniel breed also presented it, so the frequency of SP varies among healthy individuals of different breeds. The median frequency of SP also varies among different breeds, English cocker spaniel, German shepherd, Boxer and Doberman dogs, showed median SP, respectively, 3, 32.5, 366 and 141 in 24 hours. The average of greater SP is more similar: 2.5, 2.87, 3.5 and 3.3, respectively. Therefore, these preliminary studies, the median of PS varies among breeds more than the average of greater SP. The SP occurs with a wide variation among the numbers of healthy individuals and may be higher than 5 seconds at times of low HR in normal dogs.

4.5 Ventricular arrhythmias

The EC detects the same changes of the ECG ventricular depolarization, but measured by a prolonged period and it identifies the different morphologies of normal and premature ventricular complexes (PVC) and is useful in predicting the DCM and in the identification of intermittent arrhythmias.

The presence of multiple PVC, single or repeated (mainly paroxysmal ventricular tachycardia) is associated with episodes of syncope or weakness. In more than half of the Boxer dogs who have syncope, the examination identifies at least 50 PVC in 24 hours.

Also, there is variation in the amount of PVC in dogs of different breeds; the occurrence is 26% for animals of the Beagle breed, 10% for the Doberman dogs, 20% in German shepherd and 40% in English cocker spaniel. However, between 50 and 70% of healthy Boxer dogs, have presented ventricular ectopic activity.

Doberman dogs that have more than 50 PVC in 24 hours have a higher risk of developing DCM, and it is considered abnormal if the presence of PVC is greater than 100 in 24 hours, especially if PVC are repeated or multiform. Based on studies a table was made (Table 1) for the evaluation of EC exams in Doberman pinschers. And Boxer dogs that present more than 91 PVC in 24 hours may have represented dogs with arrhythmogenic right ventricular cardiomyopathy or other disease processes that could have resulted in the development of ventricular arrhythmias (Stern et al., 2010).

Frequency of premature ventricular complexes in 24 hours	Interpretation
None	Annual reassessment of the CE, if the animal is of a predisposed family
Less than 50	May be the initial DCM, annual reassessment required.
Between 50 and 100	Suspect animal, revaluation every three to six months.
More than 100	Likely to be suffering from DCM, revaluation of 3 to 6 months.

Table 1. Recommendations for interpretation of CE exam in Doberman dogs (Goodwin, 1998).

Sudden death has been reported in young dogs of German shepherd breed, very similar to the illness of sudden death in human babies. Both are characterized by paroxysmal ventricular tachycardia at moments of parasympathetic influences, as in deep sleep and after exercise. Any animal displaying episodes of ventricular tachycardia has high risk of sudden death.

This familiar arrhythmia in German shepherd dogs usually decreases after seven months of age, but can be detected in animals older than 5 years and are considered affected, the animals showing more than 240 PVC during the examination. Death was observed in 15 to 20% of affected animals and usually occurs at the age of 4-8 months (Moise et al., 1997). In Brazil, there are no reports of this hereditary disease that has been observed in North America.

4.6 CE Indication for detection of myocardial ischemia

The study of myocardial ischemia by CE was made possible by technical advances in recording systems, with enhancements to the registry of the ST segment. As the ST as a sign of low frequency response, its correct detection and registration will depend on the frequency range of recorders, which should be between 0.05 and 100 Hz. The analysis system should be able to do it automatically, exposing the behavior of the ST segment graphically in time function, with the possibility of observation and interaction of events by the analyst. In general, the electrocardiogram is superior to CE in detecting myocardial ischemia due to greater number of variations of electrode placement.

In veterinary medicine, myocardial ischemia due to obstruction of large coronary vessels is uncommon and ST segment changes may occur in cases of hypertrophic left ventricle due to obstruction of small vessels. No healthy dogs of the Boxer, English cocker spaniel, Doberman and German shepherd breeds have shown significant change in the ST segment.

The ventricular arrhythmia is a common consequence of myocardial infarction and is easily identified and quantified in dogs with induced myocardial infarction and examined by CE. Moreover, it is known that at the time of greatest oxygen demand of the heart muscle (e.g. during exercise, stress ECG) ST changes are best identified in humans and the same occurs in dogs subjected to experimental infarction and evaluated by CE, moreover, the CE has a higher ability to detect small myocardial infarction in dogs than the resting ECG.

The ST segment analysis by electrocardiography is inadequate when there is left bundle branch block or Wolff-Parkinson-White syndrome or QRS complex with very low amplitude. Moreover, in large variations in the morphology of the QRS complex, the ST segment assessment should be undertaken with caution, even if it is caused by change of posture. It is noteworthy that a normal examination does not exclude ischemic heart disease, even by the natural variability of ischemic injury.

4.7 Evaluation of therapeutic procedures results

The diagnosis of cardiac arrhythmias and evaluation of therapeutic results requires prolonged observation of the ECG; in addition, the long-term ECG enables clarity of paroxysmal symptoms that may occur after a therapeutic procedure. The aggravation of preexisting arrhythmias or the emergence of new arrhythmias (proarrhythmia) is a phenomenon that can occur during treatment, especially the drug induced treatment, and the CE exam can provide information, even in asymptomatic cases. The CE should not be routinely done in the monitoring of patients with artificial pacemakers; however, it constitutes a powerful tool in the elucidation of paroxysmal symptoms in this group of patients.

In veterinary medicine, it is considered efficient the drug that causes a decrease of at least 70% in the number of premature supraventricular complexes. In dogs of the Doberman breed with DCM and treated with tocainamida, the total number of CVPs decreased between 70 and 80%, and the ventricular tachycardia 90%, which can be defined as therapeutic success in the treatment of DCM in PVC (Calvert et al., 1996).

4.8 Heart rate variability

Several types of analysis of the variability of intervals between R waves are provided by the examination. The RR variability is a way to evaluate the cardiac autonomic system functioning, and high variability of RR indicates a good heart function.

The analysis of HR variability is achieved through the study of parameters for time domain and frequency, provided directly by the computer analysis program. Although there are certain parameters in the time domain that can be analyzed by the ECG, the best method for this analysis is the CE, because this evaluation is made considering more data and obtains more consistent results.

The following variables are evaluated in the time domain: mean NN (mean of all normal RR intervals of the exam), SDNN (standard deviation of all normal sinus R-R intervals), SDANN (standard deviation of the averaged normal sinus R-R intervals for all 5-minute segments of the entire recording), SDNN index (mean of the standard deviations of all normal R-R intervals for all 5-minute segments of the entire recording), rMSSD (root mean square of the sum of squared difference of adjacent normal RR intervals throughout the exam) and pNN50 (the percentage of adjacent R-R intervals that varied by more than 50 ms).

In the frequency domain, the following variables are assessed: HF (high frequency), which comprises 0.15 to 0.4 Hz or 2.5 to 6.6 s / cycle; LF (low frequency), consisting of 0.0033 to 0.04 Hz or 5min/ciclo the 25s, ULF (ultra low frequency), which includes values less than 0.0033 Hz, and the ratio between the LF and HF components (LF / HF).

Human beings studies of HR variability in the spectral domain have shown that the spectral power contained in certain frequency bands reflects, in part, the sympathetic and parasympathetic modulation of sinus node activity. Studies using selective autonomic blockade have shown that power in the frequencies of HR variability > 0.15 Hz can be attributed to parasympathetic modulation, whereas power in frequencies < 0.15 Hz are related to both sympathetic and parasympathetic modulation, thus spectral analyses of HR variability have been used in a variety of settings and pathologies to assess autonomic modulation. Rapid changes in sympathovagal control are known to occur in the setting of exercise and recovery from exercise. Exercise is characterized by a decrease in parasympathetic tone and an increase in sympathetic tone, resulting in an increase in heart rate. During recovery from exercise, HR gradually decreases as parasympathetic tone returns and sympathetic tone withdraws.

In human medicine the variability of RR intervals provides useful information in defining patient prognosis after acute myocardial infarction, in the person with DCM, as well as to stratify the risk of sudden death in diabetic patients. it is known that the variability decreases in the presence of heart failure in dogs with chronic myxomatous mitral valve (Crosara et al., 2010) and in advanced DCM (Calvert and Wall, 2001), and a study of HR variability in cats has not been found yet. Perhaps with more studies, HR variability will be more used in Veterinary Medicine

5. References

Calvert, C. A.; Jacobs, G. J.; Smith, D. D.; Rathbun, S. L.; Pickus, C. W. Association between results of ambulatory electrocardiography and development of cardiomyopathy during long-term follow-up of Doberman pinchers. *Journal of the American Veterinary Association*, v.216, n. 1, p.34-39, 2000.

Calvert, C. A.; Pickus, C.W.; Jacobs, G. J. Efficacy and toxicity of tocainide for the treatment of ventricular tachyarrhythmias in Doberman pinschers with occult cardiomyopathy. *Journal of Veterinary Internal Medicine*, v.4, n. 10, p.235-240, 1996.

Calvert, C. A.; Wall, M. Effect of severity of myocardial failure on heart rate variability in Doberman pinschers with and without echocardiographic evidence of dilated cardiomyopathy. *Journal of the American Veterinary Medical Association*, v.219, n. 8, p.1084-1088, 2001.

Cavalcanti, G. A. de O.; Nogueira, R. B.; Gonçalves, R.S. ; Araujo, R. B. ; Muzzi, R. A. L.; Sampaio, G. R. Influência da utilização de colar elizabetano ou colar cervical em cães submetidos a eletrocardiografia contínua (Holter). *Revista Universidade Rural. Série Ciências da Vida*, v. 27, p. 461-463, 2007.

Cavalcanti, G. A. O.; Nogueira, R. B.; Sampaio, G. R.; Araujo, R. B.; Gonçalves, R.S. Avaliação por eletrocardiografia contínua (holter) em cães da raça Pastor Alemão praticantes de atividade física regular. *Arquivo Brasileiro de Medicina Veterinária e Zootecnia*, v. 61, p. 1446-1449, 2009.

Crosara, S.; Borgarelli, M.; Perego, M.; Häggström, J.; LA Rosa, G.; Tarducci, A.; Santilli, R. A. Holter monitoring in 36 dogs with myxomatous mitral valve disease. *Australian Veterinary Journal*, v. 88, n. 10, p. 386-392, 2010.

Dae, M. W.; Lee, R. J.; Ursell, P. C. ; Chin, M.C.; Stillson, C. A.; Moise, S. Heterogeneous Sympathetic Innervation in German Shepherd Dogs With Inherited Ventricular Arrhythmia and Sudden Cardiac Death. *Circulation*, v. 96, n. 4, p. 1337-1342, 1997.

Deschryver, C.; Mertens-Strythaggen, J. Heart tissue acetylcholine in chronically exercised rats. *Experientia*, v.31, p.316– 318, 1975.

Goodwin, J. CE monitoring and cardiac event recording. *Veterinary Clinics of North America: Small Animal Practice*, v. 28, n. 6, p. 1391-1407, 1998.

Mackie, B. A.; Stepien, R. L.; Kellihan, H. B. Retrospective analysis of an implantable loop recorder for evaluation of syncope, collapse, or intermittent weakness in 23 dogs (2004-2008). *Journal of Veterinary Cardiology*, v. 12, n. 1, p. 25-33, 2010.

Martinelli, F. S.; Chacon-Mikahil, M. P. T.; Martins, L. E. B.; Lima-Filho, E. C. ; Golfetti, R.; Paschoal, M. A.; Gallo-Junior, L. Heart rate variability in athletes and nonathletes at rest and during head-up tilt. *Brazilian Journal of Medical and Biological Research*, v. 38, n. 4, p. 639-647, 2005.

Möise, N.S.; Defrancesco, T. Twentyfour-hour ambulatory eletrocardiography (CE monitoring). In: BONAGURA, J.D. (Ed). KIRK'S current veterinary therapy. XII: Small animal practice. Philadelphia: W.B. Saunders, 1995. p.792-799.

Moïse, N. S.; Gilmour, R. F.; Riccio, M. L.; Flahive, W. S. Diagnosis of inherited ventricular tachycardia in German Shepherd dogs. *Journal of the American Veterinary Medical Association*, v.210, n. 3, p.403-410, 1997.

Nogueira, R. B.; Muzzi, R. A. L.; Herrera, D. S.; Falco, I. R.; CavalcantI, G. A. O. Avaliação do ritmo cardíaco em cães da raça boxer saudáveis por meio da eletrocardiografia contínua (CE). *Arquivo Brasileiro de Medicina Veterinária e Zootecnia*, v.58, n. 1, p.133-136, 2006.

Smith M. L.; Hudson, D. L.; Graitzer, H. M..; Raven, P. B. Exercise training bradycardia: the role of autonomic balance. *Medicine Science Sports Exercise*, v.21, p. 40–44, 1989.

Sosa, E. A.; Terzi, R.; Gruppi, S.; Brito, F. S.; Paola, A. A. V.; Pimenta, J.; Lorga, A. M.; Maia, I. G.; Gizi, J. C.; Solimene, M. C.; Camargo, S. P. A. B.; Albanezi-Filho, F. Consenso SOCESP-SBC sobre Eletrocardiografia pelo Sistema CE. *Arquivo Brasileiro de Cardiologia*, v.65, n. 5, p.447-450, 1995.

Stepien, R. L.; Hinchcliff, K. W.; Constable, P. D.; Olson, J. Effect of endurance training on cardiac morphology in Alaskan sled dog. *Journal of Applied Physiology*, v. 85, n. 4, p. 1368-1375, 1998.

Stern, J. A.; Meurs, K. M.; Spier, A. W.; Koplitz, S. L.; Baumwart, R. D. Ambulatory electrocardiography evaluation of clinically normal adult Boxers. *Journal of American Veterinary Medical Association*, v. 236, n. 4, p. 430-433, 2010.

Tilley, L.P. *Essentials of canine and feline electrocardiography interpretation and treatment*. 3.ed. Philadelphia: Lea & Febiger, 1992. 470p.

Yamada, M.; Tokuriki, M. Effects of a canine Elizabethan collar on ambulatory electrocardiogram recorded by a CE recording system and spontaneous activities measured continuously by an accelerometer in beagle dogs. *Journal of Veterinary Medical Science*, v.62, n. 5, p. 549-552, 2000.

Ware, W. A. Practical use of CE monitoring. Compendium on Continuing Education for the Practicing Veterinarian, v.20, n. 2, p.167-177, 1998.

9

Arthroscopic Follow-Up After Rupture of the Cranial Cruciateligament

Cleuza Maria de Faria Rezende,
Eliane Gonçalves de Melo and Natalie Ferreira Borges
Escola de Veterinária – UFMG,
Brazil

1. Introduction

The femoro-patellar-tibial joint is one of the most complex, with different structures that, altogether, keep the joint stability and functionality (Cook, 2010; de Rooster et al., 2010, Pozzi & Kim, 2010). Due to its big functional demand, the cranial cruciate ligament (CrCL) is the structure most frequently injured. The rupture of this ligament is of the most common causes of lameness in the dog and the resulting instability leads to the development of degenerative joint disease (DJD). Still today, the search for the ideal technique to treat CrCL rupture, with joint stabilization and prevention or minimization of DJD, persists. However, would joint stabilization be the only factor to be considered for the treatment of CrCL rupture and its sequels? Won't the effects of instability on the other structures deserve more investigation? Would the inflammatory response mediators be responsible for the looseness of the intra-articular graft or is it exclusively due to improper fixation or inadequate post-operatory?

According to literature, the joint degeneration process starts as soon as seven days after CrCL rupture and, at Day 21, signals suggesting osteoarthritis can be radiographically examined. In arthroscopic evaluation, however, lesions are already significant and characteristic in this period.

Arthroscopy is a minimally invasive exam, with minimal morbidity, which allows complete visualization of structures with magnification of images and precise diagnosis of the articular condition in its natural environment. The minimal morbidity enables periodic joint evaluations (Van Ryssen & Van Bree, 1998; Hulse et al., 2010; Goldhammer et al., 2010), what permits to monitor the lesion (Borges, 2006; Bleedorn et al., 2011) and the treatment evolution (Case et al., 2008), allowing the articular function prognosis (Ljungvall & Ronéus, 2011).

Arthroscopy can be employed for evaluating and defining the DJD staging. It gives the surgeon a better understanding of the joint degeneration process, the most prevailing condition among all diseases in the dog.

The arthroscopy had a significant impact especially in our clinical practice in what concerns diagnosis and treatment of articular alterations in the dog. Non-specific arthritis, hemarthrosis, meniscus lesions, ligament partial ruptures, femoral condyle osteochondrosis, and presence of loose cartilaginous bodies not diagnosed by the radiographic exam are examples of clinical situations that nowadays are promptly diagnosed and treated through arthroscopy.

The arthroscopic monitoring of the femoro-patellar-tibial joint after CrCL rupture changed the procedures by which such condition is treated in our routine, and its stabilization began to be considered as a condition that must be treated the most precociously possible and through arthroscopy. The main challenge is the development of efficient therapies for the osteoarthritis control. The association of articular stabilization with cell therapy (stem cells, growth factors) is one of the alternatives under study at the moment. The purpose of this chapter is, in the meanwhile, presenting the articular findings, under arthroscopic vision, 21 days after arthroscope-guided CrCL rupture, taking as reference the parameters found in the arthroscopic exam right before rupture.

The rupture of the cranial cruciate ligament (CrCL) is one of the most frequent orthopedic affections in the dog (Johnston, 1997; Beale & Hulse, 2010; Van Bree et al., 2010). After rupture, the CrCL does not regenerate, leading to lost of joint stability. The resulting instability leads to the development of degenerative joint disease (DJD) (Innes, 2010), and treatment is still challenging. The unavoidable DJD progression (Vasseur & Berry, 1992; Lazar et al., 2005) is caused by enzymatic degradation of the articular cartilage (Bennett & May, 1997). As the alterations in the cartilage progress, reduction in the content of proteoglycans, hyalunoric acid, and collagen (in minor proportion) occur due to the action of catabolic enzymes released by the DJD. Once the cartilaginous lesion is installed, the subchondral bone is exposed to the synovial fluid and, when subjected to abnormal pressures and tensions, reacts forming osteophytes and subchondral sclerosis (Johnston, 1997). Periarticular osteophytes indicate articular instability and are one of the most evident radiographic signs of DJD (Schrader, 1995; Innes et al., 2004). Osteophytes are bone projections located at the peripheral region of the joint, most frequently at the osseous insertion of the synovial membrane, the perichondrium, and the periosteum, though it may occur at the central region of the joint (Johnston, 1997). As the DJD evolves, after CrCL rupture, formation of osteophytes occurs first at the osteochondral margin of the lateral and medial trochlear ridges, then at the proximal region of the tibia and the proximal and distal borders of the patella (Lewis et al., 1987; Moore & Read, 1996). Some studies attribute to the synovitis and the consequent release of inflammatory mediators by the synoviocytes the triggering factors for DJD (Lipowitz et al., 1985;). In DJD, the synovial villi are hypertrophied, with augmented mature and immature collagen in the subsynovial tissue (Jonhston, 1997). McIlwraith & Fessler (1978) classified morphologically the villi in filamentous, thin, interlaced, short, and in the shapes of polyp, staff, fringe, bush, fan, and cauliflower. In normal joints, the thin polyp-shaped, and the short, rounded, membranous and staff-shaped villi are commonly observed (McIlwraith & Fessler, 1978; Kurosaka et al., 1991), while bigger and reddish villi, with petechial hemorrhages, and in the shapes of fan, cauliflower, fringe and bush are frequently found in joints with synovitis (McIlwraith & Fessler, 1978). The follow-up of the degenerative process evolution, as well as its treatment response, is constantly challenging. Among the methods routinely used in DJD diagnosis are the clinical evaluation (which is subjective) and the radiographic and ultrasonographic examination (which rely on the evaluator's experience). The arthroscopy is another diagnostic method that, despite being an invasive procedure and requiring anesthesia, it allows direct and magnified evaluation of the articular structures, and enables serial evaluations, once tissue invasion and morbidity are minimal (Van Ryssen & Van Bree, 1998; Frizziero & Ronchetti, 2002; Arias et al., 2003; Beale et al., 2003; Melo et al., 2003; Borges, 2006; Rezende et al., 2006; Hulse, et al., 2010; Bleedorn et al., 2011). In CrCL rupture, arthroscopy offers more than a precise diagnosis of the affection, it allows evaluation through magnified visualization of degenerative alterations of the joint such as fibrillation, cartilage erosion, synovial membrane proliferation, and neovascularization, osteophytes formation, besides detecting lesions in the menisci (Siemering & Eilert, 1986; Adamiak, 2002; Borges, 2006; Case et al., 2008; Goldhammer et al., 2010; Hulse et al., 2010; Bleedorn et al., 2011.

Arthroscopy in dogs, initially employed for diagnostic purpose, has become a surgical alternative to many articular affections (Rochat, 2001) and has revolutionized the treatment of joint disease in human beings and animals (Beale & Hulse, 2010). The magnifying of the structures allows the surgeon to recognize and treat lesions that could not be distinguished through arthrotomy (Van Ryssen & Van Bree, 1998; Sams, 2000; Borges, 2006; Beale & Hulse, 2010; Hulse et al, 2010; Bleedorn et al., 2011).

2. Materials and methods

This study was submitted to and approved by the UFMG committee for ethics in research, under protocol number 14/02.

Eighteen mixed breed, adult healthy dogs (9 males and 9 females), with 18 to 25kg of body mass were used in this work. The animals were sheltered in individual cages and fed with commercial pet food and water *ad libitum.* Bilateral radiographic examination of the stifle joint was performed in craniocaudal and mediolateral incidences to confirm the radiographic normality of these joints in both hind limbs. Blood samples were collected for serum biochemical analysis, hemogram, and coagulogram. After 15 days of adaptation period, the first arthroscopy was performed to articular evaluation and CrCL section, and the second, for articular evaluation 21 days later. The animals were premedicated with atropine sulfate (0.044mg/kg), subcutaneously, and xylazine chlorhydrate (1mg/kg), intramuscularly. Anesthesia was induced with sodium thiopental (12.5mg/kg), intravenously, and maintained with isoflurane in semi-open circuit. Antibiotic prophylaxis was performed with cefalotin (30mg/kg), intravenously, 30 minutes before surgery. All animals received tramadol chlorhydrate (2mg/kg), intramuscularly, in the immediate post-operative period and during 24 hours for pain management.

The animals were submitted to arthroscopic evaluation and subsequent CrCL section. The animals were prepared for aseptic surgery and positioned in dorsal recumbence over a metal gutter. Synovial fluid was collected in heparinized syringe and the joint cavity was distended with 10 to 15 ml of Ringer's lactate solution thereafter. A 5 mm cutaneous incision was performed in the lateral parapatellar region, and the articular capsule was incisioned with a scalpel blade No. 11 for introducing the arthroscopic sheath guided by atraumatic trocar. The irrigation system was attached to the arthroscopic sheath while the atraumatic trocar was removed and replaced by a 2.7mm, 30° arthroscope, connected to the camera. Arthroscopic evaluation was conducted based on the articular compartments, as suggested by Person (1985). Articular structures were inspected: the lateral compartment with visualization of the synovial membrane, the lateral femur condyle, the long digital extensor tendon, the intercondylar fossa, the tibial plateau, the menisci, the intermeniscal and cruciate ligaments, the medial compartment, the medial femur condyle, the femoropatellar joint (articular surfaces of femur and patella, trochlear ridges), and the suprapatellar pouch. A second cutaneous and capsular incision was performed in the medial parapatellar region for introducing the arthroscopic scissor, as mentioned above. The CrCL was sectioned, remaining the intra-articular stumps. The CrCL section was confirmed by direct visualization by arthroscopy, and by the test of cranially dislocating the tibia in relation to the femur. The animals were kept in individual 4.5m^2 cages, for 21 days, and, after that period, submitted to new arthroscopic examination. Synovial fluid was collected, and the joint cavity was distented as described above. The intra-articular structures were systematically evaluated, as previously described, and the alterations found in the soft and in the hard tissues were documented . The images were analyzed based on the arthroscopic findings on Day 0, when the first arthroscopy was performed followed by the CrCL rupture, and at Day 21, when a new arthroscopy was conducted for evaluating the articular lesions. The structures were also individually evaluated, and abnormalities were registered. According to their quantity and aspect, villi were classified (Tab.1) from 1 to 4; grade 1 represented absence of lesion and 4, severe lesion. The villi were also morphologically classified as: filamentous, thin, interlaced, short, and as polyp, staff, fringe, bush, fan, and cauliflower shapes. In addition, they were categorized in respect to their localization: lateral and medial compartments, patellar ligament, and patellar tendon. The presence or absence of intra-articular fibrous cords was taken into account. Additionally, the articular cartilage was evaluated by area: patella, suprapatellar pouch, trochlear surface, intercondylar fossa, medial and lateral femur condyles, trochlear ridges of the femur, and surface of the tibia condyles. Likewise, the presence of osteophytes was classified according to the scores in Table 2.

Villi of the synovial membrane

Score	Description
1	Absent.
2	Mild: thin, filamentous, and short villi; white or rosy.
3	Moderate: thin, filamentous, short, in the shapes of polyp, fan, bush, and staff villi; reddish.
4	Severe: numerous villi, dense in the shapes of cauliflower, fringe, located mainly at the lateral and medial compartments; with hemorrhagic aspect and reddish color.

Vascularization of the synovial membrane

Score	Description
1	Absent.
2	Mild: discrete presence of ingurgitated vessel in up to two regions.
3	Moderate: vascularization, apparent hyperemia at the lateral and medial compartments, cruciate ligaments, and menisci.
4	Severe: hypervascularization, hyperemia of the lateral and medial compartments, menisci, and ligaments; articular hemorrhage.

Table 1. Scores system for the synovial membrane characteristics evaluated by stifle joint arthroscopy on Days 0 and 21 after section of the cranial cruciate ligament (CrCL).

Score	Description
1	Absent: smooth articular surface.
2	Mild: low projection irregularities.
3	Moderate: remarkable low projection irregularities, countable bone neoformation with individualized contour.
4	Severe: remarkable irregularity, neoformations with high projection individualized contour, present all over the periarticular region, including the suprapatellar pouch, the tibial plateau, and the intercondylar fossa.

Table 2. Score system for the presence of osteophytes evaluated by stifle joint arthroscopy on Days 0 and 21 after section of the cranial cruciate ligament (CrCL).

The lateral and medial menisci were evaluated according to the presence or absence of lesion. For the cruciate ligaments, the presence or absence of alterations was determined, according to neovascularization, looseness, and fiber rupture. The long digital extensor tendon was evaluated according to the presence or absence of neovascularization and villi at the insertion of the femur condyle. The variables of this study are the lesions that might have arisen or evolved, after the ligament rupture, evaluated on Day 21 by arthroscopy. The Wilcoxon test was employed for paired samples for all variables (lesions), except osteophytes, for which a unilateral test was used. For comparisons, significance was set at $P < 0.05$.

3. Results and discussion

The arthroscopic examination was possible in both evaluation moments, and, as described by Person (1985), systematic examination of the joint allowed the ordering of the results (Tables 3 and 4). All structures that were visualized on Day 0 were also seen on Day 21, when it was possible to detect, in 100% of the joints, alterations in relation to Day 0, fact that was expected given the joint instability, however, statistical significance was noticed in 83.3% (Table 5). The parapatellar approach was adequate for evaluating all the articular structures or regions in this study, as well as for sectioning the CrCL. The positioning of the animal allowed flexion, extension, *varus* and *valgus* movements, and rotation of the joint during the procedure, facilitating the detailed arthroscopic examination, as reported in literature (Van Ryssen & Van Bree, 1998; Arias et al., 2003; Beale et al., 2003; Melo et al., 2003; Rezende et al., 2006, Borges, 2006).

Structure	Region	Alteration	Wilcoxon	P-value
Synovial membrane	Lateral and medial compartments, suprapatellar region, and insertion of the patellar ligament and tendon	Villi	136.0	0.000
		Vascularization	120.0	0.001
		Fibrous cord		No variability[1]
Articular cartilage	Patella	Erosion		No variability[1]
		Fibrillation	10.00	0.100
		Vascularization	1.00	1.000

Structure	Region	Alteration	Wilcoxon	P-value
Patellar pouch		Erosion		No variability[1]
		Fibrillation	10.00	0.100
		Vascularization	45.00	0.009
Trochlea		Erosion		No variability[1]
		Fibrillation	6.00	0.181
		Vascularization	3.00	0.371
Intercondy-lar fossa		Erosion		No variability[1]
		Fibrillation	3.00	0.371
		Vascularization	10.00	0.100
Lateral femoral condyle		Erosion		No variability[1]
		Fibrillation	55.00	0.006
		Vascularization	36.00	0.014
Medial femoral condyle		Erosion		No variability[1]
		Fibrillation	21.00	0.036
		Vascularization	21.00	0.036
Tibia		Erosion		No variability[1]
		Fibrillation		
		Vascularization		No variability[2]
Lateral trochlear ridge		Erosion		No variability[1]
		Fibrillation	10.00	0.100
		Vascularization	1.000	1.000
Medial trochlear ridge		Erosion		No variability[1]
		Fibrillation	6.00	0.181
		Vascularization	1.00	1.000
Medial meniscus			10.00	0.100
Lateral meniscus				No variability[1]
Cranial cruciate ligament		Vascularization	10.00	0.100

Structure	Region	Alteration	Wilcoxon	P-value
Caudal cruciate ligament		Vascularization	6.00	0.181
Long digital extensor tendon		Neovasculariza-tion	6.00	0.181
		Villi		No variability[3]

[1] All the evaluated differences were equal to 0, thus no lesion was observed on Days 0 and 21.
[2] All the evaluated differences were equal to 0, thus no lesion was observed on Days 0 and 21, except one dog, which already presented lesion before the rupture that persisted after the 21-day period.
[3] All the evaluated differences were equal to 0, thus no lesion was observed on Days 0 and 21, except one dog, which already presented lesion before the rupture that persisted after the 21-day period.

Table 3. Wilcoxon test for the different lesions, observed between Days 0 and 21, at the stifle joint of dogs submitted to arthroscope-guided section of the cranial cruciate ligament (CrCL).

Structure	Region	Wilcoxon	P-value
Articular cartilage	Patella	1.00	0.500
	Suprapatellar pouch	10.00	0.050
	Trochlea	No variability[1]	
	Intercondylar fossa	No variability[1]	
	Lateral femoral condyle	1.00	1.000
	Medialfemoral condyle	No variability[1]	
	Tibia	No variability[1]	
	Lateral trochlear ridge	45.00	0.005
	Medial trochlear ridge	13.00	0.089

[1]All the observed differences were equal to 0, thus no lesion was observed on Days 0 and 21.

Table 4. Wilcoxon test for osteophyte presence, observed on Days 0 and 21, at the stifle joint of dogs submitted to arthroscope-guided section of the cranial cruciate ligament (CrCL).

Dog	1	2	3	4	5	6	7	8	9
Wilcoxon	120	3	113	15	105	66	66	21	21
P-value	0.001	0.371	0.003	0.059	0.001	0.004	0.004	0.036	0.036
Dog	10	11	12	13	14	15	16	17	18
Wilcoxon	28	21	91	91	105	66	10	15	55
P-value	0.022	0.036	0.002	0.002	0.001	0.004	0.100	0.059	0.006

Table 5. Individual analysis of the intra-articular lesions of dogs, evaluated on Days 0 and 21, after arthroscope-guided section of the cranial cruciate ligament (CrCL).

It was possible to precisely evaluate the menisci, the intermeniscal ligament, the synovial membrane, the articular cartilage, the cruciate ligaments, and the long digital extensor tendon by arthroscopy, as described in the literature (Van Ryssen & Van Bree, 1998; Adamiak, 2002; Beale et al., 2003; Borges, 2006). The precision offered by the arthroscopic examination is reported by distinct authors (Sams, 2000; Adamiak, 2002; Arias et al., 2003; Beale et al., 2003; Melo et al., 2003; Rezende et al., 2006; Beale & Hulse, 2010). Subcutaneous infiltration of liquid was not observed. Depending on the intensity of infiltration, capsule distension is impaired, and the examination becomes unviable (Van Ryssen et al., 1993). The arthroscopy showed itself as an effective method for evaluating the degenerative alterations after experimental CrCL section. According to Lipowitz et al. (1985) and Lewis et al. (1987), the CrCL section is accepted as the best model for the experimental induction of degenerative lesions. This result was observed in this study, with the additional advantage of direct visualization and magnification of image by arthroscopy, which allowed the evaluation of joint instability, as already reported by others (Lipowitz et al., 1985; Johnson & Johnson, 1993; Glyde et al., 2002; Hulse et al., 2010; Bleedorn et al., 2011). According to Arnoczcky & Marshall (1977), the degenerative process begins one week after CrCL rupture; thus, the articular evaluation on Day 21 is able to identify macroscopic lesions precociously, as observed in this study.

Signs of articular degeneration and inflammation could be noticed at the arthroscopic examination. According to Bennett & May (1997) and Vaughan-Scott & Taylor (1997), the unavoidable DJD progression can be mostly attributed to the enzymatic degradation of the articular cartilage.

For each dog, 14 areas or structures were evaluated by arthroscopy, in a total of 504 individual evaluations. On the day of the CrCL section, 17 dogs (94.4%) presented shining and smooth surfaces of femur and patella, without periarticular osteophytes (Fig. 1.1). Only in one animal (5.6%), irregularities at the trochlear ridge were detected.

From 18 evaluated joints, eight (44.4%) had villous proliferation of thin and filamentous types (Fig. 1.2), considered normal according to McIlwraith & Fessler (1978) and Lewis et al. (1987). In one animal, these villi were hyperemic, and, in nine, the synovial membrane was smooth, without villi. In all the cases, the villi were located at the medial compartment. McIlwraith & Fessler (1978) reported that, in normal articulations, the thin, filamentous, and polyp-shaped villi are commonly observed in equines. Similar findings were reported in dogs (Lewis et al., 1987). Joint with normal villi, but with evident hyperemia suggests inflammatory process, and once the villous type is considered normal, it is possible to infer that the process is at an initial phase.

Discrete hyperemia of the synovial membrane was observed in 17 (94.4%) dogs on Day 0 (Fig. 1.3), during the procedure that lasted 15 minutes on average. Due to the absence of dense fibrous tissue in the synovial membrane, increased local irrigation with congestion and swelling can develop during prolonged arthroscopic procedure (Lewis et al., 1987; Kurosaka et al., 1991), what may explain the findings of this work, not being, necessarily, synovial inflammation. In 14 dogs, the presence of blood vessels (Fig. 1.4) was observed along the CrCL. Despite these alterations in vascularization, the CrCL was intact and shining on Day 0, as well as the caudal cruciate ligament, the meniscofemoral ligament (Fig.1.5), and the long digital extensor tendon (Fig. 1.6). The origin of the popliteus muscle tendon was visualized in two animals (Fig. 1.7), and presented no alteration.

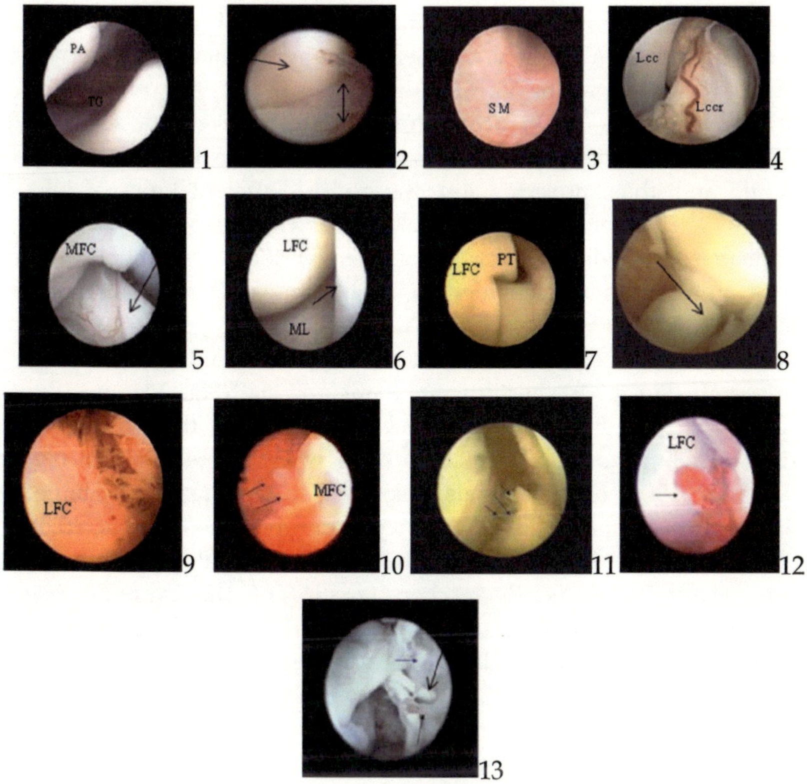

Fig. 1. Arthroscopic images of the stifle joint of dogs submitted to experimental section of the cranial cruciate ligament (Days 0 and 21) guided by arthroscopy. 1) Articular surfaces of the patella (PA) and trochlea (TC) of the femur without alterations; 2) Synovial membrane of the lateral compartment with filamentous villi (arrow), lateral condyle of the femur (arrow); 3) Synovial membrane (MS) with discrete hyperemia at the medial compartment; 4) Blood vessel along of the cranial cruciate ligament (LCCr), caudal cruciate ligament(LCCd); 5) Meniscofemoral ligament (arrow); 6) Lateral compartment showing the lateral femoral condyle (LFC), long digital extensor tendon (arrow) and lateral meniscus (LM) 7) Origin of the popliteus muscle tendon(PT); 8) Intercondylar fossa with thin villi, stump of the cranial cruciate ligament showing nodular formation on the end of the torn ligament (arrow); 9) Villi in the shape of filamentous at the lateral compartiment; 10) Villi in the shape of polyp (arrows) at the medial compartment, medial femoral condyle (MFC); 11) Villus in the shape of fringe(arrows), lateral femoral condyle (LFC); 12) Villus in the shape of fan (arrow), lateral femoral condyle (CLF); 13) Lateral compartment showing villi in the shape of staff (arrow).

On Day 21 after CrCL section, at the arthroscopic examination, alterations suggesting articular degenerative process came to evidence. The arthroscopy, as mentioned by Adamiak (2002), is effective in precisely diagnosing the CrCL rupture and evaluating the evolution of the degenerative lesions of the joint, fact that was verified in this study.

At the arthroscopic examination, in all animals, nodular formation on the end of the torn ligament were observed in the remaining stump of the CrCL on Day 21 (Fig. 1.8). In all animals, villous proliferation and synovial membrane hyperemia were observed, suggesting synovitis. As already reported by Van Ryssen & Van Bree (1998), Sams (2000), Adamiak (2002), and Beale & Hulse (2010), arthroscopy is an ideal diagnostic mean for evaluating macroscopically the synovial membrane, since the villi are kept in suspension in the irrigation liquid and, thus, projected into the cavity. Different types of villi, as well as different grades of hyperemia and vascularization were identified in detail (Table 1). Increase in the quantity of villi and new shapes were verified. Also, villous proliferations in all articular compartments were evidenced, suggesting DJD, as described in literature (Lewis et al., 1987; Kurosaka et al., 1991; Beale et al, 2003; Borges, 2006). Filamentous and thin (Fig 1.9), short, interlaced, as well as in the shape of polyp (Fig. 1.10), fringe (Fig. 1.11), fan (Fig. 1.12), staff (Fig. 1.13), short, membranous and staff(Fig. 2.1) and cauliflower villi were identified. In 21.8% of the joints, there were only short, thin, and filamentous villi, characterizing discrete synovitis. These three types of villi were found in bigger quantity and in all joints. In 55.5% of the animals, associated to the already mentioned villi, there were also those in the shape of bush, fan, polyp, and interlaced, characterizing moderate synovitis. In 22.7%, villi in the shape of fringe and cauliflower were also found, characterizing severe synovitis. Such characteristics of the synovial membrane are associated to DJD. In humans, the arthroscopic examination of the synovial membrane is employed for characterizing and diagnosing different types of pathological articular processes, as, for instance, traumatic, suppurative, tubercular, and rheumatoid arthritis (Kurosaka et al., 1991). At the insertion of the long digital extensor tendon, neovascularization was noticed in three animals (Fig. 2.2), while, in two, thin and apparently normal-colored villi (Fig. 2.3) were seen. Fibrillation in the articular cartilage and absence of erosion in the articular surfaces were verified in all evaluated animals. Fibrillation, according to Johnston (1997), is the initial microscopic finding of DJD. It may be observed as soon as one week after CrCL rupture (Johnson and Johnson, 1993), while erosive lesions are late findings, which occur with, at least, 60 days of joint instability. Statistically, fibrillation was predominant (Table 3) at the medial and lateral condyles (Fig. 2.4). Areas of fibrillation at the articular surface of the patella, the trochlear ridges of the femur, and the intercondylar fossa were also observed (Fig. 2.5). Fibrillation of the articular surface was arthroscopically evidenced as filaments from the cartilage into the articular space. In arthroscopy, the visualization of fibrillation is possible due to the amplified image and the liquid environment. By arthrotomy, it is only possible to detect areas that are apparently thickened, rugose, and opaque, corresponding to fibrillation (Sams, 2000; Adamiak, 2002). There was also fibrin at the lateral (66.%) and medial (34%) compartments (Fig. 2.6 and 2.7, respectively), and at the suprapatellar pouch (27.7%; Fig. 2.8), as well as osteophytes at the trochlear ridges, distal extremity of the patella, and suprapatellar pouch. Osteophytosis was statistically significant (Table 4) at the lateral trochlear ridge (Fig. 2.9). Muzzi (2003) detected radiographically the formation of osteophytes with, at least, 30 days of CrCL rupture. The arthroscopic findings concerning the presence of osteophytes, 21 days after articular destabilization were similar to those reported by Lewis et al. (1987). The presence of periarticular osteophytes is one of the signs of DJD (Elkins et al., 1991; Moore & Read, 1996). Thirteen animals presented vascularization of the articular cartilage, evidenced at the insertion of the patellar ligament and quadriceps tendon (Fig. 2.10 and 2.11), at the suprapatellar pouch towards the trochlea (Fig. 2.12), at the femoral condyles (2.13), and at the intercondylar fossa towards the condyles. This was statistically significant for the region of the femoral condyles. Such a finding corresponds not to the cartilage vascularization itself, but to the synovial membrane vascularization that invades the cartilage.

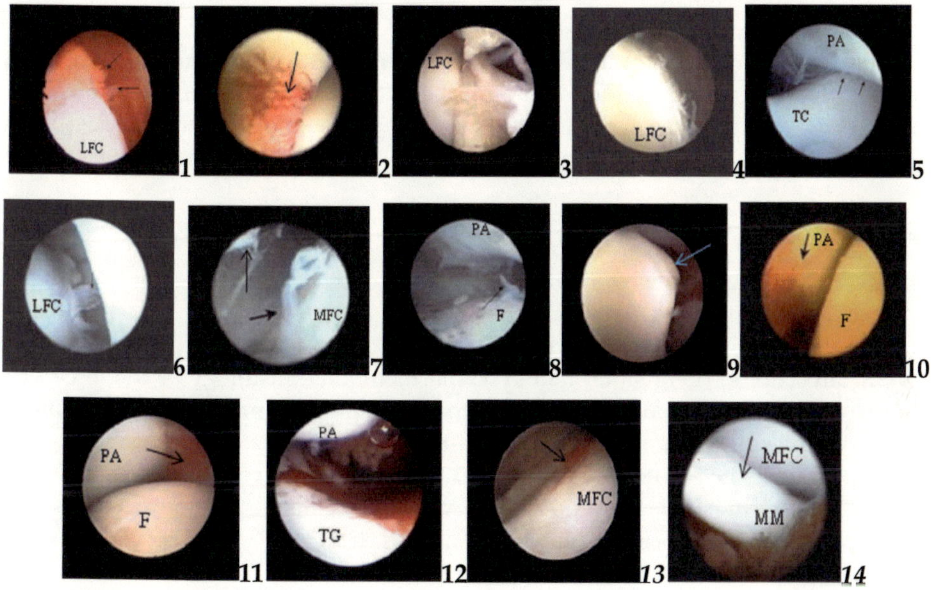

Fig. 2. Arthroscopic images of the stifle joint of dogs submitted to experimental section of the cranial cruciate ligament (Days 0 and 21) guided by arthroscopy. 1) Lateral compartment showing villi short, membranous and in the shape of staff (arrows); 2) Visualization of the long digital extensor tendon, vascularization (arrow), 3) Thin villi at the long digital extensor tendon (arrow), lateral femoral condyle (LFC), 4) Fibrillation at the lateral femoral condyle (LFC); 5) Fibrillation (arrows) and fibrin at the surface of the patella (PA), proximal trochlear groove (TC); 6) Fibrin attached to the lateral femoral condyle (LFC) (arrows); 7) Fibrin at the medial femoral condyle (arrow) and patella; 8) Supratellar pouch with vascularization and fibrin (arrow); 9) Lateral femoral condyle (LFC) with mild osteophytes (irregularities: arrow); 10) Vascularization at the insertion of the patellar ligament (arrow); 11) patellar tendon (arrow); 12) suprapatellar pouch 13); and at the medial trochlear ridge of the femoral condyle (arrow) 14) Prolapse of the caudal horn of the medial meniscus (MM).

Statistically, the results regarding the lesion in the menisci were not significant; nevertheless, prolapse of the caudal horn of medial meniscus (Fig. 2.14) was visualized in four animals (22.2%). In one of them, on Day 0, there were alterations compatible with synovitis, like hyperemia and increased synovial membrane villi. Three of them presented, on Day 21, meniscus lesion and periarticular irregularities. The medial meniscus is more susceptive to lesions due to its capsule fixation (Moore & Read, 1996). It is reported that meniscus lesions occur about the seventh week of articular instability by CrCL rupture (Johnson & Johnson, 1993). In this work, the precocious occurrence of this kind of lesion may be attributed to the eventual excited behavior of the patients. Concerning the DJD, the body mass of the animals, which ranged from 18 to 25 kg, must also be considered for this study. According to Bennett et al. (1988), the CrCL rupture in dogs with body mass under 15 kg usually causes degenerative alterations that are less severe than in heavier dogs.

4. Conclusions

As a valuable instrument for macroscopic evaluation of articular tissues, arthroscopy is a safe method for diagnosing and following-up the alterations in the stifle joint of dogs. The technique allows tracking the evolution of the degenerative lesions, as well as classifying the synovitis according to the villi shape and synovial membrane hyperemia.

5. References

Adamiak, Z. (2002). Arthroscopy in dogs with cranial cruciate ligament injuries.*Indian Vet. J.* vol.79, nº 2, pp.177-178, ISSN 0019-6479

Arias, S.A., Rezende, C.M.F., Melo, E.G.,Nunes, V.A. & Correa, J.C (2003). Radiological, arthroscopical evaluation and synovial membrane histology of the knee of dogs treated with chondroitin sulphate- sodium hialuronate association after experimental degenerative joint disease. *Brasilian J. Vet. Res. Anim. Sci.*, vol.55, nº4, pp. 421-429, ISSN 0102-0935

Arnoczky, S.P. & Marshall, J.L. (1977). The cruciate ligaments of the canine stifle: An anatomical and funcional analysis. *Am. J. Vet. Res.*, vol.38, nº 11, pp.1807-1814, ISSN 0002-9645

Beale, B. S., Hulse, D. A., Schulz, K. S.,Whitney, W. O. (2003). Arthroscopically assisted surgery of the stifle joint, In: *Small Animal arthroscopy,* 117-157, Saunders, ISBN 0-7216-8969-8, Philadelphia, Pennsylvania

Beale, B. S. & Hulse, D. A.(2010). Arthroscopy versus arthrotomy for surgical treatment, In: *Advances in the canine cranial cruciate ligament*, P. Muir,(Ed.), 145-158, Wiley-Blackwell, ISBN 978-0-8138-1852-8, Ames, Iowa

Bennett, D. & May, C. (1997). Joint diseases of dogs and cats. In: *Textbook of Veterinary Internal Medicine*. 4.ed., S.J. Ettinger & E.C. Feldman (Eds.), pp. 2032-2077, Saunders, ISBN 0721634273, Philadelphia, Pensylvania.

Bennett, D. Tennant, B. Lewis, D.G. Baughan, J. May, C. & Carter, S. (1988). A re-appraisal of anterior cruciate ligament disease in the dog. *J. Small Anim. Pract.*, vol. 29, nº5, pp.275-297, ISSN 0022-4510

Bleedorn, J. A, Greuel, E. N., Manley, P. A., Schaefer, S. L., Markel, M. D., Holzman, G., Muir, P. (2011). Synovitis in dogs with stable stifle joints and incipient cranial cruciate ligament rupture: a cross-sectional study. Vet. Surg. vol 40, nº 5, pp. 531-543, ISSN 0161 3499

Borges, N.F. (2006).*Videoarthroscopy of the femorotibiopatellar joint in dogs before and 21 daysafter cranial cruciate ligament rupture (experimental study)*. 40pp. Dissertation) Escola de Veterinária, Universidade Federal de Minas Gerais, Belo Horizonte, Brazil

Case, J. B., Hulse, D. Kerwin, S. C., Peycke, L. E. (2008). Meniscal injury following initialcranial cruciate ligament stabilization surgery in 26 dogs (29 stifles). *Vet. Comp.Orthop. Traumatol.*, vol. 21, nº 4, pp. 365-367, ISSN 0932-0814

Cook, J. L. (2010). Cranial cruciate ligament disease in dogs: biology versus biomechanics.Vet. Surg. vol. 39, n° 3, pp. 270-277, ISSN 0161-3499

de Rooster, H., de Bruin,T., Van Bree, H. (2010). Morphology and function of the cruciate ligaments, In: *Advances in the canine cranial cruciate ligament*, P. Muir,(Ed.), 5-12, Wiley-Blackwell, ISBN 978-0-8138-1852-8, Ames, Iowa

Elkins, A.D., Pechman, R. & Kearney, M.T. (1991). A retrospective study evaluating the degree of degenerative joint disease in the stifle joint of dogs following surgical repair of anterior cruciate ligament rupture. *J. Am. Anim. Hosp. Assoc.*, vol.27, September/October, pp.533-540, ISSN 0587-2871

Frizziero,L. & Ronchetti, I.P.(2002). Intra-articular treatment of osteoarthritis of the knee: an Anthroscopic and clinical comparison between sodium hyaluronate (500-730 kDa) and methilprednisolone acetate. *J.Orthopaed. Traumatol.*, vol. 3, n°2, pp.89-96, ISSN 1590-9921

Glyde, M.R. Wong, W.T., Lidbetter, D. Parry, B. & Middleton, D. (2002). Partial rupture of the cranial cruciate ligament in 13 dogs: Clinical, radiological, clinicopathological and histhopathological features. *Ir. Vet. J.*, vol. 55, n°6, pp. 271-276, ISSN 0368-0762

Goldhammer,M.A., Smith, S. H., Fitzpatrick, N., Clements, D. N. (2010). A comparison of radiographic, arthroscopic and histological measures of articular pathology in the canine elbow joint. *The Vet. Journal*, vol.186, n° 1, pp. 96-103, ISSN 1090-0233

Hulse, D., Beale, B., Kerwin, S. (2010). Second look arthroscopic findings after tibial plateauLeveling osteotomy. *Vet. Surg*. vol. 39, n° 3, pp. 350-354, ISSN 0161-3499

Innes, J. F., Costello ,M., Barr, F. J., Rudorf, H., Barry, A. R. S.(2004). *Vet. Radiol. Ultrasound*,vol. 45, n° 2, pp.143-148, ISSN 1058-8183

Innes, J. F.(2010). Progression of arthritis after stifle stabilization, In: *Advnces in the caninecranial cruciate ligament*, P. Muir,(Ed.), 229-232, Wiley-Blackwell, ISBN 978-0-8138 -1852-8, Ames, Iowa

Johnson, J.M. & Johnson, A.L. (1993). Cranial cruciate ligament rupture: pathogenesis, diagnosis, and postoperative rehabilitation. *Vet. Clin. North Am.: Small Anim. Pract.*, vol.23 n°4, pp.717-733, ISSN 0195-5616

Johnston, A.S. (1997). Ostearthritis: joint anatomy, physiology and pathobiology. *Vet. Clin. North. Am.:Small Anim. Pract.*, vol.27, n°4, pp. 699-719, ISSN 0195-5616

Kurosaka, M., Ohno, O.& Hirorat, A.K.(1991). Arthroscopic evaluation of synovitis in the knee joints. *Arthroscopy*, vol.7, n° 2, pp.162-170, ISSN 0749-8063

Lazar, T. P., Berry, C. R., de Haan, J. J., Peck,J. N. Correoa, M.(2005). Long-term radio-graphic comparison of tibial plateau leveling osteotomy versus extra-capsular stabilization for cranial cruciate ligament rupture in the dog. *Vet. Surg.*, vol.34, n° 2, pp. 133-141, ISSN 0161-3499

Lewis, D.D.; Goring, R.L.; Parker, R.B. & Curasi, P.A. (1987). A comparison of diagnostic methods used in the evaluation of early degenerative joint disease in the dog. *J. Am. Anim. Hosp. Assoc.*, vol.23, May/June, pp. 305-315, ISSN 0587-2871

Lipowitz, A.J.; Wong, P.L.; Stevens, J.B. (1985). Synovial membrane changes after experimental transecction of the cranial cruciate ligament in dogs. *Am. J. Vet. Res.*, vol. 46, n° 5, pp.1166-1170, ISSN 0002-9645

Ljungvall, K. & Ronéus, B. (2011). Arthroscopic surgery of the middle carpal joint in trotting Standardbreds: findings and outcome. Vet. Comp. Orthop. Traumatol. , vol. 24, N° 5, PMID 21792476, DOI: 10.3415/VCOT-10-12-0161, ISSN 0932-0814

McIlwraith, C.W.& Fessler, J.F. (1978). Arthroscopy in the diagnosis of equine joint disease. *J. Am. Vet. Med. Assoc.*, vol.172, nº 3, p.263-268, ISSN 0003-1488

Melo, E.G., Rezende, C.M.F., Gomes, M.G., Freitas, P.M., Arias, S.A.S. (2003) .Chondroitin sulfate and sodium hialuronate in treatment of the degenerative joint disease in dogs. Clinical and radiological aspects. *Brasilian J. Vet. Res. Anim. Sci.* vol. 55, nº 1, pp.35-43, ISSN 0102-0935

Moore, K.W. & Read, R.A. (1996). Rupture of the cranial cruciate ligament in dogs. Part I. *Compend. Cont. Educ. Pract. Vet.*, vol.18, nº3, p.223-234, ISSN 0193-1903

Muzzi, L.A.L. (2003). *Physiotherapie and temporary immobilization after arthroscopic repair of the cranial cruciate ligament in dogs.* 79pp. (Thesis) - Escola de Veterinária, Universidade Federal de Minas Gerais, Belo Horizonte, Brazil

Person, M.W. (1985). A procedure for arthroscopic examination of the canine stifle joint. *J. Am. Anim. Hosp. Assoc.*, vol.21, March/April, pp.179-186, ISSN 0587-2871

Pozzi, A. & Kim, S. E.(2010). Biomechanics of the normal and cranial cruciate ligament-deficient stifle, In: *Advances in the canine cranial cruciate ligament*, P. Muir,(Ed.), 37-42, Wiley-Blackwell, ISBN 978-0-8138-1852-8, Ames, Iowa

Rezende, C.M.F., Melo, E.G., Madureira, N.G., Freitas, P. M. (2006). Arthroscopy of the stifle joint in dogs. *Brasilian J.Vet.Res.Anim. Sci.* vol. 58, nº 5, pp. 841-848, ISSN 0102-0935

Rochat, M.C. Arthroscopy. (2001).*Vet. Clin. North. Am.:Small Anim. Pract.*, vol.31, pp.761-787, ISSN 0195-5616

Sams, A.E. (2000). Canine elbow joint arthroscopy: introduction and description of technique. *Comp. cont. Educ. Pract.Vet.* vol. 22. nº 2, pp.135-144, ISSN 0193-1903

Schrader, S.C. (1995). Joint diseases of the dog and cat. In: *Small animal orthopedics*, M.L. Olmstead, (Ed.), pp.437-471, Mosby, ISBN 0-8016-5874-8, St. Louis, Missouri

Siemering, G.B. & Eilert, R.E. (1986). Arthroscopy study of cranial cruciate ligament and medial meniscal lesions in the dog. *Vet. Surg.* vol.15, nº 3,pp.265-269, ISSN 0161-3499

Van Bree, H., de Rooster, H., Gielen, I. (2010). Stress radiography of the stifle,In: *Advances in the canine cranial cruciate ligament*, P. Muir,(Ed.), 113-116,Wiley-Blackwell, ISBN 978-0-8138-1852-8, Ames, Iowa

Van Ryssen, B., Van Bree, H. & Missinne, S. (1993).Successfull arthroscopy treatment of shoulder osteochondrosis in the dog. *J. Small Anim. Pract.*, vol.34, nº 10, pp.521-528, ISSN 0022-4510

Van Ryssen, B. & Van Bree, H. (1998). Diagnostic and surgical arthroscopy in small animals, In: *Canine Sports Medicine and Surgery*, M.S. Bloomberg, J.F.Dee & R. A.Taylor,(Eds.), pp.250-254, Saunders, ISBN 0-7216-5022-8, Philadelphia, Pennsylvania

Vasseur, P. B. & Berry, C. R.(1992). Progression of stifle osteoarthrosis followingreconstruction of the cranial cruciate ligament in 21 dogs. J. Am. Anim. Hosp.Assoc., vol. 28, March/april, pp.129-136, ISSN 0587-2871

Vaughan-Scott, T. & Taylor, J.H. (1997).The pathophysiology and medical management of canine osteoarthrtitis. *J. S. Afr. Vet. Assoc.*, vol.68, nº 1, pp.21-25,ISSN 0038-2809

10

Congenital Aplasia of the Uterine-Vaginal Segment in Dogs

Bruno Colaço, Maria dos Anjos Pires
and Rita Payan-Carreira
*CECAV, Univ. of Trás-os-Montes and Alto Douro,
Portugal*

1. Introduction

Müllerian duct abnormalities consist of a set of structural malformations that include diverse situations of agenesis or aplasia, which may evolve according to three distinct patterns: a failure of the paramesonephric ducts (Müllerian ducts) to develop in the whole or in part (uterus unicornis and segmental aplasia of uterine horn, respectively), the failure of caudal part of the paramesonephric ducts to the fuse and to form one single lumen (leading to situations such as uterus didelphus and the cranial vagina septation) and the failure of the fused caudal paramesonephric ducts to fuse with the urogenital sinus with consequent absence of anatomic continuity between the cranial vagina and the vestibule (originating an imperforate hymen or the vaginal stenosis).

Prevalence of these anomalies ranges from 0.02 to 0.05% in the canine population (Roberts, 1971; McIntyre et al., 2010). The relative frequency of the different abnormalities varies upon the region and the year of publication (Roberts, 1971; Ortega & Pacheco, 2007; McIntyre et al., 2010), and the clinical descriptions are insufficient to establish epidemiology of the defects.

In dogs, the uterine and vaginal segments can show developmental abnormalities ranging in severity from hypoplasia to complete agenesis (Romagnoli & Schlafer, 2006; McIntyre et al., 2010), and the severity of the defect influences the reproductive outcome and the existence of clinical side effects. The mildest form is the vagino-vestibular stricture, where secretions are collected in the cranial vagina and uterus, while in the more extreme form the uterus may be partially or in the whole reduced to a string, fibrous structure.

Although not uncommonly found during elective ovariohysterectomy, often they are detected co-existing with increased dimensions of the uterus and signs of mucometra or pyometra. There is a complete obstruction to fluid drainage from the uterus (as in vaginal or cervical atresia) and the owner may complaints that the female does not evidence the expected estrus vulvar discharge. With time, in obstructive diseases of the uterus or vagina, it is also possible to observe dysuria, cystitis, pyometra and renal failure (Kyles et al, 1996). However, in unicornuate uterus this is seldom the case and often the main complaint is infertility or sterility (Romagnoli & Schlafer, 2006; McIntyre et al., 2010). The animal may have been breed with success previous to the diagnosis, albeit the number of puppies in the litter is usually remarkably small comparing to the normal for the breed (Romagnoli & Schlafer, 2006; McIntyre et al., 2010). Sporadically, vulvar discharge, excessive vulvar licking and attraction of the male have been described (Kyles et al, 1996; Tsumagari et al., 2001).

The aim of this study was to discuss embryology, the gross anatomic features, clinical signs and implications, and available diagnostic approaches in cases of Müllerian duct anomalies.

2. Classification

Non-development or non-fusion (partial or complete) of Müllerian ducts may result in a variety of anomalies ranging from complete agenesis to duplication of female reproductive organs (Moore & Persaud, 2008).

In different clinical reports, a tendency for the use of different classifications for a common defect was found. Also, it is frequently found similar defects named distinctly for different species, in which develop with different patterns. These are indicative of the necessity of adopting a more uniform nomenclature. In an attempt to clarify the nomenclature used in this review, and to allow the easy identification of underlying major causes for a given defect, table 1 presents the classification for the anatomical congenital defects of the uterine-vaginal segment in dogs.

<table>
<tr><th>Main cause</th><th>Malformation</th><th colspan="2">Classification</th></tr>
<tr><td rowspan="2">Failure in organogenesis</td><td rowspan="2">Failure of segments of Müllerian ducts to develop</td><td>Complete</td><td>Uterus unicornis</td></tr>
<tr><td>Incomplete</td><td>Segmental aplasia of the uterus (uni or bilateral)</td></tr>
<tr><td rowspan="4">Failure in fusion</td><td rowspan="2">Failure of the caudal parts of the two Müllerian ducts to fuse appropriately and develop a single lumen</td><td colspan="2">Segmental aplasia of the body or the cervix</td></tr>
<tr><td colspan="2">Longitudinal septation of the uterine body*
Uterus didelphus*</td></tr>
<tr><td rowspan="2">Failure of caudal ends of the Müllerian ducts to fuse with invaginated urogenital sinus thus impairing the establishment of anatomical continuity</td><td colspan="2">Segmental stenosis or aplasia of the vagina (Also named vestibulo-vaginal constriction)</td></tr>
<tr><td colspan="2">Longitudinal septum in the cranial vagina*
Imperforate hymen*</td></tr>
</table>

* Out of the scope of this review

Table 1. Classification of developmental abnormalities of the uterine-vaginal segment in dogs according to underlying malformations.

Congenital aplasia of the uterus or vagina may occur in two variants: showing the normal development of muscular and serosal external layers, but without development of the mucosa, originating the atresia of the more internal layers; or may result from failure in development of all the layers in a more or less extended segment, which is reduced to a fibrous cordiform remnant. The former gives origin to a dense fibrous transverse partition in middle to caudal vagina, while in the later the agenesis segment miss from the normal anatomy (Gee et al., 1977; Moore & Persaud, 2008; McIntyre et al., 2010).

3. Physiopathology

Many vertebrates share a common genetic system for embryonic patterning that includes the reproductive tract. To better understand the uterus and vagina congenital anomalies we will first discuss the normal tubular genitalia development.

Normal development of the female reproductive tract involves a series of highly orchestrated, complex interactions that direct differentiation of the Müllerian ducts and urogenital sinus to form the internal female reproductive tract. This dynamic process is completed throughout mechanisms of differentiation, migration, fusion, and canalization. In the indifferent stage both male and female embryos have 2 sets of paired genital ducts (Mcgeady et al., 2006): Wolffian (mesonephric) and the Müllerian (paramesonephric). Differentiation of the Wolffian ducts occurs earlier in the male embryo, and persists after the mesonephros disintegrates (Noden & de Lahunta, 1985; Moore & Persaud, 2008). The female differentiation occurs in a later gestational age, and is characterized by regression of the Wolffian ducts due to absence of masculinisation influences from the gonads, and by stabilization of the Müllerian ducts, which is estrogen sensitive in a precise window of time (Moore & Persaud, 2008). Estrogens block the development of the Müllerian ducts if applied before the differentiation began, or cause hypertrophy of the differentiated portion and prevented further differentiation of the ducts in more caudal regions, when applied afterwards (Dood & Wibbels, 2008).

The funnel-shaped cranial region of each paramesonephric duct remains open (communicating with the coelomic cavity) and will form the uterine tube. Postnatally the communication persists from the peritoneal cavity to the exterior (exclusively in females). Caudal to this, each duct develops into an uterine horn. The bilateral paramesonephric ducts caudal portion shift medially and fuse into a single tube. The paired Müllerian ducts, initially separated by a septum, fuse and form a single Y-shaped tubular structure, the uterovaginal primordium (UVP). UVP becomes the uterine body, uterine cervix, and the cranial third of the vagina. Uterine morphology varies significantly in mammals, due to different degrees of fusion of the distinct Müllerian ducts (Noden & de Lahunta, 1985; Moore & Persaud, 2008).

Normal vaginal development requires the fusion of components that derive from 2 embryologic structures: the mesodermal Müllerian ducts and the endodermal urogenital sinus (UGS). To achieve this fusion the bilateral blind ending paramesonephric ducts enter in contact with UGS. This contact promotes the cellular proliferation of the endoderm from the urogenital sinus and the formation of the vaginal plate. The vagina is derived from both the vaginal plate and the fused ends of paramesonephric ducts (Mcgeady et al., 2006). The cranial one-third comes from fused paramesonephric ducts and the caudal two-thirds originate from the vaginal plate. Degeneration of the center of the vaginal plate creates the vaginal lumen. A hymen may persist where the vagina joins urogenital sinus although in most domestic species it tend to disappear before puberty. The urogenital sinus forms the vestibule (Noden & de Lahunta, 1985; Moore & Persaud, 2008).

Apart from some remnants of the excretory tubes and a small portion of the mesonephric duct, with varying importance according to the species, all the female mesonephric derivates atrophies (Moore & Persaud, 2008).

The cause and heritability of congenital abnormalities in dogs remain undetermined (McIntyre et al., 2010). With exception for the situations accompanying intersex conditions, the karyotype is normal. We are unaware if the condition develops in consequence of genetic, endocrine, or environmental influences. Neither it is known if it might be associated to failure of the gonad or the mesonephros development (which absence may co-exist with the *uterus unicornis*), to failure in local gene expression or, in the case of a segmental aplasia, if it might be determined by disruption of the blood supply to the affected segment (Ribeiro et al., 2009; McIntyre et al., 2010).

A *unilateral uterine aplasia* (also termed unicornuate uterus) develops when one paramesonephric duct fails to develop; this results in a uterus with one uterine horn (Moore & Persaud, 2008). The kidney and the paramesonephric ducts have the same embryologic origin therefore this anomaly is usually associated with ipsilateral kidney absence (Chang et al., 2008). Unilateral uterine aplasia has been reported in dogs and cats (Schulman & Bolton, 1997; Pinto Filho et al., 2001; Güvenç et al., 2006). However, the ovary is of a separate embryological origin and is usually present (Moore & Persaud, 2008; Thode & Johnston, 2009).

Development defects of the Müllerian duct system may cause *segmental aplasia* in several portions of the Müllerian duct system. Partial or complete fusion or occlusion of one uterine horn, of the body of the uterus, or of the most caudal segments, such as the cervix and cranial vagina, may cause fluid accumulation cranially to the occlusion (McEntee, 1990; Oh et al., 2005; Romagnoli & Schlafer, 2006; Almeida et al., 2010; McIntyre et al., 2010). Failure of canalization of the vaginal plate results in atresia (blockage) of the vagina, originating a *transverse vaginal stenosis* or a segmental aplasia, which is found between the middle and the caudal third of the vagina (Schlafer & Miller, 2007). Isolated vaginal atresia is an extremely rare finding.

Given the many variables that are involved in the female genital tract differentiation and growth, the pathogenesis for each anomaly may be multifactorial and hence of difficult identification (McIntyre et al., 2010).

4. Gross morphology

The gross morphological signs vary according to the defect and also the co-existence of secondary diseases or other concomitant uterine or vaginal disease.

4.1 Unilateral uterine aplasia (Unicornuate uterus)

In unicornuate uterus situations, aplasia of one uterine horn is the major finding, as it is reduced to a fibrous threadlike structure (Figure 1A and 1B), composed of fibrous tissue with some muscle strands (McIntyre et al., 2010) that frequently fail to present a lumen. As the contralateral uterine horn and the other genital segments retain patency, fluid accumulation does not occur unless the female develop cystic endometrial hyperplasia (CEH)/pyometra or other situations of segmental aplasia in the contralateral uterine horn are present (Figure 1A). The non-patent uterine horn may induce fluid accumulation in the oviducts that distend (Figure 1C).

Due to a common embryologic origin, this malformation might co-exist with ipsilateral tubal and renal agenesis (Chang et al., 2008). Although this situation is described with more frequency in cats than in dogs, on its survey McIntyre et al. (2010) found a relative lower frequency of 28% in cats than 45.5% in dogs. Co-existence of renal agenesis appears also on older reports on *uterus unicornis* in female beagles in research colonies (Höfliger, 1971, cited by McEntee, 1990). As those animals were euthanized after the end of the experiments (they integrated the control groups), at young ages, clinical signs were absent. Often the abnormality is detected only at necropsy or surprises the surgeon during the surgery. In contrast, due to a different embryologic origin, most frequently both ovaries are found (Romagnoli & Schlafer, 2006; McIntyre et al., 2010). Other congenital defects besides kidney agenesis were found in animals bearing unicornuate uterus: ectopic contralateral ureter, absent ipsilateral suspensory ligament and umbilical hernia (McIntyre et al., 2010).

Acquired diseases found in described conditions of unicornuate uterus include CEH of the contralateral uterine horn and polycystic ovaries. According to the descriptions, the later could correspond to cystic proliferation of the rete or the cranial uterine tube (Gu̇venc et al., 2006).

4.2 Segmental aplasia of the uterus

In cases of segmental aplasia, whether its location may be at the uterine horns, body or cervix, frequently the anomalies are grossly visible as missing segments or strictures that interrupt the normal anatomy of the uterine-vaginal segment, and that correspond to failure of the development of all the layers of that segment, which are often reduced to a streak cord-like rudiment. However, in few situations the external layers of the Müllerian duct derivates (serosal and muscle layers) are properly differentiated, but the inner layers (mucosal and sub-mucosal layers) do not differentiate (Moore & Persaud, 2008). This is more often found in the vagina than the uterus. In practical terms, the potential deleterious effects over the reproductive potential and the occurrence of secondary diseases are similar in the two situations.

As the situation usually remains undiagnosed in young animals that maintain regular reproductive activity, the normally developed portion cranial to the atresia is distended due to fluid accumulation (Figure 1D to 1H). Distension of the uterine tube (Figure 1F) is possible (McIntyre et al., 2010). Primary lesions at the endometrium are rarely reported, but CEH may develop in older animals. However, due to excessive mucous fluid pressure of the ongoing mucometra, with time reduction in the thickness of the uterine walls and compression and attenuation of the endometrial glands is commonly found (McIntyre et al., 2010).

Segmental aplasia of the uterine horns may develop at any point of the structure (Figure 1A, 1C and 1G) (Schlafer & Miller, 2007; McIntyre et al., 2010), with segmental agenesis being irregularly distributed for different segments of the uterine horns.

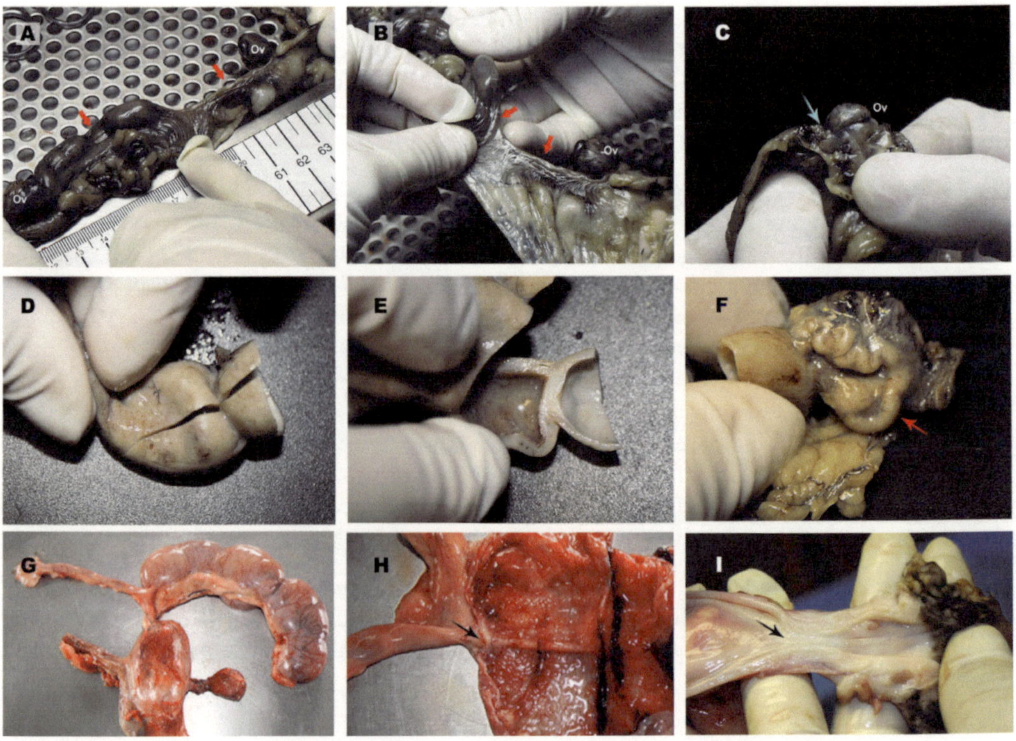

Fig. 1. Canine congenital aplasia of the uterine-vaginal segment: **A**: The external morphological evaluation showed the complete agenesis of the right uterine horn (unicornuate uterus) co-existing with segmental aplasia of the contralateral uterine horn. Both ovaries (Ov) were present. Arrows indicate areas of agenesis. **B**: Closer inspection showed that the caudal segment of the left uterine horn is patent, while the right uterine horn is reduced to a threadlike structure parallel to the uterine vessels. The ipsilateral ovary (Ov) was present, within the ovarian bursa. Arrows indicate areas of agenesis. **C**: Dilatation of the uterine tube (cyan arrow) ipsilateral to the non-developed uterine horn. **D**: Externally, segmental aplasia of the uterine horn may appear as a marked constriction of the organ. **E**: When opened, those areas showed the absence of fusion between adjacent segments, which fail to communicate. Outside undeveloped areas the uterus maintain its normal morphology. **F**: The ipsilateral uterine tube is distended (arrow), due to fluid pressure. **G**: Segmental aplasia at the basis of the right uterine horn in a 15 years-old bitch (distension of the vagina was due to a fibroma). **H**: The arrow points to the area of aplasia. Long-term effects of the secondary fluid accumulation induced a reduction in the thickness of the walls. **I**: Segmental stenosis of the vagina: the arrow points to the obstruction.

In animals with segmental aplasia of the uterine body or cervix, as communication between the two uterine horns is possible, equivalent and bilateral uterine horn distension is found (McEntee, 1990; McIntyre et al., 2010). Undeveloped segments are usually reduced to cord-like remnants. In cases of cervical aplasia, the uterine body ends blindly in a pouch, a membranous tissue separating the uterus from the vagina (McEntee, 1990).

4.3 Segmental stenosis of the vagina

Vaginal stenosis has been described as occurring as a complete misdevelopment of all the layers of the vagina, which may be reduced to an atretic segment with variable importance (Figure 1I) (Gee et al., 1977; Viehoff & Sjollema, 2003; Romagnoli & Schlafer, 2006) or as originating from the absence of internal canalization with normal developmental of the external layers (Wadsword et al., 1978). The existence of a vaginal cyst cranial to the occlusion has been referred, which has been attributed to incomplete canalization of the embryonic derivate, hence allowing the development of intermediary segments between missing portions (Gee et al., 1977; McEntee, 1990). The undeveloped segment is usually located near the vestibulo-vaginal junction (Gee et al., 1977; Kyles et al., 1996; Viehoff & Sjollema, 2003) and on gross evaluation morphology of the anatomical region may share some resemblances with imperforate hymen. The missing segment may present variable extension.

Similarly to the findings for segmental aplasia of the uterine body or cervix, fluid accumulation above the agenesis induces uterine dilatation, but contrasting to those malformations, fluid accumulation also occurs on the remainder cranial segment of the vagina.

Vaginal cysts near the point of stenosis have been described (Gee et al., 1977). It has been proposed to correspond to incomplete canalization of atrectic segments (Gee et al., 1977; McIntyre et al., 2010). Impairment of fluid drainage induces fluid accumulation above the obstruction and consequently distension of cranial reproductive tract (Gee et al., 1977; Viehoff & Sjollema, 2003; McIntyre et al., 2010). In contrast to the observed in the previously described situations, fluid accumulates within the uterus (mucometra) and the vagina (hydrocolpos). As vaginal walls are thinner than those of the uterus they distend more easily, thus reaching higher increases in size comparatively to the uterine distension. These morphological features are not exclusive for the segmental stenosis of the vagina. Imperforate hymen shares similar gross appearance (Tsumagari et al., 2001; Schlafer & Miller, 2007), except that the transverse partition is thinner in the later than in the segmental stenosis.

5. Clinical signs

Congenital abnormalities of the uterine-vaginal segment in dogs are seldom detected before puberty. Furthermore, in a large number of cases remains unnoticed until secondary diseases develop or until unsuccessful attempts for breeding the female alert the owner to the need for a detailed reproductive examination of the infertile animal. Early diagnosis of those conditions would avoid owners to maintain a bitch with severe congenital diseases in the reproductive stock. Also, the correct identification of the disease and evaluation of the importance of the defect would allow proper counselling on the therapeutic approach and establish a prognosis for a specific condition. Further, it will permit to anticipate the occurrence of secondary, associated diseases that can develop if the primary condition is not treated and that may threaten the health of the female.

5.1 Unilateral uterine aplasia (Unicornuate uterus)

The occurrence of unicornuate uterus is rare, but results from different surveys report different relative frequencies. Ortega and Pacheco (2007) in tropical regions reported a frequency of 0.3% (1:300 lesions of the genital tract) whilst McIntyre et al. (2010) found 73.3% (11:15 of the canine uterus disorders found). It is possible that the genetics of the population surveyed may have influenced, in particular the proportion of purebred and crossbred females, as well as the analysed segments of the genital tract.

Most often, in cases of complete aplasia of one uterine horn the female usually presents regular oestrous cycles (Pinto Filho et al., 2001; Güvenç et al., 2006; Almeida et al., 2010), and often the animal history include the reference to successful pregnancy and parturition, although with a smaller litter size than the expected for the breed (Romagnoli &Schlafer, 2006; McIntyre et al., 2010). It is also possible the reference to the existence of a reduced intensity in the vaginal discharge accompanying estrus, but this is not usually reported in any of the cited literature. Individuals with agenesis of one uterine horn may also have an increased incidence of premature delivery (Seyrek-Intas et al., 2004). Usually, in young dogs clinical signs other than sub-fertility are absent, the condition being detected at the necropsy or during a convenience OVH. That was the case in one of the situations we had: a small Poodle dog of 2 years of age with regular estrous cycles that was submitted to convenience OVH. The uterine defect was found during surgical procedure, in the absence of symptoms. The owner only reported the existence of a vulvar discharge of decreased intensity when compared to that of her mother. Pinto Filho et al. (2001) described a similar situation in a Poodle female with 2 years of age, without clinical signs, that was submitted to elective surgery.

However, older animals may present signs of disease located in the uterus (such as CEH or pyometra), as in the report by Güvenç et al. (2006). Nevertheless, one should be aware that those are generally of independent occurrence, as CEH/pyometra frequently develops in the uterus of old intact females. Güvenç et al. (2006), report in a 12 years-old Cocker Spaniel, the existence of a serous sanguineous vulvar discharge persisting for 5 weeks, in the absence of other clinical symptoms, which was the main reason for the consultation. However, at surgery polycystic ovaries were found. Thus it cannot be excluded that the discharge might rather be associated to the ovarian disease than to the uterine defect.

The bilateral agenesis of the uterine horns is extremely rare in dogs (Schlafer and Miller, 2007).

5.2 Segmental aplasia of the uterine(s) horn(s)

Partial or complete agenesis of a more or less extended segment of the uterine horns (segmental aplasia) is relatively common in dogs; in the survey by McIntyre and colleagues (2010) it accounted for 20% of the congenital anomalies of the uterus. Most frequently it is an unilateral condition, but some descriptions referred to bilateral situations (McIntyre et al., 2010), in which segmental agenesis was irregularly distributed for different segments of both uterine horns. This was also observed in one of our cases, in middle aged Poodle crossbred female. The agenesic segment may develop at any point of the uterine horns (McEntee, 1990; Romagnoli & Schlafer, 2006; McIntyre et al., 2010). The moment for the onset of clinical signs detection partly depends on its location and extension. Whenever it exists an obstruction to the normal outflow of the endometrial secretions, accumulation of a sterile fluid with distension of uterine walls (Figure 1 D and 1G) is the main outcome

(McEntee, 1990; Romagnoli & Schlafer, 2006; McIntyre et al., 2010). As consequence of the excessive pressure, the female may present mild signs of uterine disease (such as abdominal distension, tenesmus or dysuria) albeit usually haematological exams and clinical biochemistry are within or close to normal values. In one of the situations we had, in a Rottweiler female with 5 years of age, signs of cystic endometrial hyperplasia co-existed in the remainder segments of the uterus. Though these are unrelated conditions, as already mentioned, the uterine segment retaining patency may develop a pyometra, which may completely change the clinical scenario for the patient. In this particular case, the segmental aplasia was diagnosed after surgery.

Whenever a segmental aplasia of the cranial part of the uterine horn is found, it may also be found a distended uterine tube. This situation has been described for cats and rabbits (Thode & Johnston, 2009), but was not mentioned in the reports on dogs we have accessed. However, non-mention may also mean that the structure was not evaluated. In our Rottweiler case, gross evaluation of the excised genitalia fail to evidence distension of the uterine tubes; however, a muco-purulent fluid was found in the ovarian bursa suggesting the flow of the uterine content through the oviducts. Also in two other situations, oviductal dilatation was found during histopathological evaluation (Figures 1C and 1F).

5.3 Segmental aplasia of the uterine body and cervix

Although rare, sporadic reports on the agenesis of the uterine body have been published for the dog (Oh et al., 2005; McIntyre et al., 2010). The aplasia of the canine cervix was seldom described (McEntee, 1990).

Both conditions allow communication between the two uterine horns and completely impair fluid drainage from the uterus. Consequently, one of the main symptoms may be the existence of estrus signs (swollen vulva and male attraction) in the absence of the characteristic vulvar discharge (non-technically described as "dry" estrus). Furthermore, as none of the endometrial secretions drain from the genital tract, as the female retains her cyclicity, signs secondary to excessive fluid accumulation may develop earlier when compared to segmental aplasia, and the abdominal distension is usually noticed earlier. Uterine distension, in this situation, is bilateral and symmetric (Oh et al., 2005; McIntyre et al., 2010).

5.4 Segmental stenosis of the vagina

Animals presenting segmental stenosis of the vagina, as do those showing aplasia of the uterus or of the cervix, do not present the typical estrus vulvar discharge, despite the existence of regular cyclicity (Viehoff & Sjollema, 2003). However, chronicle, intermittent vaginitis is the most frequently described clinical symptom and the common to all the reports on segmental stenosis of the vagina (Gee et al., 1977; Kyles et al., 1996; Viehoff & Sjollema, 2003). It usually induces a more or less intense vulvar discharge, excessive licking of the vagina and dysuria. Recurrent cystitis was also found in such conditions (Kyles et al., 1996). Due to common symptoms, differential diagnosis between segmental stenosis of the vagina (Viehoff & Sjollema, 2003) and the imperforate hymen (Tsumagari et al., 2001) on the basis of the symptoms is difficult.

6. Diagnosis

Diagnosis of aplasia of the uterine-vaginal segment in dog before puberty or in young non-breeding dogs is seldom achieved, as clinical evidences are absent. Also, since the ovaries are present in congenital aplasia of the uterine-vaginal segment, normal cyclicity after puberty does not allow suspecting the existence of a defect.

Thus congenital aplasia of the uterine-vaginal is frequently asymptomatic and often missed during routine gynaecological examination, unless side effects associated to failure of genital patency develop in more severe or more prolonged situations. Further, due to unspecific and limited clinical symptoms described, congenital aplasia often requires more than one diagnostic approach to assure success.

Clinical signs or mild illness appears in older bitches, usually associated with fluid retention within the uterus or with other age-related diseases of the genital tract. The later is not taken into consideration herein when describing the diagnostic approach.

In *unilateral uterine aplasia*, few clinical signs develop even in older animals that may be related to the defect. The owner may complain on the female low prolificacy (Romagnoli & Schlafer, 2006; McIntyre et al., 2010) or on the tendency of recurrent abortion in the absence of microbiological or parasitic agents. Vaginoscopy or digital manipulation fails to evidence the primary defect. On ultrasound, the inability to find one uterine horn or the finding of a fibrous threadlike structure on its normal location are the expected findings. Both ovaries are found. Simple X-rays are not useful for diagnosis unless hysterosalpingography (a contrast X-ray) is performed. This would reveal that diffusion of the contrast is limited to one of the uterine horns, particular when using dorso-ventral projections.

A similar situation is found in the *segmental aplasia of the uterine horns*. Unless fluid accumulation in the uterine segment above the atresia becomes important and may be suspected during routine abdominal palpation or ultrasound, the defect remains undetectable in routine consultations. Complains on the female sub-fertility in the presence of regular estrous cycles may exist. The vaginoscopy and digital manipulation are not useful to evidence the primary defect. On ultrasound, segments of the uterus may show uterine distension, uterine walls of variable thickness and areas where the uterus is absent or reduced to a cord-like fibrous structure. However, often the image resembles that of a mucometra/pyometra, except that uterine distension may be limited to one portion of the uterus. On simple X-ray it may be observed dense, non-continuous masses in the normal location of the uterus, frequently in one side of the body, while the remainder parts of the genital tract are not visible. Comparison of these X-rays with those obtained by hysterosalpingography will allow to visualise the occlusion and the existence of a dense pouch cranially.

In the *segmental aplasia of the uterine body and cervix*, the most important symptom for diagnosis is the absence of the typical estrus vulvar discharge in a regularly cycling female (Oh et al., 2005; McIntyre et al., 2010). With time, also abdominal distension associated to uterine distension due to mucometra is found. During abdominal palpation, a uterus increased in size is detected. The vaginoscopy and digital manipulation fail to evidence the primary defect. On ultrasound, bilateral distension of the uterus, thin uterine walls and anechoic or hypoechoic uterine content are commonly found. The distension of the body of the uterus may be found when cervical agenesis exists. The thickness of the uterine walls is equivalent in the uterine horns and body. Radiographic signs are similar to those of mucometra or pyometra. Hysterosalpingography allows to identify the local of occlusion.

In case of *segmental stenosis of the vagina*, the major symptom is also the absence of the estrus vulvar discharge in a regularly cycling female. However, a large number of females also present clinical history of intermittent, chronic vaginitis or of recurrent, chronic lower urinary tract infection (Kyles et al, 1996; McIntyre et al., 2010), which have not being reported in the aplasia of the uterine body or cervix. Abdominal palpation shows increased size of the uterus that prolongs caudally into the pelvic brim (cranial segment of the vagina). As the vagina dilates more than the uterine body, it can appear as being more ballooned. Also transrectal digital palpation shows the distension of the vagina, that appears as a balloon at the entrance of the pelvic vault. Digital manipulation of the vagina or vaginoscopy usually allows detecting the defect (Kyles et al., 1996; Viehoff & Sjollema, 2003). On the ultrasound bilateral distension of the uterus is observed, as in the segmental aplasia of the uterine body or cervix, which is prolonged beneath the urinary bladder into the pelvic brim. The bladder may be dislocated from its normal position. Ultrasound scans allow to distinguish between the uterine and the vaginal segment, the latter having thinner walls and usually increased dilatation. On simple X-rays cranioventral displacement of the bladder and increased dimensions of the uterus, extending into the vaginal position, are observed. Hysterosalpingography allows the localization of the occlusion.

For the defects originating in the cervix or the vagina, endoscopic visualization of the vaginal cavity and of the cervical morphology may be useful in diagnosis. Furthermore, three-dimension ultrasonography and the magnetic resonance imaging may give useful information on the development of the genital structures even when secondary diseases are absent. However, these techniques are not easy to develop in the current veterinary clinics.

On table 2 we condensed the most relevant information to reach diagnosis.

7. Therapeutic management

To propose a treatment for any of the different conditions of congenital aplasia of the uterine-vaginal segment in dogs is a difficult task. For unicornuate uterus no direct negative effects arise from the defect by itself, and there are no additional risks for the health of the bitch other than those associated with age in intact females.

For the different segmental aplasia of the uterus (concerning the uterine horns, body or cervix) no effective therapeutic approach is available, and ovariohysterectomy remains the solution to avoid future health problems for the bitch, associated to fluid retention in the uterus with posterior inflammation.

In the case of segmental stenosis of the vagina, the treatment of the defect may be attempt to maintain the breeding status of the bitch, unless a large segment of the vagina is missing, which could bring potential problems during mating and delivery (Viehoff & Sjollema, 2003). Anastomosis of the vagina and vestibule may be attempt through episiotomy (Kyles et al., 1996; Viehoff & Sjollema, 2003). The success of the technique has yet to be ascertained in more severe cases of agenesis of the vaginal segment.

Defect	Unicornuate uterus	Segmental aplasia of the uterine horns	Segmental aplasia of the uterine body or cervix	Segmental vaginal stenosis
Main clinical sign	Sub-fertility (low prolificity) in the presence of regular estrous cycles		Absence of vulvar discharge at estrus ("Dry" estrus)	
Other symptoms	None, unless uterine or ovarian concurrent diseases	Uterine distension as "in pouches" None, unless uterine or ovarian concurrent diseases	Mucometra Bilateral uterine distension Abdominal distension	Intermittent, chronic vaginitis or cystitis Bilateral mucometra and hydrocolpos Abdominal distension
Physical examination	Without results	Areas of uterine dilatation alternating with areas where the uterus is not perceived or appear as a fibrous cord.	Increased dimensions of the uterus during abdominal palpation	Digital manipulation and vaginoscopy shows occlusion Increased dimensions of the uterus during abdominal palpation
Simple X-rays	Not useful	Dense, non-continuous masses in the normal location of the uterus	Visualization of bilateral increased dimensions of the uterus (similar to mucometra or pyometra)	Cranioventral displacement of the bladder and increased dimension of the uterus and cranial vagina
Ultrasound	Only one uterine horn usually detected; presence of both ovaries	Uterine wall distension similar to the observed in mucometra or pyometra	Visualization of bilateral increased dimensions of the uterus (similar to mucometra or pyometra) The distension is limited to the horns or body; image of the uterine body below the bladder	Bilateral enlargement of the uterus and vagina. Differences in the thickness of the walls may distinguish between uterus and vagina
Other complementary exams	Hysterosalpingography		Vagino-uretrocystography: the existence of an occlusion hampers the passage to the contrast medium	

Table 2. Resumed guidelines for reach a diagnosis in congenital segmental agenesis of the uterine-vaginal segment in dogs.

8. Conclusions

Congenital agenesis of the uterine-vaginal segment in dogs consists of a wide range of defects that usually progress undetected until development of associated side effects of the defect, such as mucometra and hydrocolpos, or recurrent cystitis or chronic vaginitis. Surgical correction of the congenital abnormality is seldom possible. Thus it is important to be aware of the most adequate approaches for its early diagnosis, allowing to exclude the affected animals from the reproductive stock.

9. Aknowledgements

This work was supported by the project from CECAV/UTAD with the reference PEst-OE/AGR/UI0772/2011, by the Portuguese Science and Technology Foundation.

10. References

Almeida, M.V.; Rezende, E.P.; Lamounier, A.R.; Rachid, M.A.; Nascimento, E.F.; Santos, R.L. & Valle, G.R. (2010) Aplasia segmentar de corpo uterino em cadela sem raça definida: relato de caso. *Arq. Bras. Med. Vet. Zootec.*, Vol. 62, Nº4, pp.794-800.

Chang, J.; Jung, J.H.; Yoon, J.; Choi, M.C.; Park, J.H.; Seo, K.M. & Jeong, S.M. (2008) Segmental aplasia of the uterine horn with ipsilateral renal agenesis in a cat. *J. Vet. Med. Sci.*, Vol. 70, pp. 641-643.

Dood, K.L. & Wibbels, T. (2008) Estrogen inhibits caudal progression but stimulates proliferation of developing müllerian ducts in a turtle with temperature-dependent sex determination. *Comp. Biochem. Physiol.*, Vol.150, Nº3,pp. 315-9

Gee, B.R.; Pharr, J.W. & Fuerneaux R.W. (1977) Segmental aplasia of Müllerian duct system in a dog. *Can. Vet. J.*, Vol. 18, pp. 281-286.

Güvenç, K.; Toydemir, F.S.; Sontaş, H.B. & Şenünver, A. (2006) A Cocker Spaniel bitch with uterus unicornis (unilateral cornual agenesis). *J. Fac. Vet. Med. Istanbul Univ.*, Vol. 32, nº3, 69-73.

Kyles, A.E.; Vaden,S.; Hardie, EM & Stone, E.A. (1996) Vestibulo-vaginal stenosis in dogs:18 cases[1987-1995]. *J. Am.Vet.Med. Assoc.*, Vol. 209, pp. 1889-93.

McEntee K. (1990) Reproductive pathology of domestic mammals, 1st ed. Academic Press, San Diego, CA.

Mcgeady, T.A.; Quinn, P.J.; Fitzpatrick, E.S. & Ryan, M.T. (2006) Veterinary embryology. Blackwell Publishing, Oxford.

McIntyre, R.L.; Levy, J.K.; Roberts, J.F. & Reep, R.L. (2010) Developmental uterine anomalies in cats and dogs undergoing elective ovariohysterectomy. *J. Am. Vet. Med. Assoc.*,Vol. 237, pp. 542-546

Moore, K.L. & Persaud, T.V. (2008) The urogenital system. In: The developing human clinically oriented embryology.8th ed. Saunders&Elsevier, Philadelphia, PA., pp. 243-269

Noden D.M. & de Lahunta A. (1985) The embryology of domestic animals,1st ed. Williams &Wilkins, Baltimore, MD.

Oh, K.S.; Son, C.H.; Kim, B.S.; Hwang, S.S.; Kim, Y.J.; Park, S.J.; Jeong, J.H.; Jeong, C.; Park, S.H. & Cho, K.O. (2005) Segmental aplasia of uterine body in an adult mixed breed dog. *J. Vet. Diag. Invest.*, Vol. 17, pp. 490-492

Ortega-Pacheco, A.; Segura-Correa, J.C.; Jimenez-Coello, M. & Forsberg, C.L.(2007) Reproductive patterns and reproductive pathologies of stray bitches in the tropics. *Theriogenology*, Vol. 67, pp. 382-390.

Pinto Filho, S.T.; Cunha, O.; Raiser, A.G.; Barbosa, G.S.;Portella, L.C. & Irigoyen, L.F. (2001) Unilateral uterine agenesis in a bitch- A case report. *Arq. Ciên. Vet. Zool. UNIPAR*, Vol. 4, N°1, pp.77-79.

Ribeiro, S.C.; Tormena, R.A.; Peterson, T.V.; Gonzáles, M.O.; Serrano, P.G.; Almeida, J.A. & Baracat, E.C. (2009) Müllerian duct anomalies:review of current management. *Sao Paulo Med. J.*, Vol. 127, n°2, pp. 92-6

Roberts, S.J. (1971). *Veterinary Obstetrics and Genital Diseases.* Woodstock, VT: Self published; First edition.

Romagnoli, S. & Schlafer, D.H. (2006) Disorders of sexual differentiation in puppies and kittens: a diagnostic and clinical approach. *Vet. Clin. North Am. Small Anim. Pract.*, Vol. 36, pp. 573-606.

Schlafer, D.H. & Miller, R.B. (2007) Female genital system. In: MAXIE, M.G. (Ed). Jubb, Kennedy, and Palmer's Pathology of Domestic Animals. 4th ed. Philadelphia: Elsevier Saunders:429-564.

Schulman, M.L. & Bolton, L.A. (1997) Uterine horn aplasia with complications in two mixed-breed bitches. *J.S. Afr. Vet. Assoc.*, Vol. 68, pp.150-153

Seyrek-Intaş, K.; Wehrend, A.; Nak, Y.; Tek, H.B.; Yilmazbaş, G.; Gökhan, T. & Bostedt, H. (2004) Unilateral hysterectomy (cornuectomy) in the bitch and its effect on subsequent fertility. *Theriogenology*, Vol. 61, pp.1713-1717

Thode,H.P. & Johnston, M.S. (2009) Probable congenital uterine developmental abnormalities in two domestic rabbits. *Vet. Rec.*, Vol. 164, pp.242-244

Tsumagari, S.; Takagi, S.; Takeishi, M. & Memon, M.A. (2001) A case of a bitch with imperforate hymen and hydrocolpos. *J.Vet. Med.Sci.* Vol. 63, N°4, pp. 475-477.

Viehoff, F.W & Sjollema, B.E. (2003) Hydrocolpus in dogs: surgical treatment in two cases. *J. Small Anim. Pract.* Vol. 44, pp. 404-407

Wadsworth, P.F.; Hall, J.C. & Prentice, D.E. (1978) Segmental aplasia of the vagina in the beagle bitch. *Lab. Anim.*, Vol. 12, pp.165-166.

11

Atresia Ani in Dogs and Cats

Lysimachos G. Papazoglou[1] and Gary W. Ellison[2]
[1]*Department of Clinical Sciences, Faculty of Veterinary Medicine, Aristotle University of Thessaloniki,*
[2]*Department of Small Animal Clinical Sciences, Health Science Center, University of Florida Gainesville,*
[1]*Greece*
[2]*USA*

1. Introduction

Congenital deformities of the anorectum are rarely encountered in small animals with atresia ani being the most common one. Atresia ani is a congenital defect of the anorectum, resulting in anal canal closure and /or abnormal routing of feces (Bright & Bauer, 1994). Among large animals congenital atresia ani most often occurs in pigs and calves and is considered hereditary. In these species atresia may be seen as a separate entity or in conjunction with other malformations of the distal vertebral column, urogenital tract and sometimes with intestinal atresia or colonic agenesis (Kilic & Sariepler, 2004; Maxie, 2007). In pigs atresia is associated with a high mortality rate but selective breeding decreased its occurrence (Partlow et al., 1993; Viana &Tobias, 2005).

2. Epidemiology

Atresia ani is uncommonly described in dogs and it is even less frequently reported in cats; the true incidence of this abnormality is difficult to determine and may be greater than reported because many newborn puppies and kittens are euthanatized before being evaluated based on the hypothesis that surgical correction is unsuccessful; additionally unpublished data may hide true prevalence as complications associated with surgical correction of these deformities are common (Prassinos et al.,. 2003; Mahler & Williams, 2005; Viana & Tobias, 2005). According to a review of Veterinary Medical Database atresia ani in dogs accounts for 0.0007 with females more likely to be affected than males (female/male =1.79/1) [Viana & Tobias, 2005]. Canine breeds overpresented include Finish spitz, Boston teriers, Maltese, chow chow, German shorthair pointer, toy poodle and miniature schnauzer (Viana & Tobias, 2005). In cats females are more commonly affected than males (Suess et al., 1992; van de Broek et al., 1988; Tsioli et al., 2009, Tomsa et al., 2011).

3. Normal and abnormal embryologic development

The embryologic development of the canine and feline anorectum resembles that of the human development (Greiner, 1972; Amand, 1974; Suess et al., 1992; Sadler, 1995; Viana &

Tobias, 2005). The cloaca is a common route for gastrointestinal and urogenital tracts in the canine and feline embryo. Retention of the cloaca occurs in vertebrate animals other than placental mammals; in higher mammals the cloaca separates during embryologic development. By the seventh to eighth week of development the urorectal fold, which located between the allantois and the hindgut openings in the cloaca, initially divides the cloaca into the dorsal part called the rectum and the ventral part called the urogenital sinus (Figures 1 &2). The urogenital sinus is further differentiated into the urethra and the urinary bladder. The terminal end of the hindgut forms the cranial anal canal and the anus is formed at a later time by ectodermal ingrowth of the perineum. A breakdown in physiological embryologic differentiation of the cloacal region may lead to a variety of congenital malformations of the anorectum. Failure of the urorectal fold to divide the cloaca completely or failure of the anal membrane to rupture after anal creation results in atresia ani.

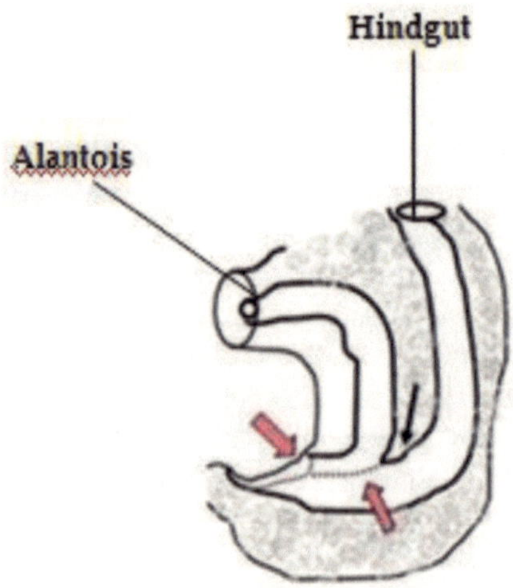

Fig. 1. Urorectal fold (arrows) located between the alantois and hindgut by 7th to 8th week of development.

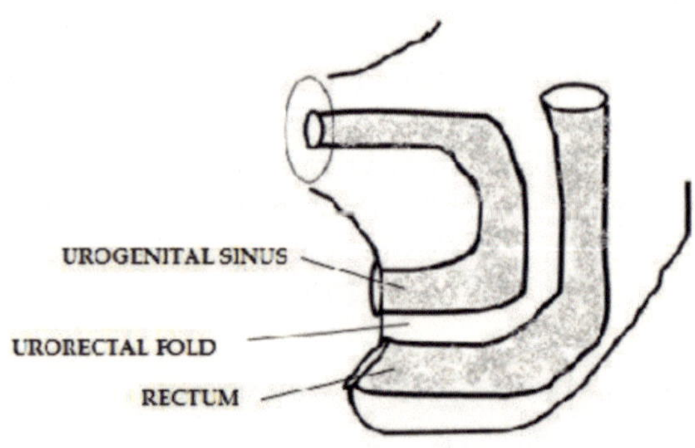

Fig. 2. The cloaca is divided into urorectal sinus and rectum.

4. Classification

Four anatomic types of atresia ani have been described in dogs and cats (Aronson, 2003; Viana &Tobias, 2005; Ellison & Papazoglou, 2011) [Figures 3-7].

External anal sphincter and anal sacs are usually develop normally in type II anomalies (Seim, 1986; Ellison & Papazoglou, 2011), while agenesis of the external anal sphincter, anal sacs or tail are reported in type III anomalies (Rawlings & Capps, 1971; Knecht & Westerfield, 1971; Loug & van Schouwenburg, 1982; Ellison & Papazoglou, 2011). Occasionally dogs and more rarely cats with type II and more uncommonly with type III atresia ani may be associated with rectovaginal, rectovestibular or urethrorectal fistulas (Holt, 1985; van den Broek et al., 1988; Chandler & MacPhail, 2001; Aronson, 2003; Ellison & Papazoglou, 2011). Animals with type III atresia ani associated with rectovaginal fistula are also reported as having an ectopic anus (Prassinos et al., 2003). However, it is unclear if type IV atresia ani has ever been reported in dogs and cats (Ellison & Papazoglou, 2011).

Atresia Type	Anomaly
Type I	Congenital stenosis of a patent anus
Type II	Persistence of a complete anal membrane alone or a combination of an anal membrane with the rectum ending as a blind pouch cranial to the membrane.
Type III	Presence of an imperforate anus with the rectum terminating further cranially.
Type IV	Normal ending of the terminal rectum and anus while the cranial rectum terminates as a blind pouch within the pelvis.

Table 1. Anatomic types of atresia ani in dogs and cats

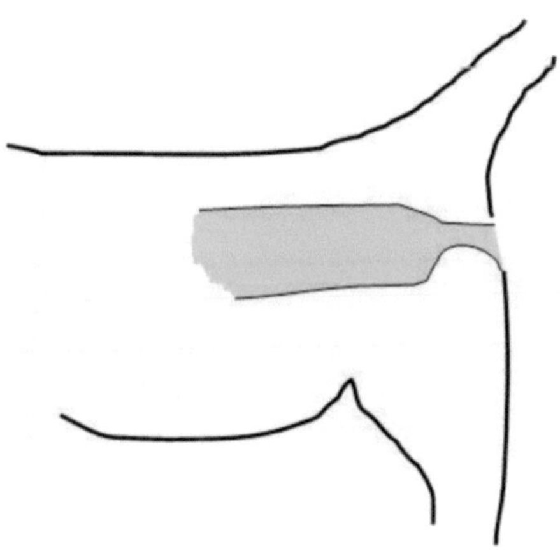

Fig. 3. Schematic representation of type I atresia ani.

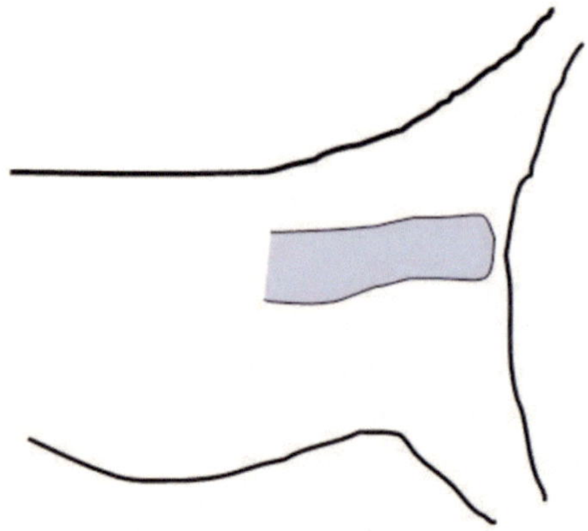

Fig. 4. Schematic representation of type II atresia ani.

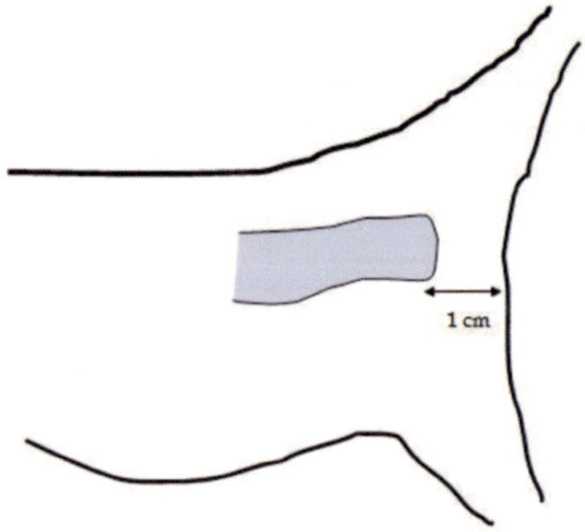

Fig. 5. Schematic representation of type III atresia ani. The blind rectal pouch is more than 1 cm away from the anal dimple.

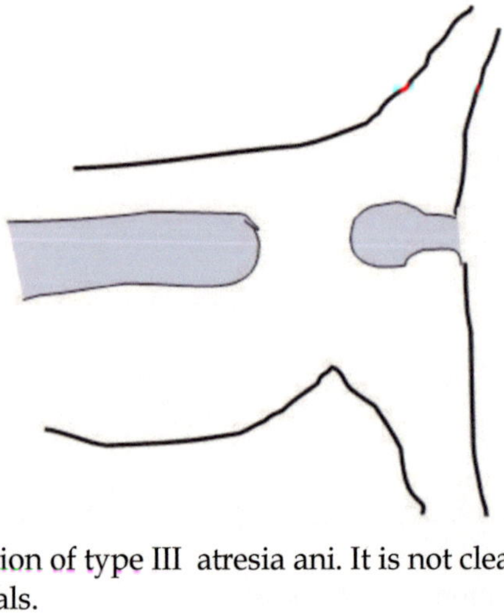

Fig. 6. Schematic representation of type III atresia ani. It is not clear if this type of atresia has been reported in small animals.

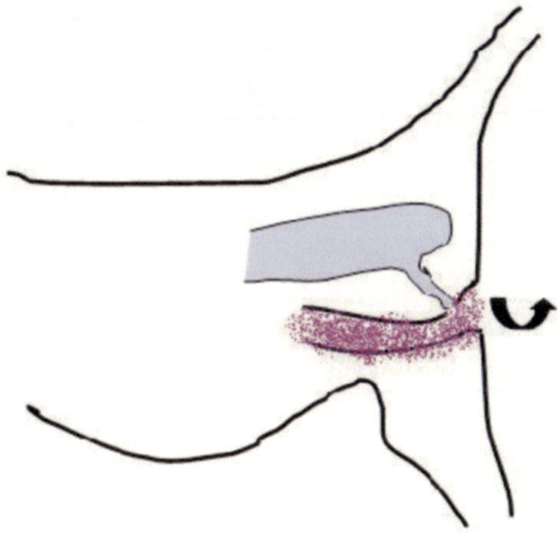

Fig. 7. Schematic representation of type II atresia ani combined with rectovaginal fistula.

5. History and clinical signs

A history of absence of defecation, which sometimes may go unnoticed, is reported. Clinical signs of atresia ani are usually evident within a few weeks of birth and depend on the type of atresia (Aronson, 2003; Viana &Tobias, 2005; Ellison & Papazoglou, 2011). Animals with type I atresia ani may exhibit constipation and tenesmus soon after weaning (Figure 8). A stenosed anal opening is evident on digital rectal palpation. Puppies and kittens belonging to types II, III and IV are clinically normal for the first 2 to 4 weeks of birth. A history of anorexia, depression, absence of defecation, tenesmus and abdominal distention are reported to follow this time frame (Figure 9). Physical examination may also reveal a dimple at the site of the closed anal opening (Figures 10 &11), abdominal enlargement, discomfort on abdominal palpation and perineal swelling depending on the type of anomaly. Abdominal enlargement may be attributed to colonic distention with feces or gas, fecal impaction or even megacolon. Longstanding cases may develop vomiting and dehydration. Animals with rectovaginal or rectourethral communications often show passage of watery to formed feces through the vagina or urethra; these animals may be in better physical condition that others with no rectovaginal or rectourethral communications (Prassinos et al., 2003; Rahal et al., 2007) [Figure 12]. Multiple congenital anomalies such as umbilical hernias, cleft palates, open fontanels, hypospadias, tail agenesis and deafness in dogs and sacrocaudal dysgenesia and hydrocepahalus in kittens may accompany atresia ani and should not escape a thorough physical or other diagnostic examination (Suess et al., 1992; Aronson, 2003; Prassinos et al., 2003; Rahal et al., 2007; Ellison & Papazoglou, 2011) [Figure 13]. Rectocutaneous fistulas associated with type II atresia ani have been recently reported in a cat (Tsioli et al., 2009) [Figure 14].

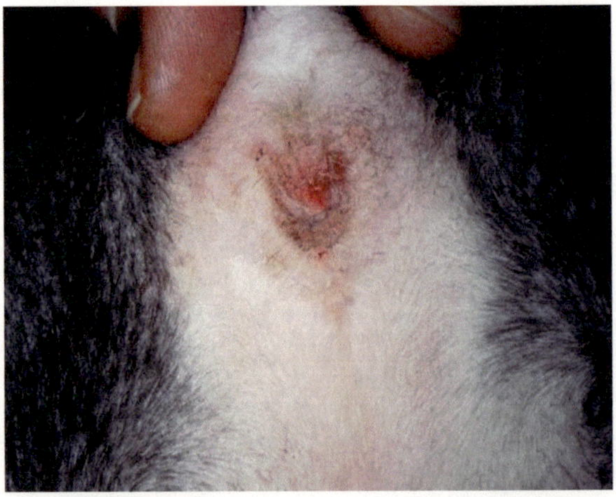

Fig. 8. Atresia ani type I in a mixed breed puppy.

Fig. 9. A mixed breed puppy with abdominal distention associated with type II atresia ani.

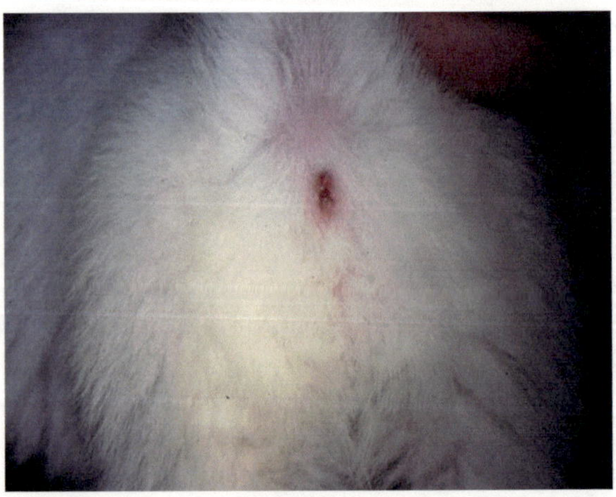

Fig. 10. Atresia ani type II of the dog of figure 9.

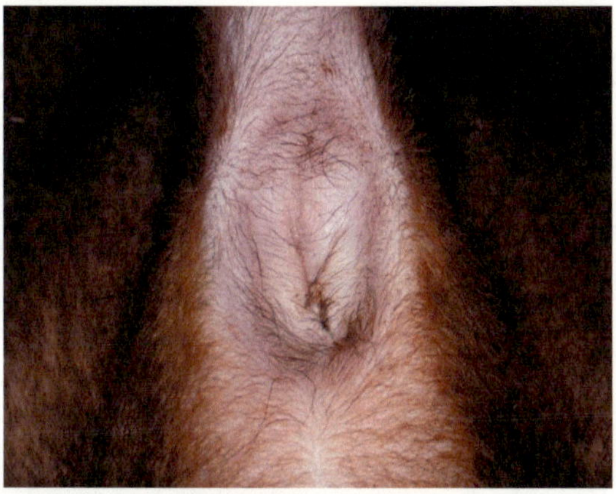

Fig. 11. An anal dimple associated with type II atresia ani in a dog.

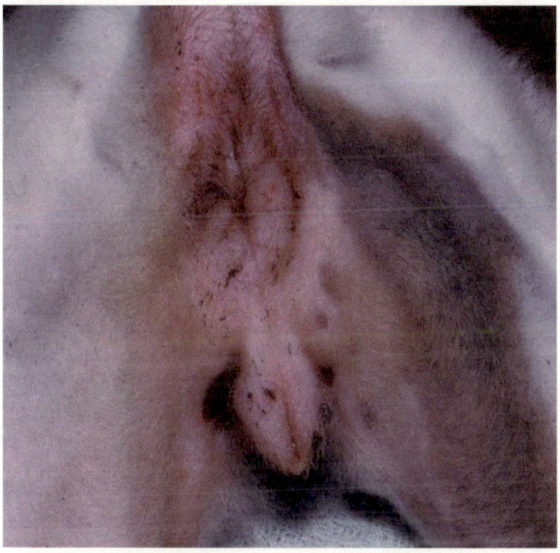

Fig. 12. Atresia ani type II associated with rectovaginal fistula in a puppy.

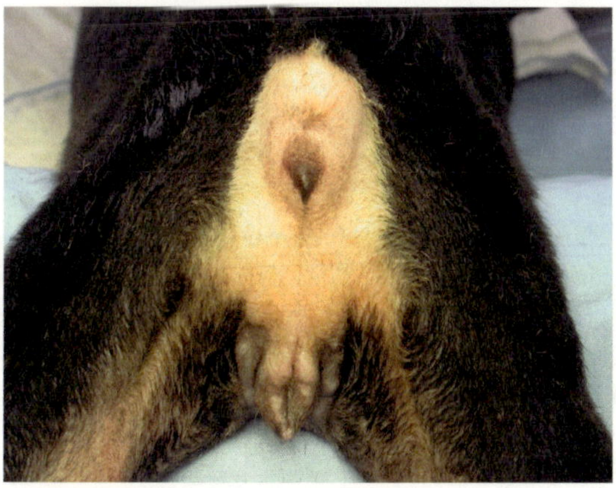

Fig. 13. Atresia ani type III associated with rectovaginal fistula and tail agenesis in a puppy.

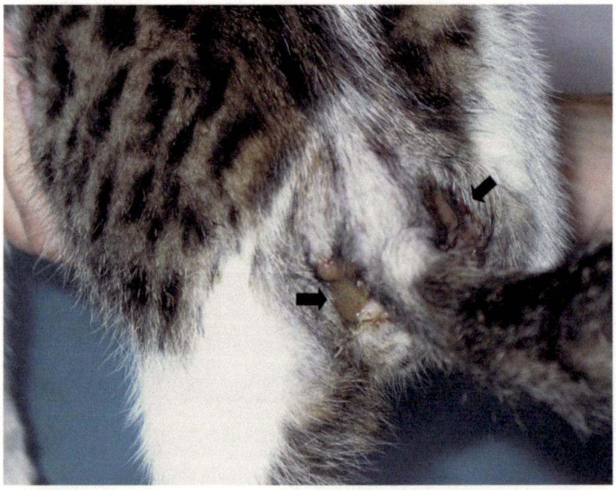

Fig. 14. Rectocutaneous fistulas ventral and lateral to the tail (arrows) in a cat with type II atresia ani

6. Diagnosis

Diagnosis of atresia ani is based in clinical signs, physical examination findings and confirmed with radiographic examination. Evaluation of external anal sphincter muscle presence and function is important for prognostic purposes.

6.1 Diagnostic tests

Penile bulb or vulva pinching may result in an anal wink (bulbourethral reflex) [Chambers, 1986; Hosgood & Hoskins, 1998]. Electro stimulation of perineal muscles using electrocautery at surgery or nerve stimulators may also be used to check for a strong anal response (Mahler & Williams, 2005). Electromyography for external anal sphincter muscle may be used if available for a more accurate evaluation of the anal tone (Ellison & Papazoglou, 2011).

6.2 Radiographic examination

Abdominal radiography is useful in atresia ani anatomic typing and ruling out colonic distention, which may lead in megacolon that affects management and prognosis (Figures 15-18). Gas accumulation in colon and rectum, as visualized in plain radiographs, may help in determining the position of the terminal rectum. Horizontal beam abdominal radiographs with the animal suspended upside down and pressure around the abdomen may help to visualize gas accumulation migrating to the terminal colon and rectum and outlining the borders of the rectal pouch (Greiner, 1972; Bright & Bauer, 1994). Additionally, a coin is placed at the anal dimple to help outline the anus (Greiner et al., 1984). In case of rectovaginal or urethrorectal fistulation positive contrast vaginography or urethrography may help determine fistula and terminal rectum location (Aronson, 2003; Prassinos et al., 2003; Viana &Tobias, 2005; Rahal et al., 2007; Ellison & Papazoglou, 2011).

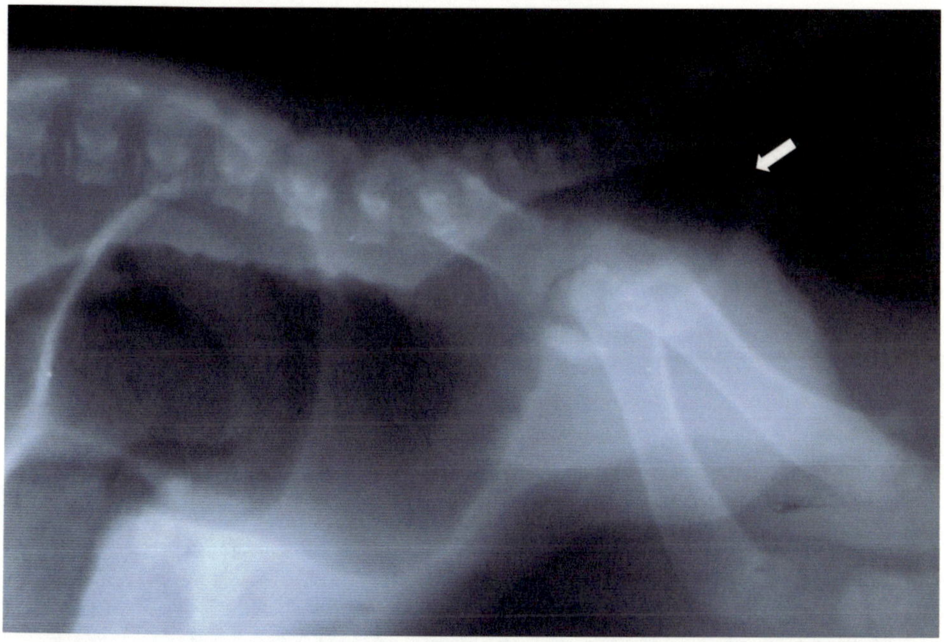

Fig. 15. Abdominal radiograph of a dog visualizing gas accumulation in the colon and rectum. The location of the terminal rectum is also shown (arrow) [Courtesy Dr. M.N. Patsikas, Aristotle University of Thessaloniki].

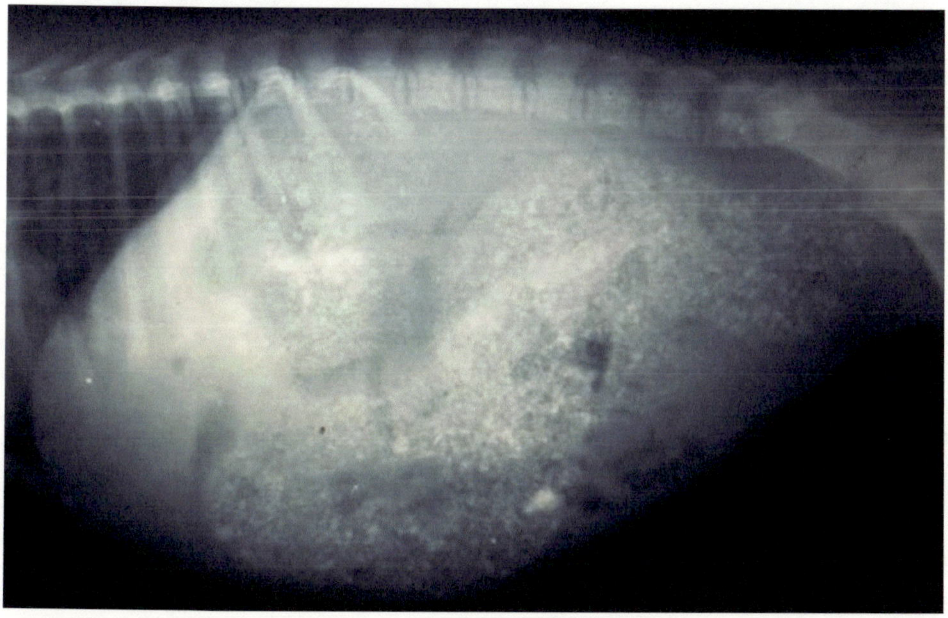

Fig. 16. Abdominal radiograph showing megacolon associated with type II atresia ani in a dog (Courtesy Dr. M.N. Patsikas, Aristotle University of Thessaloniki).

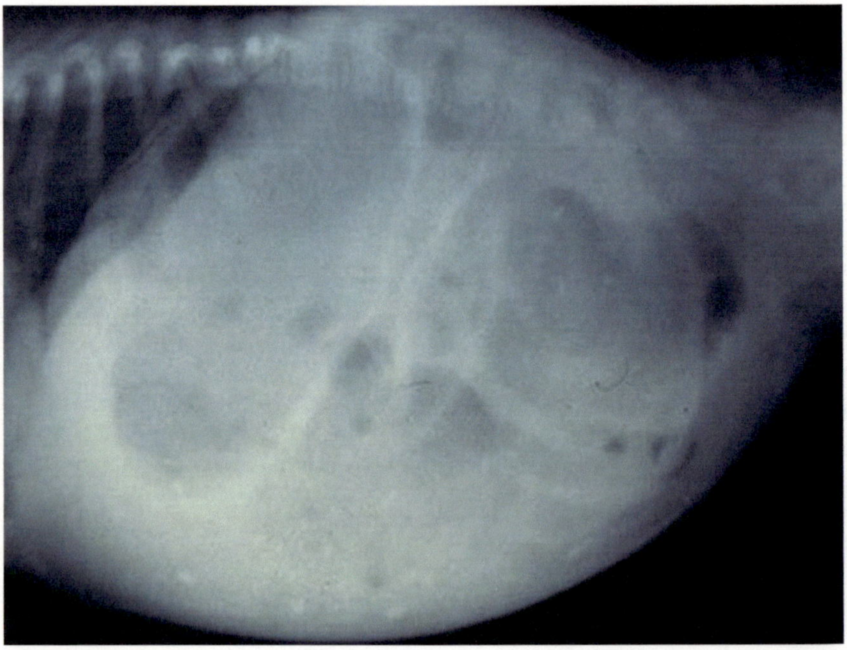

Fig. 17. Abdominal radiograph of a boxer puppy depicting colonic gas distention associated with type II atresia ani (Courtesy Dr. M.N. Patsikas, Aristotle University of Thessaloniki).

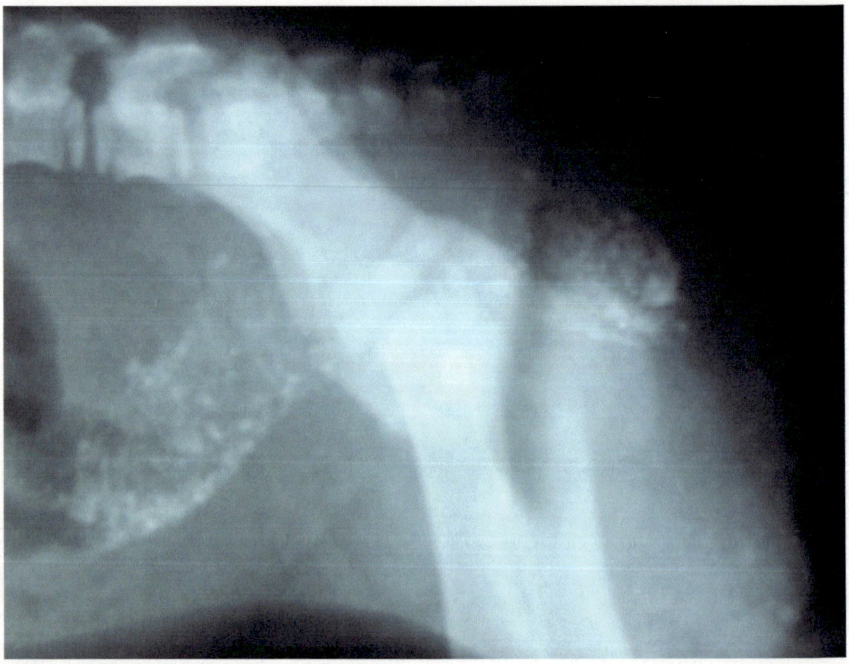

Fig. 18. Abdominal radiograph of a female rottweiler cross puppy with type III atresia ani and rectovaginal fistula showing fecal material in the colon, terminal rectum and over the vagina level. (Courtesy Dr. M.N. Patsikas, Aristotle University of Thessaloniki).

7. Treatment

Surgical correction is considered the only treatment for atresia ani. Anatomical typing of atresia ani should be performed to help determining the type of surgical correction for each case (Ellison & Papazoglou, 2011). However, non surgical management may also be applied for type I cases (Tomsa et al., 2011; Ellison & Papazoglou, 2011). Numerous case reports have appeared in the literature reporting surgical treatment of atresia ani in dogs and cats. Currently there are only 3 small and 1 larger case series reporting surgical treatment in dogs (Prassinos et al., 2003; Vianna & Tobias, 2005; Rahal et al., 2007) and 2 small case series in cats describing surgical or medical management of atresia ani (Suess et al., 1992; Tomsa et al., 2011). Anoplasty is the most common procedure performed. The aim of surgery is to restore anorectal continuity, to preserve the external anal sphincter, to preserve or restore colonic function and to eliminate any rectovaginal or urethrorectal communication. Surgical treatment should be prompt and performed before colonic atony or megacolon associated with chronic and prolonged distention or possible urinary tract infection ensues (Prassinos et al., 2003). Animals with atresia ani type II and III that are unable to defecate if not treated surgically will die because of bowel stasis. End-on temporary colostomy may be considered as an option for the treatment of cats with rectocutaneous fistulas associated with atresia ani (Tsioli et al., 2009). The perineal approach is used for all surgical corrections of atresia types I-III. Prophylactic antibiotics (cefoxitin 20 mg/kg) are administered intravenously at anesthetic induction.

7.1 Type I atresia ani

Animals with this type of atresia ani are placed in ventral decumbency and treated with resection of the strictured portion of the rectum through a 360^0 anoplasty. After excision of the stricture the rectal mucosa is brought distally so as mucosa to skin apposition is achieved with simple interrupted sutures using synthetic non absorbable monofilament 4/0-5/0 suture material (Figure 19). Care is taken during dissection to preserve the external anal

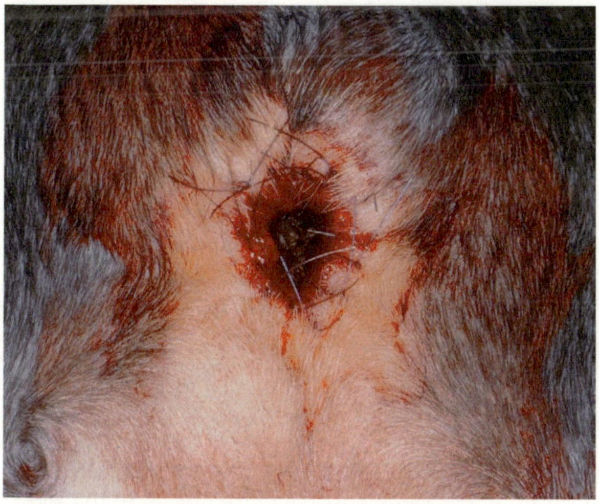

Fig. 19. Anoplasty in the dog of figure 8 with atresia ani type I.

sphincter and anal sacs (Hosgood & Hoskins, 1998; Aronson, 2003; Prassinos et al., 2003; Ellison & Papazoglou, 2011). Type I atresia ani may also be treated with bougienage or balloon dilatation in a single or multiple treatments; however, failures are not uncommon (Hosgood & Hoskins, 1998; Webb et al., 2007; Tomsa et al., 2011; Ellison & Papazoglou, 2011). Recently, a single balloon dilatation procedure alone or combined with intralesional triamcinolone injection was used to successfully treat type I atresia ani in 5 kittens and two dogs (Webb et al., 2007; Tomsa et al., 2011). Prospective studies are needed to evaluate balloon dilatation for the treatment of congenital anorectal strictures in small animals.

7.2 Type II & III atresia ani

For animals with type II or III atresia ventral recumbency is used and a cruciate, vertical or vertical elliptical incision is made over the anal dimple and medial to the ducts of the anal sacs. The triangular flaps or elliptical skin created are excised. The external sphincter and distal rectal pouch are identified and dissection is continued medial to the sphincter using fine scissors. The rectum is mobilized through the sphincter by using stay sutures, opened and sutured to the subcutaneous tissue and skin with 4/0-5/0 monofilament absorbable or non absorbable suture material respectively (Figures 20-22). In some animals with Type III atresia ani the rectal pouch is located more than 1 cm away from the anal dimple dissection for identification and mobilization of the rectum is achieved through rectal pull through procedure. The colon and rectum should be evacuated from feces before recovery, while the animal is still in anesthesia, to promote normal intestinal function (Hosgood & Hoskins, 1998; Aronson, 2003; Prassinos et al., 2006; Ellison & Papazoglou, 2011).

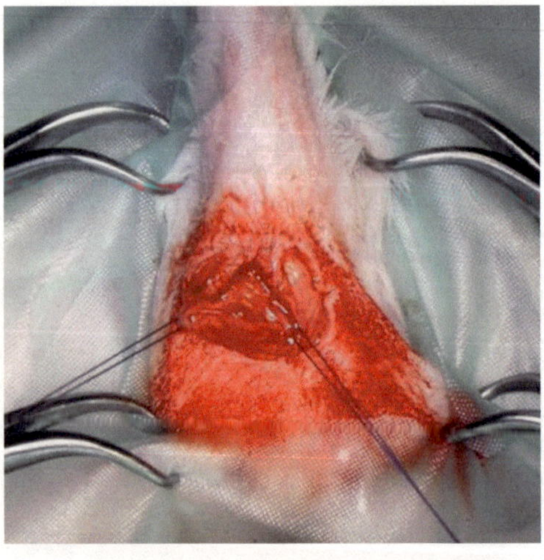

Fig. 20. The rectal pouch was identified through a cruciate incision made over the anal dimple in a puppy of figure 9. Two stay sutures were placed in the rectum to allow easy handling.

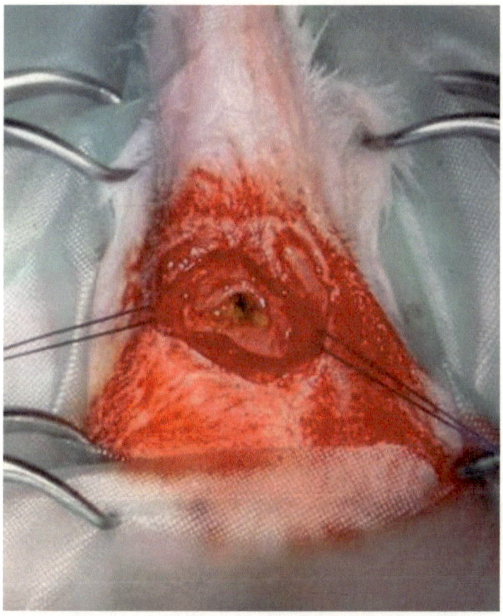

Fig. 21. The rectum of the dog of figure 9 was mobilized and opened to allow an anoplasty procedure.

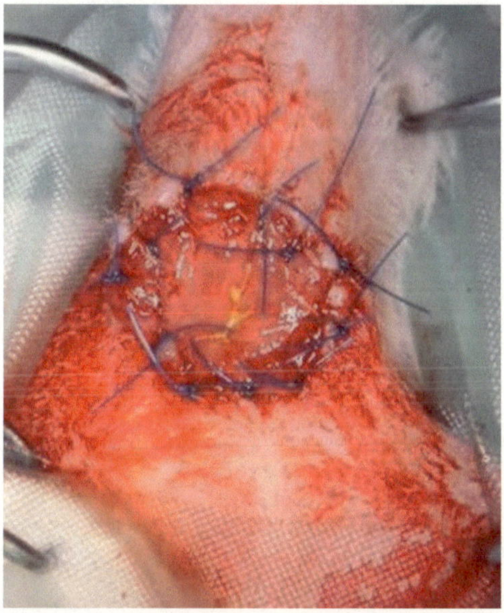

Fig. 22. The rectal mucosa of the dog of figure 9 was sutured to the skin with simple interrupted nylon sutures to complete the anoplasty.

7.3 Type IV atresia ani

Animals with this type of atresia may need an abdominal approach to isolate, mobilize and anastomose the cranial colon with the distal colon and rectum usually through a pubic symphisiotomy procedure (Hosgood & Hoskins, 1998; Aronson, 2003).

7.4 Types II-III atresia ani with rectovaginal fistula

Three surgical techniques are used for the correction of rectovaginal fistula in dogs and cats. Initial approach is through a vertical midline perineal incision extending from the ventral anus to the vulva. In one technique the fistula is isolated, excised and the rectal and vulvar lumens are ligated or oversewn separately with 3/0 -4/0 monofilament absorbable suture material or hemostatic clips followed by closure of the vertical perineal incision; An anoplasty procedure for atresia correction as previously described is performed afterwards (Chambers, 1986; Hosgood & Hoskins, 1998; Aronson, 2003; Prassinos et al., 2003; Viana &Tobias, 2005; Rahal et al., 2007; Ellison & Papazoglou, 2011). Fistula obliteration by performing numerous purse-string sutures placed along its length was also described in a dog (Louw & van Schouwenburg, 1982). According to another technique, used in 3 cats and 1 dog, the rectum is transected cranial to the fistulous opening, the affected rectal portion is excised and the terminal part of the rectum is sutured to the anus (Rawlings & Capps, 1971; Suess et al., 1992; Aronson, 2003). In the third technique the fistulous tract is preserved, isolated, mobilized through the anus and sutured to the skin at the level of the external anal sphincter; this technique was used for anal reconstruction in 2 dogs (Prassinos et al., 2003; Ellison & Papazoglou, 2011) and 1 cat (Bornet, 1990) and modified to be performed in 2 dogs through an episiotomy approach (Mahler &Williams, 2005) [Figures 23-25]. With this technique the preserved fistulus tract is considered the terminal part of the rectum and thus function of the internal anal sphincter is maintained (Prassinos et al., 2003; Mahler & Williams, 2005).

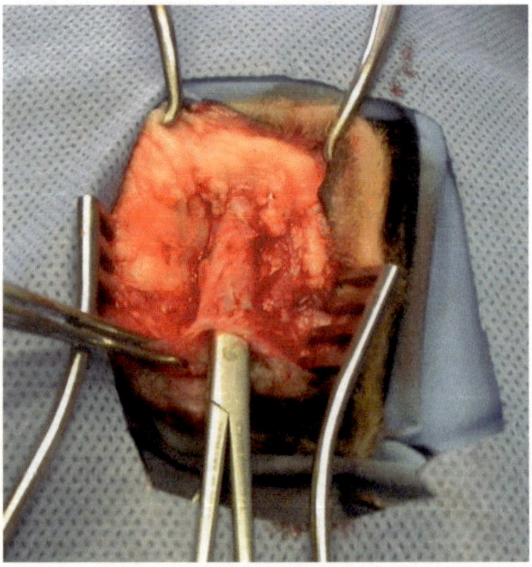

Fig. 23. Atresia ani type III associated with a rectovaginal fistula. The fistulous tract of the dog of figure 12 was isolated through a vertical midline perineal incision extending from the anus to the vulva. A pair of closed needle holders was inserted through the tract.

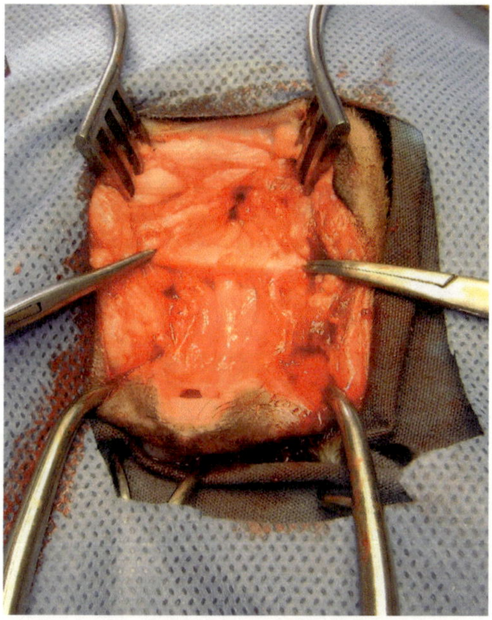

Fig. 24. Atresia ani type III associated with a rectovaginal fistula. The fistulous tract grasped with mosquito hemostats was dissected from the vagina.

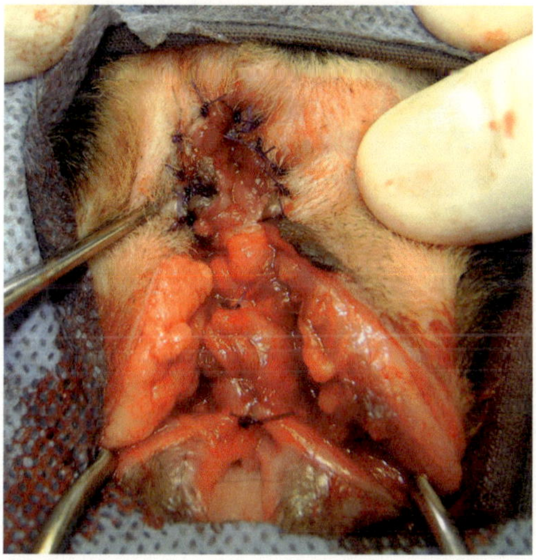

Fig. 25. Atresia ani type III associated with a rectovaginal fistula. The fistulous tract was sutured to the external anal sphincter and the vaginal mucosa was closed with simple interrupted sutures.

7.5 Temporary end-on colostomy for the management of cats with rectocutaneous fistulas associated with type II atresia ani

Temporary incontinent end-on colostomy may be performed initially for fecal diversion to help eliminate rectocutaneous fistulas. Colostomy is located in the lateroventral abdominal wall. Soon after colostomy and following resolution of fistulas a second surgery is performed for colostomy closure and reconstruction of the atresia ani by performing anoplasty (Tsioli et al., 2009) [Figures 26 & 27].

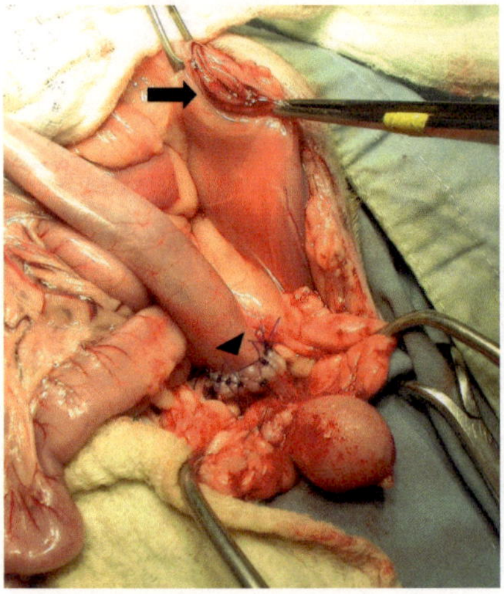

Fig. 26. Following resolution of fistulas the colostomy was closed (arrow) and an end to end colocolonic anastomosis was performed (arrowhead) through a ventral midline celiotomy in a cat.

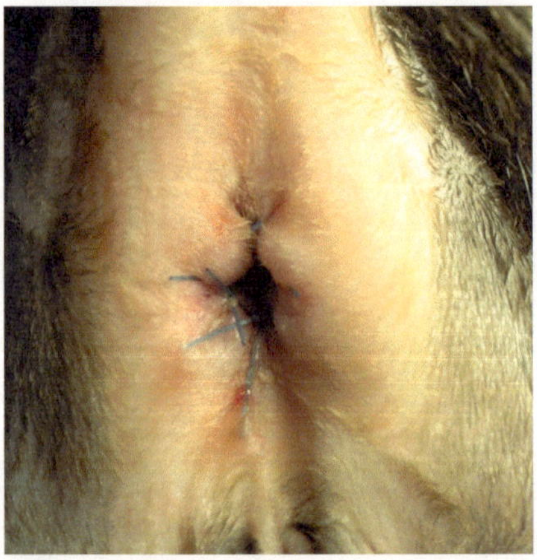

Fig. 27. Atresia ani reconstruction through anoplasty was performed following colostomy closure in a cat.

8. Postoperative care

Postoperatively analgesia through opioid administration for a week's time, cisapride to promote colonic motility and stool softeners as lactulose to combat constipation are provided. Fecal impaction may be relieved manually and multiple enemas should be avoided. Many animals start to defecate soon after surgery (Prassinos et al., 2003; Rahal et al., 2007; Ellison & Papazoglou, 2011).

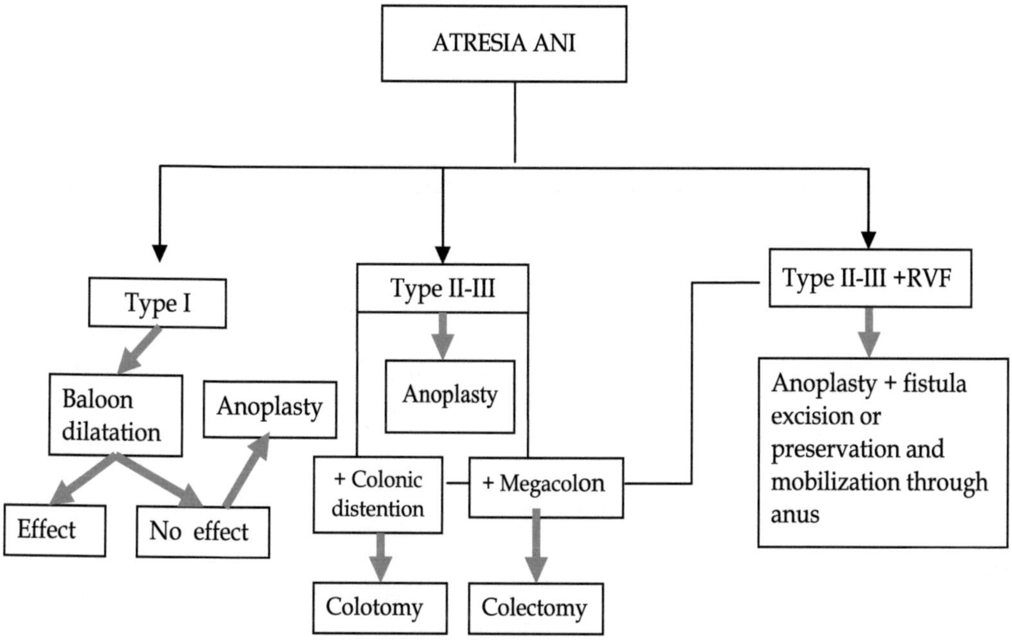

Table 2. Algorithm for atresia ani treatment methods in dogs and cats. RVF : rectovaginal fistula

9. Complications

Postoperative complications may include tenesmus, fecal incontinence, wound dehiscence, stricture of the anoplasty, colonic atony or megacolon and rectal prolapse (Suess et al., 1992; Prassinos et al., 2003; Vian & Tobias, 2005; Rahal et al., 2007; Ellison & Papazoglou, 2011). Fecal incontinence, a common complication after surgery, may be transient (Prassinos et al., 2003; Viana & Tobias, 2005, Rahal et al., 2007), intermittent or permanent (Suess et al., 1992; Ellison & Papazoglou, 2011) and related to a congenital absence of functional external anal sphincter or surgical trauma to the sphincter muscle innervation during dissection (Aronson, 2003; Prassinos et al., 2003; Viana & Tobias, 2005; Ellison & Papazoglou, 2011). Fecal incontinence secondary to surgical intervention in dogs may resolve several weeks to a year after surgery (Prassinos et al., 2003; Viana & Tobias, 2005; Rahal et al., 2007). Semitendinosus muscle flap application was proposed as an option to improve anal tone in a dog with atresia ani and rectovaginal fistula (Chambers & Rawlings, 1991). Wound dehiscence may be related to tension on the anastomosis and fecal contamination of the surgical site (Suess et al., 1992) and may be prevented by meticulous surgical technique. Stricure of the anoplasty site and fecal retention associated with colonic atony or megacolon are two common complications requiring a second surgery such as revision anoplasty, subtotal colectomy or colotomy (Viana & Tobias, 2005, Ellison & Papazoglou, 2011). Animals with stricture of the anoplasty site may develop tenesmus, constipation and fecal impaction. This complication may need a revision anoplasty or baloon dilatation to resolve. In a recent study with 12 cases of dogs and cats with atresia ani having surgical management, 5 animals with type II-III atresia ani, 4 of which combined with rectovaginal fistula, developed postoperative stricture and had initially balloon dilatation, which failed in all but one case.

Of these failures 5 had revision anoplasty which was successful in 4 animals and 3 had concurrent colotomy for fecal impaction removal, all with good results (Ellison & Papazoglou 2011). Subtotal colectomy that was performed in one dog with type II atresia ani 2 weeks after anoplasty continued to show constipation associated with megacolon 9 months after surgery (Viana & Tobias, 2005). Colotomy or subtotal colectomy might be performed at the same time with anoplasty to help improve colonic function in animals with colonic dilatation or megacolon.

10. Outcome and prognosis

In the larger study reported to date 10 of 12 dogs with type I and II atresia ani survived long term with survival times ranged from 12-96 months and 6 of the 8 animals were continent. In contrast 2 of the 3 animals having type III atresia ani were euthanized because of continuous tenesmus (Ellison & Papazoglou, 2011).

11. Comparative aspects of atresia ani

Anorectal malformations represent a wide spesctrum of disorders in boys and girls associated with urogenital tract, sacral or spinal defects. These defects are grouped in anatomic categories sharing similar prognosis and management. Treatment of these anorectal malformations aims at anatomical reconstruction, diagnosis and treatment of any associated defects and provides patients with good quality of life by addressing the functional sequelae of these malformations (Pena & Hong, 2000). The most common defect in males include imperforate anus with rectourethral fistula; some of these fistulas are associated with good quality muscles, well developed sacrum and anal dimple whereas some others are related with poor quality muscles, abnormally developed sacrum and hardly visible dimple. The most common defect in females is rectovestibular fistula and this anomaly has excellent functional prognosis (Levitt & Penna, 2005). About 75% of the patients with anorectal malformations will pass voluntary bowel motions and enjoy a good quality of life postoperatively, while constipation urinary and fecal incontinence are common complications following anorectal reconstruction (Pena & Hong, 2000).

12. References

Amand, W. (1974). Nonneurogenic disorders of the anus and rectum. *Veterinary Clinics of North America*, Vol. 4, No. 3, (August 1974), pp. 535-550

Aronson, L. (2003). Rectum and Anus, In: Textbook of Small Animal Surgery, D. Slatter, (Ed), 682-708, Saunders, ISBN 0721686079, Philadelphia, USA

Bornet, J. (1990). Fistule recto-vaginale et imporforation anale chez une chatte: traitment chirurgical. *Bulletin de l'Academie Veterinaire de France*, Vol. 63, No. 1, (January-March 1990), pp. 53-65

Bright, R. & Bauer M. (1994). Surgery of the digestive system, In: The Cat Diseases and Clinical Management, R. Sherding, (Ed), 1353-1401, Churchill Livingstone, ISBN 0443088799, New York, USA

Brown, C., Baker, D. & Barker, I. (2007). Alimentary system, In: Jubb, Kennedy and Palmer's Pathology of Domestic Animals, M. Maxie (Ed), ISBN 13978070202785, Saunders, Edinburgh, UK

Chambers, J. (1986). Surgical diseases of the anorectum, In: Canine and Feline Gastroenterology, B. Jones (Ed), 279-294, Saunders, ISBN 0721652069, Philadelphia, USA

Chandler, J. & MacPhail, C. (2001). Congenital urethrorectal fistulas. *Compendium on Continuing Education for the Practicing Veterinarian*, Vol. 23, No. 11, (November 2001), pp. 995-1002

Ellison, G. & Papazoglou, L. (2011). Long-term results of surgery for atresia ani in small animals-12 cases. *Journal of the American Veterinary Medical Association*, n.d.

Greiner, T. (1972). Surgery of the rectum and anus. *Veterinary Clinics of North America*, Vol. 2, No. 1, (January 1972), pp. 167-180

Holt, P. (1985). Anal and perianal surgery in dogs and cats. *In Practice*, Vol. 7, No. 3, (May 1985), pp. 82- 89

Hosgood, G. & Hoskins, J. (1998). Abdominal cavity and gastrointestinal disorders, In: Small Animal Paediatric Medicine and Surgery, G. Hosgood & J. Hoskins (Eds), 68-114, Butterworth Heinemann, ISBN 0750635991, Woburn, USA

Kilic, N. & Sarierler, M. (2004). Congenital intestinal atresia in calves. *Revue de Medecine Veterinaire*, Vol. 155, No. 7 (July 2004), pp. 381-384

Knecht, C. & Westerfield, C. (1971). Anorecto-urogenital anomalies in a dog. *Journal of the American Veterinary Medical Association*, Vol. 159, No. 1 (July 1971), pp. 91-92

Levitt, M. & Pena, A. (2005). Outcomes from the correction of anorectal malformations. *Current Opinion in Pediatrics*, Vol. 17, No.3, (June 2005), pp. 394-401

Louw, G. & van Schouwenburg, J. (1982). The surgical repair of atresia ani in a dobermann bitch. *Journal of the South African Veterinary Association*, vol. 53, No. 2, (June 1982), pp. 119-120

Mahler, S. & Williams, G. (2005). Prerservation of the fistula for reconstruction of the anal canal and the anus in atresia ani and rectovestibular fistula in 2 dogs. *Veterinary Surgery*, Vol.34, No. 2, (March-April 2005), pp. 148-152

Partlow, G., Fisher, K., Page, P., MacMilan, K. & Walker, A. (1993). Prevalence and types of birth defects in Ontario swine determined by mail survey. *Canadian Journal of Veterinary Research*, Vol. 57, No. 2, (April 1993), pp. 67-73

Pena, A. & Hong, A. (2000). Advances in the management of anorectal malformations. *The American Journal of Surgery*, Vol. 180, No. 11, (November 2000), pp. 370-376

Prassinos, N., Papazoglou,L., Adamama-Moraitou, K., Galatos, A., Gouletsou, P. & Rallis, T. (2003). Congenital anorectal abnormalitics in six dogs. *Veterinary Record*, Vol.153, No. 3, (July 2003), pp. 81-85

Rahal, S., Vicente, C., Mortari, A., Mamprim, M. & Caporalli, E. (2007). Rectovaginal fistula with atresia ani in 5 dogs. *Canadian Veterinary Journal*, Vol. 48, No. 8, (August 2007), pp. 827-830

Rawlings, C. & Capps, W. (1971). Rectovaginal fistula and imperforate anus in a dog. *Journal of the American Veterinary Medical Association*, Vol. 159, No. 3, (August 1971), pp. 320-326

Sadler, T. (1995). Digestive system, In: Langman's Medical Embryology, T. Sandler (Ed), 242-271, Williams and Wilkins, ISBN 068307489X, Baltimore, USA

Suess, R., Martin, R., Moon, M. & Dallman, M. (1982). Rectovaginal fistula with atresia ani in three kittens. *Cornell Vet*, Vol. 82, No. 2, (April 1992), pp. 141-153

Tomsa, K. , Major, A. & Galus , T. (2011). Treatment of atresia ani type I by balloon dilatation in 5 kittens and one puppy. *Schweizer Archiv fur Tierheilkunde*, Vol. 153, No. 6, (June 2011), pp. 277 -280

Tsioli, V., Papazoglou, L., Anagnostou, T., Kouti, V. & Papadopoulou, P. (2009). Use of a temporary incontinent end-on colostomy in a cat for the management of rectocutaneous fistulas associated with atresia ani. *Journal of Feline Medicine and Surgery*, Vol. 11, No. 12, (Dec 2009), pp. 1011-1014

Van den Broek, A., Else, R. & Hunter, M. (1988). Atresia ani and urethrorectal fistula in a kitten. *Journal of Small Animal Practice*, Vol. 29, No. 2, (February 1988), pp. 91-94

Vianna, M. & Tobias, K. (2005). Atresia ani in the dog: a retrospective study. *Journal of the American Animal Hospital Association*, Vol. 41, No. 5. (September-October 2005), pp. 317-322

Webb, G., McCord, K. & Twedt, D. (2007). Rectal stricture in 19 dogs: 1997-2005. *Journal of the American Animal Hospital Association*, Vol. 43, No. 6, (November-December 2007), pp. 332-336

12

Prospective Study of Tumor Markers as Prognostic Factors in the Histopathological Differential Diagnosis of Mammary Gland Neoplasms in Female Canines

Anna M. Badowska-Kozakiewicz
*Department of Biophysics and Human Physiology,
Medical University of Warsaw,
Poland*

1. Introduction

Cancer constitutes a major problem in animal pathology and is a subject of intensive research. The aim of this study was to gain comprehensive knowledge of cancer biology. Some animal tumors are also considered good research models in comparative oncology. Literature and own studies show that dogs often suffer from skin cancer and mammary gland tumors (Moulton, 1990). The incidence of mammary gland tumors in female canines is second only to skin cancer (Misdorp & Meuten, 2002; Ramalho et al, 2006). They most commonly appear between 6 and 10 years of age (Misdorp & Meuten, 2002; Hampe & Misdorp, 1974; Kubiak, 200). There are also documented cases of such tumors as early as at the age of two years and, very rarely, in males (Moulton, 1990). Histopathological examination in about 40-50% of cases shows malignant tumors, most of which are of epithelial origin (Rutteman, 2001). These cancers originate from epithelial follicles or ducts and take forms of papillary adenocarcinomas, simple or complex tubular and solid cancers (Misdorp & Meuten, 2002). There are also tumors derived from myoepithelial cells and mesenchymal tissue. In veterinary medicine there is a constant search for prognostic factors and predictors that allow for proper evaluation of the disease state, survival, susceptibility to treatment and risk of relapse. Recognized factors are: age of the animal, sentinel lymph node status, histological type of tumor and a histological grade of differentiation of malignancy. Beside the fundamental method in cancer diagnosis, which is the histopathological examination, additional information useful in this regard is provided by research on cancer biology, including, among others, the study of tumor markers. In histological examination of a tumor, markers of proliferative activity such as mitotic index, thymidine labeling index, percentage of cells in the S phase of cell cycle, expression of nuclear antigen Ki-67 and proliferating cells nuclear antigen (PCNA) are also taken into account. The following factors are also examined: expression of estrogen receptors, growth factor receptors such as: epidermal growth factor receptor (EGFR), insulin growth factor receptor (IGFR), growth hormone (GH),markers of high risk of metastasis: plasminogen activator and cathepsin D, oncogenes and tumor suppressor genes: c -erbB-2, c-myc, p53,

BRCA-1, BRCA-2, markers of tumor angiogenesis, and heat shock proteins (Koda et al, 2004; Niwińska, 1995; Olszewski, 1994). Diagnosis of neoplasms of the mammary gland in canines does not usually pose a problem, but in some cases there may be difficulties. Targeted diagnostic immunohistochemistry using a panel of antibodies makes it easier to determine the correct diagnosis. Difficulties encountered during routine diagnosis of neoplasms of the mammary gland in female dogs relate to assessing the origin of the tumor and its biology. Immunophenotyping can be helpful in the diagnostic process and tumor markers (expression of nuclear antigen Ki-67 and expression of estrogen receptors and the p53 protein) carry useful information, but determination of malignancy is still based on morphological criteria, such as mitotic index, which together with the histopathological criteria are the most important factors differentiating benign tumors from those of malignant nature;. In veterinary oncology there is a constant search for new tumor markers that could aid in differential diagnosis of mammary gland neoplasms in female canines. The aim of research on the mammary gland neoplasms in female dogs is to extend the panel of tumor markers by addition of new markers such as COX-2, P-gp, Hsp90 and Hsp70 and to incorporate them into routine diagnostics. Occurrence of mammary tumors in females as well as potential to use canine models in comparative pathology of these tumors makes it an interesting research material. It is known that the degree of malignancy is positively correlated with proliferative activity. Relationship between the expression of estrogen receptors and the degree of malignancy is not entirely clear, although most authors believe that expression of these receptors is greater in benign tumors. COX-2 is expressed in many canine tumors such as canine kidney, bladder, prostate, and mammary gland cancers, but there is no data on the relationship between expression of COX-2 and other markers, even though such relation would seem logical because the expression of COX-2 may be related to prognosis by indirect immunosuppressive effect and inhibition of apoptosis. Literature on the expression of heat shock protein or glycoprotein - P in breast cancer is very scarce. There is evidence of Hsp27, Hsp70 and Hsp90, as well as glycoprotein – P expression, which may influence susceptibility to therapy. One would expect a correlation between proliferative activity, degree of malignancy and expression of estrogen receptors, but no data is available on such relationship. The aim of the study was to analyze the role of classical and new tumor markers in the histopathological differential diagnosis of mammary gland neoplasms in female dogs. The objective of the present study was to investigate the expression of cyclooxygenase - 2 and heat shock proteins in canine mammary gland tumors and their correlations with histological type of tumor, degree of its histological malignancy, proliferative activity, estrogen receptor expression, infiltration by cells, as well as P-gp and p53 protein expression.

2. Materials, methods and results

2.1 Materials

Material for the investigation comprised 133 tumor samples of the mammary gland collected from female canines during surgical procedures performed at Warsaw Veterinary Clinics and Small Animal Clinic of the Department of Clinical Sciences, Faculty of Vetrinary Medicine, Warsaw University of Life Sciences – SGGW. Tumor samples were fixed in 8% formalin buffered with phosphates. After 24h fixing, material was dehydrated in a series of increasing concentrations of alcohol and embedded in paraffin. Paraffin blocks were cut into serial sections of 4μm in thickness.

2.2 Methods

Then, they were stained using proper methods. In sections stained with the routine H&E method, the following determinations were carried out: type of neoplasm (Misdorp et al, 1999), tumor grade including tubule formation, intensity of division and the degree of neoplastic cell differentiation (Misdorp & Meuten, 2002), mitotic index defined as a mean number of mitoses in neoplastic cells counted in 10 fields of vision at the lens magnification of 400x (surface field $0.17 mm^2$). Paraffin sections on slides covered with 2% saline solution in acetone, at the temperature of 42^0C were used for immunohistochemical methods. In immunohistochemical reactions, the following antibodies, properly diluted in 1% BSA (Sigma), were used: mouse monoclonal antibody against human nuclear antigen Ki-67 (Dako) diluted 1:75 (Nieto et al, 2000), mouse monoclonal antibodies against p53 (Dako) human protein, diluted at a ratio 1:25 (Gamblin et al, 1994; Rodo, 2007), mouse monoclonal antibodies against alpha (Dako) human estrogen receptor, diluted at a ratio 1:35 (Mulas et al, 2005), mouse monoclonal antibodies against COX-2 (Dako) human receptor, diluted at a ratio 1:100 (Queiroga et al, 2007; Doré et al, 2003; Soslow et al, 2000), mouse monoclonal antibodies against Hsp70 (Novocastra) human heat shock proteins, diluted at a ratio 1:40 (Romanucci et al, 2006), mouse monoclonal antibodies against Hsp90 (Novocastra) human heat shock proteins, diluted at a ratio 1:40 (Romanucci et al, 2006) and mouse monoclonal antibodies against P-glycoprotein (Sigma), diluted at a ratio 1:100 (Petterino et al, 2006). Sections were deparafinized in xylene and rehydrated in increasing concentrations of alcohol. Then, they were placed in a buffer with a pH of 6 (Dako). In order to uncover the antigen epitope, sections were warmed up in a microwave oven (1x5 min at 600 W, 2x5 min at 300 W). Then, they were cooled for 20 min. After two washings in distilled water (5 min), slides were washed in TRIS-buffered saline with pH of 8 (Sigma) and incubated with the primary antibody for 1h at room temperature. Then, the preparations were washed with TRIS buffer for 10 min. The En Vision + ™ System (Dako) was used for visualization. After 30 min of incubation with reagent, the slides were washed with TRIS buffer and a solution of diaminobenzidine (DAB, Dako), prepared according to the procedure supplied by the producer, was put on slides. The degree of slide staining was controlled. They were washed in tap water and stained with Ehrlich hematoxylin for 5 min, contrasted in 1% acid alcohol and washed again in tap water. Then, slides were dehydrated in graded alcohol concentrations, passed through xylene and fixed in DPX mounting medium (Gurr®). Computer image analysis and Lucia v. 4.21 software were used for interpretation of the results of Ki-67, ER, p53, COX-2, Hsp70, Hsp90 expression; using those facilities we could count the number of neoplastic cells featuring stained cytoplasm per 1,000 neoplastic cells. Results were analyzed using the SPSS 12.0 program. To determine whether differences for a few independent traits were significant, Kruskal-Wallis test was used. This test is an equivalent to the test of variance for traits without normal distribution. Two-sided correlations were performed using Spearman correlation test. The differences were deemed statistically significant at $P \leq 0.05$.

2.3 Results

Investigated material contained 14 adenomas (Fig.1), 66 complex carcinomas (adenocarcinomas), 47 simple carcinomas (adenocarcinomas) (Fig. 2, Fig. 3) and 6 solid carcinomas. The number of cancers with a defined grade amounted to, respectively, 1st grade - 48, 2nd grade - 39 and 3rd grade - 32. Mammary gland neoplasms were excised from female dogs belonging to 18 breeds in the ages between 3 and 16 years.

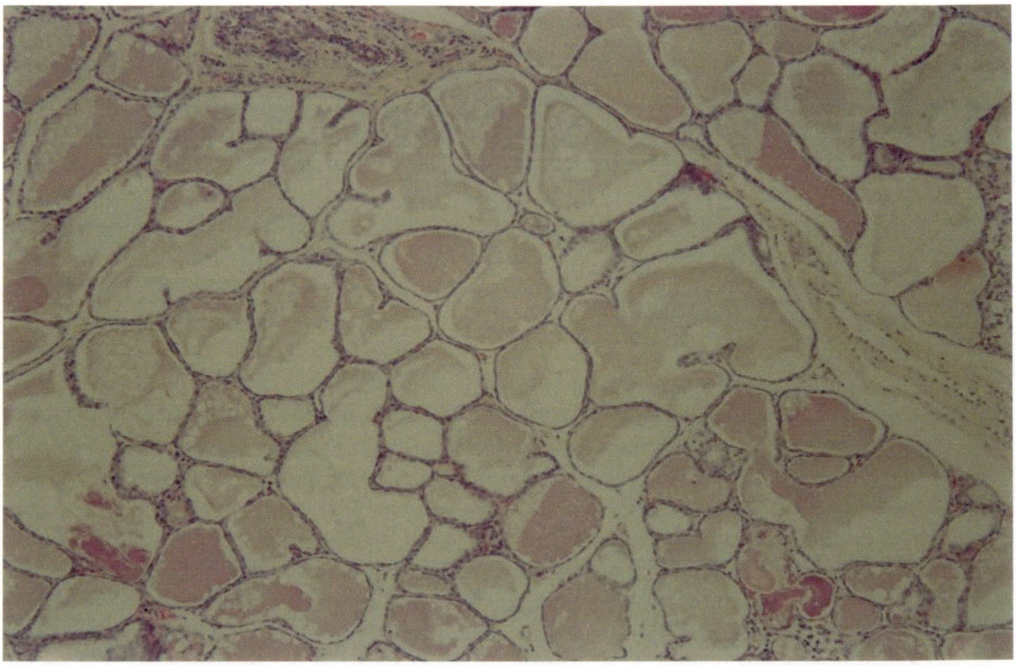

Fig. 1. Adenoma, H&E method x 20

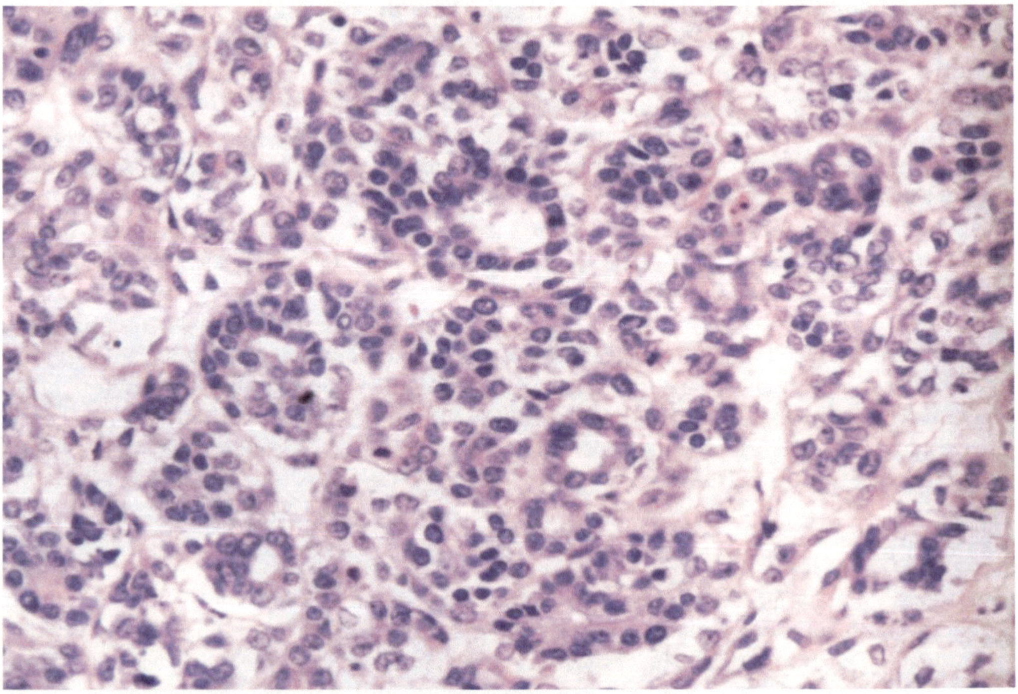

Fig. 2. Adenocarcinoma simplex, H&E method x 20

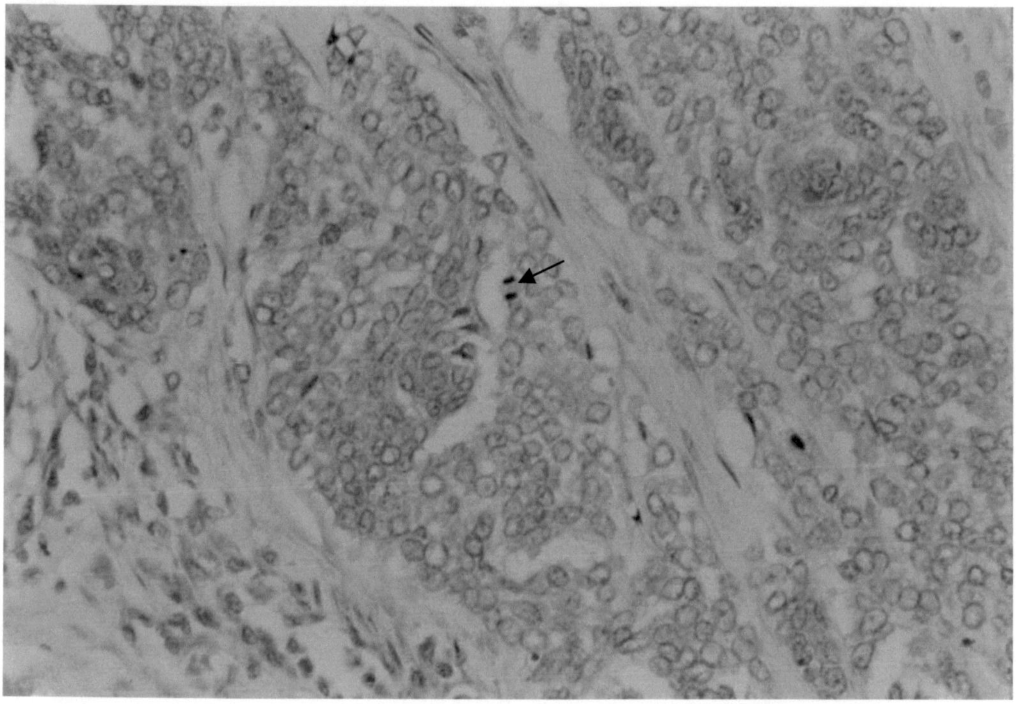

Fig. 3. Adenocarcinoma simplex, H&E (40x). The following mature tissues are visible: pathologic mitotic figures

2.3.1 Relationship between age of bitches and the grade of malignancy and histological type of tumor of epithelial origin

Dogs were divided into three age groups: <8 years, 8 -12 years and >12 years. In the group of bitches below the age of 8, majority (61.1%) consisted of tumors with the lowest histological grade of malignancy (1st). In the oldest group, 1st and 2nd grade tumors in the 1st and 2nd accounted for 77.8%. In the entire pool of studied tumors in all age groups, the largest share consisted of tumors with the lowest degree of histological malignancy (40.4%) (Table 1). Assessment of the contribution of individual types of tumors at different ages showed that in bitches younger than 8 years the most common findings were adenomas (21.7%) and complex carcinomas (56.5%) and in those over 12 years simple carcinomas occurred most often (55.0%) (Table 2).

Age of bitches	Tumor grade			Total
	I°	II°	III°	
<8 lat (n=18)	11 (61,1%)	3 (16,7%)	4 (22,2%)	18
8-12 lat (n=83)	30 (36,0%)	29 (35,0%)	24 (29,0%)	83
>12 (n=18)	7 (38,9%)	7 (38,9%)	4 (22,2%)	18
total (n=119)	48 (40,4%)	39 (32,8%)	32 (26,8%)	119

Table 1. Incidence of malignancies of various grades in bitches in different age groups

Age of bitches	Types of tumors				Total
	Adenoma	Carcinoma solidum	Adenocarcinoma simplex	Adenocarcinoma complex	
<8 lat	5 (21,7%)	1 (4,3%)	4 (17,4%)	13 (56,5%)	23
8-12 lat	7 (7,8%)	4 (4,4%)	32 (35,6%)	47 (52,2%)	90
>12 lat	2 (10,0%)	1 (5,0%)	11 (55,0%)	6 (30,0%)	20
Total	14 (10,5%)	6 (4,5%)	47 (35,3%)	66 (49,6%)	133

Table 2. Occurrence of individual types of epithelial neoplasms in different age groups in bitches

2.3.2 Results of proliferative activity

The value of mitotic index differed significantly between types of tumors. The lowest proliferative activity was observed in adenomas, the highest in simple and solid carcinomas. The highest proliferative activity was found in tumors in the 3rd grade of malignancy, the lowest in the tumors in the 1st grade (Fig.4). Expression of Ki-67 protein was observed in the nuclei of neoplastic cells that have undergone division (Fig.5). Statistical analysis shows significant differences between particular types of tumors. The lowest number of cells exhibiting Ki-67 protein expression was observed in adenomas, the highest in the solid and simple carcinomas and in tumors with the highest histological grade of malignancy (3rd).

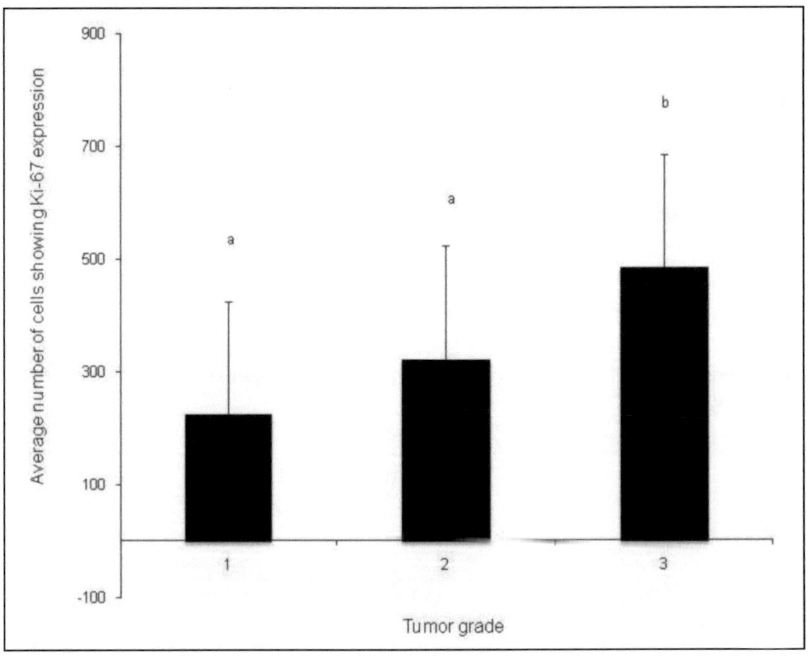

Fig. 4. Average number of cells showing Ki-67 expression depending on the tumor grade. Letters (a, b, c) above the columns show that the differences between mean values were statistically significant (P≤0.05).

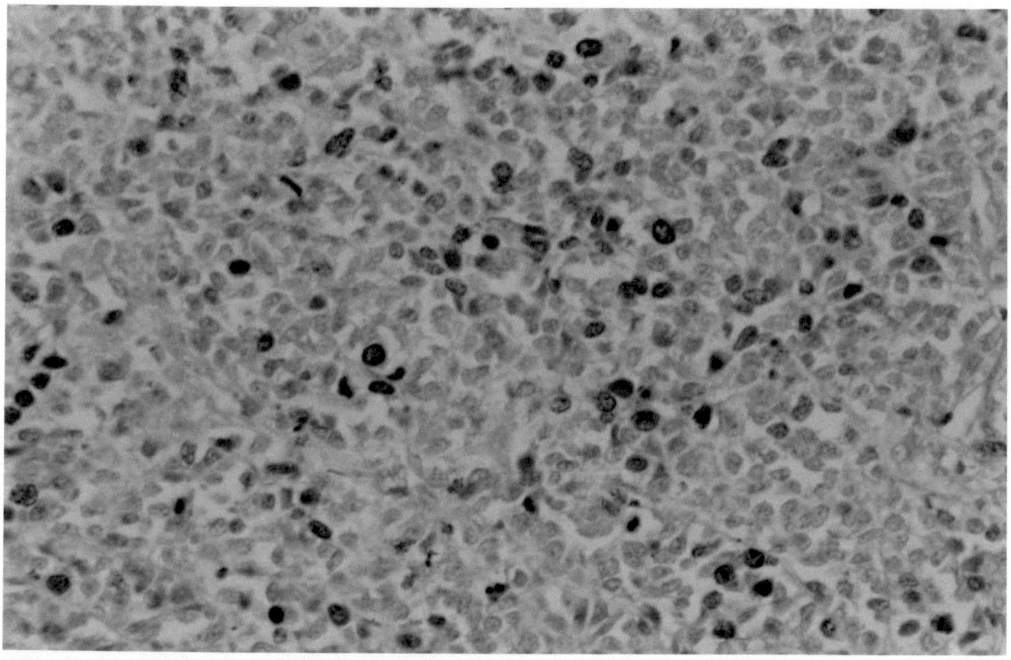

Fig. 5. Adenocarcinoma simplex, expression of Ki-67, immunohistochemical method. (40x)

2.3.3 Results of cellular infiltration in tumors of epithelial origin

In most cases, infiltrates were located predominantly in the stroma at the periphery of the tumor, and, in rare cases, occurred in the glandular alveoli. Inflammatory infiltrates consisted of lymphocytes, neutrophils and macrophages. The intensity of cellular infiltration was assessed on a four-level scale:: -/+; +; ++; +++. Statistical analysis indicates no correlation between the age of the dog and the intensity of cellular infiltration in the tumor. It is worth noting that, in bitches between the ages of 8 to 12 years, most tumors exhibited the intensity of infiltration at the first (+) level , which constituted 50% of tumors in this age group. A significant relationship was found between the intensity of cellular infiltration and the type of tumor. In 50% of solid cancers, there was no cellular infiltration. Only 33% of these tumors exhibited the presence of cellular infiltration at the first level (+), and 16.7% on the second level (++). Among the complex cancers, the largest group consisted of those, in which cellular infiltration was found at the first (+) and the second (++) level. The highest average intensity of cellular infiltration was found in complex carcinomas, followed by simple carcinomas, adenomas and solid carcinomas. Specific differences were seen between adenomas and complex carcinomas (P=0.007) and between solid carcinomas and complex carcinomas (P=0.005). While examining the relationship between intensity of cellular infiltration and histological grade of malignancy, no significant differences were revealed. But it may be noted that among tumors of all histological grades of malignancy, the largest group consisted of cancers with the first level of cellular infiltration. While analyzing the average intensity of cellular infiltration in tumors with various histological grades of malignancy no significant differences were found between groups with the exception of the tumors in the 3rd grade, where a significant difference was observed (P=0.023).

2.3.4 Results of p53 expression in neoplasms of mammary gland

Expression of p53 protein was observed in the nuclei of cancer cells (Fig.6). Positive reaction of p53 protein was observed in 70 (52.6%) of all tumors. Expression of p53 protein was found in 4 adenomas. The largest group of tumors positive for p53 protein were complex and simple cancers. Solid cancers belonged to a group, which rarely exhibited a positive reaction for nuclear p53. Analysis of the average number of cells with a positive reaction for p53 showed its highest expression in the case of complex cancers, as well as in 1st and 2nd grade tumors (Fig.7). The lowest level of expression of p53 protein was found in adenomas.

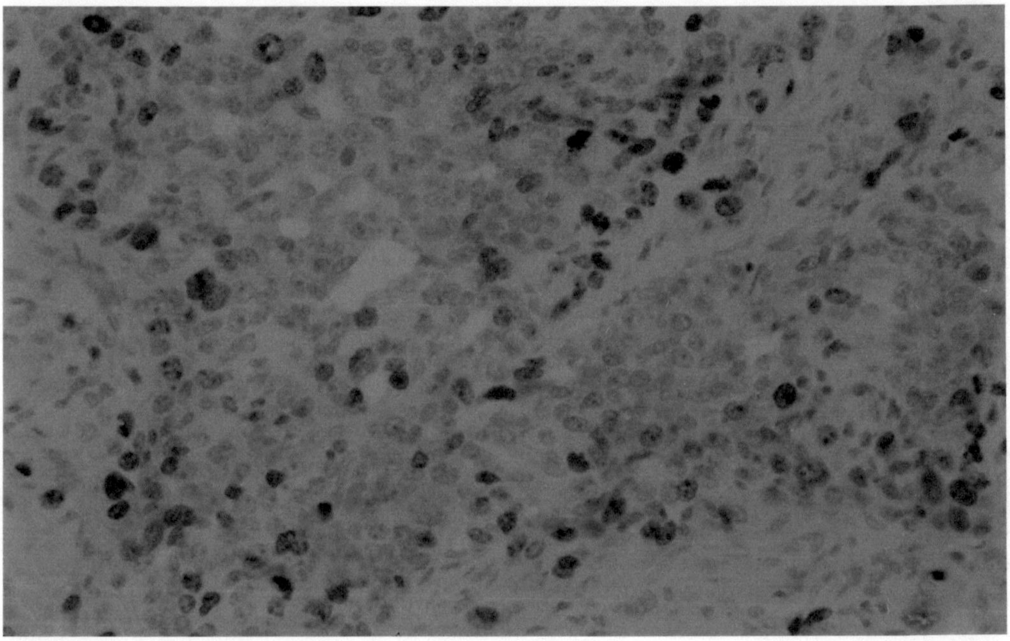

Fig. 6. Adenocarcinoma simplex, expression of p53, immunohistochemical method. (40x)

No statistically significant differences were shown between the examined groups. Comparing the ages of bitches, the highest average number of cells expressing p53 protein was found in tumors excised from younger dogs. However, there were no statistically significant differences among the age groups.

2.3.5 Results of estrogen receptor expression in neoplasms of mammary gland

Estrogen receptor expression was detected in the nuclei of tumor cells, but it was also seen in the cytoplasm. Cytoplasmic reaction was considered to be nonspecific. Among all tumors of epithelial origin, expression of estrogen receptors was found in 54 (40.6%) , and no reaction was noted in 79 (59.4%). Expression of estrogen receptors was most commonly found in complex cancers (43.9%) (Fig.8) , followed by simple cancers (42.6%) (Fig.9, Fig.10) and adenomas (28.6%), while solid cancers rarely expressed them (16.7%). The highest expression of estrogen receptors was found in simple carcinomas as well as in 3rd grade tumors, but no statistically significant differences were found between study groups (Fig. 11). Analysis of the average number of cells showing positive expression of estrogen receptors reveals that the level of expression increases with the histological grade of malignancy. It was also found that most tumors expressing estrogen receptors came from dogs younger than 8 years. A positive correlation was found between mitotic index and expression of estrogen receptors in specific types of cancer and statistically significant differences between tumor characteristics were demonstrated (P=0.042).

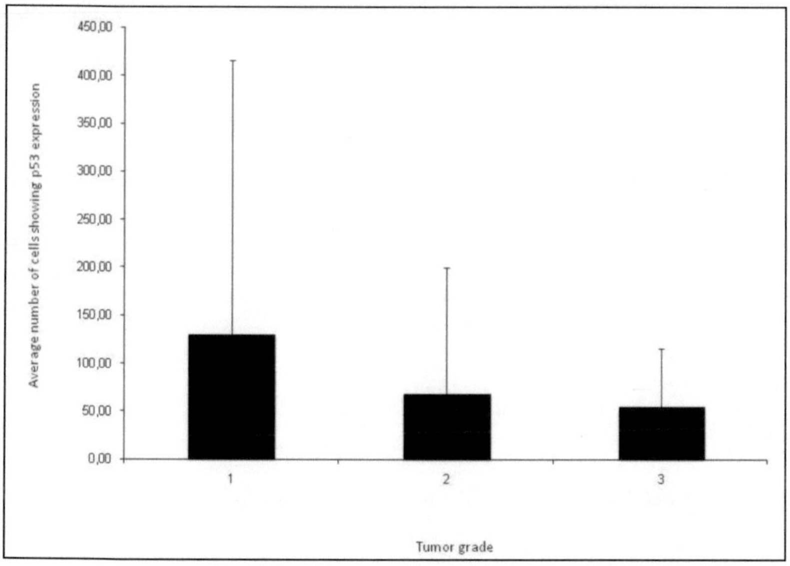

Fig. 7. Average number of cells showing p53 expression depending on the tumor grade.

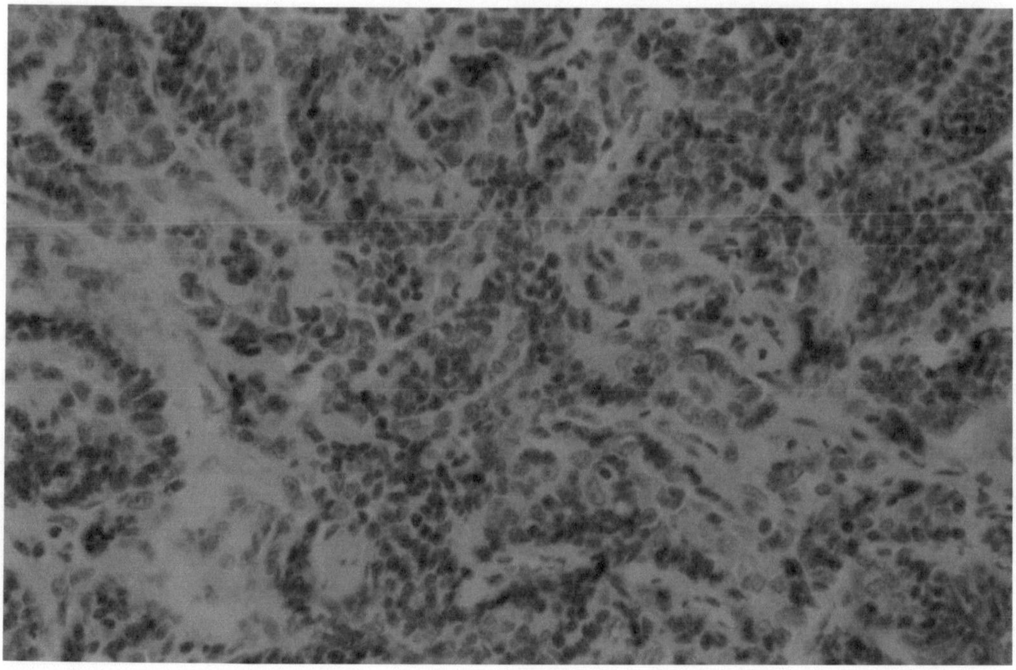

Fig. 8. Adenocarcinoma complex, expression of ER, immunohistochemical method. (40x)

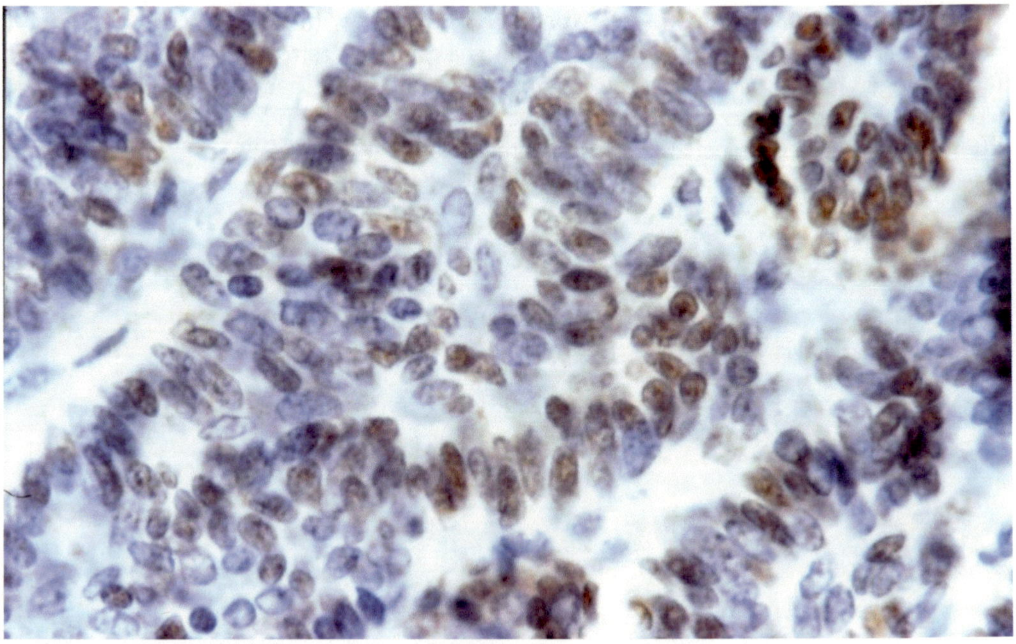

Fig. 9. Adenocarcinoma simplex, expression of ER, immunohistochemical method. (100x)

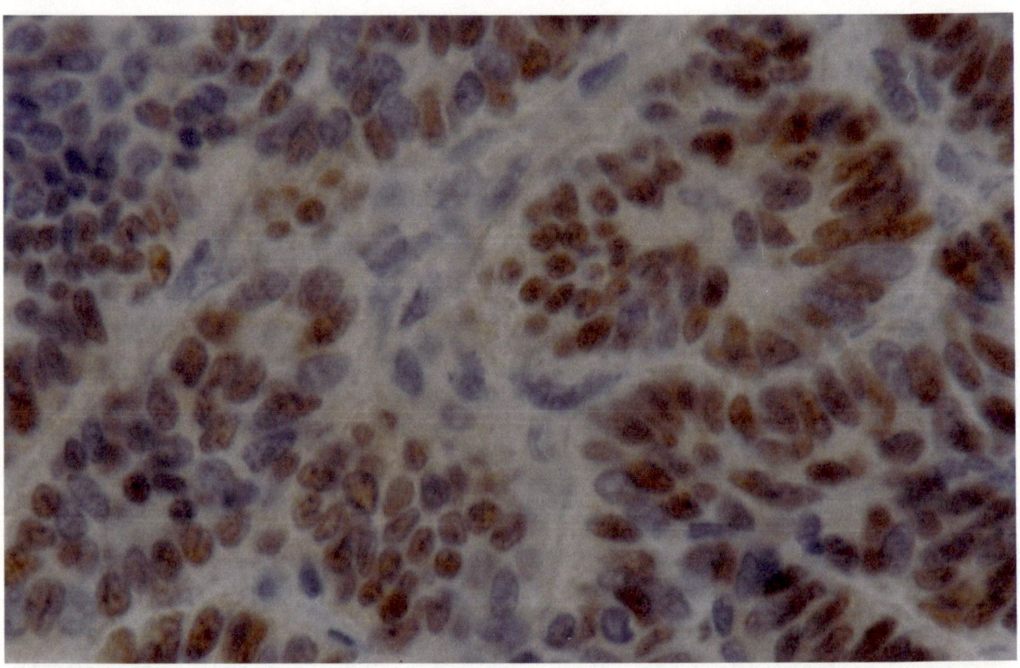

Fig. 10. Adenocarcinoma simplex, expression of ER, immunohistochemical method. (100x)

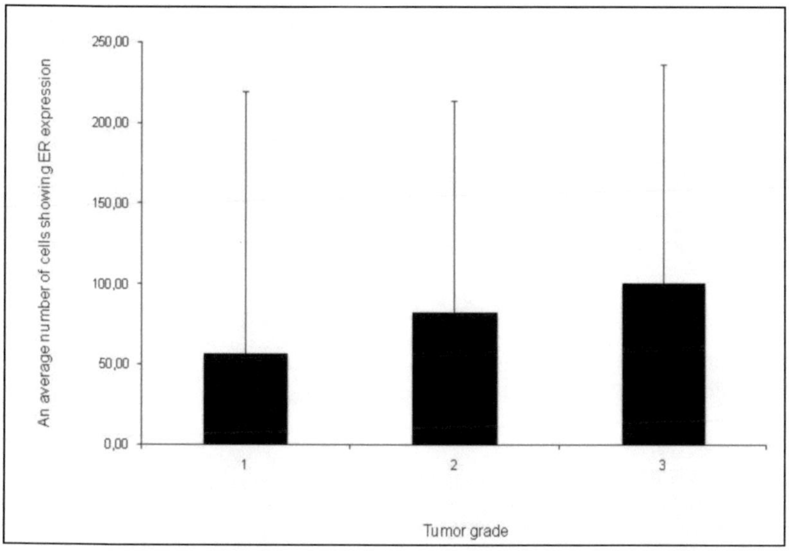

Fig. 11. An average number of cells showing ER expression depending on the tumor grade

2.3.6 Results of cyclooxygenase-2 expression in neoplasms of mammary gland

Expression of cyclooxygenase - 2 was observed in the neoplastic cell cytoplasm. It was confirmed in 13 of 14 adenomas. Similar results were obtained in solid cancers (5 of 6) (Fig. 12). Positive COX-2 reaction was found in 36.9% of simple carcinomas (Fig.13). Of all tumors, which exhibited a positive response (122), complex tumors demonstrated the highest percentage (48.4%). Statistical analysis, however, failed to demonstrate any significant differences regarding the COX-2 expression between the individual types of carcinoma (P = 0.978). According to the analysis of histological malignancy grades, the highest expression was found in 3rd grade cancers and statistical significance was confirmed for the individual histological malignancy grades (P = 0.047) (Fig. 14, Fig. 15). Furthermore, higher COX-2 expression level was demonstrated in carcinomas with higher mitotic index; statistical analysis demonstrated highly significant differences between neoplasms with high versus low mitotic index (P = 0.009) (Fig.16). The investigation into the relationship between expression of Ki-67 nuclear antigen and the COX-2 expression level did not reveal statistical significance (P = 0.614) (Fig. 17). Analysis of the relationship between the p53 protein expression and COX-2 expression level demonstrated a statistically significant correlation between these features (P = 0.034) (Fig. 18). An average number of cells expressing p53 protein was higher in carcinomas with higher COX-2 expression. In addition, experiments explored the relationship between Hsp70 protein and cyclooxygenase - 2 expression levels. The study proved a statistically high correlation between the investigated neoplastic characteristics (P = 0.006). The lowest Hsp70 protein expression was found in carcinomas exhibiting second degree COX-2 expression (Fig. 19). Analysis of the relationship between representative neoplastic features demonstrated statistically significant, positive correlations between COX-2 expression, mitotic index (P = 0.009) and p53 protein expression (P = 0.003); on the other hand, no correlation was found with regard to the Ki-67 nuclear antigen expression (P = 0.686). There was no correlation between the estrogen receptor expression and the above mentioned parameters. The highest expression of estrogen receptors was found in simple cancer types.

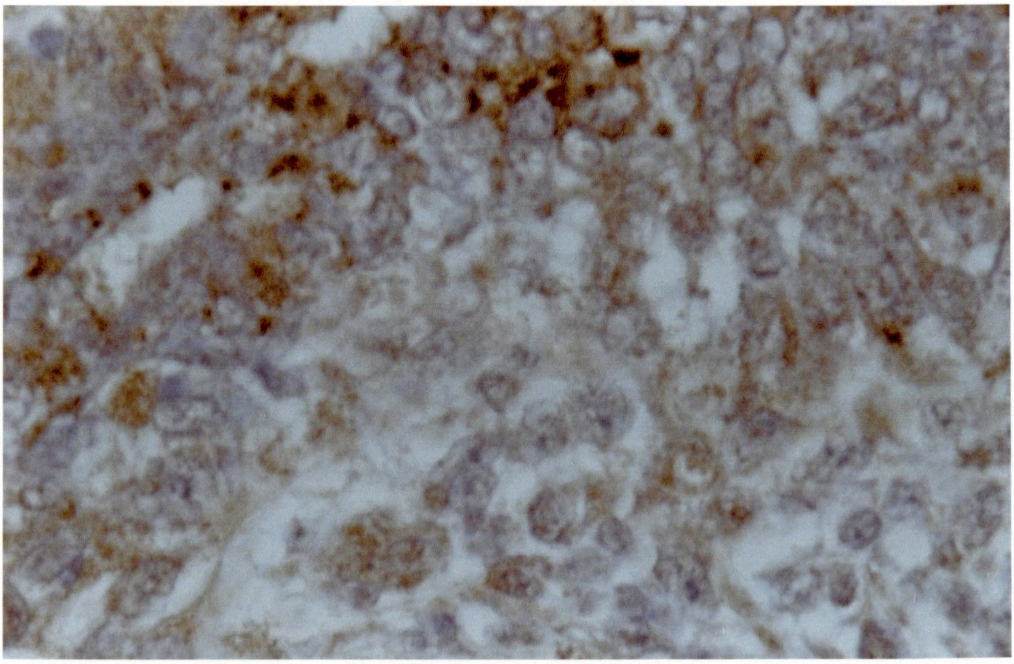

Fig. 12. Carcinoma solidum, expression of COX-2, immunohistochemical method. (100x)

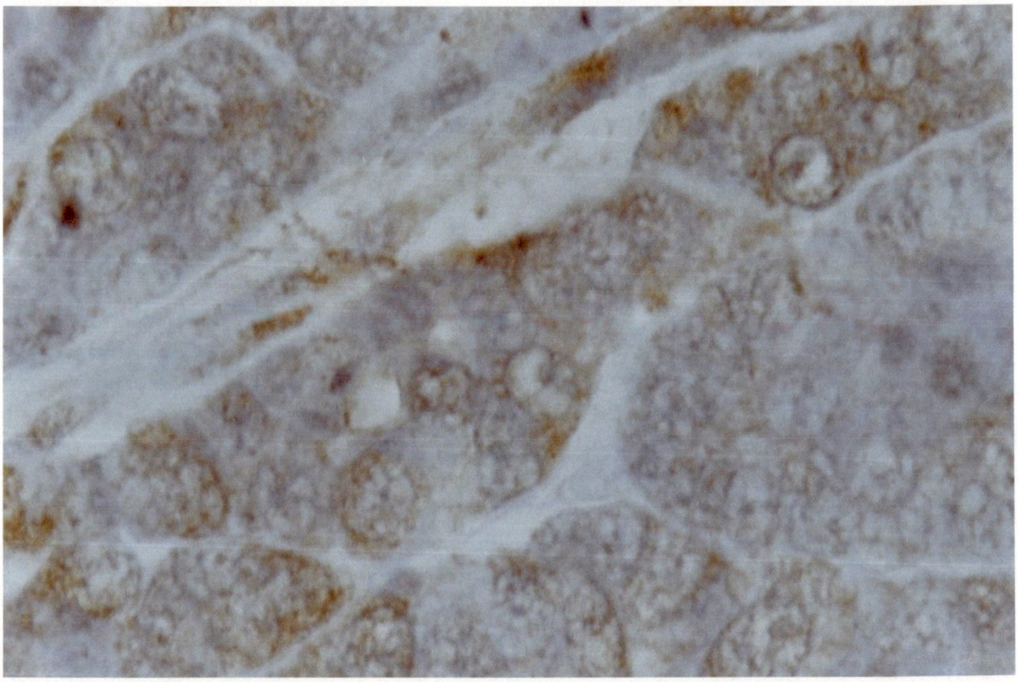

Fig. 13. Adenocarcinoma simplex, expression of COX-2, immunohistochemical method. (100x)

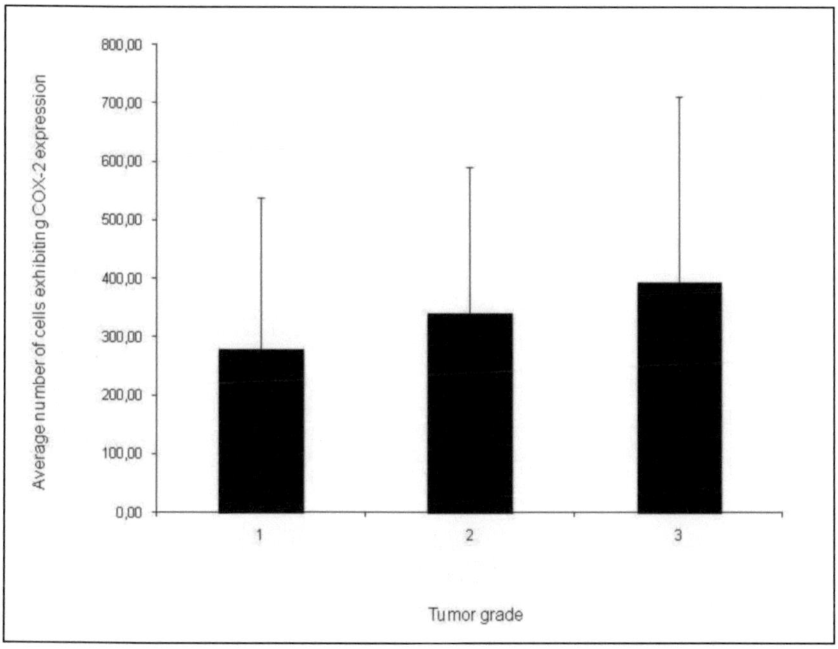

Fig. 14. Average number of cells exhibiting COX-2 expression depending on the tumor grade.

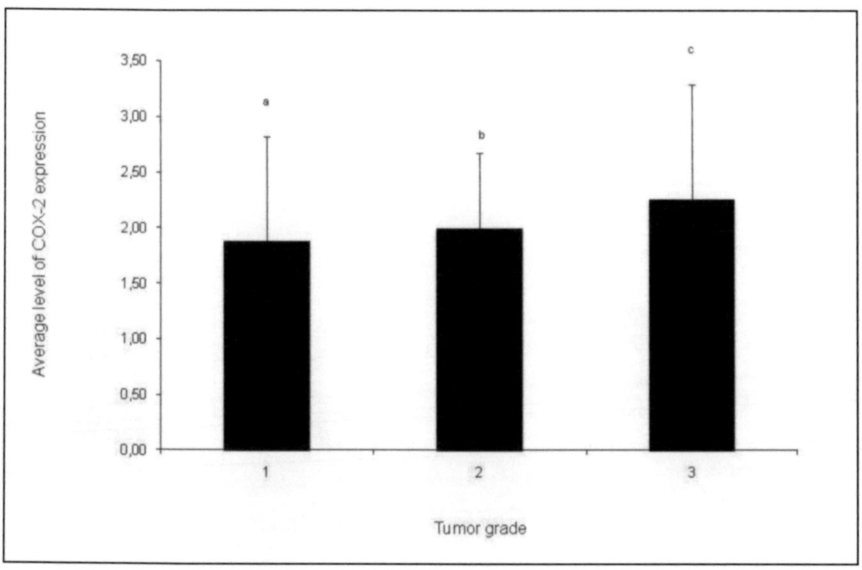

Fig. 15. Average level of COX-2 expression depending on tumor grade. Letters (a, b, c) above the columns show that the difference between means was statistically significant (P≤0,05).

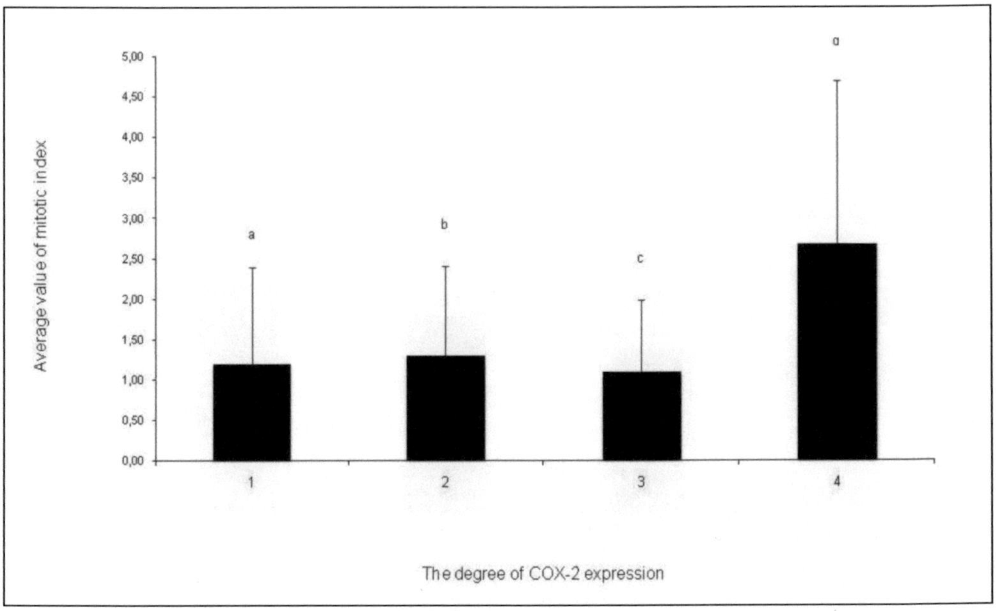

Fig. 16. Average value of mitotic index in tumors with different degrees of expression of cyclooxygenase - 2. Letters (a, b, c, d) above the columns show that the difference between means was statistically significant (P≤0.05).

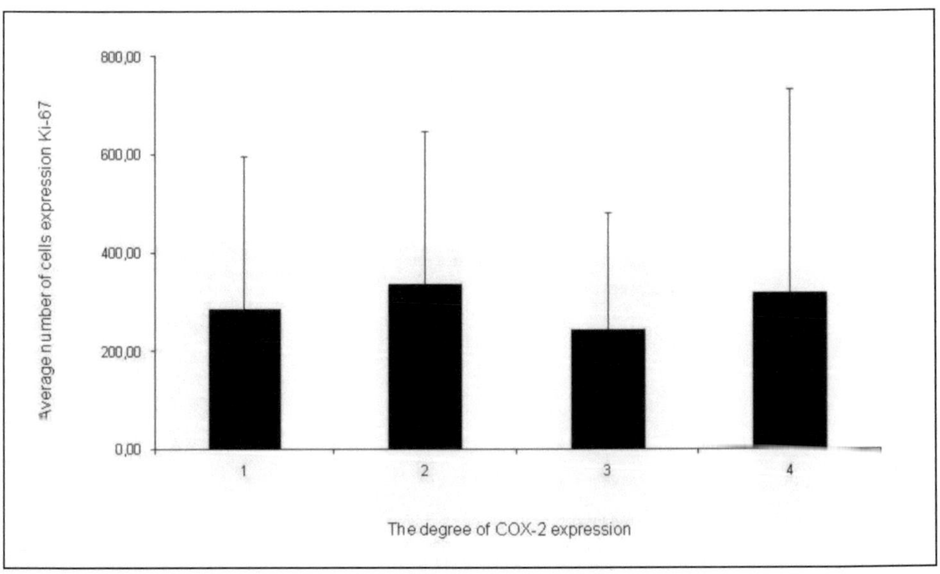

Fig. 17. Average number of cells expressing Ki-67, depending on the degree of expression of cyclooxygenase - 2.

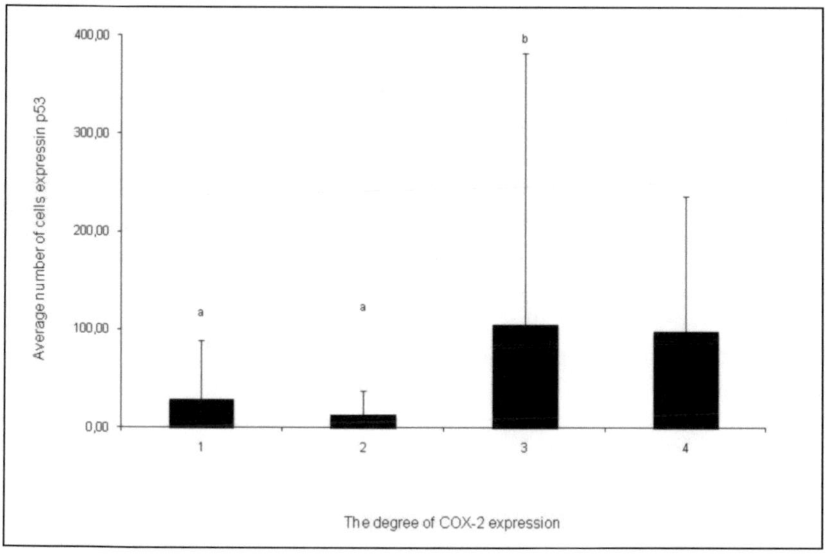

Fig. 18. Average number of cells expressing p53, depending on the degree of expression of cyclooxygenase - 2. Letters (a, b) above the columns show that the difference between means was statistically significant (P≤0.05).

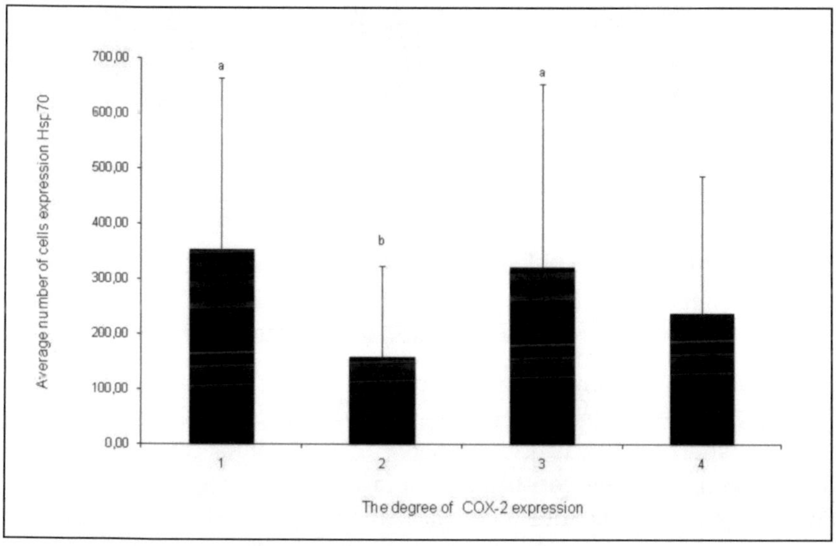

Fig. 19. Average number of cells expressing Hsp70, depending on the degree of expression of cyclooxygenase - 2. Letters (a, b) above the columns show that the difference between means was statistically significant (P≤0.05).

2.3.7 Results of P-glycoprotein expression in neoplasms of the mammary gland

Expression of P-glycoprotein was identified in cytoplasm and cell membranes of neoplastic cells. Positive reaction was found in 76% of all neoplasms. Complex carcinomas were the biggest group among cancer types, which demonstrated positive reaction to P-gp. In terms of histological malignancy grade, the most numerous were cancers featuring the lowest grade of malignancy (Fig. 20). In female canines aged 9 through 12 years, cancers exhibiting a positive P-gp reaction constituted the most numerous group (63.2%); on the other hand, this cancer type barely appeared in the oldest group (10.5%). Analysis of the average P - glycoprotein expression in different types of carcinomas did not reveal any significant differences. Correspondingly, evaluation of malignancy grade showed no statistical significance between examined features. Among different cancer types, the highest P-gp expression was demonstrated in solid carcinomas and cancers featuring the highest histological grade of malignancy. Positive correlation between investigated cancer features was found in the analysis of the relationship between cyclooxygenase-2 and P-gp expression, where P value was equal to 0.021.

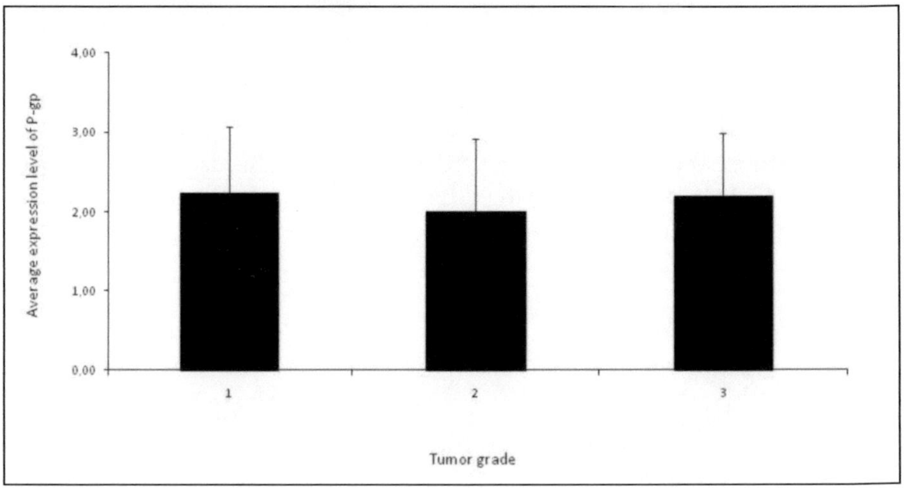

Fig. 20. Average expression level of P–gp depending on the tumor grade.

2.3.8 Results of heat shock protein expression in neoplasms of mammary gland

Heat shock proteins were found in the cytoplasm and nuclei of cancer cells (Fig. 21, Fig. 22, Fig. 23, Fig. 24). The largest group of tumors exhibiting Hsp70 and Hsp90 expression included simple and complex cancers, whereas solid tumors were the least numerous group. Grade 1 and 2 cancers constituted the largest group expressing both Hsp70 and Hsp90 (Fig. 25, Fig. 26) . Immunohistochemical analysis showed high expression of Hsp90 in simple and complex cancers, but no statistically significant differences were found between investigated types of tumors (P=0.443). High expression of Hsp90 was confirmed in solid cancers and, in this particular group, significant differences were found between types of tumor (P=0.032). As far as grading was concerned, no statistically significant differences were found between the mean number of cells exhibiting Hsp70 and Hsp90 protein expression and malignancy grade. When comparing the expression of Hsp70 to the expression of Hsp90 in particular types of cancers, we found a highly significant statistical difference (P=0.005) between the expression of both proteins. Results of a study on the expression of heat shock proteins (Hsp70 and Hsp90) were compared to nuclear antigen Ki-67 expression. It was found that the expression of nuclear antigen Ki-67 was most pronounced in solid tumors, as was the expression of Hsp90. Also, in neoplasms of the highest grade, expression of nuclear antigen Ki-67 and protein Hsp90 was the highest. High expression of Hsp70 was found in tumors with grade 1 and 3 of cyclooxygenase-2 expression, whereas the lowest Hsp70 expression was observed in tumors with grade 2 of cyclooxygenase-2 expression. Statistical significance was found between the investigated features of neoplasms (P=0.009). Increased expression of cyclooxygenase-2 was observed in tumors with a low mean number of cells showing positive immunohistochemical reaction for protein Hsp90. The highest level of expression of this protein was confirmed in tumors with grade 1 of cyclooxygenase-2 expression (Fig. 27). Between the mean number of cells showing positive Hsp90 reaction and cyclooxygenase-2 expression, there was a statistical significance observed for grade 1 and 2 of cyclooxygenase-2 expression (P=0.039). Statistical analysis proved a correlation between Hsp70 and p53 protein expression in tumors of epithelial origin. High statistical significance was shown for investigated neoplastic features (P = 0.002). Taking into account the type of tumor, expression of both proteins was highest in complex carcinomas and tumors with the lowest histological grade. High expression levels of Hsp90 protein and Ki-67 nuclear antigen was shown in cases of solid carcinomas and in carcinomas exhibiting 3rd histological grade of malignancy, described by a low apoptotic index.

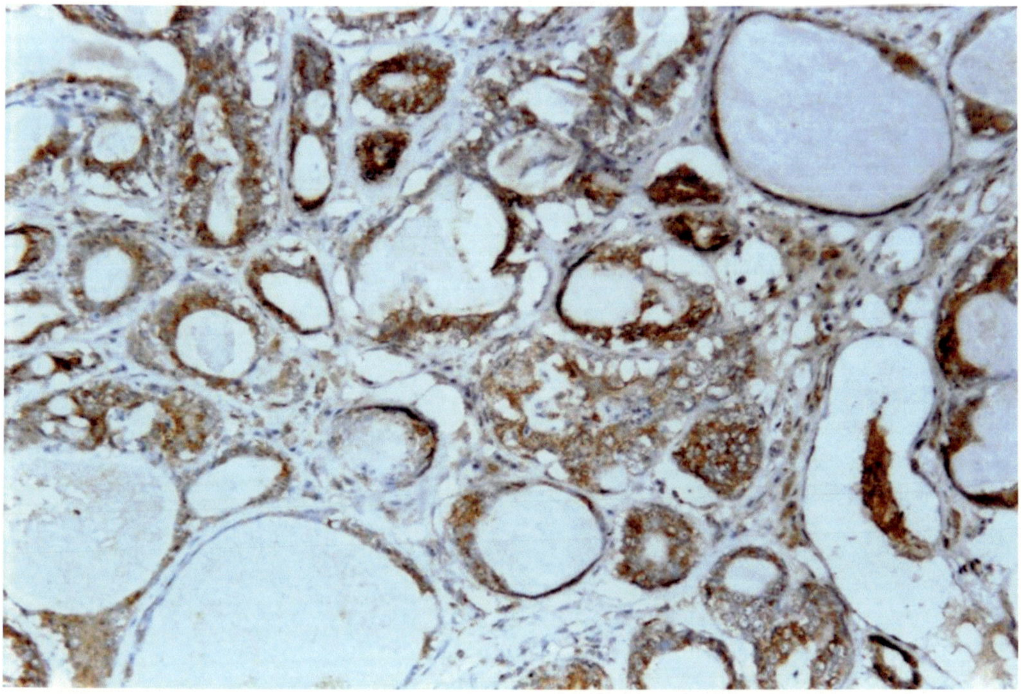

Fig. 21. Adenoma, expression of Hsp70, immunohistochemical method. (20x)

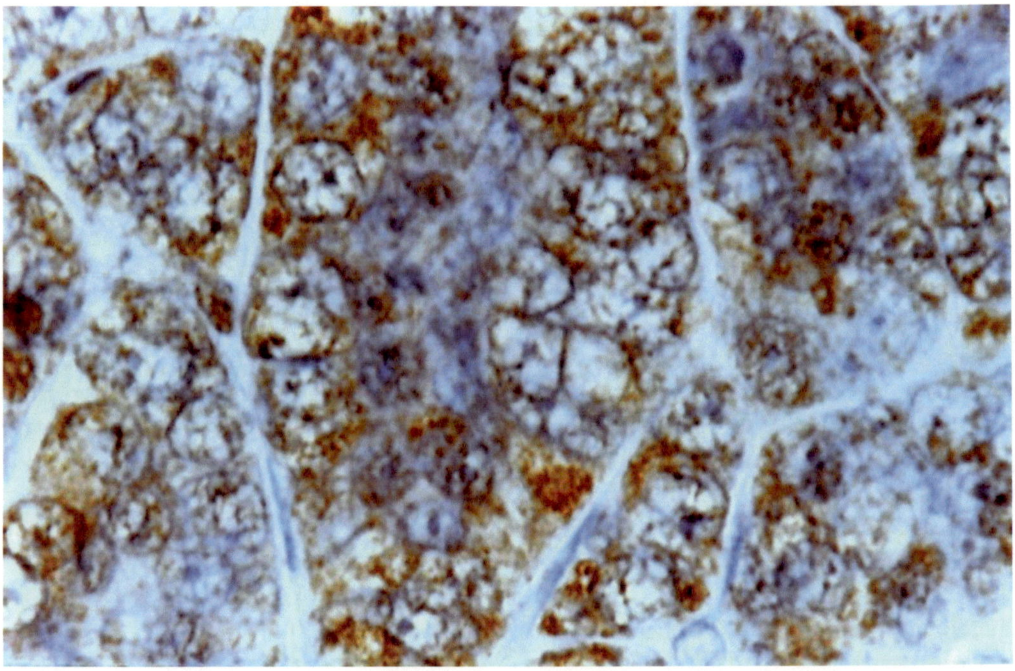

Fig. 22. Adenocarcinoma simplex, expression of Hsp70, immunohistochemical method. (100x)

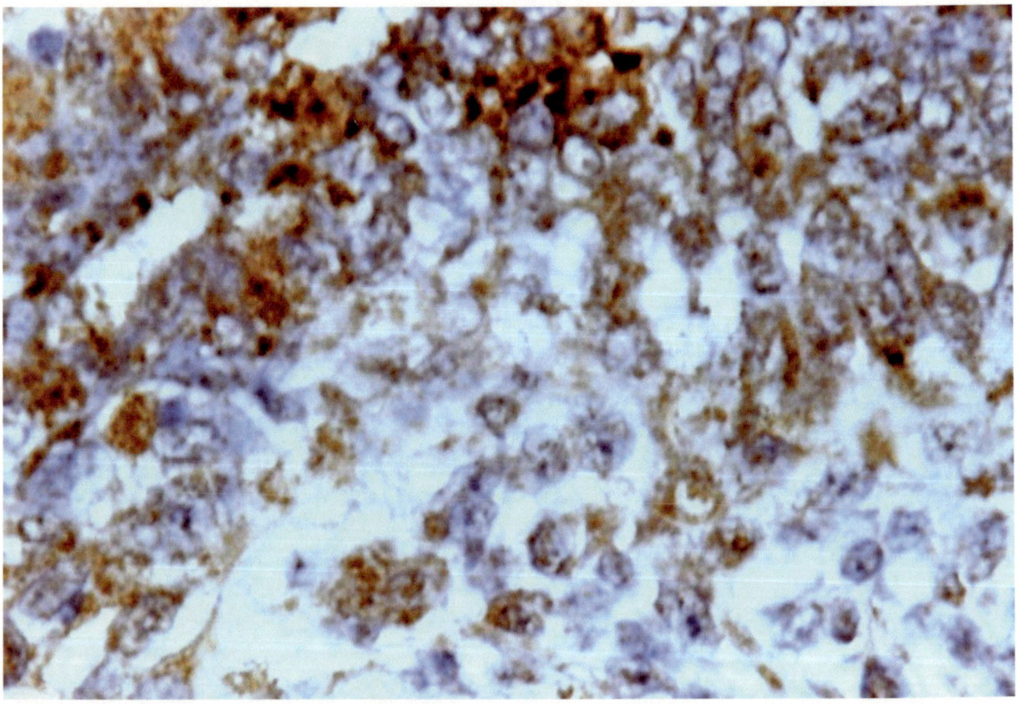

Fig. 23. Adenocarcinoma complex, expression of Hsp70, immunohistochemical method. (100x)

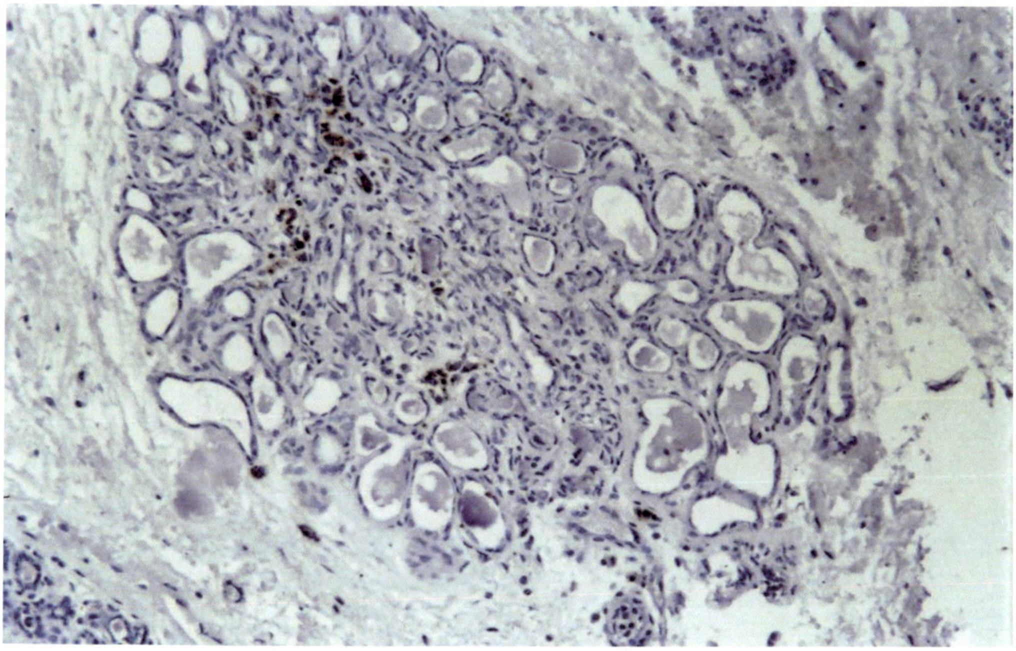

Fig. 24. Adenocarcinoma complex, expression of Hsp90, immunohistochemical method. (10x)

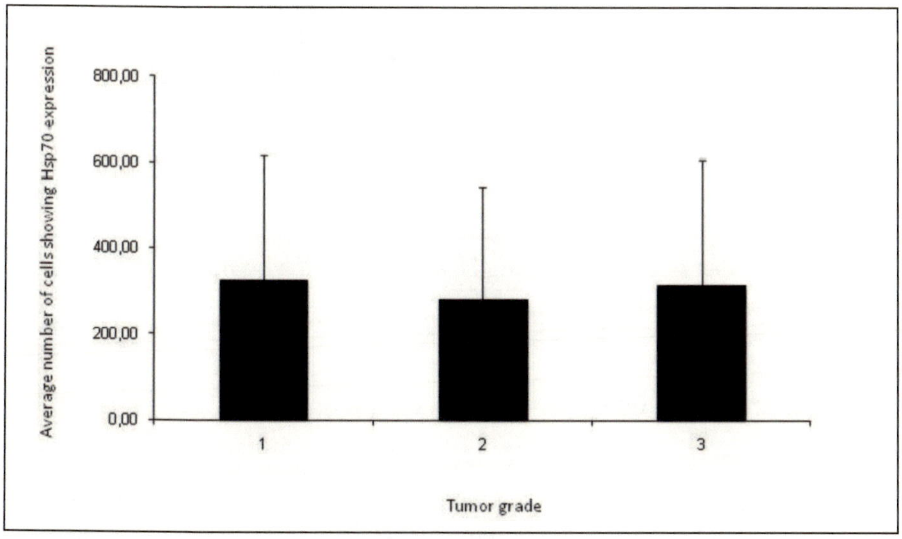

Fig. 25. Average number of cells showing Hsp70 expression depending on the tumor grade.

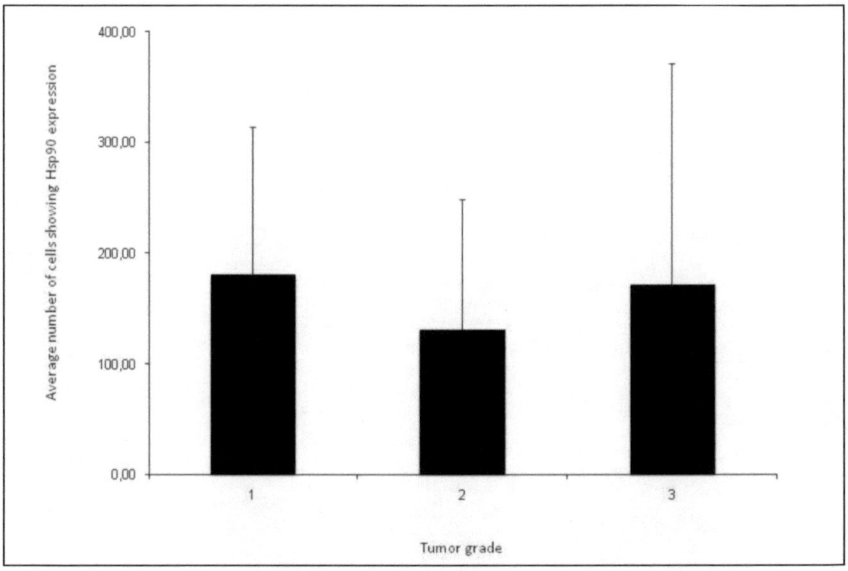

Fig. 26. Average number of cells showing Hsp90 expression depending on the tumor grade

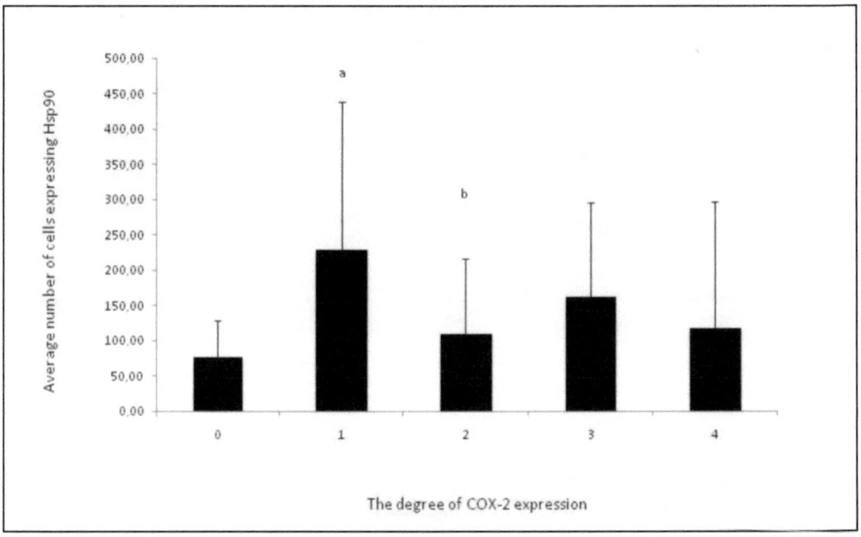

Fig. 27. Average number of cells expressing Hsp90, depending on the degree of cyclooxygenase - 2 expression. Letters (a, b) above the columns show that the difference between means was statistically significant (P≤0.05).

3. Discussion and conclusion

3.1 Discusion

In our study, cancers accounted for 89.8% of tumors of epithelial origin. This is consistent with the results of some authors, although the results of studies on the incidence and types of tumors of the mammary gland in canine females given by various authors differ.

Löhr (1997) found that 50% of the mammary gland tumors in female canines are malignant tumors (Löhr et al, 1997). Hellmén (1993) determined the incidence of malignant tumors to be 68% (Hellmén et al 1993). There are also studies showing, that about a half of tumors are benign (Bostock et al, 1992; Gilbertson et al, 1983). Moulton (1990) believes that benign tumors represent about 80% of cases and the majority of tumors studied are "benign mixed tumors" (65%) (Moulton, 1990). They are tumors in which, apart from epithelial and mesenchymal tissue, there is also cartilage and bone tissue. Their histopathological examination reveals no sign of malignancy. Nerurkar (1990) says that benign tumors represent about 27% of the mammary gland tumors in female dogs (Nerurkar et al, 1990). In our study, benign tumors of epithelial origin accounted for only 10.1% of all cases. The diversity of these results relates to the lack of uniform diagnostic criteria and lack of uniform classification of tumors of the mammary gland in dogs. In human medicine, age is a very important prognostic factor. Detection of breast cancer in women at a very young age and at the age over 60 is associated with worse prognosis (Host & Lund, 1986). In Philibert's (2003) studies there were no significant differences in survival of young and older dogs (Philibert et al, 2003). Hellmén (1993) presented completely different results (Hellmén et al, 1993). She found that age may be an important prognostic factor and showed that older bitches had a shorter survival time after surgery. Results obtained by Benjamin corresponded to that (1999) (Benjamin et al, 1999). Own results support the concept of age as an important prognostic factor, as there is very little information regarding survival or recurrence of malignancy in female dogs after surgery. Current study shows that the presence or absence of cellular inflammatory infiltrates may be a prognostic factor. Data on the incidence of cellular infiltration in mammary gland tumors in female dogs is also scarce. Gilbertson (1983) found that cell infiltration plays a role in the process of development of precancerous and invasive carcinomas (35%) (Gilbertson et al, 1983). In our study, the percentage of tumors with cellular infiltration was higher (88%). The highest intensity of cellular infiltration was observed in tumors with a high histological grade of malignancy. Skrzypczak (2004) (Skrzypczak, 2004) presented similar results in his study . Authors believe that presence of cellular infiltration plays a positive role in inhibiting tumor growth. Some studies have shown that presence of cellular infiltration is associated with good (Rilke et al, 1991) or poor (Parl & Dupont, 1982) prognosis and others, that it carries no prognostic value. In our study, no relationship was found between cellular infiltration and other tumor markers, consistent with the studies by Rodo (2007) (Rodo, 2007; Roses et al, 1982). It is worth noting the presence of necrosis in tumor foci, which may obscure the accuracy of the results. Necrosis in the tumor is the result of disparities between high proliferative activity and tumor vascularization. It may trigger cellular inflammatory reactions. We also analyzed the distribution of cellular infiltration. Our own results differ from the results of other authors, because infiltration was observed in the stroma, not scattered (Lee et al, 1996), and was more pronounced on the periphery of the tumor. The role of cellular infiltration within the tumor remains unclear and controversial. Estrogen receptors are recognized markers in the diagnosis of breast cancer in women. It is estimated that about 70-80% of breast cancers in women exhibit the expression of estrogen receptors. These tumors are characterized by slower growth, higher diversity, better prognosis with a suitable treatment regimen and correlate with the length of survival after surgical removal (Bacus et al, 1989; Barzanti et al, 2000). Study of the expression of estrogen receptors in mammary gland neoplasms in female canines was, as usual, inconclusive. Martin et. al (1984) studied 228 tumors of the mammary gland in bitches and showed the expression of estrogen receptors in only 2.1% of tumors (Martin et al, 1984). Pena et. al (1998), on the other hand, diagnosed 21 cancers with an ongoing inflammatory process and found no expression of estrogen receptors (Pena et al, 1998). According to Sartin et. al (1992) the greatest chance of long-term survival after surgery , is associated with tumors expressing ER alone or together with PgR. Indeed, in the absence of ER and PgR, researchers observed the shortest period of survival (Sartin et al, 1997). Millanta et. al (2005) study on 47 mammary gland neoplasms in bitches reported that ER and PgR expression did not correlate with survival or histological parameters of tumors (Millanta et al, 2005). Similarly, Sobczak-Filipiak and Malicka (1997) have shown no correlation between expression of ER and mitotic index (Sobczak-Filipiak & Malicka, 1997). In our study, the expression of estrogen receptors was demonstrated in 40% of the tested tumors. The highest expression of estrogen receptors was found in simple carcinomas and tumors with the highest histological grade of malignancy. There was a significant correlation between mitotic index and expression of estrogen receptors. Nowak (2007) found expression of estrogen receptors in only 6% of cancers (Nowak et al, 2007), but Mulas (2005) obtained different results- he found that estrogen receptors were expressed in benign tumors (Mulas et al, 2005). Similarly, McEwen et. al (1982) have demonstrated the expression of estrogen receptors in about 50% of cases, with a significantly higher levels of expression present in benign tumors (McEwen et al, 1982). Results obtained by Nieto et. al (2000) are different - they show a correlation between the expression of nuclear antigen Ki-67 and estrogen receptors (Nieto et al, 2000). Authors of these studies observed the highest levels of ER expression in simple and complex carcinomas, as we did in our study. Otherwise, Rutteman (1998), Sartin (1992), and Geraldes (2000), claim that high expression of ER is present in adenomas compared to adenocarcinomas, which exhibit a lower level of expression of this marker (Sartin et al, 1992; Rutteman et al, 2001; Geraldes et al, 2000). Opinions about the value of estrogen receptors as a prognostic factor are divided. Some authors consider ER expression a positive factor, but there are also voices postulating it is a negative prognostic marker. Interestingly, it seems that in our study, low expression of ER negatively correlated with high expression of nuclear antigen Ki-67. It is in agreement with the research done by Peña (1998), who led the study on the mammary gland neoplasms in canines (Peña et al, 1998). From our research we conclude, that the expression of estrogen receptors may be important in assessment of malignancy, but does not show significant correlation with other markers. An important marker of malignancy is the proliferative activity. Proper evaluation of proliferative activity of tumor cells is crucial for the evaluation of its biological activity and is used in determining the treatment of cancer. High mitotic index correlated with tumor size and presence of lymph node metastases (Niwińska, 1995; Mirecka et al, 1993). In our study, proliferative activity depended on both, the type of tumor and the degree of histological malignancy. The highest values of mitotic index were recorded in simple and solid carcinomas and in tumors with the highest histological grade of malignancy. Similar results were reported for the expression of Ki-67. The highest expression of Ki-67 was seen in solid, simple carcinomas and in 3rd grade tumors. Similar results were obtained by Nieto (2000), who confirmed that the expression of Ki-67 is high in tumors with a higher histological grade of malignancy (Nieto et al, 2000). Similar results were obtained by Peña (1998), Giziński (2003) and Szczubiał (2002) (Peña et al, 1998; Giziński et al, 2003; Szczubiał & Łopuszański, 2002). They studied the expression of nuclear antigen Ki-67 in mammary gland tumors in dogs and came to the conclusion that higher expression of Ki-67 is an important prognostic factor and is associated with a higher risk of metastasis, a shorter period to recurrence and a shorter overall survival period. The rate of tumor growth is influenced by factors related to inhibition of cell cycle and promotion of apoptosis of cancer cells., Under physiological conditions it is a function of, among others, p53 gene and its protein product. Mutation of the p53 gene plays an important role in the uncontrolled proliferation and resistance to apoptosis in cancer cells , leading to functional changes in proteins. In dogs, expression of the p53 gene was found in both benign (Muto et al, 2000) and malignant (Mayr et al, 1994; Veldhoen et al, 1999) tumors of mammary gland. Muto (2000) in his study showed no association between p53 expression and histological type of cancer, but claimed that this protein may play an important role in carcinogenesis and may be a negative prognostic factor (Muto et al, 2000). Chung-Ho (2004) came to a similar conclusion, arguing that the presence ofp53 in tumor cells is associated with their malignancy and bad prognosis in mammary gland tumors (Chun-Ho et al, 2004). In our study, p53 expression was observed in benign as well as malignant tumors. Among all tumors, more than 50% exhibited a positive reaction for 53 protein. It contradicts the data obtained in his study by Gamblin (1997), who carried out an analysis of 16 adenocarcinomas and found a positive reaction for p53 protein in only 12.5% of cases (Gamblin et al, 1994). In our study, the highest expression of p53 was observed in complex carcinomas and in 1st and 2nd grade malignant tumors. These results are consistent with the results obtained by Rungsipipata (1999), who found the highest levels of p53 expression in simple and complex carcinomas and also showed expression of this protein in adenomas (16%) (Rungspipata, 1999). Rodo (2007) obtained results on the expression of p53 , stating that there is a positive correlation between proliferative activity and the number of cells expressing p53 protein (Rodo, 2007). In our study, we found no statistically significant differences between the expression of p53 and nuclear antigen Ki-67. Obtained data in part explain why the activation of oncogenes does not always lead to uncontrolled proliferation when normal signal transduction leads to stabilization of p53 and activation of programmed cell death. There is a need for further research in this area, because the data in the literature is ambiguous and does not allow for a definite conclusion as to the importance of p53 protein and its role in mammary gland neoplasms. Varying results undermine the claim that a positive reaction for p53 protein would predict worse prognosis in cancer and that p53 mutation may be responsible for increased proliferation in tumors with advanced malignancy. Research on COX-2 expression in canine mammary cancers is scarce (Heller et al, 2005; Nowak et al, 2005; Doré et al, 2003) despite it being an attractive and motivating topic. Overexpression of COX-2 is known to occur in 56-100% of canine mammary carcinomas (Doré et al, 2003; Millanta et al, 2005; Queiroga et al, 2007; Heller et al, 2005; Mohammed et al, 2004), but there is a marked variation in the percentage of COX-2-positive tumor cells and the intensity of its expression. Doré et. al (2003) did not find any COX-2 expression in four samples of normal mammary tissue (Doré et al, 2003), whilst Mohammed et. al (2004) (Mohammed et al, 2004) and Queiroga et. al (2007) (Queiroga et al, 2007) reported expression of COX-2 in one of seven and two of four samples of normal mammary glands, respectively. Our own research demonstrated that as much as 91.7% of all carcinomas in the study displayed COX-2 expression. Doré et.al (2003) obtained similar results: he found a positive COX-2 reaction in 67% of complex cancers and in 47% of simple carcinomas (Doré et al, 2003). Our research demonstrated the highest COX-2 expression in simple cancers and the lowest in solid cancers. Similar data were presented by Heller et.al (2005), who found COX-2 expression in adenocarcinomas, whereas he failed to detect it in solid carcinomas (Heller et al, 2005). Ristimäki et. al (2002) demonstrated the relationship between COX-2 expression and certain clinical and pathological features of the tumor (Ristimäki et al, 2002). Her research into mammary gland tumours proved that COX-2 expression was positively correlated with tumour size as well as the strength of Ki-67 nuclear antigen staining and p53 protein expression. Results of our study are consistent with the results of works by Ristimäki et. al (2002) : increased COX-2 expression in carcinomas is related to higher mitotic index, i.e. with the proliferating activity (Ristimäki et al, 2002). Expression of Ki–67 nuclear antigen was demonstrated in carcinomas exhibiting increased COX-2 expression, yet research failed to prove the correlation between these markers; On the other hand, it proved the existence of statistical correlation between COX-2 expression and expression of p53 and Hsp70 proteins. Average number of cells exhibiting p53 protein expression was higher in carcinomas with higher levels of COX-2 expression. High expression level of Hsp70 protein and COX-2 was shown in carcinomas with the 3rd histological grade of malignancy, as described by low apoptotic index. Similar results were obtained by Lanza-Jacoby et. al (2004), who conducted her studies on experimental animals (Lanza-Jacoby et al, 2006), and by Liu and Rose (1996) who studied the COX-2 expression in cell cultures (Liu & Rose, 1996). Furthermore, Ristimäki et al (2002) and Dempke et al (2001) came to a conclusion that COX-2 promoted the mammary gland neoplasm growth, invasive capacity, and probability of metastasis (Ristimäki et al, 2002; Dempke et al, 2001). COX-2 expression may be relevant to a number of physiological processes within this tissue including proliferative activity, inhibition of apoptosis, increased angiogenesis and activation of matrix metalloproteinases (Dempke et al, 2001). Mammary physiology is a complex process regulated by hormones, estrogens, progesterone, growth hormone, prolactin, and epidermal growth factor (Howlin et al, 2006). As COX-2 expression is induced by different stimuli including cytokines, oncogenes, hormones and growth factors (Thomas et al, 2008), the same factors that control mammary growth and differentiation may act as trigger stimuli for COX-2 expression. The results of our work are consistent with the results obtained by other researchers. P-gp expression was discovered in epithelium and neoplasms of mesenchymal origin in dogs (Ginn, 1996). Author of these studies argues that continuation of this line of research will deliver additional prognostic data. There is a small number of works attempting to evaluate the P-gp expression in canine mammary cancers. A considerable progress in the investigation into P-gp expression in canine mammary cancers was made by Petterino et. al (2006), who attempted to define the P-gp expression in mammary gland carcinomas in female canines (Petterino et al, 2006). His study covered cases of both malignant and benign neoplasms; Petterino et al (2006) found the expression of the examined marker in two test groups and confirmed the statistical significance between the two groups (Petterino et al, 2006). Furthermore, he found that simple and complex carcinomas were the most numerous groups among the examined cancers and that P-gp expression was absent from the tissues of healthy mammary glands. We achieved similar results in our research: we found the P-gp expression in 76% of the examined canine mammary cancers. The most numerous groups exhibiting a positive P-gp immunohistochemical reaction were composed of complex carcinomas (90.9%) and simple carcinomas (73%). High expression was found in cancers with the highest histological grade of malignancy. Due to fact that there are hardly any works investigating into the P-gp expression in canine mammary cancers, we attempted not only to confirm the presence of P-gp expression and location, but also to prove the relationship between P-gp and other neoplastic markers. Statistical analysis confirmed a positive correlation between P-gp and COX-2 expression: it demonstrated a statistical significance in the case of examined characteristics ($P = 0.021$). It is worth mentioning that the expression of two markers was significant in carcinomas featuring high histological grades of malignancy. Furthermore, P-gp expression was studied in canine lymphomas; additionally, attempts were made to conduct studies on carcinomas treated by means of chemotherapy; the results were compared with the results obtained in the control group of untreated carcinomas. Higher expression was observed in the treated cancer cells. On the other hand, increased expression in untreated cancers appears to be a negative prognostic factor, given that survival rate shall decrease in such cases (Lee et al, 1996). Due to the fact that this line of research has not brought about many works, there are several questions and ambiguities concerning the role of P-gp in the process of neoplastic genesis and the importance of this factor from the clinical point of view. Additionally, the question arises whether, or not P-gp can be classified into the significant prognostic marker category. We are not aware of defensive mechanisms of the cells equipped with membrane transporters against apoptosis generated by the compounds, that are not pump substrates. There are suggestions that the phenomenon may be connected with evacuation (pumping out) of a certain important mediator of apoptosis or with the impact of P-gp on the intracellular pH (Johnson et al, 1998). Identification of a mechanism of cell resistance to apoptosis is a key instrument in selection of suitable therapy. According to studies in humans, heat shock proteins may be important predictors of breast cancer (Park & Dupont, 1982). There is not much published data in the literature on heat shock protein expression in mammary gland tumors of female dogs. The role of these proteins in carcinogenesis has not been clearly defined either. Studies have only shown that expression of heat shock proteins takes place in canine breast cancers , but the linkage between these proteins and other tumor markers was not confirmed. Seymour (1990) studied endometrial cancer in women and found that heat shock proteins were useful markers in the diagnostics of these tumors (Saymour et al, 1990). Expression of Hsp27 was also investigated in breast cancer in women (Ciocca et al, 1993). A relationship between the expression of this protein and the degree of differentiation of the tumor cells was found. Similar studies were conducted by Storm (1996), who stated that tumors with Hsp27 expression showed a higher histological grade than tumors negative for the expression of this protein (Storm et al, 1993). Kumaraguruparan (2006) studied the expression of Hsp70 and Hsp90 in breast cancer in women and found a correlation between the expression of both proteins and proliferative activity (Kumaraguruparan et al, 2006). He suggested that heat shock proteins are important prognostic factors. Presence of Hsp70 was also reported in normal gastrointestinal tissues, but an increased expression of this protein was observed in gastrointestinal tumors (Isomoto et al, 2003). Isomoto (2003) believed that Hsp70 plays an important role in the degree of cellular differentiation in tumors, which would indicate that Hsp70 is an important prognostic factor. Similar studies are conducted in veterinary medicine, but to a lesser extent (Isomoto et al, 2003). A similar study was conducted by Rommanucci (2005) in mammary gland tumors in female dogs (Romanucci et al, 2006). She found Hsp70 and Hsp90 expression in simple and complex, as well as solid-type adenocarcinomas. Expression of these proteins was observed in the cytoplasm, as well as in the nuclei of tumor cells. Rommanucci (2006), in her study of breast cancers in female dogs, has not attempted to verify the possible correlations between heat shock proteins and other tumor markers (Rommanucci et al, 2006). She focused on establishing the degree of protein expression and their localization in tumor cells. Basing on the results of her study, the author suggested that these proteins may play an important role in the process of carcinogenesis. In our study, Hsp70 protein expression was found in 86.4% of tumors, while the expression of Hsp90 was observed in 66.2% of tumors. The aim of our study was to demonstrate the relationship between the expression of heat shock proteins and other prognostic factors. When comparing the expression of Hsp70 to the expression of Hsp90 in different types of tumors, we found a high statistical significance between the expression of both proteins. Analysis of the relationship between the expression of nuclear antigen Ki-67 and the expression of heat shock proteins showed that the highest expression of Ki-67 as well as Hsp90 was present in solid tumors. The highest expression of Hsp90 protein and Ki-67 was also found in cancers of the highest histological grades. Based on these data it can be concluded that Hsp90 is an important factor, which could be considered a marker of malignancy and which may be useful in the diagnostics of cancers. Analysis of the results showed high expression of Hsp70 in cancers with grade 1 and 3 of COX-2 expression; the lowest expression of Hsp70 was found in cancers with grade 2 of COX-2 expression. Statistical significance of $P = 0.009$ was found between the expression of Hsp70 and expression of COX-2. An increased expression of COX-2 was observed in tumors with a low mean number of cells showing positive reaction to Hsp90 protein. Correlation between the expression of Hsp70 protein and p53 protein in tumors of epithelial origin was also confirmed. Based on the analysis of expression of both proteins and taking into account the mean apoptotic index, it was found that in cancers with the third (the highest) histological grade, expression of p53 protein is significantly lower than the expression of Hsp70, and the value of apoptotic index in these cancers was the lowest.

3.2 Conclusions

Occurrence of mammary tumors in female dogs and their potential use as a model in comparative pathology makes it an interesting research material. It is known that the degree of malignancy is positively correlated with proliferative activity. The relationship between expression of estrogen receptors and the degree of malignancy is not entirely clear, although most authors believe that the expression of these receptors is higher in benign tumors.

Literature concerning the expression of p53 protein in mammary tumors in canines is scarce. It deals only with the expression of p53, but leaves no answers in respect to the expression of other factors that are considered prognostic markers, such as proliferative activity, histological grade of malignancy and expression of estrogen receptors. We conclude that COX-2 is an important prognostic factor and may be applied as a marker of canine mammary neoplasm malignancy, given the fact that higher expression of COX-2 was found in adenocarcinomas and in cancers featuring the highest histological malignancy grades in comparison to simple adenomas. Obtained results suggest that cyclooxygenase-2 may be a prognostic factor, but it requires clinical confirmation. The P-gp expression is also positively correlated with the degree of histological malignancy. This suggests a prudent approach to decisions concerning chemotherapy. Expression of Hsp90 can be considered as a marker of the degree of differentiation and of histological grade of mammary gland tumors in dogs, because the highest level of expression was found in solid carcinomas and in cancers exhibiting the highest histological grades of malignancy. Hsp70 expression was confirmed, but no correlation with other factors was found. This may suggest that Hsp70 is not a useful marker in the diagnostics of breast cancers in female dogs.

4. Acknowledgments

Study was carried out at the Institute of Pathology, Department of Clinical Sciences, Faculty of Veterinary Medicine, Warsaw School of Life Sciences, 159C Nowoursynowska Street, 02-766 Warsaw. Research was conducted as a part of a doctoral dissertation, partially funded by a grant from the Ministry of Science and Information Technology, No N30800632/0667

5. References

Bacus, S.S.; Goldschmidt, R.; Chin, D.; Moran, G.; Weinberg, D.; Bacus, J.W. (1989). Biological grading of breast cancer using antibodies to proliferating cells and other markers. *The American Journal of Pathology* , Vol.135, No.5, pp. 783-792

Barzanti, F.; Dal Susino, M.; Volpi, A.; Amadori, D.; Riccobon, A.; Scarpi, E.; Medri, L.; Bernardi, L.; Naldi, S.; Aldi, M.; Gaudio, M.; Zoli, W. (2000). Comparison between different cell kinetic variables in human breast cancer. *Cell Proliferation,* Vol.33,:pp. 75-89

Benjamin, S.A.; Lee, A.C.; Saunders, W.J. (1999). Classification and behavior of canine mammary epithelial neoplasms based on life-span observations in beagles. *Veterinary Pathology,* Vol. 36, pp. 423-436

Bostock, D.E.; Moriarty, J.; Crocker, J.(1992). Correlation between histologic diagnosis mean nucleolar organizer region count and prognosis in canine mammary tumors. *Veterinary Pathology,* Vol. 29, pp. 381-385

Chung-Ho, L.; Wan-Hee, K.; Ji-Hey, L.; Min-Soo, K.; Dae-Yong, K.; Oh-Kyeong, K. (2004). Mutation and overexpression of p53 as a prognostic factor in canine mammary tumors. *Journal of Veterinary Science,* Vol.5, No.1, pp. 63-69

Ciocca, D.R.; Oesterreich, S.; Chamness, C.; Mcguire, W.L.; Fuqua, S.A. (1993). Biological and clinical implications of heat shock protein 27000 (Hsp27): a Review. *Journal of the National Cancer Institute*, Vol. 85, pp. 1558-1570

Dempke, W.; Rie, C.; Grothey, A.; Schmoll, H.J. (2001). Cyclooxygenase-2: a novel target for cancer chemotherapy? *Journal of Cancer Research & Clinical Oncology*, Vol. 127, p. 411-417

Doré, M.; Lanthier, I.; Sirois, J. (2003). Cyclooxygenase-2 expression in canine mammary tumors. *Veterinary Pathology*, Vol.40, pp. 207-212

Gamblin, R.M.; de Maria, R.; Piccoli, M. (1994). Cd44 triggering enhanced human NK cell cytotoxic function. The *Journal of Immunology,* Vol. 153, pp. 4399-4407.

Geraldes, M.; Gärtner, F.; Schmitt, F. (2000). Immunohistochemical study of hormonal receptors and cell proliferation in normal canine mammary glands and spontaneous mammary tumors. *Veterinary Record*, Vol. 146, pp. 403-406

Gilbertson, S.R.; Kurzaman, I.D.; Zachran, R.E.; Hurvitz, H.J.; Black, M.M. (1983). Canine mammary epithelial neoplasms: biologic implications of morfrologic characteristics assessed in 232 dogs. *Veterinary Pathology,* Vol. 20, pp. 127-142

Ginn, PE. (1996). Immunohistochemical detection of P-glycoprotein in formalin - fixed and paraffin-embedded normal and neoplastic canine tissues. *Veterinary Pathology,* Vol. 33, pp. 533-54

Giziński, S.; Boryczko, Z.; Katkiewicz, M.; Bostedt, H. (2003). Ki-67 protein as a prognostic indicator in breast cancer in females. *Veterinary Medicine,* Vol. 59, No. 10, pp. 888-891.

Hampe, J.F.; Misdorp, W. (1973) Tumours and dysplasias of the mammary gland. *Bull World Health Organ,* Vol. 50, pp. 111-133.

Heller, D.A,; Clifford, C.A; Goldschmidt, M.H.; Holt, D.E.; Shofer, F.S.; Smith, A.; Sorenmo, K.U. (2005). Cyclooxygenase-2 expression is associated with histologic tumor type in canine mammary carcinoma. *Veterinary Pathology,* Vol. 42, pp. 776-780.

Hellmen, E.; Bergstrom, R.; Holmberg, L.; Spanberg, I.B.; Hansson, K.; Lindgren, A. (1993). Prognostic factors in canine mammary tumors: a multivariate study of 202 consecutive cases. *Veterinary Pathology*, Vol. 30, pp. 20-27.

Host, H.; Lund, E. (2006). Age as prognostic factor in breast cancer. *Cancer*, Vol. 57, pp. 2217-2221.

Howlin, J.; McBryan, J; Martin, F. (2006) Pubertal mammary gland development: insights from mouse models. *Journal of Mammary Gland Biology and Neoplasia*, Vol. 11, pp. 283-297.

Isomoto, H.; Oka, M.; Yano, Y.; Kanazawa, Y.; Soda, H.; Terada, R.; Yasutake, T.; Nakayama, T.; Shikuwa, S.; Takeshima, F.; Udano, H.; Murata, I.; Ohtsuka, K.; Kohno, S. (2003). Expression of heat shock protein Hsp70 and Hsp40 in gastric cancer. *Cancer Letters*, Vol. 198, No.2, pp. 219-228.

Johnson, A.S.; Couto, C.G.; Weghorst, C.M. (1998). Mutation of the p53 tumor suppressor gene in spontaneously occuring osteosarcomas of the dog. *Carcinogenesis,* Vol. 19, pp. 213- 217.

Koda, M.; Reszec, J.; Sulkowska, M.; Kanczuga - Koda, L.; Sulkowska, S. (2004). Expression of the insulin-like growth factor-I receptor and proapoptotic Bax and Bak proteins in human colorectal cancer. *Annals of the New York Academy of Sciences* , Vol.1030, pp. 377-383.

Kubiak, J.Z. (2001). Cancer and cell cycle. *Advances in Cell Biology*, Vol. 28, pp. 97-307.

Kumaraguruparan, R.; Kurunagaran, D.; Balachandran, C.; Manohar, B.M.; Nagini, S. (2006). Of humans and canines: a comparative evaluation of heat shock and apoptosis-associated proteins in mammary tumors. *Clinica Chimica Acta*, Vol. 365, No.1-2, pp. 168-176.

Lanza-Jacoby, S.; Burd, R.; Rosato, F.E.J.; McGuire, K.; Little, J.; Nougbilly, N.; Miller, S. (2006). Effect of simultaneous inhibition of epidermal growth factor receptor and cyclooxygenase-2 in HER-2/neu-positive breast cancer. *Clinical Cancer Research*, Vol. 12, pp. 6161-6169.

Lee, J.; Hughes, C.S.; Fine, R.L.; Page, L.R. (1996). P-glycoprotein Expression in Canine Lymphoma a Relevant, Intermediate Model of Multidrug Resistance. *Cancer*, Vol. 77, pp. 1892-1898.

Liu, X.H.; Rose, D.P. (1996). Differential expression and regulation of cyclooxygenase-1 and –2 in two human breast cancer cell lines. *Cancer Research*, Vol. 56, pp. 5125-5127.

Lőhr C.V., Teifke, J.P.; Failing, K.; Weiss, E. (1997). Characterization the standarized AgNOR method with postfixation and immunohistologic detection of Ki-67 and PCNA. *Veterinary Pathology*, Vol. 3, No. 4, pp. 212-221.

Martin, P.M.; Cotard, M.; Mialot, J.P.; Andre, F.; Rayanaud J.P. (1984) Animal models for hormone-dependent human breast cancer. Relatioship between steroid receptor profiles in canine and feline mammary tumors and survival rate. *Cancer Chemotherapy and Pharmacology*, Vol. 12, pp. 13-70.

Mayr, B.; Schellander, K.; Schlleger, W.; Reifinger, M. (1994). Sequence of exon of the canine p53 gene-mutation in a papilloma. *British Veterinary Journal*, Vol. 150, pp. 81-84.

McEwen, E.G.; Patnaik, A.K.; Harvey, H.J.; Panko, W.B. (1092). Estrogen receptor in canine mammary tumors. *Cancer Research*, Vol. 42, pp. 2255-2259.

Millanta, F.; Calandrella, M.; Bari, G.; Niccolini, M.; Vannozzi, I.; Poli, A. (2005). Comparison of steroid receptor expression in normal, dysplastica and neoplastic canine and feline mammary tissues. *Research in Veterinary Science*, Vol. 70, No.3, pp. 225-232.

Mirecka, J.; Korabiowska, M.; Schauer, A. (1993). Correlation between the occurrence of Ki-67 antigen and clinical parameters in human breast carcinoma. *Folia Histochemica et Cytobiologica*; Vol. 31, No.2, pp. 83-86.

Misdorp, W.; Else, R.W.; Hellmen, E.; Lipscomb, T.P. (1999). Histological classification of mammary tumors of the dog and cat, *World Health Organization*, Geneva

Misdorp, W.W.; Meuten, D. (edit.). (2002). Tumors in Domestic Animals. Iowa State Press, Black Publishing Company. 4th ed., 575-606.

Mohammed, S.I.; Khan, K.N.; Sellers, R.S.; Hayek, MG.; DeNicola, D.B.; Wu, L.; Bonney, P.L.; Knapp, D.W. (2004). Expression of cyclooxygenase-1 and 2 in naturally-occurring canine cancer. *Prostaglandins Leukotrienes and Essential Fatty Acids*, Vol. 70, pp. 479-483.

Moulton, J.E. (1990). Tumors of mammary gland, W: *Tumors in Domestic Animals. 3rd ed., University of California Press*, pp. 518-549, Berkeley.

Mulas, J.M.; Millán, Y.; Dios, R. (2005). A prospective analysis of immunohistochemically determined estrogen receptor α and progesterone receptor expression and host and tumor factors as predictors of disease-free period in mammary tumors of the dog. *Veterinary Pathology*, Vol. 42, pp. 200-212.

Muto, T.; Wakui, S.; Takahashi, H.; Maekawa, S.; Masaoka, T.; Ushigome, S.; Furusato, M. (2000). p53 Gene Mutations occurring in spontaneous benign and malignant mammary tumors of the dog. *Vet Pathol*, Vol. 37, pp. 248-253.

Nerurkar, V.R.; Naik, S.N.; Lalitha, V.S.; Chitale, A.R.; Ishwad, C.S;, Jalnapurkar, B.V. (1990). Mammary tumours in dogs and their similarities with human breast cancer. W: Bamji M.S. (edit.).: *Proceedings of Symposium on Animal Information Service Center*, pp. 35-40, NIN, ICMR, India

Nieto, A.; Péna, L.; Pérez-Alenza, M.D.; Sànchez, M.A.; Flores, J.M.; Castańo, M. (2000). Immunohistologic detection of estrogen receptor alpha in canine mammary tumors: clinical and pathologic associations and prognostic significance. *Veterinary Pathology*, Vol. 37, pp. 239-247.

Niwińska, A (1995). New prognostic factors in patients with breast cancer. *Journal of Oncology*, Vol. 45, pp. 459-469.

Nowak, M.; Madej, J.A.; Dzięgiel, P. (2005). Immunohistochemical localization of cox-2 in cells of mammary adenocarcinomas in bitches as related to tumour malignancy grade. The *Bulletin of the Veterinary Institute in Pulawy*, Vol. 49, pp. 433-437.

Nowak, M.; Madej, J.A.; Dzięgiel, P. (2007). Comparison of expressions of estrogen and progesterone receptors in adenocarcinomas of the mammary gland in bitches with mitotic activity of neoplastic cells. *Veterinary Medicine*, Vol. 63, No.10, pp. 1211-1215.

Olszewski, W.(1994). Selected aspects of the pathology of breast cancer. *Journal of Oncology*, Vol. 44, No.2, pp. 10-16.

Parl, F.F.; Dupont, W.D. (1982). Retrospective cohort study of histologic risk factor in breast cancer patient. *Cancer*, Vol. 50, pp. 2410-2416.

Peńa, L.; Nieto, A.; Perez-Alenza, D. (1998). Immunohistochemical detection of Ki-67 and PCNA in canine mammary tumors: Relationship to clinical and pathologic variables. *Journal of Veterinary Diagnostic Investigation*, Vol.10, pp. 237-246.

Petterino, C.; Rossetti, E.; Bertoncello, D.; Martini, M.; Zappulli, V.; Bargelloni, L.; Castagnaro, M. (2006). Immunohistochemical detection of P-glycoprotein (clone C494) in canine mammary gland tumours. *Journal of Veterinary Medicine Series A - Physiology Pathology Clinical Medic*, Vol. 53, pp. 174-178.

Philibert, J.; Snyder, P.W.; Glickman, N.; Glickman, L.T.; Knapp, D.W.; Waters D.J. (2003). Influence of host factors in survival in dogs with malignant mammary gland tumors. *Journal of Veterinary Internal Medicine*, Vol. 17, pp. 102-106.

Queiroga, FL.; Alves, A; Pires, I; Lopes, C. (2007). Expression of Cox-1 an Cox-2 in canine mammary tumours. *Journal of Comparative Pathology*, Vol. 136, pp 177-185.

Ramalho, L.N.Z.; Ribeiro-Silva, A.; Cassali, G.D. (2006). Zucoloto S. The Expression of p63 and cytokeratin 5 in mixed tumors of the canine mammary gland provides new insights into the histogenesis of these neoplasms. *Veterinary. Pathology*, Vol. 43, pp. 424-429.

Rilke, F.; Colnaghi, M.I.; Cascinelli, N.; Andreola, S.; Baldini, M.T.; Bufalino, R.; Della Porta, G.; Menard, S.; Pierotti, M.A.; Testori, A. (1991). Prognostic significance of HER-2/neu expression in breast cancer and its relationship to other prognostic factors. *International Journal of Cancer*, Vol. 49, pp. 44-49.

Ristimäki, A.; Sivula, A.; Lundin, J.; Lundin, M.; Salminen, T.; Haglund, C.; Joensuu, H.; Isola, J. (2002) Prognostic significance of elevated cyclooxygenase-2 expression in breast cancer. *Cancer Research* , Vol. 62, pp. 632-635.

Rodo, A. (2007). Expression of HER-2 receptors. E cadherin, and p53 protein in mammary gland tumors in bitches. Doctoral thesis, Agricultural University, Warsaw.

Romanucci, M.; Marinelli, A.; Giuseppe, S.; Dell Salda, L. (2006). Heat shock protein expression in canine malignant mammary tumours. *BMC Cancer*, Vol. ,6, pp. 171.

Roses, D.F.; Bell, D.A.; Flotte, T.J.; Taylor, R.; Ratech, H.; Dubin, N. (1982). Pathologic predictors of recurrence in stage 1 (T1NOMO) breast cancer. *American Journal of Veterinary Research*, Vol. 78, pp. 817-820.

Rungsipipata, A. (1999) Immunohistochemical analysis of c-yes and c-erbB-2 oncogene products and p53 tumor suppressor protein in canine mammary tumors. The *Journal of Veterinary Medical Science*; Vol.61, No.1, pp. 27-32.

Rutteman, G.R.; Withrow, S.J.; MacEwen, E.G. (2001). Tumors of the mammary gland. W: *Small Animal Clinical Oncology*. S. J. Withrow; E. G. MacEwen, (Ed.), 455-477, 3rd ed., Philadelphia

Sartin, E.A.; Barnes, S.; Kwapien, R.P.; Wolfe, L.G. (1992). Estrogen and progesterone receptor status of mammary carcinomas and correlation with clinical outcome in dogs. *American Journal of Veterinary Research*, Vol. 53, pp. 2196-2200.

Saymour, L.; Bezwoda, W.R.; Meyer, K. (1990) Tumor factors predicting for prognosis in metastatic breast cancer. The presence of P24 predicts for response to treatment and duration of survival. *Cancer*; Vol.66, No.11, pp. 2390-2394.

Skrzypczak, M. (2004). Angiogenesis in mammary gland tumors in bitches. Doctoral thesis, Agricultural University, Warsaw.

Sobczak-Filipiak, M.; Malicka, E.(1997). Diagnosis of mammary gland tumors including immumocytochemistry methods. *Materials Conference. Veterinary Oncology*. T. Rotkiewicz, (Ed.), 100-107, Olsztyn

Soslow, R.A.; Dannenberg, A.J.; Rush, D.; Werner, B.M.; Khan, K.N.; Masferrer, J.; Koki, A.T. (2000). COX-2 is expressed in human pulmonary, colonic and mammary tumours. *Cancer*, Vol. 89, pp. 2637-2645.

Storm, F.K.; Mahvi, D.M.; Gilchrist, K.W. (1993). Heat shock protein 27 overexpression in breast cancer lymph node metastasis. *Annals of Surgical Oncology*; Vol. 3, No.6, pp. 570-573

Szczubiał, M.; Łopuszański, W. (2002). Prognosis in mammary gland tumors in bitches. *Veterinary Medicine*, Vol. 58, No. 40, pp. 261-264.

Thomas, W.; Caiazza, F.; Harvey, B.J. (2008). Estrogen, phospholipase A and breast cancer. *Frontiers in Bioscience*, Vol. 13, pp. 2604-2613.

Veldhoen, N.; Watterson, J.; Brash, M. (1999). Identification of tumor-associated and germ line p53 mutation in canine mammary cancer. *British Journal of Cancer* , Vol.81, pp. 409-415.

Part 3

Veterinary Care in Livestock

13

Local Anesthesia for Husbandry Procedures and Experimental Purposes in Farm Animals

Maria Laura Marongiu
Department of Scienze Zootecniche,
University of Sassari,
Italy

1. Introduction

Surgical husbandry practices carried out on farm animals in many management systems are: castration, docking, disbudding, dehorning, spaying, ear notching, ear tagging, branding-hot, branding-freeze, mulesing, teeth grinding, teeth clipping, tusk grinding, nose ringing, tail nicking, beak trimming. The management of farm animals is based on the ability to control them and many aspects of their environments.

The standards of husbandry and welfare during animal production are becoming important factors influencing consumer perceptions in many markets. Welfare attention on surgical husbandry practices for farm animals focuses on the fact that many of the procedures involve innervated tissues.

One of the approaches available for addressing welfare concerns surrounding these procedures is performing them without pain by the use of local anesthesia.

Food-animal practitioners commonly perform local-anesthetic techniques due to the dangers associated with general anaesthesia. Local anesthetic techniques usually are simple, cheap and have relatively few side effects.

The problems involved in anaesthetizing farm animals for husbandry procedures are usually less complicated than those encountered when local anesthetics have to be used in these animals for experimental purposes where it is important that anesthesia should have little or no influence on the result of the experiment.

Blood serial sampling is required for many applied reproduction and nutrition trials with farm animals. The experimental design of these researches very often includes repeated samplings of large volumes of blood for metabolites and hormones analysis. In addition, the harvested blood can be used for analysis of biochemical, toxicological or immunological parameters, for examination or culture of micro-organisms, for production of antibodies.

The use of topically applied local anesthetics is very useful in removing sensation from the vessel puncture.

There are several features of local anesthesia which render it particularly helpful in veterinary practice. Local anesthetics provide a reversible regional loss of sensation, reduce pain, thereby facilitating surgical procedures. Many surgical procedures can be carried out satisfactorily using different local anesthetics.

Whether or not sedation is necessary as an adjunct will depend on the species, temperament and health of the farm animal, and on the magnitude of the procedure.

Many operations, including some husbandry procedures and practices for experimental purposes, are performed on standing animals. Local anesthesia, enabling protracted operations in standing animals, eliminates the dangers associated with forcible casting and restraint, and prolonged recumbency.

The techniques are not difficult to learn and do not involve the use of expensive or complicated equipment.

2. Local anesthetics

In 1860, cocaine, the oldest anesthetic, was extracted from the leaves of the *Erythroxylon coca* bush. In 1884, Sigmund Freud and Karl Koller were the first to use it as an anesthetic agent during ophthalmologic procedures.

Procaine, a synthetic alternative to cocaine, was not developed until 1904. Procaine is an ester of para-aminobenzoic acid (PABA). As procaine is metabolized, PABA, a known allergen, is released as a metabolic product. The potential for severe allergic reactions limits the use of procaine and other ester-type anesthetic agents.

Tetracaine, another ester-type anesthetic, was introduced in 1930. Tetracaine is more potent than procaine, and it causes similar allergic reactions.

In 1943, an alternative class of anesthetics was discovered when Lofgren developed lidocaine. This agent is an amide derivative of diethylaminoacetic acid, not PABA; therefore, it has the benefit of a low allergic potential.

Since then, multiple amide-type anesthetics have been introduced into clinical use. Slight chemical alterations to the compounds have imparted beneficial characteristics, including increased duration and potency, to each. These compounds offer the surgeon more choices, and anesthetics can be appropriately matched to different procedures.

The lidocaine is the most widely used general-purpose local anesthetic in veterinary use. It possesses reasonably rapid onset of action, with good spreading properties. It may cause some local irritation and swelling. It is available in a variety of concentrations or injection; with and without epinephrine; and in the form of solutions, creams, jellies, sprays etc.

Duration of action is variable (depending on uptake) but will be around one hour without epinephrine, and two hours with epinephrine.

The bupivacaine has a prolonged duration of action: up to eight hours when combined with epinephrine. It is therefore used whenever long action is required.

The mepivacaine causes very little swelling and edema in the area of injection, possibly as it lacks vasodilatory action. Onset of action is faster and reliability of block greater than with prilocaine.

The prilocaine has slower onset of action, and spreads less well compared to lidocaine. The unique ability of prilocaine to cause dose-dependent methemoglobinemia limits its clinical usefulness.

Other local anesthetics are:

The proparacaine is used to anesthetize the cornea of the eye. When dropped on the cornea it has a rapid onset of action (within 1 minute) and lasts for about 15-30 minutes. It is nonirritant, and does not affect the size of the pupil.

The amethocaine is well absorbed by surfaces and is used on mucous membranes.

The procaine, an older drug with slow onset of action and poor spreading powers has been superseded by the more modern drugs.

The cocaine is the only one to cause vasoconstriction. It is now not used as a local anesthetic because of its potential for abuse.

Newer long acting local anesthetics with less cardiotoxicity, e.g. ropivacaine or lovobupivacaine, are now available for man, but are currently very expensive for veterinary use.

3. Physiology of nerve conduction

Reviewing the physiology of nerve conduction is important before any discussion of local anesthetics. Nerves transmit sensation as a result of the propagation of electrical impulses; this propagation is accomplished by alternating the ion gradient across the nerve cell wall, or axolemma.

In the normal resting state, the nerve has a negative membrane potential of -70 mV. This resting potential is determined by the concentration gradients of 2 major ions, Na^+ and K^+, and the relative membrane permeability to these ions (also known as leak currents).

The concentration gradients are maintained by the sodium/potassium ATP pump (in an energy-dependent process) that transports sodium ions out of the cell and potassium ions into the cell. This active transport creates a concentration gradient that favors the extracellular diffusion of potassium ions.

In addition, because the nerve membrane is permeable to potassium ions and impermeable to sodium ions, 95% of the ionic leak in excitable cells is caused by K^+ ions in the form of an outward flux, accounting for the negative resting potential. The recently identified 2-pore domain potassium (K2P) channels are believed to be responsible for leak K^+ currents.

When a nerve is stimulated, depolarization of the nerve occurs, and impulse propagation progresses. Initially, sodium ions gradually enter the cell through the nerve cell membrane. The entry of sodium ions causes the transmembrane electric potential to increase from the resting potential. Once the potential reaches a threshold level of approximately -55 mV, a rapid influx of sodium ions ensues. Sodium channels in the membrane become activated, and sodium ion permeability increases; the nerve membrane is depolarized to a level of +35 mV or more.

Once membrane depolarization is complete, the membrane becomes impermeable to sodium ions again, and the conductance of potassium ions into the cell increases. The process restores the excess of intracellular potassium and extracellular sodium and reinstates the negative resting membrane potential.

Alterations in the nerve cell membrane potential are termed the action potential. Leak currents are present through all the phases of the action potential, including setting of the resting membrane potential and repolarization.

4. Mechanism of action of local anesthetics

Local anesthetics inhibit depolarization of the nerve membrane by interfering with both Na^+ and K^+ currents. The action potential is not propagated because the threshold level is never attained.

Although the exact mechanism by which local anesthetics retard the influx of sodium ions into the cell is unknown, two theories have been proposed.

The membrane expansion theory postulates that the local anesthetic is absorbed into the cell membrane, expanding the membrane and leading to narrowing of the sodium channels. This hypothesis has largely given way to the specific receptor theory.

This theory proposes that the local anesthetic diffuses across the cell membrane and binds to a specific receptor at the opening of the voltage-gated sodium channel. The local anesthetic affinity to the voltage-gated Na^+ channel increases markedly with the excitation rate of the neuron. This binding leads to alterations in the structure or function of the channel and inhibits sodium ion movement.

Blockade of leak K^+ currents by local anesthetics is now also believed to contribute to conduction block by reducing the ability of the channels to set the membrane potential.

On the basis of their diameter, nerve fibers are categorized into 3 types. Type A fibers are the largest and are responsible for conducting pressure and motor sensations. Type B fibers are myelinated and moderate in size. Type C fibers, which transmit pain and temperature sensations, are small and unmyelinated. As a result, anesthetics block type C fibers more easily than they do type A fibers.

All local anesthetics have a similar chemical structure, which consists of three components: an aromatic portion, an intermediate chain, and an amine group. The aromatic portion, usually composed of a benzene ring, is lipophilic, whereas the amine portion of the anesthetic is responsible for its hydrophilic properties.

The degree of lipid solubility of each anesthetic is an important property because its lipid solubility enables its diffusion through the highly lipophilic nerve membrane. The extent of an anesthetic's lipophilicity is directly related to its potency.

Local anesthetics are weak bases that require the addition of hydrochloride salt to be water soluble and therefore injectable. Salt equilibrates between an ionized form and a nonionized form in aqueous solution.

Equilibration is crucial because, although the ionized form is injectable, the nonionized base has the lipophilic properties responsible for its diffusion into the nerve cell membrane. The duration of action of an anesthetic or the period during which it remains effective is determined by its protein-binding activity, because the anesthetic receptors along the nerve cell membrane are proteins.

The intermediate chain, which connects the aromatic and amine portions, is composed of either an ester or an amide linkage. This intermediate chain can be used in classifying local anesthetics.

5. Nerve block for husbandry procedures in farm animals

Lidocaine is the most common local anaesthetic drug in worldwide veterinary use. It is used to alleviate the acute pain experienced by animals during and for 1–2 hours after a number of painful procedures, including some husbandry practices in farm animals (Mellor & Stafford, 2000; Stafford & Mellor, 2005a, b).

It is a short-acting local anaesthetic that is usually cleared from the site of injection quickly enough for its effects to last for about 60–120 minutes.

Lidocaine block of the corneal nerve prior to dehorning of calves virtually eliminates the plasma cortisol response indicative of pain for about 2 hours (Stafford and Mellor, 2005a).

There are other local anaesthetics such as bupivacaine and mepivacaine, with different characteristics, that may be useful in husbandry procedures. For instance, bupivacaine blockade of the cornual nerve of calves virtually eliminates the cortisol response to dehorning for about 4 hours (Stafford & Mellor, 2005a). However, most such local anaesthetics are not licensed for livestock in many countries.

Epidural nerve block is produced by injecting local anaesthetic into the epidural space of the spinal cord (Flecknell & Waterman-Pearson, 2000).

Effective epidural analgesia can also be achieved by injecting alpha-2 agonists such as xylazine epidurally and this is used when castrating adult cattle because it has the added advantage of being accompanied by sedation.

A lidocaine-xylazine mixture produces effective epidural analgesia in cattle castrated using a castration clamp (Burdizzo®) and extends the duration of analgesia.

6. Systemic and toxic effects of local anesthetics

Accidental intravenous injection of local anesthetics is the most common cause of adverse reaction associated with local anesthetic administration. In severe cases it can cause cardiac arrest.

When the plasma concentration of local anesthetics is excessive, sufficient cardiac sodium channels become blocked so that conduction and automaticity become adversely depressed. For example, excessive plasma concentration of lidocaine may slow conduction of cardiac impulses through the heart, manifesting as increased PR interval and widened QRS complex on the ECG.

Effects of local anesthetics on calcium and potassium ion channels and local anesthetic induced inhibition of cyclic adenosine monophosphate (cAMP) production may also contribute to cardiac toxicity.

Bupivacaine is more cardiotoxic than lidocaine. It is always important to draw back on syringe to check to be not in vein before injecting local anesthetics.

General overdose depends on blood levels, therefore is influenced by total dose and speed of uptake from the tissues.

As a very rough guide, the toxic dose of lidocaine would be 8 mg/kg and 4 mg/kg of bupivacaine.

In very small animals such as small mammals (goat kids) and birds this amount can be easily exceeded using solutions of standard concentration, so it must be diluted carefully and used with caution.

Signs of overdose are initial sedation, followed with increasing dosage by twitching, convulsions, coma and death.

Reports implicate prilocaine, benzocaine, lidocaine and procaine as causative agents to produce methemoglobinemia in some animals.

7. Methods of producing local anesthesia

Common methods of producing local anesthesia are:

- Surface (topical) anesthesia
- Intrasynovial anesthesia
- Infiltration anesthesia
- Spinal anesthesia
- Intravenous regional local anesthesia
- Regional anesthesia.

7.1 Surface (topical) anesthesia

This refers to the use of local anesthetics in solution sprays as well as in various creams and ointments; drops into the eye; sprays or brush in laryngeal area, infuses into the nostrils, urethra, or rectum.

7.2 Intrasynovial anesthesia

Used in joints, bursa, and tendon sheaths. Useful for both diagnosis of lameness, and for general pain relief. The local anesthetic chosen must cause minimal irritation, and great care in sterility is necessary as infection in these sites occurs easily.

7.3 Infiltration anesthesia

By this method the nerve endings are affected at the actual site of operation. Most minor surgery can be done this way, excluding surgery on teats in cattle or small animals digits.

Problems occur through infection (never inject local analgesic through infected tissues), irritation, distortion of the wound, swelling and some delay in post-operative healing.

A variation of infiltration anesthesia designed to minimize these effects is field anesthesia. Here, walls of anesthesia are made by infiltrating the tissues around (rather than at) the surgical site.

Advantages include absence of distortion of the anatomical features in the line of incision; muscle relaxation and no interference to healing. An example of a field anesthesia technique which is widely used in cattle is the Inverted L or 7 block for anesthesia of the abdominal fossa.

Ring blocks whereby the tissue all around a distal organ is infiltrated with local anesthetic, is another form of field anesthesia: examples of where this is used is on the teats of cattle (epinephrine must not be used here, as vasoconstriction could lead to ischemic necrosis and sloughing of tissue) or around the limb of cattle.

7.4 Spinal anesthesia

Spinal anesthesia is the injection of local anesthetic around the spinal cord. When local anesthetics such as lidocaine or bupivacaine are used, all the segmental nerves (sensory and motor) which pass through the anesthetic are paralyzed, although when opioids are used only sensory block occurs.

Spinal anesthesia is divided into two types: epidural and true spinal.

Epidural (or extradural) anesthesia refers to depositing of local anesthetics into the extradural space. The needle enters the spinal canal, but does not penetrate the meninges. The anesthetic is therefore limited to the canal outside the dura mater.

True spinal anesthesia refers to the subarachnoid access in which the needle penetrates the dura mater, and the analgesic is injected into the cerebrospinal fluid.

7.4.1 Epidural anesthesia in cattle

For caudal and epidural anesthesia the injection site used is between coccygeal C1 and C2. For a 500 kg bovine 5-10 ml 2% lidocaine will give caudal anesthesia without causing hind limb ataxia or paralysis.

Onset of paralysis of the tail should occur in 1-2 minutes. The block will last 1-2 hours. Larger doses will produce increasingly anterior effects.

By the time 100-150 ml 2% lidocaine is injected, the block will be sufficiently anterior to allow surgery of the hindlimbs, mammary tissue, flanks and abdominal wall. However, the bovine will be recumbent. Injection of local anesthetics can be carried out at the lumbosacral junction in order to produce an anterior block with less anesthetic.

There is a danger of accidental subarachnoid injection. Segmental epidural anesthesia, where the anesthetic is injected into the epidural space at the region required can be used for analgesia of any 'segment' with less overall side effects. It is more difficult to perform; penetration of the meninges is likely, but in skilled hands it is a very useful technique.

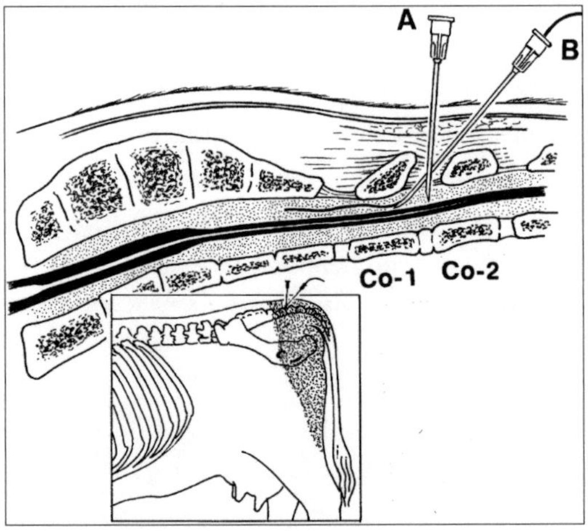

Fig. 1. For caudal and epidural anesthesia in cattle the injection site used is between coxygeal C1 and C2 (Modified from Thurmon *et al*, 1996)

7.4.2 Epidural anesthesia in the sheep and goat

In both sheep and goats, anterior epidural anesthesia, induced by injection at the lumbosacral junction is easily performed and provides excellent analgesia and muscle relaxation for abdominal surgery.

Recumbency may occur but is not a problem in these small animals. As in cattle, there is a risk of subarachnoid injection.

7.4.3 Epidural anesthesia in the horse

Hind limb ataxia is a serious problem, so only caudal epidural techniques are used. These are useful for various obstetrical manipulations and surgery on the rectum, vagina and tail.

The technique is less reliable than in cattle. Site of injection is usually sacrococcygeal junction but can be between C1 and C2. For a 500 kg horse, a mixture of 50 mg of xylazine and 6 ml of 2% mepivacaine (may be repeated for another dose) may prove very effective.

7.5 Intravenous regional local anesthesia

In this technique, a limb vein is catheterized. The limb is then exsanguinated (Esmarchs bandage), and a tourniquet placed around the limb, at a pressure adequate to prevent arterial circulation (> 150 mmHg).

Local anesthetic (preferably without epinephrine) is then injected into the vein. After a period of 15 minutes the area distal to the tourniquet is anesthetized until the tourniquet is removed.

Potential problems are: a) Difficulty in finding the vein once the limb is exsanguinated (this is why it is best to have a catheter in place first). c) Cardiac arrhythmias or even arrest (this is due to an inadequate tourniquet). d) Failure to take effect. (common reasons are inadequate tourniquet, inadequate time, and lack of exsanguination). e) Collapse when tourniquet is removed (this is because of anoxic waste products re-entering circulation); it is preferable if the animal is recumbent at this time.

7.6 Regional anesthesia

This term is used where specific nerves to the area concerned are blocked. Examples include specific nerve blocks to the limbs; paravertebral blocks; cornual block (for dehorning) and many others.

7.6.1 Paravertebral anesthesia

Paravertebral anesthesia refers to the perineural injection of local anesthesia about the spinal nerves as they emerge from the vertebral canal through the intervertebral foraminae.

The technique may theoretically be carried out in any species, and at any level of the spinal cord but in practice, its main use is to provide anesthesia of the lumbar region in ruminants.

Its advantage is that it provides analgesia and muscle relaxation of the whole area covered by the segmental nerves blocked.

Several different methods of achieving paravertebral anesthesia have been described.

All methods approaching from the dorsal surface are equally effective.

The method described whereby the needle is inserted ventral to the transverse processes of the spine has the disadvantage that the dorsal branches of the segmental nerves are not blocked, thus some skin sensitivity remains.

Paravertebral anesthesia is easy to carry out, and almost always effective, except in the very large beef breeds where it may be very difficult to locate the necessary landmarks.

7.6.2 Proximal paravertebral block (Farquharson, Hall or Cambridge Technique)

Indicated for standing laparotomy surgery such as C-section, rumenotomy, cecotomy, correction of gastrointestinal displacement, intestinal obstruction and volvulus.

The dorsal aspect of the transverse processes of the last thoracic (T-13) and first and second lumbar (L-1 and L-2) vertebrae is the site for needle placement.

The dorsal and ventral nerve roots of the last thoracic (T-13) and 1st and 2nd lumbar spinal nerves emerging from the intervertebral foramina are desensitized. 10-20 ml of 2% lidocaine is injected to each site and onset of anelgesia occurs usually within 10 minutes after injection

Increased skin temperature due to vasodilation (paralysis of cutaneous vasomotor nerves) indicates effective block. Duration of analgesia is approximately 90 minutes.

7.6.3 Distal paravertebral block (Magda, Cakala, or Cornell technique)

Indication is the same as the proximal paravertebral block.

The dorsal and ventral rami of the spinal nerves T13, L1 and L2 are desensitized at the distal ends of L-1, L-2 and L-4. A 7.5-cm, 18-gauge needle is inserted ventral to the tips of the respective transverse processes in cows where approximately 10-20 ml of a 2% lidocaine solution are injected in a fan-shaped infiltration pattern.

The needle is completely withdrawn and reinserted dorsal to the transverse process, where the cutaneous branch of the dorsal rami is injected with about 5 ml of the analgesic.

The procedure is repeated for the second and fourth lumbar transverse processes.

10-20 ml 2% lidocaine is used per site and onset and duration of analgesia are similar to the proximal technique.

7.6.4 Local nerve blocks of the head

Cornual nerve block - The cornual nerve block in cattle is indicated for dehorning and treating horn injury. Ophthalmic division of the fifth cranial nerve is desensitized by the injection of the anesthetic (3-5 ml of 2 % lidocaine) on the upper third of the temporal ridge, about 2.5 cm below the base of the horn. The nerve is relatively superficial, about 0.7-1 cm deep.

Onset of analgesia occurs 10-15 minutes following injection and duration of analgesia is approximately one hour.

In adult cattle with well-developed horns, a ring block around the base of the horn may be necessary.

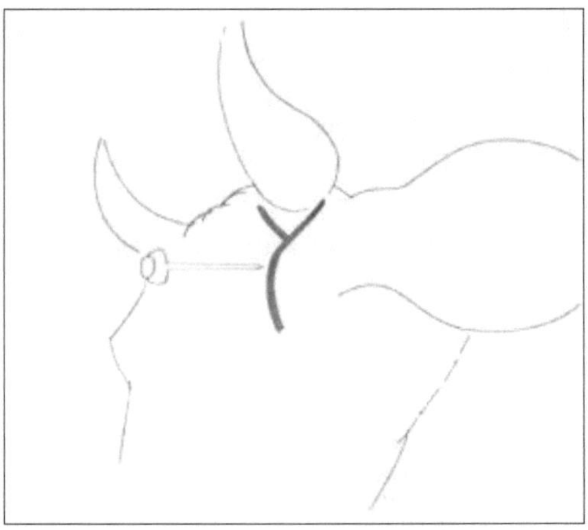

Fig. 2. Needle placement site for cornual nerve block in cattle

The cornual nerve block in goats is indicated for dehorning, not as easy as in the cow as there are two branches of the nerve (lacrymal and intratrochlear branches).

It must not be used in kids, as tend to use a total overdose of local anesthetics in these very small animals.

Auriculopalpebral block in cattle and horses - Auriculopalpebral nerve supplies motor fiber to the orbicularis oculi muscle. It runs from the base of the ear along the facial crest, past and ventral of the eye, giving off its branches on the way.

Fig. 3. Needle placement sites for cornual nerve block in the goat (Modified from Thurmon *et al*, 1996)

The needle is inserted in front of the base of the ear at the end of the zygomatic arch and is introduced until its point lies at the dorsal border of the arch. 10-15 ml of 2% lidocaine are injected.

Onset of analgesia occurs 10-15 minutes and duration of analgesia is approximately one hour. This block does not produce analgesia of the eye or the lids. In conjunction with topical analgesia (2% lidocaine), it is useful for the removal of foreign bodies from the cornea and conjunctival sac. Also used, but less frequently, in other species.

This block has no sensory effects but paralyses the muscles of the eyelid. It is used to keep the eye open, for example, as an adjunct in ophthalmic surgery.

This is one of the most commonly used techniques to block the motor function of the upper eye lid for ophthalmic surgery, preventing eyelid closure during examination of the eyeball.

Retrobulbar nerve block in cattle - Widely used in practice to enucleate the eye. May also be used in many other species.

Peterson's eye block in cattle - Requires more skill and specific anatomic knowledge than retrobulbar nerve block, but involves less risk in damaging surrounding anatomic structure around the eye globe, and less volume requirement reducing potential for systemic toxicity and expense.

The point of injection is the notch formed by the supraorbital process cranially, the zygomatic arch ventrally, and the coronoid process of the mandible caudally.

An one inch, 14 gauge needle is inserted through a desensitized skin as far anterior and ventral as possible in the notch.

A 4-5 inch, 18 gauge straight or slightly curved needle is inserted at the point of injection mentioned above in a horizontal and slightly posterior direction until it hits the coronoid process of the mandible.

The needle must be gently manipulate anteriorly until its point passes medially around the coronoid process, then advanced to the pterygopalatine fossa rostral to the solid bony plate that is in close proximity of the orbitorotundum foramen. Following aspiration, 7–15 ml of local anesthetics are injected.

Oculomotor, trochlear, abducens, and three branches of the trigeminal nerve (ophthalmic, maxillary, and mandible) are desensitized in 10–15 minutes following injection.

7.6.5 Brachial plexus block

Brachial plexus block is suitable for inducing analgesia for the surgery on the front limb, any area below distal part of humerus.

The technique should be performed in a well-sedated or anesthetized animal.

This block can be used in small ruminants, calves, and foals.

Brachial plexus block is relatively simple and safe to perform and produces selective anesthesia and relaxation of the limb and analgesia to the forelimb.

This technique places a local anesthetic in close proximity to brachial plexus nerves that include the radial, ulnar, median, musculocutaneous and axillary nerves.

The brachial plexus nerves derived from C-6, C-7, C-8 and T-1 spinal nerves roots.

Gradual sensation and loss of motor function occurs within 10-30 minutes depending on the type of drugs used. Anesthesia lasts for approximately 2-6 hours, and total recovery requires approximately 6-9 hours.

A 20-22 gauge spinal needle is inserted medial to the shoulder joint and directed parallel to the vertebral column toward the costochondral junction.

In large size animal, if no blood is aspirated into the syringe as the needle is withdrawn, approximately 10-15 ml of 2% lidocaine or 0.5 % bupivacaine at 3 mg/kg with dilution up to 20-30 ml is slowly injected.

Local anesthetics are injected to brachial plexus which would diffuse into the area to effect.

8. Local anesthesia for husbandry procedures in farm animals

Some of the husbandry objectives are met by practices that are painful for the animals concerned. The behavioural and physiological responses, and by inference the pain, caused by particular husbandry procedures may be reduced by choosing methods that cause less pain, by carrying out the procedures in young rather than older animals and by using pain-relieving drugs, such as local anesthetics.

For each species several such practices may be carried for different reasons: to minimize the risk of injury to animals and people (e.g. dehorning); to reduce aggressive behaviour and make male animals easier to handle (e.g. castrating oxen); to prevent carcass damage such as bruising (e.g. dehorning); to enhance carcass quality (e.g. castration); to prevent damage to the environment (e.g. nose ringing in pigs); to aid in identification (e.g. ear marking or notching, branding);

Cattle can be subjected to the following painful husbandry procedures: disbudding, dehorning, castration, branding and ear notching or tagging. In addition, some heifers may be spayed, and dairy cattle may have their tails docked.

Disbudding is used to mean the prevention of horn growth before it has become advanced, and dehorning means the amputation of horns at any stage after their growth has progressed beyond the early budding stage. Disbudding of calves at a very young age is recommended.

Cautery disbudding is preferable, but extreme escape behaviour of the calf shows that application of the hot iron is painful. This behaviour, and the often less obvious response that occurs with surgical disbudding, can both be eliminated using local anaesthetic blockade of the corneal nerve or a ring block around the base of each horn bud.

Removal of horns from mature cattle without anaesthetics is very painful. Dehorning mature animals may be more painful than most other procedures.

The behavioural response to the pain caused by the act of amputating the horn can be eliminated using local anaesthetic blockade of the corneal nerve or by a ring block around the horn base.

Effective cornual nerve blockade using local anaesthetic eliminates the initial peak of the plasma cortisol response and cortisol concentrations remain at pre-treatment levels for about 2 hours. Thereafter the concentrations increase for about 6 hours before returning to pre-treatment levels (McMeekan *et al.*, 1998a).

This is interpreted to indicate that effective corneal nerve blockade eliminates pain for about 2 hours, after which some pain is experienced. This delayed pain is probably dull and not as acutely sore as the initial pain of horn amputation. To eliminate the acute cortisol response for at least 12 hours the dehorned animal needs to be given a systemic analgesic along with the local anaesthetic; a combination of an NSAID (non-steroidal anti-inflammatory drug) with a local anaesthetic is effective in this regard (McMeekan *et al.*, 1998b; Stafford *et al.*, 2003; Milligan *et al.*, 2004; Stewart *et al.*, 2009).

When local anaesthetic is given before dehorning, cautery of the wound to control haemorrhage also reduces the plasma cortisol response, and by inference the pain experienced by cattle, for at least 24 hours (Sutherland *et al.*, 2002).

Castration is a standard husbandry practice in most cattle production systems and it is painful. The cortisol response to clamp castration is less than the response to rubber-ring or surgical castration, which suggests that the clamp is less painful (Stafford *et al.*, 2002).

Local anaesthesia placed in the distal scrotum and testicles eliminates or reduces the pain-related behaviours seen at castration by these methods. It eliminates the plasma cortisol response to castration by rubber ring or band. This is because the tight ring or band stops blood and lymph perfusion of the testicles and scrotum so that the local anaesthetic remains in those tissues beyond the time required for anoxic death of the pain receptors and associated nerves.

Heifers are spayed to prevent mis-mating and to prevent mating of cull cows in extensive farming systems. The ovaries are removed either through the vagina in larger cows and heifers following epidural anaesthesia or in smaller heifers through an incision in the flank. A flank incision is certainly painful and warrants local anaesthesia.

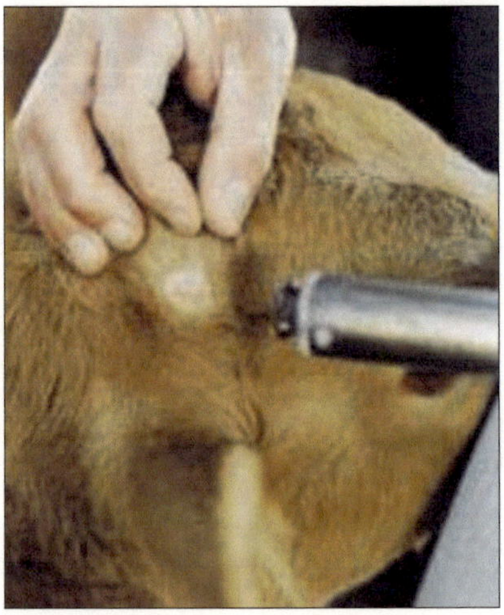

Fig. 4. Cautery disbudding in the calf

Tail docking young calves by rubber ring or cautery is not especially painful, and local anaesthetic either given as a tail ring block or an epidural can be used to alleviate the pain.

Ear tagging and is a common way of identifying individual animals. Ear tags or ear notches are used also to indicate the ownership of cattle. Little work has been done on reducing the pain caused by these procedures, and as it is not easy to anaesthetize the ears using local anaesthetic, systemic analgesia would be preferable.

Branding by hot iron or freeze branding are common means of identifying cattle both for ownership and for recognizing individual animals. Little work has been done on reducing the pain caused by these procedures.

As multiple injections of local anaesthetic in the area to be branded are impractical, systemic analgesia would probably reduce but not eliminate the pain experienced following branding.

The most common painful husbandry procedures used in sheep are castration, tail docking and ear tagging or notching. These procedures are generally carried out on lambs.

Local anaesthetic injected into the scrotum and testicles eliminates the plasma cortisol response and by inference the acute pain caused by rubberring castration (Dinniss *et al.*, 1997). Kent *et al.* (1998) found that lidocaine injected into the neck of the scrotrum was more effective than injection into the testicles, but Dinniss *et al.* (1997) showed that intratesticular injections were just as effective but practically much easier to execute than scrotal neck injections.

Local anaesthetics have less impact on responses to surgical or clamp castration, and with these techniques a systemic analgesic such as an NSAID is required together with local anaesthesia to eliminate the pain caused by castration.

Squirting local anaesthetic into the scrotal neck and on to the spermatic cords before they are severed towards the inguinal area may reduce the pain experienced by lambs following surgical castration where the spermatic cords are drawn out and broken by traction.

Tail docking is carried out on woolly sheep. Local anaesthesia may be injected as a ring block around the tail (subcutaneously), but this is difficult and time consuming. The use of local anaesthetics greatly reduced behavioural indicators of pain (Kent *et al.*, 1998).

Ear tagging or ear notching in sheep are similar to those carried out in cattle.

Disbudding is carried out on young goats before the horn bud becomes too large. Local anaesthetic is often used when disbudding goats.

Castration and ear tagging or ear notching and the use of analgesia are conducted in goats in a manner similar to those used in cattle and sheep.

The management of pigs ranges from intensive indoor to extensive outdoor approaches. Piglets are subject to a number of painful procedures including castration, teeth clipping, tail docking and, if they are going to live outdoors, rings may be inserted into their noses.

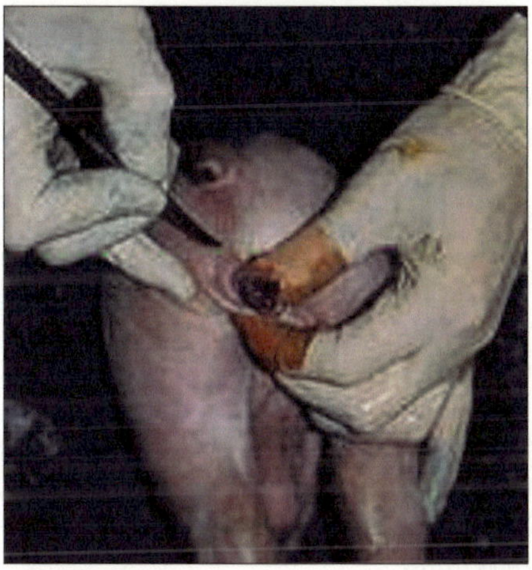

Fig. 5. Piglet castration

Traditionally, pigs have been castrated surgically. Local anaesthetics prevented pain-related behaviours in 2-week-old piglets, but not in 7- week-old piglets (McGlone & Heilman, 1988).

Gaseous anaesthesia technology suited to pig farms is being developed and it may soon become normal to give piglets a general anaesthetic and then to castrate them, clip teeth and dock tails at one time.

Local anaesthesia and sedation could be used to reduce the pain caused by ring placement.

Horse castration is carried out in many countries under general anaesthesia, or standing under sedation and with local anaesthesia (Ohme & Prier, 1974).

Beak trimming is the most important amputation carried out on chickens and turkeys. Beak trimming is acutely painful and food intake levels decrease following it (Glatz, 1987).

Behavioural and physiological evidence suggests that pain and beak sensitivity persists for weeks or months after trimming (Craig and Swanson, 1994). A local analgesic (bupivacaine and diethyl sulfoxide) administered to the cut beak of 6-week-old chicks after trimming prevents at least some of the acute pain and the reduction in feed intake usually seen in the first 24 hours after trimming.

9. Using farm animals for experimental purposes

In many countries, laboratory animals (mainly rodents) but also farm animals (mainly horses, cattle, sheep and goats) are used in research, testing and teaching.

Experiments conducted on farm animals may cause pain which ought to be eliminated or reduced to minimum for ethical and scientific reasons.

The suffering associated with surgical procedures may be prevented with the use of local anaesthetics. Most of them, if injected or inhalated, have an effect on physiological functions of animal body which may distort the results of an experiment, but this is also true for the pain which triggers stress reaction. The experiments may be performed without anaesthesia only in exceptional cases, when it is necessary from the scientific point of view.

Reliable results of experiments on farm animals are largely dependent on the standardisation of the factors affecting physiological reactions of these animals and on the broad idea of their wellbeing. The main factors influencing the quality of experiments conducted on farm animals include: their biological status (sex, age, body mass); health condition; nutrition; maintenance conditions; animal headquarters (ventilation, temperature, humidity, lighting, noise); exposure to stressogenic stimuli; proper care; the choice of appropriate experimental techniques. Meeting these requirements allows to obtain repeatable, reliable results of experiments and to create proper living conditions.

Experiments conducted on animals often require blood sampling. It is important to safeguard the welfare of farm animals from which blood is harvested for research purposes. Contemporary animal welfare requirements are more stringent than they have been in the past, and it is appropriate that there should be guidelines to protect the welfare of animals used in blood harvesting operations.

For the research purposes, blood harvesting usually implicates the removal of a relatively large volume of blood over a short period of time, i.e. more than would usually be required or routine diagnostic tests.

It is in the interests of good science as well as the welfare of the animals used that stress on the animals is kept to a minimum. If the process of harvesting is stressful for the animal because of handling, pain or discomfort, physiological changes occur which may compromise the quality of the scientific result obtained.

10. Topical anesthetics

Although local areas of anaesthesia can be produced by subcutaneous injection of local anaesthetics, the process of injection may itself cause pain. A number of attempts have been made to obtain suitable formulations for effective topical application to the skin.

Local anaesthetic creams are generally ineffective when applied to intact animal skin because they are poorly absorbed. This has limited their use to topical anaesthesia of mucous membranes. To diffuse through nerve membranes the local anaesthetic must be in the form of the unchanged base, but formulations in which the lipid soluble free base dissolves easily achieve poor penetration of unbroken skin.

This problem has been overcome with a eutectic combination of lidocaine and prilocaine. When mixed together at room temperature crystals of these drugs form an oily liquid. No additional solvent is necessary, so droplets of an oil in water emulsion of lidocaine and prilocaine each have a concentration of 80% of active drug compared with the 20% of conventional formulations. The total concentration of local anaesthetic remains low at 5%.

Lidocaine-prilocaine cream produces effective topical dermal anaesthesia. The advantages of avoiding painful needle punctures are selfevident. Not only is animal discomfort reduced but also the procedure is easier to carry out.

11. Local anesthesia for vessel puncture

Pain associated with vessel puncture has long been accepted as an unavoidable consequence of blood collection and intravenous drug administration. The degree of pain associated with the procedure is believed to be minimal, provided that a fine gauge needle is used and that the animal is expertly restrained and the operator skilled in the technique.

In many instances, however, it is necessary to use larger gauge needles and this may produce a detectable reaction from the animal. Such responses are also noted more frequently when over-the-needle type cannulae are inserted to provide long-term vascular access.

When vessel puncture is undertaken by relatively inexperienced staff, repeated attempts may be necessary before the procedure is carried out successfully. Under these circumstances, the animal may experience greater pain or discomfort than when the technique is undertaken by a skilled operator.

There has been considerable debate as to the nature of pain in animals, since it is unclear what form animal experiences may take. It has been suggested that one constructive approach is to accept that animals are capable of sensory experiences which, given the opportunity, they avoid and which modify their biochemistry, physiology, and behaviour in a species-specific way. Many of the stimuli that produce these reactions in animals would produce the sensation described as pain in man, and hence nociceptive reactions in animals such as withdrawal of the limb, vocalizations and aggressive behaviour may be interpreted as one type of response to pain.

In human anaesthetic practice, it has been demonstrated that pain caused by the intravessel insertion of needles and cannulae can be alleviated by the topical application of local anaesthetic cream. The feasibility of using a similar technique in animals has been shown.

A commercially available local anaesthetic cream (EMLA®) can be used for application prior to vessel puncture in farm animals. EMLA cream consists of a eutectic mixture of lignocaine base (25 mg/ml) and prilocaine base (25 mg/ml), together with an emulsifier, a viscosity increasing agent and water and with the pH of the cream adjusted to pH 9,4 with sodium hydroxide.

12. Blood sampling in farm animals

The farm animals used for blood collection should be more than 6 months old and preferably fully grown adults. They must be in good body condition and in good health.

Horses should be at least 3 years old and at least 400 kg body weight. Horses with a body condition score of 3 or less must not be bled.

Sheep and goats should be at least 12 months old. Sheep should weigh at least 40 kg. Goats vary greatly in size according to breed, so a minimum weight cannot be given for the species, but they should be in good body condition.

Cattle should be at least 18 months old and should weigh at least 250 kg if 18 to 24 months of age and at least 300 kg if older.

If growing animals are used their diet must be nutritious, supplemented with iron, and fed at appropriate above-maintenance levels to allow for replacement of blood as well as steady growth.

The animals to be bled must be accustomed to handling so that they are relaxed and calm throughout the harvesting procedure. Excitement and fear can cause splenic contraction which results in altered blood parameters.

The design of the bleeding facilities and methods of restraint must be such that the procedure can be carried out efficiently with the minimum of stress to the animals.

Cattle and horses must be standing when bled.

It is preferable to bleed sheep and goats in the standing position, but where large numbers of sheep and goats are bled, it may be acceptable to strap them in lateral recumbency.

The harvesting procedure should begin as soon as the animal is restrained, and animals should be kept under close supervision during the bleeding process to guard against inhalation of ruminal contents or the development of ruminal bloat. After release they should be allowed to return to their paddock at their own pace.

The skin over the sampling site may be clipped or shaved to facilitate placement of the needle and the site may be cleaned with disinfectant such as alcohol. It is important that time be taken to locate the vessel accurately and that it be distended by gentle pressure before the needle is inserted.

A needle with as large a bore size as possible should be used to ensure efficient blood withdrawal without collapsing the vessel, without causing haematoma formation and without causing blood pressure to drop too rapidly.

Immediately after removal of blood, all animals must have unrestricted access to water.

As concerning the volume and frequency of bleeding for adult animals, not more than 15% of the estimated circulating blood volume should be removed in any 4-week period, i.e. 0.9% liveweight in cattle and sheep and 1.1% liveweight in goats and horses.

Circulating blood volume (litres) can be estimated from body weight (kg) using a conversion factor of 0.06 for cattle and sheep, 0.07 for goats and 0.075 for horses. As a guide,

1% of body weight is the weight of 16 to 17% of the circulating blood volume in sheep and cattle; about 13% in horses and about 14% in goats.

If more than 15% of blood volume is removed, consideration should be given to fluid replacement using lactated Ringer's solution with 5% dextrose.

For young animals, the volumes removed should be relatively less. For animals 6 months old, not more than 10% circulating blood volume should be removed, with incremental increases to the maximums above when fully grown (more than 12 months old for sheep and goats, 18 months old for cattle and 3 years old for horses).

If too much blood is drawn too quickly or too frequently without replacement, animals may develop hyperpnoea (deep and rapid breathing) and may go into a state of hypovolaemic shock. In the longer term the removal of to much blood causes anaemia, muscle weakness, increased susceptibility to cold, reduced exercise tolerance and ill-thrift, particularly if management and nutrition are suboptimal.

If 15% to 20% of the blood volume is removed, cardiac output and blood pressure will be reduced. Removal of 30% to 40% can induce shock.

Inexpert bleeding techniques can result in bruising around the vein, haematoma formation and/or inflammation at the site.

12.1 Bleeding technique in cattle

Blood can be collected from the jugular vein in cattle of all ages or from the tail (coccygeal) vein of older cattle.

A variety of collection devices may be used: vacutainers, bleeding tubes, syringe and needle. Restraint should ensure quick, easy and safe collection of the sample causing minimal distress.

This may involve use of a bail, race, or crush for tail bleeding. For jugular bleeding the animal may require minimal restraint (e.g. halter) or may need to be restrained in a crush with head bail and the employment of a halter or nose grips. Use of nose grips should be avoided wherever possible.

Operators should use gloves and disinfect them between animals to prevent the transmission of blood-borne diseases. Equipment such as vacutainer holders should also be cleaned between animals.

Jugular bleeding - The animal must be restrained with the head elevated and the jugular groove exposed. The jugular vein is raised by pressure at the base of the jugular groove. The needle is passed through the skin and into the vein by a firm thrust directed an angle of 20° to the skin surface and the blood sample is withdrawn.

Tail bleeding - Restraint should prevent the animal from moving away during the procedure. The tail must be raised vertically with one hand until it is horizontal with the ground.

Approximately 150 mm from the base of the tail, the groove lying in the ventral midline of the tail is located. Midway along the body of a coccygeal vertebra, the needle must be inserted perpendicularly to the surface of the skin to a depth of a few millimetres and the blood sample is withdrawn. Pressure must be applied to the venipuncture site after withdrawal of the needle until the bleeding stops.

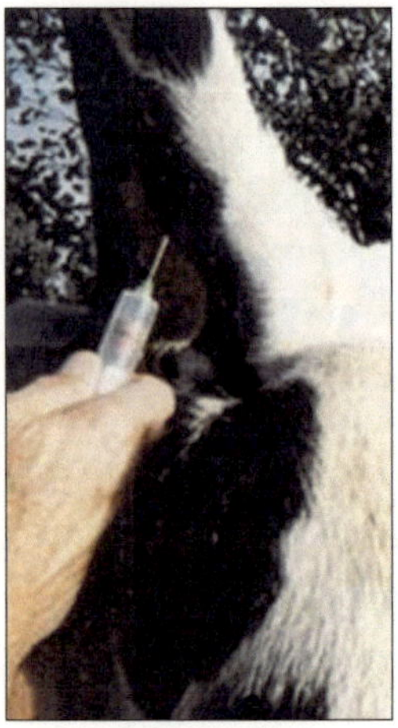

Fig. 6. Tail bleeding in cattle

13. Conclusion

Most of the husbandry procedures used today were developed many years ago. They were usually selected to be carried out quickly, easily, cheaply and with inexpensive equipment, and to be generally safe for the animals and people involved.

At the time of their development, little significance was attributed to any associated animal pain. Since the mid-1990s, many advances have been made in the field of animal pain research but much remains to be done.

It is necessary a full analysis of matters that should be considered when deciding whether or not, and how, to undertake particular painful husbandry procedures.

Public attention on the welfare consequences of surgical husbandry procedures used on farm animals is increasing. This suggest that alternative approaches must be considered.

The best way to reduce the pain and distress caused by horn injuries or horn removal would be to select for polled cattle. At present the common dairy breeds, Friesian and Jersey, are horned. It may be possible for such breeds to be bred polled if an effort were made to do so and this would completely remove the necessity for disbudding or dehorning.

A possible alternative to castration in cattle could be the use of immunocastration. The vaccines used for this are now available, and it is a matter of seeing whether this practice is acceptable to producers and consumers.

Also spaying heifers could be avoided by rendering the cows infertile: this is now possible immunologically.

Tail docking of dairy cows is prohibited in some countries and in many natural and organic programmes. It is a procedure that should be phased out, because there is no scientific justification for it.

Microchips to identify cattle are becoming a feasible option to replace branding for identification and their placement will be less painful than branding. However, they are not really suited to everyday identification and it will take some time before hand-held identity chip readers become a widely used tool on cattle farms.

Another alternative for animal identification is DNA testing. The cost of DNA testing is steadily falling and it may become economically feasible in the future.

Immunocastration vaccines for lambs are likely to become available in the near future and they could be used to eliminate the pain caused by castration if their use is acceptable throughout the production and consumption chain and provided that they are cheap enough.

It is possible to breed sheep with reduced wool in the perineal area and for shorter and less woolly tails. This would reduce the necessity of docking to reduce flystrike, but shearing would still be impeded by the presence of the tail.

Identity microchips may reduce the need for ear marking and tagging in the future, but they are economically and practically difficult to use on large commercial sheep farms at present.

The Australian wool industry aims to phase out mulesing by 2010. This will be facilitated by breeding Merino sheep with fewer wrinkles in the perineal area and the use of anti-fly vaccination or long-lasting insecticides.

Immunocastration is a possible alternative to surgical castration in pigs and may become more a widely accepted practice.

A combination of using specific genetic lines of laying hens that have a low tendency to feather peck, and the use of particular environmental features, might reduce cannibalism to the degree that beak trimming is not necessary.

Although it is generally seen as preferable for the future not have to perform surgery to achieve husbandry outcomes, there are several reasons why the use of local anesthesia should not be discounted.

Firstly, genetic improvement takes time. Even for a trait such as polledness in cattle, where only a few genes control the polled condition, there would be a significant time lag following the identification of the gene(s) and development of a test before homozygous breeding animals were disseminated throughout the cattle population. There is also likely to be some time delay before the development and implementation of some of the non surgical approaches currently under investigation.

Removal of blood is one of the most common procedures performed on animals for experimental purposes. The choice of a technique appropriate for the purpose and the species, by a trained staff, and the use of local anesthesia are essential for ensuring that any pain, distress or discomfort is kept to a minimum.

Minimisation of such adverse effects is important for scientific as well as ethical and legal reasons, since they can cause biological changes which may affect the blood sample, and hence the validity of the research results and the number of animals used to achieve the scientific objective.

14. References

Dinniss, A.S, Mellor, D.J., Stafford, K.J., Bruce, R.A. & Ward, R.N. (1997). Acute cortisol responses to lambs to castration using rubber ring and/or a castration clamp with or without local anaesthetic. *New Zealand Veterinary Journal,* Vol.45, pp. 114–121

Flecknell, P. & Waterman-Pearson, A. (2000). *Pain Management in Animals,* W.B. Saunders, London, UK

Glatz, P. (1987). Effects of beak trimming and restraint on heart rate, food intake, body weight and egg production in hens. *British Poultry Science,* Vol.28, No.4, pp. 601–611

Hall, L.W. & Clarke, K.W. (1991). *Veterinary Anaesthesia,* Baillière Tindall, London, UK

Kent, J.E., Molony, V. & Graham, M.J. (1998). Comparison of methods for the reduction of acute pain produced by rubber ring castration of week-old lambs. *Veterinary Journal,* Vol.155, pp. 39–51

McGlone, J.J. & Heilman, J.M. (1988). Local and general anesthetic effects on the behaviour and performance of two and seven week old castrated and uncastrated piglets. *Journal of Animal Science,* Vol.66, pp. 3049–3058

McMeekan, C.M., Mellor, D.J., Stafford, K.J., Bruce, R.A., Ward, R.N. & Gregory, N.G. (1998a). Effects of local anaesthesia of 4 or 8 hours duration on the acute cortisol response to scoop dehorning in calves. *Australian Veterinary Journal,* Vol.76, pp. 281–285

McMeekan, C.M., Stafford, K.J., Mellor, D.J., Bruce, R.A., Ward, R.N. & Gregory, N.G. (1998). Effect of regional analgesia and/or a non-steroidal antiinflammatory analgesic on the acute cortisol response to dehorning in calves. *Research in Veterinary Science,* Vol.64, pp. 147–150

Mellor, D.J. & Stafford, K.J. (2000) Acute castration and/or tailing distress and its alleviation in lambs. *New Zealand Veterinary Journal,* Vol.48, pp. 33–43

Milligan, B.N., Duffield, T. & Lissemore, K. (2004). The utility of ketoprofen for alleviating pain following dehorning of young dairy calves. *Canadian Journal of Veterinary Research,* Vol.45, pp. 140–143

Ohme, F.W. & Prier, J.E. (1974). *Large Animal Surgery,* Williams and Wilkins, Baltimore, Maryland

Stafford, K.J., Mellor, D.J., Todd, S.E., Bruce, R.A. & Ward, R.N. (2002). Effects of local anaesthesia or local anaesthesia plus a non-steroidal anti- inflammatory drug on the acute cortisol response of calves to five different methods of castration. *Research in Veterinary Sciences,* Vol.73, pp. 61–70

Stafford, K.J., Mellor, D.J., Todd, S.E., Ward, R.N. & McMeekan, C.M. (2003). Effects of different combinations of lignocaine, ketoprofen, xylazine and tolazoline on the acute cortisol response to dehorning in calves. *New Zealand Veterinary Journal,* Vol.51, pp. 219–226

Stafford, K.J. & Mellor, D.J. (2005a). Dehorning and disbudding distress and its alleviation in calves. *Veterinary Journal,* Vol.169, pp. 337–349

Stafford, K.J. & Mellor, D.J. (2005b). The welfare significance of the castration of cattle: a review. *New Zealand Veterinary Journal,* Vol. 53, pp. 271–278

Stewart, M., Stookey, J.M., Stafford, K.J., Tucker, C.B., Rogers, A.R., Dowling, S.K., Verkerk, G.A., Schaefer, A.L. & Webster, J.R. (2009). Effects of local anaesthetic and a nonsteroidal anti-inflammatory drug on pain responses of dairy calves to hot-iron dehorning. *Journal of Dairy Science,* Vol.92, pp. 1512–1519

Sutherland, M.A., Bryer, P.J., Krebs, N. & McGlone, J.J. (2008). Tail docking of pigs: acute physiological and behavioural responses. *Animal,* Vol.2, pp. 292–297

Thurmon, J.C., Tranquilli W.J. & Benson, G.J. (1996). *Veterinary Anesthesia,* Williams & Wilkins, Baltimore, USA

14

Artificial Insemination and Its Role in Transmission of Swine Viruses

Tanja Opriessnig, Luis G. Giménez-Lirola and Patrick G. Halbur
Department of Veterinary Diagnostic and Production Animal Medicine, College of Veterinary Medicine, Iowa State University, Ames, Iowa, USA

1. Introduction

Artificial insemination (AI) in swine is not a new technique and reports as early as the 1930s (Lush, 1925) describe collecting semen for AI. However, because of farm structure changes, increasing farm sizes and separation of production stages, interest in intensive pig production is growing and AI has become a critical component in modern pig production. In 2001, nearly 60% of North American swine herds utilized AI (Singleton, 2001), a drastic increase from the estimated 5% in the 1990's (Flowers & Esbenshade, 1993). This is still relatively low compared to the 90% or greater use of AI in Western Europe (Madsen, 2005; Maes et al., 2008). The extensive use of AI in pig reproduction in the last decade has facilitated the exchange of desirable genetic characteristics at an international level, allowing producers to make greater use of superior genetics at a lower cost than some natural-service systems (Gerrits et al., 2005). However, the growth in use of AI has increased the risk of quick and widespread transmission of venereally transmissible pathogens (Thacker et al., 1984). It has been reported that the porcine male reproductive tract is highly susceptible to viral infections (Phillips et al., 1972; Spradbrow, 1968). This, coupled with the ability of boars to produce tens to thousands of insemination doses per week and the widespread distribution of the processed semen (both nationally and internationally), further increases the risk of wide transmission of viral pathogens by semen.

Many viruses have been reported to be present in boar semen (Yaeger et al., 1993; Lucas at al., 1974, Madson et al., 2008) and have the potential of being transmitted to susceptible breeding animals during AI, particularly when the boar is viremic or clinically sick. Viral shedding may also continue long after clinical signs have abated. In addition, infertility or reduced reproductive performance have been reported in boars with detectable virus in their semen (Guerin & Pozzi, 2005; Larsen et al., 1980). The potential economic impact and liability associated with transmission of diseases through semen has created great interest and investment in testing boars for viral diseases prior to entry and while at AI centers.

In many countries, foot-and-mouth disease virus (FMDV), porcine reproductive and respiratory syndrome virus (PRRSV), Japanese B encephalitis virus (JBEV), pseudorabies virus (PRV), classical swine fever virus (CSFV) and African swine fever virus (ASFV) are of particular importance. Accurate monitoring of boars in AI stations is essential to reduce the risk of virus transmission. Monitoring for the presence of other ubiquitous viruses that can be found in semen, such as porcine circovirus type 2 (PCV2), torque teno sus virus (TTSuV), porcine parvovirus (PPV) and porcine adenovirus (PAV) is commonly not a high priority in most countries despite the fact that some of these viruses have been associated with sporadic reproductive failure in breeding animals.

Due to differences in pathogenicity, economic consequences, geographic localization, and epidemiological parameters, different viruses are commonly assigned different levels of importance. The aims of this review are to summarize the information on swine viruses that can be transmitted via AI, their shedding characteristics in boars, effects of these viruses on naïve breeding animals, testing that is currently done to reduce or prevent the risk of transmission, and current best practices being used to reduce or eliminate virus shedding in semen.

2. Impact of viral contamination of semen

Microbial contamination of boar semen not only reduces semen quality but can also reduce conception rate resulting in considerable economic loss. It is not uncommon that a single boar stud serves multiple breeding herds with an average of 200 to 10,000 sows on each site. The risk for breeding herds becoming infected via AI increases with the number of pathogens present in the semen, the dose of the pathogen(s) in the semen, the number of sows inseminated with the contaminated semen, and the level of protective immunity in the breeding herd. From a practical point of view, most economically important viral diseases are associated with clinical signs, so semen collection from sick boars will not always take place, and consequently the risk of pathogen transmission into susceptible breeding herds by this route may be lower than expected. However, when clinical signs are mild or absent the risk of viral transmission via contaminated semen is a concern.

Artificial insemination is now widely recognized as a route for spread of swine diseases (Thacker et al., 1984) and continues to be a major concern to pig breeders and regulatory authorities in countries where AI is practiced. Many viruses known to contaminate semen have been shown to be highly infectious via the uterine route, and wide dissemination of porcine pathogens by semen could occur when semen is shipped globally. Transmission of viruses via the semen route to breeding females has been experimentally proven for CSFV (de Smit et al., 1999), PRRSV (Yaeger et al., 1993), and PPV (Lucas et al., 1974). It is of importance to note that, although virus-contaminated semen indeed constitutes a serious risk for transmission, it does not guarantee tthat transmission to the sow or gilt by AI will consistently occur (Swenson et al., 1994b; Yaeger et al., 1993). The conditions required for establishment of infection in the breeding female are complex, and lack of transmission might be explained by sow immunity or failure to reach the minimum infectious dose. In this regard, much research has been conducted concerning the risk of transmission of PRRSV by semen, and the minimum dose necessary to establish infection in the sow (Benfield et al., 2000; Prieto et al., 1996b).

Contamination of semen with virus can occur through three possible routes: 1) fecal contamination of semen during collection, 2) systemic viral infection, or 3) local viral infection (testes, accessory gland, etc.). Once virus-contaminated semen is utilized to inseminate a breeding female, the outcome can vary depending on the time of fetal infection. Early embryonic death may result from direct invasion of the embryo by the pathogen or by uterine epithelial alterations in response to the pathogen (Wrathall & Mengeling, 1979). Until six to seven days after conception, an embryo is surrounded and protected by an impervious barrier called the zona pellucida, which helps the embryo avoid pathogen invasion (Mateusen et al., 2004). However, after entering the blastocyst stage, embryos may become susceptible to viral infections. Fetuses that are infected prior to 70 days of gestation usually die (mummies) and contribute to smaller litter sizes or early termination of pregnancy. Fetuses that are infected around or after 70 days of gestation are able to mount an active immune response against the pathogen and survive intrauterine infection.

As the risk of rapid virus spread via semen is well known, sanitation protocols in AI centers are constantly being reviewed to assure customers that current best practices are in place. A new categorization of viruses has been proposed to further address different risk levels (Guerin & Pozzi, 2005): Category 1 includes viruses for which there is scientific proof of transmission by semen, but without any risks for AI due to official national eradication programs such as FMDV, PRV, CSFV, and ASFV. Category 2 includes viruses for which there is scientific proof of transmission by semen and which can be commonly encountered such as PRRSV. Category 3 includes viruses for which additional scientific proof is needed to better assess the risk of transmission by semen (PCV2, porcine rubulavirus). In addition, some viruses have not been shown to be present in semen and thus are considered non-hazardous to AI.

3. Swine viruses that can be present in boar semen

3.1 Porcine Reproductive and Respiratory Syndrome Virus

Porcine reproductive and respiratory syndrome virus (PRRSV) is a single-stranded, positive-sense, enveloped RNA virus that belongs to the family *Arteriviridae*, genus *Arterivirus* (Cavanagh, 1997). PRRSV was first recognized in the late 1980's (Cavanagh, 1997; Meulenberg et al., 1993) and today is found globally in swine producing countries. As the name of the virus implies, PRRSV infection is associated with reproductive failure in pregnant sows and respiratory disease in pigs of all ages. It is less commonly associated with neonatal diarrhea (Albina et al., 1994; Bierk et al., 2001; Neumann et al., 2005; Rossow, 1998; Wensvoort et al., 1991). Although clinical disease associated with PRRSV in growing pigs can be quite severe, no or mild symptoms are typically seen in boars (Wensvoort et al., 1991). One of the main characteristics of PRRSV is its high transmissibility, making it difficult to maintain pig populations free of PRRSV (Prieto & Castro, 2005). The 50% tissue culture infective dose ($TCID_{50}$) for exposure via oral, intranasal and AI routes was determined to be $10^{5.3}$, $10^{4.0}$ and $10^{4.5}$, respectively (Benfield et al., 2000). Experimental infection in boars has demonstrated seminal shedding of PRRSV (Prieto et al., 1996b; Swenson et al., 1994a), and epidemiological evidence confirms that transmission of PRRSV from fresh semen of acutely infected boars into breeding herds is possible (Robertson, 1992; Yaeger et al., 1993). However, successful transmission is not always achieved (Swenson et al., 1994b; Yaeger et al., 1993) and appears to depend largely on the amount of PRRSV present in the semen (Benfield et al., 2000). In semen of adult boars, PRRSV can persist for variable periods (Christopher-Hennings et al., 1995a; Swenson et al., 1994) suggesting that the virus continues to replicates in the one or more tissues of the reproductive tract. Several studies have since been carried out to determine the temporal localization of the virus in different organs and tissues (Prieto et al., 2003; Sur et al., 1997) and vascular dissemination and replication of PRRSV in tissues of the reproductive tract was confirmed (Sur et al., 1997). In addition, migration of infected monocytes and macrophages directly from blood and lymph of the reproductive tract into the semen has also been suggested as a mechanism for PRRSV contamination of semen (Christopher-Hennings et al., 1998; Prieto et al., 2003). PRRSV has been detected in semen samples of experimentally infected boars ranging from 4 to 92 days post infection (Christopher-Hennings et al., 1995a, 1998). This marked variability may be due to individual host factors (Christopher-Hennings et al., 1995a, 2001; Swenson et al., 1994a), the breed of the boars (Christopher-Hennings et al., 2001), and the strain of virus used in the experimental inoculation. European and North American PRRSV strains differ genetically and antigenically (Bautista et al., 1993; Meng et al., 1995). A bioassay has been used to determine the presence of infectious PRRSV in semen samples for a period of time lasting from 4 to 42 days post infection (Swenson et al., 1994a). In semen, PRRSV is sometimes associated with drastic changes in quality (reduced motility, abnormal acrosomes, morphological alterations, etc.) and volume of the semen (Teuffert et al., 1998). Conversely, some researchers have found that the quality of semen remains within normal limits after infection with PRRSV (Swenson et al., 1994a; Yaeger et al., 1993). The etiological role of PRRSV in reproductive failure of swine is firmly established (Christianson et al., 1993; Lager & Mengeling, 1995; Mengeling et al., 1994; Prieto et al., 1996a; Terpstra et al., 1991). The effect of PRRSV on reproductive parameters was found to be highly related to strain pathogenicity (Prieto et al., 1997). Sows and gilts bred by infected boars (Gradil et al., 1996; Yaeger et al., 1993) or with experimentally contaminated semen (Prieto et al., 1997; Swenson et al., 1994b) typically show seroconversion to PRRSV, even in the absence of detectable viremia (Christopher-Hennings et al., 1995a, 2001). It has been reported that insemination of seronegative or pre-immunized gilts with boar semen containing PRRSV often has little to no effect on conception rates, but may result in early embryonic infection and death (Prieto et al., 1997). In addition, PRRSV can be a major cause of prenatal death in commercial swine herds and often preweaning mortality is also increased.

3.2 Porcine Parvovirus

Porcine parvovirus (PPV) is a small, single-stranded, non-enveloped DNA virus (Molitor et al., 1984) that belongs to the family *Parvoviridae*, genus *Parvovirus*. PPV is considered ubiquitous in the pig population. The main routes of transmission of PPV are oronasal and transplacental. In general, infection of growing pigs and mature boars with PPV alone is not associated with clinical disease (Allan et al., 2000; Brown, Jr. et al., 1980; Kennedy et al., 2000; Krakowka et al., 2000). There is evidence that PPV-isolates vary in pathogenicity (Mengeling & Cutlip, 1976; Oraveerakul et al., 1993). PPV-isolates have been classified as non-pathogenic (Mengeling & Cutlip, 1976; Paul et al., 1979), pathogenic to non-immunocompetent fetuses leading to death (Mengeling, 1979), pathogenic to immunocompetent fetuses and inducing dermatitis (Choi et al., 1987; Kresse et al., 1985; Lager et al., 1992; Lager & Mengeling, 1994), and enteric PPV-strains (Dea et al., 1985; Duhamel et al., 1991). PPV has previously been isolated from semen of naturally infected boars (Cartwright & Huck, 1967). Boars can shed the virus in semen during the acute phase of infection (Gradil et al., 1990); shedding beyond this phase has not been demonstrated, but the possibility of immunotolerant carriers of PPV as a result of early *in utero* infection has been suggested (Cartwright et al., 1971). Semen may also become contaminated with PPV from feces or from the male reproductive organs (Biront & Bonte, 1983; Lucas et al., 1974). Changes in sperm output, ejaculate volume, motility, or morphologic defects have not been observed when the semen quality was evaluated after experimental inoculation of boars with PPV (Thacker et al., 1987a). Acute infection of breeding females with PPV is usually subclinical (Johnson et al., 1976). During initial infection, PPV replicates extensively and can be found in many tissues including lymphoid tissues (Cutlip & Mengeling, 1975a). The major and often only clinical response to PPV infection is reproductive failure. Return to estrus, fewer pigs per litter, and increased numbers of mummified fetuses are often observed (Cartwright & Huck, 1967; Joo et al., 1976). The consequence of PPV infection depends on timing of infection of the fetus: Death and reabsportion are usually observed in 10-30 gestation day embryos, death and mummification are seen when 30 to 70 day gestation fetuses are infected, and an active immune response and survival *in utero* is seen in fetuses when infected after 70 days of gestation (Bachmann et al., 1975; Cutlip & Mengeling, 1975b). The role of semen contamination with PPV leading to clinical reproductive problems has not been clearly established (Lucas et al., 1974; Mengeling & Paul, 1986; Thacker et al., 1987b; Wrathall & Mengeling, 1979).

3.3 Pseudorabies Virus

Pseudorabies virus (PRV) is a double-stranded DNA virus that belongs to the subfamily *Alphaherpesvirinae* of the family *Herpesvirus* (Mettenleiter, 2000). PRV is distributed worldwide; however, in recent years eradication efforts have been successful to eliminate PRV from the domestic pig population in parts of Europe, Canada, New Zealand and the United States. PRV replication typically occurs in the nasal and pharyngeal mucosa and primary transmission occurs via the nasal route. Viral replication has also been reported to take place in the genital tract and transmission by copulation is therefore possible (Hall, Jr. et al., 1984a; Vannier & Gueguen, 1979). Differences in pathogenicity and duration of virus shedding exist among PRV strains (Maes et al., 1983). Clinically affected boars are often unable to mount a dummy (Guerin & Pozzi, 2005) and can additionally develop respiratory disease (Hall, Jr. et al., 1984b). PRV can be isolated infrequently from urine, preputial membranes, and semen after either natural (Medveczky & Szabo, 1981) or experimental infection (Vannier & Gueguen, 1979). Very high viral concentrations have been reported in semen ranging from 10^8 to 10^9 $TCID_{50}$ per ml (Medveczky & Szabo, 1981) to $10^{3.7}$ to $10^{4.9}$ $TCID_{50}$ per ml (Vannier & Gueguen, 1979). Transient viral excretion for up to 12 days usually occurs during the acute phase of the disease (Ressang, 1973) but viral excretion has been observed for long periods of time after natural infection (Wittmann, 1985). After experimental PRV infection by the intratesticular route, testicular degeneration and transient elevation in sperm abnormalities have been reported (Hall, Jr. et al., 1984a; Larsen et al., 1980). Sows inseminated with contaminated semen may develop vaginitis or endometritis resulting in embryonic death (Maes et al., 2008). Sows infected in the first trimester of pregnancy often reabsorb their fetuses and return to estrus. If sows get infected in the later stages of pregnancy they may abort or have larger numbers of stillborn and weakborn piglets. If susceptible females are infected with PRV close to parturition, there are increased numbers of weakborn pigs with signs of nervous system disease.

3.4 Porcine Rubulavirus

Porcine rubulavirus (PoRV), also known as La-Piedad-Michoacan paramyxovirus (LPMV), is associated with blue eye disease in pigs. This virus was first isolated in Mexico in the early 1980's. PoRV belongs to the genus *Rubulavirus* and family *Paramyxoviridae.* Only one serotype has been recognized to date. Subclinically infected pigs are the primary source for PoRV transmission which occurs primarily via oronasal route. Boars, like other adult animals infected with PoRV, generally do not show clinical signs except for epididymitis and orchitis and in some cases loss of libido (Maes et al., 2008). Transmission of the virus through semen has not been proven experimentally; however, PoRV has been recovered from semen, testes, and other tissues of the reproductive tract for up to 49 days after inoculation (Solis et al., 2007). Based on semen evaluation in herds naturally infected with PoRV, approximately 30% of boars develop temporary or permanent infertility. Semen abnormalities include a decrease in concentration, increased morphologically abnormal sperm, decreased sperm motility and viability, and azoospermia (Maes et al., 2008). Most sows are clinically normal with a few animals developing corneal opacity. Infected breeding herds may experience increased returns to estrus and a reduction in farrowing rates.

3.5 Porcine Enteric Picornaviruses

Porcine enterovirus (PEV) and porcine teschovirus (PTV) are single-stranded positive-sense RNA viruses in the family *Picornaviridae.* PEV and PTV infections are commonly transmitted between pigs by oral exposure to contaminated feces; however, contamination of semen via aerosol during semen collection cannot be excluded (Guerin & Pozzi, 2005). PTV has been isolated from the male genital tract (Phillips et al., 1972); however, insemination of gilts with PTV contaminated semen had no effect on their fertility (De Meurichy & Pensaert, 1977). PEV can cause seminal vesiculitis, sperm abnormalities, and decrease libido (Phillips et al., 1972). Usually, there are no clinical signs in the sows. Although evidence is limited, there are reports that suggest that semen contaminated with PEV and PTV could cause embryonic and neonatal death (Dunne et al., 1969).

3.6 African Swine Fever Virus

African swine fever virus (ASFV) is a double-stranded, linear DNA virus in the genus *Asfivirus* of the family *Asfarviridae* (Tabares et al., 1980). ASFV replicates in monocytes and macrophages of the lymph nodes near the site of viral entry and subsequently spreads through the blood system, the lymphatic system, or both (Maes et al., 2008). ASFV transmission is generally vector-borne via tick bites (Parker et al., 1969); however, infections can also occur through direct interaction of sick and naïve animals (Plowright et al., 1970). Clinical signs range from subclinical to severe acute systemic disease that can resemble other hemorrhagic diseases in pigs. In naïve pigs, acute disease is characterized by high fever, loss of appetite, and hemorrhages in the skin (Moulton & Coggins, 1968). ASFV has been experimentally isolated from semen (Schlafer & Mebus, 1987; Thacker et al., 1984). Effect on semen quality and volume has to the authors knowledge not been reported. Under experimental conditions, ASFV has been shown to induce abortion when sows were infected between 38 and 92 days of gestation (Schlafer & Mebus, 1987). It was determined that abortion resulted from the effect of the virus on the dam (clinical illness) rather than direct effect on the fetuses (Schlafer & Mebus, 1987).

3.7 Classical Swine Fever Virus

Classical swine fever virus (CSFV) is a small, enveloped, positive-sense, single-stranded RNA virus in the genus *Pestivirus* of the family *Flaviviridae* (Becher et al., 1999). CSFV has been eradicated from many countries including Australia, New Zealand, North America, and Western Europe; however, it has been periodically reintroduced into domestic pigs that are in contact with wild boars. During the CSFV epizootic of 1997-1998 in the Netherlands (de Smit et al., 1999), two AI centers became infected and more than 100,000 sows from approximately 1,700 farms were affected (Hennecken et al., 2000) highlighting the economic importance of disease spread though AI centers. The main transmission route of CSFV is oronasally with primary virus replication in tonsils.Data from natural infection(Hennecken et al., 2000) and experimental inoculations (Choi & Chae, 2003; Floegel et al., 2000) have demonstrated that CSFV can be excreted in semen from infected boars. Boars experimentally infected with CSFV shed the virus in semen for up to 53 days post-infection (Choi & Chae, 2003). The virus does not affect the semen quality and motility and concentration are within normal range. Sows that were inseminated with contaminated semen seroconverted; however, the virus was shown to cross the placental barrier and infected the fetuses causing embryonic mortality (de Smit et al., 1999). CSFV continues to be important, especially in areas where CSFV is endemic, because the virus is highly contagious and infection of pregnant sows, in contrast to acute infections in piglets, may not be apparent (Hare et al., 1985; Ressang, 1973). If clinically normal CSFV immunotolerant piglets are born, they can spread virus for months without showing signs of disease or developing an antibody response (de Smit et al., 1999).

3.8 Foot and Mouth Disease Virus

Foot and mouth disease virus (FMDV) is a small single-stranded positive-sense RNA virus which belongs to the genus *Apthovirus* in the family *Picornaviridae* (Carrillo et al., 1990). FMDV is endemic in Asia, some areas in South America and in Africa. FMDV can be isolated from all excretions and secretions including urine, feces, milk and semen (Lubroth & Brown, 1995). Infected pigs develop fever, lethargy and lameness which is associated with vesicular lesions in the hoof area (Mebus, 1978). Infection with FMDV leads to viremia, with subsequent systemic dissemination of the virus, including the genital tract and the skin around the preputial orifice (Alexandersen et al., 2001). FMDV has been detected in boar semen before and during manifestation of clinical signs of the disease. The viral concentration in semen has been found to be low (Guerin & Pozzi, 2005). After natural infection, FMDV has been isolated from infected boar semen for up to 9 days, but AI with contaminated semen failed to transmit the disease to sows (McVicar et al., 1977). To the authors' knowledge, an effect on semen quality has not been described. Abortion storms can be observed and are mainly due to the high fever and clinical illness of the dams

3.9 Japanese B Encephalitis Virus

Japanese B encephalitis virus (JBEV) is a member of the *Flavivirus* genus of the family *Flaviviridae* (Solomon et al., 2003). JBEV is a mosquito-borne pathogen affecting humans and animals. This virus represents an economically important reproductive pathogen of breeding pigs, especially in Asia and Northern Australia, and is a common cause of infertility in Japanese pigs (Habu et al., 1977). Infection of susceptible boars resulted in edematous, congested testes and semen with numerous abnormal spermatozoa and significantly decreased total and motile sperm counts (Habu et al., 1977). These changes are usually temporary and most boars recover completely. JBEV has been isolated from the testicles of boars with orchitis and also can be shed in the semen for 5 weeks (Habu et al., 1977). JBEV infection can easily be transmitted if gilts are inseminated with infected semen (Guerin & Pozzi, 2005; Habu et al., 1977).

3.10 Porcine Circovirus type 2

Porcine circovirus type 2 (PCV2) is a small, single-stranded, ambi-sense DNA virus that belongs to the family *Circoviriade*, genus *Circovirus* (Tischer et al., 1982). PCV2 is a ubiquitous virus and most herds worldwide are seropositive for PCV2 (Dulac & Afshar, 1989; Edwards & Sands, 1994; Segalés et al., 2008; Tischer et al., 1986). When first described in 1998, PCV2 was linked to disease mainly characterized by wasting and generalized lymphadenopathy in growing pigs (Allan et al., 1998; Morozov et al., 1998). Since that time, PCV2 has been associated with several disease manifestations in pigs commonly referred to as PCV-associated disease (PCVAD) which includes systemic disease, respiratory disease, enteric disease, porcine dermatitis and nephropathy syndrome (PDNS) and reproductive failure in dams (late term abortions and stillbirths) (Choi & Chae, 2001; Harms et al., 2002; Kim et al., 2004a; Kim et al., 2004b). PCV2-associated reproductive disease under field conditions is rare. The main route of transmission has been postulated to be the fecal-oral route (Segalés et al., 2005); however, due to the rapid spread and the extensive use of AI, semen transmission has been suggested as a potentially significant route of dissemination of PCV2 (Horlen et al., 2007; Lawton et al., 2004; Rose et al., 2003; Sibila et al., 2004; West et al., 1999). Mature boars infected with PCV2 generally lack clinical signs and lesions (Larochelle et al., 2000; Madson et al., 2008). The virus has been detected in semen of naturally and experimentally infected boars, even after the appearance of antibodies in the serum (Larochelle et al., 2000). In the acute phase of infection, PCV2 DNA was detected in serum and semen at two and five days post inoculation, respectively (Larochelle et al., 2000; Madson et al., 2008). Detection of PCV2 viremia commonly precedes the detection of semen-associated virus (Madson et al., 2008), but semen shedding has been reported in the absence of viremia (Larochelle et al., 2000). Anti-PCV2 antibodies typically develop by two weeks post inoculation (Larochelle et al., 2000; Madson et al., 2008). Following intranasal inoculation, intermittent semen shedding of PCV2 DNA was observed during a 47 day observation period (Larochelle et al., 2000). Intermittent semen shedding was confirmed by a different study (Grasland et al., 2008) after testing semen samples for 56 days in four inoculated boars. In contrast, continuous semen shedding was observed following intranasal and intramuscular inoculation following a 90 day observational period (Madson et al., 2008). In addition, naturally infected boars were found to sporadically shed PCV2 DNA in semen for up to 27.3 weeks in a PCV2 positive boar stud (McIntosh et al., 2006). Peak PCV2 shedding in semen occurs between nine and 20 days post inoculation (Grasland et al., 2008; Madson et al., 2008). Changes in morphology, motility, or quantity are not commonly associated with natural or experimental PCV2 infection (Madson et al., 2008; McIntosh et al., 2006). Clinical signs of PCV2 infection in the dam are typically absent or unapparent; however, a low percentage of females may abort due to systemic illness.

Pyrexia and anorexia are frequently observed in aborting dams (Park et al., 2005). The virus is also able to replicate in the zona pellucida-free embryos, leading to embryonic death (Mateusen et al., 2004, 2007). Delayed farrowing (>118 days of gestation) (Ladekjær-Mikkelsen et al., 2001) or pseudopregnancy (Josephson & Charbonneau, 2001) is less frequently observed with PCV2-associated reproductive failure. PCV2 affected litters are commonly composed of increased numbers of non-viable fetuses (mummified and stillborn) at parturition (Madson & Opriessnig, 2011).

3.11 Other viruses that are of minor importance

3.11.1 Swine Vesicular Disease Virus

Swine vesicular disease virus (SVDV) is a small, non-enveloped single-stranded positive-sense RNA virus, in the family *Picornaviridae,* genus *Enterovirus* (Nardelli et al., 1968). Documented outbreaks of this disease have been limited to selected countries in Asia, Europe and Central America. Clinical signs include appearance of vesicles around the coronary bands, snout, tongue and lip (Kanno et al., 1995) making this disease an important differential for FMDV. After natural infection, SVDV has been isolated from infected boar semen for up to four days, but AI with contaminated semen failed to transmit the disease to sows (McVicar et al., 1977).

3.11.2 Non-Classical Swine Fever Virus pestiviruses

Pigs are also susceptible to non-CSFV pestiviruses, including bovine viral diarrhea virus (BVDV) and border disease virus (BDV), which are associated with disease in cattle and sheep, respectively. Pigs congenitally infected with these viruses may shed large amounts of virus. Previously, BVDV has been isolated from oropharyngeal fluid, urine and semen of a congenitally infected, infertile boar (Terpstra & Wensvoort, 1997).

3.11.3 Porcine retroviruses

Retroviruses are RNA viruses that exist in two main groups: endogenous or exogenous retroviruses. Endogenous retroviruses are thought to be present in all vertebrates accounting for approximately 8% of their genomes (Tucker et al., 2006). All pigs carry PERV in their genome (Wilson, 2008) and three porcine endogenous retrovirus (PERV) subtypes have been identified based on their envelope sequence and tropism in cell culture (Le et al., 1997; Wilson et al., 2000): PERV-A, PERV-B and PERV-C (Martin et al., 2000a, 2000b; Patience et al., 1997; Specke et al., 2001; Takeuchi et al., 1998). More recently, the possible existence of exogenous porcine retrovirus in pigs was proposed (Scobie et al., 2004; Wood et al., 2004). It was demonstrated that a human-tropic recombination between PERV-A and PERV-C (designated as PERV-A/C) was not a product of *in vitro* recombination; instead PERV-A/C appeared to exist *in vivo* as an exogenous virus (Wood et al., 2004). Furthermore, a PERV-A/C recombinant was isolated from porcine peripheral blood mononuclear cells and was not present in a proviral form in the miniature swine genome (Scobie et al., 2004a; Wood et al., 2004). The PERV A/C virus is a recombination within the *env* region and is thought to arise from exogenous recombination of mRNA (Martin et al., 2006). Shedding via the semen route occurs, since PERVs are embedded in the genome.

3.11.4 Swine Influenza Virus and transmissible gastroenteritis virus

Swine influenza virus (SIV), a negative-sense, single-stranded RNA virus in the genus *influenzavirus A* of the family *Orthomyxoviridae* (Lamb & Krug, 2001; Wright & Webster, 2001), and transmissible gastroenteritis virus (TGEV), a positive-sense, single-stranded RNA virus in the *Coronaviridae* family (Lai & Cavanagh, 1997) have not been isolated from semen. Contamination of semen and transmission of SIV and TGEV have not been demonstrated to date; however, these viruses are highly contagious and could be easily transmitted via aerosol at the time of collection and preparation of semen. Boars in the acute stages of SIV or TGEV infection are often clinically ill on examination (Guerin & Pozzi, 2005).

3.11.5 Porcine Cytomegalovirus

Porcine cytomegalovirus (PCVM), a double-stranded DNA virus, is classified in the genus *Cytomegalovirus* in the subfamily *Betaherpesvirinae* of the family *Herpesviridae* (Roizman, 1982). Infection with PCMV is usually subclinical in adults including boars. Following infection in boars, the virus was detected in the testis and the epididymis (Shirai et al., 1985); however, semen shedding of the virus has not been determined.

3.11.6 Porcine Adenovirus

Porcine adenovirus (PAV) is a double-stranded, linear DNA virus which belongs to the family *Adenoviridae.* PAV is distributed worldwide and associated with gastroenteric disease. As such, its importance in AI is due to the potential of fecal contamination of semen.

4. Diagnostic approaches to prevent virus transmission by the semen route

The concern of spreading diseases on a global scale via semen has placed restrictions on international movement of semen. From the early 1990s onwards, major improvements in detection methods particularly in terms of sensitivity have better enabled diagnostic laboratories to detect viruses in boar semen. Today, various methods are available for demonstration of viruses in boars and include monitoring for presence of clinical signs, demonstration of virus (virus isolation, bioassay), nucleic acids, and demonstration of antibodies against the virus. However, it needs to be considered that the results of any of these methods may vary substantially, depending on individual sample type and the sensitivity and specificity of a chosen diagnostic method. Accurate diagnosis of semen associated viruses (correctly identified positive or negative status of any given sample) is extremely important as false positive as well as false negative test results can cause substantial economic damage. For example, with PRRSV, any false positive sample out of a boar stud has the following consequences: (1) Immediate hold on all semen samples from the stud to breeding herds resulting in missed cycles and associated production losses in hundreds to thousands of sows supplied by the boar stud. (2) Retesting of boars. (3) Culling of suspect positive boars.

4.1 Importance of sample type

Samples that are commonly collected in boar studs for routine surveillance include semen (raw or extended), blood swabs, serum and more recently oral fluids. Blood swabs have become a popular sample type in recent years. For blood swab collection, the ear is pricked with a needle during semen collection and a cotton swab is used to catch the resulting blood drop (Reicks et al., 2006). Pathogens that cause systemic disease are typically identified in serum or blood first before they enter the semen and this should be taken into consideration when collecting appropriate samples. A negative result on semen only means that the tested sample does not contain virus, and that the particular semen sample is likely to be virus-free. It does not provide certainty that there will be no risk of contamination in the future. On the other hand, a negative result on serum does not guarantee that the semen is free of the virus as the viremic stage may have been very short.

4.2 Pooling of samples

Economic aspects are important to consider. For interpretation of diagnostic results at the boar stud level, in addition to the detection limit of the diagnostic method used, it is extremely important to test a representative number of semen samples, and to include evaluation of more than one parameter, such as presence of antibodies, viremia, or clinical symptoms. Because of the potential high cost of this approach, producers commonly rely on testing pooled samples. This strategy allows testing a larger number of animals while running the same number of tests, thus decreasing the cost per boar stud. Pooling can be successfully applied if the assay utilized has a high analytical sensitivity; however, because of a dilution effect, the sensitivity of the test when run on pooled samples is lower than its sensitivity when run on individual samples (Munoz-Zanzi et al., 2006).

4.3 Detection methods

4.3.1 Clinical evaluation of the boar

As a very basic and broad approach, in order to avoid spread of disease via contaminated semen, the health status of the animals should be checked daily. In case of any abnormality or clinical disease signs, semen collection should be halted until the animal has recovered. However, as indicated before, clinical examination alone is insufficient for most viral infections, since clinically normal boars can shed pathogens in their semen for extended periods of time.

4.3.2 Detection of viable virus

For demonstration of viable virus in semen two options exist including virus isolation and conducting bioassays. Virus isolation is often difficult as bacterial contamination of semen can be substantial. Virus isolation can be further complicated by the existence of cytotoxic factors in semen that destroy cell culture systems, and antiviral factors that nonspecifically neutralize virus (Christopher-Hennings et al., 1995b, 1997; Prieto et al., 1996b, 2003). In many cases swine bioassays have been found to be more sensitive than cell culture systems for demonstrating viable viruses in semen. However, animal inoculations cannot be justified to be used routinely for large numbers of samples and turnaround time for test results is poor.

4.3.3 Direct detection of RNA/DNA

Much progress has been made in recent years to improve the quality of molecular diagnostic assays used on semen. Numerous PCR techniques are now available for detecting pathogens in boar semen. Many of these are highly sensitive, specific and rapid (Christopher-Hennings et al., 1995b). In the case of PRRSV, reverse transcriptase-PCR is today considered to be the most sensitive diagnostic technique. It allows for the detection of as little as 10^0 $TCID_{50}$ per seminal dose (Christopher-Hennings et al., 1995b; Gradil et al., 1996), 20 times less virus than what has been shown experimentally to result in transmission of PRRSV in gilts (Shin et al., 1997).

4.3.4 Detection of antibodies

Serology is a simple and relatively inexpensive way to survey for pathogens that should not be present in the boar stud. The main disadvantage of serology is the time period between pathogen exposure and detectable levels of antibodies in a boar which may range from a few days to weeks during which time the boar is not identified as being positive and may shed the virus via semen.

5. Prevention of introduction of viruses into boar studs

The best way to prevent disease transmission via semen is to assure pathogens are not introduced to the boar studs by maintaining very strict biosecurity measures (Madec et al., 1999) and routine surveillance. Due to the high risk of dissemination of disease via AI, the most important goal is to provide pathogen-free semen, which is feasible with adequate control measures.

5.1 Biosecurity

Strict biosecurity measures should be maintained, involving all of the following: (1) *Location of the boar stud:* To avoid introduction of pathogens via aerosol routes, boar studs are often located in remote areas away from pig dense populations. (2) *Protection from aerosol contamination:* High traffic areas and location next to heavily travelled roads have shown to be a risk factor for exposure to airborne viruses (Otake et al., 2002). Boar stud owners often attempt to protect the boars by implementing filter systems for incoming air in AI centers to safeguard boars against the entry of airborne pathogens (Dee et al., 2005, 2006). (3) *Incoming replacement boars:* Replacement boars are typically isolated for 30 to 60 days and housed in a separate off-site facility. During the isolation time, the health status of the boars is evaluated. Animals that test positive for any disease of concern are typically not allowed to enter. During the isolation time, incoming boars are often vaccinated against diseases present in the resident boars. The annual boar replacement rate in AI centers is typically 60% (Singleton, 2001). (4) *Employees and visitors:* Strict regulations are typically in place for people entering AI centers. Often, there is 24 to 72 hour down-time required during which time people are not allowed to have contact with other live pigs. In addition, there is often a shower in/shower out requirement for all visitors and employees. (5) *Deliveries:* Any deliveries (clothes, mail, supplies, feed, tools, and equipment) are brought to a special entrance with no direct contact with the boar stud and are fumigated prior to entry into the boar studs. (6) *Rodent and insect control:* Rodents and insects have been shown to be capable of carrying several pig viruses (Lorincz et al., 2010; Otake et al., 2003a, 2003b; Plowright et al., 1970) and appropriate prevention measures need to be in place. (7) *Implementation of a regular and appropriate cleaning and disinfectant procedure.*

5.2 Routine surveillance testing

Commercial AI centers need be regularly checked for conformity with specific criteria, assuring that their products are free of certain pathogens and contain a minimal or acceptable number of microorganisms (Guerin & Pozzi, 2005; Prieto et al., 2004). Many veterinary diagnostic laboratories took this need into account and started to offer daily service for boar studs.

5.3 General semen handling and storage

To reduce the unavoidable presence of bacteria in the ejaculate and to prolong *in vitro* longevity of sperm, use of antimicrobials in semen is a common part of most semen extenders. Apart from a possible dilution effect of pathogens, semen processing and addition of antimicrobials does not eliminate viruses. The use of effective antiviral agents to render semen virus-free has so far not been adopted in the swine AI industry. However, several best practices for handling, storage conditions, and use of fresh semen have been described to reduce the potential for transmission of viral pathogens (Guerin & Pozzi, 2005). For boars, the immediate use of fresh semen increases the risk of pathogen transmission (Prieto et al., 2004).

5.4 Vaccination

The use of modified-live and inactivated-virus vaccines in boars can be highly effective in eliminating or decreasing shedding of viruses and with that decreasing the risk of virus transmission by AI. For example, vaccination against PPV may help to reduce shedding of the virus following infection. This has also been shown for PCV2 under experimental conditions (Opriessnig et al., 2011). In many countries, vaccination is done as part of a PRV eradication program (Siegel & Weigel, 1999). In the case of PRRSV, the use of a modified live virus vaccine shortened or eliminated virus shedding in boars challenged with wild-type virus for 50 days after vaccination (Christopher-Hennings et al., 1997; Nielsen et al., 1997). However, based on other studies it appears that PRRSV vaccination provides only partial protection (Nielsen et al., 1997). Moreover, semen shedding has been demonstrated after vaccination with a modified live vaccine virus (Christopher-Hennings et al., 1997). In addition, the modified live PRRSV vaccine virus has been shown to be shed in the semen in low levels (Christopher-Hennings et al., 1995b). In contrast, an inactivated vaccine, did not clearly reduce subsequent shedding of wild-type virus in semen (Christopher-Hennings et al., 1997; Stevenson et al., 1994). The presence of PRRSV in semen was demonstrated by PCR in most of the vaccinated boars during an interval of 7-21 days post-vaccination, although some boars sporadically shed the virus for longer periods of time (Christopher-Hennings et al., 1997). Similarly, when a swine bioassay was used, the presence of infectious virus in semen samples of vaccinated boars was confirmed 14 days post-vaccination (Nielsen et al., 1997).

6. Summary

Virus contamination of boar semen poses a great risk for breeding herds worldwide due the possibility of fast introduction of viruses into large naïve and susceptible populations.

Producers in some countries have led the way in establishing and maintaining specific pathogen-free AI centers in which incoming and resident boars are systematically screened for specific viruses to prevent introduction of viruses and if viruses are introduced to prevent spread of the virus within the boar studs and to breeding herds that receive semen from the studs. Serious attention is being given to appropriate location of boar studs and major investments are being made in air filtration equipment and building design and strict adherence is given to biosecurity protocols to further decrease the risks of introduction of diseases, particularly PRRSV, to boar studs in North America.

7. References

Albina, E., Madec, F., Cariolet, R., & Torrison, J. (1994). Immune response and persistence of the porcine reproductive and respiratory syndrome virus in infected pigs and farm units. *Veterinary Record,* Vol. 134, No. 22, pp. 567-573.

Alexandersen, S., Oleksiewicz, M.B., & Donaldson, A.I. (2001). The early pathogenesis of foot-and-mouth disease in pigs infected by contact: a quantitative time-course study using TaqMan RT-PCR. *Journal of General Virology,* Vol. 82, No. 4, pp. 747-755.

Allan, G.M., McNeilly, F., Ellis, J., Krakowka, S., Meehan, B., McNair, I., Walker, I., & Kennedy, S. (2000). Experimental infection of colostrum deprived piglets with porcine circovirus 2 (PCV2) and porcine reproductive and respiratory syndrome virus (PRRSV) potentiates PCV2 replication. *Archives of Virology,* Vol. 145, No. 11, pp. 2421-2429.

Allan, G.M., McNeilly, F., Kennedy, S., Daft, B., Clarke, E.G., Ellis, J.A., Haines, D.M., Meehan, B.M., & Adair, B.M. (1998). Isolation of porcine circovirus-like viruses from pigs with a wasting disease in the USA and Europe. *Journal of Veterinary Diagnostic Investigation,* Vol. 10, No. 1, pp. 3-10.

Bachmann, P.A., Sheffy, B.E., & Vauhan, J.T. (1975). Experimental in utero infection of fetal pigs with a porcine parvovirus. *Infection and Immunity,* Vol. 12, No. 3, pp. 455-460.

Bautista, E.M., Goyal, S.M., & Collins, J.E. (1993). Serologic survey for Lelystad and VR-2332 strains of porcine respiratory and reproductive syndrome (PRRS) virus in US swine herds. *Journal of Veterinary Diagnostic Investigation,* Vol. 5, No. 4, pp. 612-614.

Becher, P., Orlich, M., Kosmidou, A., Konig, M., Baroth, M., & Thiel, H.J. (1999). Genetic diversity of pestiviruses: identification of novel groups and implications for classification. *Virology,* Vol. 262, No. 1, pp. 64-71.

Benfield, D., Nelson, C., Steffen, M., & Rowland, R. (2000). Transmission of PRRSV by artificial insemination using extended semen seeded with different concentrations of PRRSV. *Proceedings of the American Association of Swine Practitioners,* Vol. 31, pp. 405-408.

Bierk, M.D., Dee, S.A., Rossow, K.D., Otake, S., Collins, J.E., & Molitor, T.W. (2001). Transmission of porcine reproductive and respiratory syndrome virus from persistently infected sows to contact controls. *Canadian Journal of Veterinary Research,* Vol. 65, No. 4, pp. 261-266.

Biront, P. & Bonte, P. (1983). Porcine parvovirus (P.P.V.): infection in boars. I. Possibility of a genital localization in the boar after oronasal infection. *Zentralblatt fürVeterinärmedizin Reihe B,* Vol. 30, No. 7, pp. 541-545.

Brown, T.T., Jr., Paul, P.S., & Mengeling, W.L. (1980). Response of conventionally raised weanling pigs to experimental infection with a virulent strain of porcine parvovirus. *American Journal of Veterinary Research,* Vol. 41, No. 8, pp. 1221-1224.

Carrillo, C., Plana, J., Mascarella, R., Bergada, J., & Sobrino, F. (1990). Genetic and phenotypic variability during replication of foot-and-mouth disease virus in swine. *Virology,* Vol. 179, No. 2, pp. 890-892.

Cartwright, S.F. & Huck, R.A. (1967). Viruses isolated in association with herd infertility, abortions and stillbirths in pigs. *Veterinary Record,* No. 81, pp. 196-197.

Cartwright, S.F., Lucas, M., & Huck, R.A. (1971). A small haemaggultinating porcine DNA virus. II. Biological and serological studies. *Journal of Comparative Pathology,* Vol. 81, No. 1, pp. 145-155.

Cavanagh, D. (1997). Nidovirales: a new order comprising Coronaviridae and Arteriviridae. *Archives of Virology,* Vol. 142, No. 3, pp. 629-633.

Choi, C. & Chae, C. (2001). Colocalization of porcine reproductive and respiratory syndrome virus and porcine circovirus 2 in porcine dermatitis and nephrology syndrome by double-labeling technique. *Veterinary Pathology,* Vol. 38, No. 4, pp. 436-441.

Choi, C. & Chae, C. (2003). Detection of classical swine fever virus in boar semen by reverse transcription-polymerase chain reaction. *Journal of Veterinary Diagnostic Investigation,* Vol. 15, No. 1, pp. 35-41.

Choi, C.S., Molitor, T.W., Joo, H.S., & Gunther, R. (1987). Pathogenicity of a skin isolate of porcine parvovirus in swine fetuses. *Veterinary Microbiology,* Vol. 15, No. 1-2, pp. 19-29.

Christianson, W.T., Choi, C.S., Collins, J.E., Molitor, T.W., Morrison, R.B., & Joo, H.S. (1993). Pathogenesis of porcine reproductive and respiratory syndrome virus infection in mid-gestation sows and fetuses. *Canadian Journal of Veterinary Research,* Vol. 57, No. 4, pp. 262-268.

Christopher-Hennings, J., Holler, L.D., Benfield, D.A., & Nelson, E.A. (2001). Detection and duration of porcine reproductive and respiratory syndrome virus in semen, serum, peripheral blood mononuclear cells, and tissues from Yorkshire, Hampshire, and Landrace boars. *Journal of Veterinary Diagnostic Investigation,* Vol. 13, No. 2, pp. 133-142.

Christopher-Hennings, J., Nelson, E.A., Hines, R.J., Nelson, J.K., Swenson, S.L., Zimmerman, J.J., Chase, C.L., Yaeger, M.J., & Benfield, D.A. (1995a). Persistence of porcine reproductive and respiratory syndrome virus in serum and semen of adult boars. *Journal of Veterinary Diagnostic Investigation,* Vol. 7, No. 4, pp. 456-464.

Christopher-Hennings, J., Nelson, E.A., Nelson, J.K., & Benfield, D.A. (1997). Effects of a modified-live virus vaccine against porcine reproductive and respiratory syndrome in boars. *American Journal of Veterinary Research,* Vol. 58, No. 1, pp. 40-45.

Christopher-Hennings, J., Nelson, E.A., Nelson, J.K., Hines, R.J., Swenson, S.L., Hill, H.T., Zimmerman, J.J., Katz, J.B., Yaeger, M.J., Chase, C.C., & .Benfield, D.A. (1995b). Detection of porcine reproductive and respiratory syndrome virus in boar semen by PCR. *Journal of Clinical Microbiology,* Vol. 33, No. 7, pp. 1730-1734.

Christopher-Hennings, J., Nelson, E.A., Nelson, J.K., Rossow, K.D., Shivers, J.L., Yaeger, M.J., Chase, C.C., Garduno, R.A., Collins, J.E., & Benfield, D.A. (1998).

Identification of porcine reproductive and respiratory syndrome virus in semen and tissues from vasectomized and nonvasectomized boars. *Veterinary Pathology,* Vol. 35, No. 4, pp. 260-267.

Cutlip, R.C. & Mengeling, W.L. (1975a). Experimentally induced infection of neonatal swine with porcine parvovirus. *American Journal of Veterinary Research,* Vol. 36, No. 8, pp. 1179-1182.

Cutlip, R.C. & Mengeling, W.L. (1975b). Pathogenesis of in utero infection: experimental infection of eight- and ten-week-old porcine fetuses with porcine parvovirus. *American Journal of Veterinary Research,* Vol. 36, No. 12, pp. 1751-1754.

De Meurichy, W. & Pensaert, M. (1977). Effect of an enterovirus in gilts inseminated with a semen-virus mixture. *Zentralblatt für Veterinärmedizin Reihe B,* Vol. 24, No. 2. pp. 97-103.

De Smit, A.J., Bouma, A., Terpstra, C., & Van Oirschot, J.T. (1999). Transmission of classical swine fever virus by artificial insemination. *Veterinary Microbiology,* Vol. 67, No. 4, pp. 239-249.

Dea, S., Elazhary, M.A., Martineau, G.P., & Vaillancourt, J. (July 1985). Parvovirus-like particles associated with diarrhea in unweaned piglets. *Canadian Journal of Comparative Medicine,* Vol. 49, No. 3, pp. 343-345.

Dee, S., Batista, L., Deen, J., & Pijoan, C. (2005). Evaluation of an air-filtration system for preventing aerosol transmission of porcine reproductive and respiratory syndrome virus. *Canadian Journal of Veterinary Research,* Vol. 69, No. 4, pp. 293-298.

Dee, S.A., Deen, J., Cano, J.P., Batista, L., & Pijoan, C. (July 2006). Further evaluation of alternative air-filtration systems for reducing the transmission of Porcine reproductive and respiratory syndrome virus by aerosol. *Canadian Journal of Veterinary Research,* Vol. 70, No. 3, pp. 168-175.

Duhamel, G.E., Bargar, T.W., Schmitt, B.J., Molitor, T.W., & Lu, W. (1991). Identification of parvovirus-like virus particles in intestinal crypt epithelial cells of pigs with diarrhea. *Journal of Veterinary Diagnostic Investigation,* Vol. 3, No. 1, pp. 96-98.

Dulac, G.C. & Afshar, A. (1989). Porcine circovirus antigens in PK-15 cell line (ATCC CCL-33) and evidence of antibodies to circovirus in Canadian pigs. *Canadian Journal of Veterinary Research,* Vol. 53, No. 4, pp. 431-433.

Dunne, H.W., Wang, J.T., Clark, C.D., Hokanson, J.F., Morimoto, T., & Bubash, G.R. (1969). The effects of in utero viral infection on embryonic, fetal, and neonatal survival: a comparison of SMEDI (porcine picorna) viruses with hog cholera vaccinal virus. *Canadian Journal of Comparative Medicine,* Vol. 33, No. 4, pp. 244-252.

Edwards, S. & Sands, J.J. (1994). Evidence of circovirus infection in British pigs. *Veterinary Record,* Vol. 134, No. 26, pp. 680-681.

Floegel, G., Wehrend, A., Depner, K.R., Fritzemeier, J., Waberski, D., & Moennig, V. (2000). Detection of classical swine fever virus in semen of infected boars. *Veterinary Microbiology,* Vol. 77, No. 1-2, pp. 109-116.

Flowers, W.L. & Esbenshade, K.L. (1993). Optimizing management of natural and artificial matings in swine. *Journal of Reproduction and Fertility Supplements,* Vol. 48, pp. 217-228.

Gerrits, R.J., Lunney, J.K., Johnson, L.A., Pursel, V.G., Kraeling, R.R., Rohrer, G.A., & Dobrinsky, J.R. (2005). Perspectives for artificial insemination and genomics to improve global swine populations. *Theriogenology,* Vol. 63, No. 2, pp. 283-299.

Gradil, C., Dubuc, C., & Eaglesome, M.D. (1996). Porcine reproductive and respiratory syndrome virus: seminal transmission. *Veterinary Record,* Vol. 138, No. 21, pp. 521-522.

Gradil, C., Molitor, T., Harding, M., & Crabo, B. (1990). Excretion of porcine parvovirus through the genital tract of boars. *American Journal of Veterinary Research,* Vol. 51, No. 3, pp. 359-362.

Grasland, B., Blanchard, P., Jan, B., Oger, A., Rose, N., Madec, F., Jestin, A., & Cariolet, R. (2008). Transmission of porcine circovirus of type 2 (PCV2) with semen. *Proceedings of the 20th International Pig Veterinary Society Congress,* Vol. 2, p. 56.

Guerin, B. & Pozzi, N. (2005). Viruses in boar semen: detection and clinical as well as epidemiological consequences regarding disease transmission by artificial insemination. *Theriogenology,* Vol. 63, No. 2, pp. 556-572.

Habu, A., Murakami, Y., Ogasa, A., & Fujisaki, Y. (1977). [Disorder of spermatogenesis and viral discharge into semen in boars infected with Japanese encephalitis virus (author's transl)]. *Uirusu,* Vol. 27, No. 1, pp. 21-26.

Hall, L.B., Jr., Kluge, J.P., Evans, L.E., Clark, T.L., & Hill, H.T. (1984a). Testicular changes observed in boars following experimental inoculation with pseudorabies (Aujeszky's) virus. *Canadian Journal of Comparative Medicine,* Vol. 48, No. 3, pp. 303-307.

Hall, L.B., Jr., Kluge, J.P., Evans, L.E., & Hill, H.T. (1984b). The effect of pseudorabies (Aujeszky's) virus infection on young mature boars and boar fertility. *Canadian Journal of Comparative Medicine,* Vol. 48, No. 2, pp. 192-197.

Hare, W.C., Mitchell, D., Singh, E.L., Bouillant, A.M., Eaglesome, M.D., Ruckerbauer, G.M., Bielanski, A., & Randall, G.C. (1985). Embryo transfer in relation to bovine leukemia virus control and eradication. *Canadian Veterinary Journal,* Vol. 26, No. 8, pp. 231-234.

Harms, P.A., Halbur, P.G., & Sorden, S.D. (2002). Three cases of porcine respiratory disease complex associated with porcine circovirus type 2 infection. *Journal of Swine Health and Production,* Vol. 10, pp. 27-30.

Hennecken, M., Stegeman, J.A., Elbers, A.R., van, N.A., Smak, J.A., & Verheijden, J.H. (2000). Transmission of classical swine fever virus by artificial insemination during the 1997-1998 epidemic in The Netherlands: a descriptive epidemiological study. *Veterinary Quarterly,* Vol. 22, No. 4, pp. 228-233.

Horlen, K.P., Schneider, P., Anderson, J., Nietfeld, J.C., Henry, S.C., Tokach, L.M., & Rowland, R.R. (2007). A cluster of farms experiencing severe porcine circovirus associated disease: Clinical features and association with the PCV2b genotype. *Journal of Swine Health and Production,* Vol. 15, No. 5, pp. 270-278.

Johnson, R.H., Donaldson-Wood, C., & Allender, U. (Feb. 1976). Observations on the epidemiology of porcine parvovirus. *Australian Veterinary Journal,* Vol. 52, No. 2, pp. 80-84.

Joo, H.S., Donaldson-Wood, C.R., & Johnson, R.H. (1976). Observations on the pathogenesis of porcine parvovirus infection. *Archives of Virology,* Vol. 51, No. 1-2, pp. 123-129.

Josephson, G. & Charbonneau, G. (2001). Case report of reporductive problem in a new startup operation. *Swine Health and Production,* Vol. 9, No. 5, pp. 258-259.

Kanno, T., Inoue, T., Wang, Y., Sarai, A., & Yamaguchi, S. (1995). Identification of the location of antigenic sites of swine vesicular disease virus with neutralization-resistant mutants. *Journal of General Virology,* Vol. 76, pp. 3099-3106

Kennedy, S., Moffett, D., McNeilly, F., Meehan, B., Ellis, J., Krakowka, S., & Allan, G.M. (2000). Reproduction of lesions of postweaning multisystemic wasting syndrome by infection of conventional pigs with porcine circovirus type 2 alone or in combination with porcine parvovirus. *Journal of Comparative Pathology,* Vol. 22, No. 1, pp. 9-24.

Kim, J., Ha, Y., Jung, K., Choi, C., & Chae, C. (2004a). Enteritis associated with porcine circovirus 2 in pigs. *Canadian Journal of Veterinary Research,* Vol. 68, No. , pp. 218-221.

Kim, J., Jung, K., & Chae, C. (2004b). Prevalence of porcine circovirus type 2 in aborted fetuses and stillborn piglets. *Veterinary Record,* Vol. 155, No. 16, pp. 489-492.

Krakowka, S., Ellis, J.A., Meehan, B., Kennedy, S., McNeilly, F., & Allan, G. (2000). Viral wasting syndrome of swine: experimental reproduction of postweaning multisystemic wasting syndrome in gnotobiotic swine by coinfection with porcine circovirus 2 and porcine parvovirus. *Veterinary Pathology,* Vol. 37, No. 3, pp. 254-263.

Kresse, J.I., Taylor, W.D., Stewart, W.W., & Eernisse, K.A. (1985). Parvovirus infection in pigs with necrotic and vesicle-like lesions. *Veterinary Microbiology,* Vol. 10, No. 6, pp. 525-531.

Ladekjær-Mikkelsen, A.S., Nielsen, J., Storgaard, T., Bøtner, A., Allan, G., & McNeilly, F. (2001). Transplacental infection with PCV-2 associated with reproductive failure in a gilt. *Veterinary Record,* Vol. 148, No. 24, pp. 759-760.

Lager, K.M. & Mengeling, W.L. (1994). Porcine parvovirus associated with cutaneous lesions in piglets. *Journal of Veterinary Diagnostic and Investigation,* Vol. 6, No. 3, pp. 357-359.

Lager, K.M. & Mengeling, W.L. (1995). Pathogenesis of in utero infection in porcine fetuses with porcine reproductive and respiratory syndrome virus. *Canadian Journal of Veterinary Research,* Vol. 59, No. 3, pp. 187-192.

Lager, K.M., Mengeling, W.L., & Liu, W. (1992). Comparison of the virulence of two isolates of porcine parvovirus in 72-day-old porcine fetuses. *Journal of Veterinary Diagnostic and Investigation,* Vol. 4, No. 3, pp. 245-248.

Lai, M.M. & Cavanagh, D. (1997). The molecular biology of coronaviruses. *Advances in Virus Research,* Vol. 48, pp. 1-100.

Lamb, R.A. & Krug, R.M. (2001). Orthomyxoviridae: The viruses and their replication, In: *Field's Virology,* Knipe, D.M., Howley, P.M., Griffin, D.E., Lamb, R.A., Martin, M.A., Roizman, B., & Straus, S.E., pp. 1487-1531, Lippincott-Raven Publishers, ISBN Philadelphia

Larochelle, R., Bielanski, A., Müller, P., & Magar, R. (2000). PCR detection and evidence of shedding of porcine circovirus type 2 in boar semen. *Journal of Clinical Microbiology,* Vol. 38, No. 12, pp. 4629-4632.

Larsen, R.E., Shope, R.E., Jr., Leman, A.D., & Kurtz, H.J. (1980). Semen changes in boars after experimental infection with pseudorabies virus. *American Journal of Veterinary Research,* Vol. 41, No. 5, pp. 733-739.

Lawton, D.E., Morris, R.S., & King, C.M. (2004). PMWS in New Zealand part 2: Epidemiological evidence for a novel agent. *18th International Pig Veterinary Society Congress,* Vol. 1, p. 128.

Le, T.P., Stoye, J.P., Takeuchi, Y., Patience, C., & Weiss, R.A. (1997). Two sets of human-tropic pig retrovirus. *Nature,* Vol. 389, No. 6652, pp. 681-682.

Lorincz, M., Csagola, A., Biksi, I., Szeredi, L., Dan, A., & Tuboly, T. (2010). Detection of porcine circovirus in rodents - short communication. *Acta Veterinaria Hungarica,* Vol. 58, No. 2, pp. 265-268.

Lubroth, J. & Brown, F. (1995). Identification of native foot-and-mouth disease virus non-structural protein 2C as a serological indicator to differentiate infected from vaccinated livestock. *Research in Veterinary Science,* Vol. 59, No. 1, pp. 70-78.

Lucas, M.H., Cartwright, S.F., & Wrathall, A.E. (1974). Genital infection of pigs with porcine parvovirus. *Journal of Comparative Pathology,* Vol. 84, No. 3, pp. 347-350.

Lush, J.L. (1925). The possibility of sex control by artificial insemination with centrifuged spermatozoa. *Journal of Agricultural Research,* Vol. 30, No. 10, pp. 893-913.

Madec, F., Eveno, E., Morvan, P., Hamon, L., Morvan, H., Albina, E., Truong, C., Hutet, E., Cariolet, R., Arnauld, C., & Jestin, A. (1999). La maladie de l'amaigrissement du porcelet (MAP) en France 1 - Aspects descriptifs, impact en élevage. [Porcine wasting disease (PWD) in France: 1 - Decription of the disease and impact in affected herds]. *Journées Recherche en Porcine France,* Vol. 31, No. 347-354.

Madsen, K.S. (2005). Management of disease control and epidemics in AI in Denmark. *Theriogenology,* Vol. 63, No. 2, pp. 585-594.

Madson, D.M. & Opriessnig, T. (2011). Effect of porcine circovirus type 2 (PCV2) infection on reproduction: disease, vertical transmission, diagnostics and vaccination. *Animal Health Research Reviews,* Vol. 12, No. 1, pp. 47-65.

Madson, D.M., Ramamoorthy, S., Kuster, C., Pal, N., Meng, X.J., Halbur, P.G., & Opriessnig, T. (2008). Characterization of shedding patterns of porcine circovirus types 2a and 2b in experimentally inoculated mature boars. *Journal of Veterinary Diagnostic Investigation,* Vol. 20, No. 6, pp. 725-734.

Madson, D.M., Ramamoorthy, S., Kuster, C., Pal, N., Meng, X.J., Halbur, P.G., & Opriessnig, T. (2009). Infectivity of porcine circovirus type 2 DNA in semen from experimentally-infected boars. *Veterinary Research,* Vol. 40, No. 1, p. 10.

Maes, D., Nauwynck, H., Rijsselaere, T., Mateusen, B., Vyt, P., de, K.A., & Van, S.A. (2008). Diseases in swine transmitted by artificial insemination: an overview. *Theriogenology,* Vol. 70, No. 8, pp. 1337-1345.

Maes, R.K., Kanitz, C.L., & Gustafson, D.P. (1983). Shedding patterns in swine of virulent and attenuated pseudorabies virus. *American Journal of Veterinary Research,* Vol. 44, No. 11, pp. 2083-2086.

Martin, S.I., Wilkinson, R., & Fishman, J.A. (2006). Genomic presence of recombinant porcine endogenous retrovirus in transmitting miniature swine. *Virology Journal,* Vol. 3, p. 91.

Martin, U., Winkler, M.E., Id, M., Radecke, H., Arseniev, L., Groteluschen, R., Simon, A.R., & Steinhoff, G. (2000a). Transmission of pig endogenous retrovirus to primary human cells. *Transplantation Proceedings.*, Vol. 32, No. 5, p. 1157.

Martin, U., Winkler, M.E., Id, M., Radeke, H., Arseniev, L., Takeuchi, Y., Simon, A.R., Patience, C., Haverich, A., & Steinhoff, G. (2000b). Productive infection of primary human endothelial cells by pig endogenous retrovirus (PERV). *Xenotransplantation.*, Vol. 7, No. 2, pp. 138-142.

Mateusen, B., Maes, D.G., Van, S.A., Lefebvre, D., & Nauwynck, H.J. (2007). Effect of a porcine circovirus type 2 infection on embryos during early pregnancy. *Theriogenology,* Vol. 68, No. 6, pp. 896-901.

Mateusen, B., Sanchez, R.E., Van, S.A., Meerts, P., Maes, D.G., & Nauwynck, H.J. (2004). Susceptibility of pig embryos to porcine circovirus type 2 infection. *Theriogenology,* Vol. 61, No. 1, pp. 91-101.

McIntosh, K.A., Harding, J.C., Parker, S., Ellis, J.A., & Appleyard, G.D. (2006). Nested polymerase chain reaction detection and duration of porcine circovirus type 2 in semen with sperm morphological analysis from naturally infected boars. *Journal of Veterinary Diagnostic Investigation,* Vol. 18, No. 4, pp. 380-384.

McVicar, J.W., Eisner, R.J., Johnson, L.A., & Pursel, V.G. (1977). Foot-and-mouth disease and swine vesicular disease viruses in boar semen. *Proceedings of the Annual Meeting of the United States Animal Health Association,* Vol. 81, pp. 221-230.

Mebus, C.A. (1978). Ulcerative diseases of animals with an infectious etiology. *Journal of Oral Pathology,* Vol. 7, No. 6, pp. 365-371.

Medveczky, I. & Szabo, I. (1981). Isolation of Aujeszky's disease virus from boar semen. *Acta Veterinaria Hungarica,* Vol. 29, No. 1, pp. 29-35.

Meng, X.J., Paul, P.S., Halbur, P.G., & Lum, M.A. (1995). Phylogenetic analyses of the putative M (ORF 6) and N (ORF 7) genes of porcine reproductive and respiratory syndrome virus (PRRSV): implication for the existence of two genotypes of PRRSV in the USA and Europe. *Archives of Virology,* Vol. 140, No. 4, pp. 745-755.

Mengeling, W.L. (1979). Prenatal infection following maternal exposure to porcine parvovirus on either the seventh or fourteenth day of gestation. *Canadian Journal of Comparative Medicine,* Vol. 43, No. 1, pp. 106-109.

Mengeling, W.L. & Cutlip, R.C. (1976). Reproductive disease experimentally induced by exposing pregnant gilts to porcine parvovirus. *American Journal of Veterinary Research,* Vol. 37, No. 12, pp. 1393-1400.

Mengeling, W.L., Lager, K.M., & Vorwald, A.C. (1994). Temporal characterization of transplacental infection of porcine fetuses with porcine reproductive and respiratory syndrome virus. *American Journal of Veterinary Research,* Vol. 55, No. 10, pp. 1391-1398.

Mengeling, W.L. & Paul, P.S. (1986). Interepizootic survival of porcine parvovirus. *Journal of the American Veterinary Medical Association,* Vol. 188, No. 11, pp. 1293-1295.

Mettenleiter, T.C. (2000). Aujeszky's disease (pseudorabies) virus: the virus and molecular pathogenesis--state of the art, June 1999. *Veterinary Research,* Vol. 31, No. 1, pp. 99-115.

Meulenberg, J.J., Hulst, M.M., de Meijer, E.J., Moonen, P.L., den, B.A., de Kluyver, E.P., Wensvoort, G., & Moormann, R.J. (1993). Lelystad virus, the causative agent of porcine epidemic abortion and respiratory syndrome (PEARS), is related to LDV and EAV. *Virology,* Vol. 192, No. 1, pp. 62-72.

Molitor, T.W., Joo, H.S., & Collett, M.S. (1984). Porcine parvovirus DNA: characterization of the genomic and replicative form DNA of two virus isolates. *Virology,* Vol. 137, No. 2, pp. 241-254.

Morozov, I., Sirinarumitr, T., Sorden, S.D., Halbur, P.G., Morgan, M.K., Yoon, K.J., & Paul, P.S. (1998). Detection of a novel strain of porcine circovirus in pigs with postweaning multisystemic wasting syndrome. *Journal of Clinical Microbiology,* Vol. 36, No. 9, pp. 2535-2541.

Moulton, J. & Coggins, L. (1968). Comparison of lesions in acute and chronic African swine fever. *Cornell Vet,* Vol. 58, No. 3, pp. 364-388.

Munoz-Zanzi, C., Thurmond, M., Hietala, S., & Johnson, W. (2006). Factors affecting sensitivity and specificity of pooled-sample testing for diagnosis of low prevalence infections. *Preventive Veterinary Medicine,* Vol. 74, No. 4, pp. 309-322.

Nardelli, L., Lodetti, E., Gualandi, G.L., Burrows, R., Goodridge, D., Brown, F., & Cartwright, B. (1968). A foot and mouth disease syndrome in pigs caused by an enterovirus. *Nature,* Vol. 219, No. 5160, pp. 1275-1276.

Neumann, E.J., Kliebenstein, J.B., Johnson, C.D., Mabry, J.W., Bush, E.J., Seitzinger, A.H., Green, A.L., & Zimmerman, J.J. (2005). Assessment of the economic impact of porcine reproductive and respiratory syndrome on swine production in the United States. *Journal of the American Medical Association,* Vol. 227, No. 3, pp. 385-392.

Nielsen, T.L., Nielsen, J., Have, P., Baekbo, P., Hoff-Jorgensen, R., & Botner, A. (1997). Examination of virus shedding in semen from vaccinated and from previously infected boars after experimental challenge with porcine reproductive and respiratory syndrome virus. *Veterinary Microbiology,* Vol. 54, No. 2, pp. 101-112.

Opriessnig, T., Madson, D.M., Schalk, S., Brockmeier, S., Shen, H.G., Beach, N.M., Meng, X.J., Baker, R.B., Zanella, E.L. & Halbur, P.G. (2011). Porcine circovirus type 2 (PCV2) vaccination is effective in reducing disease and PCV2 shedding in semen of boars concurrently infected with PCV2 and *Mycoplasma hyopneumoniae. Theriogenology,* Vol. 76, No. 2, pp. 351-360.

Oraveerakul, K., Choi, C.S., & Molitor, T.W. (1993). Tissue tropisms of porcine parvovirus in swine. *Archives of Virology,* Vol. 130, No. 3-4, pp. 377-389.

Otake, S., Dee, S.A., Jacobson, L., Torremorell, M., & Pijoan, C. (2002). Evaluation of aerosol transmission of porcine reproductive and respiratory syndrome virus under controlled field conditions. *Veterinary Record,* Vol. 150, No. 26, pp. 804-808.

Otake, S., Dee, S.A., Moon, R.D., Rossow, K.D., Trincado, C., Farnham, M., & Pijoan, C. (2003a). Survival of porcine reproductive and respiratory syndrome virus in houseflies. *Canadian Journal of Veterinary Research,* Vol. 67, No. 3, pp. 198-203.

Otake, S., Dee, S.A., Moon, R.D., Rossow, K.D., Trincado, C., & Pijoan, C. (2003b). Evaluation of mosquitoes, Aedes vexans, as biological vectors of porcine reproductive and respiratory syndrome virus. *Canadian Journal of Veterinary Research,* Vol. 67, No. 4, pp. 265-270.

Park, J.-S., Kim, J., Ha, Y., Jung, K., Choi, C., Lim, J.-K., Kim, S.-H., & Chae, C. (2005). Birth abnormalities in pregnant sows infected intranasally with porcine circovirus 2. *Journal of Comparative Pathology,* Vol. 132, No. 2-3, pp. 139-144.

Parker, J., Plowright, W., & Pierce, M.A. (1969). The epizootiology of African swine fever in Africa. *Veterinary Record,* Vol. 85, No. 24, pp. 668-674.

Patience, C., Takeuchi, Y., & Weiss, R.A. (1997). Infection of human cells by an endogenous retrovirus of pigs. *Nature Medicine,* Vol. 3, No. 3, pp. 282-286.

Paul, P.S., Mengeling, W.L., & Brown, T.T., Jr. (1979). Replication of porcine parvovirus in peripheral blood lymphocytes, monocytes, and peritoneal macrophages. *Infection and Immunity,* Vol. 25, No. 3, pp. 1003-1007.

Phillips, R.M., Foley, C.W., & Lukert, P.D. (1972). Isolation and characterization of viruses from semen and the reproductive tract of male swine. *Journal of the American Veterinary Medical Association,* Vol. 161, No. 11, pp. 1306-1316.

Plowright, W., Perry, C.T., Peirce, M.A., & Parker, J. (1970). Experimental infection of the argasid tick, Ornithodoros moubata porcinus, with African swine fever virus. *Archiv für die Gesamte Virusforschung,* Vol. 31, No. 1, pp. 33-50.

Prieto, C. & Castro, J.M. (2005). Porcine reproductive and respiratory syndrome virus infection in the boar: a review. *Theriogenology,* Vol. 63, No. 1, pp. 1-16.

Prieto, C., Garcia, C., Simarro, I., & Castro, J.M. (2003). Temporal localization of porcine reproductive and respiratory syndrome virus in reproductive tissues of experimentally infected boars. *Theriogenology,* Vol. 60, No. 8, pp. 1505-1514.

Prieto, C., Garcia, C., Simarro, I., & Castro, J.M. (2004). Temporal shedding and persistence of porcine reproductive and respiratory syndrome virus in boars. *Veterinary Record,* Vol. 154, No. 26, pp. 824-827.

Prieto, C., Sanchez, R., Martin-Rillo, S., Suarez, P., Simarro, I., Solana, A., & Castro, J.M. (1996a). Exposure of gilts in early gestation to porcine reproductive and respiratory syndrome virus. *Veterinary Record,* Vol. 138, No. 22, pp. 536-539.

Prieto, C., Suarez, P., Bautista, J.M., Sanchez, R., Rillo, S.M., Simarro, I., Solana, A., & Castro, J.M. (1996b). Semen changes in boars after experimental infection with porcine reproductive and respiratory syndrome (PRRS) virus. *Theriogenology,* Vol. 45, No. 2, pp. 383-395.

Prieto, C., Suarez, P., Simarro, I., Garcia, C., Martin-Rillo, S., & Castro, J.M. (1997). Insemination of susceptible and preimmunized gilts with boar semen containing porcine reproductive and respiratory syndrome virus. *Theriogenology,* Vol. 47, No. 3, pp. 647-654.

Reicks, D.L., Munoz-Zanzi, C., & Rossow, K. (2006). Sampling of adult boars during early infection with porcine reproductive and respiratory syndrome virus for testing by polymerase chain reaction using new blood collection technique (blood swab method). *Journal of Swine Health and Production,* Vol. 14, pp. 258-264.

Ressang, A.A. (1973). Studies on the pathogenesis of hog cholera. I. Demonstration of hog cholera virus subsequent to oral exposure. *Zentralblatt für Veterinärmedizin Reihe B,* Vol. 20, No. 4, pp. 256-271.

Robertson, I.B. (1992). Transmission of blue-eared pig disease. *Veterinary Record,* Vol. 130, No. 21, p. 478.

Roizman, B. (1982). The family Herpesviridae: General description, taxonomy, and classification, In: *The Herpesviruses,* Vol. 1, pp. 1-23, Plenum Press, ISBN New York

Rose, N., Larour, G., Le, D.G., Eveno, E., Jolly, J.P., Blanchard, P., Oger, A., Le, D.M., Jestin, A., & Madec, F. (2003). Risk factors for porcine post-weaning multisystemic wasting syndrome (PMWS) in 149 French farrow-to-finish herds. *Preventive Veterinary Medicine,* Vol. 61, No. 3, pp. 209-225.

Rossow, K.D. (1998). Porcine reproductive and respiratory syndrome. *Veterinary Pathology,* Vol. 35, No. 1, pp. 1-20.

Schlafer, D.H. & Mebus, C.A. (1987). Abortion in sows experimentally infected with African swine fever virus: pathogenesis studies. *American Journal of Veterinary Research,* Vol. 48, No. 2, pp. 246-254.

Scobie, L., Taylor, S., Wood, J.C., Suling, K.M., Quinn, G., Meikle, S., Patience, C., Schuurman, H.J., & Onions, D.E. (2004). Absence of replication-competent human-tropic porcine endogenous retroviruses in the germ line DNA of inbred miniature Swine. *Journal of Virology,* Vol. 78, No. 5, pp. 2502-2509.

Segalés, J., Allan, G.M., & Domingo, M. (2005). Porcine circovirus diseases. *Animal Health Research Reviews,* Vol. 6, No. 2, pp. 119-142.

Segalés, J., Olvera, A., Grau-Roma, L., Charreyre, C., Nauwynck, H., Larsen, L., Dupont, K., McCullough, K., Ellis, J., Krakowka, S., Mankertz, A., Fredholm, M., Fossum, C., Timmusk, S., Stockhofe-Zurwieden, N., Beattie, V., Armstrong, D., Grassland, B., Baekbo, P., & Allan, G. (2008). PCV-2 genotype definition and nomenclature. *Veterinary Record,* Vol. 162, No. 26, pp. 867-868.

Shin, J., Torrison, J., Choi, C.S., Gonzalez, S.M., Crabo, B.G., & Molitor, T.W. (1997). Monitoring of porcine reproductive and respiratory syndrome virus infection in boars. *Veterinary Microbiology,* Vol. 55, No. 1-4, pp. 337-346.

Shirai, J., Narita, M., Iijima, Y., & Kawamura, H. (1985). A cytomegalovirus isolated from swine testicle cell culture. *Nippon Juigaku Zasshi,* Vol. 47, No. 5, pp. 697-703.

Sibila, M., Calsamiglia, M., Segalés, J., Blanchard, P., Badiella, L., Le, D.M., Jestin, A., & Domingo, M. (2004). Use of a polymerase chain reaction assay and an ELISA to monitor porcine circovirus type 2 infection in pigs from farms with and without postweaning multisystemic wasting syndrome. *American Journal of Veterinary Research,* Vol. 65, No. 1, pp. 88-92.

Siegel, A.M. & Weigel, R.M. (1999). Herd factors affecting the selection and success of intervention strategies in the program for eradication of pseudorabies (Aujeszky's disease) virus from Illinois swine farms. *Preventive Veterinary Medicine,* Vol. 40, No. 3-4, pp. 243-259.

Singleton, W.L. (2001). State of the art in artificial insemination of pigs in the United States. *Theriogenology,* Vol. 56, No. 8, pp. 1305-1310.

Solis, M., Ramirez-Mendoza, H., Mercado, C., Espinosa, S., Vallejo, V., Reyes-Leyva, J., & Hernandez, J. (2007). Semen alterations in porcine rubulavirus-infected boars are related to viral excretion and have implications for artificial insemination. *Research in Veterinary Science,* Vol. 83, No. 3, pp. 403-409.

Solomon, T., Ni, H., Beasley, D.W., Ekkelenkamp, M., Cardosa, M.J., & Barrett, A.D. (2003). Origin and evolution of Japanese encephalitis virus in southeast Asia. *Journal of Virology,* Vol. 77, No. 5, pp. 3091-3098.

Specke, V., Tacke, S.J., Boller, K., Schwendemann, J., & Denner, J. (2001). Porcine endogenous retroviruses: in vitro host range and attempts to establish small animal models. *Journal of General Virology,* Vol. 82, pp. 837-844

Spradbrow, P.B. (Sept. 1968). The isolation of infectious bovine rhinotracheitis virus from bovine semen. *Australian Veterinary Journal,* Vol. 44, No. 9, pp. 410-412.

Stevenson, G.W., Van Alstine, W.G., & Kanitz, C.L. (1994). Characterization of infection with endemic porcine reproductive and respiratory syndrome virus in a swine herd. *Journal of the American Medical Association,* Vol. 204, No. 12, pp. 1938-1942.

Sur, J.H., Doster, A.R., Christian, J.S., Galeota, J.A., Wills, R.W., Zimmerman, J.J., & Osorio, F.A. (1997). Porcine reproductive and respiratory syndrome virus replicates in testicular germ cells, alters spermatogenesis, and induces germ cell death by apoptosis. *Journal of Virology,* Vol. 71, No. 12, pp. 9170-9179.

Swenson, S.L., Hill, H.T., Zimmerman, J.J., Evans, L.E., Landgraf, J.G., Wills, R.W., Sanderson, T.P., McGinley, M.J., Brevik, A.K., & Ciszewski, D.K. (1994a). Excretion of porcine reproductive and respiratory syndrome virus in semen after experimentally induced infection in boars. *Journal of the American Medical Association,* Vol. 204, No. 12, pp. 1943-1948.

Swenson, S.L., Hill, H.T., Zimmerman, J.J., Evans, L.E., Wills, R.W., Yoon, K.J., Schwartz, K.J., Althouse, G.C., McGinley, M.J., & Brevik, A.K. (1994b). Artificial insemination of gilts with porcine reproductive and respiratory syndrome (PRRS) virus-contaminated semen. *Swine Health and Production,* Vol. 2, No. 6, pp. 19-23.

Tabares, E., Marcotegui, M.A., Fernandez, M., & Sanchez-Botija, C. (1980). Proteins specified by African swine fever virus. I. Analysis of viral structural proteins and antigenic properties. *Archives of Virology,* Vol. 66, No. 2, pp. 107-117.

Takeuchi, Y., Patience, C., Magre, S., Weiss, R.A., Banerjee, P.T., Le, T.P., & Stoye, J.P. (1998). Host range and interference studies of three classes of pig endogenous retrovirus. *Journal of Virology,* Vol. 72, No. 12, pp. 9986-9991.

Terpstra, C. & Wensvoort, G. (1997). A congenital persistent infection of bovine virus diarrhoea virus in pigs: clinical, virological and immunological observations. *Veterinary Quarterly,* Vol. 19, No. 3, pp. 97-101.

Terpstra, C., Wensvoort, G., & Pol, J.M. (1991). Experimental reproduction of porcine epidemic abortion and respiratory syndrome (mystery swine disease) by infection with Lelystad virus: Koch's postulates fulfilled. *Veterinary Quarterly,* Vol. 13, No. 3, pp. 131-136.

Teuffert, J., Schluter, H., & Muller, T. (1998). [Boar semen--a possible risk factor in infection occurrence of porcine reproductive and respiratory syndrome]. *Deutsche Tierärztliche Wochenschrift,* Vol. 105, No. 9, pp. 340-345.

Thacker, B.J., Joo, H.S., Winkelman, N.L., Leman, A.D., & Barnes, D.M. (1987a). Clinical, virologic, and histopathologic observations of induced porcine parvovirus infection in boars. *American Journal of Veterinary Research,* Vol. 48, No. 5, pp. 763-767.

Thacker, B.J., Joo, H.S., Winkelman, N.L., Leman, A.D., & Barnes, D.M. (May 1987b). Clinical, virologic, and histopathologic observations of induced porcine parvovirus infection in boars. *American Journal of Veterinary Research,* Vol. 48, No. 5, pp. 763-767.

Thacker, B.J., Larsen, R.E., Joo, H.S., & Leman, A.D. (Sept. 1984). Swine diseases transmissible with artificial insemination. *J Am.Vet Med Assoc.,* Vol.185, No.5, pp.511-516.

Tischer, I., Gelderblom, H., Vettermann, W., & Koch, M.A. (1982). A very small porcine virus with circular single-stranded DNA. *Nature,* Vol. 295, No. 5844, pp. 64-66.

Tischer, I., Mields, W., Wolff, D., Vagt, M., & Griem, W. (1986). Studies on epidemiology and pathogenicity of porcine circovirus. *Archives of Virology,* Vol. 91, No. 3-4, pp. 271-276.

Tucker, A.W., Mellencamp, M.M., Donadeu, M., & Scobie, L. (2006). Retroviremia in commercial pigs and its preliminary association with poor health. *Journal of Clinical Microbiology,* Vol. 44, No. 10, pp. 3846-3847.

Vannier, P. & Gueguen, B. (1979). Excrétion du virus de la maladie d'Aujeszky par les voies génitales mâles du porc. *Journee Recherche Porcine,* pp. 4001-406

Wensvoort, G., Terpstra, C., Pol, J.M., ter Laak, E.A., Bloemraad, M., de Kluyver, E.P., Kragten, C., van, B.L., den, B.A., Wagenaar, F., & . (1991). Mystery swine disease in The Netherlands: the isolation of Lelystad virus. *Veterinary Quarterly,* Vol. 13, No. 3, pp. 121-130.

West, K.H., Bystrom, J.M., Wojnarowicz, C., Shantz, N., Jacobson, M., Allan, G.M., Haines, D.M., Clark, E.G., Krakowka, S., McNeilly, F., Konoby, C., Martin, K., & Ellis, J.A. (1999). Myocarditis and abortion associated with intrauterine infection of sows with porcine circovirus 2. *Journal of Veterinary Diagnostic Investigation,* Vol. 11, No. 6, pp. 530-532.

Wilson, C.A. (2008). Porcine endogenous retroviruses and xenotransplantation. *Cellular and Molecular Life Sciences,* Vol. 65, No. 21, pp. 3399-3412.

Wilson, C.A., Wong, S., VanBrocklin, M., & Federspiel, M.J. (2000). Extended analysis of the in vitro tropism of porcine endogenous retrovirus. *Journal of Virology,* Vol. 74, No. 1, pp. 49-56.

Wittmann, G. (1985). Aujeszky's disease: factors important for epizootiology and control. *Review of Science Technology OIE,* Vol. 4, pp. 5-44.

Wood, J.C., Quinn, G., Suling, K.M., Oldmixon, B.A., Van Tine, B.A., Cina, R., Arn, S., Huang, C.A., Scobie, L., Onions, D.E., Sachs, D.H., Schuurman, H.J., Fishman, J.A., & Patience, C. (2004). Identification of exogenous forms of human-tropic porcine endogenous retrovirus in miniature Swine. *Journal of Virology,* Vol. 78, No. 5, pp. 2494-2501.

Wrathall, A.E. & Mengeling, W.L. (1979). Effect of inseminating seropositive gilts with semen containing porcine parvovirus. *British Veterinary Journal,* Vol. 135, No. 5, pp. 420-425.

Wright, P.F. & Webster, R.G. (2001). Orthomyxoviruses, In: *Field's Virology,* Knipe, D.M., Howley, P.M., Griffin, D.E., Lamb, R.A., Martin, M.A., Roizman, B., & Straus, S.E., pp. 1533-1579, Lippincott-Raven Publishers, ISBN Philadelphia.

Yaeger, M.J., Prieve, T., Collins, J., Christopher-Hennings, J., Nelson, E., & Benfield, D. (1993). Evidence for the transmission of porcine reproductive and respiratory syndrome (PRRS) virus in boar semen. *Swine Health and Production,* Vol. 1, No. 5, pp. 7-9.

15

Porcine Herd Health Management Practices for the Control of PRRSV Infection

V.G. Papatsiros
*Clinic of Medicine, Faculty of Veterinary Medicine,
University of Thessaly,
Greece*

1. Introduction

Porcine reproductive and respiratory syndrome (PRRS) is a highly contagious viral disease that was first recognized almost simultaneously in Western Europe (Wensvoort et al., 1991) and North America in the late 1980s (Keffaber, 1989). The causative agent is PRRS virus (PRRSV), a small single-stranded positive sense RNA virus, classified in the order Nidovirales, family Arteriviridae and genus Arterivirus. Since its appearance, PRRS has devastated the worldwide swine industry with tremendous economic losses (Neumann et al., 2005). PRRSV causes reproductive failure in breeding stock (e.g. premature farrowings, late term abortions, poor farrowing rate, mummified fetuses and stillborn piglets), as well as respiratory disease, elevated mortality and reduction of growth performance in piglets and growing / finishing pigs (Cho & Dee, 2006). Generally, after an acute outbreak of a PRRSV infection, herds may undergo a chronic loss of production in growing / finishing pigs and an endemic infection of breeding stock characterized by several outbreaks throughout the year (Stevenson et al., 1993).

The severity of PRRS may result from a number of factors such as differences in virulence among the PRRSV isolates, probable recombination between the different isolates within the same farm, immune status, host susceptibility and concurrent infections (other viruses and bacteria) and hygiene monitoring programme (Goldberg et al., 2000). PRRSV infected pigs are more susceptible to some bacterial (e.g. *Mycoplasma hyopneumoniae, Actinobacillus pleuropneumoniae, Bordatella bronchiseptica, Pasteurella multocida, Haemophilus parasuis, Streptococcus suis*) and viral diseases (e.g. swine influenza virus, Aujezky's disease virus, porcine respiratory coronavirus, porcine circovirus 2 - PCV2) (Brockmeier et al., 2002). One of the main pathogens involved in the porcine respiratory disease complex (PRDC) is PRRSV, as it has an additive or synergistic effect with the above other bacteria or viruses, that leads to a more severe and chronic respiratory disease in growing / finishing pigs (Thacker, 2006).

Before trying to control diseases at the farm level, it is very important to get information about what we really have to control. For instance, it is important to understand the pathogenesis, epidemiology and clinical forms of diseases. Therefore, the more scientific knowledge is known about PRRS, the better are the chances that this disease will be kept under control at relatively low losses. Table 1 shows some basic information of PRRSV infection at farm level (Cho & Dee 2006; Zimmerman et al., 2006).

Infection source	• Direct contact, nasal secretions, saliva, urine, feces, milk, semen, blood, aerosol, transplacental, fomites (e.g. boots, coveralls), equipment, insects, vehicles, human
Incubation time	• 3 days and over
Shedding period	• 99 days
Survival in the environment	• Less than 24 hours at 25°C on solid material • 9 to 11 days in water at 25°C • 8 days in lagoon water at 4°C • Stable at pH 6.5–7.5, but infectivity is rapidly lost at below 6 > pH < 7.5 • Survival >4 months at -70 to -20°C / its viability decreases with increasing temperature • 90% of PRRSV infectivity is lost within 1 week at 4°C, but low titers of infectious virus can still be detected for at least 30 day • In solution, PRRSV infectivity persists for 1–6 days at 20–21°C, 3–24 hours at 37°C, and 6–20 minutes at 56°C • Shedding of PRRSV in saliva, urine, and feces are high risk factors for the environmental contamination
Other properties	• High biological, antigenic, genetic and pathogenic heterogeneity exists among PRRSV strains / genetically diverse PRRSV strains may coexist in the same farm • Partial and variable cross protection between isolates • Frequent genetic changes or recombination • Fetuses are most susceptible to active infection after 60 days of conception • Pigs may become carriers until at least 150 days of age (PRRSV persistence) • PRRSV is associated with outbreaks of other pathogens • Lack of complete heterologous protection by commercial vaccines

Table 1. Main information about PRRSV infection for a herd health management programme

The above information should not be used as definitive, as many factors may interact with field situations that can result in potential timeline changes. Veterinary medicine, just as human medicine, is not a static science. So, it is possible that the scientific data for PRRS could change year by year. The data of Table 1 are very basic and representative of the scientific literature regarding PRRSV infection in order to take decisions on what herd health management practices should be applied in a farm. For this reason, the above data should be interpreted, since the goal at farm level is the herd and not individual animals.

2. Principles of a porcine herd health management programme in PRRSV infected farms

2.1 Biosecurity measures

Biosecurity measures are always a fundamental part of a porcine herd health management programme. Generally, the risk of a herd becoming PRRSV-positive increases with: a) density of PRRSV-positive neighboring herds, within 500 meters; b) increasing herd size; c) high number of animal introductions; d) purchase of semen for artificial insemination; e) absence of quarantine for replacement boars and gilts (Weigel et al., 2000). The aim of porcine herd health management programmes for the prevention and control of PRRSV infection is either to stop the introduction of PRRSV into negative herds or limit the introduction of new strains into already PRRSV-infected herds. However, the control of PRRS in a herd, continuously exposed to exogenous infection, is a loss of time and money. Biosecurity measures should be reviewed with attention to both internal and external biosecurity. External biosecurity deals with the risk of introduction of new PRRSV strains or other co-infections from outside the farm. Internal biosecurity deals with the spread of PRRSV within a farm after the exogenous introduction of the virus. Table 2 shows the basic elements of a biosecurity programme for a commercial farrow-to-finish farm or a sow unit.

Gilts source	• PRRSV negative or at least non-shedding • Site of gilt production should be away from any other pig facilities
Semen source	• PRRSV negative boar stud • Unit under air filtration or far from any other pig facilities
Quarantine	• For a minimum of 30 days before the introduction into the herd • Separate building than the main unit
AIAO	• Strictly applied in farrowing rooms, nursery and finisher • Washing/disinfecting/drying between batches • Consider batch farrowing to facilitate AIAO
Environment/Feed, water, air	• Mechanical air filtration • Water sanitation • Feed free of mycotoxins / balanced energy and amino acids
Equipment/ instruments	• One set of processing equipment should be disinfected while the other is in use • Boots, hands and coveralls should be kept clean • Disposable gloves can be used between litters • Separate equipment (shovels, brooms, scrapers) should be used for the manure passage and the feed alley at all times in order to reduce the risk of PRRSV spread • Needles: *Sows and Boars: discard after one injection / Piglets: discard after each litter or pen* • Equipment / instruments for castration, tooth-clipping or tail-docking: *Washing and heating (propane burner) between litters; Disinfection after using*
Transport	• Transport of pigs in cleaned vehicles (washing, disinfection, drying)
Hygiene	• Disinfectants • Washing of boots at the end of the day with a brush and disinfect in a bath of disinfectant (new disinfectant solution should be used every day) • Scrape sow's manure each day in farrowing room with a shovel (1 shovel per room) • Carcass disposal: compost or incineration
Personnel/ Visitors	• Minimize the numbers of visitors • Wash hands • Fomites belonging to the farm • Forbid to enter equipment, tools or materials in the farm that have been in contact with pigs or pig manure of other farms

Table 2. Basic elements of a biosecurity programme for a commercial farrow-to-finish farm or a sow unit

Direct routes of PRRSV transmission within and between pig populations include infected animals and contaminated semen. For this reason, all replacement boars and gilts should originate from PRRSV negative farms and be properly isolated in quarantine facilities for a minimum of 30 days, including serological tests and PRRSV vaccinations prior to their introduction into the herd. Ideally, isolation facilities should be located on another farm site and visited at the end of the working day. Moreover, semen for artificial insemination should come from PRRSV-negative boar studs. Indirect transmission involves transmission by fomites (boots and coveralls), contaminated equipment (e.g. needles), farm personnel and visitors, transport vehicles (contaminated trailers, coolers, containers), substances (e.g., water, food), insects (e.g. houseflies and mosquitoes), or aerosols (Dee et al., 2002; Cho & Dee, 2006; Desrosiers, 2011). In general, biosecurity efforts should focus on all inputs and outputs of the farms, such as pigs, supplies and materials, feed, water, personnel, removal of manure, and reclaims. In addition, the entry of pests such as rodents, insects, and birds from all buildings should be avoided (Zimmerman et al., 2006). Finally, All-in/All-out (AIAO) pig flow is effective in controlling a variety of respiratory pathogens in weaned pigs. AIAO consists of dividing buildings into individual rooms, allowing thorough cleaning and disinfection of facilities between groups of pigs. This method is very effective in reducing the horizontal spread of PRRSV from older, infected pigs to those recently placed in the finishing stage. Although, AIAO does not directly control the transmission of PRRSV, it reduces the impact of secondary bacterial co-infections.

The disinfection is crucial for a biosecurity programme in a PRRSV-infected farm. The survival of PRRSV outside a living host is affected by factors that include the substrate, pH, temperature, relative humidity, and exposure to ultraviolet light. PRRSV is inactivated by lipid solvents, such as chloroform and ether. In addition, at "room temperature" the virus can be inactivated completely with the use of chlorine (0.03%) in 10 minutes, iodine (0.0075%) in 1 minute, and a quaternary ammonium compound (0.0063%) in 1 minute (Shirai et al., 2000). PRRSV is also relatively labile in the environment and particularly susceptible to heating and drying (Pirtle & Beran, 1996). Leaving a room to completely dry with or without supplemental heating is one of the most effective ways of killing viruses and bacteria. PRRSV can remain infectious for an extended time under specific conditions of temperature, moisture, and pH. It is stable at pH 6.5–7.5, but infectivity is rapidly lost at pH below 6 and above 7.5 (Bloemraad et al., 1994). At temperatures ranging from -70 to -20°C, PRRSV can survive more than 4 months, but when temperature increases, its survivability decreases. Approximately 90% of PRRSV infectivity is lost within 1 week at 4°C, but low titers of infectious virus can still be detected for at least 30 days. In solution, the virus infectivity persists for 1–6 days at 20–21°C, 3–24 hours at 37°C, and 6–20 minutes at 56°C (Zimmerman et al., 2006). At 25–27°C, infectious PRRSV is not detected on plastic, stainless steel, rubber, alfalfa, wood shavings, straw, corn, swine starter feed, or denim cloth (Pirtle & Beran, 1996). Cleaned and disinfected pens should be left to dry for a minimum of 24 hours before pigs are placed, while barns should be allowed to dry for a minimum of 7 - 14 days between batches. Since PRRSV persists in cold and wet conditions, all equipment and material used at the farm or for transport of pigs must be cleaned and dried (Dee et al., 2002).

Herd health management has a significant impact on disease expression, as some management parameters like environmental conditions, water and feed quality may compromise affect the animal resistance to diseases. In addition, the above management parameters if are not in compliance with the rules of animal welfare can also increase the stress and favor virus / bacteria transmission in pigs or have a negative impact on immunity against several diseases.

Water: PRRSV survives in water up to 11 days, but drying quickly inactivates it (Zimmerman et al., 2006). It is important to use chlorine or hydrogen peroxide for the water sanitation, because water sources and delivery systems may be contaminated with pathogens. However, the use of chlorine should be monitored by testing the levels of free chlorine. Well capacity, water flow rates, pressure, drinker number, type and placement are very important for the optimal water intake. Finally, factors such as mineral content, hardness, total dissolved solids, and pH should also be considered.

Feed: it is unlikely that PRRSV can be transmitted through contaminated feed, but feed suppliers present a risk acting as a vector from infected farms to other farms. However, energy and amino acids must be balanced in diet, in order to provide the required energy for the enhancement of the immune system and the basic building blocks of antibodies. Minerals and vitamins are also required in sufficient quantities for the optimal function of immune system. According to the rules of animal welfare, feeders must be managed appropriately, so that the feeder provides adequate access for each pig. Moreover, the feed needs to be free of mould and mildew, because mycotoxins can cause immune suppression, resulting in increased incidences of clinical signs of PRRS, such as pneumonia.

Environment: It plays also a main role, as PRRSV shedding in saliva, urine, and feces is a high risk factor for the environmental contamination, creating the potential for transmission via fomites (e.g. boots, coveralls), personnel and vehicles. Air exchange rates have an impact on control of PRRS. The filtering of air entering pig housing has been proposed as a means to reduce the risk of airborne transmission of PRRSV from infected herds to at-risk populations (Pitkin et al., 2009). Therefore, air filtration is an effective means to reduce the risk of external PRRSV introduction to large breeding herds located in areas with high pig population density (Dee et al., 2010). Moreover, increased relative humidity may increase the survival time of respiratory pathogens in the room environment and increased ventilation rates may increase air speed, causing chilling. Chilling, due to the wide daily and rapid small temperature fluctuations, contributes significantly to the increased prevalence of disease by increasing stress levels in affected pigs. For this reason, ventilation and temperature controllers should be adjusted so as to ensure that they are set to control temperature fluctuation and daily variability. The use of simple environmental testing equipment, such as humidity monitors, data loggers, air speed and gas testers are very important. Moreover, PRRSV can be inactivated through the process of composting or incinerating carcasses, so only these methods should be applied. Finally, PRRSV can survive in lagoon effluent for up to 3 days at 20 °C and for 7 days at 4 °C. Contact with PRRSV-positive effluent can be a source of infection to naïve pigs. Therefore, producers that utilize recycled lagoon water in their waste management protocols may be at higher risk for external PRRSV introduction than those who use deep pits.

2.3 Control strategies

PRRSV tends to circulate within a herd indefinitely after the initial infection of a herd. For example, PRRSV was isolated from nursery pigs up to 3.5 years after the initial PRRS outbreak (Larochelle et al., 2003). Several parameters related to the herd (size, management and type), or the vaccine (type, management) may determine whether PRRS can be successfully controlled in a farm. Unfortunately, the ability of the virus to persist in herds and its wide biologic, antigenic and genetic variability may further complicate control plans (Zimmerman et al., 2006). In endemically PRRSV infected herds, the virus circulates, because, in any given time, animals are in different stages of infection and immunity. Control and prevention of the virus should be based on: a) minimizing PRRSV circulation at farm level, b) limiting the effects of the virus circulation and secondary bacterial co-infections and c) maximizing and stabilizing the herd immunity (Dee, 2003).

2.3.1 Vaccination programmes

Vaccinations against PRRSV with both modified (or attenuated) live vaccines (MLV) and inactivated (or killed) vaccines have frequently been reported by many studies. The absence of complete protection has frequently been attributed to antigenic differences and the limited cross-reactivity between strains of commercial vaccines and challenge strains. Such variability is largely attributed to genetical and antigenical heterogeneity between isolates mainly among European and American isolates, but also within the same area or the same farm. The high heterogeneity among PRRSV strains is likely to be the main obstacle to effective control of PRRSV infection using current commercial vaccines (MLV and inactivated), since the immunity induced by one strain may be only partial against a different strain, even within the same genotype (Mateu & Diaz, 2008; Kimman et al., 2009). However, vaccine efficacy may be associated with an efficient cell-mediated immunity and it is not only related with its immunological properties, but also with the characteristics of the challenging strain to trigger an immune response (Martelli et al., 2009). Therefore, the ability of each strain to induce a strong cell-mediated immune response is more important than the genetic similarity between the vaccine strain and the field strains for inducing clinical protection (Mateu & Diaz, 2008). The complexity of the immune response to PRRSV and the ability of the virus to escape or modulate the host's immune system make difficult the development of an effective vaccine for control and eradication of PRRS.

MLV vaccines have been widely used in breeding stock and young piglets. The results of many studies have shown beneficial effects on PRRS clinical disease occurrence and severity, the duration of viremia and virus shedding (Scortti et al., 2006; Martelli et al., 2007; Kimman et al., 2009), as well as improvement of health status and performance of gilts/sows and their litters (Alexopoulos et al., 2005). However, the use of MLV vaccines is questionable. Virus-neutralizing (VN) antibodies against PRRSV protect against viremia, virus replication in lungs, transplacental spreading of the virus and reproductive failure (Labarque et al., 2003, 2004; Lopez & Osorio, 2007). MLV vaccination induce VN antibodies and protect against viremia, virus replication in lungs and virus induced respiratory and reproductive disorders (Labarque et al., 2003; Scortti et al., 2007; Zuckermannet al., 2007). The protective immune response induced by current commercial MLV vaccines is influenced by genetic diversity, as these vaccines do not always sufficiently protect (or only partially) against re-infection and transplacental infections caused by heterologous strains (Scortti et al., 2006; Prieto et al., 2008; Kimman et al., 2009). However, Martelli et al. (2009) reported that vaccination of piglets at 5 weeks of age with a commercial MLV vaccine induced a partial clinical protection, associated with an efficient cell-mediated immune response, when the above vaccinated pigs were exposed to a heterologous field strain. Moreover, there are major concerns about the safety of current vaccination programmes with MLV vaccines. Experimental and field studies reported that MLV strains can cause viremia, revert to virulence and spread transplacentally and horizontally not only within the vaccinated herds (transmission and detection in non-vaccinated pigs), but also to neighbouring non-vaccinated herds (Bøtner et al., 1997; grosse Beilage et al., 2009; Kimman et al., 2009). It is possible that farmers using an MLV vaccine for the first time may experience a decrease in the herd productivity. MLV vaccinations in sows reported to cause acute PRRS-like clinical signs, characterized by increased late term abortions, increased numbers of stillborns and mummified piglets, as well as reduced numbers of live born and weaned piglets (Bøtner et al., 1997; Dewey et al., 1999). An additional problem is that the MLV vaccination has been shown to decrease the efficacy of *Mycoplasma hyopneumoniae* vaccines (Thacker et al., 2000; Le Roitha et al., 2011).

Based on the Greek experience, in farms with history of endemic PRRSV infection, the MLV vaccination of breeding stock (gilts once at the age of 180 days and sows 10 days post-partum) can have beneficial effects on their health and performance. Alexopoulos et al. (2005) reported reduction of premature farrowing rate, return-to-oestrus rate, number of dead and mummified born piglets as well as the increase of farrowing rate, the number of piglets born alive and weaning pigs per litter. Furthermore, in farms affected by both PRRSV and PCV2, the MPV vaccinations in piglets (at roughly 5 weeks old) can lead to a reduction of morbidity and mortality of growing pigs, as well as an improvement of the growth performance in vaccinated pigs (Alexopoulos et al., 2005; Kritas et al., 2007). However, recent evidences based on personal experience and field observations in endemic PRRSV-infected farms suffering by significant reproductive failure, indicate that the MLV vaccination of breeding stock can improve; a) the reproductive performance, b) the viraemic status of piglets, c) the morbidity and mortality of piglets, d) the growth performance of piglets. The vaccination protocol depends on the average breeding herd size and the current clinical expression of PRRS. In cases of; a) large scale farms with a capacity of over 700 sows under production, b) farms suffering from acute outbreaks of PRRS and c) endemic PRRSV-infected farms with 4-5 outbreaks every year, the preferable vaccination schedule includes a vaccination at 60th day of gestation and booster at 6th day of lactation. Vaccination at 2-4 weeks before mating and revaccination before each consecutive gestation is proposed for; a) farms with a capacity of over 100-200 sows under production and b) endemic PRRSV-infected farms with 4-5 outbreaks every year. Moreover, the MLV vaccination of piglets at 2-3 weeks of age can have beneficial effects on their health and growth performance, including reduction of morbidity and mortality rate due to PRRSV infection, as well as improvement of average daily gain and feed conversion ratio (Papatsiros, 2011).

Inactivated vaccines are considered safer than MLV vaccines, as the vaccine virus cannot transmit to other pigs and cannot revert to virulence. However, their efficacy has been seriously questioned. Field studies reported that the inactivated vaccines did not induce reproductive failure in vaccinated sows and improved efficiently the reproductive parameters at a farm level (Plana-Duran et al., 1997; Papatsiros et al., 2006). However, their capacity to induce a protective immunity against challenge with wild-type virus has been questioned, as these vaccines induce poor immune responses in naïve pigs and provide weak memory responses with sequential challenge without any obvious active immune responses in the vaccinated pigs (Kim et al., 2011). On the one hand, commercially available inactivated vaccines do not induce VN antibodies and do not sufficiently protect against viremia or prevent from the clinical signs associated with PRRSV infection, i.e., post-challenge viremia and transplacental infection of the piglet (Nilubol et al. 2004; Scortti et al., 2007; Zuckermannet al., 2007). On the other hand, experimental inactivated PRRSV vaccines can induce VN antibodies and reduce the duration of viremia (Misinzo et al., 2006, Vanhee et al., 2009). However, without VN antibodies induction, a commercial inactivated vaccine reported to induce significant improvement of sow reproductive performance and litter characteristics (Papatsiros et al., 2006). In addition, some types of adjuvants were used as effective vaccine adjuvants to enhance the humoral and cellular responses of piglets against PRRSV (Linghua et al., 2007). Hence, the effectiveness of vaccination programme based on inactivated vaccine depends on the vaccine (virus strain and used cells to prepare the vaccine) and vaccination strategy (Misinzo et al., 2006).

Based on the Greek experience, the vaccination of gilts and sows with inactivated vaccines can have beneficial effects on their health and performance, as well as on their litters. The long term use of a commercial inactivated vaccine for a period of 1.5 years in breeding stock of a closed single commercial farm with persistent PRRSV infection and high seroprevalence, proved to reduce the negative effects of the virus on the breeding herd (Papatsiros et al., 2006). The vaccination scheme included a primary vaccination of all gilts/sows of the herd by administering two doses 3–4 weeks apart, except those being 1 week prior to 2 weeks post-service. The skipped females were subjected to primary vaccination, but starting 3 weeks later. All previously vaccinated animals received a booster vaccination between 55 and 60 days of next gestation, and thereafter at each gestation for a period of 1.5 years. The gilts were vaccinated twice prior to breeding (primary vaccination) and boostered in each pregnancy as described previously. Vaccinations resulted in a significant improvement of sow reproductive performance and their litters' characteristics, as are shown in Table 3. This study indicated also that under practical conditions, the use of an inactivated vaccine should be administered on a regular basis for obtaining the maximum of its effect, as it has been observed that the higher the degree of immunization of sows, the better the improvement of their reproductive parameters. It is interesting to note that while the number of booster vaccinations improved several performance parameters, it did not improve the level of immunity as measured by IPMA, suggesting that the levels of IPMA-antibodies may not always reflect protection. Moreover, the vaccination led to a significant reduction of culling rate due to reproductive failure, resulting in an improvement of longevity, herd age distribution and number of non productive days in the breeding stock (Table 4). Finally, the long-term vaccination of boars with the same inactivated vaccine was safe and no significant changes in semen characteristics after each vaccination were noticed (Papatsiros et al., 2008). The above vaccination schedule is proposed to be applied on a regular basis in breeding stock of endemic PRRSV-infected farms, in order to achieve a stabilization of the immunity status of all breeding herd and prevent the losses due to the yearly outbreaks of PRRS.

	Period relative the start of vaccination				
	1 year prior (Non vaccinated sows)	1.5 years after (Vaccinated sows)		1 year prior (Non vaccinated sows)	1.5 years after (Vaccinated sows)
Reproductive parameters			**Number of piglets per litter**		
Returns to oestrus rate (%)	8.9^{a}	6.8^{b}	*Totally born*	11.4^{a}± 3.1	11.6^{a} ± 3.1
Abortion rate (%)	2.2%^{a}	1.3%^{b}	*Live born*	10.5^{b} ± 3.1	10.9^{a} ± 3.1
Farrowing rate (%)	87.7%^{b}	90.3%^{a}	*Stillborn*	0.7^{a} ± 1.4	0.5^{b} ± 1.2
«Empty» sows rate (%)	0.4%^{a}	0.5%^{a}	*Mummified*	0.2^{a} ± 0.6	0.1^{b} ± 0.5
Premature farrowing rate (%)	22.8%^{a}	1.4%^{b}	*Light weight*	1.0^{a} ± 1.6	0.1^{b} ± 0.6
Farrowing rate (%)	87.7%^{b}	90.3%^{a}	*Weak*	1.5^{a} ± 2.0	0.6^{b} ± 1.5
Culling rate (%)	24.5%^{a}	21.9%^{b}	*Splay-legs*	0.7^{a} ± 1.0	0.1^{b} ± 0.4
Gestation length (days)	114.3^{b} ± 2.3	115.2^{a} ± 1.4	*Alive first 24h*	10.1^{b} ± 3.0	10.5^{a}± 3.0
Lactation length (days)	25.0^{a} ± 3.3	21.6^{b} ± 2.8	*Weaned*	8.9^{b} ± 1.5	9.3^{a} ± 1.4
Weaning-to-service (days)	6.7^{a} ± 6.3	5.6^{b} ± 3.7			

[a.b.] Percentages and Means (± SD) in a row with different superscripts differ ($P \geq 0,05$).

Table 3. Reproductive and litter characteristics prior (non vaccinated sows) and after (vaccinated sows) the start of vaccination with inactivated vaccine: Source: Papatsiros 2006.

	Period relative the start of vaccination	
Causes of culling (%)	1 year prior (Non vaccinated sows)	1.5 years after (Vaccinated sows)
Reproductive failure	40.5%^{a}	26.6%^{b}
Deaths	13.7%^{a}	10.4%^{a}
Old age	14.8%^{a}	33.9%^{b}
Locomotor problems	12.5%^{a}	12.7%^{a}
Other	18.5%^{a}	16.3%^{a}

[a, b] Percentages in a row with different superscripts differ ($P \leq 0.05$).

Table 4. Causes of culling in female breeding stock prior (non vaccinated animals) and after (vaccinated animals) the start of vaccination with inactivated vaccine. Source: Papatsiros 2006.

2.3.2 Gilt development & isolation / acclimatization

The first step in breaking virus circulation in the breeding herd is to use replacement animals that have been exposed to PRRSV and developed immunity prior to their introduction into the herd (Dee, 2003). Gilt introduction in breeding population is the key for PRRS control, consisting of 3 periods: the isolation period, the acclimatization period, and the recovery period (Pesente et al., 2006; Vashisht et al., 2008). The length of each period may range from 30 to 60 days, depending on the age of the replacement animal, the PRRS history of intra and inter farms, and the type and size of farm. The isolation period consists of serological tests (day 1 to 2) in order to determine the PRRSV infection status of the incoming animals. The PRRSV vaccination of the incoming breeding stock should be done shortly after arrival. The acclimatization period starts 30 days after the introduction of replacement gilts, aiming to expose the ''new'' gilts to the specific PRRSV strain of the farm. Finally, a period of recovery (at least 30 days) is implemented to reduce the risk of introducing actively infected gilts into the breeding herd. By the end of acclimatization period, the replacement gilts could be vaccinated against other diseases, such as Aujeszky's disease. Moreover, the acclimatization should be also applied in farms with grandparent nucleus of sows for producing its own gilts, which are separately housed (McCaw, 1995).

Several acclimatization programmes were described, including methods such as: a) feedback with tissues from weak-born piglets and stillbirths in the face of outbreaks, b) use of MLV and inactivated vaccines, c) contact exposure between gilts and weaning pigs that are used as donor sources of virus and d) inoculation of negative replacement animals with serum collected from viremic pigs from the same farm (Dee, 2003; Vashisht et al., 2008). The use of serum containing viable field virus has recently increased due to the increased genetic heterogeneity among PRRSV strains and the perception that commercial vaccines do not induce enough protective cross-protection against newly identified PRRSV strains. This method has inherent risks and requires thoughtful application and high quality control standards (Zimmerman et al., 2006). Consistent acclimatization of incoming breeding stock to PRRSV results in: a) the stabilization of clinical signs, b) the production of PRRSV negative piglets at weaning, c) the prevention of PRRSV outbreaks associated with endemic PRRSV strains, d) the development of specific immunity to the homologous herd strains, and e) the improvement of production parameters (Pesente et al., 2006; Vashisht et al., 2008). It is generally accepted that early exposure (2–4 months of age) can result in the protection of the exposed animals and the introduction of the replacement animals at a time when shedding has stopped. Therefore, it is recommended that gilts can be purchased and acclimatized at an early age (Vashisht et al., 2008).

In a closed-herd system, replacements are produced internally and are introduced into the sow herd directly from the grower or finisher stage, regardless their PRRSV infection status. Generally, closed herd systems do not eliminate PRRSV infection, because replacement gilts usually have previous exposure to PRRSV strains circulating in the herd. For this reason, gilt developer facilities are very important for introduction of gilts into an infected farm. Gilt developer facilities should be located in the sow site or preferably, in an alternative location far from the other unites. Gilts may be introduced at the age of weaning piglets or 2 to 5 months of age, under AIAO pig flow practices.

A field study in a commercial farrow-to-finish farm with 1,100 sows, where the gilts were housed with finishing pigs, indicated that even if the breeding stock were vaccinated with a PRRSV inactivated vaccine for a period of 24 months, the non-vaccinated gilts show high (93,3 - 100%) seroprevalence (Figure 1). This was an indication of high virus circulation in fattening buildings, but also was important evidence that some infected gilts re-introduced PRRSV into the breeding stock (Papatsiros, 2006).

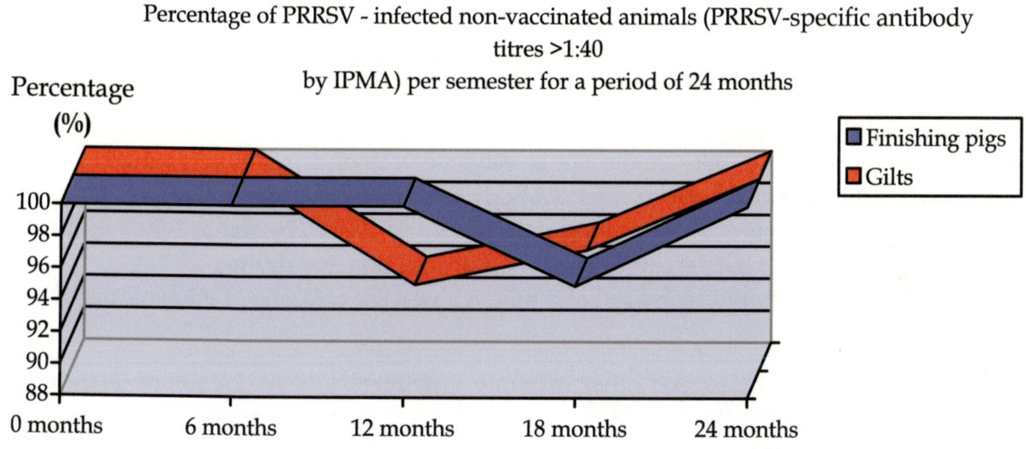

Fig. 1.

Finally, the strategy of "herd closure" could also be applied in order to eliminate the virus circulation by the reduction of viral shedding and elimination of carrier pigs. The basis of herd closure is the cessation of replacement gilt introduction for an extended period (4 to 8 months), depending on the herd health status and pig flow. During the period of herd closure, the introduction of replacement gilts and boars is discontinued, and only the introduction of semen for artificial insemination is allowed. In addition, the vaccination of the breeding herd could be applied at this period. However, herd closure can result in the production of an improper parity distribution and the development of a PRRSV-negative breeding population over time. These effects can be minimized through the use of off-site breeding projects for replacement gilts (Cho & Dee, 2006; Zimmerman et al., 2006).

2.3.3 Other management practices

McCaw (1995) proposed the McREBEL (Management Changes to Reduce Exposure to Bacteria to Eliminate Losses) system for PRRS control. The aim of McREBEL is to maximize the number of piglets remaining with their birth mother and, secondly, to maximize the number of piglets remaining with the colostrum mother. According to the author, the McREBEL system reduces considerably the economic losses, without eliminate successfully the virus circulation in the farm. This system is recommended to be applied in farms with separate premises for the different ages both at farrowing and in newly weaned piglets, using the AIAO system as following:

i. Limitation or interruption of the cross-fostering of newborn piglets for resizing or saving sick pigs, fall-behinds, and runts. The cross-fostering of piglets should be performed only within the first 24 hours of age.
ii. Piglets should be moved within farrowing rooms at birth and not between different rooms to "nurse sows" (especially poor-doing or possibly diseased piglets to younger age groups attempting to save them). In addition, sows or piglets should not be moved between rooms.
iii. Interruption of using "nurse sows" for weak-born PRRSV-infected pigs, fall-behinds, and runts.
iv. Minimization of piglets' handling, especially routine antibiotic treatments. The effect of each handling or treatment procedure should be evaluated on clinical disease levels.
v. Immediate euthanization of all very sick and debilitated piglets those are non-responsive to therapy.
vi. Sick or lightweight piglets should not be retained with or mixed with younger piglets. Euthanization of small piglets with poor body condition at weaning.
vii. Suckling and weaning piglets should only be moved AIAO, by room. 2-3 days should be allowed between batches for cleaning and disinfection and drying.
viii. Weaning piglets should be loaded at a time with earlier weaning of a few of the oldest and biggest litters.

Co-infections of PRRS by several other respiratory pathogens have the ability to increase both the severity and duration of PRRS associated disease. For this reason, a porcine herd health management for the prevention and control of PRRSV infection should also include measures for the control of bacterial co-infections (e.g. *Haemophilus parasuis, Streptococcus suis, Actinobacillus pleuropneumoniae, Pastereula multocida*) and viral diseases such as swine influenza, Aujeszky's disease and PCV2 associated diseases. Appropriate medication and vaccination protocols need to be applied for the individual infections. Water soluble antimicrobials or feed medications can be selected based on the sensitivity pattern of the secondary bacteria. Oxytetracycline, chorotetracycline, trimethoprim/sulpha, or synthetic penicillins are the medicines of choice, but tiamulin, tylosin or lincomycin may be also used. If *Actinobacillus pleuropneumonia* is active, ceftiofur or florfenicol could be medicines of choice for individual treatments. Feed medications (e.g. 500 to 800g of tetracycline or 400g of trimethoprim/sulpha per tonne in-feed) must be provided at the earliest possible post infection time in order to maximise efficacy. Anti-fever drugs may be used to decrease the negative effects of fever, such as it depresses appetite, abortion in sows, and decreases milk production. Finally, vaccination protocols should be applied, including vaccines against viral pathogens such as PCV2, swine influenza virus and Aujeszky's disease virus, as well as bacterial pathogens such as *Mycoplasma hyopneumoniae, Actinobacillus pleuropneumoniae* and *Haemophilus parasuis*. The most important vaccinations of weaning pigs in farms suffering from PRDC are considered those against *Mycoplasma hyopneumoniae* and PCV2 (Alexopoulos et al., 2004).

3. Monitoring serology

The preliminary diagnosis of PRRSV infection is most often based on a review of herd history and clinical signs. Gross post mortem findings can help to confirm the suspicions and finally serological, histological, immunohistochemistry and PCR examinations can confirm the diagnosis. Because of the PRRSV genetic diversity, it is preferable to isolate and characterize the farm specific PRRSV strains, using genetic typing such as restriction fragment length polymorphism (RFLP) and gene sequencing. This allows also determining any new PRRSV strain (Zimmerman et al., 2006).

Before starting a vaccination programme, it is essential to investigate when the PRRSV infection took place. Routine serological monitoring tests for PRRSV serum antibodies are very useful diagnostic tools for the determination of PRRSV herd exposure status. Therefore, the vaccination programme can be applied prior to the usual infection time in order to provide efficient protection in vaccinated animals. The serological diagnostics of PRRS include enzyme-linked immunosorbent assay (ELISA), blocking ELISA, serum virus neutralization (SVN), indirect fluorescent antibody (IFA) and immunoperoxidase monolayer assay (IPMA). The IFA has high specificity (99.5%), but unknown sensitivity for individual animals, while the IPMA is also considered to be a highly specific and sensitive test (Wensvoort et al., 1991). The ELISA is also reported to be sensitive and specific (O'Connor et al., 2002). The specificity of a commercial PRRS ELISA (HerdChek® PRRS ELISA, IDEXX Laboratories Inc., Westbrook, Maine), has been estimated to be between 99.3 and 99.5 % (O'Connor et al., 2002). The ELISA is well-suited for monitoring since it is sensitive, specific, standardized, simple to perform, relatively inexpensive, easy to implement, able to detect both American and European strains, and potential for rapid analysis of numerous serum samples. Antibodies can be detected as early as 9 DPI (days post infection), peak at 30–50 DPI and then decline to negative levels 4–12 months after infection. ELISA results are interpreted as positive when a sample-to - positive (S/P) ratio ≥ 0.4 (indicates presence of antibody to PRRSV) or negative (S/P <0.4). The demonstration of seroconversion (negative to positive), using acute and convalescent serum samples, is the most definitive method to diagnose PRRSV infection serologically. Increasing titers of PRRSV specific antibody demonstrated by rising ELISA S/P ratios in a group of infected animals can also indicate PRRSV infection. Finally, sequencing of PRRSV isolates followed by phylogenetic analysis is a powerful tool to monitor the spread of virus intra and inter farms and allows a fuller understanding of the success or failure of the control programme (Pesente et al., 2006). The ELISA in all of its current forms is the best practical diagnostic method to detect exposure events and confirm vaccinations, but it is not a reliable indicator of protection against PRRSV.

The number of serum blood samples needed to identify an infected herd depends on the seroprevalence. Shortly after an acute outbreak, the prevalence of seropositive pigs in infected herds is high. For this reason, it is possible to identify infected herds by testing only a small number of samples. In case of a herd with unknown serostatus, more samples from breeding herd than from a finishing herd should be tested, because seroprevalence in breeding herd is usually lower compared to seroprevalence in finishing herd. A porcine herd health management programme for the control of PRRS should include a herd serological monitoring, once or twice per year. The herd serostatus is predicted by testing a representative number of animals. The number and ages of animals to be tested is determined by the assumed seroprevalence in each age group. Seroprevalence to PRRSV in the breeding herd can vary widely depending on how long the herd has been infected.

Herds with a recent history of reproductive failure have often been infected less than 1 year and tend to have a high seroprevalence (approximately 50%) in breeding animals. To detect seropositive animals with 95% confidence (assume 50% seroprevalence in the breeding herd) at least 10-20 blood samples from breeding stock is recommended. In endemic PRRSV-infected herds, which have probably been infected more than 1 year, the seroprevalence is usually low in the breeding herd (20% or less) and high in the finishing pigs (50% or greater) (Stevenson et al., 1993). For example, in a breeding herd of 100 animals, to detect seropositive animals with 95% confidence (assume 10% seroprevalence in breeders and 50% seroprevalence in finishing pigs) at least 25 blood samples from breeding stock and 7 from finishing pigs are recommended. However, herd serology only demonstrates previous exposure to the virus and does not provide a definitive diagnosis of PRRSV as a cause for the clinical problems on the farm. It is generally recommended that young pigs (10-15 blood samples from pigs at 3, 5, 7, 9 weeks and 16 of age), rather than breeding stock, should be tested to determine PRRSV circulation and infection status in a herd. In single-site, farrow-to-finish herds, the seroprevalence of PRRSV infection is usually considered to be highest in the grow-finish unit. Moreover, in herds without clinical signs of PRRS, testing 12 blood samples in both PCR and ELISA, (6 samples in pigs 9 weeks of age and 6 samples in pigs 16 weeks of age), is reported as a cost-efficient first evaluation of the PRRSV infection-status (Duinhof et al., 2011). In order to detect an increasing seroprevalence over time is preferable to perform serology in nursery pigs, including ear tagging and bleeding 10-20 pigs at 3-4 weeks of age and bleeding the same pigs at 7-8 weeks of age (Van Alstine et al., 1993). A greater percentage of seropositive pigs at the second bleeding would indicate that the virus is actively spreading among pigs on the farm. This may be confounded by passively acquired maternal antibody that usually lasts for 6-8 weeks, but can last up to 16 weeks of age.

4. Conclusion

PRRS remains to have a great negative economic impact on the the global swine industry. The control of PRRS economic losses in commercial swine farms is a challenge for farmers and swine veterinarians. Many factors including management practices and pig flow, the level of risk associated with local pig density, and the inherent characteristics of the specific PRRSV strains on farm level contribute to a successful a porcine herd health management programme development. The goal of a porcine herd health management to control PRRSV infection is based mainly on: a) minimizing the virus transmission and circulation in the farm, b) maximizing the herd immunity against PRRSV and c) limiting the losses due to secondary co-infections. A good knowledge of PRRS disease at the farm level, the establishment of strict biosecurity measures, acclimatization of replacement gilts by exposure to the specific PRRSV circulating strains in farm as well as the application of appropriate vaccination programme are commonly considered the most sound strategies to control PRRS. Furthermore, a completed herd history, assessment of clinical signs and proper diagnostic tests are very important tools for a swine veterinarian practitioner to control successfully PRRSV infections. Among the diagnostic tests, ELISA is the best practical diagnostic method to detect exposure events and confirm vaccinations, but it is not a reliable indicator of protection against PRRSV.

Finally, the establishment of a porcine herd health management programme to control PRRS is an excellent investment for swine farmers in order to improve the herd health status and reduce the cost of pork meat production.

5. References

Alexopoulos, C.; Kritas, S.K.; Papatsas, I.; Papatsiros, V.G.; Tassis, P.D. & Kyriakis, S.C. (2004). Efficacy of one and two shot vaccines for the control of Enzootic Pneumonia (EP) in a pig unit suffering from respiratory syndrome due to EP, PRRS and PMWS. *Proceedings of the 18th International Pig Veterinary Society congress*, Vol. 1, pp. 449, Hamburg, Germany, 27 June 1 July, 2004.

Alexopoulos, C.; Kritas, S.K.; Kyriakis, C.S.; Tzika, E. & Kyriakis, S.C. (2005). Sow performance in an endemically porcine reproductive and respiratory syndrome (PRRS)-infected farm after sow vaccination with an attenuated PRRS vaccine. *Veterinary Microbiology*, Vol. 111, No. 3-4, pp. 151-157, ISSN 0378-1135.

Bloemraad, M; de Kluijver, E.P.; Petersen, A.; Burkjardt, G.E. & Wensvoort, G. (1994). Porcine reproductive and respiratory syndrome: temperature and pH stability of Lelystad virus and its survival in tissue specimens from viraemic pigs. *Veterinary Microbiology*, Vol. 42, No. 4, pp. 361–371, ISSN 0378-1135

Bøtner, A.; Strandbygaard, B.; Sorensen, K.J.; Have, P.; Madsen, K.G.; Madsen, S. & Alexandersen, S. (1997). Appearance of acute PRRS-like symptoms in sow herds after vaccination with a modified live PRRS vaccine. *Veterinary Record*, Vol. 141, No. 19, pp. 497-499, ISSN 0042-4900.

Brockmeier, S.L.; Halbur, P. & Thacker, E.L. (2002). Porcine Respiratory Disease Complex. In: *Polymicrobial Diseases,* Brogden, K.A. Guthmiller, & J.M. (eds), Washington (DC), ASM Press, USA, ISBN-10: 1-55581244-9.

Cho, J.G. & Dee, S.A. (2006). Porcine reproductive and respiratory syndrome virus. *Theriogenology*, Vol. 66, No.3, pp. 655–662, ISSN 0093-691X.

Dee, S.A.; Deen, J.; Rossow, K.; Wiese, C.; Otake, S.; Joo, H.S. & Pijoan, C. (2002). Mechanical transmission of porcine and respiratory syndrome virus throughout a coordinated sequence of events during cold weather. *Canadian Journal of Veterinary Research*, Vol. 66, No. 4, pp. 232–239, ISSN 0830-9000

Dee, S.A. (2003). Approaches to prevention, control, and eradication. In: *The PRRS Compendium* (2nd ed). Zimmerman, J.& Yoon, K-J (eds), pp. 119–130, National Pork Board, Des Moines, Iowa, USA.

Dee, S.; Spronk, G.; Reicks, D.; Ruen, P. & Deen, J. (2010). Further assessment of air filtration for preventing PRRSV infection in large breeding pig herds. *Veterinary Record*, Vol. 167, pp. 976-977, ISSN 0042-4900.

Desrosier, R. (2011). Transmission of swine pathogens: different means, different needs. *Animal Health Research Reviews*, Vol. 12, No. 1, pp. 1-13, ISSN 1466-2523.

Dewey, C.E.; Wilson, S.; Buck, P. & Leyenaar, J.K. (1999). The reproductive performance of sows after PRRS vaccination depends on stage gestation. *Preventive Veterinary Medicine*, Vol. 40, pp. 233-241, ISSN 0167-5877.

Duinhof, T.F.; van Schaik, G.; van Esch, E.J. & Wellenberg, G.J. (2011). Detection of PRRSV circulation in herds without clinical signs of PRRS: comparison of five age groups to assess the preferred age group and sample size. *Veterinary Microbiology,* Vol. 150, No. 1-2, pp. 180-184, ISSN 0378-1135

Goldberg, T.L.; Hahn, E.C.; Weigel, R.M. & Scherba G. (2000). Genetic, geographical and temporal variation of porcine reproductive and respiratory syndrome virus in Illinois. *Journal of General Virology*, Vol. 81, No. 1, pp. 171–179, ISSN 0022-1317.

Grosse Beilage, E.; Nathues, H.; Meemken, D.; Harder, T.; Doherr, M.; Grotha, I. & Greiser-Wilke, I. (2009). Frequency of PRRS live vaccine virus (European and North American genotype) in vaccinated and non-vaccinated pigs submitted for respiratory tract diagnostics in North-Western Germany. *Preventive Veterinary Medicine,* Vol. 92, No. 1-2, pp. 31-37, ISSN 0167-5877.

Keffaber, K.K. (1989). Reproductive failure of unknown etiology. *American Association of Swine Practitioners Newsletter*, Vol. 1, No. 2, pp. 1-10,

Kim, H.; Kim, H.K.; Jung, J.H.; Choi, Y.J.; Kim, J.; Chang, G.U.; Hyun, S.B.; Shin, S.; Lee. B.; Jang, G.; Kang, B.K.; Moon, H.J. & Song, D.S. (2011). The assessment of efficacy of porcine reproductive respiratory syndrome virus inactivated vaccine based on the viral quantity and inactivation methods *Virology Journal*, Vol. 8, pp. 323, ISSN 1743-422X.

Kimman, T.G.; Cornelissen, L.A.; Moormann, R.J.; Rebel, J.M.J.; & Stockhofe-Zurwieden, N. (2009). Challenges for porcine reproductive and respiratory syndrome virus (PRRSV) vaccinology. *Vaccine*, Vol. 27, No. 28, pp. 3704-3718, ISSN 0264-410X.

Kritas, K.; Alexopoulos, C.; Kyriakis.; CS, Tzika, E. & Kyriakis, S.C. (2007). Performance of Fattening Pigs in a Farm Infected with Both Porcine Reproductive and Respiratory Syndrome (PRRS) Virus and Porcine Circovirus Type 2 Following Sow and Piglet Vaccination with an Attenuated PRRS Vaccine. *Journal of Veterinary Medicine A, Physiology, pathology, clinical medicine*, Vol. 54, No. 6, pp. 287-291, ISSN 0931-184X.

Labarque, G.; Van Gucht, S.; Van Reeth, K.; Nauwynck, H. & Pensaert M. (2003). Respiratory tract protection upon challenge of pigs vaccinated with attenuated porcine reproductive and respiratory syndrome virus vaccines, *Veterinary Microbiology*, Vol. 95, pp. 187–197, ISSN 0378-1135.

Labarque, G.; Van Reeth,V.; Nauwynck, H.; Drexler, C.; Van Gucht, S. & Pensaert M. (2004). Impact of genetic diversity of European-type porcine reproductive and respiratory syndrome virus strains on vaccine efficacy. *Vaccine,* Vol. 22, No. 31-32, pp. 4183–4190, ISSN 0264-410X.

Larochelle, R.; D'Allaire, S. & Magar, R. (2003). Molecular epidemiology of porcine reproductive and respiratory syndrome virus (PRRSv) in Quebec. *Virus Research,* Vol. 96, No. 1-2, pp. 3-14, ISSN 0168-1702.

Le Roitha, T.; Hammonda, S.; Todda, S.M.; Nia, Y.; Cecerea, T. & Pelzerb, K.D. (2011). A modified live PRRSV vaccine and the pathogenic parent strain induce regulatory T cells in pigs naturally infected with Mycoplasma hyopneumoniae. *Veterinary Immunology and Immunopathology*, Vol. 140, No. 3-4, pp. 312–316, ISSN 0165-2427.

Linghua, Z.; Xingshan, T. & Fengzhen, Z. (2007). Vaccination with porcine reproductive and respiratory syndrome killed virus vaccine and immunostimulatory oligodeoxynucleotides induces specific immunity in piglets. *Vaccine*, Vol. 25, No. 10, pp. 1735-1742, ISSN 0264-410X.

Lopez, O.J.; Oliveira, M.F.; Garcia, E.A.; Kwon, B.J.; Doster, A.; & Osorio, F.A. (2007). Protection against porcine reproductive and respiratory syndrome virus (PRRSV) infection through passive transfer of PRRSV-neutralizing antibodies is dose dependent. *Clinical and Vaccine Immunology*, Vol. 14, No. 3, pp. 269–275, ISSN 1556-6811.

Madsen, K.G.; Hansen, C.M.; Madsen, E.S.; Strandbygaard, B.; BøtnerA. & Sørensen, K.J. (1998). Sequence analysis of porcine reproductive and respiratory syndrome virus of the American type collected from Danish swine herds. *Archives of Virology*, Vol. 143, No. 9, pp. 1683-1700, ISSN 0304-8608.

Martelli, P.; Cordioli, P.; Alborali, L.G.; Gozio, S.; De Angelis, E.; Ferrari, L., Lombardi, G. & Borghetti, P. (2007). Protection and immune response in pigs intradermally vaccinated against porcine reproductive and respiratory syndrome (PRRS) and subsequently exposed to a heterologous European (Italian cluster) field strain. *Vaccine* 25, No. 17, pp. 3400-3408, ISSN 0264-410X.

Martelli, P.; Gozio, S.; Ferrari, L.; Rosina, S.; De Angelis, E.; Quintavalla, C.; Bottarelli, E. & Borghetti, P. (2009). Efficacy of a modified live porcine reproductive and respiratory syndrome virus (PRRSV) vaccine in pigs naturally exposed to a heterologous European (Italian cluster) field strain: Clinical protection and cell-mediated immunity. *Vaccine*, Vol. 27, No. 28, pp. 3788-3799, ISSN 0264-410X.

Mateu, E. & Diaz, I. (2008). The challenge of PRRS immunology. *Veterinary Journal*, Vol. 177, No. 3, pp. 345-35, ISSN 1090- 0233.

McCaw, M. (1995). PRRS control: whole herd management concepts and research update. Proceedings of the North Carolina Healthy Hogs Seminar, pp. 57-64, Greenville, North Carolina, U.S.A.

Misinzo, G.; Delputte, P.L.; Meerts, P.; Drexler, C. & Nauwynck, H.J. (2006). Efficacy of an inactivated PRRSV vaccine: induction of virus-neutralizing antibodies and partial virological protection upon challenge. *Advances in Experimental Medicine and Biology*, Vol. 581, pp. 449–454, ISSN 0065-2598.

Neumann, E.J.; Kliebenstein, J.B.; Johnson, C.D.; Mabry, J.W.; Bush, E.J.; Seitzinger, A.H.; Green, A.L. & Zimmerman, J.J. (2005). Assessment of the economic impact of porcine reproductive and respiratory syndrome on swine production in the United States. *Journal of the American Veterinary Medical Association*, Vol. 227, No. 3, pp. 385-392, ISSN 0003-1488.

Nilubol, D.; Platt, K.B.; Halbur, P.G.; Torremorell M.; & Harris D.L. (2004) The effect of a killed porcine reproductive and respiratory syndrome virus (PRRSV) vaccine treatment on virus shedding in previously PRRSV infected pigs. *Veterinary Microbiology*, Vol. 102, No. 1-2, pp. 11–18, ISSN 0378-1135.

O'Connor, M.; Fallon, M. & O'Reilly, P.J. (2002). Detection of antibody to porcine reproductive and respiratorysyndrome (PRRS) virus: reduction of cut-off value of an ELISA, with confirmation by immunoperoxidasemonolayers assay. *Irish Veterinary Journal*, Vol. 55, No. 2, pp. 73-75, ISSN 0368-0762.

Papatsiros, V.G. (2006). Porcine Respiratory and Reproductive Syndrome (PRRS); A field study on the effect of long-term use of an inactivated vaccine on the health status of boars, sows and piglets. *PhD Thesis*, Faculty of Veterinary Medicine, Aristotle University of Thessaloniki, Greece.

Papatsiros, V.G. (2011). PRRSV and PCV-2 Infections in Greek Swine Farms: Clinical forms and vaccination programmes. *VetScan*, Vol. 6, No. 1, Article 73, ISSN 0973-6980.

Papatsiros, V.G.; Alexopoulos, C.; Boscos, C.; Joisel, F. & Kyriakis, S.C. (2008). Effect of a PRRSV inactivated vaccine on health status and semen characteristics of boars. *Reproduction in Domestic Animals*, Vol. 43, No. 3, pp. 121-122, ISSN 0936-6768.

Papatsiros, V.G.; Alexopoulos, C.; Kritas, S.K.; Koptopoulos, G.; Nauwynck, H.J.; Pensaert, M.B/, & Kyriakis, S.C. (2006). Long-term administration of a commercial porcine reproductive and respiratory syndrome virus (PRRSV)-inactivated vaccine in PRRSV-endemically infected sows. *Journal of Veterinary Medicine B, Infectious Diseases and Veterinary Public Health,* Vol. 53, No. 6, pp. 266-272, ISSN 0931-1793.

Pesente, P.; Rebonato, V.; Sandri, G.; Giovanardi, D.; SperatiRuffoni, L. & Torriani, S. (2006). Phylogenetic analysis of ORF5 and ORF7 sequences of porcine reproductive and respiratory syndrome virus (PRRSV) from PRRS-positive Italian farms: A showcase for PRRSV epidemiology and its consequences on farm management. *Veterinary Microbiology* Vol. 114, No. 3-4, pp. 214-224, ISSN 0378-1135.

Pirtle, E.C. & Beran, G.W. (1996). Stability of porcine reproductive and respiratory syndrome virus in the presence of fomites commonly found on farms. *Journal of American Veterinary Medical Association*, Vol. 208, No. 3, pp. 390–392, ISSN 0003-1488.

Pitkin, A.; Deen, J. & Dee, S. (2009). Use of a production region model to assess the airborne spread of porcine reproductive and respiratory syndrome virus. *Veterinary Microbiology*, Vol. 136, No. 1-2, pp. 1-7, ISSN 0378-1135.

Plana-Durán, J.; Bastons M.; Urniza A.; Vayreda M.; Vila X. & Mane H. (1997). Efficacy of an inactivated vaccine for prevention of reproductive and respiratory failure induced by porcine reproductive and respiratory syndrome virus. *Veterinary Microbiology*, Vol. 55, No. 1-4, pp. 361-370, ISSN 0378-1135.

Prieto, C.; Álvarez, E.; Martínez-Lobo, F.J.; Simarro, I. & Castro, J.M. (2008). Similarity of European porcine reproductive and respiratory syndrome virus to vaccine strain is not necessarily predictive of the degree of protective immunity conferred. *Veterinary Journal*, Vol. 175, No. 3, pp. 356-363, ISSN 1090-0233.

Scortti, M.; Prieto, C.; Simarro, I. & Castro, J.M. (2006). Reproductive performance of gilts following vaccination and subsequent heterologous challenge with European strains of porcine reproductive and respiratory syndrome virus. *Theriogenology*, Vol. 66, No.8, pp. 1884-1893, ISSN 0093-691X.

Scortti, M., Prieto, C., Alvarez, E., Simarro, I., & Castro, J.M. (2007). Failure of an inactivated vaccine against porcine reproductive, respiratory syndrome to protect gilts against a heterologous challenge with PRRSV. *Veterinary Research,* Vol. 161, No. 24, pp. 809–813, ISSN 0830-9000.

Shirai, J.; Kanno, T.; Tsuchiya, Y.; Mitsubayashi, S. & Seki, R. (2000). Effects of chlorine, iodine, and quaternary ammonium compound disinfectants on several exotic disease viruses. *Journal of Veterinary Medical Science,* Vol. 62, No. 1, pp. 85–92, ISSN 0916-7250.

Stevenson, G.W.; Van Alstine, W.G.; Kanitz, C.L.; & Keffaber, K.K. (1993). Endemic porcine reproductive and respiratory syndrome virus infection of nursery pigs in two swine herds without current reproductive failure. *Journal of Veterinary Diagnostic Investigations,* Vol. 5, No. 3, pp. 432-434, ISSN 1040-6387.

Thacker, E.L. (2006). *Mycoplasmal diseases.* In: Diseases of Swine (9th ed.), Straw, B.E.; Zimmerman, J.J.; D'Allaire, S. & Taylor, D.J. (Eds.), pp. 701-717, ISBN-13: 978-0-8138-1703-3, Blackwell Publishing, Ames IA, USA.

Thacker, E.L.; Thacker, B.J.; Young, T.F. & Halbur, P.G. (2000). Effect of vaccination on the potentiation of porcine reproductive and respiratory syndrome virus (PRRSV)-induced pneumonia by Mycoplasma hyopneumoniae. *Vaccine,* Vol. 18, No. 13, pp. 1244–1252, ISSN 0264-410X.

Van Alstine, W.G.; Stevenson, G.W. & Kanitz C.L. (1993). Diagnosis of porcine reproductive and respiratory syndrome. *Journal of Swine Health and Production,* Vol. 1, No. 4, pp. 24-28, ISSN 1066-4963.

Vanhee, M.; Delputte, P.L.; Delrue, I.; Geldhof, M.F. & Nauwynck H.J. (2009). Development of an experimental inactivated PRRSV vaccine that induces virus-neutralizing antibodies. *Veterinary Research,* Vol. 40, No. 6, pp.63, ISSN 0830-9000.

Vashisht, K.; Erlandson, K.R.; Firkins, L.D.; Zuckermann, F.A. & Goldberg, T.L. (2008). Evaluation of contact exposure as a method for acclimatizing growing pigs to porcine reproductive and respiratory syndrome virus. *Journal of the American Veterinary Medical Association,* Vol. 232, No. 10, pp. 1530-1535, ISSN 0003-1488.

Weigel, R.M.; Firkins, L.D.; & Scherba, G. (2000). Prevalence and risk factors for infection with Porcine Reproductive and Respiratory Syndrome Virus (PRRSV) in swine herds in Illinois (USA). *Veterinary Research,* Vol. 31, No. 1, pp. 87-88, ISSN 0830-9000.

Wensvoort , G.; Terpstra, C.; Pol, J.M.; ter Laak, E.A.; Bloemraad, M.; De Kluyver, E.P.; Kragten, C.; van Buiten, L.; Den Besten, A.; Wagenaar, F.; Broekhuijsen, J.M.; Moonen, P.L.J.M.; Zetstra, T.; De Boer, E.A.; Tibben, H.J.; De Jong, M.F.; Van' t Veld, P.; Groenland, G.J.R.; van Gennep, J.A.; Voets, M.T.H.; Verheijden, J.H.M. & Braamskamp J. (1991). Mystery swine disease in the Netherlands: The isolation of Lelystad virus. *Veterinary Quarterly,* Vol. 13, No. 3, pp. 121-130, ISSN 0165-2176.

Zimmerman, J.J.; Benfield, D.A.; Murtaugh, M.P.; Osorio, F.A.; Stevenson, G.W. & Torremorell, M. (2006). *Porcine Reproductive and Respiratory Syndrome Virus (Porcine Arterivirus).* In: Diseases of Swine (9th ed.), Straw, B.E.; Zimmerman, J.J.; D'Allaire, S. & Taylor, D.J. (Eds.), pp. 387-417, ISBN-13: 978-0-8138-1703-3, Blackwell Publishing, Ames IA, USA.

Zuckermann, F.A.; Garcia, E.A.; Luque, I.D.; Christopher-Hennings, J.; Doster, A.; Brito, M.; & Osorio, F. (2007). Assessment of the efficacy of commercial porcine reproductive and respiratory syndrome virus (PRRSV) vaccines based on measurement of serologic response, frequency of gamma-IFN-producing cells and virological parameters of protection upon challenge. *Veterinary Microbiology*, Vol. 12, pp. 69–85. ISSN 0378-1135.

16

Values of Blood Variables in Calves

Martina Klinkon and Jožica Ježek
Clinic for Ruminants,
Veterinary Faculty,
University of Ljubljana,
Slovenia

1. Introduction

Haematological and biochemical variables are most widely used medical decision making tool. Haematological and biochemical analyses of blood are very useful to get an insight in metabolic and health status of animal. During diagnostic procedure it is very useful to compare the values obtained from ill animal with normal values in healthy animals. Therefore specific reference intervals are needed for each animal species for appropriate interpretation of results of haematological and biochemical analyses. The reference values of different blood variables are well established for adult cattle, but for calves there is not a lot of data available. The values of different blood variables in calves and other young animals are changing with age. The values of haematological variables change after birth because of colostrums intake, short life time of erythrocyte and decrease of concentration of foetal haemoglobin. In growing calves the feeding and rearing system has an important influence on the values of different blood variables. Influence of feeding becomes more apparent after 5th week when consummation of dry food (hay, starter) increases

In our contribution we would like to present the results of our research about different blood variables (haematological, biochemical) in calves and their dependence from age (from birth till age of 6 months). Further we will discuss about factors which influence the values of different blood variables like; laboratory methods, feeding and rearing systems, health status, etc. We would like to explain how the knowledge about values of blood variables could be used for proper interpretation of results of laboratory analyses and for assessment of health or metabolic status in an animal.

2. Haematological variables in calves

The number of erythrocytes (E) in blood is lover in young animals and indicates more signs of regeneration (higher number of reticulocytes) as in adult animals (Kraft, 1999a). The life span of E in cattle lasts 160 days. Different authors studied values of haematological variables in calves and observed that the values are changing with age (Greatorex, 1954; Scheidegger, 1973; Hanschke & Schulz, 1982; Steinhardt & Thielscher, 2000a; Steinhardt & Thielscher, 2000b; Klinkon et al., 2000;Brun-Hansen et al., 2006; Moosavian et al., 2010;Mohri et al., 2010). Immediately after parturition the values of (PCV), haemoglobin (Hb) and number of E are higher and they decreased in first days of life what could be associated with colostrum intake, which increase the plasma volume due to osmotic effect (Scheidegger, 1973; Harvey, 1997). To decreased values contribute also lover production of erythrocytes in first days after parturition and shorter life span of intrauterine produced E (Harvey, 1997). The findings of Kurz and Willet (1991) are in agreement with these establishments. They studied the dynamics of Hb and PCV values and number of E in calves from the 1st to the 6th day after calving and established continuous decrease of mentioned variables from birth till the 6th day when the Hb concentration was 9.3 g/dl, value of PCV was 27.2 % and number of E was 7.1 x 10^{12}/L. The different time of first colostrum intake did not influence the dynamics of these variables. Similar dynamics till 5th day after birth observed also Muri et al. (2005) in calves of Simmental and Holstein Friesian breed only they established slightly lover values of Hb (6.6 g/dl), PCV (24 %) and E (7.8 x 10^{12}/L) as were established in other studies (Scheidegger, 1973; Kurz & Willet, 1991; Egli & Blum, 1998). Mohri et al. (2007) observed a decrease of PCV and Hb from birth to the age of 28 days later the values increased with age. Heidarpour Bami et al. (2008) found lower values of Hb and PCV to the age of 28 d in calves, which did not received iron and higher values in those, which received iron injections. In veal calves the highest values of some variables were established immediately after birth E (9.35 x 10^{12}/L), Hb (12.86 g/dl), PCV (41 %), MCV (43.2 fL) in L (13.99 x 10^{9}/L). The values decreased to the age of 48 hours and at the age of 3 weeks they observed higher values except MCV value which still decreased (Adams et al., 1992). Higher values of E, Hb and PCV are response of calf's organism to the hypoxia which could appear at the end of the pregnancy (Steinhardt et al., 1993). The decrease of haematological variables in the first weeks after calving was observed by majority of authors and is influenced also by decline of foetal erythrocytes which contain foetal Hb (HbF), they are replacing with E containing haemoglobin of adult animals HbA (Scheidegger 1973; Harvey, 1997) and have smaller volume. Smaller MCV is compensated with higher number of E to retain normal Hb amount which is nearly on the level of adult animals (Brun-Hansen et al. 2006).

In growing calves feeding and rearing system has an important influence on the values of haematological variables (Reece & Hotchkiss, 1987; Scheidegger, 1973). Influence of feeding regime becomes more apparent after the 5th week when consummation of dry food (hay, starter) increases. In this period the values of RBC, Hb and Ht increase. In veal calves fed predominantly with milk the values of mentioned variables decreased and calves become anaemic due to iron deficiency. In calves fed exclusively with milk replacer (feeding for white meat) the number of E decreased from birth to the age of 15 weeks when it was only 3.96 ± 0.66 x 10^{12}/L. Similar dynamics was observed by Hb concentration which was 7.65 ± 1.28 g% at the age of 15 weeks and PCV value which decreased from 41.40 ± 5.86 % in the 1st week to 26.82 ± 5.03 % in the 15th week (Jazbec et al., 1973).

The number of leucocytes (WBC) in blood of calves is higher in comparison with adult animals and is more variable as values of other haematological variables. Different types of leucocytes have different life spans so their number can change rapidly and blood serves only as transport medium from the place of origin to the place of inflammation (Kraft, 1999a). The number of WBC usually increases at the presence of disease, especially in association with inflammatory processes and possibly at stress. The number of WBC increased from birth to the age of 10 weeks, than it oscillated more or less to the age of 1 year and is slightly higher as in adult animals the most of values are between 6.5-11.5 x 10^9/L (Greatorex, 1954). Terosky et al. (1997) observed the oscillation of WBC number in the period from birth to the age of 18 months; the number was between reference values for adult animals. In calves fed exclusively with milk replacer the number of WBC decreased after the 1st week (10.42 ± 2.29 x 10^9/L) and was 7.30 ± 1.21 x 10^9/L at the age of 21 weeks (Jazbec et al., 1973).

The number of platelets (PLT) in calves is around 400 x 10^9/L in the first three days of life, then it increase rapidly to 900 x 10^9/L at 10 days of age. Then after the PLT number increase slowly when it amount 1100 x 10^9/L, later it decrease slightly (Knowles et al. 2000). In Norwegian research performed on calves of Norwegian Red breed the increasing of PLT number to the age of 2 weeks was observed when it was 987 x 10^9/L. Then the PLT number decreased slowly to the age of 27-29 weeks, when it was 518 x 10^9/L (Brun-Hansen et al., 2006). In Simmental calves the PLT number increased rapidly from the 1st to the 7th day of life, later it almost did not change to the age of 84 days (Egli & Blum, 1998). The observations of different studies indicate that PLT number increase rapidly in the first days of life, and for later the establishments of studies about PLT dynamics differ.

The results of various studies on age dependent dynamics of haematological variables are different what established also the other researchers (Jain, 1986; Knowles et al., 2000). The differences are influenced by the fact that blood samples were taken in different age periods, breeds, rearing systems, geographic regions, and different feeding system.

By analysing of results of different studies was established, that the most marked differences are between calves which are fed with milk or milk replacer and receive starter and hay and veal calves which receive predominantly or exclusively milk or milk replacer and no hay. The differences in values of haematological variables are partly influenced with the use various analysing methods. In older studies the blood cells were counted manually with Neubauer chambers, PCV values were estimated with centrifugation in microhaematocrit tubes, and Hb concentration was measured photometrical (Greatorex, 1954; Hanschke & Schulz, 1982). In later studies the blood cell numbers were counted with automated counters from different producers, PCV value was still estimated by microhaematocrit method (Scheidegger, 1973; Terosky, 1997; Steinhardt & Thielscher, 2000a; Steinhardt & Thielscher, 2000b; Reece & Hotchkiss, 1987). In most of recent researches, which were published in last years the haematological variables were measured with various automated counters (Egli & Blum, 1998; Muri et al., 2005; Knowles et al., 2000; Klinkon et al., 2000; Mohri et al., 2007, Heidarpour Bami et al., 2008).

3. Biochemical variables in calves

The knowledge about normal values of biochemical variables in blood serum and other physiological variables is important for assessment of damage of organs and tissues in different diseases and for assessment of development from the welfare aspect (Steinhardt & Thilescher, 2000c; Terosky et al., 1997). The values of biochemical variables in calf's serum differ from the values in adult animals. Different authors ascertained that there are deficient data available about physiological values of biochemical variables in calves and that results of various studies differ (Hanschke & Schulz, 1982; Bouda & Jagoš, 1984; Steinhardt et al., 1993; Knowles et al., 2000 Mohri et al., 2007). The majority of data available is for the calves in the first days after birth, and for the calves of few weeks or months of age there is very few data available. The changing of values of biochemical variables in the first days of life is the consequence of adaptation on the extra uterine life and is importantly influenced by maturation of organs and intake of nutrients.

3.1 Aspartat Aminotransferase (AST)

The enzyme AST is present in different tissues and is a sensitive indicator of soft tissue damage. In heart and skeleton musculature as in liver there is high activity of AST. The AST is present in cytoplasm and in mitochondria so its activity is increased chiefly by cell necrosis in smaller amount also by damage of the cell membrane (Kraft & Dürr, 1999a). Measuring of AST activity in combination with CK is used for diagnostics of muscle damage (Kaneko, 1997). High activity of AST is also in liver and in the case of liver damage AST activity in serum increase.

After first colostrum intake the AST activity in serum increased from 23 U/L before intake, to 38 U/L at the age of 3 hours, what is most likely due to absorption from colostrum or because of activation of enzymes in calf intestine as consequence of colostrum intake (Kurz & Willet, 1991). However Hammon and Blum (1998) established that in calves which received only milk replacer instead of colostrum, activity of AST increased on the second day after birth so they are of the opinion that also other factors influence on the increased activity of some enzymes. The activity of AST decreased after the first week, and from 42nd to 84th day of life it increased slowly (Egli & Blum, 1998). Mohri et al. (2007) observed the increase of AST activity from the 14th to the 84th day of age.

3.2 Lactate Dehydrogenase (LDH)

Enzyme LDH catalyse reversible oxidation of pyruvate to lactate. The enzyme is present in numerous organs and tissues.

The activity of LDH in calves increased slowly in the first 24 hours of life, from 421 U/L immediately after birth to 759 U/L at the age of 24 hours and it is more likely this is a physiological event than it is due to absorption from colostrum (Kurz & Willet, 1991). The LDH activity increased slowly to the age of 56 days, later it remains on the same level to the age of 84 days (Egli & Blum, 1998).

3.3 Creatin Kinase (CK)

The highest activity of CK is in the skeleton and heart musculature. Measuring of the activity of CK in serum is first of all used for diagnostics of skeleton musculature damage. The activity of CK could be increased also after effort, long-lasting lying of the animal or convulsions. Miopathias as the consequence of vitamin E and Selenium deficiency which usually appear in veal animals (calves, lambs), sometimes also in adult animals, cause increased activity of CK (Smith et al., 1994).

The activity of CK in calves is high after birth later it decreased rapidly and almost did not change to the age of 60 days, at the age of 80 days it increased slightly (Knowles et al., 2000).

The mean activity of CK in calves of Simmental breed was 11.2 ± 2 µkat/L (671.8 ± 119.9 U/L) at birth then it decreased to the age of 7 days and remained on this level to the age of 42 days, later it increased to the 84th day when it was 21.3 ± 10.7 µkat/L (1277.7 ± 641.8 U/L) (Egli & Blum, 1998). Increased activity of CK after birth could be associated with the parturition and adaptation to the extra uterine life which represent for the newborn calf an effort on which it was not used in intrauterine life. The increasing of CK activity with age could be attributed to the growth of calves and gaining of muscle mass, partly also to the increased activity of calves, at this age they are in group pens where they have enough space for movement.

3.4 Alkaline Phosphatase (ALP)

For a long time the enzyme ALP is used in diagnostics as indicator of liver damage. The ALP is important also by diseases of skeleton. The enzyme was found in the intestine, liver, kidney and bones. In serum of young fast growing animals predominates isoenzyme from bones, in older animals which grow slower its activity decrease (Kaneko, 1997). Serum activity of ALP is higher in young animals than in adult ones and it decrease with age. After first colostrum intake serum activity of ALP increased from 235 U/L before intake to 364 U/L at the age of 3 hours, what is most likely due to absorption from colostrum and activation of enzymes in the calf's intestine because of colostrum intake (Kurz & Willet, 1991). The activity of ALP was high in calves after birth then it decreased and remained stable to the age of 60 days, later it decreased slightly more (Knowles et al., 2000). In calves to the age of 6 months the activity of ALP can reach 1800 U/L, in young cattle to the age of 3 years it decrease to 500 U/L (Kraft & Dürr, 1999a). In adult animals activity of ALP can increase at increased activity of osteoblasts. Activity of ALP is increased at acute and chronic liver diseases (especially cholestatic hepatopathias) and in diseases of bones (rachitis, periostitis)

3.5 Gamma Glutamyl Transferase (GGT)

The highest activity of GGT is in bile ducts epithelium and in kidney. The enzyme is located in membrane structures of the cells. The increased serum activity of GGT is usually associated with cholestasis and bile ducts damage. Very high activity of GGT is also in colostrum of cattle, sheep and goats. Hammon and Blum (1998) measured in colostrum of cows mean activity of GGT 22.432 U/L. After colostrum intake the enzyme is absorbed through intestinal wall, consequently the GGT activity is very increased in this period and can be used for indirect estimation of colostrum supply (Bostedt, 1983). The GGT activity in newborn calves was 10-31 U/L, after colostrum intake it increased to 370-5000 U/L, then it slowly decreased to the age of 20 days when it stabilised (Braun et al., 1982). In calves which received only milk or milk replacer instead of colostrum, the GGT activity did not increase (Boediker, 1991). In the first week of life GGT activity was high later it decreased rapidly (Knowles et al., 2000; Egli & Blum, 1998). The GGT activity below 100 U/L at the age of 2 days indicates insufficient colostrum supply or disturbed absorption (Klee, 1985). Tyler et al. (1999b) claimed that activity of GGT above 50 U/L in the calves serum indicate sufficient colostrum supply, Perino et al. (1993) stated 200 U/L for the boundary value.

By comparison of various data from the literature in some cases considerable differences in the activity of enzymes were established. There are also considerable differences between reference values of enzyme activity claimed by different authors. Enzymes are very sensitive indicators of cell damage, their activity change by tiny alternations, so it is understandable that enzyme activity could rather differ between single animals. Consequently there are different results between studies. Reasons for these differences are partly differences between breeds and breeding conditions where the results were obtained. Great influences on the measurements have also analytical procedures and temperatures at which the activity of enzymes was measured. In majority of literature the mentioned analytical procedures are deficiently described, so it is very difficult to compare the data, because it is not clear at which temperature the activity of enzyme was measured. Some sources (Egli & Blum, 1998) claimed twice as high activity of LDH, as was measured by us. The reason is most likely in measuring procedure because the activity of LDH is measured with reaction of transformation pyruvate to lactate which can expire in both directions. With experiences in our laboratory was established that if the LDH activity is measured with reaction from lactate to pyruvate almost for a half lower values are obtained, than when LDH activity is measured with reaction from pyruvate to lactate.

3.6 Total serum bilirubin

Bilirubin is formatted by enzymatic decay of hem at disintegration of mature erythrocytes. These processes are taking course in spleen, liver and bone marrow. In blood plasma the unconjugated (indirect) bilirubin is bound to the albumin and is carried to the liver. With help of special protein called ligandin, bilirubin pass into hepatocytes. In the liver of newborns there is very low concentration of ligandin, so their ability to excrete the bilirubin is decreased. They can have for 50-100% higher serum concentrations as the adult animals (Jazbec, 1990; Tennant, 1997). At the end phase conjugated (direct) bilirubin pass into bile and than in the intestine. Increased concentration of total serum bilirubin in cattle is usually associated with cholestasis, fatty liver or haemolytic anaemia (Tennant, 1997), but can also be in connection with decreased appetite (Kraft & Dürr, 1999a).

The total serum bilirubin concentration in one day old calves was higher (9.58 μmol/L) than in adult animals. Later it decreased slowly and at 31-60 days of age it was 4.45 μmol/L (Hanschke & Schulz, 1982; Egli & Blum, 1998). Kurz and Willett (1991) observed increase of total serum bilirubin concentration in the first 12 hours of life for approximately five times and until 6th day the concentration decreased approximately to the value at birth. In the research of Mohri et al. (2007) bilirubin concentration decreased from 1st to the 14th day of age and later it remain stable. Higher concentration of total serum bilirubin after birth is associated with destruction of foetal haemoglobine and slower excretion of bilirubin because of lower concentration of transport protein ligandin. Ligandin enables passing of indirect bilirubin into hepatocytes where it is transformed to direct bilirubine which can be excreted (Tennant, 1997). After the second week of age the concentration of total serum bilirubin is inside reference values for adult animals. What can be associated with improvement of glomerular filtration and liver function with age.

3.7 Serum iron

Iron is present in the milk in very low concentration, but it is absorbed very efficiently from the gut (Underwood & Suttle, 2001). Efficiency of iron absorption depends from the needs of the organism and is controlled by iron amount in the intestine mucosa. The majority of iron is absorbed in the duodenum. The absorption is more efficient in animals with iron deficiency (Underwood & Suttle, 2001). The iron is very important for synthesis of haemoglobin. Approximately 60-75% of iron in the body is bound to haemoglobin (Kraft, 1999a). Haemoglobin is composed from four molecules of hem and each of them contains one atom of iron. The main occupation of iron is transportation of oxygen. The rest of iron in the body is bound to transferine and feritine, small amount also to mioglobin. Iron is important for normal functioning of numerous enzymes. Iron deficiency in calves cause decreasing of haemoglobin and mioglobin concentration (Underwood & Suttle, 2001). In milk the iron is in relatively small quantity for all that in this period intensive synthesis of haemoglobin is expiring in calves (Kraft, 1999a). Long lasting iron deficiency influence on lost of appetite, growth retardation, whitening of mucous membranes due to progressive hypochromic anaemia (Underwood & Suttle, 2001). In fast growing suckling calves mild anaemia can appear, but usually it does not come to iron deficiency in ruminants, because there is enough iron in forage.

The calves were born with iron concentration around 27.7 ± 9.6 μmol/L, then it decreased on 18.0 ± 3.0 μmol/L at the age of 4 days. After the 2nd week of life iron concentration started to increase and at the age of 6 weeks it reached similar level as at birth (28.0 ± 7.8 μmol/L) (Bostedt et al., 1990). Knowles et al. (2000) established gradually increasing of iron concentration from 30th to 80th day of age, but at the age of 80 days the concentration was still on the lower limit of reference values for adult cattle. Other studies observed gradual increase of iron concentration to the age of 60 days (Steinhardt & Thielscher, 2000a; Steinhardt & Thielscher, 2000b).

Iron dynamics with age is influenced by feeding regime of calves. By suckler calves the iron concentration was low in the first weeks (around 9 μmol/L) later it increased to the age of 14 weeks (approximately 36 μmol/L). The calves which received milk replacer, have already at the age of 1-2 weeks, higher iron concentration as suckler calves. In calves which received starter and hay in addition to milk replacer the iron concentration increased to the 3rd week of age later it slightly oscillated but it did not decrease to the age of 13 weeks. In calves fed exclusively with milk replacer the iron concentration decreased after 2nd week of age and at the age of 14 weeks it was only 4 μmol/L (Reece & Hotchkiss, 1987; Scheidegger, 1973). In the calves fed only with milk replacer the iron concentration decreased from birth to the 10th week when it was 5.19 ± 4.29 μmol/L, later it increased slightly, but the lowest concentration was established at the age of 22 weeks when it was only 1.79 ± 1.43 μmol/L (Jazbec et al., 1975).

3.8 Urea

The concentration of urea in blood depends from nutrition, diagnostically is important also at diseases of kidneys (Kraft & Dürr, 1999b; Jazbec, 1990). Increased concentration of urea in calves' serum indicates increased catabolism of proteins and appears at long lasting diarrhoeas (Jazbec, 1990).

The colostrum intake did not influence the urea concentration (Steinhardt et al., 1993). In calves the urea concentration slightly decreased from birth to the age of 60 days when it was 2.7 mmol/L (Steinhardt & Thielscher, 2000d). Knowles et al. (2000) observed increasing of urea concentration from 40th to 80th day of age. Hanschke and Schulz (1982) established higher values of urea in the age 31-60 days in calves in subtropical climate, where the concentration of urea was 5.14 mmol/L. In calves with diarrhoea at the age 4-15 days twice as high mean urea concentration (7.98 mmol/L) in plasma was established as in healthy calves of the same age (3.89 mmol/L) (Maach et al., 1992). The authors claimed that measuring of urea concentration is very helpful for assessment of dehydration and disturbances of acid-base balance in calves with diarrhoea. Hugi et al. (1997) ascertained association between protein amount in forage and serum urea concentration in calves at the age 8 to 15 weeks.

3.9 Creatinin

Creatinin is synthesised at endogen metabolism in muscles. Creatinin is excreted with urine; its concentration in serum does not depend from the nutrition. Diagnostically it is important for the assessment of functioning of the glomerular system in the kidneys, but it concentration increase only at serious damage (Kraft & Dürr, 1999b). In calves after birth very high serum concentration of creatinin was established (256 ± 106 μmol/L), the value normalised to the 4th day of age (108 ± 28 μmol/L) (Klee, 1985). Similar was established by Maach et al. (1991), and after 15th day of age they observed increase of creatinin concentration to the age of 60 days when it was 146.7 ± 23.9 μmol/L. In the research of Mohri et al. (2007) the concentration of creatinin in Holstein calves decreased from 1st to the 70th day of age.

3.10 Total Serum Protein (TSP)

The absorption of proteins which are degraded into amino acids takes place in small intestine. The main source of proteins in ruminants is microbe synthesis in the rumen. In liver the majority of body's own proteins are synthesised. Albumins (Alb) are proteins in blood plasma which are synthesised in the liver. Other part of plasma proteins are globulins which are produced by the immune system. Albumins carry 75% of osmotic activity in plasma and they serve as transport proteins in metabolic processes. Globulins can be divided into three fractions with help of electrophoresis. In α fraction there are proteins of acute phase of inflammation (haptogobin, serum amyloid A). In β fraction are parts of complement (C_3 and C_4), transferin, C reactive protein and partly also some immunoglobulin (IgA, IgM). Gamma fraction consists of immunoglobulin (Kaneko, 1997). Measuring of concentration of TSP and quantity of albumin and globulin is very important in diagnostics of numerous diseases and disturbances in functioning of organism.

The concentration of TSP and proportion between albumin and globulin is changing with age. Usually the calves have lower concentration of TSP (50-70 g/L) as adult animals (60-80 g/L) (Kraft & Dürr, 1999c). Pregnancy, lactation, nutrition and inflammation can influence on TSP concentration (Kaneko, 1997). In calves after birth the TSP concentration is almost for a half lower than in cows (45.8 g/L), after colostrum intake the concentration increases (54.5 g/L), but is still lower than in adult animals. Numerous studies found correlation between TSP and concentration of immunoglobulin in calves' serum. Measuring of TSP concentration in the 1st week of age can be used as indirect indicator of colostrum supply (Tyler et al., 1998; Tyler et al., 1996a; Selim et al. 1995).

The concentration of albumin decreased after colostrum intake (27.5 g/L), and was approximately on the lower limit of reference values for adult cattle (Steinhardt et al., 1993; Kurz & Willet, 1991). In calves, which received colostrum, higher concentration of TSP was established, as in calves which received only milk replacer, what is associated with immunoglobulin absorption in the first ones (Muri et al., 2005). The concentration of TSP and albumin is influenced with nutrition of calves and functioning of the liver. The albumins are predominantly synthesised in the liver, so their amount depends on maturity and functional ability of the liver (Steinhardt & Thilescher, 2000c). Hypoalbuminemia in calves could be the consequence of liver damage or protein catabolism at long lasting diarrhoeas (Jazbec, 1990). In calves at the age from 8 to 15 weeks statistically significant differences in albumin concentration were established when they received meals with different amount of proteins (Hugi et al. 1997). Hammon et al. (2002) compared the calves which received limited or unlimited amounts of milk and established significantly lower albumin concentration in calves, to the age of 28 days, which were fed with unlimited amount of milk. Lower concentration of albumin in healthy animals could be the consequence of insufficient supply with amino acids (Whitaker, 1997). In calves from the 5th to the 40th day of age the concentration of TSP decreased slightly, after the 60th day it increased again and was 55.7 g/L. The concentration of albumin gradually increased from the 5th to the 60th respectively the 80th day of age (Steinhardt & Thielscher, 2000d; Knowles et al., 2000). In sucker calves gradual increase of TSP and albumin concentration was established from birth to the age of two months. The values of both variables were higher than in calves which were fed with limited amounts of milk (Steinhardt & Thielscher, 2000b).

3.11 Inorganic Phosphate (iP)

The absorption of iP is taking place in the forestomacs and in the front part of the small intestine. The iP is excreted with faeces and urine in cows in lactation also with milk. Phosphorus is important for normal growth and mineralization of bones. In the skeleton of adult animals is stored approximately 80% of all phosphorus in the body, which could be mobilised at need. The rest 20% of phosphorus is in soft tissues and body fluids where it collaborates in numerous important processes. It is the ingredient of deoxi- and ribonuclein-acids, as phospholipids it is the part of cell membranes, as phosphate is important by regulation of osmotic and acid-base balance in the organism. Phosphorus has an important role in metabolism of energy, where it collaborates by transport of energy and fatty acids, synthesis of amino acids and proteins and working of Na/K pump (Underwood & Suttle, 2001). The serum concentration of iP is higher in young animals because the growth hormone increases the reabsorption of phosphate in the kidney (Rosol & Capen, 1997). In the rumen the phosphate is working as buffer for volatile fatty acids and as substrate for microorganisms. In ruminants with phosphorus deficiency, lower microbe synthesis of proteins in rumen is established. Phosphorus is important for the control of appetite and efficiently use of nutrients (Underwood & Suttle, 2001). By decrease of iP concentration in serum, increase the activity of ALP. Phosphorus deficiency in young animals causes the lost of appetite and retarded growth. In rachitic calves and lambs the concentration of iP is decreased for 30-50 % (Jazbec, 1990). Normal plasma concentration of iP in calves should be 1.3-1.9 mmol/L and 1.0-1.5 mmol/L in adult cattle (Underwood & Suttle, 2001). Kraft (1999b) claimed slightly higher values in calves serum namely; at the age until 2 months 2.6-3.5 mmol/L, from 2 to 6 months it should be 2.5-3.1 mmol/L and from 12 to 18 months the concentrations are on the level of adult animals (1.6-2.3 mmol/L). The colostrum intake did not influence the iP concentration in calves' serum (Steinhardt et al., 1993). Kurz and Willett (1991) established decrease of iP concentration in first 24 hours of age. In sucker calves the increase of iP concentration in the first 14 days was established later it remained stable. The values were all the time higher as in adult animals (Egli & Blum, 1998). At the age of 60 days the concentration of iP in calves' serum was 2.6 mmol/L (Steinhardt & Thielscher, 2000d).

3.12 Calcium (Ca)

The majority of calcium (99 %) in organism is stored in bones and teeth. Calcium is important for activation of numerous enzymes and hormones. The calcium collaborates in processes of blood coagulation, nerves stimulation and muscles contraction. In blood serum approximately 55% of Ca is in ionised form and this is biologically active. The part of ionised Ca depends from blood pH, by decrease of pH the part of ionised Ca increase. A part of Ca in serum is bound to the albumins (40%) and smaller part (5%) to organic acids (Jazbec, 1990; Kraft, 1999b). The most of Ca is absorbed in small intestine.

In newborn calves the mean serum concentration of Ca was 3.35 ± 0.27 mmol/L. Six hours after birth the Ca level decreased to 2.41 ± 0.18 mmol/L and in the next days and weeks almost did not change (Bostedt & Schramel, 1982). Kurz and Willett (1991) observed decrease of Ca concentration in the first 24 hours of life. In the first two months of life the concentration of Ca was around 2.7 mmol/L and did not change a lot. After the 3rd month the Ca concentration started to decrease and at the age of 6 months it was 2.53 ± 0.10 mmol/L (Bouda & Jagoš, 1984). Similar dynamics was established in other study in calves which were fed with milk replacer, only they have slightly higher Ca concentration at the age of 5 days (3.02 ± 0.2 mmol/L) and from the age of 15 days to 2 months it was 2.8 ± 0.1 mmol/L (Steinhardt & Thielscher, 2000d). In sucker calves of Simmental breed the decrease of Ca concentration from birth to the age of 28 days was established (2.6 mmol/L), later the concentration almost did not change to the age of 84 days (Egli & Blum, 1998). Mohri et al. (2007) established a decrease of Ca concentration in the first two weeks of life, later the Ca concentration slowly increased.

3.13 Potassium (K)

In ruminants the potassium is absorbed from rumen and small intestine and is excreted over the kidney and with faeces. The K is main cation in milk (36 mmol/L). High concentrations of K in organism have citotoxical effect. Potassium is the main intracellular ion, its concentration in the cells is 25 to 30 times higher than in plasma (Ward, 1966), 96-98 % of total amount of K is in the cells (Wirth, 1999). Potassium is important for making electrical potential for transport of nervous impulses and for maintenance of muscle tonus. Hypokalemia increase the membrane potential and cause hyper polarisation block, what influence on lower muscle tonus and paralysis (Carlson, 1997). Potassium is important for regulation of acid-base balance in the body. The forage of ruminants usually contains enough potassium; more often comes to the surplus than to potassium deficiency in the meal. Tucker et al. (1991) and Weil et al. (1988) established association between potassium amount in the meal and concentration of K in calves' serum, and between mean daily weight gains. The mean K concentration in plasma of 8 weeks old calves was from 5.13 mmol/L, at 0.35% K in the meal, to 5.60 mmol/L at 0.53 % K in the dry matter of the meal. The authors claimed that the measuring of K concentration in serum and control of daily weight gain is a good indicator of supply with this element in calves. In calves with diarrhoea comes to hyperkalemia which is connected with grade of acidosis (Maach et al., 1992).

The colostrum intake influence on increase of K concentration in calves' serum, what is most likely, the consequence of higher amount of this mineral in the colostrum (Steinhardt et al., 1993). In the age from 1 week to 2-3 months the K concentration in calves' serum almost did not change and was around 5 mmol/L (Maach et al. 1991). Bouda and Jagoš (1984) measured slightly higher values, around 5.4 ± 0.4 mmol/L, at the age of 6 month K concentration slightly decreased to 4.7 ± 0.4 mmol/L. Reece (1980) Established a decrease of K concentration from the 1st week of age when it was 7.2 mmol/L to the age of 15 weeks when it fell to 4.4 mmol/L.

3.14 Sodium (Na)

The sodium is the most important cation in extracellular fluid, where it is responsible for maintenance of osmotic pressure. Together with chlorine (Cl) collaborates in metabolism of water and regulation of acid-base balance in the organism (Jazbec, 1990). In ruminants a big part of Na which comes in to digestive tract originates from saliva (rotation in the organism). The rumen could contain up to 50% of whole amount of Na which is available for the organism. From the body is excreted with the urine, faeces and milk (Underwood & Suttle, 2001). At diarrhoeas the calves lose higher amounts of Na. Maach et al. (1992) established significantly lower concentration of Na in serum of calves with acute diarrhoea (131.2 ± 7.2 mmol/L) in comparison to healthy calves of same age (140.0 ± 9.9 mmol/L). In newborn calves after colostrum intake the Na concentration increased what was attributed to absorption from the colostrum (Steinhardt et al., 1993). But Maach et al. (1991) established higher concentration of Na before colostrum intake when it was 145.7 ± 3.7 mmol/L as then after (137.8 ± 6.8 mmol/L), later it almost did not change. The concentration of Na did not change a lot in the first three months, it was about 145 mmol/L, and at the age of 6 months it was slightly lower, about 136.6 ± 5.1 mmol/L (Bouda & Jagoš, 1984). Reece (1980) established higher concentration of Na in serum of calves which received milk replacer in comparison to the calves which were fed with milk.

3.15 Chlorine (Cl)

Chlorine is the most important anion in the extracellular fluid, inside the cells there is only 12% of total amount in organism. Together with Na it is responsible for maintenance of osmotic pressure in extracellular fluid. Chlorine is absorbed in small intestine, and is excreted mainly through the kidney (Wirth, 1999). The calves with acute diarrhoea, in which the volume of faeces can increase for 40 times, with the fluid lose also electrolytes. In such calves significantly lower concentration of Cl was established (95.6 ± 6.9 mmol/L) in comparison to healthy calves (103.3 ± 6.9 mmol/L) (Maach et al., 1992).

In calves immediately after birth higher concentration of Cl was established (107.3 ± 12.3 mmol/L), then it decreased to the 7th day of age to 95.9 ± 6.6 mmol/L, later it increased slightly to 102.3 ± 6.2 mmol/L at the age of 2 months (Maach et al., 1991). In suckler calves the concentration of Cl increased with age from 98.0 mmol/L in the 1st week of age to 102.4 mmol/L at the age of 14 weeks (Reece, 1984). In calves which were fed with limited amounts of milk the increase of Cl concentration was established to the age of 5 weeks, later it decreased slightly and oscillated between 97.7 and 99.3 mmol/L (Reece, 1980).

4. Results of our study

The aim of the study was to define the physiological pattern of haematological and biochemical variables for different age groups of dairy calves, which should help in interpretation of laboratory results.

The study was performed on 65 Holstein-Friesian calves (31 females, 34 males) from two dairy farms. After calving, the calves received 1-1.5 l of colostrum from their mother *via* nipple bottle. The colostrum and milk were supplied three times a day for the first 4 day; later they received milk two times a day. After 10 day of age, they had free access to commercial starter and hay. They were weaned at the age of 16 weeks.

Blood samples were taken from the jugular vein into evacuated tubes containing K_3EDTA (Venoject, Terumo, Belgium) for haematology and in tubes without additives for biochemical analysis. They were taken once weekly till the age of 6 weeks, and then at the age of 8, 12, 16, 20, and 24 weeks. During the research period, health status of calves was monitored regularly. When the calf was found to be sick at sampling time, the sample was excluded from the study. Some calves were lost from the study before the end because they were sold.

Age (weeks)	n	RBC x 10^{12}/L	Hb g/L	MCV fL	PCV L/L	WBC x 10^9/L	PLT x 10^9/L
1	61	7.63 ± 1.49 [8,12,16,20,24]	104.1 ± 22.1 [20]	42.8 ± 3.5 [2,3,4,5,6,8,12,16,20,24]	0.3 ± 0.07 [6]	9.6 ± 3,4 [16]	451 ± 212 [2,3,4,5,6,8,12, 16]
2	64	8.35 ± 1.70 [12,16,20,24]	108.8 ± 23.7 [5,6,8]	40.5 ± 2.6 [1,4,5,6,8,12,16,20,24]	0.34 ± 0.07 [5,6,8]	10.9 ± 4.1	723 ± 218 [1,16,20,24]
3	58	8.22 ± 1.55 [12,16,20,24]	103.7 ± 22.7 [16,20]	38.9 ± 2. [1,5,6,8,12,16, 20,24]	0.32 ± 0.07	9.9 ± 3.2	730 ± 221 [1,16,20,24]
4	62	8.11 ± 1.54 [12,16,20,24]	100.1 ± 22.5 [16,20,24]	37.7 ± 2.9 [1,2,6,8,12,16, 20,24]	0.31 ± 0.07	10.1 ± 3.1	746 ± 201 [1,16,20,24]
5	63	7.98 ± 1.47 [12,16,20,24]	94.9 ± 20.4 [2,12,16,20,24]	36.4 ± 2.7 [1,2,3]	0.29 ± 0.07 [2,12,16,20,24]	10.9 ± 3.1	745 ± 194 [1,16,20,24]
6	63	8.05 ± 1.33 [12,16,20,24]	94.5 ± 18.3 [2,12,16,20,24]	35.7 ± 2.5 [1,2,3,4]	0.29 ± 0.06 [1,2,12,16,20,24]	9.7 ± 3.1 [16]	682 ± 206 [1,24]
8	65	8.53 ± 1.01 [12,16,20,24]	98.3 ± 12.9 [2,12,16,20,24]	35.1 ± 2.5 [1,2,3,4]	0.30 ± 0.05 [2,16,20]	9.8 ± 3.3 [16]	681 ± 252 [1,24]
12	61	9.33 ± 0.89 [1,2,3,4,5,6,8]	109.7 ± 9.0 [5,6,8]	35.2 ± 2.7 [1,2,3,4]	0.33 ± 0.04 [5,6]	10.8 ± 2.6	659 ± 198 [1,24]
16	62	9.72 ± 0.76 [1,2,3,4,5,6,8]	114.6 ± 8.6 [3,4,5,6,8]	35.0 ± 2,0 [1,2,3,4]	0.34 ± 0.03 [5,6,8]	11.8 ± 3.1 [1,6,8]	585 ± 201 [1,2,3,4,5]
20	56	9.85 ± 0.81 [1,2,3,4,5,6,8]	116.2 ± 10.1 [1,3,4,5,6,8]	34.8 ± 2.2 [1,2,3,4]	0.34 ± 0.03 [5,6,8]	11.4 ± 3.0	573 ± 168 [2,3,4,5]
24	38	9.59 ± 1.11 [1,2,3,4,5,6,8]	113.8 ± 12.3 [4,5,6,8]	35.3 ± 2,1 [1,2,3,4]	0.34 ± 4.0 [5,6]	11.5 ± 4.1	481 ± 213 [2,3,4,5,6,8,12]

The age group differs significantly ($P<0.05$) from age groups in superscript

Table 1. Haematological variables in calves (Ježek, 2007)

Age (weeks)	n	AST U/L	LDH U/L	CK U/L	ALP U/L	GGT U/L
1	64	38.7 ± 16.1 [3,12,16,20,24]	483.2 ± 108.8 [8,12,16,20,24]	168.8 ± 211 [2]	262.2 ± 185 [2,3,4,5,12,16,20,24]	329.8 ± 358.1 [2,3,4,5,6,8,12,16,2024]
2	64	32.7 ± 11.1 [8,12,16,20,24]	451.3 ± 105.2 [5,6,8,12,16,20,24]	99.6 ± 69.7 [1]	143.1 ± 61.3 [1,6,8]	79.8 ± 72.1 [1]
3	59	31.5 ± 6.2 [1,8,12,16,20,24]	459.8 ± 88.9 [5,6,8,12,16,20,24]	121.9 ± 48.4 [24]	132.8 ± 57.0 [1,5,6,8,12]	45.1 ± 34.5 [1]
4	62	33.3 ± 11.0 [8,12,16,20,24]	503.1 ± 160.9 [12,16,20,24]	106.3 ± 71.8 [16,20,24]	157.8 ± 79.7 [1,8]	30.5 ± 18.3 [1]
5	63	33.9 ± 12.1 [12,16,20,24]	532.9 ± 125,9 [2,3,16,20,24]	113.3 ± 69,9 [16,24]	189.5 ± 93.4 [1,3]	24.6 ± 11.2 [1]
6	63	35.7 ± 7.4 [12,16,20,24]	544.9 ± 115.3 [2,3,12,16,20,24]	106.9 ± 46.9 [16,20,24]	208.1 ± 104.3 [2,3]	20.2 ± 7.7 [1]
8	64	40.3 ± 10.2 [2,3,4,12,16,20,24]	575.8 ± 122.4 [1,2,3,16,20,24]	140.4 ± 70.5	216.2 ± 97.7 [2,3,4,20]	16.1 ± 3.5 [1]
12	61	47.6 ± 9.2 [1,2,3,4,5,6,8]	633.6 ± 100.9 [1,2,3,4,6,20,24]	154.5 ± 103.1	196.9 ± 67.9 [1,4]	13.4 ± 3.0 [1]
16	61	48.4 ± 8.1 [1,2,3,4,5,6,8]	671.1 ± 102.7 [1,2,3,4,5,6,8]	181.9 ± 94.2 [2,4,5,6]	175.6 ± 65.2 [1]	13.0 ± 2.6 [1]
20	56	50.2 ± 10.3 [1,2,3,4,5,6,8]	722.6.,3 [1,2,3,4,5,6,8,12]	166.5 ± 96.3 [2,4,6]	158.9 ± 55.4 [1,8]	13.9 ± 3.4 [1]
24	39	53.6 ± 10.1 [1,2,3,4,5,6,8]	770.7 ± 125.5 [1,2,3,4,5,6,8,12]	177.0 ± 115.5 [2,3,1,5,6]	162.4 ± 58.5 [1]	14.2 ± 4.9 [1]

The age group differs significantly ($P<0.05$) from age groups in superscript

Table 2. Activity of enzymes in calves (Ježek, 2007)

Age (weeks)	n	Bilirubin µmol/L	Fe µmol/L	Urea mmol/L	Creatinin µmol/L	TSP g/L	Alb g/L
1	64	16.08 ± 13.16 [2,3,4,5,6,8,12,1620,24]	16.91 ± 10.95 [8,12,16,20,24]	3.64 ± 1.33 [2,3]	113.76 ± 36.20 [3,4,5,6,8,12,16,20,24]	52.09 ± 7.09 [8,12,16,20,24]	26.35 ± 2.10 [2,3,4,5,6,8,12,1620,24]
2	64	8.23 ± 3.41 [1,12,16,20,24]	14.52 ± 7.16 [5,8,12,16,20,24]	5.90 ± 2.01 [1,4,5,6,8,12,16, 20,24]	116.53 ± 19.41 [3,4,5,6,8,12,16, 20,24]	52.92 ± 5.48 [8,12,16,20,24]	28.94 ± 2.79 [1,4,5,6,8,12,16, 20,24]
3	59	6.99 ± 1.94 [1]	16.61 ± 7.88 [8,12,16,20,24]	5.12 ± 3.42 [1,4,5,6,8,12]	100.79 ± 21.66 [1,2,5,6,8,12,16,20,24]	52.71 ± 4.85 [8,12,16,20,24]	30.24 ± 2.48 [1,16,20,24]
4	62	6.96 ± 1.52 [1]	20.97 ± 12.16 [12,16,20,24]	4.01 ± 1.21 [2,3]	93.55 ± 16.66 [1,2,8,12,16,20,24]	52.41 ± 3.91 [8,12,16,20,24]	30.99 ± 2.53 [1,2,20,24]
5	63	6.88 ± 2.11 [1]	22.84 ± 13.47 [2,12,16,20,24]	3.85 ± 1.34 [2,3]	89.35 ± 16.67 [1,2,3,12,16,20]	53.77 ± 4.47 [8,12,16,20,24]	31.67 ± 2.48 [1,2]
6	63	6.79 ± 3.05 [1]	20.93 ± 11.81 [12,16,20,24]	3.79 ± 1.26 [2,3]	87.89 ± 15.67 [1,2,3,12,16,20]	54.95 ± 5.21 [12,16,20,24]	31.79 ± 2.91 [1,2]

Age (weeks)	n	Bilirubin µmol/L	Fe µmol/L	Urea mmol/L	Creatinin µmol/L	TSP g/L	Alb g/L
8	64	5.82 ± 1.72 [1]	25.67 ± 12.53 [1,2,3,16,20]	3.90 ± 1.19 [2,3]	82.34 ± 15.10 [1,2,3,4,16]	56.94 ± 5.19 [1,2,3,4,5,12,16,20,24]	31.76 ± 2.68 [1,2]
12	61	5.19 ± 1.32 [1,2]	30.43 ± 10.48 [1,2,3,4,5,6]	4.07 ± 1.21 [2,3]	75.31 ± 10.60 [1,2,3,4,5,6]	61.09 ± 4.98 [1,2,3,4,5,6,8,16,20,24]	31.71 ± 2.82 [1,2]
16	61	4.82 ± 0.98 [1,2]	34.08 ± 13.49 [1,2,3,4,5,6,8]	4.49 ± 1.29 [2]	69.92 ± 8.79 [1,2,3,4,5,6,8]	64.69 ± 5.17 [1,2,3,4,5,6,8,12]	32.52 ± 2.40 [1,2,3]
20	56	4.42 ± 1.15 [1,2]	32.90 ± 9.27 [1,2,3,4,5,6,8]	4.52 ± 1.36 [2]	75.96 ± 11.06 [1,2,3,4,5,6]	67.55 ± 4.37 [1,2,3,4,5,6,8,12]	33.21 ± 2.91 [1,2,3,4]
24	39	4.47 ± 1.52 [1,2]	30.12 ± 9.88 [1,2,3,4,5,6]	4.48 ± 1.04 [2]	78.38 ± 9.25 [1,2,3,4]	66.92 ± 5.03 [1,2,3,4,5,6,8,12]	32.89 ± 2.84 [1,2,3,4]

The age group differs significantly ($P<0.05$) from age groups in superscript

Table 3. Biochemical variables in calves (Ježek, 2007)

Age (weeks)	n	iP mmol/L	Ca mmol/L	K mmol/L	Na mmol/L	Cl mmol/L
1	64	2.94 ± 0.48 [2]	2.97 ± 0.30 [2,3,4,5,6,8,12,16,20,24]	5.72 ± 0.60 [2,3,4]	144.9 ± 4.1 [2,20,24]	100.0 ± 2.6 [5,6,8,12,16,20,24]
2	64	3.27 ± 0.42 [1,4,6,8,12]	2.0 ± 0.23 [1]	6.49 ± 0.69 [1,4,5,6,8,12,16,20,24]	142.6 ± 3.9 [1,4,5,6,8,12,16,20,24]	98.6 ± 3.1 [3,4,5,6,8,12,16, 20,24]
3	59	3.11 ± 0.42	2.63 ± 0.23 [1]	6.17 ± 0.65 [1,12,16,20,24]	143.8 ± 4.4 [16,20,24]	100.5 ± 2.9 [2,6,8,12,16,20,24]
4	62	3.02 ± 0.42 [2]	2.66 ± 0.25 [1]	6.08 ± 0.53 [1,2,16,24]	144.6 ± 3.2 [2,16,20,24]	101.3 ± 2.6 [2,20]
5	63	3.04 ± 0.49	2.66 ± 0.23 [1]	6.02 ± 0.53 [2,16,24]	144.9 ± 3.0 [2,16,20,24]	101.8 ± 2.6 [1,2]
6	63	2.94 ± 0.45 [2]	2.60 ± 0.19 [1,16,20]	5.95 ± 0.48 [2,16,24]	144.6 ± 2.9 [2,16,20,24]	102.3 ± 2.7 [1,2,3]
8	64	2.88 ± 0.33 [2]	2.65 ± 0.20 [1]	5.85 ± 0.37 [2]	145.1 ± 2.3 [2,20,24]	102.7 ± 2.5 [1,2,3]
12	61	3.01 ± 0.35 [2]	2.73 ± 0.18 [1]	5.79 ± 0.54 [2,3]	145.2 ± 2.5 [2,20,24]	102.3 ± 2.6 [1,2,3]
16	61	3.06 ± 0.36	2.74 ± 0.20 [1,6]	5.61 ± 0.57 [2,3,4,5,6]	146.8 ± 2.5 [2,3,4,5,6]	102.2 ± 2.1 [1,2,3]
20	56	3.07 ± 0.34	2.76 ± 0.20 [1,6]	5.76 ± 0.48 [2,3]	148.4 ± 2.4 [1,2,3,4,5,6,8,12]	102.9 ± 1.9 [1,2,3,4]
24	39	2.99 ± 0.33	2.67 ± 0.21 [1]	5.52 ± 0.57 [2,3,4,5,6]	148.6 ± 2.8 [1,2,3,4,5,6,8,12]	102.3 ± 1.7 [1,2,3]

The age group differs significantly ($P<0.05$) from age groups in superscript

Table 4. Mineral concentration in calves' serum (Ježek, 2007)

The data were processed with the statistical software SPSS (Ver 15.0). For investigated haematological and biochemical variables the descriptive statistics were calculated regarding to the age. Bonferroni test was used to investigate significant difference between age groups.

The age of calves influenced the investigated haematological and biochemical variables what should be considered by interpretation of laboratory results.

5. Conclusion

The age of calves influence the haematological and biochemical variables what should be considered by interpretation of laboratory results.

The results of various studies on age dependent dynamics of haematological and biochemical variables in calves are different. The differences are influenced by the fact that blood samples were taken in different age periods, breeds, rearing systems, geographic regions, and were analysed with different methods.

For proper interpretation of laboratory results it is the best to use the reference values from the laboratory which performed the analyses of blood.

6. Acknowledgements

This work was supported by the Slovenian Ministry of Higher Education, Science and Technology, programme group "Animal Health, Environment and Food Safety" (P4-0092).

Special thanks go to author's collegues who contributed to the research work, presented in this chapter: Jože Starič, DVM, PhD and Marija Nemec, DVM, Msc.

7. References

Adams, R.; Garry, F.B.; Aldridge, B.M.; Holland, M.D.& Odde, K.G. (1992). Hematologic values in newborn beef calves. *American Journal of Veterinary Research*, Vol.53, No.6, (June 1992), pp. 944-950, ISSN 0002-9645

Boediker, R. (1991). Die Bestimung der GGT im Serum als Indikator für die Kolostralmilchversorgung des Kalbes. *Tierärztliche Umschau*, Vol.46, No.4 (April 1991), pp. 190-194, ISNN 0049-3864

Bostedt, H. & Schramel, P. (1982). Zur dynamik der Blutserumkonzentration von Kalzium und Magnesium sowie der Spurenelemente Eisen, Kupfer und Zink in den ersten Lebenswochen des Kalbes. *Tierärztliche Umschau*, Vol.37, No.7 (July 1982), pp. 471-476, ISNN 0049-3864

Bostedt, H. (1983). Vergleichende Untersuchung über die Entwicklung des Enzymprofils im Blut von Kälbern und Lämern in der neonatalen Adaptationsperiode. *Berliner und Münchener Tierärztliche Wochenschrift*, Vol.96, No.12 (December 1983), pp. 431-438, ISNN 0005-9366

Bostedt, H.; Jekel, E. & Shramel, P. (1990). Zur entwicklung der Eisen und Kupferkonzentration im Blutplasma von Kälbern in den ersten Lebenstagen und Wochen, gleichzeitig ein Beitrag zur larvierten neonatalen Eisenmangelanämie.

Deutsche Tierarztliche Wochenschrift, Vol.97, No.10, (october 1997), pp. 400 - 403, ISSN 0341-6593

Bouda, J.& Jagoš, P. (1984). Biochemical and hematological reference values in calves and their significance for health control. *Acta Veterinaria Brno*, Vol.53, No.3-4. (September-December 1984), pp. 137-142, ISNN 0001-7213

Braun, J.P.; Tainturier, D.; Laugier, C.; Benard, P.; Thouvenot, J.P. & Rico, A.G (1982). Early variations of gamma-glutamyl transferase in newborn calves - a test of colostrum intake. *Journal of Dairy Science*, Vol.65, No.11 (November 1982), pp. 2178 - 2181, ISSN 0022-0302

Brun-Hansen, H.C.; Kampen, A.H. & Lund, A. (2006). Hematologic values in calves during the first 6 months of life. *Veterinary Clinical Pathology*, Vol.35, No.2 (June 2006), pp. 182 - 187, ISSN 0275-6382

Carlson, G.P. (1997). Fluid, electrolyte, and acid - base balance, In : *Clinical biochemistry of domestic animals*, J.J. Kaneko, J.W. Harvey & M.L. Bruss, (Ed.), pp.485-516, Academic Press , ISBN 0-12-396305-2, San Diego, California

Egli, C.P. & Blum, J.W. (1998). Clinical, hematological, metabolic and endocrine traits during the first three months of life of suckling simmentaler calves held in a cow-calf operation. *Journal of Veterinary Medicine: Series A*, Vol.45, No.2 (March 1998), pp. 99-118, ISSN 0514-7158

Greatorex, J.C. (1954). Studies on the haematology of calves from birth to one year of age. *British Veterinary Journal*, Vol.110, No.4, (April 1954), pp. 120-138, ISNN 0007-1935

Hammon, H.M. & Blum, J.W. (1998). Metabolic and endocrine traits of neonatal calves are influenced by feeding colostrum for different durations or only milk replacer. *The Journal of Nutrition*, Vol.128, No.3 (march 1998), pp. 624 - 632, ISSN 0022-3166

Hammon, H.M.; Schiessler, G.; Nussbaum, A. & Blum, J.W. (2002). Feed intake paterns, growth performance and metabolic and endocrine traits in calves fed unlimited amounts of colostrum amd milk by automate, starting in the neonatal period. *Journal of Dairy Science*, Vol.85, No.12 (december 2002), pp. 3352-3362, ISSN 0022-0302

Hanschke, G. & Schulz, C. (1982). Blutuntersuchungen bei klinisch gesunden Kälbern im subtropischen Klima (Marokko). *Tierärztliche Umschau*, Vol.37, No.8 (August1982), pp. 554-563, ISNN 0049-3864

Harvey, J.W. (1997). The erytrocyte: physiology, metabolism, and biochemical disorders. In: *Clinical biochemistry of domestic animals*, J.J. Kaneko, J.W. Harvey & M.L. Bruss, (Ed.), pp. 157-203, Academic Press , ISBN 0-12-396305-2, San Diego, California

Heidarpour Bami, M.; Mohri, M.; Seifi, H.A. & Alavi Tabatabaee, A.A. (2008). Effects of parenteral supply of iron and cooper on hematology, weight gain, and health in neonatal dairy calves. *Veterinary Research Communications*, Vol.32, (May 2008), pp. 553-561, ISNN 0165-7380

Hugi, D.; Gut, S.H. & Blum, J.W. (1997). Blood metabolites and hormones - especially glucose and insulin - in veal calves: Effects of age and nutrition. *Journal of Veterinary Medicine: Series A*, Vol.44, No.7 (september 1997), pp. 407-416, ISSN 0514-7158

Jain, N.C. (1986). Cattle: Normal Hematology with Comments on Response to Disease. In: Schalm´s veterinary hematology, N.C. Jain, (Ed.) pp. 178-207, Lea & Febiger, ISBN 0-8121-0942-2, Philadelphia, United States of America

Jazbec, I.; Gregorović, V. & Skušek, F. (1973). Dinamika krvne slike pri teletih v pitanju na belo meso. *Zbornik Biotehniške Fakultete Univerze Ljubljana, Veterina,* Vol.10, No.1, (January 1973), pp. 96-107, ISNN 1408-3442

Jazbec, I.; Gregorović, V. & Skušek, F. (1975). Gibanje koncentracije serumskega železa in bakra pri teletih ob pitanju na belo meso. *Zbornik Biotehniške Fakultete Univerze Ljubljana, Veterina,* Vol.12, No.1, (January 1973), pp. 23-34, ISNN 1408-3442

Jazbec, I.(1990). Klinično laboratorijska diagnostika, I. Jazbec, (Ed.) pp. 82-206, Veterinarska fakulteta, Ljubljana, Slovenija

Ježek,J. (2007). The dynamics of serum immunoglobulin concentrations and hematological and biochemical parameters in the period to the age of 24 weeks in diferently reared calves, pp. 172, *Univerza v Ljubljani, Veterinarska fakulteta,* Ljubljana, Slovenija

Kaneko, J.J. (1997). Serum proteins and the disproteinemias. In: *Clinical biochemistry of domestic animals,* J.J. Kaneko, J.W. Harvey & M.L. Bruss, (Ed.), pp. 117-138, Academic Press , ISBN 0-12-396305-2, San Diego, California

Klee, W. (1985). Untersuchungen über die Nierenfunktion bei gesunden und bei an akutem Durchfall erkrankten Kälbern. In: *Klinische Labordiagnostik in der Tiermedizin,* W. Kraft, U.M. Dürr, (Ed.), pp. 194-195, Schattauer, ISBN 978-3794519422, Stuttgart, Germany

Klinkon, M.; Zadnik, T.; Mesarič, M. & Nemec, M. (2000). Haematological profile of neonatal white & black calves with regard to age and sex. *Comparative Haematology International,* Vol.10, No.4, (December 2000), pp. A4, ISSN 0938-7714

Knowles, T.G.; Edwards, J.E.; Bazeley, K.J.; Brown, S.N.; Butterworth, A. & Warriss, P.D. (2000). Changes in the blood biochemical and haematological profile of neonatal calves with age. *Veterinary Record,* Vol.147, No.21, (November 2000) pp. 593-598, ISSN 0042-4900

Kraft, W. (1999a). Hämatologie. In: *Klinische Labordiagnostik in der Tiermedizin,* W. Kraft, U.M. Dürr, (Ed.), pp. 43-77, Schattauer, ISBN 978-3794519422, Stuttgart, Germany

Kraft, W. & Dürr, U.M. (1999b). Leber. In: *Klinische Labordiagnostik in der Tiermedizin,* W. Kraft, U.M. Dürr, (Ed.), pp. 112-133, Schattauer, ISBN 978-3794519422, Stuttgart, Germany

Kraft, W. & Dürr, U.M.(1999c). Harnapparat. In: *Klinische Labordiagnostik in der Tiermedizin,* W. Kraft, U.M. Dürr, (Ed.), pp. 169-200, Schattauer, ISBN 978-3794519422, Stuttgart, Germany

Kraft, W. & Dürr U.M. (1999d). Serum-Protein. In: *Klinische Labordiagnostik in der Tiermedizin,* W. Kraft, U.M. Dürr, (Ed.), pp. 148-155, Schattauer, ISBN 978-3794519422, Stuttgart, Germany

Kraft, W. (1999). Skelettmuskulatur, Knochen, Kalzium-, Phosphor-, Magnesiumstoffwechsel. In: *Klinische Labordiagnostik in der Tiermedizin,* W. Kraft, U.M. Dürr, (Ed.), pp. 251-258, Schattauer, ISBN 978-3794519422, Stuttgart, Germany

Kurz, M.M. & Willett, L.B. (1991). Physiology and management; carbohydrate, enzyme, and hematology dynamics in newborn calves. *Journal of Dairy Science*, Vol.74, No.7 (July 1991), pp. 2109 - 2118, ISSN 0022-0302

Maach, L.; Gründer, H.D. & Boujija, A. (1992). Klinische und hämatologische Untersuchungen bei schwarzbunten an Durchfall erkrankten neugeborenen Aufzuchtkälbern in Marokko. *Deutsche Tierarztliche Wochenschrift*, Vol.99, No.4, (April 1992), pp. 133-140, ISSN 0341-6593

Maach, L.; Gründer, H.D. & Faio, A. (1991). Hämozytologische und hämobiochemische Untersuchungen bei schwarzbunten, klinisch gesunden Aufzuchtkälbern in Marokko. *Deutsche Tierarztliche Wochenschrift*, Vol.98, No.3, (March 1991), pp. 77-116, ISSN 0341-6593

Mohri, M.; Sharifi, K. & Eidi, S. (2007). Hematology and serum biochemistry of Holstein dairy calves: Age related changes and comparison with blood composition in adults. *Research in Veterinary Science*, Vol.83, No.1 (August 2007), pp. 30-39, ISSN 0034-5288

Mohri, M .; Poorsina, S. & Sedaghat, R. (2010). Effect of parenteral supply of iron on RBC parameters, performance, and health in neonatal dairy calves. *Biological Trace Element Research*, Vol.136,No1(July 2010), pp.33-39, ISSN 0163-4984

Moosavian H.R.; Mohri, M. & Seifi H.A. (2010). Effect of parenteral over-supplementation of vitamin A and iron on haematology, iron biochemistry, weight gain, and health of neonatal dairy calves. *Food and Chemical Toxicology*, Vol.48, No.5 (May 2010), pp.1316-1320, ISSN 0278-6915

Muri, C.; Schottstedt. T.; Hammon, H.M.; Meyer, E. & Blum, J.W. (2005). Hematological, metabolic and endocrine effects of feedning vitamin A and lactoferin in neonatal calves. *Journal of Dairy Science*, Vol.88, No.3 (March 2005), pp. 1062-1077, ISSN 0022-0302

Perino, L.J.; Sutherland, R.L. & Woollen, N.E. (1993). Serum γ- glutamiltransferase activity and protein concentration at birth and after suckling in calves with adequate and inadequate passive transfer of immunoglobulin G. *American Journal of Veterinary Research*, Vol.54, No.1 (January 1993), pp. 56-59, ISSN 0002-9645

Reece, W.O .(1980). Acid-base balance and selected hematologic, electrolyte, and blood chemical variables in calves: milk-fed vs conventionally fed. *American Journal of Veterinary Research*, Vol.41, No.1, (January 1980), pp. 109-113, ISSN 0002-9645

Reece, W.O. (1984). Acid-base balance and selected hematologic, electrolytic, and blood chemical variables in calves nursing cows: one week through fifteen weeks. *American Journal of Veterinary Research*, Vol.45, No.4, (April 1984), pp. 666-669, ISSN 0002-9645

Reece, W.O. & Hotchkiss, D.K. (1987). Blood studies and performance among calves reared by different methods. *Journal of Dairy Science*, Vol.70, No.8 (August 1987), pp. 1601-1611, ISSN 0022-0302

Rosol, T.J. & Capen, C.C. (1997). Calcium regulating hormones and diseases of abnormal mineral (calcium, phpsphorus, magnesium) metabolism. In: *Clinical biochemistry of domestic animals*, J.J. Kaneko, J.W. Harvey & M.L. Bruss, (Ed.), pp. 619-702, Academic Press , ISBN 0-12-396305-2, San Diego, California

Scheidegger, H.R. (1973). Veränderungen des Roten Blutbildes und der Serumeisen-Konzentration bei Simmentaler Kälbern. *Schweizer Archiv für Tierheilkunde*, Vol.115, No.11, (November 1973), pp. 483-497, ISNN 0036-7281

Selim, S.A.; Smith, B.P. & Cullor, J.S. (1995). Serum immunoglobulins in calves: their effects and two easy reliable means of measurement. *Veterinary Medicine*, Vol.90, No.4, (April 1995), pp. 387-404, ISNN 8750-7943

Smith, G.M.; Fry, J.M.; Allen, J.G. & Costa, N.D. (1994). Plasma indicators of muscle damage in a model of nutritional myopathy in weaner sheep. *Australian Veterinary Journal*, Vol.71, No.1 (January 1994), pp. 12-17, ISSN 0005-0423

Steinhardt, M.; Gollnast, I.; Langanke, M.; Bünger, U. & Kutschke, J. (1993). Klinisch-chemische Blutwerte bei neugeborenen Kälbern. Tierärztliche Praxis, Vol.21, No.4 (August 1993), pp. : 295-301, ISNN 0303-6286

Steinhardt, M. & Thielscher, H.H. (2000a). Reaktionen von Milchrindkälbern auf ACTH-Applikation und Flüssignahrungsaufnahme an verschiedenen Alterspunkten vor und während der Aufzucht am Tränkeautomaten - Plasmacortisol, Speichelcortisol, hämatologische, metabolische Variablen und Herzschlagfrequenz. *Deutsche Tierärztliche Wochenschrift*, Vol.107, No.5 (May 2000), pp. 180-187, ISNN 0341-6593

Steinhardt, M. & Thielscher, H.H. (2000b). Physiologische Variablen und Wachstumsleistung bei Saugkälbern der Mutterkuhhaltung in den ersten beiden Lebensmonaten. *Tierärztliche Umschau*, Vol.55, No.7, (July 2000), pp. 380-389, ISNN 0049-3864

Steinhardt, M. & Thielscher, H.H. (2000c). Tiergerechte Haltung und physiologische Funktionen von Tieren. *Tierärztliche Umschau*, Vol.55, No.4, (April 2000), pp. 189-198, ISNN 0049-3864

Steinhardt, M. & Thielscher, H.H. (2000d). Reaktionen von Milchrindkälbern auf Flüssignahrungaufnahme an verschiedenen Alterspunkenten vor und während der Aufzucht. *Tierärztliche Umschau*, Vol.55, No.12, (December 2000), pp. 663-673, ISNN 0049-3864

Tennant, B.C. (1997). Hepatic function. In: *Clinical biochemistry of domestic animals*, J.J. Kaneko, J.W. Harvey & M.L. Bruss, (Ed.), pp. 327-352, Academic Press , ISBN 0-12-396305-2, San Diego, California

Terosky, T.L.; Wilson, L.L.; Stull, C.L. & Stricklin, W.R. (1997). Effects of individual housing design on special fed Holstein veal calf growth performance, hematology and carcass characteristics. *Journal of Animal Science*, Vol.75, No.7, (July 1997), pp. 1697-1703, ISSN 0021-8812

Tucker, W.B.; Jackson, J.A.; Hopkins, D.M. & Hogue, J.F. (1991). Influence of dietary sodium bicarbonate on the potassium metabolism of growing dairy calves. *Journal of Dairy Science*, Vol.74, No.7, (July 1991), pp. 2296-2302, ISSN 0022-0302

Tyler, J.W.; Hancock, D.D. & Parish, S.M. (1996a). Evaluation of 3 assays for failure of passive transfer in calves. Journal of Veterinary Internal Medicine, Vol.10, No.5, (September-October 1996), pp. 304-307, ISSN 0891-6640

Tyler, J.W.; Hancock, D.D.; Wiksie, S.E.; Holler, S.L.; Gay, J.M. & Gay, C.C .(1998). Use of serum protein concentration to predict mortality in mixed-source dairy replacement heifers. Journal of Veterinary Internal Medicine, Vol.12, No.2, (March-April 1998), pp. 79-83, ISSN 0891-6640

Tyler, J.W.; Parish, S.M.; Besser, T.E.; Van Metre, D.C.; Barrington, G.M. & Middleton, J.R. (1999b). Detection of low serum immunoglobulin concentrations in clinicaly ill calves. Journal of Veterinary Internal Medicine, Vol.13, No.1, (January-February 1999), pp. 40-43, ISSN 0891-6640

Underwood, E.J. & Suttle, N.F. (2001). *The mineral nutrition of the livestock*. pp.105-397, CABI Publishing , ISBN 0-85199-128-9,Wallingford, United Kingdom

Ward, G.M. (1966). Potassium metabolism of domestic ruminants: a review. *Journal of Dairy Science*, Vol.49, No.3 (March 1966), pp. 268-276, ISSN 0022-0302

Weil, A.B.; Tucker, W.B. & Henken, R.W (1988). Potassium requirement of dairy calves. *Journal of Dairy Science*, Vol.71, No.7 (July 1988), pp. 1868-1872, ISSN 0022-0302

Whitaker, D.A. (1997). Interpretation of metabolic profiles in dairy cows. *Cattle Practice*, Vol.5, No.1, (January 1997), pp. 57-60, ISNN 0969-1252.

Wirth, W. (1999). Elektrolyte und Säure - Basen - Haushalt. In: *Klinische Labordiagnostik in der Tiermedizin*, W. Kraft, U.M. Dürr, (Ed.), pp. 159-168 Schattauer, ISBN 978-3794519422, Stuttgart, Germany

17

The Growth Performance Evaluation of Cattle Breeds in the South Western Agro-Ecological Zone (SWAEZ) of Uganda

Charles Lagu[1], Robert Nsubuga Mutaka[1], James Oluka[2], Steven Byenkya[1], Betty Laura Ayoo[1], Immaculate Nabukenya[1] and Proscovia Ntakyo[1]
[1]*Mbarara Zonal Agriculture Research and Development Institute, Mbarara,*
[2]*National Livestock Resources Research Institute (NaLIRRI), Tororo, Uganda*

1. Introduction

Agriculture is Uganda's most important sector, contributing 48% of gross domestic product (GPD) and directly supporting 85% of the population in rural areas. Livestock which is concentrated along 'the cattle corridor' runs southwest to northeast across Uganda, encompassing 29 districts represents 7.5% of the GDP and 17% of the agricultural GDP. The south-western agro-ecological zone (SWAEZ) of Uganda, comprising of the districts of Bushenyi, Ibanda, Isingiro, Kiruhura, Lyantonde, Mbarara, Ntungamo, Rakai and Sembabule lies within the corridor with a population of 3,085,900 people (UBOS 2006).

The cattle population in the SWAEZ number to 1,689,605 (UBOS and MAAIF, 2009). There are a number of different cattle breeds on farms in the south western agro-ecological zone (SWAEZ) which include the following: Friesian and Friesian crosses 50% (F1) and 75%(F2), Boran and Boran crosses and the Ankole cattle. The Ankole cattle used to be the most predominant breed of cattle but they have been extensively crossed with the Friesian cattle.

There are generally no guidelines on the optimum productivity and cost-effective level at which to stop cross-breeding. It is known that the more one up-grades the more susceptible the crossbred animals become to common diseases requiring extra cost on feeding and disease control. Studies by Kiwuwa *et al* 1983; Kugonza (2005) and Ndumu (2007) indicate similar findings about the production and husbandry constraints affecting other local breeds, the Ankole cattle in addition to extensive nature of cross breeding of the Ankole cattle. But these evaluations have been limited in scope and approaches while covering limited agro-ecological environments and unique production systems and furthermore, there have been a lot of changes in the grazing systems, feed resources availability and supply as a result of continued adverse changes in climatic conditions which impacts on the overall performance of the animals. It is, therefore, necessary that up-to date information is constantly generated to guide farmers in the best practices with regards to adoption of appropriate breeding and production technologies for improving production and productivity of various cattle breeds in Uganda.

The overall objective of the study was to evaluate and document the performance of existing breeds of cattle in the south western agro-ecological zone (SWAEZ) for improved livestock productivity. The specific objectives include: (i) to evaluate the growth performance of existing cattle breed under on-station and on-farm conditions.

The study is meant to generate up to-date and accurate information on performance of selected cattle populations in the SWAEZ in terms of their growth, maturity and production parameters. This would help in packaging of appropriate information on better management practices to guide farmers in partnership with National Agriculture Advisory Services (NAADS) and extension workers as conduits for effective uptake pathways for technology adoption.

2. Materials and methods

A longitudinal study involving growth data collection from Ankole, Friesian, Friesian cross, Boran and Boran crosses was conducted. Growth data on-station and on-farm from the experimental animals was collected over a period of three years (July 2006 to June 2009) and 9 months from October 2008 to June 2009 respectively.

The on farm study was conducted in the districts of Kiruhura, Ibanda and Sembabule. In Kiruhura district, the sub-counties of Kikatsi and Sanga were involved. In Ibanda district, Nyabuhikye and Kikenkye sub-counties were involved. In Sembabule the studies were conducted in Lugusulu and Migwala sub-counties.

On-station the studies were conducted at Mbarara Zonal Agriculture Research and Development Institute (ZARDI) found in Mbarara district. It involved taking bi- weekly (i.e. fortnightly) weights using the heart girth tape. The heart girth tape was graduated in both cm and kg. Both measurements were taken but for purposes of analysis, only weight values were used. There was a high and positive correlation between heart girth and body weight of cattle (Oluka, 2006). Individual study animals were identified using the ear tag numbers for effective continuous monitoring. Animals were restrained in the crush and weighed using the heart girth tape.

The sample sizes varied from district to district. In Kiruhura district, the study animals were 141, Ibanda district, 182 and in Sembabule district 189. Calves below the age of three months were identified and randomly selected and tagged for the study. Only farmers who were willing to participate in the study were recruited and involved in the study. The farmers willingly gave the number of animals requested for the study. More Female animals were preferred because they were normally kept for longer periods by the farmers and were not easily sold. In contrast the male were sold off at any age to meet the financial needs of the farmers e.g. paying school fees, meet family health care needs or provide food.

Under on-station management, a total of 30 Ankole cattle and 35 Friesian cattle were recruited for the study. Unlike the on-farm cattle, parameter measurement and data collection started immediately after birth of calves by taking and recording birth weights and there after bi-weekly.

The heart girth tape was placed around the girth for reading off the weights. Weights of cattle were immediately recorded in data sheets and transferred for entry into Excel programme as back up and for further collation for analysis; all weights in kilograms (kg) were entered in Excel worksheet

2.1 Description of on-station data

On-station data involved only two cattle breeds i.e., the pure (100%) Friesian and the pure (100%) Ankole cattle. The weights were taken immediately after birth. The sample sizes were 30 Ankole cattle and 35 Friesian cattle. Continuous recording of heart girth and body weight measurements was done every two weeks for a period of 3 years. The data was, however entered in Excel at interval of 30 days i.e. monthly 1 for period 1-30 days, period 2 for 31-60 days, etc. All calves were managed under similar conditions allowing suckling up to 6 months while their dams were on open free range grazing management under the care a herdsman. Comparison was based on fixed effects under a single management of one on-station site focussing on breed differences.

Prior to the analysis of variance (ANOVA), a simple descriptive statistical analysis was done to examine the distribution of weight data structured at 30-day intervals for distribution of the means. Those values showing gross errors and outlying were excluded from the data set to avoid adversely affecting the means and hence, standard deviations, variances and standard errors. This was achieved using VIEWTABLE, GPLOT and MEANS procedures of SAS, version 9 (2002). The GPLOT showed the distribution of observations against the different time intervals of 30 day means.

2.1.1 Analysis models

ANOVA using General Linear Model (GLM) (SAS, 2000) for calf body weight was performed to determine the fixed effects of various factors on body weight at 30-day intervals from the first month of birth to 12th months of age (360 days of age). The data was classified according to the experimental variables covering the twelve periods 1-30, 31-60, 61-90, 91-120, 121-150, 151-180, 181-210, 211-241, 241-270, 271-300, 301-330 and 331-360 days (Table 1).

Various analyses were run for different combination of the fixed effects and effects with non-significance levels of P>0.05 dropped from the next test model until the final model was developed. After final adjustments for the various differences and exclusions of non-significant effects such as year of birth and multiple births and interaction of cattle breed with season of birth, the final model (i) below was arrived at. This includes only the effect of breed, sex of calf and season of birth (model i) below.

$$Y_{ijkl} = \mu + \mathbf{B}_i + \mathbf{X}_j + \mathbf{S}_k + e_{ijkl} \qquad \text{(i)}$$

Where:

Y_{ijklm} = the live weight of the lth calf born under ith calf cattle breed, jth sex of calf and kth season of birth.

μ = the common parameter (konstant)
$\mathbf{B}_i$ = the fixed effect of breed of cattle (i = 1, 2)

$\mathbf{X}_j$ = the fixed effect of sex of calf (j = 1, 2)
$\mathbf{S}_k$ = the fixed effect of season of birth (k =1, 2)
$\mathbf{e}_{ijkl}$ = the random residual effect

2.2 Description of on farm cattle data

Cattle handling varied from farmer to farmer depending on availability of handling structures. In some cases the animals were put in a crush, restrained and weight measurement taken using the heart girth tape. In other cases, individual animals were restrained manually by field assistants using ropes. The heart girth and weight estimates were taken using the tape when the animal was calmed down and standing in an upright and straight position. The measurements were immediately recorded in field data sheets. Thereafter, the data was transferred into Excel for back up, collation and analysis.

On-farm data involved 4 genetic breeds of cattle including pure Ankole and Boran crossbred Friesian and Boran. It was, however difficult to determine the exact level of the crossbred animals. At the start of the field study, the calves that were 3 months old were recruited for measurements. However, with time, calves were born among the herds of the participating farmers. The measurements of such calves were recorded within the first month of birth. The heart girth and weights were again taken at the time of the next visit. This was done on monthly intervals though, in some cases, delays could occur due to late release of operational funds. The sample sizes were however larger than on-station with Kiruhura district offering 141 animal records, 182 in Ibanda district and 189 in Sembabule . Continuous recording of heart girth and body weight estimates was done for only 9 months due to the heavy expense and limited funding. All data was entered in Excel at 30 days intervals i.e. period 1 for 1-30 days, period 2 for 31-60 days, etc. All calves were managed under similar conditions allowing suckling up to 8-12 months or at the time when the farmers observe that the cow is nearing to calve down again. All experimental animals from the selected participating farmers were on free range management system under herding. The comparison was based on fixed effects under a free range management focussing on four cattle breed differences, that is Ankole, Boran, Boran x Ankole , and Friesian x Ankole.

Prior to the ANOVA was employed to examine the distribution of live body weight data structured at 30-day intervals. Those values whose means showed gross errors and outlying were excluded from the data set to avoid adversely affecting the means, standard deviations, variances and standard errors. This was achieved again by applying the VIEWTABLE, GPLOT and MEANS procedures of SAS (2002). The GPLOT showed the distribution of observations against the different time intervals of 30 day means.

2.2.1 Analysis models

ANOVA using General Linear Model (GLM) (SAS, 2000) for calf body weight was tested. This was done to determine the fixed effects of various factors on body weight at 30-day intervals from the first month of birth to 12 months of age (360 days). The data was classified according to the experimental variables covering the twelve months periods 1-30, 31-60, 61 90, 91 120, 121-150, 151-180, 181-210, 211-241, 241-270, 271-300, 301-330 and 331-360 days (Table 1).

Various analyses were run for different combinations of the fixed effects. Those effects with non-significance levels at P>0.05 were dropped from the next test model until the final model was developed. After final adjustments for the various differences and exclusions of non-significant effects such as year of birth, farmer herd and interaction of cattle breed with season of birth, the final model (i) below was arrived at; that included the fixed effects of breed, sex of calf, season of birth and district of birth (model ii) below.

$$Y_{ijklm} = \mu + \mathbf{B}_i + \mathbf{X}_j + \mathbf{S}_k + \mathbf{D}_l + e_{ijklm} \quad \text{(ii)}$$

Where:

Y_{ijklm} = the live weight of the mth calf born under i[th] calf cattle breed, j[th] sex of calf, k[th] season of birth and l[th] district of birth

μ = the common parameter (konstant)
$\mathbf{B}_i$ = the fixed effect of breed of cattle (i = 1, 2)
$\mathbf{X}_j$ = the fixed effect of sex of calf (j = 1, 2)
$\mathbf{S}_k$ = the fixed effect of season of birth (k =1, 2)
$\mathbf{D}_l$ = the fixed effect of district of birth (l =1, 2, 3)
$\mathbf{e}_{ijklm}$ = the random residual effect

3. Results and discussions

3.1 Genetic group effect on body weight and daily gain for on-station and on-farm calves

On station there was significant effect of breed on body weight of the cattle between month 1 and 2, then between the later months from month 7 to 12 (Tables 1a and 1 b). The birth weights of the Friesian calves were higher than that of the Ankole (Table 2a). There was no significant difference in growth weight of both the Friesians and the Ankole up to six months. After the age of 6 months, the Ankole grew faster than the Friesian. This was attributed to the weaning shock that was experienced by the Friesian calves versus the Ankole that continued to suckle their dams. This is clearly illustrated in Fig. 1. This therefore explained the weight variation between the Friesian and Ankole after six months as illustrated in figure1. Animal used on the station were composed of an old stock that had been breeding for along time without replacement of bulls. This apparently introduced some element of inbreeding depression on the herd. The inbreeding could be an important factor that could be used to explain the variation in the growth rate. Furthermore, disease prevalence and worm burden were present on station; however, the Friesian calves were more susceptible to the diseases than the Ankole calves. The Friesian calves were intensively kept in paddocks compared to the Ankole calves which were moving with their dams. This increased the chances for the Friesian calves to pick worm eggs. These findings are similar to the results by Said at al., 2001 and Kugonza *et al.*, 2005.

On farm the study found out that there was significant relationship between breed and growth rate ($p<0.05$) from the first month to the ninth months as illustrated in table 3a and 3b. The growth weight of the Boran was much higher than for the Boran x Ankole crosses and Friesian x Ankole crosses respectively. This could be explained by the fact that the Borans had great potential for growth intensity traits, have high feed conversion efficiency,

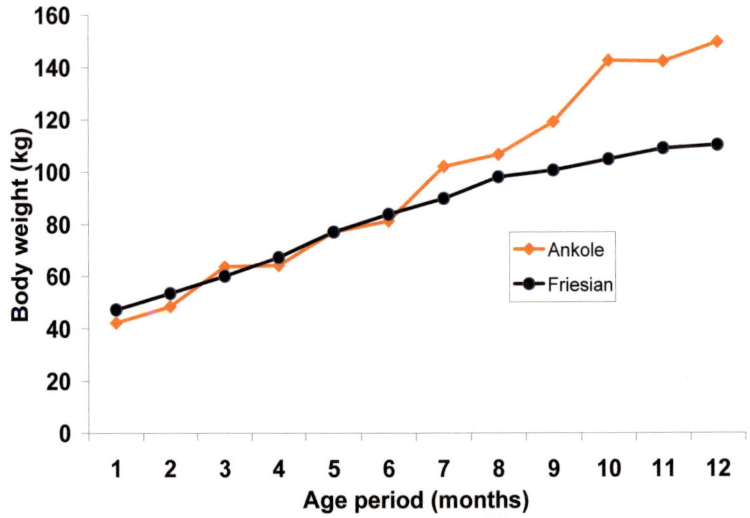

Fig. 1. On station graph for weights of Ankole and Friesian Crosses

were well adapted to the environment and resistant to harsh conditions (Ndumu, 2007). The dams of the Borans were not milked, therefore the calf got as much milk as it required for growth. Ankole on the other hand were dual purpose i.e. they were for both meat and milk production (Kugonza et al., 2005). This had effect on their growth pattern in terms of functional roles. The Ankole had low genetic potential were also well adapted to prevailing environmental conditions and exhibited better abilities to utilize low quality forages (Tuah and Yaa Nyamaa Danso, 1985; LRSP, 1999; Scerf, 2000 in Kugonza, 2005). The growth pattern of the Boran and Friesian crosses were lower compared to the pure breeds. This could be because of the following: high susceptibility to diseases, harsh environmental conditions, insufficient amounts of feeds during dry seasons, management related issues e.g. prophylaxis, deworming, and tick control (Moran, 2002; Asimwe and Kifaro, 2007). It was however difficult to determine the level of crossing since farmers had no records to substantiate the level of crossing among the breeds. Farmers had poor recording system and culture, information on the breeds was mainly got from memory and oral communication. This had challenges related to inaccuracy in records and hence human memory failure. These results were in agreement with those documented by Kugonza *et al.,* 2004.

3.2 Effect of sex on the growth weights of calves on station and on farm

On station, there was a significant relationship between sex and body weight of Ankole and Friesian calves ($p<0.05$). The birth weight of both male and female were similar (Figure 2). However, females' calves after the third month were heavier than the males. Similar studies by Vial, V. E. and More O'Ferrall G. J. (1965) had similar results. This could be explained as follows: large value in error variance, less managemental factors that were considered in the model during analysis, another reason could be castration stress that affects the males. This finding agrees with Fisher *et al.,* 2001 where they established that banding or branding and surgical castration negatively affects growth rates of castrated bull calves.

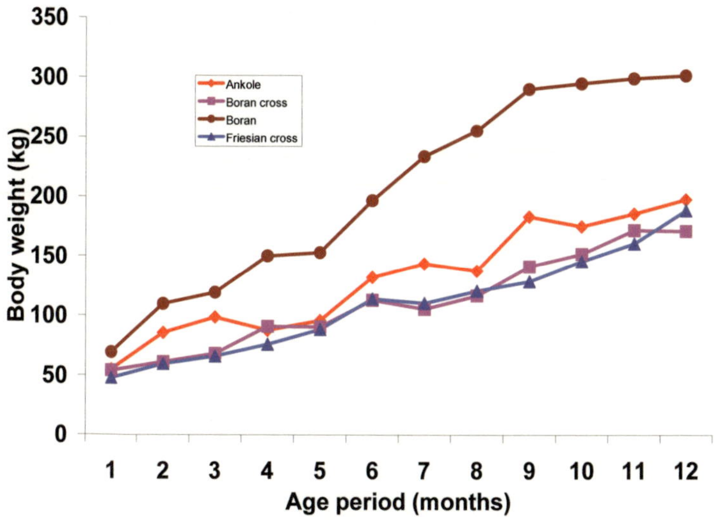

Fig. 2. Growth Patterns on breeds of Cattle in Kiruhura, Ibanda and Sembabule

On farm there was no significant relationshp between sex and growth weights p>0.05 among the breeds (table 3a and 3b). This could be explained by the fact that some factors in the model were not captured. The values taken were not large enough to distinctively give differences betweeen female and male animals, and probably sample size was small. However, from field experience, male calves were usually restricted from suckling unlike the females, this therefore counteracts hormonal effects. It is only when the male calves are intended for breeding that they get adequate milk. It becomes easy to distinguish between male calves intended for breeding to those left for slaughter. Female calves on the other hand were given preferential treatment because they were intended for breeding. Generally, female and male calves had competitive growth. These findings were in variation with the study by Krupá *et al.,* 2005. They found out that calves sex and age of dam at calving jointly explained the highest proportion variability (56 -75%). Riha *et al* 1999 as cited in Krupá *et al* 2005 found out that birth weight of calves both male-singles and male twins had higher birth weights than female-singles and female-twins. This particular study found out that twinning of cattle was a rear occurrence in Kiruhura, Ibanda and Sembabule. Riha *et al.,* 1999 in Krupá *et al.,* 2005 reported higher birth weights, weights at 120 days, weaning weights, and average daily gains for males and singles. Higher growth intensity in twins could be due to either a smaller number of twins involved in the analysis or to milk stealing behaviour of twin calves. In the study areas of Kiruhura, Ibanda and Sembabule, twinning hardly ever occurs and milk stealing behaviour was common among the Friesian crosses. In the Ankole, calf attachment to the dams is very strong as such; milk stealing behaviour was a rare phenomenon.

Goyache *et al.,* 2003 reported the highest weaning weights for calves descending from seven to eleven years old dams. On the other hand, Pribyl *et al.,* 2003 as cited in Krupá *et al.,* 2005 reported highest weights in fastest growth for calves of seven to nine year old dams. Lowest weights were reported for calves of first primiparas (2 year olds or younger dams). This could be explained by the fact that young dams needed more nutrients to complete their own growth (Kifaro, 1984 and Katyega 1988). In comparison with mature dams, cows of higher age usually produce calves of lower birth weights as production abilities decreases along with the increasing age of the dam. The results in this study concur with the observations made by Krupá *et al.*, 2005. In this study, insignificant differences between the male and female calf weights could be explained by the fact that sample sizes were small, higher variance ratios, high herd variability in both husbandry practices and geographical locations figure 5.

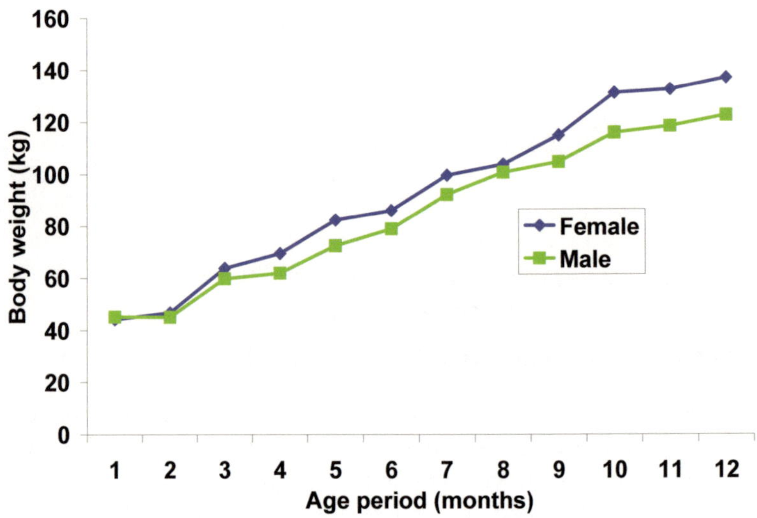

Fig. 3. On-station comparison of weights of Female and Male cattle

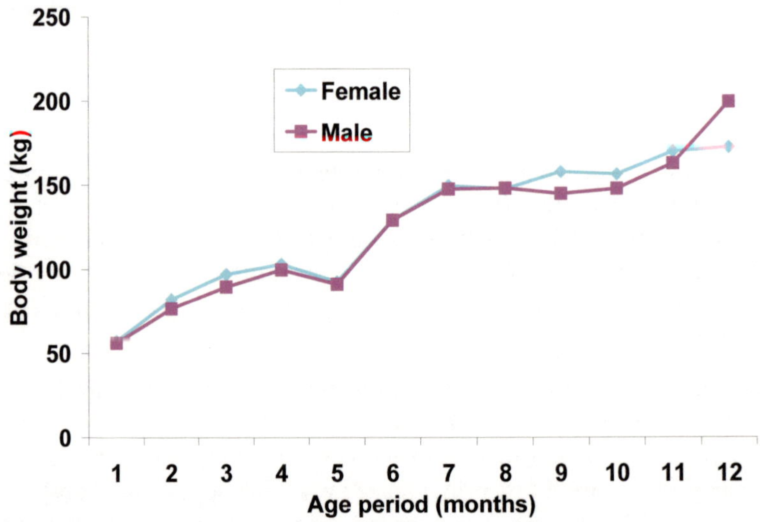

Fig. 4. On-farm comparison of weights of female and male cattle compared

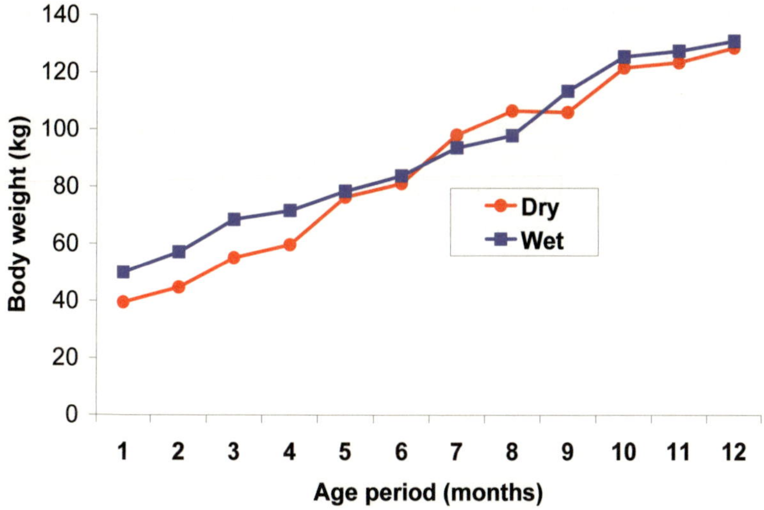

Fig. 5. On-station comparison of seasonal effects on cattle growth

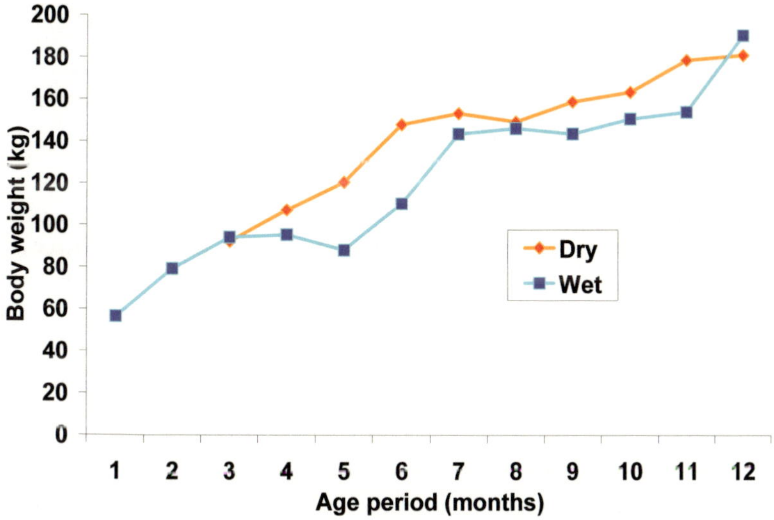

Fig. 6. Graph for weight Patterns of on farm Cattle and seasonal Variation

3.3 Effect of season on the growth weights of calves on station and on farm

On station, there was a significant relationship between season and body weight among Friesian and Ankole calves. In the first five months of neonatal life, calves weighed higher during the wet season than the dry season. This could be explained by fact that there was fodder availability during the wet season for the dam and the calf. The variation in body weight after the fifth month tends to fluctuate between the wet and dry season. During these periods, the calves had grown faster and had better adaptability and tolerance to fodder fluctuation. From six months onwards, there was increased resistance towards worm burden and other diseases hence explaining the fluctuation between the body weights during the wet and dry season. This finding agrees with observation made by Kugonza *et al*, 2005 and Twinamatsiko, 2001.

On farm there was significant relationship between season and the growth rate among the breeds in period four (p=0.0059) table 3a. Calves born during the dry season weighed heavier than the calves born in the wet season. This was due to adequate forage availability for the dams during the rainy season (Twinamasiko, 2001). This therefore met the gestational needs of dams and the foetus especially during the last trimester. There was inadequate forage during the dry season, as such, the dams had inadequate amount of forage availability. However, calves born during the wet season were weaker than those born during the dry season. The ability of these calves to cope with the environmental conditions was compromised. This explained why calves born during the dry season would perform much better than those born during the wet season. Further still, calves born in the wet season would perform less compared to the ones born in the dry season. This could be explained by the fact that there was coldness predisposing calves to pneumonia, high worm burden, high tick challenge, presence of bacterial and protozoan diseases e.g. collibacillosis, coccidiosis, etc. all these factors negatively affect the growth of the calves during the wet season. This agreed with the report of LRSP, 1999. The lack of environmental modification impacts negatively on growth pattern of calves. During the dry season, grazing ruminants often show signs of distress with only short periods of grazing from mid morning to late afternoon. The restricted grazing is usually attributed to the direct effects of temperature and solar radiation on the animal. But this is not necessarily the case as climate forage interactions also contribute to animal distress (Anonymous, 1981).

High ambient temperatures ushered in rapid maturation of forages leading to a rise in cell wall content. The particular parts affected were the stems and leaves of the pasture grasses. There was a direct relationship between plant maturation and temperature rises, as such, forages matured fast and led to increase in cell wall content. The increase in cell wall content decreased the digestibility of the cell wall. As the ambient temperature rose, the digestibility of the dry matter of the forage decreased due to a rise in the cell wall content and decrease in digestibility of the cell wall (Minson and McLeod, 1970 in Anonymous 1981). On the other hand, high light intensity led to increase in the content of water soluble carbohydrates, whereas high temperatures decreased water soluble carbohydrates (Anonymous, 1981). This therefore meant there was low fodder availability, poor nutritional value of the fodder. The dams during dry season were deprived of requisite nutritional needs to meet the functional roles of maintenance and reproduction (Mc Donald *et al.*, 2002).

3.4 Effects of location on the weights of calves on-farm

Generally, the study found that there was no significant differences between body weight of calves among the different district (Kiruhura, Ibanda and Sembabule) (p>0.05) but there was significant relationship in the months 6 (p=0.001) and months 8 (p=0.0007). This could be explained by the fact that the three districts were located in the same agro ecological zone, and therefore, the, management husbandry related practices were similar basically extensive systems with minimum investment. The similar practices included grazing, deworming, spraying, etc. It was however noted that water was more available in Ibanda than in Kiruhura and Sembabule. In the dry season, animals in Kiruhura and Sembabule had to trek long distances in search for water for drinking. This finding agrees with study conducted by Nsubuga, 1996.

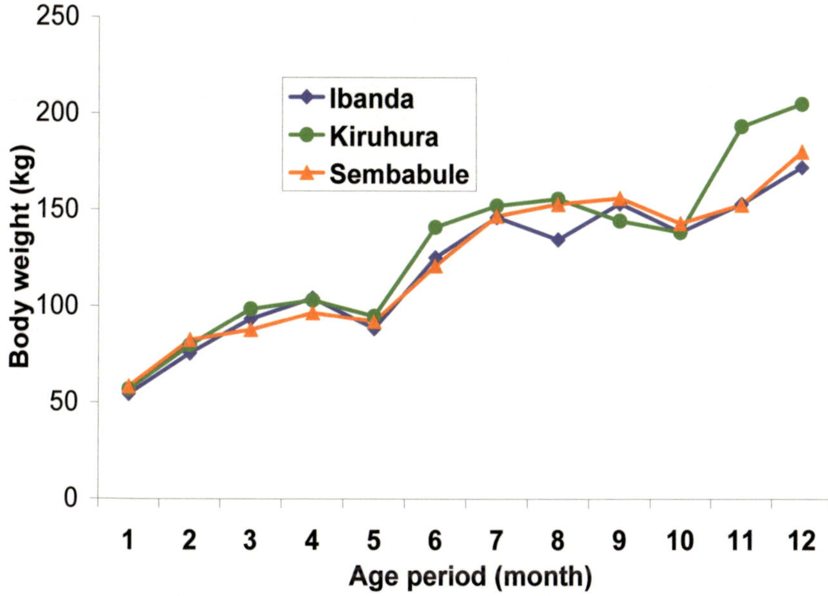

Fig. 7. Comparison of Growth patterns of cattle in Kiruhura, Ibanda and sembabule locations

3.5 Results of Analysis of Variance ann least square means

Source		Variance for live body weights (kg)									
	df	W1		W2		W3		W4		W5	
		MS	P	MS	P	MS	P	MS	P	MS	P
Breed	1	79.55	0.0453*	2339.10	0.0001***	275.29	0.2102	237.75	0.2959	17.01	0.7982
Sex	1	8.75	0.4973	43.64	0.3624	381.95	0.1406	1475.94	0.0102	0.79	0.0005***
Season	1	1029.00	0.0001***	5171.60	0.0001***	2856.87	0.0001***	2991.21	0.0003***	157.99	0.4362
Overall		354.26	0.0001***	2257.33	<.0001***	1144.84	0.0004***	1392.36	0.0005***	1101.84	0.0066**

Key: *=p<0.005, **=p <0.001, ***=p<0.0001, MS= Mean square

Table 1(a). Analysis of variance of On-station MBAZARDI Ankole and Friesian up to 5 months of age

Source		Variance for live body weights (kg)											
		W6		W7		W8		W9		W10		W11	
	Df	MS	P	MS	P	MS	P	MS	P	MS	P	MS	P
Breed	2	0.038	0.3723	5419.67	0.0003***	2527.08	0.0118***	12810.69	<.0001***	46552.13	0.0001***	32136.75	0.0001***
Sex	1	1889.28	0.0081	2309.09	0.0171	354.40	0.3410	4048.35	0.0178	8473.26	0.0017***	6430.67	0.0004***
Season	1	281.75	0.3016	791.83	0.1604	2626.85	0.0102	2037.26	0.0913	530.25	0.4267	449.60	0.3376
Overall		755.78	0.0378	3395.64	<.0001***	2639.55	0.0003***	5081.26	0.0001***	17461.11	<.0001***	13080.93	<.0001***

Table 1(b). Analysis of variance of On-station MBAZARDI Ankole and Friesian Cattle from 6-11 months of age

Source	Variance for live body weights (kg)												
		W12		W13		W14		W15		W16		W17	
	Df	MS	P	MS	P	MS	P	MS	P	MS	P	MS	P
Breed	2	48504.95	<.0001***	18953.47	<.0001***	35975.63	<.0001***	58680.75	<.0001***	27175.23	<.0001***		
Sex	1	6166.57	0.0003***	3594.66	0.0066***	5943.41	0.0002***	6781.15	0.0025***	2353.32	0.0867	3067.20	0.0778
Season	1	6166.57	0.5615	1366.74	0.0902	3240.59	0.0058***	6365.49	0.0034***	3024.48	0.0529	563.33	0.4436
Overall		18118.15	<.0001**	8791.74	<.0001***	13195.86	<.0001***	28507.09	<.0001***	13967.80	<.0001***	1557.70	0.2029

Table 1(c). Analysis of variance of On-station MBAZARDI Ankole and Friesian Cattle from 12-17 months of age

Source	Variance for live body weights (kg)												
		W18		W19		W20		W21		W22		W23	
	Df	MS	P	MS	P	MS	P	MS	P	MS	P	MS	P
Breed	2												
Sex	1	6.67	0.9214	1410.67	0.2397	29040.00	0.0011***	13005.00	0.0548	8836.00	0.0015***	13689.00	0.0057***
Season	1	308.17	0.5033	1.25	0.9718	3.33	0.9695	333.33	0.7484				
Overall		214.58	0.7298	975.65	0.3818	18859.29	0.0012***	9023.00	0.0790	8836.00	0.0015***	13689.00	0.0057***

Table 1(d). Analysis of variance of On-station MBAZARDI Ankole and Friesian Cattle from 18-23 months of age

Effect		Live body weights (kg)										
		W1	W2	W3	W4	W5	W6	W7	W8	W9	W10	W11
Overall			47.22	56.64	64.95	69.00	78.00	83.29	94.11	101.13	108.15	117.85
Breed	Ankole		42.25 ±2.24	38.41 ±2.23	63.69 ±2.67	64.12 ±2.57	77.02 ±2.57	81.21 ±2.42	102.01 ±2.63	106.57 ±2.48	119.17 ±3.34	142.49 ±4.08
	Friesian		47.25 ±0.71	53.47 ±1.09	60.03 ±1.83	67.29 ±1.86	77.02 ±1.62	83.79 ±1.56	89.81 ±1.92	98.01 ±2.15	100.55 ±2.71	104.75 ±2.81
Sex	Female		44.25 ±1.54	46.77 ±1.78	63.85 ±2.33	69.46 ±2.13	82.34 ±1.93	86.01 ±1.89	99.60 ±2.23	103.83 ±2.31	114.98 ±2.99	131.26 ±3.43
	Male		45.25 ±1.14	45.1 ±1.44	59.87 ±2.11	61.94 ±2.26	72.47 ±2.18	78.99 ±1.99	92.22 ±2.22	100.75 ±2.24	104.74 ±3.02	115.98 ±3.39
Season	Dry		39.50 ±1.55	34.74 ±2.23	55.17 ±3.029	59.72 ±2.76	76.33 ±2.11	81.12 ±1.99	98.14 ±2.17	106.67 ±2.38	106.16 ±3.02	121.66 ±3.15
	Wet		50.00 ±1.10	57.14 ±1.09	68.56 ±1.59	71.69 ±1.75	78.48 ±1.99	83.87 ±1.92	93.68 ±2.34	97.92 ±2.27	113.56 ±3.06	125.59 ±3.74

Table 2(a). Least Square Means (LSM±STDERR) for live body weights of On-station MBAZARDI Ankole and Friesian Cattle

Effect		Live body weights (kg)								
		W12	W13	W14	W15	W16	W17	W18	W19	W20
Over--all		120.28	125.39	129.23	141.21	161.38	182.23	207.50	224.67	235.71
Breed	Anko-le	142.19 ±3.36	149.56 ±3.04	143.36 ±3.26	156.77 ±3.05	183.32 ±4.34	193.25 ±4.33			
	Friesi-an	108.89 ±2.27	110.17 ±2.31	116.52 ±2.82	116.28 ±3.31	127.34 ±4.50	140.07 ±7.44			
Sex	Fem-ale	132.62 ±2.89	137.09 ±2.90	136.04 ±3.40	144.82 ±3.20	164.63 ±4.66	172.96 ±5.33	220.58 ±8.87	222.92 ±7.87	244.58 ±10.71
	Male	118.47 ±2.69	122.64 ±2.56	123.83 ±2.79	128.23 ±3.19	146.04 ±4.07	160.36 ±5.62	201.67 ±5.61	223.92 ±5.31	229.25 ±7.01
Sea-son	Dry	123.55 ±2.46	128.68 ±2.38	133.84 ±2.60	143.24 ±2.43	165.21 ±3.66	174.18 ±5.55	206.79 ±5.25	227.00 ±5.04	236.67 ±6.39
	Wet	127.54 ±3.26	131.05 ±3.16	126.03 ±3.65	129.81 ±4.06	145.46 ±5.33	159.14 ±5.66	215.46 ±9.51	219.83 ±8.39	237.17 ±11.80

Table 2(b). Least Square Means (LSM ±STDERR) for live body weights of On-station MBAZARDI Ankole and Friesian Cattle

Effect		Live body weights (kg)			
		W21	W22	W23	W24
Overall		240.57	235.40	277.60	298.80
Breed	Ankole				
	Friesian				
Sex	Female	268.50 ±16.77	263.00 ±23.52	287.00 ±5.84	310.50 ±8.58
	Male	202.50 ±12.99	212.00 ±15.39	240.00 ±11.67	252.00 ±17.16
Season	Dry	236.00 ±9.19	242.50 ±12.57	263.50 ±6.53	281.25 ±9.59
	Wet	235.00 ±23.12	232.50 ±28.10		

On-farm growth data

Table 2(c). Least Square Means (LSM ±STDERR) for live body weights of On-station MBAZARDI Ankole and Friesian Cattle

Sour-ce	Variance for live body weights (kg)												
		W1		W2		W3		W4		W5		W6	
	Df	MS	P	MS	P	MS	P	MS	P	MS	P	MS	P
Breed	3	2276.64	<.0001***	1994.67	0.0003***	6595.22	<.0001***	19748.96	<.0001***	130.52	0.7571	1447.86	0.0221**
Sex	1	38.14	0.5969	425.06	0.2236	498.43	0.1838	315.32	0.3924	49.34	0.7459	1.07	0.9606
Season	1	.	.	.	.	10.79	0.8441	3337.52	0.0059***	1062.89	0.1341	5964.76	0.0003
Distr-ict	3	156.65	0.3181	338.84	0.3069	426.42	0.2212	651.41	0.2220	401.14	0.4267	4726.64	<.0001***
Over-all		52.01		64.49		74.94		82.23		87.68		101.56	

Table 3(a). Analysis of variance of On-farm Kiruhura, Ibanda and Sembabule Ankole, Boran and Friesian crosses up to 6 months of age

Source	Variance for live body weights (kg)												
		W7		W8		W9		W10		W11		W12	
	Df	MS	P	MS	P	MS	P	MS	P	MS	P	MS	P
Breed	3	43394.75	<.0001***	18738.28	<.0001***	5941.96	0.0123**	3739.08	0.0140	249.60	0.6822	344.70	0.7232
Sex	1	67.89	0.7715	6.69	0.9257	1718.59	0.2481	446.099	0.4987	202.58	0.7122	2870.00	0.1073
Season	1	1604.24	0.1594	268.49	0.5548	2258.80	0.1861	1322.24	0.2462	2078.51	0.2411	686.45	0.4248
Distri-ct	3	524.30	0.5217	5809.68	0.0007***	514.69	0.6677	158.45	0.8484	2796.13	0.1622	3032.60	0.0687
Over-all		120.12		124.24		136.39		145.88		157.20		170.09	

Table 3(b). Analysis of variance of On-farm Kiruhura, Ibanda and Sembabule Ankole, Boran and Friesian crosses from 7-12 months of age

Effect	Live body weights (kg)								
		W1	W2	W3	W4	W5	W6	W7	W8
Over-all		52.01	64.49	74.94	82.23	87.68	101.56	120.12	124.24
Breed	Ankole	54.97±2.69	85.77 ±6.89	118.75 ±9.85	87.79 ±6.13	95.94 ±9.90	132.33 ±8.92	143.66±8.69	137.74±10.72
	Boran cross	54.09±3.66	61.36 ± 4.35	68.35 ±9.12	91.04 ±7.44	90.56 ±5.83	113.11 ±8.21	105.49±10.93	116.85 ±7.67
	Boran	69.38±2.94	110.01 ±17.19	119.51 ±8.47	150.23 ±6.05		156.71 ±22.11	233.66 ±9.27	215.16±11.17
	Friesian cross	47.86±1.10	59.78 ±2.63	66.15 ±4.65	76.14 ± 6.05	88.64 ±2.87	114.12 ±5.19	110.48 ±4.12	120.90 ±3.21
Sex	Female	57.10 ±1.53	82.04 ±4.87	96.95 ± 5.88	102.97 ±3.05	92.45 ±4.15	129.21 ±7.87	149.33 ±4.76	147.41 ±4.65
	Male	56.05 ±2.04	76.42 ±6.30	89.43 ± 6.95	99.63 ±4.42	90.97 ±5.27	128.92 ±9.24	147.32 ±7.32	147.92 ±6.33
Sea-son	Dry			92.19 ±9.78	107.24 ±4.60	95.34 ±5.29	147.87 ±11.94	153.18 ±7.13	149.25 ±5.94
	Wet	56.58 ±1.50	79.23 ±5.14	94.19 ±4.75	95.36 ±3.03	88.09 ±4.27	110.27 ±6.31	143.46 ±5.02	146.08 ±5.02
Distr-icts	Ibanda	54.48±2.49	75.57 ±6.09	93.36 ±7.19	104.21 ±4.14	88.34 ±5.33	125.22 ±9.31	146.09 ±6.62	134.48 ±6.19
	Kiruhura	56.86±1.86	79.72 ±6.16	98.53 ±6.67	103.16 ±4.27	94.82 ±5.35	140.96 ±8.76	152.09 ±6.46	155.60 ±6.35
	Sembabule	58.39±1.67	82.41 ±4.99	87.68 ±6.12	96.54 ±4.17	91.98 ±4.27	121.02 ±7.75	146.77 ±6.07	152.91 ±5.29

Table 4(a). Least Square Means (LSM ±STDERR) for live body weights of On farmKiruhura, Ibanda and Sembabule Ankole, Boran and Friesian crosses

Effect	Live body weights (kg)				
		W9	W10	W11	W12
Overall		136.39	145.88	157.20	170.09
Breed	Ankole	183.23 ±18.07	175.09 ±16.39		197.58 ±26.92
	Boran cross	141.27 ±14.34	101.89 ±19.07	171.90 ±25.35	171.18 ±25.36
	Boran		144.89 ±32.93		
	Friesian cross	128.89 ±6.39	145.68 ±10.44	160.69 ±12.20	188.42 ±8.86
Sex	Female	157.72 ±9.38	136.22 ±10.75	169.83 ±14.28	172.32 ±13.03
	Male	111.55 ±12.48	147.57 ±19.53	162.77 ±19.97	199.12 ±18.48
Season	Dry	158.75 ±13.03	133.19 ±17.53	178.69 ±20.64	181.05 ±14.55
	Wet	143.52 ±8.66	150.59 ±12.69	153.89 ±14.61	190.39 ±15.33
Districts	Ibanda	153.22 ±13.51	138.32 ±15.62	152.99 ±17.05	172.06 ±17.05
	Kiruhura	144.25 ±13.25	138.32 ±14.78	193.29 ±24.31	204.89 ±19.62
	Sembabule	155.94 ±8.50	142.97 ±15.27	152.62 ±18.55	180.22 ±13.72

Table 4(b). Least Square Means (LSM ±STDERR) for live body weights of On farm Kiruhura, Ibanda and Sembabule Ankole, Boran and Friesian crosses

4. Conclusion and recommendations

The study concludes that the factors that affect growth performance among on-station and on-farm cattle breeds include: breed, sex of the animal and seasonal variation. The study revealed that twinning in cattle was a rare occurrence. The study demonstrated that Boran cattle performed much better than the rest of other breeds in terms of growth followed by Ankole, Friesian cross and Boran cross respectively. It was found that the performance of cattle breeds did not vary significantly ($p>0.005$) among the different geographical areas of Mbarara, Kiruhura, Ibanda and Sembabule except months six ($p<005$) and months eight ($p<0.05$). The genotypes of the breeds were very important in performance evaluations. It was however noted that, husbandry practices related to feeding, deworming, spraying, mineral supplementation and other disease control measures were paramount in promoting the full potential of the breeds. Other important factors to consider included environmental conditions e.g. temperature, humidity etc. The study recommended genetic characterisation of the cattle breeds. Guidelines for appropriate Ankole and Friesian breeding and management practices developed and disseminated.

5. Acknowledgement

The following are acknowledged for their contributions in getting this work done. National Agriculture Research Organization (NARO) for funding this field based research activities. The farmers contributed immensely by availing cattle to be used in the research study and willingly accepting to cooperate during the data gathering process which most times was strenuous. Extension workers from Kiruhura, Ibanda and Sembabule districts are applauded.

6. References

Anonymous (1981). Dairy cattle. Retrieved from http://www.nap.edu/openbook.php?isbn

Asimwe L and Kifaro G C 2007. Effect of breed, season, year and parity on reproductive performance of dairy cattle under smallholder production system in Bukoba district, Tanzania. Livestock Research for Rural Development. Volume 19, article #152 Retrieved June 30, 2010 from http://www.lrrd.org/lrrd19/10/asim1952.htm

Fisher, A.D., Knight, T.W., Cosgrove, G.P., Death, A.F., Anderson, C.B., Duganzich, D.M., Matthews, L.R (2001). Effects of surgical or banding castration on stress responses and behaviour of bulls. *The Journal of the Australian Veterinary Association.* Volume 79, Issue 4 , Pages279 - 284 2001

Katyega P M 1998.Performance of Jerseys on the slopes of Mount Meru. World Animal Review 65: 24-30

Kifaro G C 1984. Production efficiency of *Bos taurus* cattle in Mbeya region. Msc. Thesis . Sokoine University of Agriculture. Morogoro, Tanzania. pp 52-95.

Kiwuwa G H, Trail J C M, Kurtu M Y, Worku G, Anderson F M and Durkin J 1983. Crossbred dairy cattle productivity in Arsi region, Ethiopia (Research report No. 11, ILCA, Addis Ababa, Ethiopia). 1-29.

Krupa, E., Oravcová, M., Polák, P., Huba, J.,Krupová., (2005). Factors affecting growth traits of beef cattle breeds raised in Slovakia. *Czech J. Anim.Sci.*,50, 2005 (1): 14-21

Kugonza, D.R., Mwai, O.A., *et al.* 2005. Ankole Cattle Breed of Uganda: Functions and Criteria Used in Identification, Selection and Parentage Assignment by Herdsmen. *Rare Breeds International (RBI) Conservation of Domestic Animal Genetic Resources Global Conference Proceedings* 6:18 (Abstract).

Mc Donald P, Edward R A, Greenhalgh J F D and Morgan C A 2002 Animal Nutrition. 6th Edition. Pearson Education Limited, Edinburgh Gate. Harlow 693 pp

Moran J 2002 . Calf rearing. A practical guide. 2nd edition. CSIRO publishing. Collingwood Victoria. 228 pp

Ndumu, D., Baumung, R., *et al.* 2006. Performance and fitness traits versus phenotypic appearance: a novel approach to identify selection criteria for indigenous breeds. *Deutscher Tropentag, Conference on International Agricultural Research for Development*, Bonn-University of Bonn, October 11-13, 2006. www.tropentag.de/2006/abstracts/ full/151.pdf

Ndumu, D., Baumung, R., *et al.* 2008. Performance and fitness traits versus phenotypic appearance: A novel approach to identify selection criteria for indigenous breeds. *Livestock Science* 113: 234–24.

Nsubuga, H.S.K. 1996. Small Ruminants: Goats and sheep in Uganda. In: Lebbie, H.B. and Kagwini,E.(Eds).*Small Ruminant Research Development in Africa.* Proceedings of the third biennial Conference of the African small ruminant research network.UICC, Kampala,Uganda 5-9 December 1994. Pp. 1-4.

Oluka, J (2006). On-farm Comparative Perfomance evaluation of selected indigenous, Exotic and Crossbred Goat Populations in Uganda. PhD thesis.The Royal Veterinary and Agricultural University Copenhagen, Denmark.

Petersen, P.H., Ndumu, D.B., *et al.* 2003. Characteristics of Ankole Longhorn cattle and their production environments in South Western Uganda: milk offtake and body measurements. *Animal Genetic Resources Information* 34: 1–9.

Said R, Bryant M J and Msechu J K K (2001). Growth and survival of crossbred beef cattle in Tanzania. *Tanzania society for Animal production* proceedings, volume 28, 2001.

Tuah A K and Yaa Nyamaa Danso 1985. Preliminary studies on the performance and productivity indices of N'Dama and West African shorthorn cattle in Ghana. *Tropical Animal Health production*, 17, 114- 120.

Twinamasiko, N.I. 2001. Dairy Production. In: *Agriculture in Uganda.* Livestock and Fisheries. Mukiibi, J.K. (ed). Fountain Publishers / CTA / NARO 4: 18–42.

Vial, V. E. and More O'Ferrall G. J. (1965). The Effects of Breed, Sex, Season and Age of Dam on the Birth Weight of Dairy and Crossbred Calves and their Growth during the First Seven Days of Age *Irish Journal of Agricultural Research*, Vol. 4, No. 2 (Oct., 1965), pp. 223-230

Wurzinger, M., Ndumu, D., *et al.* 2007. Comparison of production systems and selection criteria of Ankole cattle by breeders in Burundi, Rwanda, Tanzania and Uganda. *Tropical Animal Health and Production* 38: 571–581.

18

Clinical Approach to the Repeat Breeder Cow Syndrome

Carlos C. Perez-Marin, Laura Molina Moreno
and Guillermo Vizuete Calero
University of Cordoba,
Spain

1. Introduction

1.1 Importance of RBC in dairy cattle

Dairy cattle production requires great intensification, which has been demonstrated to affect negatively on the reproduction. One calf by cow and year is the reproductive objective in these animals. It means that cows must get pregnant after AI, maintain the pregnancy, have parturition after 270 days approx., and wait for a period of 40-50 days to be successfully inseminated again. Nevertheless, this is not always attained and cows must be re-inseminated during several consecutive cycles. In this context appears the Repeat Breeder Cow syndrome (RBC), comprising a heterogeneous group of subfertile cows, without anatomical abnormalities nor infections, that exhibit a variety of reproductive disturbances in a consistent pattern during the course of 3 or more consecutive estrous cycles of normal duration (17-25 days). Any of these disturbances may affect the delicate interplay of estrous behavior, hormone patterns, and ovarian dynamics, which in synchrony with the uterine functions finally determines the outcome of mating or artificial insemination (AI) (Bage et al., 2002). Epidemiological studies of RBC prevalence have shown disparate results, ranging from a 5% described by Ayalon (1984) in Jordan, to a 36% observed by Zambrano et al. (1982) in Cuba. Nevertheless, considering the great demands on dairy cow production (which ideally requires obtaining a calf per cow per year), RBC has an important impact on dairy cattle economics (Figure 1).

1.2 Importance of RBC in beef cattle

Beef cattle reared under extensive system usually have a deficient data record, and then it is difficult to identify this syndrome. In this case cows are considered as RBC when they do not get pregnant after being exposed to bulls during the established reproductive periods. The bull-cow interaction usually ranges from 45 to 90 days, although sometimes it lasts 6 months or it may also be permanent. Few reports on RBC exist in beef cattle (Maurer & Echternkamp, 1985; Ferreira et al., 2008).

In contrast with dairy cows, to obtain a high reproductive efficiency in beef cattle, it is necessary to get a calving interval of 365 days and an overall calving rate of 95% (Duffy et al., 2004). The cow should get pregnant around day 85 after parturition. But how many cows really get this objective? Studies carried out in US indicate that only between 50-70% of cows reach this goal. In Spain, the production system presents notable differences with respect to reproductive indexes. Under intensive farming, beef cows usually calve for the first time at 28 months old, while in dairy cows the first parturition occurs at 26 months old. In beef cattle exists a tendency to reduce the calving interval, in contrast with dairy cows.

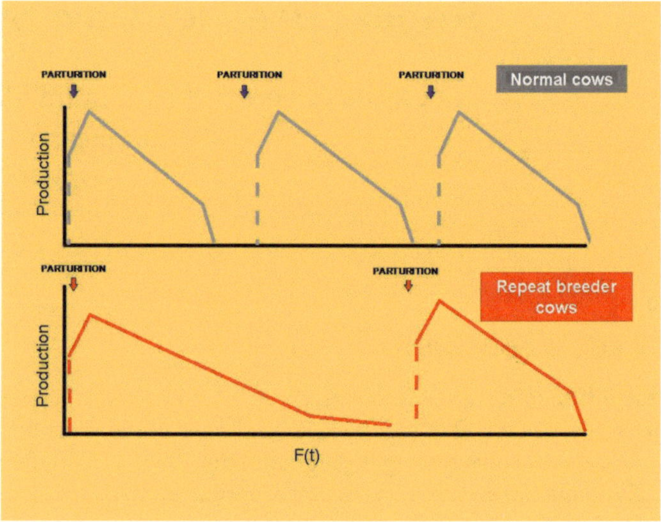

Fig. 1. Evolution of the lactation curves in normal and repeat breeder cows. Note that RBCs have fewer calves per cow and therefore milk yield is also reduced.

The preovulatory follicle size is correlated with fertility in beef cattle, and it has been postulated that the implementation of management to optimize the size of the ovulatory follicle could improve the fertility (Perry & Perry, 2008). For getting good reproductive results it is recommended to select those animals that:

- calve one alive and health calf per year.
- wean a calf that will reach 60% of the cow body weight at 7 months old.
- get pregnant after weaning.
- have a good maternal condition.

Then, the reproductive efficiency can be measured throughout the number of weaned calves/number of cows in reproduction x 100. This percentage ranges between 60-70%. The majority of cows that failure to get pregnant (30-40% of cows in reproduction) has infertility problems during the breeding season, and many of these cows are defined as RBC. This syndrome should be treated in beef cattle farms, since expenses may rise due to the decrease of the number of calves per season and a prolonged breeding season. If beef cows are mated for a long time (for a long season), a low production will be obtained: calves will be less heavy and then its price will be lower, the health management of different ages calves will be more difficult, and the calving interval will be longer. A higher conception rate means that a higher number of calves will be weaned, and then, a higher beef yield is obtained. The costs are the same in pregnant or open cows, but profits are different. It has been reported that dystocia provokes higher economical expenses associated with perinatal mortality of calves, veterinary costs and worse reproductive indexes (longer parturition-estrus interval or lower conception rate).

2. Aethiology of RBC syndrome

The aethiology of RBC syndrome is unclear and multifactorial. The cow, the bull and several environmental/handling factors are incriminated. All of them are often overlapped and it is difficult to determine the primary origin.

2.1 Maternal factors involved in RBC syndrome

Causes provoking repeat estrus in cattle are often related to maternal defects, which used to be at individual-level and make more difficult the clinical assessment and diagnosis. Age, genetic defects, genital tract infections, conformational defects, hormonal disorders, embryo mortality and nutritional defects have been reported. But usually there is no clear aethiology and several concomitant causes often appear, what makes difficult to characterize the problem.

2.1.1 Influence of maternal age

It has been widely reported that age impacts negatively on fertility (Hodel et al., 1995), and higher RBC rates have been described in old cows (Hewett, 1968). It is attributed to alterations in hypothalamic or pituitary hormonal levels or to inability of the ovary response (Bullman & Lamming, 1978). It has also been demonstrated the relationship between old age and low oocytes viability, what explains the fertility decline (Lanman, 1968). Studies in RBCs showed that age and breed affect the FSH and LH levels. FSH was higher in cows with six or more lactations (1.03 ± 0.12 ng/ml) and LH ranged from 1.31 ± 0.21 ng/ml in heifers and 2.19 ± 0.28 ng/ml in cows after three calves, falling to 0.94 ± 0 25 ng/ml in six or more calves (Santana et al. 2000). It is supported that fertility in dairy cattle improves after the 1st or 2nd parturition, and decreases from the 4th or 5th, but it should be taken into account the time required for uterine involution or problems associated to puerperium (Dominguez, 1989).

2.1.2 Genetic factors in repeat breeder cows

Individuals inherit their parent´s genetic merit, and then chromosomal or genetic abnormalities of parent, or those that occurred during the differentiation process may compromise fertility. RBC syndrome has been described in cows with chromosomal abnormalities as translocation 1/29 or trisomy X (Roberts, 1971; Lafi & Kaneene, 1988; Bruyas et al., 1993). A study regarding Robertsonian translocation 1/29 in cattle (Rodriguez et al., 2000) informed that cows diagnosed with this chromosomal defect were classified as RBC. Few embryos (less than 10%) exhibit chromosomal or genetic abnormalities, and are usually associated with high inbreeding or aged gametes. Humblot (1986) observed higher RBC rate in Holstein and Charolais breeds than in Norman or Frisian breeds, considering breed as a risk factor for this syndrome.

The utilization of bulls to produce bigger calves could be considered as a genetic- or management-based problem when dystocia or postpartum disease appear, and metritis, infertility or repeat estrus are developed.

2.1.3 Uterine infection and repeat estrous cycles

The uterine environment promotes the normal embryonic development. So, any disorder compromises the survival of the embryo and induces the RBC syndrome. Some authors have observed a correlation between repeat estrus and endometrial abnormalities (Roine & Saloniemi, 1971; Francos, 1979; Santana et al., 1998b). In fact, reproductive failure appears after metritis, and Francos (1979) observed that from 3.5 to 5.7% of cows with metritis had repeat estrus. Arguably, uterine infections (specific and nonspecific) will adversely affect the reproductive indexes by enlargement of the uterine and cervical postpartum involution, by alteration of follicular development (Lewis, 1997), and by increased embryo mortality and repeat estrus rates (Santana et al., 1998).

Subclinical endometritis should be considered when pregnancy failure or repeat estrus is observed. However, clinical signs are difficult to detect; it is not easy to do the diagnoses by rectal palpation and the bacteriological analysis of cervical mucus does not reflect the endometrium status. Endometrial biopsies and uterine microbiological culture can enhance the diagnosis. Leukocyte infiltration in moderate degree with lymphocytes, neutrophils, plasmocits, eosinophils and macrophages are histopathological findings that have been reported in the endometrium of RBCs

Many authors have identified the germs located in the reproductive tract of RBCs:

- Sagartz y Hardenbrook (1971): Stafilococos, Corinebacterium.
- Hartigan et al. (1972): Stafilococos aureus, Gram+, Streptococos β haemolitic, Streptococos microaerofiles.
- Murthy et al. (1974): Pseudomonas, Aerobacter sp, Klebsiella sp, Paracolobactrum sp, Proteus sp, Micrococos sp, Stafilococos sp, Corinebacterium sp, Bacillus sp.
- Palangala et al. (1978): Streptococos sp, Escherichia coli, Bacillus sp, Corynebacterium sp; others less frequent as Proteus sp, Klebsiella sp, Pasteurella sp, Neiseria sp, Branhamella sp, Actinobacter sp, Haemophilus sp, Kurthia sp.
- Harvey (1993): Nocardia
- Singla et al., (1993): Stafilococos aureus, Escherichia coli, Klebsiella pneumoniae
- Vasconcelos et al. (1995): Streptococos sp, Enterobacterias, Stafilococos, Yeast.
- Santana et al. (1998a): Escherichia coli, Stafilococos non-haemolitc sp, Acinetobacter sp, Streptococos β-haemolitic, Enterobacter cloacae.

Barbato et al. (1994) studied histopathological changes in repeat breeder zebus, describing focal chronic endometritis, adenomyosis, endometrial hyperplasia, glandular hyperplasia of the "rete ovarii", hyperplasia of the ovarian serous membrane, oophoritis and granulosa cells tumour. It was concluded that uterine infections are the main causes of RBC in zebus.

2.1.4 Anatomical defects of the genital tract

The reproductive tract of cow provides a suitable environment for oocyte growth, as well as for sperm transport, fertilization and implantation. Uterus is a suitable habitat for the embryonic and fetal development. A complex communication between hormones, proteins, etc. will be necessary for obtaining reproductive success. Anatomical or functional changes of these structures can drive to gestational failure and infertility. Therefore, it is essential to carry out a proper reproductive assessment for discarding animals with congenital or acquired defects.

Oviductal abnormalities, that complicate and frequently inhibit the reproduction, are present in 6-15% of adult cows and can reach up to 80% in those with a history of infertility or repeat breeding. Adhesions between the ovary, fallopian tubes or ovarian bursa, unilateral or bilateral obstructions, moderate degree of hydrosalpinx and inflammation (perisalpingitis, peritonitis) have been described in RBC syndrome (Figure 2). López-Gatius (1995), based on previous studies, proposes the use of intraperitoneal insemination in RBC to avoid the detrimental effect of uterine diseases on the sperm transit.

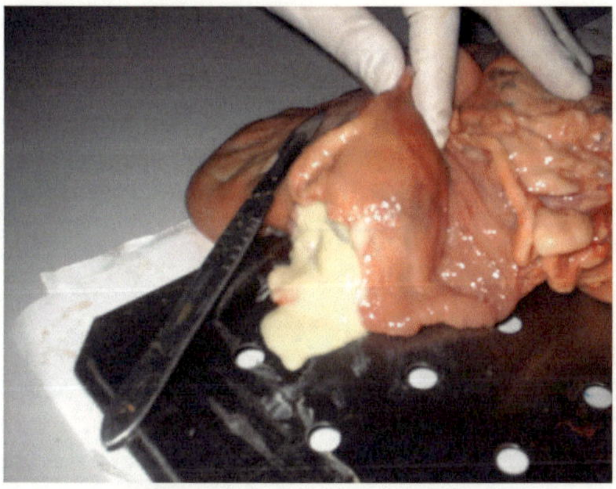

Fig. 2. Piosalpingitis observed at slaughterhouse.

Acquired uterine alterations, as metritis, are critical to the resumption of the normal cyclicity during postpartum period, provoking RBCs (Shresta et al., 2004). Other non-infectious abnormalities, as uterine degeneration and neoplasia, could also be involved in this syndrome, although their incidence is low.

Cervix is a defensive barrier and a sperm reservoir, and may undergo structural changes associated with inflammation. Cervical traumatic stenosis and obstruction, prolapse of cervical rings, adhesions or functional incompetence can be detected associated with RBCs.

Vagina acts as receptacle for semen and is one of the uterine defensive barrier. Infectious disorders alter vaginal pH and bacterial flora, allowing the infection and reducing the sperm vitality. Congenital anomalies, conformation defects (urovagina and pneumovagina) and infections (vaginitis) could be diagnosed. Vulvitis or vestibulitis may also modify the normal reproductive function.

2.1.5 Hormonal dysfunctions

Following are showed some alterations that affect the reproductive hormonal function and produce RBCs, although their diagnosis is usually difficult and uncertain. Perez-Marín & España (2007) reported hormonal dysfunctions found in RBCs in Spanish farms (Table 1).

Hypofunctional CL provokes a decrease of progesterone and affects negatively the fertility. Perez-Marín & España (2007) reported that hypofunctional CL was diagnosed in a 17,3% of RBCs, and Gonzalez-Stagnaro et al. (1993) affirm that this pathology appears in 23,7% of cows. CLs are small and poorly developed, with low progesterone production and LH peak asynchrony. Therefore, inadequate uterine environment is formed and this increases the abnormalities and the loss of embryos. As Kimura et al. (1987) suggested, a delayed increase in progesterone levels may indicate late or insufficient corpus luteum formation following ovulation or short luteal phase. Shelton (1997) argues that luteal inadequacy, due to a diminished response to circulating luteotrophic hormones, may contribute to embryo mortality in subfertile cows. A delayed and diminished post-ovulation progesterone curve has been associated with low conception rates in cattle, and a low progesterone curve has been shown to be related to significantly reduced production of interferon-tau by bovine embryos recovered on Day 16 of pregnancy (Figure 3).

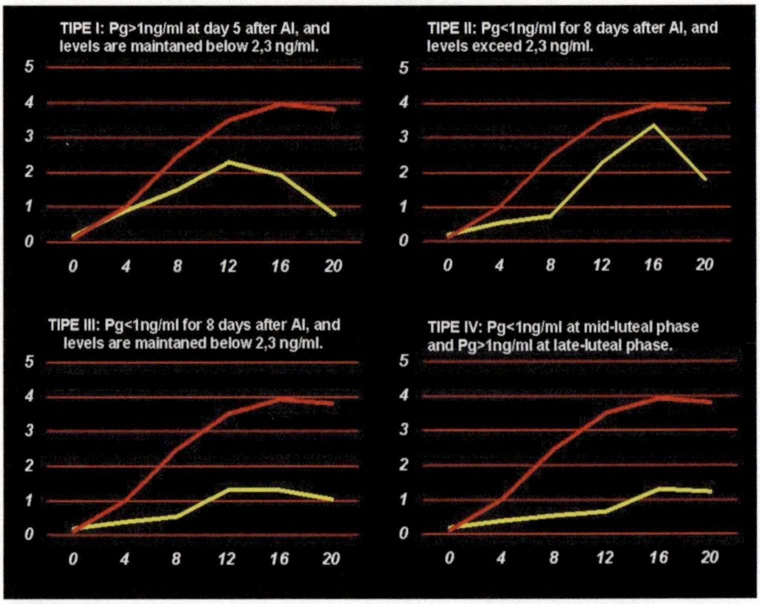

Fig. 3. Luteal phase deficiency in cow. Red line indicates normal progesterone levels, and yellow line indicates the progesterone level in cows with luteal hypofunction.

Suprabasal progesterone level around estrus has been described in RBCs (Duchens et al., 1995; Bage et al., 1997; Perez-Marín & España, 2007). It is associated with low gonadotrophin levels and with incomplete luteal regression after luteolysis, which prolongs the follicular growth and damages the oocyte. A longer estrus-ovulation interval usually appears and premature insemination is then carried out. Duchens et al. (1995, 1996) have experimentally induced this pathology using retard release progesterone. It has been informed that these levels reduce the number of blastomeres around day 3, accelerate the zygote progress though the oviduct and affect negatively the fertility.

Anovulation has been reported in 2-16% of RBCs and it is characterized by prolonged basal progesterone after estrus (Gonzalez-Stagnaro et al., 1993; Perez-Marín & España, 2007). The LH release pattern is modified and follicle does not get the stimulus for ovulation to occur. The follicle continues growing and releases estradiol, which induces the formation of persistent follicles and delayed ovulation. Also, defective follicle recruitment during the middle- and late-luteal phases has been suggested as a cause of anovulation. An abnormal hormone environment may promote continuous development in the dominant follicle, impairing follicular function and oocyte quality, and thus reducing fertility (Odde, 1990; Stock & Fortune, 1993).

Irregular and delayed ovulations have been associated with asynchrony between estrus and ovulation (Duchens et al., 1994), asynchrony of LH peak and ovulation (Lee et al., 1983), or incapacity for LH release (Duchens et al., 1995). The LH peak characteristics are altered in RBCs and estrous signs are less intense than in normal cows. As described above, suprabasal progesterone levels can be involved.

RBC	Diagnosis	n	%	Progesterone on day 0	Pre-ovulatory follicle size
Pregnant		9	24,32	0,37 ± 0,20	1,79 ± 0,33
Reproductive failure					
	Prolonged luteal phase	3	8,11	0,34 ± 0,10	1,60 ± 0,31
	Ovarian cysts and anoestrus	2	5,41	0,18 ± 0,10	1,33 ± 0,04
	Luteal dysfunction type II	1	2,70	0,12	1,65
	AI mis-timmed	1	2,69	2,47	---
	Suprabasal levels of progesterone	1	2,71	1,78	3,2
	Late ovulation	2	5,41	0,53 ± 0,01	2,03 ± 0,25
	Anovulation	1	2,70	0,49	2,15
	Luteal dysfunction type II	3	8,11	0,24 ± 0,16	1,40 ± 0,26
	WITHOUT diagnosed aethiology	**14**	**37,84**	0,28 ± 0,23	1,92 ± 0,32

Table 1. Endocrine dysfunctions and reproductive management defects detected in RBC in Spain (Pérez-Marín & España, 2007).

2.1.6 Early Embryonic Death (EED) causes RBC syndrome

The bovine embryo releases a substance of trophoblastic origin (interferon tau) into the uterus around day 16-18 that prevents luteolysis (Sheldon, 1997; Kerbler et al., 1997) and maintains the luteal function and pregnancy. EED has been attributed to irregular LH and progesterone profiles that induce failures in the maintenance of CL (Swanson & Young, 1990). EED is associated with poor quality of gametes and zygotes, uterine alterations, hormonal imbalances and defects in the immunitary mechanisms (Bruyas et al., 1993). The EED occurs between days 8 and 16 post-mating (Diskin & Sreenan, 1980), before cow returns to estrus. As a result, no variation at interestrus interval is observed and clinicians cannot differentiate between embryonic resorption and other pregnancy failures. The incidence of EED is highly variable, from 10.6 to 39.7%.

Kimura et al. (1987) consider that EED and fertilization failure are recurrent causes of RBC, and genetic, environmental and endocrine influences are risk factors involved. De la Fuente et al. (1988) observed a lower embryo production in RBCs. In this study, FSH-p was administrated as superovulatory treatment and the good results suggested that non-ovarian factors (hormonal, uterine...) directly influence the reproductive failure in RBCs. Results obtained after embryo transfer in RBC suggest that uterine environment is the most important factor in the EED, reporting abnormal concentrations of ions and proteins in the uterus. Intrinsic embryo factors are also related with this syndrome. Linares (1982) observed that blastocysts collected in RBCs had more abnormalities than those of control cows (3:1), and the oocyte collection rate was also lower, possibly due to a degenerative process initiated after fertilization. It was determined that the reduction of embryo viability could be associated with the reduction of nutrients exchange capacity and other regulatory substances. They could disrupt the process of cell differentiation, initiate degenerative processes, and slightly reduce the number of cilliar cells in the endometrium. High and early endometrial progesterone receptors have been detected 3 days after estrus in RBC that suggest hormonal and cellular changes in the uterus (Almeida et al., 1987, 1995).

Abnormal embryonic development by hormonal asynchrony during estrus and metaestrus have been involved in the RBC syndrome. Inadequate levels of estrogen and abnormal interactions with gonadotrophins injure the oocyte maturation, resulting in abnormal embryonic development. The altered hormonal pattern in RBCs could cause the incompetence of oocytes that will suffer abnormal embryonic development (Gustafsson et al., 1986; Albihn et al., 1991). However, other authors consider that the majority of embryonic abnormalities occur during the way through the oviducts, although those abnormalities are not evident until 6-7 days post-AI (i.e. in blastocyst stage). As oocytes of RBCs are competent to reach the blastocyst stage and to continue the development, embryo production (IVM-IVF-IVC-ET) could improve the reproductive success in these animals.

EED has usually no effect on the normal length of the estrous cycle. In occasions, anamnesis in RBC shows normal estrous cycle and fertilization rate, and the "apparent infertility" is due to earlier embryo mortality (day 7). No significant differences have been detected in these animals regarding estrus duration, ovulation time and the incidence of anovulation or silent estrus (Linares et al., 1984).

2.1.7 Inadequate follicular growth

It has also been proposed that failure of pregnancy in RBCs could be associated to dysfunction in the recruitment of large follicles during the second half of the estrous cycle, resulting in absent or abnormal ovulation, or early cow-embryo asynchrony, occurring EED and RBC. The critical period in the follicular recruitment occurs as from 10 days before ovulation (Staigmiller & England, 1982; Thibier et al., 1985). Dominant follicle continues its growth when progesterone levels are subluteal (or suprabasal), in which case follicular function is compromised and oocyte quality is reduced, affecting negatively the fertility (Odde, 1990; Stock & Fortune, 1993).

2.1.8 Effect of nutrition on RBC syndrome

The importance of nutrition in all vital processes is indisputable, and the qualitative and quantitative differences in the ration in dairy cattle may cause reproductive dysfunctions (Roberts, 1971; Casida, 1961; Lagneau, 1981; Pedroso & Roller, 1996; Dovensky et al., 1996). Nutritional deficiency and increment of services per cow are linked, because the decrease in food intake, weight and body condition causes endocrine imbalances that affect fertility and other organs or systems. The mentioned nutritional deficiency can also affect the postpartum period, causing delay in the uterine involution and increasing the number of open days. McClure (1995) reported that reproductive disorders could occur at 3 levels: the synthesis and release of LH from hypothalamus, at ovarian function, or at ovulation, fertilization and development of the fertilized egg, embryo and fetus.

Homeostasis of nutritional elements can be maintained through correct animal diet (deficit or excess depends on the production status). Nutritional variations may be due to an excess or deficiency of certain elements or imbalances of their concentrations in diet, and provoke alterations in their absorption or utilization. Nutritional deficiencies are more significant in high production animals, as occur in dairy cattle. In the 70's, Payne et al. (1973) developed the theory of metabolic profiles, which aimed to monitor the metabolic status and health of the herd.

2.2 Bull factors involved in the RBC syndrome

Factors related to the bull and sperm quality must also be taken into account when RBCs are evaluated. It is essential to evaluate the sperm function in both natural mating and AI. Frozen semen straws should be carefully stored and managed. Repeat pregnancy failure could be linked to the mentioned bull factors, in which estrus is repeated and interestrus interval has normal duration. Some relevant aspects are mentioned below.

2.2.1 Influence of bull fertility and semen quality on repeat breeding

Optimal bull fertility (by natural breeding or AI) is necessary to achieve high pregnancy rate and normal calving interval. Semen doses for AI must contain at least 6 million of motile sperm after thawing, and fertility drops if sperm concentration is reduced (Foote & Parks, 1993). Currently, frozen semen doses are packed with 15-25 million of motile sperm pre-freezing, because around 50% of spermatozoa recover motility after thawing. However, despite semen doses may fulfill all requirements, there are donors of sperm with erratic fertility. Eid et al. (1994) found that embryos formed from high fertile bulls reached earlier the S phase of DNA synthesis and the 2-cell phase, showing high blastocyst viability. Embryos from low fertile bulls showed a longer G2 phase associated with sperm DNA damage or DNA replication failure during the S3 phase (Eid et al., 1995). The strict selection of semen donors based on the semen quality after thawing and sexual behavior, among other variables, has reduced defects in bulls used for the creation of frozen semen banks.

At natural breeding, it is necessary to assess the reproductive performance of the bulls at least twice a year, carrying out semen assessment (macroscopic and microscopic) and physical evaluations. It is also interesting to evaluate libido and the sexual behavioral pattern for mating to diagnose reproductive failures in bulls. On the other hand, it is essential to maintain an appropriate male:female ratio for mating. For guaranteeing optimal ejaculates, bulls should do 10 mating throughout a week.

2.2.2 Site of semen deposition and estrus return

Sperm that has just been ejaculated into the vagina by the bull or placed into the uterus by the inseminator should reach the oviducts, where fertilization occurs. Sperm begins its upward through the tubular tract of the female and reaches, within minutes, the fallopian tubes to find the oocyte. It has been reported that the utero-tubal isthmus acts as a reservoir of sperm, making these cells to go up gradually towards the ampulla, preventing polyspermy and ensuring that sperm finds the oocyte into the oviduct (Hunter, 2002). The interaction of sperm with the reproductive female tract results in a drastic decline of spermatozoa. Thus, the female reproductive tract may alter the mechanism of sperm transport and induces infertility. In addition, defective inseminations (e.g. deposition of semen at the entrance of the cervix) may provoke RBC and affect negatively the fertility rates. Higher conception rates are obtained when sperm is deposited into the uterus (Gwazdauskas et al., 1986), but the incidence of uterine disease in RBCs ranges between 36 and 89% (Ellington & Schlafer, 1993). It has been suggested that intraperitoneal insemination may be an alternative to uterine insemination (López-Gatius, 1995) in cows with uterine diseases.

2.2.3 Time of semen deposition

Time of semen deposition is closely related with the detection of estrus and with all those factors that may affect the preovulatory LH. Other cows or a bull mounting the cow are recognized as signs of the cow heat. This behavior could last between 13-17 h. on average, and ovulation will take place approximately 30 h. after starting the heat. It must be ensured that semen is deposited in the female tract as close as possible to that time. The AM/PM rule is the most common practice in dairy cattle farms: it consists of making insemination at morning if the cow was in heat the previous evening, or being inseminated in the afternoon if heat was detected in the morning. If the technician palpates ovaries before insemination and considers that the follicle texture is far to ovulation, double insemination is recommended separated by 12 hours. Similar pregnancy rates have been reported after one or two inseminations per day (Wahome et al., 1985). As noted above, semen is normally deposited into the uterus, and the uterine environment will be adequate depending on the stage of the estrous cycle. This is one of the most common problems in RBCs, in which any change of pH, endometrial stroma and glands, endometrial flora, etc. could affect the reproduction (Ohtani & Okuda, 1995).

2.3 Environmental and management factors affecting the RBC syndrome

These factors are very miscellaneous. Hewett (1968) suggests a correlation between RBC and the following variables: lactational stress, size of livestock (the number of cows with repeat estrus increases in bigger livestocks), milk yield or calving season (a greater number of RBCs are detected in autumn calves). Similarly, Gustafsson & Emmanuelson (2000) considered as risk factors for RBC syndrome the first lactation, dystocia or calving difficulties, the first insemination at winter, reproductive disorders before the first insemination, high milk yield or clinical history of repeat estrus. Therefore, environmental and animal management factors should be considered in order to reduce the incidence of RBC syndrome.

2.3.1 Season

Barlett et al. (1986) reported that there is no association between breeding season and RBC syndrome. However, ovarian follicular growth and development of the dominant follicle can be altered during the summer months, and heat stress exerts a depressant effect on the endocrine mechanisms (reduced intensity of estrus, decreased preovulatory LH peak, etc) and/or the embryo, reducing the fertility.

However, other authors (Gonzalez, 1981) observed that higher progesterone concentrations was present in fertile and anestrus females during warm periods, attributing it to increased activity of the CL or the adrenals by heat stress. Photoperiod length and temperature variations are linked to the season, and could influence the endocrine regulation of the estrous cycle. Other factors as body condition at calving, feeding level and calving-insemination interval are also season-related, and can be modified through management practices (Fulkerson & Dickens, 1985). GnRH-based treatments used in RBCs reveal a marked seasonal response, and it has been demonstrated that RBCs and infertility increase during summer (hot) months (BonDurant et al., 1991; Gonzalez-Stagnaro et al., 1993).

2.3.2 Estrus detection

The poor heat detection is one of the main causes of RBC and infertility in cattle (Lafi & Kaneene, 1988; Perez-Marin & España, 2007). Computer simulation models have been implemented to improve the estrus detection and it has been showed that they increase the frequency of heat detection and reduce the services/conception and days open (Bailie, 1982). Reproductive failure may be observed after mis-timed inseminations or in cows without evident estrus. Pregnant cows could show estrous signs and they could be inseminated. This intrauterine procedure could induce early or late embryonic death, or abortion in pregnant cows. Between 3 and 10% of pregnant cows can be mounted by other cows, and early pregnancy diagnosis reduces these mis-timed insemination. Some authors recommend making intracervical insemination if reproductive status of cows in heat is doubtful (Lopez-Gatius & Camón-Urgel, 1991; Bullman & Lamming, 1978).

2.3.3 Hygiene at artificial insemination and parturition

It is essential to act carefully and prevent infections at reproductive tract. During artificial insemination, the catheter is inserted into the uterus, and may act as a disseminator of pathogenic bacteria that alter the normal uterine environment and provoke subacute, acute or chronic infectious diseases, with harmful effects on reproduction.

On the other hand, is essential to care the peripartum period. In fact, many authors consider this period, and particularly the postpartum period, the most critical in the reproductive activity, observing that animals showing problems during this period are finally RBCs (Erb et al., 1985).

2.3.4 Stress

RBCs have a higher predisposition to be stressed (Bage, 2002), as evidenced by the increased capacity to produce cortisol from adrenal glands after induced stress. Adrenal progesterone in induced-stress cows reaches suprabasal levels. Thus, stress can be considered as a potential factor in the RBC syndrome.

Heat is one of the stressful causes in cattle. It has been shown that heat-stressed cows reduce the duration and intensity of estrus, and alter hormone levels by increasing circulating corticosteroids and by reducing the concentration of progesterone. Uterine environment is altered and viability of the pre-implanted embryo is reduced. It is recommended to use sprinklers, fans cooling, etc., to handle this stress (Ferreira et al., 2011).

3. Diagnosis of RBC syndrome

It is necessary to diagnose the aethiology of reproductive failures in cows having an apparently normal clinical history and then, reduce the economical impact. However, although many diagnostic tools are available, it is usually difficult to get a correct diagnostic because sometimes it is unprofitable.

First of all, a complete clinical history should be obtained at herd and individual level. Age, parity, milk yield, previous diseases, reproductive indexes, estrous cycles characteristics, insemination schedule, bulls, estrus detection, hormones, food and farm hygiene should be registered. Now, anatomy, morphology and function of cows should be inspected. The reproductive status of animals must be according to their production. Sexual behavior must be evaluated to detect disorders, as muscle or claw lameness. Similarly, it is necessary to examine the behavior of bull and bull-cow interactions when natural breeding is carried out. Vulva, vagina, cervix, uterus, fallopian tubes and ovaries must be evaluated to diagnoses reproductive defects.

3.1 External inspection and vaginal evaluation

External inspection can identify congenital or acquired anatomical defects as pneumovagina, vulvar defects, tumors or injuries. The anatomy of the area, secretions around vulva or tail, and vulvar and vaginal coloration should be evaluated.

In vaginal palpation, vulvar and vestibular functionality are evaluated though the pressure of these structures around the clinician's hand. It also assesses the presence of adhesions, abnormal structures and cervical defects. The vaginal contents should be inspected to detect urine (if urovagina), pus (if endometritis, vaginitis ...), blood (post-ovulation or some diseases) or clear and clean mucus (associated with heat) (Figure 4).

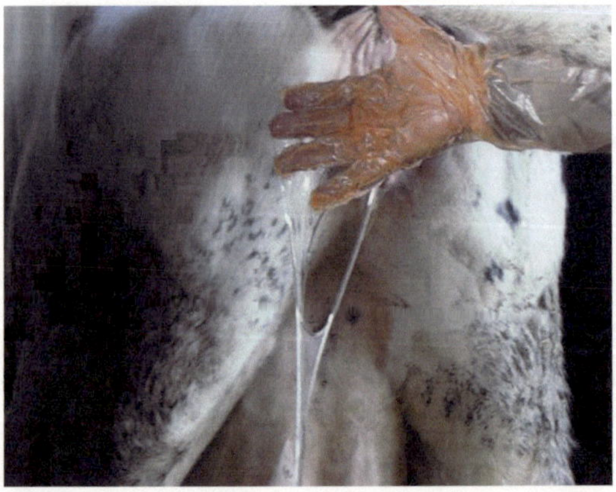

Fig. 4. Transparent and abundant vaginal mucus indicates optimal heat and good uterine environment.

Vaginoscopy is helpful to visualize the vaginal cavity and cervix. Feces are removed from rectum before the perineal area, vestibule and clitoral fossa were cleaned and finally dried. The speculum -which must be clean, dry, sterile, and lubricated- is inserted in direction of the pelvic canal. Once crossed the vestibule, the negative pressure inside allows air to penetrate. Hyperemia may result in 30-60 sec, which hampers the assessment of the vaginal mucosa coloration. The vestibule acts as a defensive barrier of the female genital tract, and hinders the entry of the speculum. If this does not occurs, dysfunctions can be present (e.g. in pneumovagina). Similarly, if material is accumulated into the vagina, speculum helps to localize its origin.

The examination of internal organs (uterus, oviducts and ovaries) demands other techniques, as manual palpation and ultrasound.

3.2 Rectal palpation in cow

Rectal palpation is a widely used diagnostic method in cattle with high accuracy, easy to be implemented and at low cost in comparison with other sophisticated techniques. Plastic gloves are lubricated and then feces are withdrawn. Air should not be present into the rectum to get a more relaxed mucosa and easily manipulate the structures beneath. The cervix is presented as a solid structure, tubular, fibrous, with 3-4 folds projected inside and localized on pelvis floor in normal non-pregnant cows. It is cylindrical, with a length of 5-10 cm and a diameter of 1,5-7,0 cm. Cranially the uterus can be palpated. At heat, uterus is turgid, erect and coiled. However, it is soft and flaccid during luteal phase and palpation is a bit more difficult; it is a consequence of the progesterone action, released from CL. Palpation helps to diagnose anomalies such as uterine infections.

After that, it is interesting to palpate the ovaries. They are located ventrolaterally to the pelvis floor, and sometimes placed under the bone. During anestrus, ovary size ranges from 2 to 3 cm approximately. Follicles (at different stages of growth) and CLs (hemorrhagic, mature or/and albicans) are developed at the ovaries and its size could suggest some diseases.

3.3 Ultrasonography (US)

The first description of cattle pregnancy by real-time ultrasound was conducted by Chaffeux et al. (1982). Later, Pierson & Ginther (1984) showed ultrasound images of normal ovarian structures in superovulated heifers. Also Reeves et al. (1984) described the echogenic cystic CL. More recently, ultrasound technology has been used for developing more effective superovulation, embryo collection and recipient´s synchronization. Foetal sex can be determined by ultrasonography (US) from day 50 onward, emphasizing that it can be accurately established around day 60 of pregnancy. Ultrasonographic anatomy of foetal sex organs, from genital tubercle (GT) to fully developed organs, has been extensively described in their entire progression. The external genitalia are initially formed on the caudoventral surface of the abdominal wall, between the hind limbs. These primordial structures are seen as poorly defined elevations and they constitute the GT, the urogenital fold and the genital swelling, which are developed into the male or female gonads. In males, the GT becomes elongated to form penis, the urogenital folds enclose the penis to form the prepuce and the genital swelling become enlarged to develop into the scrotum. In females, the GT forms the clitoris, the genital swellings completely disappear and the urogenital folds develop to enclose the GT, forming the labia (Quintela et al., 2011).

Other ultrasound application is the aspiration of oocytes, or "ovum pick-up" (OPU) (Pieterse et al., 1988). It is less traumatic and invasive than laparoscopy, does not affect the ovarian activity and can be given many other utilities. Among them the treatment of RBCs syndrome is the most notable, since it allows collecting oocytes that will be in vitro maturated, fertilized and cultured.

Transrectal ultrasound diagnosis has improved our ability to assess the reproductive organs in cattle and to follow the dynamic interactions between ovarian follicular cohorts. Even 2-3 mm follicles can be seen, quantified and sequentially monitored, allowing the development of superovulation regimens, an essential practice for the embryo transfer industry. US practical uses include routine assessment of follicular and luteal development, and differential diagnosis of cystic ovaries, ovarian abscesses and tumors, which can be considered factors associated with RBC syndrome. US helps to evaluate the uterus by detecting changes in shape and echotexture related with circulating hormone concentrations during the estrous cycle, and thereby detecting hormonal imbalances. It is also useful for detecting pathological conditions such as metritis, pyometra, maceration or mummification, and it is an important tool for diagnosing ovarian cystic disease (Rajamahendran, 1994).

3.4 Hormonal function tests

Sex hormones, neurotransmitters and other substances are involved in the regulation of the sexual cycle in cow. Synthesis, release, actuation or interaction imbalance of these substances could be related to reproductive alterations as repeat estrus and reduced fertility. Progesterone can be considered as a sensor of the reproductive capacity, both for its information about the estrous cycle and for its easy determination. Progesterone assay is an objective and accurate test to evaluate the ovarian function and to diagnose certain diseases that otherwise could not be correctly determined, such as delayed ovulation, persistent luteal activity, ovarian cysts or suprabasal progesterone levels (Lamming et al., 1989; Waldmann et al., 2000). When cows are in estrus, this assessment could confirm that the hormonal environment is within the correct hormonal framework. Then, if progesterone is measured around day 19-21 after AI, it could be an effective method to detect females with pregnancy failure. Radioimmunoassay (RIA) and enzyme immunoassay (ELISA) are the usual analytical techniques for determining steroids in biological fluids.

3.5 Oviductal patency

The determination of oviductal permeability is interesting, although it is difficult to carry out. An injection of 500 ml of sterile solution containing 30 gr of starch has been described by Kessy & Noakes (1979). It is intraperitoneally injected and starch reaches the oviduct and descends to the cervical mucus if the fallopian tubes are permeable. It takes about 12 hours to arrive at the cervical mucus, where it will remain between 2 and 4 days. A sample of mucus is collected, stained with lugol and observed under the microscope. Other procedure to study the oviducts individually is infusing phenolsulphonfthalein into the uterine horn. A Foley catheter is inserted through the uterus and the balloon is inflated on the end of the horn before dye is infused. If the oviduct is normal, the dye will cross it, reach the abdominal cavity, and is eliminated by urine. Bladder is catheterized 20 minutes later and urine turns to reddish if oviduct is normal. After 4 hours, the test can be repeated in the other oviduct (Figure 5).

Another diagnostic technique to check the oviductal patency is collecting oocytes or embryos, either with or without superovulation treatments. In addition, if they are collected, it is possible to transfer them to cows without reproductive problems, in which case it also becomes a therapeutic tool to overcome this syndrome.

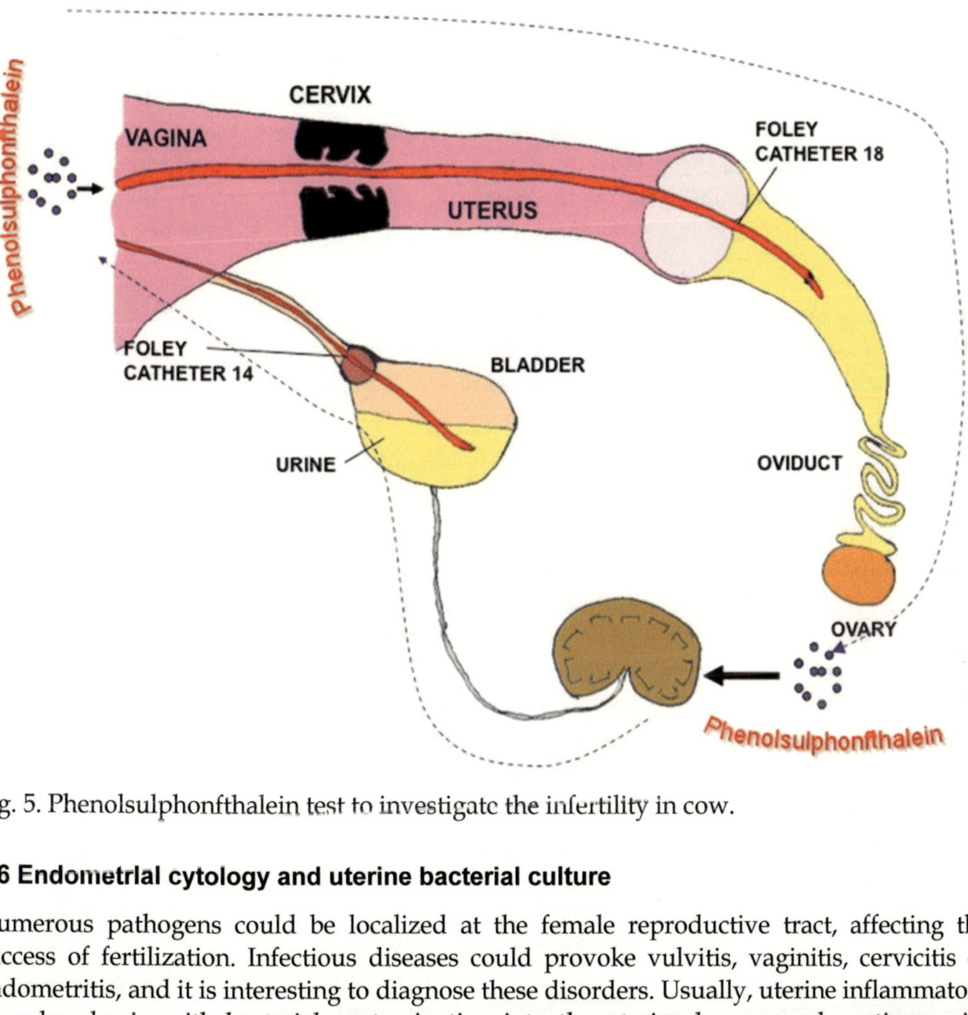

Fig. 5. Phenolsulphonfthalein test to investigate the infertility in cow.

3.6 Endometrial cytology and uterine bacterial culture

Numerous pathogens could be localized at the female reproductive tract, affecting the success of fertilization. Infectious diseases could provoke vulvitis, vaginitis, cervicitis or endometritis, and it is interesting to diagnose these disorders. Usually, uterine inflammatory disorders begin with bacterial contamination into the uterine lumen, and continue with adhesion of pathogens to the mucosa, colonization or penetration of the epithelium, and/or release of endotoxins. Uterine inflammation, even in the absence of active bacterial infection, may disrupt embryonic survival and provoke RBC syndrome. The endometrial bacteriological diagnosis is interesting to detect pathogens implicated in infertility. In cattle, especially due to the cervical anatomy, samples can be taken using a catheter connected to a syringe containing 30-60 ml of sterile saline. It is deposited into the uterus and then it is removed and cultured. Clinical or subclinical endometritis could be diagnosed. The sample can also be taken by cytology brushes (65 cm, 4 mm Ø) protected by a sterile metal tube (50 cm, 5 mm Ø). Endometrial cytology is a practical technique to diagnose subclinical endometritis, when clinical signs are absent. The number of neutrophils indicates the type and grade of endometrial inflammation.

4. Therapy and control of RBC syndrome to reduce the negative impact on farm profitability

To reduce the negative effects of repeat breeding on the farm profitability, an effective therapy must be established after a proper diagnosis. Nutritional supplements have been used to restore certain imbalances at herd level. The assisted reproductive techniques, such as in vitro production or embryo transfer, have provided new therapeutic options for resolving this syndrome. Alternative medicines have also developed protocols to overcome the syndrome, as moxibustion or acuapunture. However, hormonal treatments, with progestins, GnRH, exogenous gonadotrophins and prostaglandins, have been traditionally used.

As previously mentioned, RBC syndrome causes considerable economical losses in beef and dairy cattle associated to reproductive failure, drop in milk yield and calves, and treatment cost. Numerous treatments have been studied to solve and prevent this syndrome at herd or individual level.

4.1 Nutritional treatments

Nutritional deficiencies have been described as causes of RBC. Diets containing higher concentration of inorganic iodine from 8-12 days before estrus improve the stimulation of the pituitary gland, reducing at the same time the RBC rate (McDonald et al., 1961). Herds with problems of repeated estrus were supplemented with copper and magnesium, minimizing fertility problems (Ingraham et al., 1987). Deficiencies of these minerals have been associated with infertility, anemia or immune suppression. Concentrations of these minerals can be modified in the diet as a result of the season or agriculture practices, or as a consequence of mineral availability alteration associated with some dietary components (e.g. diets containing high level of protein reduce the copper plasma concentration). Beta-carotene, precursor of vitamin A, has recently been investigated for its involvement in the formation and function of CL. Although no conclusive results exist, it is suggested that beta-carotene improves the progesterone synthesis and reduces the luteal hypofunction (Wang et al., 1988). Further experiments are required in order to assert its therapeutic role or positive effect on reproduction. In 1992, Marai et al. reported that repetitive administration of 40 g/animal/day of sodium phosphate monobasic (in diet) plus 500 ppm zinc (in water) improves the fertility. Deficiencies of phosphorus and zinc are linked to low levels of progesterone, which could be the cause of failures in fertilization or early embryonic death. Other consequences associated with deficient diets are hypoglycemia, reduced coenzyme NAD (important for progesterone synthesis), decreased carbohydrate concentration or increased creatinine. Unsaturated fatty acid diets improve the milk yield and embryo development rates when administered at dry and postpartum periods, also being beneficial in RBCs.

In conclusion, it is important to consider the nutritional imbalance in RBCs, moreover in dairy cattle with higher nutritional requirements. However, it is not practical to analyze the mineral or chemical composition in blood. It is more reasonable to monitor the food rations in order to avoid disruptions to the reproductive function.

4.2 Assisted reproduction techniques

Abnormal implantation and transport of gametes are associated with endometrial defects, resulting in RBC syndrome. Certain assisted reproduction techniques, as in vitro production of embryos or intraperitoneal insemination, have been proposed to solve this syndrome.

Intraperitoneal insemination could be an alternative procedure to the normal deposition of semen in the genital tract of the cow (Lopez-Gatius, 1995). It is considered that between 36 and 89% of RBCs show uterine diseases but are often difficult to detect in field conditions. A bypass from vagina to peritoneal area (around the ovary) is possible to avoid the negative effects of altered uterine environment on the sperm quality (Figure 6).

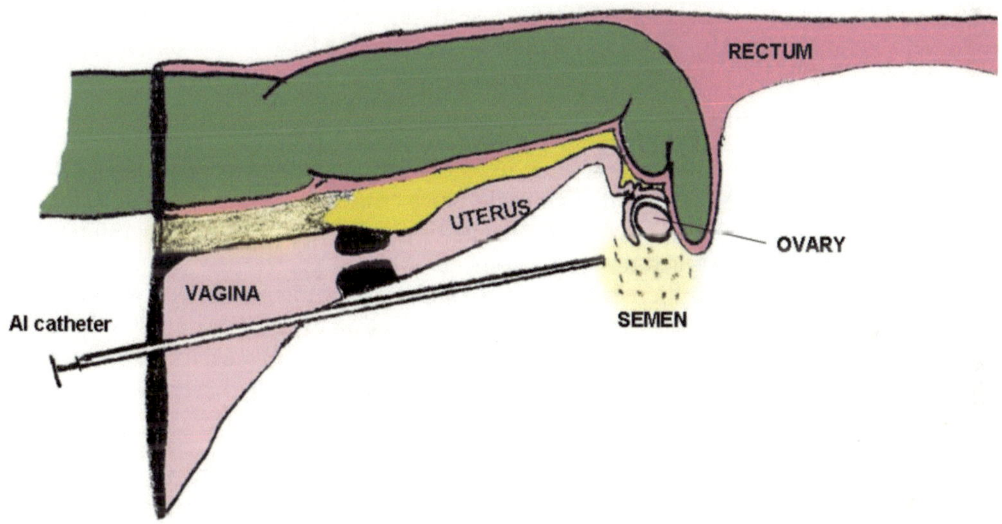

Fig. 6. Intraperitoneal insemination.

Other authors (Tanaka et al., 1994) consider that the majority of embryonic abnormalities occur in the oviducts, but are not apparent until 6-7 days post-breeding. Then, oocytes of RBCs are competent to reach the blastocyst stage. IVM, IVF, IVC and ET techniques are proposed to improve the reproductive success in these animals.

4.3 Intrauterine treatment

Disorders related to reproductive tract infections may be related to RBC syndrome and very often overlooked by clinicians. Prophylactic practices have been used, as the uterine administration of antiseptic solutions (lugol) 24 h. after mating/AI (Tanaka et al., 1994), although fertility results are poor (Huszenicsa et al., 1994).

Studies on bacteriology and histology of the uterus concluded that the non-specific genital infection is one of the main causes of RBC, and it is suggested that antimicrobial treatments (chloramphenicol, gentamicin, enrofloxacin, tetracycline, or nitrofurantine) could improve the reproductive indexes (Santana et al., 1998a).

4.4 Alternative medicine

Scientific contributions report the utility of some alternative medicines, such as acupuncture or moxibustion, to reduce the incidence of RBC syndrome.

Moxibustion has been used to reduce reproductive failure in RBCs (Hosaha & Nakama, 2002). Moxas (balls of about 3 cm in diameter) containing 2 g of Artemisia spp. are applied on nine points of the skin and burned for 15 min. The treatments are applied for 3 consecutive days, and then repeated during the heat. Authors describe an increase in blood flow at the uterine arteries, which is considered to improve fertility.

We also find references in the literature about the positive effect of aquapuncture therapy (Lin et al., 2002), i.e. acupuncture combined with injections. A volume of 5-10 ml of glucose 50% was injected at certain points set by traditional acupuncture in cattle. This study reported a good fertility rate, although many gestations failed later.

4.5 Hormonal treatments

Numerous hormonal treatments have been used to improve results in RBCs. The literature about treatments in RBCs is difficult to compile, since criteria to define animals included on this syndrome are variable. It is reported the use of progesterone, alone or combined with other substances, at different times of the estrous cycle to protect the pregnancy and to improve the maternal recognition of pregnancy. Numerous studies describe the influence of GnRH (and analogues) to improve fertility in RBCs. HCG, an exogenous gonadotrophin, has been administered during luteal phase and at AI time. References exist about the use of pregnant mare serum for the RBC therapy. Prostaglandins have also been widely used to improve heat detection.

4.5.1 Progesterone (Pg)

Progesterone is essential for implantation and maintenance of pregnancy. CL dysfunction decreases Pg concentrations and then negatively affects fertility (Gustafsson et al., 1986; Kimura et al., 1987). In contrast, some authors have not found differences between Pg concentrations in normal or repeat breeder cows (Ayalon, 1984).

It has been noted that the administration of progesterone as of 3 to 5 days after insemination and for 2-3 weeks (or more) improves conception rates in RBCs. Umakanthan (1995) administered 250 mg of retard progesterone on days 4, 14, 24, 34, 44 and 54 after insemination. The pregnancy rate was 96% in RBCs versus 20% in control cows. Kimura et al. (1987) argue that the delayed formation of the CL induces estrus repetition, and then suggest that progesterone therapy should be started on 4-5th day after insemination.

Bulman & Lamming (1978) inserted intravaginal devices (PRID) for 14 days in RBCs, previously removing the estrogen capsule. The treatment improved the ovulation and shortened the calving-conception interval. Villaroel et al. (2004) have used this disposal from day 5 after AI to day 19. The suggested fertility improvement in RBCs is based on the rise of the progesterone level above the minimum threshold necessary to maintain pregnancy and the increment of progesterone receptors. Intravaginal progestagen treatments are effective some days after the insertion, while progesterone injection (200 mg on days 5, 7, 9 and 11 post-IA) fails to increase plasma concentrations of progesterone in RBC (Walton et al., 1990).

4.5.2 GnRH (gonadotrophin-releasing hormone)

The administration of GnRH around the insemination time aims to accelerate and ensure ovulation in cows, acting directly on the pituitary, stimulating the secretion and release of gonadotrophins, such as LH and FSH, and promoting the preovulatory LH peak, which is essential for follicular dehiscence. In many cases, RBC syndrome is associated with ovulatory defects, as anovulation, delayed ovulation or gonadotrophin release failure.

If luteal deficiency is suspected as reproductive failure in cows, GnRH (100 μg) could be used on day 5 post-AI (Gonzalez-Stagnaro, 1993). GnRH induces LH and FSH discharge pituitary, promotes increased levels of progesterone and exerts a luteoprotector (luteotrophic), increasing fertility in RBCs associated with subfunctional CL or ovulation defects. The mentioned treatment has also been combined with intrauterine antibiotic (infusion of 30-50 ml containing 300000 IU penicillin and 0,5 g streptomycin), increasing fertility in RBCs. This treatment could be more appropriate when abnormal vulvar conformation is present.

A comprehensive study carried out in RBCs indicates that the administration of GnRH at the AI time may be beneficial for improving fertility in these cows (Stevenson et al., 1990). These authors do not recommend the use of double insemination in RBCs, in contrast with Singh et al. (2005) who suggest that pregnancy rates can be improved by increasing the insemination frequency.

4.5.3 Exogenous gonadotrophin

These hormones have been used in RBCs to induce ovulation and exert the luteotrophic effect on the CL.

In 1965, Roussel et al. administered intramuscularly pregnant mare serum (PMS) on day 15 or 16 of the estrous cycle, obtaining a conception rate of 73,9% in RBCs compared with 44,4% of control cows.

However, HCG is the most used exogenous gonadotrophin for treating RBC syndrome.

Treatment with hCG on day 5 after insemination can achieve higher levels of progesterone for at least 2 weeks, due to the development of accessory CL (Walton et al., 1990)

4.5.4 Prostaglandins

The luteolytic effect of prostaglandins has been used to treat RBCs. In this case, treatment aims to achieve better heat detection and to increase the number of cows in heat. Numerous protocols are used in cow, e.g. two PGF2alpha injections apart 11 days and insemination 80 hours later. Intravenous PGF2alpha (0,2 ml cloprostenol) has also been reported in RBCs at AI time. However, the most frequent use of this hormone has been combined with other substances, serving as a pretreatment and with the ultimate goal of improving the reproductive management.

4.6 Reproductive management improvement

Treatments are being used recently for fixed-timed artificial insemination, without heat detection. These protocols allow the treatment of cows with silent heats or ovulation problems. Ovsynch is a protocol based on the administration of GnRH, PGF2alpha and GnRH (Pursley et al., 1995) to schedule the insemination time.

It is also essential to improve all aspects related to heat detection, since it has been demonstrated that estrus detection mistakes involve very significant losses in reproduction and production cattle. Therefore, it could be interesting to implement different methods for estrus detection as Kamar or Bovine Heat-beacon devices and pedometers.

5. References

Albihn A, Gustafsson H, Hurst M & Rodriguez-Martinez H. Embryonic ability to prolong the interoestrous interval in virgin and repeat breeder heifers. *Anim Reprod Sci* 1991;26:193-210.

Almeida AP, Ayalon N &, Bartoov B. Progesterone Receptors in the endometrium of normal and repeat-breeder cows. *Anim Reprod Sci* 1987;14:11-19.

Almeida AP. Early embryonic mortality in "repeat-breeder" cows. *Ars Veterinaria* 1995;11(2):18-34.

Ayalon N. The repeat breeder problem. *Procceding 10th International Congress Animal Reproduction and AI.* Urbana, III, USA 1984;4:111-141.

Bage R, Gustafsson H, Forsberg M, Larsson B & Rodriguez-Martinez H. Suprabasal progesterone levels in repeat breeder heifers during the pro- and oestrous period. *Theriogenology* 1997;47:141.

Bage R. On repeat breeding in dairy heifers. With special focus on follicular dynamics, ovulations, and oocyte quality. SLU, Doctoral thesis 2002.

Bailie JH. Management and economic effects of different levels of oestrus detection in dairy herd. *Vet Rec* 1982;110:218-221.

Barbato AL, Marqués AP & Fagundes E. Alteracoes histopatologicas na genitalia de femeas Zebus cruzadas com sindrome "repeat breeder". *Rev Bras Reprod Anim* 1994;18:31-36.

Barlett PC, Kirk JH & Mather EC. Repeated insemination in Michigan Holstein-Friesian cattle: incidence, descriptive epidemiology and estimated economic impact. *Theriogenology* 1986;26:309-322.

BonDurant RH, Revah Y, Franti C, Harman RJ, Hird D, Klingborg D, McCloskey M, Weaver L, & Wilgenberg B. Effect of gonadotrophin-relesing hormone on fertility in repeat-breeder california dairy cows. *Theriogenology* 1991;35(2):365-373.

Bruyas JF, Fieni F & Tainturier D. Le syndrome "repeat-breeding": analyse bibliographique 1° partie: Étiologie. *Revue med vet* 1993;144:385-398.

Bulman DC & Lamming GE. Milk progesterone levels in relation to conception, repeat breeding and factors influencing acyclity in dairy cows. *J Reprod Fert* 1978;54:447-458.

Casida LE. Present status of the repeat-breeder cow problem. *J Dai Sci* 1961;44:2323-2329.

Chaffaux S, Valon F & Martinez J. Evolution de l´image echographique du produit de conception chez la vache. *Bull Acad Vet* 1982;55:213-221.

De la Fuente J, Monge A, Cocero MJ & Barragán C. Evaluación de la mortalidad embrionaria en vacas repetidoras de celos a través de su producción embrionaria. *Invest. Agr.: Prod Sanid anim* 1988;3(2-3):183-193.

Diskin MG & Sreenan JM. Fertilization and embryonic mortality rates in beef heifers after artificial insemination. *J Reprod Fertil* 1980;59:463-468.

Domínguez JC. Fisiopatología puerperal de la vaca. *Bovis* 1989;29:11-20.

Dovensky T, Popovsky K, Mickousky G, Kocosky Lj, Trojacanec P, Petkov V & Popovska F. Serum concentration of several biochemical parameters in cows with reproductive disorders. *13th Intern. Congress on Animal Reprod*, Australia 1996.

Duchens M, Forsberg M, Edqvist L, Gustafsson H & Rodríguez-Martínez H. Effect of induced suprabasal progesterone levels around estrus on plasma concentrations of progesterone, estradiol-17beta and LH in heifers. *Theriogenology* 1994;42:1159-1169.

Duchens M, Maciel M, Gustafsson H, Forsberg M, Rodríguez-Martínez H & Edqvist L. Influence of perioestrous suprabasal progesterone levels on cycle length, oestrous behaviour and ovulation in heifers. *Anim. Reprod. Sci.* 1995;37:95-108.

Duchens M, Rodríguez-Martínez H, Forsberg M & Gustafsson H. Ultrastructure of bovine ovarian follicles induced to extended growth by perioestrous suprabasal progesterone levels. *Acta vet. scand.* 1996;37:239-250.

Duffy P, Crowe MA, Austin EJ, Mihm M, Boland MP, Roche JF. The effect of eCG or estradiol at or after norgestomet removal on follicular dynamics, estrus and ovulation in early post-partum beef cows nursing calves. *Theriogenology* 2004;61:725-734.

Eid LN & Parrish JJ. Duration of the G2-phase and onset of M-phase during the first cell cycle of the bovine embryo is dependent on bull in vivo fertility. *Theriogenology* 1995;43:205.

Eid LN, Lorton SP & Parrish JJ. Paternal influence on S-phase in the first cell cycle of the bovine embryo. *Biol Reprod* 1994;51:1232-1237.

Erb HN, Smith RD, Oltenacu PA, Guard CL, Hillman RB, Powers PA, Smith MC & White MA. Path model of reproductive disorders and performance, milk fever, mastitis, milk yield, and culling in Holstein cows. *J Dai Sci* 1985;68:3337-3349.

Ferreira R, Coelho de Oliveira JF, Antoniazzi AQ, Pimentel CA, Ferrugem JC, Henkes LE, Bordignon V, Dias Gonçalves PB. Relationship between clinical and *postmortem* evaluation in repeat breeder beef cows. Cienc. Rural 2008; 38

Ferreira R.M., H. Ayres, M.R. Chiaratti, M.L. Ferraz, A.B. Araújo, C.A. Rodrigues, Y.F. Watanabe, A.A. Vireque, D.C. Joaquim, L.C. Smith, F.V. Meirelles, P.S. Baruselli. The low fertility of repeat-breeder cows during summer heat stress is related to a low oocyte competence to develop into blastocysts. *J Dai Sci* 2011;94:2383-2392.

Foote RH & Parks JE. Factors affecting preservation and fertility of bullsperm: a brief review. *Reprod Fertil Develop*. 1993;5:665-673.

Francos G. The relationships between the incidence of endometritis and repeat breeders in dairy herds. *Refuah Veterinarith* 1979;36:131-134.

Fulkerson WJ & Dickens AJ. The effect of season on reproduction in dairy cattle. *Australian Vet Journal* 1985; 62(11):365-367.

Gonzalez F. Características del comportamiento de los niveles séricos de progesterona y LH en vacas anéstricas. Doctoral thesis, CENSA. La Habana (Cuba), 1981.

González-Stagnaro C, Madrid-Bury N, Morales J & Marín D. Efecto luteoprotector del tratamiento GnRH en vacas mestizas repetidoras con cuerpo lúteo sub-funcional. *Revista científica FCV-LUZ* 1993;1:14-20.

Gustafsson H & Emanuelson U. Has a repeat breeder cow in the present lactation a higher risk to become a repeat breeder in the next lactation?. *14th International Congress on Animal Reproduction,* Stockholm 2000;(15:26):100.

Gustafsson H, Larsson K, Kindahl H & Madej A. Sequential endocrine changes and behaviour during oestrus and metoestrus in repeat breeder and virgin heifers. *Ani Reprod Sci* 1986;10:261-273.

Gwazdauskas FC, Whittier WD, Vinson WE & Pearson RE. Evaluation of reproductive efficiency of dairy cattle with emphasis on timing of breeding. *J Dai Sci* 1986,69:290-297.

Hartigan PJ, Murphy JA & Nunn WR. An investigation into the causes of reproductive failure in dairy cows. II.-Uterine infection and endometrial histopathology in clinically normal repeat breeder cows. *Irish Vet J* 1972;26:245-247.

Harvey D. Nocardia asteroides in the uterus of 2 repeat breeder cows. *Canadian Vet J-Revue Vet Canad* 1993;34(12):709.

Hewett CD. A survey of the incidence of the repeat breeder cows in Sweden with reference to herd size, season, age and milk yield. *British Vet J* 1968;124:342-352.

Hodel F, Moll J & Kunzi N. Factor affecting fertility in cattle. *Schweizer Fleckvieh* 1995;4:14-24.

Hosaha T & Nakama S. Effects of moxibustion for reproductive failure in dairy cattle. *XXII World Buiatrics Congress*, Hannover, Germany 2002, p 212

Humblot P. La mortalité embryonnaire chez les bovins. *Colloque de la Societe Francaise pour l`etude de la Fertilite.* "Recherches récentes sur l`epidemiologie de la fertilite". Ed. Masson, París, 1986:213-242.

Hunter RHF. Vital aspects of Fallopian tube physiology in pigs. *Reprod Dom Anim* 2002;37:86-190.

Huszenicza G, Kulcsar M, Nagy P, Mihály K, Fekete S & Zöldág L. Ovarian and metabolic characteristics of repeat breeding dairy cows and heifers and the effect of intrauterine lugol treatment on their fertility. *Proceeding 18th World Buiatrics Congress: 26th Congress of the Italian Association of Buiatrics,* Bologna, Italy, August 29-Sept 2, 1994.

Ingraham RH, Kappel LC, Morgan EB & Srikandakumar A. Correction of subnormal fertility with copper and magnesium supplementation. *J. Dairy Sci*. 1987;70:167-180

Kerbler TL, Buhr MM, Jordan LT, Leslie KE & Walton JS. Relationship between maternal plasma progesterone concentration and interferon-tau synthesis by the conceptus in cattle. *Theriogenology* 1997;47:703-714.

Kessy BM & Noakes DE. The use of the starch grain and phenolsulphophthalein test to investigate infertile cows. *Vet Rec* 1979;105:489-491

Kimura M, Nakao T, Moriyoshi M & Kawata K. Luteal phase deficiency as a possible cause of repeat breeding in dairy cows. *Br. vet. J.* 1987;143:560-566.

Lafi SQ & Kaneene JB. Risk factors and associated economic effects of the repeat breeder syndrome in dairy cattle. *Vet. Bull.* 1988;58:891-903.

Lagneau F. Infertilite des vaches a chaleurs normales. *Rec Méd vét.* 1981;157:117-131.

Lamming GE, Darwash AO & Back HL. Corpus luteum function in dairy cows and embryo mortality. *J Reprod Fertil* 1989;37(Suppl):245-252.

Lanman JT. Delays during reproduction and their effects on the embryo and fetus. *New England J Med* 1968;278:993-1092.

Lee CN, Maurice E, Ax RL, Pennington JA, Hoffman WF & Brown MD. Efficacy of gonadotrophin-releasing hormone administered at the time or artificial insemination of heifers and postpartum and repeat breeder dairy cows. *Am J Vet Res* 1983;44:2160-2163.

Lewis GS. Uterine health and disorders. *J Dai Sci* 1997;80:984-994.

Lin JH, Wu LS & Wu YL. Aquapuncture therapy or repeat breeding in dairy cattle. *Am J Chin Medic 2002,* 30;2-3:397-404

Linares T, Ploen L, Ekwall H & Bane A. The ultraestructure of blastocysts collected from virgin and repeat breeder heifers. *Nord Vet Med* 1984;42:585-598.

Linares T. Embryonic development in repeat breeder and virgin heifers seven days after insemination. *Anim. Reprod. Sci.* 1982;4:189-198.

Lopez-Gatius F & Camón-Urgel J. Confirmation of estrus rates by palpation per rectum of genital organs in normal repeat dairy cows. *J Vet Med A* 1991;38:553-556.

López-Gatius F. Intraperitoneal insemination in repeat-breeder cows: a preliminary report. *Theriogenology* 1995;44:153-158.

Lotthammer KH. Trastornos de la fertilidad de origen ambiental. In: *Infertilidad en la vaca.* Ed. Hemisferio Sur. 1988:351-391.

Marai IFM, Daader AH & El-Darawany AA. Some physiological aspects of repeat breeding in Holstein Friesians and its improvement under Egyptian environment. *Beitr trop Landwirtsch vet med* 1992;30 (H.2):199-209.

Maurer RR & Echternkamp SE. Repeat breeder females in beef cattle: influences and causes. *J Anim Sci* 1985;61(3):624-636.

McClure TJ. Infertilidad nutricional y metabólica de la vaca. Ed. Acribia, Zaragoza, España, 1995.

McDonald RJ, McKay GW & Thomson JD. The use of organic iodine in treatment of repeat breeder cows. *Proceeding of the IVth International Congress on Animal Reproduction* 1961:679-681.

Murthy KG, Nanjiah RD & Keshava MBS. Bacteriological flora of cervical mucus in repeat breeder bovines. *Indian Vet J* 1974,51:264-268.

Odde KG. A review of synchronization of estrus in postpartum cattle. *J Anim Sci* 1990;68:817-830.

Ohtani S & Okuda K. Histological observation of the endometrium in repeat breeder cows. *J Vet Med Sci* 1995;2:283-286.

Panangala VS, Fish NA & Barnum DA. Microflora of the cervico-vaginal mucus of repeat-breeder cows. *Can Vet J*. 1978,19:83-89.

Payne JM, Rowlands GJ, Manston R & Dew SM. A statistical appraisal of the results of metabolic profile test on 75 dairy herds. *Brit Vet J* 1973;129:370-381.

Pedroso R & Roller F. Problemática de las repeticiones del celo en la hembra bovina. Revisión bibliográfica. *Rev Cub Reprod Anim* 1996;22(1):1-19.

Perez-Marin CC & España F. Estrus expression and ovarian function in repeat breeder cows, monitored by ultrasonography and progesterone assay. *Reprod Dom Anim* 2007; 42, 449-456.

Perry GA, Perry BL. Effect of preovulatory concentrations of estradiol and initiation of standing estrus on uterine pH in beef cows. *Dom Anim Endoc* 2008; 34 333-338.

Pierson RA & Ginther OJ. Ultrasonography of the bovine ovary. *Theriogenology* 1984;21:495-504.

Pieterse MC, Kappen KA, Kruip TAM & Taverne MAM. Aspiration of bovine oocytes during transvaginal ultrasound scanning of the ovaries. *Theriogenology* 1988;30:751-762.

Pursley JR, Mee MO & Wiltbank MC. Synchronization of ovulation in dairy cows using PGF2α and GnRH. *Theriogenology* 1995;44:915-923.

Quintela, L.A., Becerra, J.J., Pérez-Marín, C.C., Barrio, M., Cainzos, J., Prieto, A., Díaz, C. & Herradón, P.G. Fetal gender determination by first-trimester ultrasound in dairy cows under routine herd management in Northwest Spain. *Anim Reprod Sci* 2011 10.1016/j.anireprosci.2011.02.022

Rajamahendran R. Clinical and research applications of real-time ultrasonography in bovine reproduction: A review. *Can Vet J* 1994;35:563-572.

Reeves JJ, Rantanen NW & Hauser M. Transrectal real-time ultrasound scanning of the cow reproductive tract. *Theriogenology* 1984;21:485-494.

Roberts SJ. *Veterinary Obstetrics and genital diseases* (Theriogenology). Published by the author, Ithaca New York, 1971, 496-506.

Rodriguez LAM, Lozano CB & Pérez VJE. La translocación robertsoniana 1/29 en ganado vacuno: sus consecuencias en la reproducción. *Salud Públ .Nutrición* 2000, 2.

Roine K & Saloniemi H. Incidence of infertility in dairy cattle. Acta Vet Scand 1971;19:354-367.

Roussel JD, Stallcup OT, Rakes JM & Ward RF. Treatment of repeat breeders with pregnant mare serum. *J Dai Sci* 1965;48:808.

Sagartz JW & Hardenbrook HJ. A clinical, bacteriologic and histologic survey of infertile cows. *J Amer Vet Med Assc* 1971;158:619-622.

Santana AO, Oba E, Langoni H & Uribe-Velasquez LF. Avalia□ao bacteriologica do endometrio em femeas bovinas repetidoras de cio. *Arch Reprod Ani* 1998;6;62-71.

Santana AO, Oba E, Lopes J & Uribe-Velasquez LF. Avalia□ao histopatologica do endometrio em femeas bovinas repetidoras de cio. *Arch Reprod Ani* 1998;6;58-61.

Santana AO, Oba E, Ramos AA & Uribe-Velasquez LF. Concentracao serica de FSH, LH, Estradiol e Progesterona P4 em femeas bovinas repetidoras de cio. *Arch Reprod Ani,* 2000;12:26-32.

Sheldon M. Bovine fertility-practical implications of the maternal recognition of pregnancy. *In practice* 1997;Nov-Dec:546-556.

Shrestha HK, Nakao T, Higaki T, Suzuki T & Akita M. Resumption of postpartum ovarian cyclicity in high-producing Holstein cows. *Theriogenology.* 2004 Feb;61(4):637-49.

Singh B, Savavia F, Bage R & Rodríguez-Martinez H. Pregnancy rates in repeat-breeder heifers following multiple artificial inseminations during spontaneous oestrus. *Acta Vet Scand* 2005;46:1-12.

Singla VK, Singh G & Dwived PN. In vivo evaluation of drug sensitivity pattern of bacterial isolates in repeat breeders cows. *Indian J Anim* 1993;4:425-426.

Staigmiller RB & England BC. Folliculogenesis in the bovine. *Theriogenology* 1982;17:43-52.

Stevenson JS, Call EP & Scoby RK. Double insemination and gonadotropin-releasing hormone treatment of repeat-breeding dairy cattle. *J Dai Sci* 1990;73:1766-1772.

Stock AE & Fortune JE. Ovarian follicular dominance in cattle: relationship between prolonged growth of the ovulatory follicle and endocrine parameters. *Endocrinology* 1993;132:1108-1114.

Swanson LV & Young AJ. Failure of gonadotripin-releasing hormone or human chorionic gonadotropin to enhance the fertility of repeat-breeder cows when administered at the time of insemination. *Theriogenology* 1990; 34(5):955-963.

Tanaka H, Hishinuma M, Takahashi Y & Kanagawa H. Developmental competence of oocytes from the ovaries of repeat-breeding cows after in vitro fertlization. *J Vet Med Sci* 1994;56(3):547-548.

Thibier M, Gouffe D, Jean O, Valognes J, Daunizeau A & Humblot P. Enhancing the rate recovery and quality of the embryos in repeat breeding cows by using a GnRH analogue injection at mid-luteal phase prior to breeding. *Theriogenology* 1985;24(6):725-733.

Umakanthan T. A field trial on progesterone treatment in repeat breeding cows. *Ind Vet J* 1995;72:1308.

Vasconcelos JLM, Chande CG & Silva AV. Microorganismos isolados de material uterino de vacas e sua sensibilidade "in vitro" aos antiioticos, nas bacias leiteiras de passos (MG) e Sao Jose dos Campos (SP). In: *Congresso Brasileiro de Reproducao Animal,* Belo Horizonte, 1995;11:438.

Villaroel A, Martino A, BonDurant RH, Deletang F & Sischo WM. Effect of post-insemination supplementation with PRID on pregnancy in repeat-breeder Holstein cows. *Theriogenology* 2004;61:1513-1520

Wahome JN, Stuart MJ, Snuth AE, Hearne WR & Fuquay JW. Insemination management for a one-injection PGF2α synchronization regimen. II.One versus two inseminations following detection estrus. *Theriogenology* 1985;24:501.

Waldmann A, Reksen O, Landsverk K, Ropstad E, Kommisrud E, Dahl E & Refdal AO. Progesterone at first insemination-effects on non-return and repeat-breeding. *14th International Congress on Animal Reproduction,* Stockholm 2000;(15:13):87.

Walton JS, Halbert GW, Robinson NA & eslie KE. Effects of progesterone and human chorionic gonadotrophin administration five days potsinsemination on plasma and milk concentrations of progesterone and pregnancy rates of normal and repeat breeder dairy cows. *Can J Vet Res* 1990;54:305-308

Wang JY, Owen FG & Larson LL. Effect of beta-carotene supplementation on reproductive performance of lactating Holstein cows. *J.Dairy Sci*. 1988;71:181-186.

Zemjanis R. "Repeat-breeding" or conception failure in cattle. In: *Current Therapy in Theriogenology*. Ed. D.A.Morrow, Philadelphia, WB Saunders, 1980:205-213.

Zemjanis, R. The problem of repeat breeding in cattle, *New England Vet. Meeting*, 1963.

19

Bovine Respiratory Syndrome (BRD) Etiopathogenesis, Diagnosis and Control

Dariusz Bednarek, Monika Szymańska-Czerwińska
and Katarzyna Dudek
National Veterinary Research Institute,
Department of Cattle and Sheep Diseases,
Poland

1. Introduction

1.1 Etiopathogenesis

Bovine respiratory disease (BRD) complex is a major disease, classically in the indoor calves and feedlot young cattle. The etiopathogenesis of BRD is multifactorial and complex. In this complex etiology an equivalent role plays both the infectious agents as well as the environmental factors which are called also as environmental stressors (inappropriate livestock management like mistakes in animal nutrition, transport, handling, veterinary interventions etc.) (tab. 1). The most significant pathogens which are involved in the etiopathogenesis of BRD, i.e. suitable species of viruses such as (bovine respiratory syncytial virus (BRSV), parainfluenza virus type 3 (PI3V), bovine herpes virus type 1 (BHV1), bovine viral diarrhea virus (BVDV), are usually associated with concurrent bacterial infections represented by: *Mannheimia haemolytica, Pasteurella multocida, Histophilus somni* and others (tab. 1) (Jared et al., 2010a, Kita et al., 1994, Klimentowski et al., 1995) and also mycoplasmal factors such as: *Mycoplasma bovis, Mycoplasma bovirhinis, Mycoplasma dispar, Ureaplasma diversum* and even *Mycoplasma canis* (Szymańska et al., 2010). Among bacteria *M. haemolytica* and *P. multocida* have been traditionally considered as the most common bacterial infectious factors in the BRD etiology (Fulton et al.; Reggiardo et al., 2005). On the other hand the one of the most often isolated mycoplasmal factors from BRD cases is *Mycoplasma bovis*. Its adaptative ability to a host organism increases owing to different versions of the same *vsp* gene family which encode particular adhesive factors of the mycoplasma, i.e. variable surface proteins (Vsps; Razin et al., 1998). The variability of *vsp* gene based on DNA transposition (Lysnyansky, 1996). Therefore the high Vsp variation determines the phenotypic variability of the antigen (Rosengarten et al., 1994) which further increases the virulence of the pathogen. *M. bovis* possess the ability to immunomodulate host defence against infection (Razin et al., 1998), for example inducing a synthesis of proinflammatory agents, i.e. TNF-α and nitric oxide (Jungi, et al., 1996) or stimulation of some acute phase protein production, such as haptoglobulin and serum amyloid A, which represent one of the most important components of acute phase response for cattle (Dudek, 2010). It is known that *M. bovis* acts as a stimulant on major subsets of T-lymphocytes (Vanden Bush, 2003). In contrast, some data show inhibitory properties of the pathogen which come down to its suppressive effect on lymphocytes (Vanden Bush & Rosenbusch, 2002; Thomas et al., 1990) and some neutrophil functions (Thomas, et al. 1991). The inhibitory effect of *M. bovis* on the mononuclear cells is not connected with arginine depletion or cytotoxic activity (Thomas et al., 1990). Recently, *M. bovis* is considered as primary pathogen in BRD, not only acting in coinfection conditions. The significance of *M. bovis* in etiopathogenesis of the syndrome is also important with regard to synergistic effect with some infectious factors involved in BRD. For example, *M. bovis* exhibits the colonising synergism with *Mannheimia haemolytica* (Houghton & Gourlay, 1983), whereas coinfection with bovine viral diarrhea virus (BVDV) may increases a pathogenicity of concomitant infectious agents mainly considering suppression of host immune response (Potgieter, 1997; Haines et al., 2001). The disease course during infection with *M. bovis* is generally acute and relates to young cattle (Wikse, 1985). Clinical symptoms mainly limited to dysfunction of the respiratory tract of animals, including cases of fatal diseases, and their intensity decreases with the development of chronic phase of infection (Nicholas & Ayling, 2003). However, complicated infections with *M. bovis* have usually severe course and may take the form of necrosuppurative bronchopneumonia. Pathological changes of the disease are characterized by consolidation areas with nodular lesions containing necrosuppurative material surrounded with fibrous capsule, whereas in histology they are rich in foci infiltrated by degenerating leukocytic cells (Radaelli et al., 2008).

Stress factors	Viruses	Bacteria
Heat Cold Dust Dampness Injury Fatigue Dehydration Hunger Anaxiety Irritant gases Nutrional deficiencies Surgery	*Bovine respiratory syncytial virus (BRSV)* *Parainfluenza virus type 3* (PI3) *Bovine herpes virus type 1 (BHV))* *Bovine viral diarrhoea virus (BVDV)* *Adenoviruses* *Rhinovirus* *Enteroviruses* *Bovine respiratory coronavirus*	*Mannheimia haemolytica* *P. multocida* *Histophilus somni* *Arcanobacterium pyogenes* *Streptococcus pneumonie* *Staphylococcus aureus* *Chlamydiales spp.* *Fusobacterium necrophorum* *Corynebacterium bovis* *Streptococcus spp.* *Micrococcus spp.* Mycoplasmas *M. bovis* *M. bovirhinis* *M. dipsar* *M. alkalesscens* *M. canis* *M. bovigenitalium* *Ureaplasma spp.* *Ureaplasma diversum*

Table 1. The most important etiological causes of BRD

In the complex etiology of BRD also other mycoplasmas, such as *M. dispar, M. bovirhinis* or *U. diversum* can play an important role (Autio et al., 2007). These mycoplasma species were concurrently present in 50% of examined herds and bacterial agents of the syndrome, i.e. *P. multocida, Arcanobacterium pyogenes* or *M. haemolytica* coexisted with these cases.

Additionally, main viruses of bovine respiratory system were concomitant factors of the infections. However, the highest correlation between the prevalence and clinical respiratory signs of disease were observed for *U. diversum* when compared with *M. dispar* and *M. bovirhinis* (Autio et al., 2007).

In the chronic form of the respiratory syndrome should be also mention about chlamydia, particularly *Chlamydophila psittaci* and *Ch. pecorum* apart from that anaerobes bacteria such as *Clostridium* sp. are often engaged in the etiopathogenesis of the disease. Moreover, the different endogenous factors play also an important etiological role in the BRD etiology. First of all young cattle before one - year - old have an insufficient developed respiratory system for an efficient gas exchange and fully effective and constructive mucociliary clearance mechanisms. Calves have almost by a half lower of the lung surface for gas exchange and higher primary activity of pulmonary ventilation, which in connection with the increased of metabolic demand in these animals is the main cause of low supply of oxygen in the body. The low partial pressure of oxygen causes a significant weakening of the movement of cilia snapshot, phagocytic activity of pulmonary macrophages and a reduction in lung clearance of harmful microorganisms. When an affectivity of pulmonary clearance and local immune response of the lungs are reduced, a variety of infectious agents (viruses, bacteria, mycoplasmas) have better access to lower parts of the respiratory tract and they can initiate the development of pathological lesions (Lekeux et al., 1995).

While the different pathogens involved in etiology of BRD are presented in the heard throughout the years, the outbreak of the disease and an appearance of clinical symptoms are seasonal. Usually, there are two seasonal peaks of enzootic pneumonia occurred: the first was between October and December and the second from February to May (Andrews, 2004). The sigs of the disease depend on its clinic form i.e. chronic or acute. The chronic form of BRD generally shows no signs. The calves are bright, eats well but may have a slight mucoid or mucopurulent oculonasal discharge, the body temperature is normal or slightly raised at 38.5-39.5°C and dry, explosive cough usually is produced singly. During the acute case of the disease inappetance, pyrexia (40-42°C), dull sweaty coat, mucoid and mucopurulent oculo-nasal discharge, tachypnoea (respiratory rate- RR over 40 breaths per minute) are presented. Moreover, there is a tendency to a persistent coughing. The cough can be harsh, dry, hacking type but sometimes is moist. During the auscultation of the thorax under both inspiration or expiration there are loud, harsh sounds or whistling, wheezing or squeaking (Andrews, 2004).

At present, BRD is a great economic problem in cattle husbandry, especially in young and feedlot animals all over the world. It accounts for approximately 30% of the total cattle deaths worldwide and is associated with an annual economic losses of over one billion dollars in North America alone (Adamu, 2007). However, the most often outbreak of acute disease occurs when there are calves from different sources in herds. The congestion and inadequate ventilation are the predisposing factors for the development of BRD. Despite that the different pathogens cause of BRD its initial clinical symptoms are very similar i.e. bronchitis and bronchiolititis. In the first phase of the disease the mucus production in diseased calves increases and its stand over in respiratory tract. This is cause a main cause for handicap of defence mechanism in the course of the respiratory disease and the spread of bacterial infections in lungs, which resulted in the intensification of the primary viral infections. The studies recently conducted in Poland have been found the mixed viral infections in field outbreaks of BRD (Kita et al., 1994). The highest specific antibody titres were detected particularly in relation to: BRSV, BHV-1, BVDV, PI-3 and Adeno-3.

During the necropsy the anatomicopathologic changes are observed in the ventral part of the lung lobes and involve, in the decreasing the apical, cardiac and cranial part of the caudal lobes. The area can involved 5 - 40% of the lung tissue. Histologically, the changes concerning intensive accumulations of lymphocytes in the peribronchiolar tissue are usually found, macroscopically it is seen as a mottling of the lesion's cut surface. In the acute pneumonia there are three types of pathological changes. Type 1- pulmonary tissue consolidation - is noted in the cranial lobes of lungs and the tissue is dark red, friable and there is no necrosis changes. The type 2 – in a form of the marked consolidation is very often observed in cranial lobes too with red/grey hepatisation, necrosis and suppuration. The extensive consolidation and suppuration are seen during *A. pyogenes* and *F. necrophorum* infections. The 3th type of pathological entity is characteristic for calves that suddenly developed respiratory diseases. In this case there is interstitial emphysema, pulmonary oedema and congestion with alveolar epithelia hyperplasia and hyaline membrane formation (Andrews, 2004).

Viruses are believed to predispose to bacterial infections in two distinct ways. The first, viral agent can cause direct damage to the respiratory clearance mechanism and translocation of bacteria from the upper respiratory tract. The second way, viral infection can interfere with the immune system's ability to respond to bacterial infections. The viruses can affect the leukocytes causing impairment of their function which result in increased susceptibility to infection of *Mannheimia haemolytica* (Jared et al., 2010a). The virus which seen to be the most often responsible for the appearance of BRD is BVD/MD. The virus may suppress the immune system in the affected animals. The infection of BVD/MD leads to inhibition the production of interferon, a decline in the number of leukocytes and the weakening of humoral immunity by reducing production of antibodies it causes bacterial infections (Polak, M., 2008).

The clinical cases which confirmed the presence of *Mannheimia haemolytica* have a sever character and finishing quick death (Bednarek, 2010). In sick animals are usually found high fever, mucopulurent or pulurent nasal discharge, lacrimation, incidence of painful cough with symptoms of severe shortness of breath, weakness and apathy. In some animals may have been presented watery diarrhea and ill animals not shown willingness to foraging. As a result of this pathogen infection in affected animals occur the extensive damage and inflammation of lung tissue. *M. haemolytica* produces many other potentially virulent factors, among them leukotoxin (Lkt; Hinghlander, 2001; Whitley et al., 1992). The leukotoxin izoform produced by *M. haemolytica* biotype A, serotype 1, has the most visible cytotoxic proprieties in relation to bovine leukocytes. It has been discovered that bovine leukocytes exposed to low doses of exotoxin show reduced phagocytic and killing activity engulfed bacteria. On the other hand, higher concentration of the agent causes complete destruction of the leukocytes leading to their swelling and bursting (Clinkenbeard et al., 1989; Bednarek et al., 2009). The Lkt binds specific leukocyte adhesion molecules, signal-inducing trans-membrane pore formation is generated, leading to efflux of K^+, influx of Ca^{2+}(Clinkenbeard et al., 1989). Many potentially profitable reactive substances (free radicals, lizosomal enzymes, proteases) in relation to phagocytes (netrophils, monocytes) are realised from the destroy cells and then they stimulate different pathological lesions in host affected lung tissue. The process manifested with an acute lobar fibrionecrotising pneumonia. This histopathological picture is characteristic for acute BRD. All the changes are consequences of Lkt action and the development of lung inflammatory cascade regulated additionaly by some pro-inflamatory cytokines (Bednarek et al., 2009). The clinical cases in which the *M. haemolityca* is included have severe clinical courses. According to the literature the intravenous administration of *M. haemolytica* A1 leukotoxin to clinically healthy calves caused the occurrence of the leukopenia manifested with the significant decrease of total peripheral WBC count with lower values of polymorphonuclear leukocyte and MID cell percentages, the last is a total value of all blood peripheral monoctes, eosinophils and basophils. Moreover, there were significant changes regarding some lymphocyte subpopulations such as: $CD2^+$ (T lymphocytes), $CD4^+$ (T helper lymphocytes) and $CD8^+$ (supresor/cytotoxic lymphocytes). The leukopenia resulted from the toxic influence of *M. haemolytica* A1 Lkt, demonstrating a species-specyfic depletion effect with respect to bovine leukocytes (Bednarek et al., 2008). Additionally, the concentrations of some acute phase proteins (CRP, Cp, Tf, Hp, SAA and also eicosanoids (PGE2, $PGF_{2\alpha}$, LTB4) had also significantly changes, because there were their higher values after administration of leukotoxin (Bednarek et al., 2009, 2010).

2. Diagnosis

The initial viral or mycoplasmal diseases are usually mild and are clinically distinguishing. The syndrome can be form subclinical to acute; most are somewhere in between. The initial viral/mycoplasmal diseases causes a moderate fever, sometimes accompanied by constipation. This is followed by rhinitis with a serous-to-mucopulurent discharge and pneumonia with a harsh, hacking cough, tachypnea, dyspnea and diarrhea. The calves very often are depressed, listless and anorectic. The bacterial infection causes intensity these sigs, with higher fevers, more severe dyspnea and depression and sigs of toxaemia. Calves are particularly difficult to auscult and abnormal lungs sounds may be hard to detect. In cases with severe consolidation, the normal breath sounds are replaced by harsh, high-pitched, large airway noises in the anterior-ventral lung fields. When secondary infection with *Pasteurella multocida* occurs the temperature rise to 41 -41.5°C, the area of lung affected is much increased, and increased breath sounds due to congestion are followed by pleuritic friction rub. The acute course is 10 to 14 days. The differential diagnoses should include aspiration pneumonia from improper tubing or feeding practices and purely viral pneumonias such as those caused by IBR, BVD and bovine respiratory syncytial virus (Blood et al., 1983; Smith, 1990).

The laboratory diagnostics of BRD is directly connected with an isolation and identification of suitable species of viruses, bacteria or mycoplasmas presented in a sample tested. These methods are in the various cases the same or similar, but some are specific to a given agent. However, at present the routine diagnostics is divided into two parts. The first one mainly consist of serological methods and the second is nowadays dominated by the molecular biology (PCR and real time-PCR). In the intravital diagnostic process an usual material collected is the nasal swabs or lung lavages and sera samples. On the other hand, post-mortem there are collected tissue samples from lungs and parenchymatous organs (liver, kidneys, spleen). The random amplified polymorphic DNA polymerase chain reaction

(RAPD-PCR) found excellent correlation between lung and nasal isolates. The nasal passages of sick animals may provide clues as to what strain is present in the lung (Jared et al., 2010b).

In the diagnosis of mycoplasmal infection there are used both microbiological, serological and molecular biology methods. It is worth mentioning that mycoplasmas need specific media (Eaton's or Hayflik's medium) and suitable conditions to grow, i.e. 37°C and 5% of CO_2. Mycoplasma culture methods were also described by Autio et al. (Autio et al., 2007). A characteristic feature of mycoplasmas is their growth on solid media in the form of „fried eggs" (Miles, 1998). Some species of mycoplasmas such as *M. bovis* have the ability to create spot and film reactions and the latest increase their resistance to adverse environmental conditions during culture. In order to diversify the presence of mycoplasmas and bacteria in the material from culture microscope method by Diens's was applied. In this method, mycoplasmal colonies are visible due to their ability to absorb dye (Malinowski & Kłossowska, 2002). Polymerase Chain Reaction (PCR) and its modification - real-time PCR (rt-PCR) are the techniques for identifying mycoplasmas from biological material (Sachse et al. 2010; McAuliffe et al., 2005; Miles et al., 2004; Vasconcellos et al., 2000) but they have some limitations. Technique which is devoid of these limitations is denaturing gradient gel electrophoresis (DGGE) that allows differentiation of sixty seven mycoplasma species from one sample, including thirteen bovine pathogens in this *M. bovis, M. dispar, M. bovirhinis* and *M. canis* (McAuliffe et al., 2005). However, to detect the presence of anti-mycoplasma antibodies in sera samples there were applied ELISA tests (Ghadersohi, 2005; Bansal et al., 1995).

The diagnosis of bacterial infections involved in BRD is based on many species-specific methods, such as conventional bacterial cultivation (Autio et al. 2007, Angen et al., 1998), phenotyping characterization (Angen et al., 2002), indole reaction (Autio et al., 2007) or molecular biology techniques. From the latest a PCR method was applied for main bacterial agents of the syndrome, such as *P. multocida* (Miflin & Blackall 2001), *M. haemolytica* (Angen et al., 2009) or *H. somni* (Angen et al., 1998).

In order to identify viral infections in the respiratory syndrome there are used methods detecting the presence of both antigens and specific antibodies. An antigen of some virus species, such as BVDV, BHV1, PIV-3 or BRSV can also be identified using isolation test or Elisa methods (Uttenthal et al., 1996; Autio et al., 2007). From molecular biology techniques to identify the viruses species-specific PCR and rt-PCR methods were applied (Autio et al., 2007; Vilcek et al., 1994). However, the presence of specific anti-viral antibodies in sera samples is possible to detect using Elisa techniques (Anderson et al., 2011) and others.

3. Control

The control of Bovine respiratory disease (BRD) mainly based on therapeutic management because of its multifactorial etology past prophylactic measures, including different vaccination programmes with using both mono- and polyvalent inactivated or modified live vaccines are not sufficiently effective till now. Therefore, at present new advanced, mostly complex - adjunctive therapeutic strategies are widely recommended in order to minimise the economic impact of the respiratory syndrome. According to the generally accepted opinion preferred by Pierre Lekeux (2006), i.e. a very known international expert in this discipline, the fully effective BRD system of treatment should be included three independent steps and the system could be called as "a three-pillar therapeutic strategy of BRD". The first is an elimination of infectious agents using an appropriate antibacterials, the second is modulation of the pulmonary inflammatory reaction and the third - correction of mechanical and secretolitic lung disorders.

However, directly before undertaking of the treatment due to the economic considerations and potencial reduction of therapeutic costs, the field cases of the syndrome should be classified into four grades: Grade 1, subclinical disease (therapy is usually not necessary); Grade 2, compensated clinical disease (at this stage, the inflammatory reaction generated tends to limit the impact of the disease on the animal, this clinical form of the disease needs mainly antibacterial therapy); Grade 3, noncompensated clinical disease (at this stage, the inflammatory reaction is excessive and must be controlled by additional use of anti-inflammatory drugs); Grade 4, irreversible clinical disease (which threatens the animal's survival, this BRD form is not treated because conceivable profitable effects not compensate costs, and affected animals most often die).

The first element of complex therapeutic strategy of BRD aimed quick pathogenic bacteria elimination, particularly these originated from *Pasteurellacae* family (*M. haemolytica, P. multocida, H. somni*) which as important infectious factors participate in the development of pulmonary lesions and dysfunction associated with the syndrome. These bacteria play a crucial role in the pathologic cascade: therefore, the antibiotic must be administered as soon as possible after the induction of the infection, which is most often clinically characterized by hyperthermia, reduced appetite, and nasal discharge. Antibiotic treatment must be initiated before irreversible damage (characterized by oral breathing, orthopnea, lactatemia, and cyanosis) occurs. Among antibiotics presently used most often are administered long acting antibacterials such as some tetracyklines (oxyteracykline), macrolides (florfenicol, tulathromycin, gamythromycin) and fluoroquinolones (enrofloxacin, marbofloxacin, danofloxacin) with a wider antibacterial spectrum included also mycoplasmas (*M. bovis, M. bovirhinis, Ureaplasma diversum*). Significant role of mycoplasma in BRD etiology is not now questioned, and it is especially important because their effective control is very difficult. It is generally known that the mycoplasmas are resistant to beta-lactames and cephalosporins because of the lack of cell wall. The same resistance can be observed to nalidixic acid, polymyxin, rifamycin, tylosin, lincomycin, tylmicosin, trimethoprim and to sulfonamides (Poumarat *et al.*, 1996; Ayling *et al.*, 2007). Therefore, in this case there are intensively searching new more effective generations of antibiotics against mycoplasma infection. The most important mycoplasmal etiological agent in BRD i.e. *M. bovis* as other mycoplasmas is sensitive to antibiotics, which inhibit the protein or nucleic acid synthesis. Recently, in cattle respiratory treatment there are recommended antibiotics which were only applied in swine medicine i.e. pleuromutilins (tiamulin, valnemulin). At present it is known too, that tiamulin has also excellent activity against cattle mycoplasmas like *M. bovis*. In addition an analog compound of tiamulin is valnemulin, which has proven to be affective in the control *M. bovis* infection under field conditions (Stipkovits *et al.*, 2001; Tenk, 2005).

Recently, new antibiotic-treatment conceptions of BRD have been presented during the first European Buiatrics Forum in Marseille (2009). There were described three independent conceptions which in shortening forms are called as SISAAB, SILAAB and MILAAB. The first of them means: Single Injection Shot Acting AntiBiotic among others things represented by new formulations of fluoroquinolones like Marbocyl S (marbofloxacin 100 mg/ml, administered at a single dose of 8 mg/kg b.w. *i.m.*), Baytril One or Enroxil Max (enrofloxacin 100 mg/ml, administered at a single dose of 7.5 mg/kg b.w. *s.c.*). Moreover, also long acting tetracyclines (Tetradur) could be administered here in a form of single intramuscular injections. The second conception is Single Injection Long Acting AntiBiotic represented by new generations of macrolides such as tultromycin (Draxxin) and gamytromycin (Zactran), and also the third one i.e. Multiple Injection Long Acting AntiBiotic which however is considered to be a little controversial conception due to take real risks of antibiotic-resistance increasing.

In the complex of advanced therapeutic strategy of BRD the second its component is the modulation of pulmonary inflammatory reaction. In this aspect, at present there are use anti-inflammatory medicines originated from both steroidal (SAIDs) and non-steroidal anti-inflammatory drugs (NSAIDs) (Lekeux, 2006). Steroids are powerful anti-inflammatory agents, but their effects on the animal's defensive machanisms reduce the value of their use in syndromes of infectious origin unless they have a short duration of action or are administered locally. Generally the drugs stabilise cellular and lysosomal membranes and thereby inhibit release of the chemical pro-inflammatory mediators and proteolytic enzymes. Other steroidal effects include inhibition of antibody synthesis; suppression of activity of fibroblasts; elevation of circulating neutrophil count and reduction in eosinophil count. Steroids also increase microvascular tone and decrease permeability thus reducing exudation and oedema formation. In BRD adjunctive therapy there were used different kinds of corticosteroids include betamethasone (2-10 mg/animal), dexamethasone (2-5 mg/animal), prednisolone (up to 20 mg/animal), cortisone (up to 500 mg/animal), hydrocortisone (up to 300 mg/animal), flumethasone (0.5 mg/animal) and trimcinolone (up to 5 mg/animal). The drugs because of their strong immunosupresive character were usually use only in a single administration throughout of the therapy.

In contrast to steroids, which reduce the yield of all products of arachidonic acid cascade, nonsteroidal anti-inflammatory drugs (NSAIDs) have a narrower anti-inflammatory spectrum, acting as inhibitors of cyclooxygenase. However, NSAIDs have a wider safety margin, which largely compensates for their narrower spectrum. NSAIDs act principally to inhibit the biosynthesis of prostaglandins. They also inhibit kallikrein activity and kinin formation and at the same time pharamacologically antagonise the tissue effects of kinins, prostaglandins and slow-reacting substance of anaphylaxis (SRS-A). NSAIDs are, with some exceptions, analgesic and antipyretic: effects not shown by steroids. Apart from that NSAIDs have also the ability to improve gas exchange what it has been shown in pneumonic calves in experimental conditions (Van de Weerdt *et al*, 1999). The benefits of NSAIDs therapy in bovine respiratory disease have also been demonstrated multiple in a field study (Bednarek *et al*, 2003; Bednarek *et al*, 2004; Lockwood *et al* 2003; Weingarten, 2009).

NSAIDs should be administered to animals at grade 3 of severity, which is mainly characterized by hyperthermia, anorexia, and dyspnea. NSAIDs act rapidly in the pneumonic lung, so the modulation of inflammation and the resulting improvement in the clinical status should follow quickly in the absence of irreversible damages (Lekeux, 2006). The drugs most commonly used in BRD therapy in Europe are flunixin meglumine (2.2

mg/k b.w. *i.v.*), carprofen (1.4 mg/kg b.w. *i.v., s.c.*), ketoprofen (3 mg/kg b.w. *i.v., i.m.*), meloxicam (0.5 mg/kg b.w. *s.c., i.v.*), tolfenamic acid (2 mg/kg b.w. *i.m.*), metamisole sodium (10 - 50 mg/kg b.w. i.m., i.v.).

In the most severe cases of BRD, a short-acting steroidal anti-inflammatory drug, bronchodilator, and diuretic could be added to the NSAID when pulmonary fulminating infalammation, bronchospasm, and edema are present, respectively. These additional supparative drugs i.e. bronchodilators and diuretics are included into the third component of the complex therapeutic strategy of BRD. Among bronchodilators mostly atropine sulfate (cholinolitics) and theophylline (methylxanthines) are administered. The first of them is classified as an anticholinergic drug (parasympatholytic). In general, atropine lowers the parasympathetic activity of all muscles and glands regulated by the parasympathetic nervous system. This occurs because atropine is a competitive antagonist of the muscarinic acetylcholine receptors (acetylcholine being the main neurotransmitter used by the parasympathetic nervous system). Therefore, it may cause bronchictasis and reduced secretions salivary, sweat, and mucus glands. As adjuncts in BRD therapy atropine sulfate 1% is recommended by Lekeux (2002) once a day for three consecutive days at a dose of 0.06 mg/kg b.w. s.c. Theophylline, also known as dimethylxanthine, as a methylxanthine drugs is rarely administered in cattle practice against bronchoconstriction in course of BRD mostly intramuscularly or slowly intravenously at a dose of 1-10 mg/kg b.w. However, it is recommended in veterinary medicine as bronchodilators in adjuntive therapy for respiratory diseases such as BRD, COPD in horses and feline asthma under a variety of brand names. Because of its numerous side-effects, the drug is now rarely administered for clinical use. As a member of the xanthine family, it bears structural and pharmacological similarity to caffeine.

On the other hand within the veterinary diuretics mostly in calves suffering from severe form of BRD caused with pulmonary edema is furosemide (1 mg/kg b.w. i.v.). Furosemide (INN) or frusemide (former BAN) is a loop diuretic used in the treatment of congestive heart failure and edema. It is most commonly marketed by Sanofi-Aventis under the brand name Lasix. It has also been used to prevent Thoroughbred and Standardbred race horses from bleeding through the nose during races.

It should be remember that very important is also to use expectorants (mucolitic drugs) like bromhexine hydrochloride (bromhexine HCl) in order to inflammatory mucose evacuation from the obturated respiratory airways. Bromhexine is a mucolytic agent used in the treatment of respiratory disorders associated with viscid or excessive mucus. In addition, bromhexine has antioxidant properties (Morton, 1999). Bromhexine supports the body's own natural mechanisms for clearing mucus from the respiratory tract. It is secretolytic: that is, it increases the production of serous mucus in the respiratory tract and makes the phlegm thinner and less sticky. This contributes to a secretomotoric effect: it helps the cilia - tiny hairs that line the respiratory tract - to transport the phlegm out of the lungs. For this reason at present it is generally recommended as a adjunct in the complex therapy of BRD. In clinical studies, bromhexine showed secretolytic and secretomotoric effects in the bronchial tract area which facilitates expectoration and eases cough. It is indicated as "secretolytic therapy in bronchopneumonia of calves associated with abnormal mucus secretion, impaired mucus transport and bromhexine has also antiinflammatory properties"

(Bednarek & Kondracki, 2002). Bromhexine also enhances mucus transport by reducing mucus viscosity and by activating the ciliated epithelium. Bromhexine is contained in various formulations (high and low strength syrups, tablets intended mainly for companies animals), however in cattle practice the most useful form is injectable preparation (Eres, Bisolvon, Bisolvomicin, Flegamina) recommended in affected calves at a dose of 0.5 mg/kg b.w. *i.m.* for 5 to 7 consecutive days. Bromhexine is a well established and well tolerated product in its indication.

Other expectorants have been used too in chronic cases of coughing in course of BRD (Andrews, 2004). These include a mixture of strychnine hydroxide, arsenic trioxide and ferric ammonium citrate given at a dose of about 5 ml orally twice daily, or diphenhydramine hydrochloride, ammonium chloride, sodium citrate and menthol at 5-10 ml orally two or three times daily. There is limited benefit form the antihistaminic action of diphenhydramine hydrochloride in cases of calf pneumonia. This resent indication has been accepted too but many clinicians consider that using antihistamines is a little effective in calf bronchopneumonia. This is probably because the main proinflammatory mediator of cattle is not histamine like and human and other mammals but 5-HT (5-hydroxytryptamine). The histamine that is released occurs very quickly following the antibody-antigen reaction so that antihistamines can only be of use in the early stages of the inflammatory response. Among of the drugs in USA recently for the large animal medicine (horses, cattle) are used tripelennamine hydrochloride at a dose of 1 mg/kg b.w. *i.m.* once a day (Divers, 2011). The dose may be repeated in 6 to 12 hours if necessary.

Presented above the basic principles and drugs used for the optimal therapeutic strategy of BRD especially within Grade 3 of the disease are very important to maintain the profitability of cattle breading. This strategy is a combination of an antibiotic acting against the relevant pathogens (eg. florfenicol) and an NSAID acting against the deleterious effects of inflammation (eg, flunixin) additionally supported by the correctors of mechanical and secretolitic lung disorders (Expectorants, bronchodilators). This strategy were confirmed by several experimental and field tests showing an improvement of clinical signs and a reduction of pulmonary dysfunction and lung consolidation in animals receiving such a combined therapy.

In the complex control measures of BRD should be added too about prevention using vaccination programmes. There are various vaccines available both live or attenuated consist of only one or a few bacterial or/and viral antigens. Dead vaccines are used to provide immunity against *P. multocida* and septicemic and pneumonic strains of *M. haemolytica* and *H. somni* (eg, Hiprabovis pneumos, Pastobov, Bovilis Bovipast). Killed or modified live polyvalent vaccines are used in many countries and also in Poland, and contain antigens such as BRSV, PI3 (Rispoval, Bovilis Bovipast), BVDV (Mucosiffa, Bovilis BVD) and BHV1. Recently the last presented antigen is widely utilized in the construction of so-called marker vaccines both live (eg, Rispoval-IBR marker vivum, Hiprabovis-IBR marker live) and inactivated (eg, Ibraxion, Rispoval-IBR-marker inactivatum, Bovilis IBR - marker inactivatum) applied in IBR eradication based on DIVA system i.e. **D**ifferentiating **I**nfected from **V**accinated **A**nimals. The vaccines are usually administered parenterally and usually require two injections to produce immunity. Recent developments have included combined live and dead viral components of the four main antigens which are injected intramuscularly (eg, Rispoval 3). When live vaccines are used, the integrity of the vaccine must ensure absence of other potential contaminats, both viral and other pathogens. Subsequently, modified live intranasal vaccines have been available (eg. Rispoval RS+PI3 IntraNasal). Till now numerous studies have demonstrated that immunization calves via the intranasal route not only is most effective and it gives active immunity in very young animals despite maternal antibodies, it also generates a significant systemic response and interferon induction (Stokes, 2006). Given the relative ease of antigen delivery and the organization of the local lymphoid tissue (MALT), intranasal immunization offers attractive possibilities. However, to successfully achieve this there are a number of obstacles that have to be overcome. For example in the context of IBR vaccination, sometimes, although good immunity was conferred to animals by modified live vaccination, it was shown that some cattle became carriers after exposure to field strains of BHV.

In multifactorial etiology of BRD at present there is also valued the role of mycoplasmal infectious agents. Therefore, many scientific centers all over the world try to produce suitable vaccines mainly against the most important mycoplasma i.e. *M. bovis*. Till now in Europe the saponin inactivated vaccine has been produced in the UK (by The Mycoplsma Group, AHVLA in Webridge), but it has had only limited success experimentally. The inactivated vaccine containing saponin-killed cells was shown to be safe, highly immunogenic and protective against a strong experimental challenge with virulent *M. bovis* (Nicholas *et al.*, 2002). Vaccinated calves showed few respiratory signs while all unvaccinated calves developed signs of pneumonia. Moreover, vaccination gave a statistically significant degree of protection against pyrexia, lung lesions and loss of body weight. The vaccine also reduced the spread of *M. bovis* to internal organs, including the joints. This vaccine is now under commercial development and AHVLA (Weybridge) has a licence to produce this as an autogenous vaccine in the UK. Therefore, practically no commercial vaccines currently exist for *M. bovis* in Europe. Various commercial vaccines, including autogenous preparations, are available in the USA; however there is no published evidence to support their effectiveness.

Recapitulating, the control of the Bovine respiratory syndrome is essential to maintain the profitability of most intensive bovine ventures, and veterinary practitioners have the opportunity to play an important role in this effort. Preventive and therapeutic measures must be adapted to the type of production operation, the specific features associated with the individual animal, the environment and the pathogens, the availability of drugs, and the state of the art in science.

4. References

Adamu, J.Y. (2007): *Mannheimia haemolytica*: phylogeny and genetic analysis of its major virulence factors. *Isr J Vet Med*, 62, pp. 6-13

Anderson, S.; Wakeley, P.; Wibberley, G.; Webster, K. & Sawyer, J. (2011). Development and evaluation of a Luminex multiplex serology assay to detect antibodies to bovine herpes virus 1, parainfluenza 3 virus, bovine viral diarrhea virus, and bovine respiratory syncytial virus, with comparison to existing ELISA detection methods. *J Immunol Methods*, 366, pp. 79-88

Andrews A.H. (2004): Calf respiratory diseases. In: *Bovine medicine*, edited by Andrews A. H., ISBN 0-632-05596-0, Oxford, UK, pp. 239-248

Angen, Ø.; Ahrens, P. & Bisgaard, M. (2002). Phenotypic and genotypic characterization of *Mannheimia (Pasteurella) haemolytica*-like strains isolated from diseased animals in Denmark. *Vet Microbiol,* 84, pp. 103-114

Angen, O.; Ahrens, P. & Tegtmeier, C. (1998). Development of a PCR test for identification of *Haemophilus somnus* in pure and mixed cultures. *Vet Microbiol,* 63, pp. 39-48

Angen, Ø.; Thomsen, J.; Larsen, L.E.; Larsen, J.; Kokotovic, B.; Heegaard, P.M. & Enemark, J.M. (2009). Respiratory disease in calves: microbiological investigations on trans-tracheally aspirated bronchoalveolar fluid and acute phase protein response. *Vet Microbiol,* 137, pp. 165-171

Autio, T.; Pohjanvirta, T.; Holopainen, R.; Rikula, U.; Pentikäinen, J.; Huovilainen, A.; Rusanen, H.; Soveri, T.; Sihvonen, L. & Pelkonen, S. (2007). Etiology of respiratory disease in non-vaccinated, non-medicated calves in rearing herds. *Vet Microbiol,* 119, pp. 256-265

Ayling, R. D.; Godinho, K. & Nicholas, R. A. J. (2007). Comparative studies on the *in vitro* antimicrobial sensitivities of *Mycoplasma mycoides* subsp. *mycoides* small colony type and *Mycoplasma bovis*. Proceedings of the FAO-AU/IBAR-IAEA Consultative Group Meeting on CBPP in Africa, FAO, Rome pp. 51-61

Bansal, P.; Adegboye, D.S. & Rosenbusch, R.F. (1995). Immune responses to the capsular polysaccharide of *Mycoplasma dispar* in calves and mice. *Comp Immunol Microbiol Infect Dis,* 18, pp. 259-268

Bednarek, D. & Kondracki, M. (2002). Explorations on anti-inflammatory effects of bromhexine (Bisolvon®) on the cellular immunity of calves with experimentally induced local lung inflammation. XXII World Buiatrics Congress, Hannover, pp. 37

Bednarek, D.; Kondracki, M.; Friton, G.; Trela, T. & Niemczuk K. (2004). Effect of steroidal and non-steroidal anti-inflammatory drugs on inflammatory markers in with experimentally-induced bronchopneumonia. The XXIII World Buiatrics Congress, Québec, pp. 77

Bednarek, D.; Szymańska-Czerwińska, M. & Dudek, K. (2010). The effect of leukotoxin *Mannheimia haemolytica* A1 on inflammatory response in calves. *Medycyna Wet,* 66, pp. 400-404

Bednarek, D.; Urban-Chmiel, R. & Dudek, K. (2008). Protective effect of Pastobov and alternations of peripheral blood leukocytes subpopulations in calves experimentally challenged with *Mannheimia haemolytica* A1 leukotoxin. XXV WBC M Allator LAPJA 2008, 130, supp. 2, pp. 231-232

Bednarek, D.; Urban-Chmiel, R.; Dudek, K. & Szymańska-Czerwińska, M. (2009). Evaluation of peripheral blood leukocyte subpopulation by flow cytometry in calves treated with *Mannheimia haemolytica* leukotoxin. *Bull Vet Inst Pulawy,* 53, pp. 199-20

Bednarek, D.; Zdzisińska, B.; Kondracki, M. & Kandefer-Szerszeń, M. (2003). Effect of steroidal and non-steroidal anti-inflammatory drugs in combination with long-acting oxytetracycline on non-specific immunity of calves suffering from enzootic bronchopneumonia. *Vet Microbiol,* 96, pp. 53-67

Bednarek, D.; Zdzisińska, B.; Kondracki, M.; Rzeski, W.; Paduch, R. & Kandefer-Szerszeń, M. (2003). A comparative study of the effects of meloxicam and flunixin meglumine (NSAIDs) as adjunctive therapy on interferon and tumor necrosis factor production in calves suffering from enzootic bronchopneumonia. *Pol J Vet Scienc*, 6, pp. 109-115

Blood, D.C.; Radostitis, O., M; Henderson, J.A. (1983). Veterinary Medicine. ISBN 0-7020-0987-3, London, pp. 794-795

Clinkenbeard, K.D.; Mosier, D.A. & Confer, A.W. (1989). Transmembrane pore size and role of cell swelling in cytotoxicity caused by *Pasteurella haemolytica* leukotoxin. *Infect Immun*, 57, pp. 420-425

Divers, T.J. (2011). Bovine respiratory diseases. In: Diseases of dairy cattle. Divers T.J. & Peek S.F. Polish Edition by Twardoń J & Fabisiak M., Elsevier, pp. 120

Dudek, K.; Bednarek, D. & Szymańska-Czerwińska, M. (2010). Acute chase response in calves as a result of experimental challenge with *Mycoplasma bovis. Bull Vet Inst Pulawy*, 54, pp. 517-520

Fulton, R.W.; Cook, B.J. & Step, D.L. at all (2002). Evaluation of health status of calves and the impact on feedlot performance: Assessment of a retained owner-ship program for postweaning calves. *Can J Vet Res* , 66, pp.173-180

Ghadersohi, A.; Fayazi, Z. & Hirst, R.G. (2005). Development of a monoclonal blocking ELISA for the detection of antibody to *Mycoplasma bovis* in dairy cattle and comparison to detection by PCR. *Vet Immunol Immunopathol*, 104, pp. 183-193

Haines, D.M.; Martin, K.M.; Clark, E.G.; Jim, G.K. & Janzen, E.D. (2001). The immunohistochemical detection of *Mycoplasma bovis* and bovine viral diarrhea virus in tissues of feedlot cattle with chronic, unresponsive respiratory disease and/or arthritis. *Can Vet J*, 42, pp. 857-860

Highlander, S.K. (2001). Molecular genetic analysis of virulence in *Mannheimia* (*Pasteurella*) *haemolytica. Front Biosci*, 6, pp. 1128-1150

Houghton, S.B. & Gourlay, R.N. (1983). Synergism between *Mycoplasma bovis* and *Pasteurella haemolytica* in calf pneumonia. *Vet Rec*, 113, pp. 41-42

Jared, D.; Fulton, R.W.; Lehenbauer, T.W.; Douglas, L.S. & Confer, A.W. (2010a) The epidemiology of bovine respiratory disease: What is the evidence for predisposing factors? *CVJ*, 51, pp. 1095-1102

Jared, D.T.; Fulton, R.W.; Lehenbauer, T.W.; Douglas, L. S. & Confer, A.W. (2010b). The epidemiology of bovine respiratory disease: what is the evidence for preventive measures? *CVJ* , 51, pp. 1351-1359

Jungi, T.W.; Krampe, M.; Sileghem, M.; Griot, C. & Nicolet J. (1996). Differential and strain-specific triggering of bovine alveolar macrophage effector functions by mycoplasmas. *Microb Pathog*, 21, pp. 487-498

Kita, J.; Ochmańska - Hecold, M. & Peryt, T. (1994). Mixed viral infections of calves in bronchopneumonia out-breaks. *Medycyna Wet*, 51, 459-461

Klimentowski, S.; Folwarczyn, J. & Repuła, K. (1995). Infections of the respiratory tract caused by viruses -serological survey. *Medycyna Wet*, 51, pp. 459-461

Lekeux P., (1995) Bovine Respiratory disease Complex: An European Perspective. *Bovine Pract*, 29, pp. 71-75

Lekeux, P. ; Borceux, J. ; Boutet, P. ; Bureau, F. ; Coghe, J. & Uystepruyst, C. (2002). Recent advances in bovine pneumology. In : Recent developments and perspectives in bovine medicine. Ed. by Kaske M et al. XXII WBC Hanover pp. 144-149

Leukeux, P. (2006) A therapeutic strategy for treatment of the bovine respiratory disease complex: the rationale for the combination of a nonsteroidal antiinflammatry drug with an antibiotic. In: BRDC – setting a new therapeutic standard: a multimodal approach. XXIV World Buiatrics Congress Nice France, pp. 8-11

Lockwood, P.W.; Johnson, J.C. & Katz, T.L. (2003). Clinical efficacy of flunixin, carprofen and ketoprofen as adjuncts to the antibacterial treatment of bovine respiratory disease. *Vet Rec*, 152, pp. 392-394

Lysnyansky, I.; Rosengarten, R. & Yogev, D. (1996). Phenotypic switching of variable surface lipoproteins in *Mycoplasma bovis* involves high-frequency chromosomal rearrangements. *J Bacteriol*, 178, pp. 5395-5401

Malinowski, E. & Kłossowska, A. (2002). *Diagnostyka zakażeń i zapaleń wymienia*, PIWet-PIB, ISBN 83-907862-6-5, Puławy, Polska

McAuliffe, L.; Ellis, R.J.; Lawes, J.R.; Ayling, R.D. & Nicholas, R.A. (2005). 16S rDNA PCR and denaturing gradient gel electrophoresis; a single generic test for detecting and differentiating *Mycoplasma* species. *J Med Microbiol*, 54, pp. 731-739

Miflin, J.K. & Blackall, P.J. (2001). Development of a 23S rRNA-based PCR assay for the identification of *Pasteurella multocida*. *Lett Appl Microbiol*, 33, pp. 216-221

Miles, K.; McAuliffe, L.; Ayling, R.D. & Nicholas, R.A. (2004). Rapid detection of *Mycoplasma dispar* and *M. bovirhinis* using allele specific polymerase chain reaction protocols. *FEMS Microbiol Lett*, 241, pp. 103-107

Miles, R.J. & Nicholas, R.A.J. (1998). *Mycoplasma Protocols*, Humana Press, ISBN 0-89603-525-5, Totowa, USA

Morton, I. & Hall J. (1999). *Concise Dictionary of Pharmacological Agents*. Springer. p. 55. ISBN 0751404993. Retrieved 2009-06-03

Nicholas, R. A. J., Ayling, R. D. & Stipkovits, L. (2002). An experimental vaccine for calf pneumonia caused by *Mycoplasma bovis*. *Vaccine*,. 20, pp. 3569-3575

Nicholas, R.A. & Ayling, R.D. (2003). *Mycoplasma bovis*: disease, diagnosis, and control. *Res Vet Sci*, 74, pp. 105-112

Polak M. (2008). Zakażenie wirusem BVD-MD i jego rola w etiopatogenezie syndromu oddechowego bydła. In: *Najważniejsze czynniki etiologiczne, patogeneza i najnowsze trendy w profilaktyce i terapii syndromu oddechowego bydła (BRD)*, edited by Bednarek, D., pp. 22-30

Potgieter, L.N. (1997). Bovine respiratory tract disease caused by bovine viral diarrhea virus. *Vet Clin North Am Food Anim Pract*, 13, pp. 471-481

Poumarat, F.; LeGrand, D. & Bergonier, D. (1996). Propriétés générales des mycoplasmes et hypervariabilité antigénique. *Point Vet*, 28, pp. 761-767

Radaelli, E.; Luini, M.; Loria, G.R.; Nicholas, R.A. & Scanziani, E. (2008). Bacteriological, serological, pathological and immunohistochemical studies of *Mycoplasma bovis* respiratory infection in veal calves and adult cattle at slaughter. *Res Vet Sci*, 85, pp. 282-290

Razin, S.; Yogev, D. & Naot, Y. (1998). Molecular biology and pathogenicity of mycoplasmas, *Microbiol Mol Biol Rev*, 62, pp. 1094-1156

Reggiardo, C. (2005). Role of virus in shipping fever of feedlot cattle. Case studies and diagnostic consideratiuons. 22nd Annual Preceedings Amer Assn Veterinary diagnosticians, pp. 315-320

Rosengarten, R.; Behrens, A.; Stetefeld, A.; Heller, M.; Ahrens, M.; Sachse, K.; Yogev, D. & Kirchhoff, H. (1994). Antigen heterogeneity among isolates of *Mycoplasma bovis* is generated by high-frequency variation of diverse membrane surface proteins. *Infect Immun,* 62, pp. 5066-5074

Sachse, K.; Salam, H.S.; Diller, R.; Schubert, E.; Hoffmann, B. & Hotzel, H. (2010). Use of a novel real-time PCR technique to monitor and quantitate *Mycoplasma bovis* infection in cattle herds with mastitis and respiratory disease. *Vet J,* 186, pp. 299-303

Smith, B.P. (1990). Large animal internal medicine. ISBN 0-8016-5062-3, Toronto, pp. 580

Stipkovits, L.; Ripley, P.H.; Varga, J. & Palfi, V; (2001). Use of valnemulin in the control of *Mycoplasma bovis* infection under field conditions. *Vet Rec,* 148, pp. 399-402

Stokes, Ch. (2006). Intranasal vaccination against respiratory disease in animals: a review. In: Intranasal BRSV and PI3 vaccination: innovation for the prevention of respiratory disease in cattle .XXIV World Buiatrics Congress Nice France, pp. 7-9

Szymańska-Czerwińska, M.; Dudek, K. & Bednarek, D. (2010). Occurrence and diagnosis of mycoplasma infections in ruminants. *Medycyna Wet,* 66, pp. 597-599

Tenk, M. (2005). Examination of *Mycoplasma bovis* infection in cattle. Doctoral Thesis, Szenti István University, Budapest, pp. 16-17

Thomas, C.B.; Mettler, J.; Sharp, P.; Jensen-Kostenbader, J. & Schultz, R.D. (1990). *Mycoplasma bovis* suppression of bovine lymphocyte response to phytohemagglutinin. *Vet Immunol Immunopathol,* 26, pp. 143-155

Thomas, C.B.; Mettler, J.; Sharp, P.; Jensen-Kostenbader, J. & Schultz, R.D. (1990). *Mycoplasma bovis* suppression of bovine lymphocyte response to phytohemagglutinin. *Vet Immunol Immunopathol,* 26, pp. 143-155

Thomas, C.B.; Van Ess, P.; Wolfgram, L.J.; Riebe, J.; Sharp, P. & Schultz, R.D. (1991). Adherence to bovine neutrophils and suppression of neutrophil chemiluminescence by *Mycoplasma bovis. Vet Immunol Immunopathol,* 27, pp. 365-381

Uttenthal, A.; Jensen, N.P. & Blom, J.Y. (1996). Viral aetiology of enzootic pneumonia in Danish dairy herds: diagnostic tools and epidemiology. *Vet Rec,* 139, pp. 114-117

Van de Weerdt, M.L.; Coghe, J. & Uystepruyst, C. (1999). Ketoprofen and phenylbutazone attenuation of PAF-induced lung inflammation in calves. *Vet J* 157, pp. 39-49

Vanden Bush, T.J. & Rosenbusch, R.F. (2002). *Mycoplasma bovis* induces apoptosis of bovine lymphocytes. *FEMS Immunol Med Microbiol,* 32, pp. 97-103

Vanden Bush, T.J. & Rosenbusch, R.F. (2003). Characterization of the immune response to *Mycoplasma bovis* lung infection. *Vet Immunol Immunopathol,* 94, pp. 23-33

Vasconcellos, C.M.; Blanchard, A.; Ferris, S.; Verlengia, R.; Timenetsky, J. & Florio Da Cunha, R.A. (2000). Detection of *Ureaplasma diversum* in cattle using a newly development PCR-based detection assay. *Vet Microbiol,* 72, pp. 241-250

Vilcek, S.; Elvander, M.; Ballagi-Pordány, A. & Belák, S. (1994). Development of nested PCR assays for detection of bovine respiratory syncytial virus in clinical samples. *J Clin Microbiol,* 32, pp. 2225-2231

Weingarten, A.J. (2009). Mechanisms of action and the role of anti-pyretic and anti-inflammatory intervention in the treatment of bovine respiratory disease. In: Getting it right the first time: best practices in BRD treatment. EBF Marseille, pp. 6-15.

Whitleley, L.O.; Maheswaran, S.K.; Weiss, D.J.; Ames, T.R. & Kannan, M.S. (1992). *Pasteurella haemolytica* A1 and bovine respiratory disease: Pathogenesis. *J Vet Int Med,* 6, pp. 11-22

Wikse, S.E. (1985). Feedlot cattle pneumonia. *Vet Clin North Am Food Anim Pract,* 1, pp. 289-310

20

Epidemiology, Zoonotic Aspect and Current Epidemiological Situation of Q Fever in Poland

Krzysztof Niemczuk and Monika Szymańska-Czerwińska
National Veterinary Research Institute, Department of Cattle and Sheep Diseases, Pulawy, Poland

1. Introduction

Q fever (for query fever) is a zoonosis caused by *Coxiella burnetii,* a small gram-negative (0.2 to 0.4 µm wide, 0.4 to 1 µm long), obligate-intracellular bacterium. Historically, it was considered as *Rickettsia* but gene-sequence analysis classifies *Coxiella* genus in the order *Legionellale,* family *Coxiellaceae* (Seheradi, 2003).

Pleomorphic bacteria are classified in three groups: large cell variants (LCV), small-cell variants (SVC) and small dense cells (SDC). The SCVs and the SDCs are the most frequent forms in the host and certainly (the most) resistant forms in the environment. The LCV form of *Coxiella burnetii,* which shares features common with gram-negative bacteria, diffuse chromatin and process clearly distinguishable outside and cytoplasmic membranes with LPS exposed on the surface. The LCVs are larger, more pleomorphic and metabolically active than the SCVs and the SDCs. The SDCs have been visualised in LCV as endospores, and they may be liberated upon the lysis of LCV or binary transverse fission with unequal cell division. The SDCs alone have never been isolated, free-living amoeba can provide an intra-cellular niche for SDC formation and survival of *Coxiella burnetii* in the environment. Binary transverse fission was observed both in the SCVs and the LCVs in cells cultured. The form changes within lifecycle of *Coxiella burnetii* are strategy to survive (Arricau-Bouvery & Rodolakis, 2005).

Coxiella burnetii occurs in two phases. In the culture, it transforms from a virulent phase I type to an avirulent phase II. In the phase II, bacteria have alerted expression of cell wall lipopolysaccharide (LPS), do not occur in nature and are killed by macrophages (Hotta et al., 2002). The survival strategy of the bacteria is based on multiplication in mature phagosomes of monocytes/macrophages. The virulence is connected to the LPS expression. The virulent *Coxiella burnetii* entered into monocytes through $\alpha_v\beta_3$ integrin and survived inside the cells (Dellacasagrande et al., 2000). Avirulent variants were more easily ingested than virulent bacteria but they were eliminated by monocytes. Their phagocytosis was mediated by $\alpha_v\beta_3$ integrin-IAP complex and CR3, suggesting that the efficiency of *Coxiella burnetii* phagocytosis mainly results from the activation and CR3-dependent phagocytosis, thus preventing CR3 engagement. It is possible that *Coxiella burnetii* induced impairment of CR3 function results from uncoupling $\alpha v\beta_3$ integrin from IAP (Capo et al., 1999).

The heterogeneity among strains of *Coxiella burnetii* is low degree (Vodkin et al., 1986). However, when DNA from 38 *Coxiella burnetii* isolates was examined by using restriction fragment length polymorphism (RFLP) analysis, six genomic groups (I to VI) were detected (Hendrix et al., 1991). The genome size of *Coxiella burnetii* nine Mile strain amounts to 2.1 Mb. The size of genome is highly variable among different isolates and ranging from 1.5 to 2.4 Mb (Willems et al., 1998). Genomic groups (I, II and III) are associated with animal, tick or acute Q fever in human, whereas group IV and V are isolated from human Q fever endocarditis cases. Group VI isolates were obtained from feral rodents in dungway, and their pathogenicity is unknown (Maurin & Raoult, 1999).

Coxiella burnetii is the well-known causative agent of Q fever, but neglected as zoonosis. Q fever has spread worldwide both in humans and in animals (Kruszewska et al., 1996). *Coxiella burnetii* can infect many animal species including mammals, birds and arthropods such as ticks. Generally, Q fever is asymptomatic, but it can lead to abortions and stillbirths in mammals. In these animals, *Coxiella burnetii* can induce pneumonia as well as abortion, stillbirth and delivery of weak lambs, calves or kids, which are the most frequent clinical sings of the disease. *Coxiella burnetii* in humans causes highly variable clinical manifestations ranging from acute to fatal chronic infections. However, about 60% of the infections are asymptomatic seroconversions. Acute Q fever is mainly a flu-like disease, or atypical pneumonia or hepatitis. Q fever is essentially an airborne disease. The infections occur after inhalation of aerosols generated from infected placentas, body fluids or contaminated manure. Transmission of this pathogen is generally associated with abortion of domestic ruminants, particularly sheep (Arricau-Bouvery & Rodolaski, 2005).

Tetracyclines are the best for treating Q fever. Although, Q fever endocarditis has been treated by the use of the combination of doxycyline with chloroquinolone. The time of treatment is very long and it takes for 18 to 36 months in order to cure chronic Q fever (Maurin & Raoult, 1999).

2. Entry and survival *Coxiella burnetii* in the cells

Coxiella burnetii after the entry into the organism, is phagocytosed by macrophages. The entry of *Coxiella burnetii* into monocytes and macrophages is lakely to be critical for its adaptation to host cells and development of Q fever (Capo et al., 1999). The bacterial cells multiply in phagolyosome and next cells burst, spreading the pathogens throughout the body. In animals, they are mostly located in the lung, mammary gland, testis, lymph nodes, uterus and placenta. The disease process may be also extended to the liver and cardiovascular system. There are views that the animal can be infected lifelong but it takes place in a latent form. The amplification of larger number of bacteria and prevalence of typical symptoms of Q fever occurs during pregnancy or by the action of immunosuppressive agents. *Coxiella burntii* survives and divides in a phagolisosome (Fig.1). The small and large forms express different genes that permit the bacterium to survive in the acidified environment of phagolisosome (Arricau-Bouvery & Rodolaski, 2005).

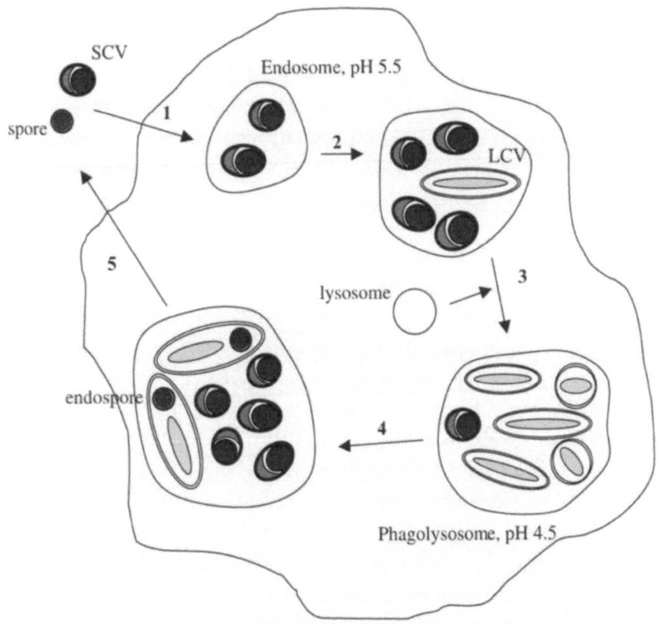

Fig. 1. Life cycle of the bacteria (Arricau-Bouvery & Rodolakis, 2005) (1). Entry of the spore in the eukaryotic cell and acidification of the endosome of the phagosome. (2) Multiplication of small-cell variants (SCV) by transverse binary fission and differentiation to large-cell variants (LCV). (3) Fusion of the endosome with the lysosome, acidification of the phagolysosomal. (4) Multiplication of LCV by transverse binary fission, differentiation of LCV to SCV and development of the polar endospore in LCV. (5) Release of the spore and SCV out of the cell.

3. Epidemiology

3.1 Zoonotic aspects

In humans, Q fever occurs the most often in direct contact with infected animals in slaughterhouses, as well as fur and veterinarians. Infections from human to human are very rare, and may have place especially in hospitals where patients infect the others through sputum released during coughing. Sporadic human infections of Q fever occurred after contact with infected parturient women, via transplacental transmission resulting in congenital infections, during autopsies, intradermal inoculation or blood transfusion. It has been postulated that transmission of *Coxiella burnetii* among mammals occurs mainly via inhalation of infected dust and aerosols (Maurin & Raoult, 1999). It is very difficult to explain some cases of *Coxiella burnetii* infections. The literature dates showed that there is possibility of sexual transmission of Q fever among mammals. The studies showed that heifers infected intravaginally with a suspension of *Coxiella burnetii* shed bacteria in their urine few days later. Moreover studies on nine Polish patients, who were employed in Spain during sheep shearing season showed occurrence of antibodies to *Coxiella burnetii* antigens in their serum. The antibodies were detected also in their spouses but they did not occur in sera from the other family members. Moreover, scanning electron microscopy of spermatozoa cells from patients with Q fever has revealed the presence of attached bacteria. (Kruszewska et al., 1993, 1996).

The incidence of Q fever in human occur very often during spring and early summer when it is warm, dry and windy because the environmental survival allows it to be transported by wind far away from its original source. Such environmental conditions in Jenna (German) in June 2005 caused that *Coxiella burnetii* - infected sheep, grazing and lambing on the meadow bordering a residential brought about infection in 331 human (Gildsdorf et al., 2007).

The source of human infections is often unknown, although sheep and goats are more frequently related to Q fever outbreaks in humans (Glisdorf et al., 2005). The largest Q fever outbreaks occurred in the Netherlands, involving 3,921 human cases in the years 2005-2009. The dairy goats and sheep were considered the source of the human Q fever outbreak. The disease was found in about 60 farms these animals (Roest et al., 2011).

The infection by ingestion (mainly drinking raw milk) is probably a minor factor of risk, and at present it constitutes even a point of controversy (Maurin & Raoult, 1999; Krumbiegel & Wiśniewski, 1996). Drinking contaminated milk has induced seroconversion in human volunteers but without clinical signs (Benson, 1963). However, some studies have reported clinical diseases linked to the ingestion of cheese (Hatchette et al, 2001; Fishbein & Raoult, 1992) but infection by inhaling contaminated dust or aerosols cannot be excluded. The serological and molecular studies performed in the herds, where antibodies were present in blood of lactating cows, showed that *Coxiella burnetii* DNA was detected in the milk (8.7%). The antibodies were detected only in the serum of 1 % of young cattle selected from these herds (Muskens, 2011).

Coxiella burnetii in humans causes highly variable clinical manifestations ranging from acute to fatal chronic infections. Initially, there is typically a noticeable slight increase in internal body temperature lasting for several days. In addition, eye inflammation is stated, as manifested by greater amounts of fluid seromucinous flowing from the conjunctiva sac and nose. These general symptoms do not give rise to suspicion of Q fever. Sometimes only the prevalence of many cases of people with the acute symptoms of flu can indicate the occurrence of the disease in animals. It comes more often to premature births and miscarriages. Swelling and local haemorrhages can be observed in the placenta of infected animals. The further course of disease depends on the degree of virulence of the agent and the immune status of animals. Acute form of Q fever is mainly a flu-like disease, atypical pneumonia or hepatitis. Atypical pneumonia is characterised by fever, headache and myalgia. Despite the prevalence of pneumonia, cough can be absent. The hepatitis could be asymptomatic, only with the raise of transaminase levels, or the infectious hepatitis - rarely with jaundice (Fournier et. al., 1998).

In about 5% of the cases, the disease may become chronic form and it can lead to the endocarditis, chronic fatigue syndrome or repeated abortions. Endocarditic is the most frequent. It represents 1.5 - 2.5% all human cases of endocarditis. Chronic fatigue syndrome is characterised by an inappropriate fatigue, myalgia, arthralagia, night sweats and changes in mood. This chronic forms occur very often in Australia (Arricau-Bouvery & Rodolaski, 2005). In pregnant women, *Coxiella burnetii* can cause placentitis leading to abortion, neonatal death, premature birth and low birth weight. The risk of chronic Q fever leading to repeating miscarriages, is very high when the infection occurs during pregnancy (Arricau-

Bouvery & Rodolaski, 2005). Despite the isolation of *Coxiella burnetii* in the placenta and milk sick of mothers, children are born healthy (Lepe et al., 1999). Mortality in humans is relatively low at 1-2%, except for tropical countries where it amounts to up to 9% (Moquin et al., 2002).

3.2 Reservoirs

Cattle, goats and sheep are considered the primary reservoirs that human contamination comes from. Epidemiological data indicated that dairy cows are more frequently chronically infected than sheep, and they are the most important source of human infections. The duration time of *Coxiella burnetii* excretion in infected animals depends on species and type of samples tested (Arricau-Bouvery N. & Rodolakis, 2005) (Tab. 1).

Ticks have been considered a reservoir and a vector of *Coxiella burnetii*. In the nature, there are about 40 species of ticks transmitting the infections including *Rihicephalus sanguineus, Amblyomma triuguattum* (Maurin & Raoult, 1999). The infection is transmitted from one stage to the next in the ticket cycle, and transovarian infection is observed. Ticks become infected after ingestion of infected mammal blood that in the time of the bacteremia. Microorganisms multiply mainly in the epithelial cells of the middle tick gut, then penetrate the intestinal wall, hemocytes and connective tissue of internal organs. Hence an infected tick is a carrier of the *Coxiella burnetii* for life, and its saliva may contain very high concentration of agent (10^8 - 10^{11}) (Niemczuk, 2010). *Coxiella burnetii* in ticks, as in mammals, are in phase I and they are highly infectious. Morover, ticks are not considered essential in the natural cycle of *Coxiella burnetii* infections in livestock because animals can have many other opportunities to become infected. Ticks may play a significant role in the transmission of coxiellosi among the wild birds and vertebrates such as rodents and lagomorphs (Maurin & Raoult, 1999).

Cats and dogs may be reservoirs of *C. burnetii*. Dogs may be infected by tick bite, consumption of placentas or milk from infected ruminants, and the aerosol route. In the literature, the cases of Q fever human infections after contact with parturient cats are described in Nova Scotia (Kosatsky, 1984). Moreover, wild mammals, including horses, rabbits, swine, camels, water buffalo, rats and mice constitute the reservoirs of *Coxiella burnetii*. The serological studies of rats in the UK have shown anti-phase II antibody seroprevalences ranging from 7 to 53% among wild brown rat populations (Webster et al., 1995).

Birds, including pigeons, chickens, ducks, geese and turkeys can be infected with coxiellosis. Infected domestic poultry can be the sources of infections in human. People become infected by consumptions of raw eggs or inhalation of infected fomites (Maurin & Raoult, 1999).

According to world experts on counter-terrorism, the agent that causes Q fever, is considered as an effective component of biological weapon (Arricau & Rodolaski, 2005). It is a potential agent of bioterrorism because of its accessibility, low infection dose, resistance to environmental degradation, and aerosol route of transmission. The minimum number of pathogens is required for human infection is 1-10 microorganisms. According to WHO, 50 kilograms of powder containing *Coxiella burnetii* can cause the comparable number of cases to tularemia or anthrax (Spicer, 1994; Moquin 2002). Q fever, although not characterised by high mortality, is important in wartime because the disease is severe, and the patient requires long-term care. Furthermore, the high resistance of the bacteria to drying, moisture, high or low temperature, promotes long-term being in the air, on skin, wool and straw

(Prusakowski, 2001), for example experiencing up to 2 years in human faeces, to 6 years in the ticks' feces, 2-3 years in the water, 1 year in the fur, 1 month in meat, and 1-2 months in dairy products (Little, 1999; Maurin & Raoult, 1999). Physical factors and especially pasteurisation kills germs in 15 seconds, while ultraviolet radiation is not an effective way to get rid of this microbe.

4. Prevention

Prevention of *Coxiella burnetii* shedding in cattle is critical to control the spread of the pathogen between animals and from animals to humans. A herd or a flock should be considered clinically infected with Q fever when abortion or stillbirths have occurred, positive PCR results confirming the presence of *Coxiella burnetii* on specimens from affected animals, positive results in the serology test. In cattle herds, the occurrence of a series of abortion: for herds less than 100 animals 2 abortion or more in the month or 3 abortions in the year; for herds more than 100 animals more than 4% of cows with abortions in the year, is the major warning sing to be taken into account (Sidi-Boumedine et al., 2010).

Proceedings in the environment of healthy animals include a serological test of animals fur, quarantine of imported animals, the prohibition on animal breeding farms from areas where infection occurred, the serological tests of crossing animals, the fight against ticks and rodents. If Q fever occurs, the following activities should be performed: the district veterinarian must notify diagnostic laboratory indicating a positive result, sporadic cases should be eliminated, the animals should be selected to serologically positive and negative in the case of numerous infections in the herd, the dynamics of antibodies should be studied in animals showing the presence of antibodies, disinfection of rooms should be performed, farrowing should be separated, placenta and the stillborn foetuses should be protected after each delivery, the direction of Q fever should be examined in newborn animals.

The newest study made in Denmark reported an increased risk of becoming infected with *Coxiella burnetii* for herds where farmers had bought cattle from other herds. Therefore, as a general rule, the purchase of animals should be avoided as much is possible, even if tests before introduction are implemented (EFSA Panel on Animal Health and Welfare (AHAW), 2010).

Prevention of *Coxiella burnetii* shedding in cattle or sheep and goat is critical to control the spread the pathogen. Q fever vaccination of animals is not widely used because it is protective and safe only for animals that are uninfected at the time of vaccination. Moreover, extensive evaluation of the protective effect of such a vaccination on animal and human health compared to the cost of such prophylactic measures is lacking. There are two types of vaccines. The first one contains formalin-killed whole-cell vaccine preparations (WCV) and the second one - chloroform methanol-extracted bacterial residue (CMR). A WCV from the Henzerling strain (Q vax, CSL Limited, Parkville, Victoria, Australian) has been commercially available and used from human vaccination in Australia since 1989 (Arricau-Bouvery & Rodolaski, 2005). Among 942 no immune abattoir workers in Australia who were vaccinated, no Q fever case was diagnosed during 18 month whereas 34 cases were recorded among 1,349 unvaccinated controls (Marmion et al., 1984). This vaccination is very useful in the protection of exposed populations such as veterinarians or breeders of animals. A soluble vaccine containing trichloroacetic acid-extracted antigen from phase I *Coxiella burnetii* nine mile strain was successfully used in Czechoslovakia. In animal, the vaccines composed of inactivated whole phase I bacteria are the most effective. Bacterial shedding in milk and placentas was strongly reduced in infected animals vaccinated with phase I vaccines. Available vaccines are prepared with a single strain and therefore, they are monovalent. The recent works have shown that the vaccines reduce the shedding of *Coxiella burnetii*. These results are very important because of limiting the clinical disease in animals and human (Arricau-Bouvery & Rodolaski, 2005).

5. Laboratory diagnosis

The diagnosis of Q fever remains difficult and epidemiological studies. Regarding the diagnosis of Q fever, several tools are available for direct and indirect detection of *Coxiella burnetii*. The studies are often based on the serological investigations. Although, the serological methods are useful. They do not allow for the identification of *Coxiella burnetii* shedding animals. Some infected animals can be seropositive without sheding *Coxiella burnetii*, and the others can shed the bacteria and remain seronegative, what is of great concern and could have an important impact both on animal and public health. Moreover, the serological test can distinguish vaccinated and naturally infected animals. The complement fixation test (CFT) was proposed as an alternative test for international trade by the Office International des Epizooties (OIE). Additionally, alternative tests such as enzyme-linked immunosorbent assay (ELISA) and indirect immunofluorescence assay (IFA) can be used for diagnosis Q fever within a local context, and it can also be used during importing and exporting animals following bilateral agreements. At present, there are three ELISA commercial kits for diagnosis Q fever in domestic ruminants whereas commercial kits using the IFA for veterinary investigation are not available. The diagnosis of Q fever is based on the serological test and the immunofluorescence assay. The ELISA and IFA tests are more sensitive than the CFT for detection of antibody response in animals or humans with acute Q fever but not for the detection of antibody response in humans with chronic Q fever and in aborting cows (Sidi-Boumedine et al., 2010). The ELISA and IFA detect antibody in acute-phase sera, and they are quite efficient in binding IgM (these antibodies predominante in the acute stage), in contrast to the CF test. Recent infections are difficult to demonstrate serologically by the CFT, because antibodies detected by this test can persist longer than acute stage. However, chronic and persistent infections of *Coxiella burnetii* are difficult to diagnose by the ELISA, because this test is more accurate in the acute stage of Q fever, when the IgM predominate in the response (Jóźwik et al., 2007). Cross-reactions are the biggest source of false negative results during interpreting serological results. Cross-reactions have been described between *Coxiella burnetii* and *Legionella pneumophila* (Deyer et al., 1988; Finidori et al., 1992), *Legionella micdadei* (Musso & Raoult, 1997), and *Bartonella quintana* and *Bartonella henselae* (La Scola & Raoult, 1996).

Other techniques used for Q fever diagnosis include dot immunobloting, Western blotting, indirect hemolysis test and radioimmunoassay. Western blotting has been reported to be sensitive and specific. The molecular masses of antigens detected vary from 10 to 100 kDa. Antibodies reacting with 50, 80, 160 kDa antigens are considered indicative of chronic Q fever (Blondeau et al, 1990). However Western blotting is very time consuming technique and the results obtained by this technique are not reproducible. Dot blotting and radioimmunobloting are sensitive and specific technique but radioimmunoblotting can be performed only in radioactivity equipped laboratories (Cowley et al., 1992, Doller et al., 1984).

PCR is rapid and the most sensitive technique for direct diagnosis of *Coxiella burnetii* and identification of shedding animal. Several PCR diagnostic assays were developed to detect *Coxiella burnetii* DNA including conventional PCR, nested PCR and real-time PCR (Klee et al., 2006). The PCR is adapted to a wide range of samples. The results of conventional PCR are presented in figure 2. The development of real-time PCR technology has recently allowed the quantification of *Coxiella burnetii,* and commercial kits are now available. The most suitable sample use in the PCR are: a vaginal swab, a placenta swab focusing the zones of necrosis or cotyledons presenting lesions, the organs (spleen, lung, liver) or the stomach contents of the aborted foetus. Besides *Coxiella burnetii* is also detected in the shed in faeces and milk. For direct detection of *Coxiella burnetii,* sampling should be targeted at pregnant animals either giving birth normally or aborting. The sampling should be carried out as soon as after abortion or parturition because the shedding level of bacteria decrease very fast. *Coxiella burnetii* genes such as 16S rDNA, sodB, and gltA may be amplified. The threshold for quantitative PCR is not officially approved at the international level. A group of French experts has suggested that abortion in ruminants should be suggested to be caused by *Coxiella burnetii* when at least 10^4 bacteria per gram of placenta or vaginal swabs are detected. Testing tissues or stomach contents from aborted foetuses, the same group considered that a positive result by quantitative PCR is sufficient to diagnose Q fever as the origin of abortion. For pooled samples, the proposed threshold is 10^3 bacteria per pool (EFSA Panel on Animal Health and Welfare (AHAW), 2010). Conventional PCR is the most commonly used technique in diagnosis. This technique cannot be completely automated, due to the time between the amplification and revelation of the DNA. Moreover, conventional PCR does not distinguish animals shedding huge amounts of bacteria and those with only few bacteria. The real-time PCR is quantitative method. The fusing media that *Coxiella burnetii* was detected and identified by RT-PCR which, presented bacterial load greater than 10^4. This concentration reflects active infection by this bacterium. Considering the frequency of secretion accompanying normal parturition and the heavy bacterial load present in the placenta or vaginal mucus of aborting females, a definitive diagnosis of infection with *Coxiella burnetii* as the cause of abortion can only be made if bacterial tissue concentrations are greater than 10^4 bacteria per gram (Alsaleh et al., 2011).

In human, rapid nested PCR assay may be used for early diagnosis of acute Q fever, using serum as the template and Light-Cycler (Roche Diagnostic, Basel, Switzerland) as the thermal cycler, named the LightCycler nested PCR assay (LCN-PCR). In this techniques the two primer pairs had different hybridisation temperatures, reaction tubes were not opened during amplification. Moreover, it is very fast techniques because both amplification and reamplification were performed within 90 minutes. However, the presence of inhibitors in the serum and usually small amount of bacterial DNA limit the efficiency of PCR for this assay. For early diagnosis of acute Q fever, where LCN-PCR is used, the serology tests should be performed together with serology in the first 2 weeks of the diseases but they should be reserved for seronegative patients in the next 2 weeks and not used later than 4 weeks following onset, when serology is highly sensitive (Fournier & Raoult, 2003).

The direct detection of *Coxiella burnetii* without using PCR or real-time PCR is very difficult because this bacteria does not grow on the standard bacteriological media, and isolation is time consuming, difficult and hazardous to perform. Moreover, the biosafety level 3 condition is required for cultivation of *Coxiella burnetii* in embryonated eggs, eukaryotic cell culture and laboratory animals. Although, *Coxiella burnetii* has been successfully isolated in guinea pig, mice and embryonated eggs, such techniques have been abandoned because they are more hazardous than in vitro cell cultures (Maurin & Raoult, 1999). *Coxiella burnetii* can infect embryonated eggs and multiply within the yolk sac. The growth and multiplication of the intracellular bacterium leads to death of the embryo, usually within 14 day postinoculation (Maurin & Raoult, 1999). The most often the Buffalo Green Monkey (GMB) cells is use for culturing this microorganism.

Although, possessing a membrane similar to gram-negative bacterium is usually stainable by the Gram techniques. The Gimenze method is usually used to stain *Coxiella burnetii* in laboratory cultures or clinical specimens (Gimenez, 1964).

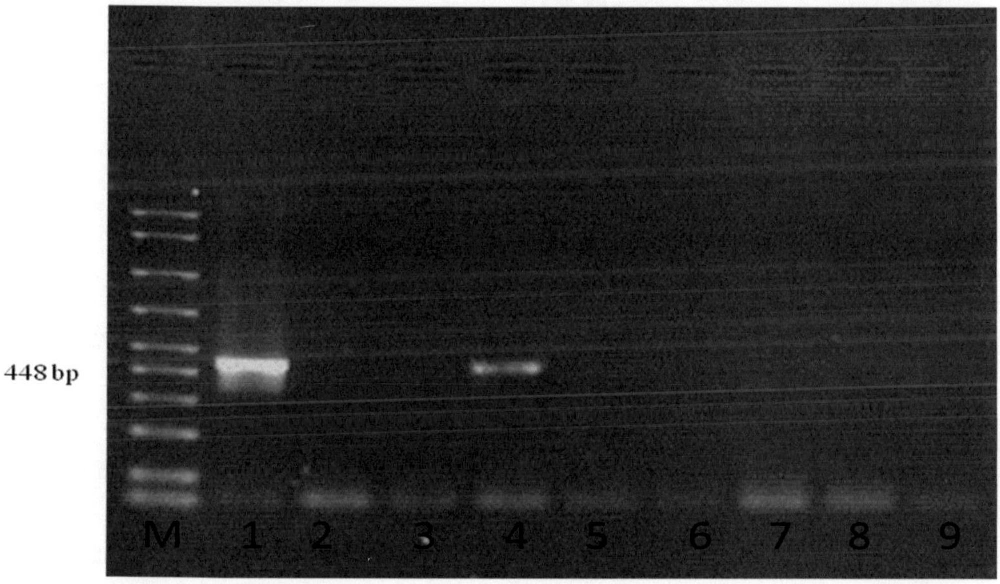

Lines:
M - marker

1 - positive control,
2 - negative control,
4 – positive sample,
3, 5 - 9 – negative sample

Fig. 2. Electrophoresis separation of amplification products.

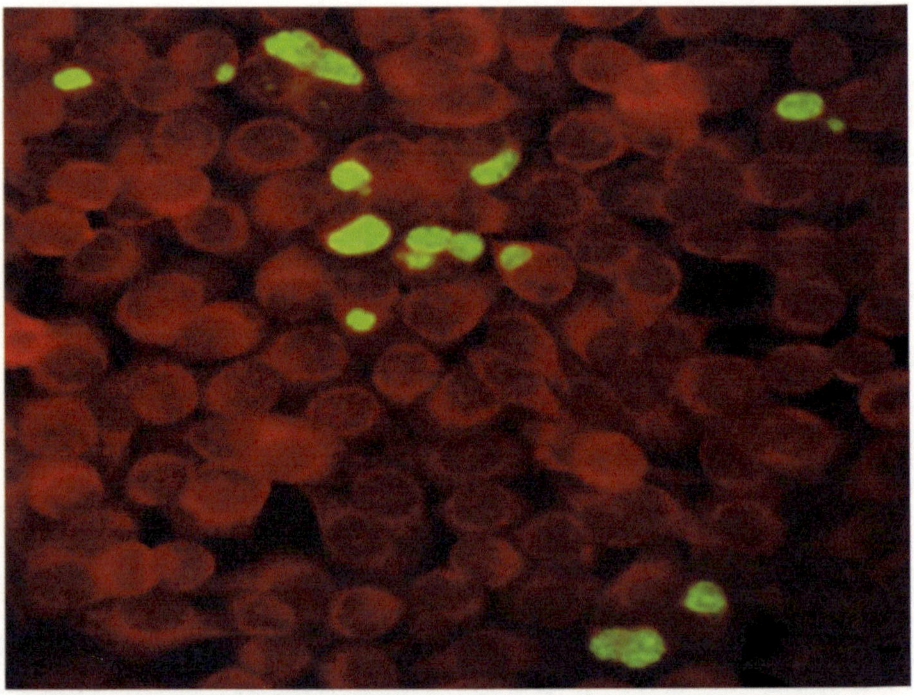

Fig. 3a. The results of immunofluorescence assay (Friedrich-Loeffler-Institute, Jena)

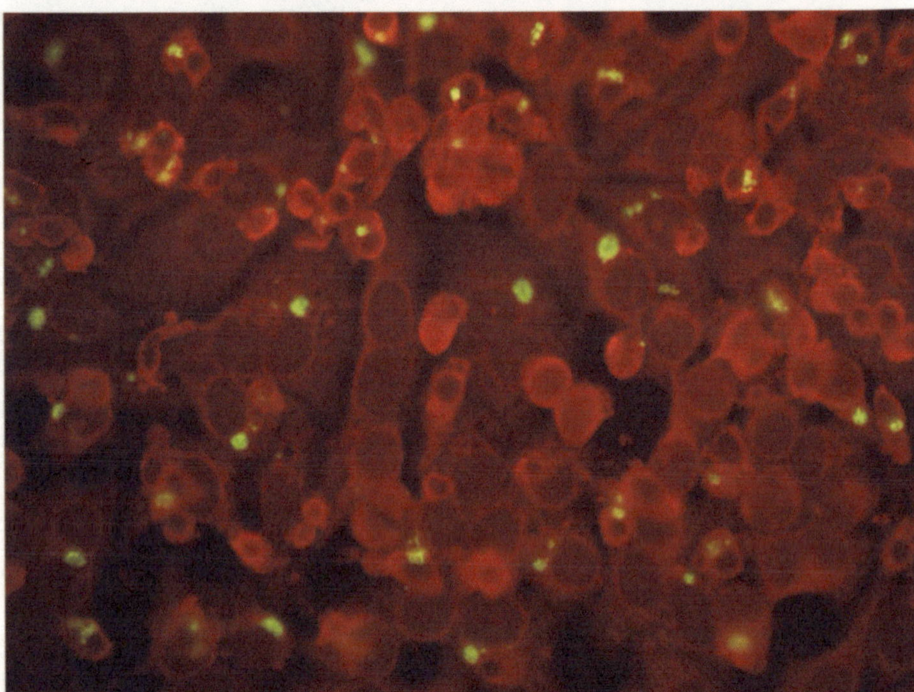

Fig. 3b. The results of immunofluorescence assay (Friedrich-Loeffler-Institute, Jena)

6. Prevalence of O fever infection in Poland

Human infection with *Coxiella burnetii* have been reported from many countries on all continents, including Poland (Lutyński, 1956). The first outbreak of Q fever was diagnosed in Poland in 1956. Since then, few outbreaks of the disease both in humans and animals were recorded in the country. Several cases of human Q fever occurred in Poland, however it is really not known whether this relatively low number of cases of illness is caused by the facts of its occurrence, whether or not all cases are detected and diagnosed. Data on these cases in human are presented in Table 1. Moreover, in recent times at the turn of 2008-2009, seven outbreaks of the disease were detected in cattle. These cases have occurred in the region of south-eastern Poland. In total, the disease was confirmed in more than 340 heads of cattle, the disease has been confirmed in humans handling these animals and in their family members. The other studies carried out in eastern Poland present that cattle may play an important role in epidemiology of Q fever, in addition to sheep and goats which are considered the main reservoirs of *Coxiella burnetii* (Cisak, 2003).

species of animals	Duration of *Coxiella burnetii* shedding		
	vaginal mucus	**feces**	**milk**
cow	not determined	14 days	13 months
goat	14 days	20 days	52 days
ewe	71 days	8 days after lambing	8 days

Table 1. The longest observed duration of *Coxiella burnetii* excretion in ruminant vaginal mucus, milk and feces during the follow-up of naturally or experimentally infected herds (Arricau-Bouvery & Rodolakis, 2005).

The European Community system for the monitoring and collecting information on zoonoses is established by Directive 2003/99/EC on the monitoring of zoonoses and zoonotic agents. According to this directive, reporting information on Q fever takes place on the basis of the epidemiological situation in the country, what means that the Member States should report the information whether those zoonotic agents are considered to be of importance in their country. For the reporting year 2006, 10 Member States reported information on Q fever in animals. Based on the Regulation of the Minister of Agriculture and Rural Development of 24 June 2010, monitoring tests in the direction of Q fever are conducted in Poland. In order to control Q fever, the blood samples from cattle or sheep and goats located in the different area of country, are tested annually. Studies conducted in the framework of this monitoring in 2010 showed the presence of seropositive units. The first study of blood serum is performed by the field laboratory. If the test result for antibodies to *Coxiella burnetii* is a positive sample, it is sent to the National Reference Laboratory (NRL) for re-implementation of the study. In 2010, the research review in the NRL of Q fever tested a total of 99 samples. The presence of antibodies was confirmed in 81 samples tested. Tests were conducted using the method of CFT. The samples which are confirmed as positive are also tested using real-time PCR. The positive results in this study allows the unambiguous confirmation of a case of Q fever.

Years	The source of infection	Cases of infection in human
1956	sheep imported from Romania	63
1956	sheep wool	20
1957-1958	guinea pigs - laboratory infection	17
1963	basils imported from America	20
1963 - 1983	no data	
1983	cattle	1000
1985	bisons, cattle	35
1985 - 1986	cattle and small ruminants	12
1985 -1986	basils	5
2005 - 2006	cattle	70
2008	cattle	21
2009-2010	cattle	20

Table 2. The infection of Q fever in human in Poland

7. References

Alsaleh, A.A; Pellerin J.L; Rodolakis, A.; Larrat ,M.; Cochonneau, D.; Bruyas J.F. & Fieni, F. (2011). Comparitive Immunology, *Microbiol & Infectious diseases*, 34, pp. 355-360

Arricau-Bouvery, N. & Rodolakis, A. (2005). Is Q fever an emerging or re-emerging zoonozis? *Vet Res* 36, pp. 327-349

Benson, W.W.; Brock, D.W. & Mather, J. (1963): Serological analysis of a penitentiary group using raw milk from a Q fever infected herd. *Public Health Rep.* 78, pp. 707-710

Blondeau, J. M; Williams, J. C. & Marrie T. J. (1990). The immune response to phase I and II *Coxiella burnetii* antigensas measured by western immunobloting. *Ann N Y Acad Sci*, 590, pp. 187-202

Capo, Ch.; Lindberg, F.P.; Meconi, S.; Zaffran, Y.; Tardei, G.; Brown, E. J.; Raoult, D. & Mege, J. L. (1999): Subversion of monocyte functions by *Coxiella burnetii*: Impairment of the cross-talk between $\alpha_v\beta_3$ integrin and CR3. *J Immunol*, 163, pp. 6078-6085

Cisak, E.; Chmielewska-Badora, J.; Mackiewicz, B. & Dudkiewicz, J. (2003) Prevalence of antibodies to *Coxiella burnetii* among farming population in eastern Poland. *Ann Agric Environ Med*, 10, pp. 265-267

Cowley, R.; Fernandez F., Freemantle W. & Rutter, D. (1992). Enzyme immunoassay for Q fever: comparison with complement fixation and immunofluorescence tests and dot immunobloting. *J Clin Microbiol*, 30, pp. 2451-2455

Dellacasagrande, J.; Ghigo E.; Machergui-El, S.; Hammami, T. R.; Raoult, D.; Capo C. & Mege, J. L. (2000) alpha(v) beta(3) integrin and bacterial lipolysaccharide are involved in *Coxiella burnetii*-stimulated production in tumor necrosis factor by human monocytes. *Immun,* 68, pp. 5673-5678

Directive 2003/99/EC of the European Parliament and of the Council of 17 November 2003 on the monitoring of zoonoses and zoonotic agents, amending Council Decision 90/424/EEC and repealing Council Directive 92/117/EEC (OJ L 325, 12.12.2003, pp. 31)

Doller, G.; Doller, P.C. & Gerth H. J. (1984) Early diagnosis of Q fever: detection of immunoglobulin M by radiommunoassay and enzyme immunoassay. *Eur J Clin Microbiol Infect Dis,* 3, pp. 550-553

Dyer, D. E.; Gibbson, V. L.; Brady, L. M. & Cunningham, A. L. (1988). Serological reaction to *Legionella pneumophila* group 4 in a patient with Q fever. *J Infect Dis,* 158, pp. 499-500

EFSA Panel on Animal Health and Welfare (AHAW), 2010. Scientific Option on Q fever. EFSA J., 114

Finidori, J. P.; Raoult, D., Bornstein, N.W & Fleurette, J. (1992). Study of cross reactions between *Coxiella burnetii* and *Legionella pneumophila* using indirect immunofluorescence assay and immunoblotting. *Acta Virol* , 36, pp. 459-465

Fishbein, D.B. & Raoult D. (1992) A cluster of *Coxiella burnetii* infections associated with exposure to vaccinated goats and their unpasteurized dairy products. *Am J Trop Med Hyg,* 47, pp. 35-40

Fournier, P. E. & Raoult D. (2003). Comparison of PCR and serology assays for early diagnosis of acute Q fever. *J Clin Microbiol,* 41, pp. 5094-5098

Furnier, P.E.; Casalta, J.P.; Piquet P.; Tournigand, P.; Branchereau, A. & Raoult D. (1998). *Coxiella burnetii* infections of aneurysms or vascular grafts: report of seven cases and revive. *Clin Infect Dis,* 26, pp. 116-121

Gilsdorf, A.; Kroh, C.; Grimm, S.; Jensen, E.; Wagner-Wening, C. & Alpers, K. (2008) Large Q fever outbreak due to sheep farming near residential areas, Germany 2005. *Epidemiol Infect,* 136, pp. 1084-1087

Gimenez, D. F. (1964). Staining rickettsiae in yolk sac cultures. *Stain Technol,* 30, 135-137.

Hatchette, T.F.; Hudson, R.C.; Schlech, W.F.; Campbell, N.A.; Hatchette, J.E.; Ratnam, S.; Raoult, D.; Donovan, C. & Marrie, T.J. (2001). Goat associated Q fever: a new disease in newfound-land. *Emerg Infect Dis,* 7, pp. 413-419

Hendrix, L. R; Samuel J.E. & Mallavia L. P. (1991). Differentiation of *Coxiella burnetii* isolates by analysis of restriction-endonuclease-digested DNA separated by SDS-PAGE. *J. Gen Microbiol,* 137, pp. 269-276

Hotta, A.; Kawamura, M.; To, H.; Andoh, M.; Yamaguchi, T. Fukushi; H. & Hirai; K. (2002). Phase variation analysis of *Coxiella burnetii* turning serial passage in cell culture by use of monoclonal antibodies. *Infect Immun,* pp. 4747- 4749

Jóźwik, A.; Jakubowski, T.; Kaba, J.; Jurkowski, W.; Witkowski, L., Nowicki, M. & Frymus, T. (2007). Evaluation of agreement of ELISA and complement fixation test in the diagnostics of Q fever in cattle. *Medycyna Wet,* 63, pp. 655-657

Klee, S.R.; Tyczka, J.; Ellerbrok, H.; Franz, T.; Linke, S.; Baljer, G. & Appel, B. (2006). Highly sensitive real-time PCR for specific detection and quantification of *Coxiella burnetii. BMC,* pp. 6:2

Kosatsky, T. (1984). Household outbreak of Q fever pneumonia related to a parturient cat. *Lancet,* pp. 1447-1449

Krumbiegel, E.R. & Wiśniewski, H.J. (1970). Q fever in Milwaukee II consumption of infected raw milk by human volunteers. *Arch Environ Health,* 21, pp. 63-65

Kruszewska, D. & Tylewska-Wierzbowska, S. (1993). *Coxiella burnetii* penetration into the reproductive system of male mice, promoting sexual transmission of infection. *Infect Immun*, 10, pp. 4188-4195

Kruszewska, D.; Lembowicz, K. & Tylewska-Wierzbanowska, S. (1996): Possible sexual transmission on Q fever among human. *CID*, 22, pp. 1087-1088

La Scola, B. & Raoult, D. (1996). Serological cross reactions between *Bartonella quintana, Bartonella henselae, and Coxiella burnetii*. *J Clin Microbiol*, 34, pp. 2270-2274

Lepe, J.A. & Guerro, F.J.; Del Castillo, E. (1999): The epidemiology of Q fever in the northern area of Huelva, Spain. *Enferm Infecc Microbiol Clin*, 17, pp. 65-68

Little, T.W.A. (1999). Q fever. *Br Vet J*, 139, pp. 277-283

Lutyński, R. (1956). First focus of Q fever on the territory of Poland. *Przegl Lek* 12, pp. 187-188

Marmion, B. P.; Ormsebee R. A.; Kyrkou, M. & Wright, D.; Worswick, A.; Cameron, S.; Esterman, A.; Feery B.; Collins W. (1984). Vaccine prophylaxis of abattoir-associated Q fever. *Lancett*, pp. 611-6161

Maurin, M. & Raoult, D. (1999). Q fever. *CMR*, pp. 518-553

Moquin, R.R. & Moquin, M.E (2002). Weapons of mass destruction: biological. *Neurosurg Focus*, 12, pp. 1232-1236

Muskens, J.; Engelen, E.; Maanen van, C.; Bartels, C. & Lam, T. J. G., M. (2011). Prevalence of *Coxiella burnetii* infection in Dutch dairy herds on the testing bulk tank milk and individual samples by PCR and ELISA. *Vet Rec*, 168, pp. 79

Musso, D.& Raoult. (1997). Serological cross-reactions between *Coxiella burnetii* and *Legionella micdadei*. *Clin Diagn Lab Immunol*, 4, pp. 208-212

Niemczuk, K. (2010) Q Fever - epidemiology, zoonotic aspects and administrative proceedings. *PIWet-PIB*, Pulawy, pp. 1-35

Prusakowski, M. (2001). Bioterror. Jak się nie dać zabić. Tower Press, Gdańsk, pp. 66-68.

Roest, I.J.H.; Ruuls, R.C.; Tilburg, J.J.H.C; Nabuurs-Franssen, M.H. & Klaassen C.H.W. at all. *Emerg Infect Dis*, 4, 668-675

Seshardi, R.; Paulsen, I.T.; Elisen, J.A.; Read, T.D.; Nelson, K.E.; Nelson, W.C.; Ward, N.L.; Tettelin, H.; Davidsen, T.M.; Beanan, M.J.; Deboy, R.Y.; Daugherty, S.C.; Brinkac, L.M.; Madupu, R.; Dodson, R.J.; Khouri, H.M.; Lee, K.H.; Carty, H.A.; Scanlan, D.; Heinzen, R.A.; Thompson, H.A.; Samuele, J.E.; Fraser, C.M. & Heidelberg, J.F. (2003). Complete genome sequence of the Q-fever pathogen *Coxiella burnetii*. *Proc Natl Acad Sci USA*, pp. 5455-5460

Sidi-Boumedine, K.; Rousset, E.; Henning K.; Ziller, M.; Niemczuk, K.; Roest H.I.J. & Thiéry R. (2010). Development of harmonised schemes for the monitoring and reporting of Q fever in animals in the European Union. Scientific. Scientific report submitted to EFSA, www.efsa.europa.eu/de/supporting/pub/48e.htm

Spicer A.J. (1998) Military significance of Q fever: a review. *J Roy Soc Med*, 71, pp. 762-765

Vodkin, M. H.; Williams, J. C. & Stephenson E. H. (1986). Genetic heterogeneity among isolates of *Coxiella burnetii*. *J Gen Microbiol*, 132, pp. 455

Webster, J. P.; Lloyd, G. & Macdonald D. W. (1995). Q fever (*Coxiella burnetii*) reservoir in wild brown rat (*Rattus norvegicus*) populations in the UK. *Parasitology*, 110, pp. 31-35

Willems, J. C.; Jäger C. & Baljer, G.: (1998). Physical and genetic map of the obligate intracellular bacterium *Coxiella burnetii*. *J Bacteriol*, 180, pp. 3816-3822

21

Risk (Predisposing) Factors for Non-Infectious Claw Disorders in Dairy Cows Under Varying Zero-Grazing Systems

J. Nguhiu-Mwangi, P.M.F. Mbithi,
J.K. Wabacha and P.G Mbuthia
Department of Clinical Studies,
Faculty of Veterinary Medicine,
University of Nairobi,
Kenya

1. Introduction

Lameness in cattle is one of the major causes of economic losses in dairy production systems (Hernandez et al., 2005; Kossaibati & Esslemont, 1997). These losses occur through various negative impact directly on cattle and indirectly on the dairy production system. These include reduced milk yield (Hernandez et al., 2005), discarding of milk due to withdrawal period of drugs used to treat some of the lameness conditions, cost of veterinary drugs and professional services in managing the conditions (Enting et al., 1997), lowered conception rate and increased calving interval (Melendez et al., 2003; Sogstad et al.,2006), reduced ovarian activity during early postpartum period (Garbarino et al., 2004), as well as premature culling and occasional mortalities (Enting et al., 1997). Lameness has also been identified as a major welfare determinant in cattle because of discomfort and pain that it causes (Offer et al., 2000).

Claw lesions account for between 60% and 90% of all lameness incidences in cattle in various countries of the world (Bergsten et al., 1994; Manske et al., 2002; Weaver, 2000). More than 60% of lameness in cattle is caused by lesions and disorders affecting the horn of the claw such as sole ulcers, heel erosion, sole bruising, white line separation and underrun (double) soles. All these claw disorders and lesions have a direct or indirect effect on the dermis (corium) of the claw and are associated with laminitis (Belge & Bakir, 2005; Manske et al., 2002; Nocek, 1997). They are common in cattle raised under intense systems and feedlots (Smilie et al., 1991). Claw horn disorders in cattle are discernible at clinical level by lameness symptoms or at subclinical level by hoof trimming to reveal non-painful lesions within or under the horn (Clarkson et al., 1996; Nocek, 1997).

However, adoption of confined housing in dairy cattle husbandry as is the practice in smallholder dairy production systems particularly in developing countries has led to higher incidences of claw disorders. This is mainly due to cattle spending long hours standing on confined hard floors that exposes claws to higher pressures which predispose them to circumscribed excessive local loading, thus stimulating more horn production and enlarging of the claws. A positive correlation exists between the time cows spend on hard floors and development of claw horn lesions. This confinement further adversely affects behavior and welfare of cattle (Bergsten, 1994: Somers et al., 2003; van der Tol et al., 2003). These practices synergistically with high levels of concentrate diets exacerbate the occurrence of claw-horn disorders (Cook et al., 2004; Donovan et al., 2004; Nocek, 1997; Somers et al., 2005; Vermunt, 2004). Feeding cattle with diets that lead to low-grade prolonged lowering of rumen pH could result in an up-surge of lameness due to increased incidences of laminitis. Excessive feeding of rapidly fermented carbohydrates and finely chopped fodder silage enhance development of laminitis due to their tendency of initiating subacute ruminal acidosis (SARA), which is also aggravated by a fall in the ratio of concentrate to forage (C:F) (Vermunnt, 2004). However, supplementing cattle with minerals and vitamins tends to improve claw health and reduce likelihood of developing laminitis and sole lesions (Bergsten et al., 2003; Tomlinson et al., 2004). Previous publications based on retrospective data show that improvement of cattle housing in the zero-grazing units from earthen slurry-laden floors to concrete floors, has led to a shifting from high prevalence of infective claw lesions and low prevalence of non-infectious laminitis related disorders to high prevalence of laminitis related disorders and low prevalence of infective claw lesions (Gitau, 1994; Mbithi et al., 1991; Nguhiu-Mwangi et al., 2009).

In developing Kenya where this research on association between non-infective claw disorders and risk factors was carried out, increased human population has led to inevitable reduction in per capita land-holdings. This has invariably led to diminished grazing land and subsequently triggering an increase in smallholder zero-grazing dairy production units particularly for the low income households (Mutugi, 2004). Rearing of dairy cows in smallholder zero-grazing units in developing countries is inevitable since they are part of the livelihood of the high population of low class smallholder farmers. For example in Kenya, smallholder units contribute 80% of the national commercial dairy herd (Wanyoike & Wahome, 2004). These smallholder dairy production units in Kenya have varied designs and management practices. They are so varied unit to unit and sometimes within individual units that they can be further classified as subunits. They completely lack standardization of housing designs, nutritional regimen and management protocols (Nguhiu-Mwangi, et al., 2008). Prevalence of claw disorders in dairy cows raised under smallholder zero-grazing systems in Kenya and possible risk factors leading to development of these claw disorders has not been documented previously. Therefore, the study was carried out in Kenya to verify the types of claw disorders in dairy cows that are raised in the smallholder subunits under varied zero-grazing systems and to find out through statistical analysis whether there is any association between these subunit-level factors, animal-level factors and the claw disorders.

2. Study procedures

2.1 Geographical study area

This study was carried out in the peri-urban areas of Nairobi, Kenya in a period of six months (partly during the long rainfall season and partly during the short rainfall season). This is a fertile agricultural zone with an altitude of 1798 metres above sea level and has an annual rainfall estimated at a maximum of 765 mm and a minimum of 36 mm in two distinct seasons, March to June (long rainfall season) and October to December (short rainfall season) respectively.

2.2 Zero-grazing unit and animal selection

Evaluation of 32 smallholder zero-grazing dairy units purposively selected was done, in which cows were also examined for lameness and claw lesions. The zero-grazing units selected were those with 5-20 adult dairy cows. The farmers owning most of the zero-grazing units were skeptical and hesitant to allow their dairy units evaluated or their cows examined. This was the reason for purposive selection of the zero-grazing units used in this study. The cows that were included in the study of claw lesions were those that had calved at least once with parities ranging between first and fourth. Those with 3rd and 4th parities were 60% and those with 1st and 2nd parities were 40% of the total population examined. All the selected cows whether lame or non-lame were examined. All the cows that had the study criteria were tagged with serial numbers 1, 2, 3, to n, where n was the serial number of the last cow being selected in the zero-grazing unit. To avoid biased numbering of the cows, assigning of serial numbers was randomly done by an independent worker employed in the zero-grazing unit. Then the author did the systematic selection of individual cows from the serialized groups by picking every second cow from the serial numbers. For example in the serial n_1 to $n_{10,}$ if the first cow selected was n_1, the next serially selected would be those with odd serial numbers (n_1, n_3, n_5, n_7, and n_9). But if the first cow selected was n_2, the next serially selected would be those with even serial numbers (n_4, n_6, n_8 and n_{10}). The serial numbers of the first cow selected was alternated between n_1 and n_2 from one zero-grazing unit to the next. Therefore, the cows selected in any zero-grazing unit were all those with either odd or even serial numbers.

2.3 Study design and data collection

The data were collected in a cross-sectional study in which each smallholder zero-grazing dairy unit was visited only once. During the single visit, the zero-grazing unit was evaluated and data recorded on possible risk factors of claw conditions. The selected cows were also evaluated for claw lesions as well as animal-level risk factors of claw lameness conditions. Data on animal-level risk factors were collected by the author administering questionnaires to the farmers or to the persons employed to manage the zero-grazing units as respondents before examination of each cow. These data included breed of the cow, parity, milk yield per day and lactation stage. The questionnaires were structured with simple "Yes" or "No" and "I do not know" responses to minimize variations and information bias from the respondents. From the 32 smallholder zero-grazing units, 300 dairy cows were examined for claw lesions and animal-level risk factors. Out the 300 cows, Friesians were 76% (228), Ayrshires 20% (60) and the remaining 4% (12) were a mixture of Guernsey and Jersey. Data from unit-level risk factors were collected during visitation to each of the 32 smallholder zero-grazing units. Some of the data (housing and stall design, number of cubicles, type of cubicle bedding, type of floor, presence or absence of a curb (kerb) at the rear end of the cubicles, lunging space, bob zone, presence or absence of neck-bar over the feeding bunk/trough and adequacy of feeding space) were collected through observational (qualitative) method. Other data such as curb height and height of neck-bar from the top surface of feeding trough were collected through actual measurements, while the rest (frequency and amount of concentrate feeding , mineral supplementation, type of fodder, and frequency of manure removal from the walk-alleys were collected by the author interviewing the farmers or the persons managing the zero-grazing units. All the data both for animal-and unit-level risk factors were entered into the questionnaires using codes adopted at the pre-coding stage during an initial pilot study in 5 sample smallholder zero-grazing dairy units.

2.4 Examination of the cows for claw lesions

Each cow was examined for general lameness using universally described locomotion scoring system (Sprecher et al., 1997). Data on claw disorders were collected by examining each claw on each cow. Each cow was restrained in a standing posture in a restraining area (crush). The limbs were lifted one at a time by tying with a rope and fastening the rope to an overhead pole or cross-bar. In that raised position, each claw was thoroughly washed with soap and water and examined for any lesions particularly on the weight-bearing surface. About 1-2 mm thickness of the horn of the sole was trimmed-off using a sharp quitter knife to expose any underlying lesions particularly sole haemorrhages. Trimming did not reach the level of corium and therefore no pain was caused on the cows. In cases with painful claw conditions, foot anaesthesia using 2% lignocaine hydrochloride was applied to allow painless manipulation. The lesions found on each cow were recorded.

2.5 Data management and analysis

The data were entered and stored in Microsoft Office Excel 2003. After verification and validation of data, they were imported into SAS© 2002-2003 (SAS Institute Inc., Cary, NC, USA). Descriptive statistics were computed for cow-level and unit-level risk factors. The prevalence rate of each claw disorder was calculated independent of other claw disorders. The prevalence of each claw disorder was calculated as the number of cows affected by the specific claw disorder or lesion (nCL) divided by the total number of cows (300) examined, and then multiplied by 100.

$$\text{Prevalence } (\%) = \frac{nCL}{100} x100 \quad (1)$$

Chi-square (χ^2) statistics were used to determine unconditional associations between all risk factors and the claw lesions. An association was considered significant at the level of $p<0.05$. Chi square values were determined using 2 x 2 contingency tables constituting 2 rows and 2 columns. In these associations the Chi square calculations were determined by evaluating each risk factor (variable) against each claw disorder (outcome) on the cow. The degrees of freedom (df) in each case was standard, being calculated by:

$$[(\text{row-1}) \times (\text{columns-1})], \text{ hence } [(2\text{-}1) \times (2\text{-}1)] \quad (2)$$

Therefore, df was 1 for each association test. Multiple logistic regressions were done through a step-down regression in which the risk factors that made the least variation to the occurrence of the claw lesions were eliminated one at a time through consideration of their odds ratios. Only the factors that were found to influence the occurrence of claw lesions significantly were retained in the model. The effects of confounding risk factors were dealt with in the analysis but they were minimal because of some similarities in the management between the smallholder zero-grazing units.

3. The study findings

3.1 Prevalence of claw disorders found in the study

The results of this study revealed that cows from the varied smallholder zero-grazing dairy units had a high (88%) prevalence of claw lesions, 69% of which did not manifest lameness but had subclinical disorders and only 31% were lame. Some of the cows had more than one claw disorders concurrently. Laminitis had the highest prevalence (70.3%), with subclinical laminitis at 49.3% and chronic laminitis at 21%. Other non-infectious claw disorders with high prevalence were sole bruising (erosion) at 45%, claw overgrowth at 30%, heel erosion at 27.3%, white line separation at 18% and underrun (double) soles at 17%. Infective lesions had very low prevalence of 4.7% (Table 1).

Claw lesions	Number of cows (n = 300)	Prevalence (%)
Subclinical laminitis	148	49.3
Chronic laminitis	63	21.0
Laminitis-related disorders		
Sole bruising	135	45.0
Heel erosion	82	27.3
White line separation	54	18.0
Underrun (double) soles	51	17.0
Sole ulcer	8	2.7
sole foreign bodies	1	0.3
Laminitis-related claw deformities		
Claw overgrowth	91	30.3
Claw horn deep horizontal grooves	53	17.8
Flattened claws (reduced toe angle)	36	12.0
Splayed toes	26	8.7
Concaved dorsal claw wall	15	5.0
Corkscrewed claws	14	4.7
Non-laminitis lesions		
Claw infective lesions	14	4.7
Claw traumatic lesions	2	0.7
Growths in the foot soft tissues	2	0.7

(NB: *Some of the cows had more than one claw disorders concurrently, hence percentage totals of more than 100*).

Table 1. Prevalence of claw disorders and lesions and their prospective risk factors as found during a cross-sectional study in dairy cows in varied smallholder zero-grazing dairy units in Kenya.

Among the cows that had chronic laminitis, 85% had more than two other concurrent claw disorders such as underrun (double) soles, sole bruising, heel erosion and white line separation. In addition, all the cases of chronic laminitis invariably had one or more ways in which the claw was misshaped such as excessive elongation, abnormal widening, excessive flattening, concavity of the dorsal wall and deep horizontal grooves (ridges) on the dorsal and the lateral walls. When chronic laminitis was found in any cow, it was always bilateral affecting all the claws but more often the hind limbs. Such cows were invariably lame.

3.2 Description of the non-infectious claw disorders found in this study

Lesions on the weight-bearing surface of the claw were described mainly according to the zones they affected. The six zones on the weight-bearing surface of the claw were universally described previously denoted as zone 1 (white line at the toe), zone 2 (abaxial white line), zone 3 (abaxial wall-bulb junction), zone 4 (sole-bulb junction), zone 5 (apex of the sole) and zone 6 (the bulb) (Greenough & Vermunt, 1991).

3.2.1 Subclinical laminitis

Cows with subclinical laminitis had normal gait. The only way of confirming subclinical laminitis was by trimming-off a thin layer of the horn of the sole. The main sign of subclinical laminitis was sole haemorrhages seen as reddish or yellow-waxy discolourations after trimming of the superficial layers of the horn of the sole. The haemorrhages in subclinical laminitis appeared in one or at most two zones of the weight-bearing surface of the claw (Fig. 1). In a few instances, the haemorrhages were visible on the surface of the horn of the sole even before trimming. Sole haemorrhages were scored according a universally suggested scale denoted as score 0 (no haemorrhages), score 1 (slight haemorrhages), score 2 (moderate haemorrhages), score 3 (severe haemorrhages), and score 4 (exposed corium) (Greenough & Vermunt, 1991). The haemorrhages were highly prevalent in zone 4 (34.7%), moderately prevalent in zones 3(14.3%), 6 (14.1%) and 2 (13.6%), and least prevalent in zones 5 and 1 at 8.7% and 0.7% respectively (Nguhiu-Mwangi et al., 2007). Haemorrhage score and the number of haemorrhagic zones had a strong positive correlation ($r = 0.546$, O.R. = 21.24) with subclinical laminitis and this correlation was highly significant ($\chi^2 = 89.45$, $p < 0.05$). Subclinical laminitis was mainly associated with haemorrhages in zone 4 and the abaxial whilte line (Fig. 1) and these associations were significant ($\chi^2 = 22.83$, $p < 0.05$). Occurrence of sole haemorrhages was significantly ($\chi^2 = 86.5$, $p <= 0.001$) higher in the hind lateral claws than the medial claws. Among the cows with subclinical laminitis, 32.4% manifested with only sole haemorrhages but 67.6% concurrently manifested sole haemorrhages as well as one or two of the other claw lesions such as sole bruising, heel erosion and white line separation.

3.2.2 Chronic laminitis

Claws with chronic laminitis had severe and widespread haemorrhages affecting several zones of the weight-bearing surfaces of the claws (Fig. 2). Haemorrhages in chronic laminitis could be visible on the surface of the sole before trimming. Haemorrhage score and the number of haemorrhagic zones had a positive correlation ($r = 0.41$) and the correlation was highly significant ($\chi^2 = 50.16$, $p < 0.05$). Haemorrhages in chronic laminitis significantly affected mainly zones 3 ($\chi^2 = 23.45$, $p < 0.05$), and 2 ($\chi^2 = 18.23$, $p < 0.05$) and occasionally but weakly zone 6 ($\chi^2 = 4.48$, $p < 0.05$). In addition to haemorrhages, 84.8% of the cows with chronic laminitis also had more than two other claw lesions such as double soles, sole bruising, heel erosion and white line separation. All cases of chronic laminitis, almost

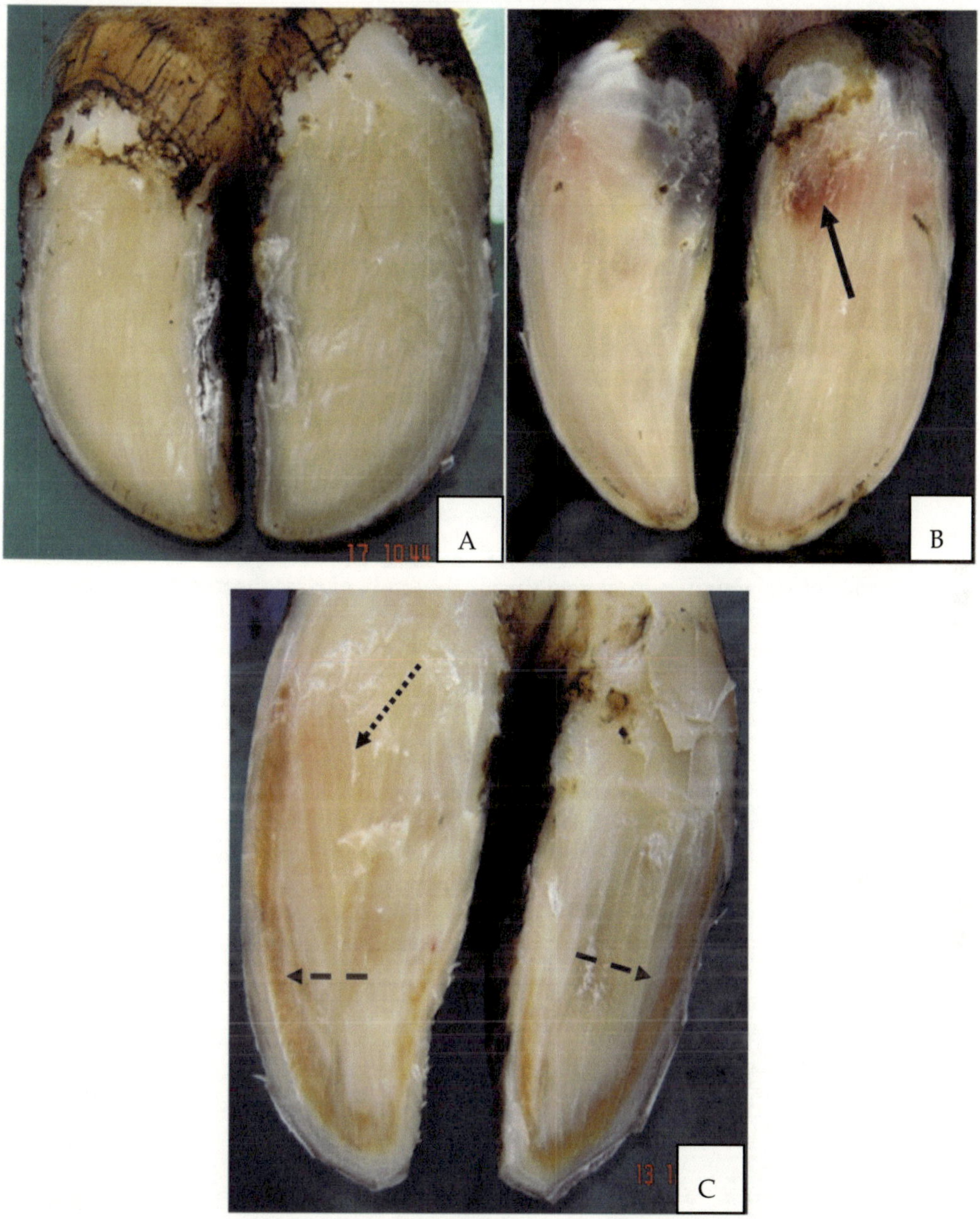

Fig. 1. Sole haemorrhages exposed after trimming of the horn of the sole signifying presence of subclinical laminitis in the claws of dairy cows examined in smallholder zero-grazing dairy units in the peri-urban areas of Nairobi, Kenya. A: normal claw with no haemorrhages; B: slight haemorrhages in zone 4 (bold arrow); C: moderate haemorrhages in the abaxial and a bit of axial white line (dashed arrows), as well as on zones 4 and 5 (dotted arrow).

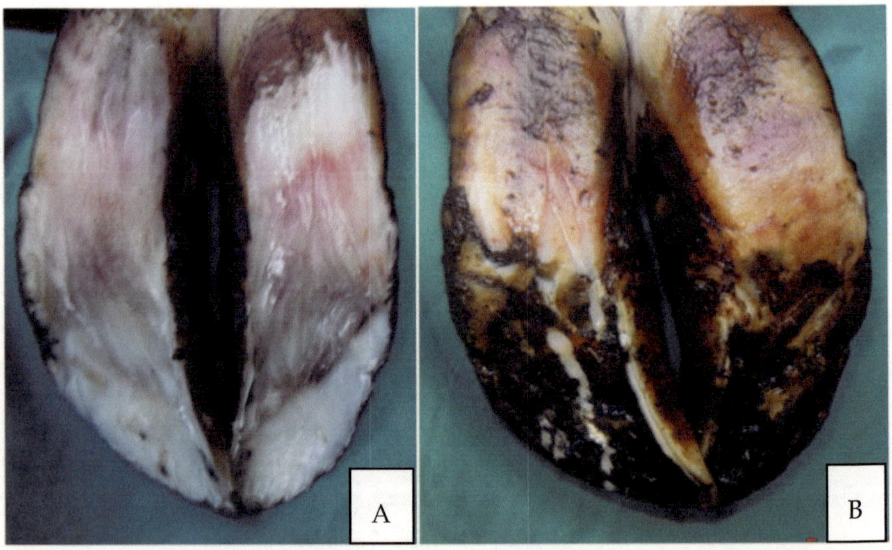

Fig. 2. Severe widespread haemorrhages affecting most zones of the weight-bearing surfaces of the claws in one of the cows examined during the study of claw disorders and their prospective risk factors in the smallholder zero-grazing dairy units in the peri-urban areas of Nairobi, Kenya. The haemorrhages are seen as reddening of the horn of the sole. A: haemorrhages seen after trimming of the horn of the sole; B: haemorrhages seen superficially on the weight-bearing surface before trimming.

invariably had misshapen claws resulting from either overgrown, elongated, widened, flattened or concaved claws and all of them with deep prominent horizontal ridges of the dorsal and lateral walls (Fig. 3). Chronic laminitis in each cow was invariably bilateral, which together with misshaped claws exacerbated poor locomotion and hence the animals were found to always have moderate degree of lameness. Some of the claws were extremely misshapen to the extent that re-shaping them back to normal would be impossible. On trimming some of the claws with chronic laminitis, they were found to have weak whitish crumbling horn (Fig. 4).

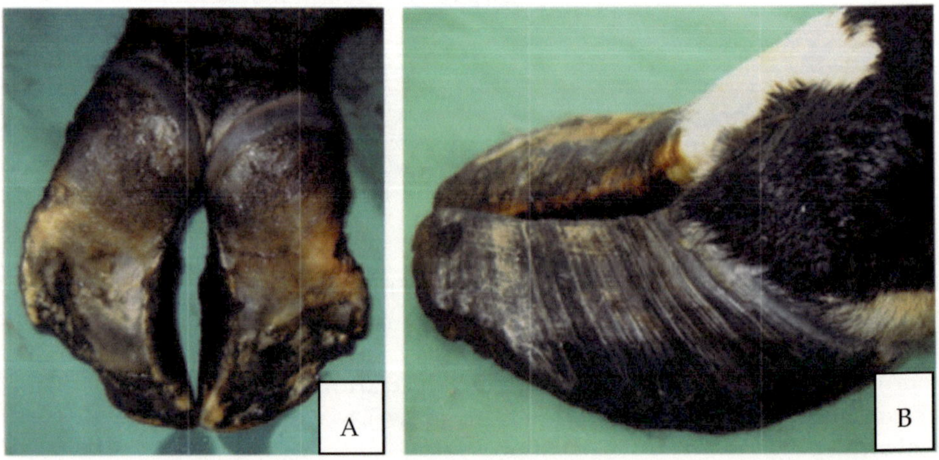

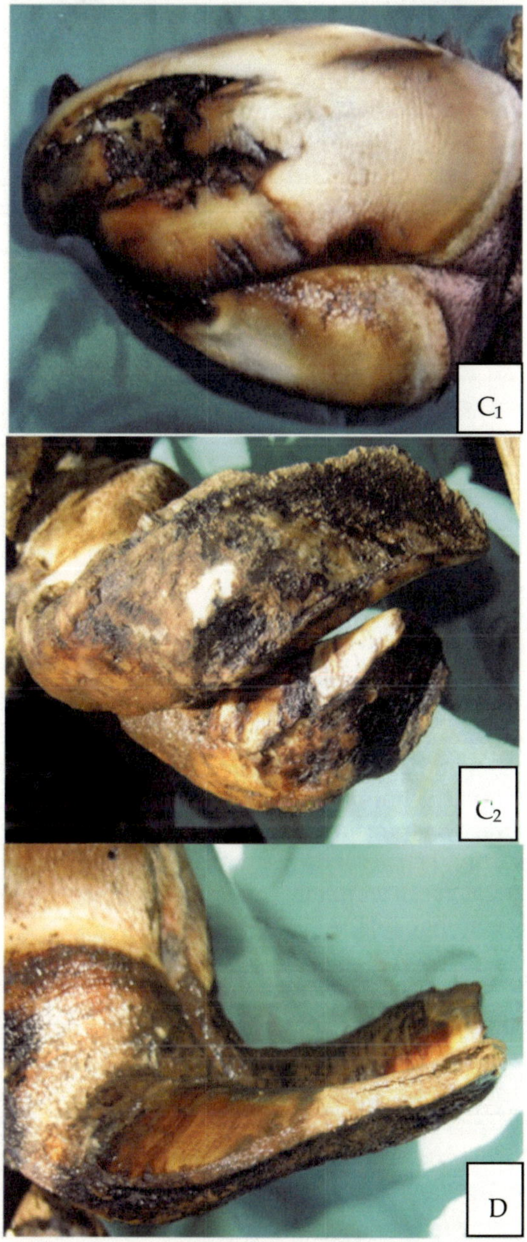

Fig. 3. The different forms of misshaping of the claws with chronic laminitis observed in dairy cows during a study on prevalence of claw disorders and the possible risk factors in smallholder zero-grazing dairy units in the peri-urban areas of Nairobi, Kenya.
A: irregularly widened claws; B: elongated flattened claws, reduced toe angle and heel height and presence of deep prominent horizontal ridges on the dorsal and lateral walls; C_1 and C_2: elongated overlapping claws with extreme concavity of dorsal wall and convex weight-bearing surface; D: elongated flattened claw, toe angle greatly reduced and concavity of dorsal walls of the claw.

Fig. 4. Softened whitish crumbling horn of the sole in a claw with chronic laminitis found in one of the cows during the study of prevalence and risk factors of claw lesions in smallholder zero-grazing dairy units in the peri-urban areas of Nairobi, Kenya.

3.2.3 Sole bruising

Sole bruising was seen as erosive areas of black necrotic-like horn. In the cows with less erosive sole bruising, it involved zone 5 only. The more invasive erosions of the sole involved zones 1, 4 and 5 simultaneously. In the less severe sole bruising, there was erosion of only the superficial parts of the horn of the sole, but in the severe sole bruising, there was invasive erosion deeper into the horn of the sole, which when trimmed revealed that only a very thin layer of the horn was left intact (Fig. 5). Sole bruising was always found to be bilateral affecting either only the lateral claws, or only the medial claws. It very occasionally involved lateral and medial claws simultaneously. It was observed to affect the claws of all the feet, but particularly common on the hind feet claws.

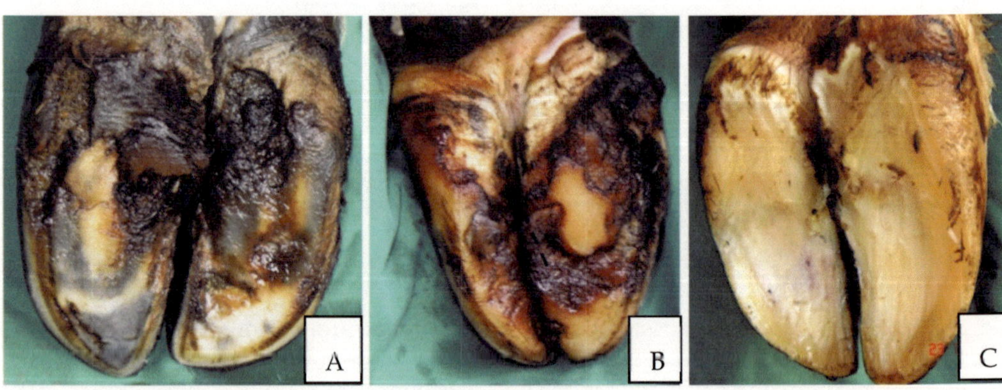

Fig. 5. Sole bruising in claws of cows observed during a study of prevalence and risk factors of claw disorders in smallholder zero-grazing dairy units in Kenya. A: superficial non-invasive sole bruising mainly in zone 5; B: deeper erosive and invasive sole bruising involving zones 1, 4 and 5, with only a small part at the centre remaining not eroded; C: thin layer of the horn left in a section of the sole after trimming areas that had sole bruising

3.2.4 Heel erosion

Heel erosion like sole bruising appeared as dark necrotic areas of the horn predominantly in zone 6. However, when deeply invasive, it occasionally extended to zones 3 and 4 (Fig. 6). Heel erosion was invariably bilateral and in the hind limbs it affected either lateral claws only, or medial claws only, but occasionally it affected both lateral and medial claws concurrently. When present in any herd, it was observed to affect all the cows in the zero-grazing unit without any exception. Some of the claws with heel erosion had some degrees of sole bruising at the same time.

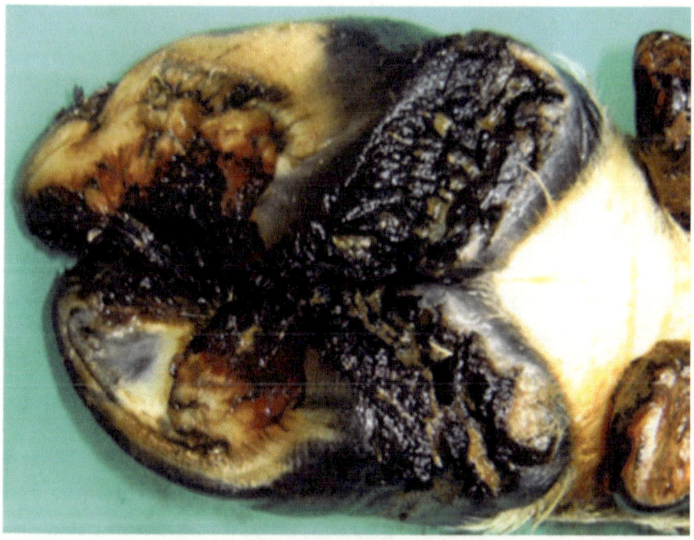

Fig. 6. Severely erosive and invasive heel erosion involving zones 6, 4 and 3 in both claws of a cow seen during the study of prevalence and risk factors of claw disorders in smallholder zero-grazing dairy units in Kenya. There is also sole bruising affecting the axial part of zone 5.

3.2.5 Underrun (double) soles

Underrun sole involved separation of the horn of the sole into two distinct layers constituting a superficial thicker harder sole and a thinner softer underlying sole. It was diagnosed by thoroughly washing the claws, which made it possible to see dung and debris stuck underneath the superficial sole. In some of the cases, it was only diagnosed during trimming by incidental finding of a superficial horn layer completely detached from a thinly forming underlying horn layer (Fig. 7). A claw that had double soles appeared grossly wider than both the normal ipsilateral and contralateral claws. Double sole phenomenon was common in zones 4 and 5. In some of the cows it was unilateral while in others it was bilateral affecting the contralateral claw on the opposite limb. Superficially traumatized horn of the sole was encountered in two cows in which the top sliced horn still attached to the underlying remaining horn at some point (Fig. 8), which erroneously can be diagnosed as underrun sole. The margins of the traumatized horn of the sole cleared appeared as distinct cut edges compared to those of underrun sole, and the underlying horn in the case of trauma was hard compared to the thin horn of the underrun sole.

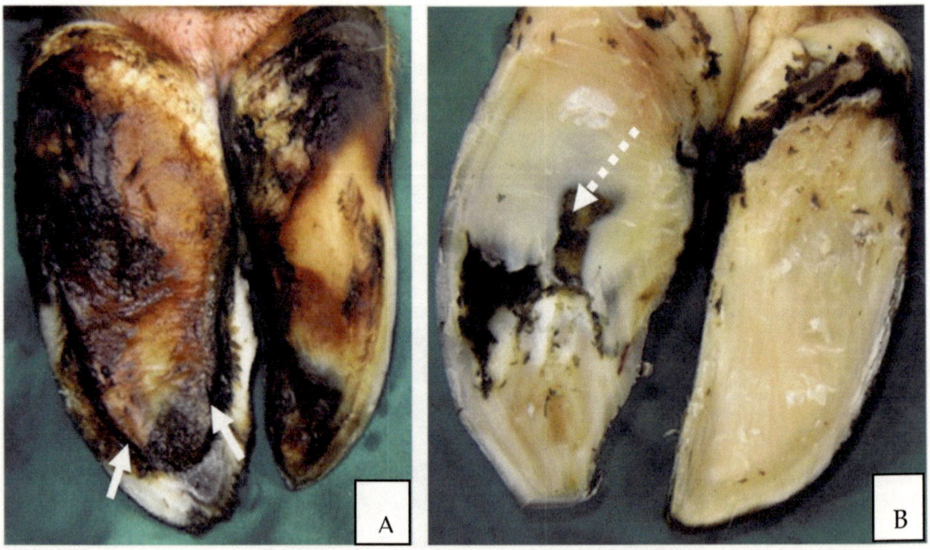

Fig. 7. The left claw in A and B has underrun (double) soles and is wider than the normal ipsilateral claw. This was seen in a cow examined during a study on prevalence and risk factors of claw disorders in smallholder zero-grazing dairy units in the peri-urban areas of Nairobi, Kenya. A:underrun sole before trimming with the margins of the superficial sole clearly demarcated (arrows); B: underrun sole after trimming, the dung matting underlying thin sole can be seen (dashed arrow).

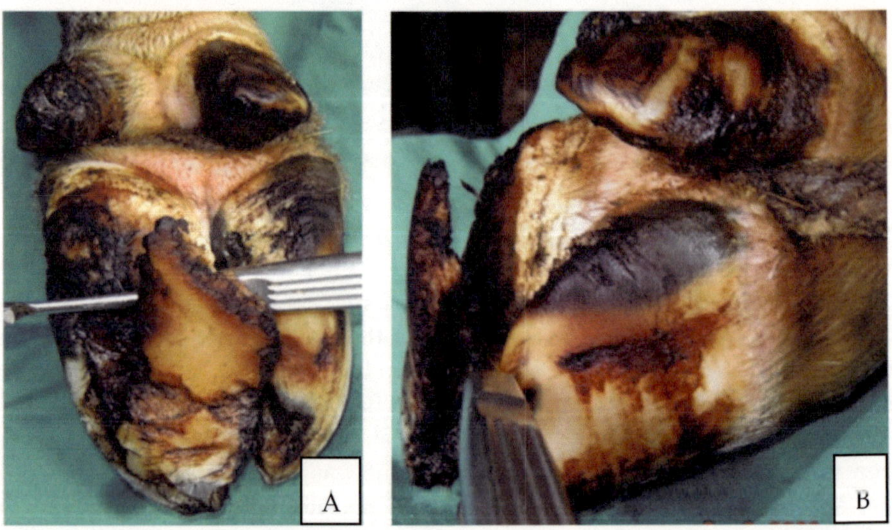

Fig. 8. Traumatized horn of the claw observed in 2 cows during the study of prevalence and risk factors of claw lesions in smallholder zero-grazing dairy units in the peri-urban areas of Nairobi, Kenya. A: plantero-dorsal view with the metallic quitter knife pointer between the detached horn and underlying horn; B: lateral view with the metallic quitter knife pointer between the detached horn and underlying horn.

3.2.6 White line separation

White line separation was seen as clear separation of the horn of the sole and the wall. It was more discernible after thorough washing of the claws and trimming the horn of the sole (Fig. 9). The gap between the avulsed horn of the sole and the wall was usually filled with dung and debris, and in addition it could also oozed infective exudates if sepsis had set into the separated white line. White line separation involved mainly zone 2, but also very occasionally the axial part of zone 4 and 5, and sometimes zone 3. It was invariably unilateral involving only one claw.

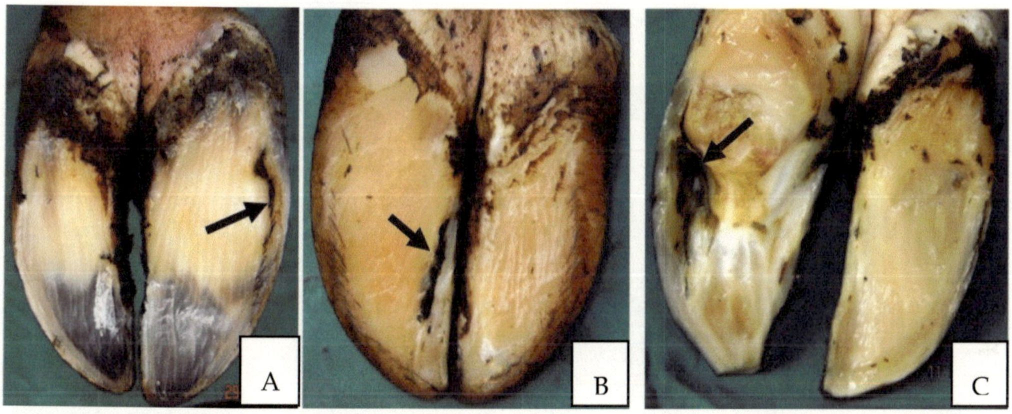

Fig. 9. Various degrees of white line separation seen after trimming of the horn of the sole in some of the cows examined during the study of prevalence and risk factors of claw lesions in smallholder zero-grazing dairy units of the peri-urban areas of Nairobi, Kenya. A: moderate separation on the abaxial white line (arrow); B: moderate separation on the axial white line (arrow); C: extensive avulsion of the white line with invasive infection (arrow).

3.2.7 Sole ulcer

Occurrence of sole ulcers exposes the corium. In this study, it occurred in 3% of the cows and invariably resulted in severe lameness. All the sole ulcers occurred at the inner part of zone 4 and were unilateral affecting only one claw in each case. Each sole ulcer had a mass of protruding granulation tissue growing out of the site of the exposed corium (Fig. 10).

3.2.8 Claw deformity disorders

In this study, 43.7% of the cows examined had one or more types of claw deformities. There were several types of claw deformities and a number of them occurred concurrently in the same cow. Cows with claw overgrowth had entire toe, sole and walls overgrown ranging from slight to extreme overgrowth and with subsequent misshaping or deformity of the claw (Fig. 11) and consequently abnormal gait. The deformities observed were upward (dorsal) turning or crookedly shaped toe (Fig. 12), irregularly widened claw (Fig. 3 and 11), concaved claw (Fig. 3), elongated flattened claw (Fig. 3 and 11), lateral claw wall turned sole-ward (ventral) to the treading surface (Fig. 13), which was observed in 3 cows, crossed toes or scissor feet (Fig. 14) and prominent deep horizontal ridges on the dorsal wall of the claw (Fig. 3). Claws that had these characteristics in most cases also had chronic laminitis as evidenced by underlying sole haemorrhages on trimming of the horn of the sole and some even before trimming (Fig. 2). Splaying of the claws was found independent of other deformities but occasionally concurrent with claw overgrowth (Fig. 15). Corkscrew claw was not common but when encountered, it was bilateral involving only medial claws or only lateral claws of fore or hind limbs (Fig. 16).

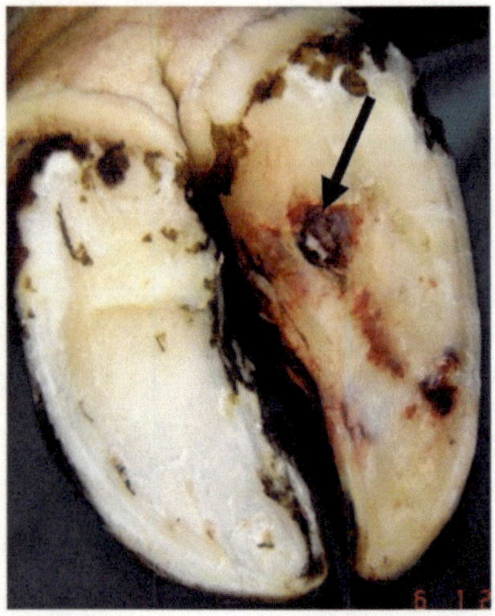

Fig. 10. Sole ulcer with protruding mass of granulation tissue (arrow) seen clearly after trimming of the horn of the sole in some cows during the study of prevalence and risk factors of claw lesions in smallholder zero-grazing dairy units in the peri-urban areas of Nairobi, Kenya. The affected claw also has sole haemorrhages, thus indicating presence of laminitis.

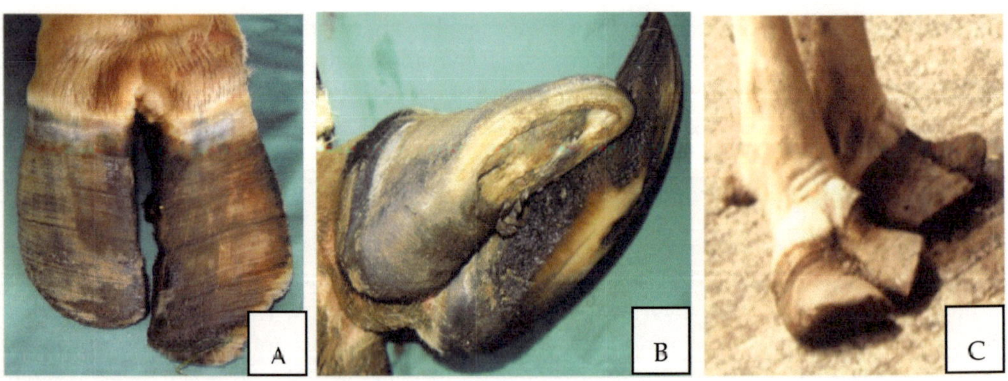

Fig. 11. Regular overgrowth showing claw widening (A), toe elongation (B), as well as widening, flattening and change of treading angle of the claw to more proximal parts toward the heel (C). These were observed in some of the cows examined during the study of prevalence and risk factors of claw lesions in smallholder zero-grazing dairy units in the peri-urban areas of Nairobi, Kenya.

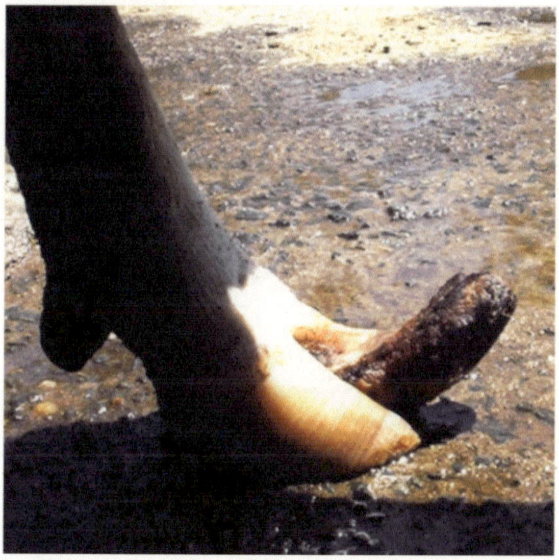

Fig. 12. Crookedly twisted and upward (dorsal) turning of the claw of a cow observed during the study of prevalence and risk factors of claw lesions in smallholder zero-grazing dairy units in the peri-urban areas of Nairobi, Kenya.

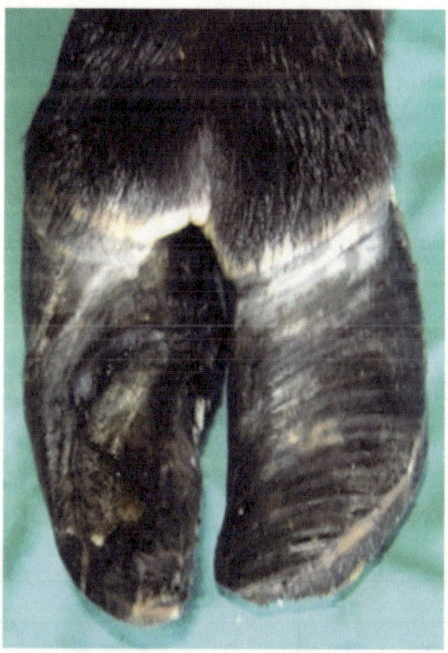

Fig. 13. Spiraling of the lateral wall of the claw of a cow with abaxial aspect turning to face ventral toward the treading surface and the axial aspect turning to face dorsally. This was observed during the study of prevalence and risk factors of claw lesions in smallholder zero-grazing dairy units in the peri-urban areas of Nairobi, Kenya.

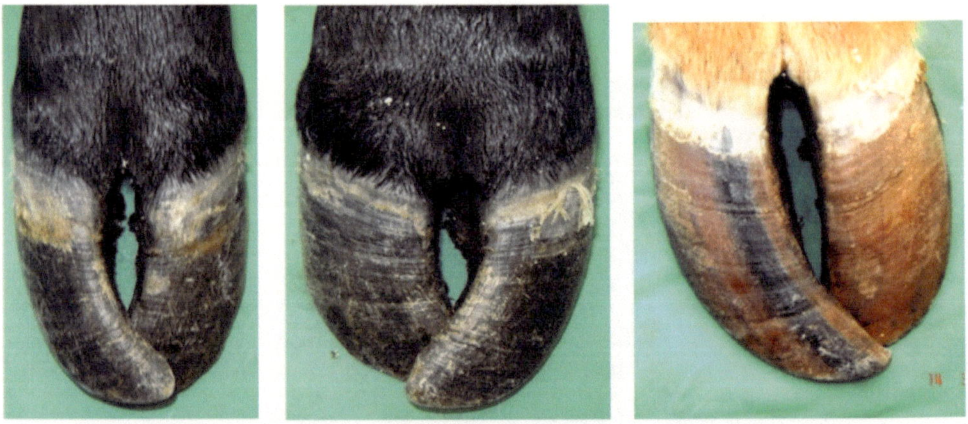

Fig. 14. Bilateral crossed toes or scissor feet in cows due to overgrown claws overlapping over each other. These were observed during the study of prevalence and risk factors of claw lesions in smallholder zero-grazing dairy units in the peri-urban areas of Nairobi, Kenya.

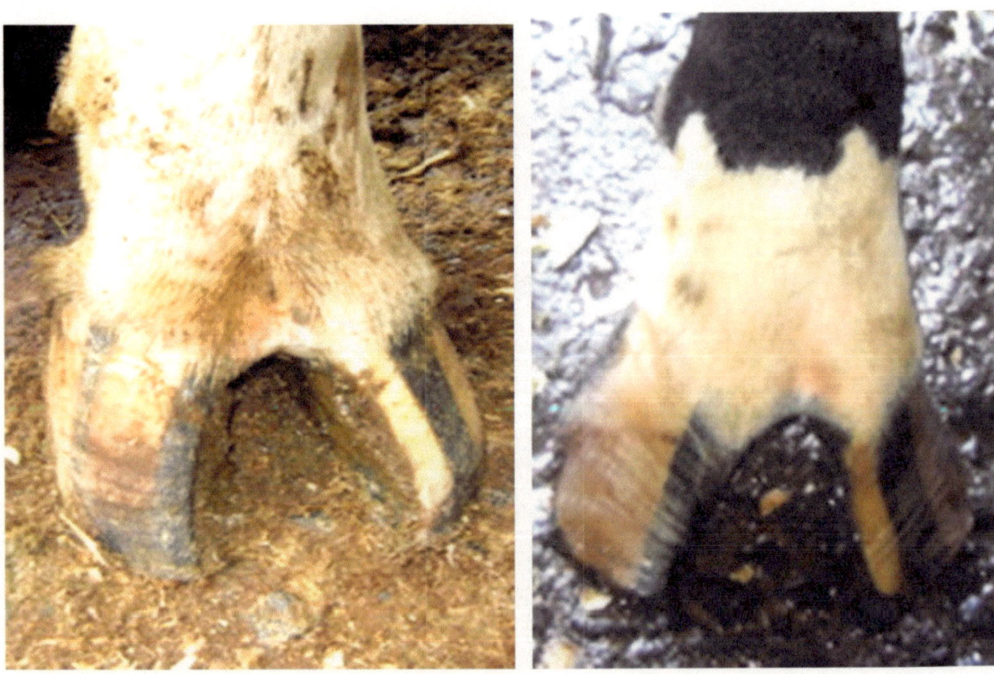

Fig. 15. Splayed toes observed in cows examined during a study of prevalence and risk factors of claw lesions in smallholder zero-grazing dairy units in the peri-urban areas of Nairobi, Kenya.

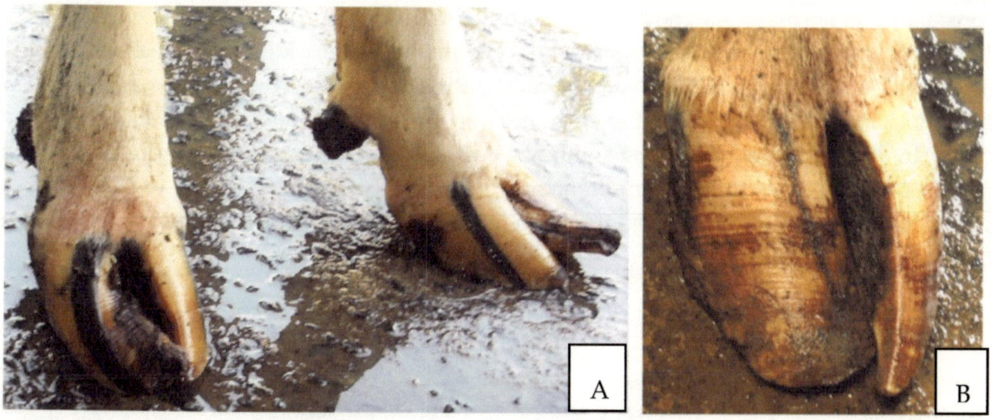

Fig. 16. Two stages of corkscrew claw seen in the cows examined during a study of prevalence and risk factors of claw lesions in smallholder zero-grazing dairy units in the peri-urban areas of Nairobi, Kenya. A: shows severe bilateral corkscrew on both lateral claws, while both medial claws are normal and, B: mild corkscrew on one of the claws spiraling toward the normal claw.

3.3 Description of risk factors

3.3.1 Cow housing factors

Good concrete floors (with firm ground-grip, non-slip and without pot-holes) were present in 68.8% of the zero-grazing units. The rest (31.2%) of the units had floors that were concreted but with loose stones, over smoothened and slippery, grossly worn out with small to large pit-like patches, or earthen (bare soil). Among these floors, 62% had slight to moderate floor slope (< 3% gradient), while 9% had steep slope (> 3% gradient) and 29% had no slope at all but were flat. A total of 59% of the zero-grazing units were stocked to capacity (cubicle to cow ratio 1:1). The other 41% were over-stocked (more cows than cubicles). The proportion of the zero-grazing units without cubicle bedding was 38%, in which the animals lay directly on bare concrete and in some of them there were loose stones present on the cubicle floor (Fig. 17a). In those with cubicle bedding, the bedding was wood shavings, sawdust, soil, sand, rice husks, or rubber mats (Fig. 17b).

The neck-bar height of 0.6 m from the upper-level of the feeding bunk was considered to be optimal. Only in 57% of the zero-grazing units were the neck-bars ≥ 0.6 m high, but in the other 43% of the units the neck-bars were < 0.6 m high. In 63% of the zero-grazing units, the lunging space (for the forward head movement during the act of standing) was ≥ 0.4 m, but in 37% of the units, it was < 0.4 m and the cow's head was in contact with the wall of the cubicle. In 72% of the units, the bob zone (for up-down movement of the head during the act of standing) was adequate between 0.4 m and 0.6 m from the ground to the cubicle cross-bar, while in 28% of the zero-grazing units, the bob zone was inadequate at < 0.4 m in which the cows struggled during the act of standing with the dorsal part of the neck coming into contact with the cross bar before fully standing (Fig. 18).

Presence of a curb (kerb) on the rear end of a cubicle was considered as a possible risk factor for development of non-infective claw disorders. Cubicle curb-height of 10-15 cm was considered as normal. Only 50% of the zero-grazing units had cubicle curbs among which 19% had cubicle curbs that were between 10 cm and 15 cm tall, while 31% had cubicle curbs that were more than 15 cm tall that presented difficulties for the cows to walk backwards off the cubicles. The other 50% of the zero-grazing units had cubicles that did not have any curbs. In 88% of the zero-grazing units, the feeding area was adequate space (≥ 0.8m per cow) for all the cows to feed simultaneously. In most of them the older and the younger stock fed separately. In 12% of the units, the feeding area was inadequate space (< 0.8m per cow) due to overstocking.

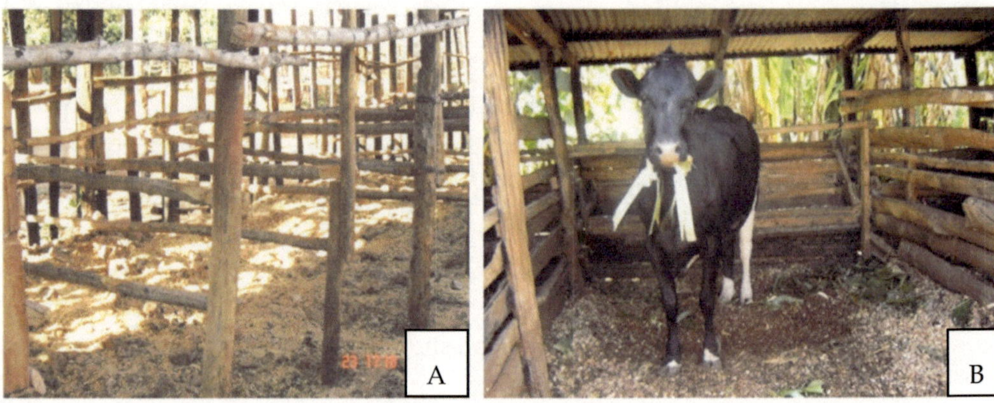

Fig. 17a. A: bare cubicle without bedding but with loose stones, B: cubicle with wood shaving bedding. These were seen during the study of prevalence and risk factors of claw lesions in cows in smallholder zero-grazing dairy units in the peri-urban areas of Nairobi, Kenya.

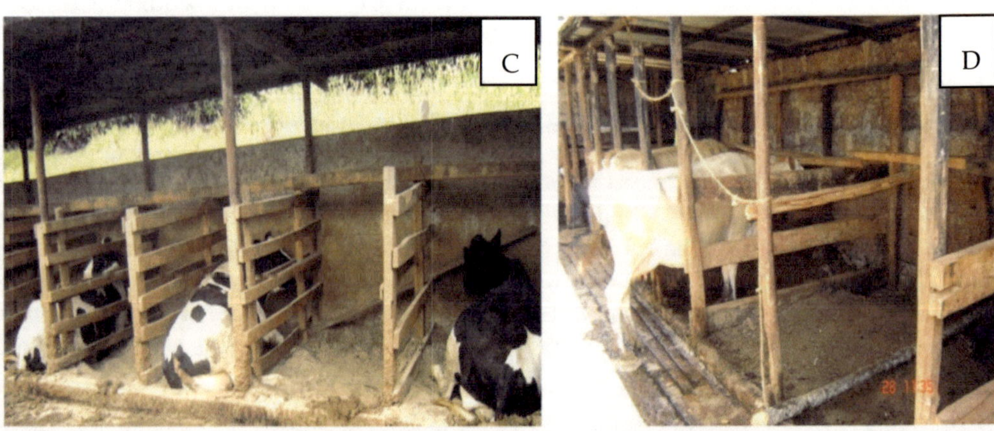

Fig. 17b. C: cubicles with sand bedding, D: cubicles with soil bedding and

Fig. 17b. E: cubicles with partial rubber mat covering. These were seen during the study of prevalence and risk factors of claw lesions in cows in smallholder zero-grazing dairy units in the peri-urban areas of Nairobi, Kenya

Fig. 18. Inadequate lunging space and bob zone causing difficulties for the cow during standing due to low cross-bar in the cubicle and short length of the cubicle. These were found in some of the zero-grazing units during the study of prevalence and risk factors of claw lesions in the peri-urban areas of Nairobi, Kenya.

3.3.2 Feeding regime

In most of these zero-grazing units, the forage types were mainly mature-wilted or mature non-wilted Napier (*Pennisetum purpureum*), Rhodes (*Cloris gayana*) or mixed wild grasses. Only in one of the units were the cows fed on succulent forages. The Napier grass was fed to the cows in chopped small pieces but the Rhodes and wild grasses in whole lengths without chopping. In addition to forages in 16% of the zero-grazing units, there were supplementations with succulent maize stover or Napier grass silage.

In 81% of the zero-grazing units, concentrate feeding was regular at every milking time which was at least 2-3 times a day and in some there was additional concentrates given to the cows between the intervals from one milking time to the next. In the remaining 19% of the units, concentrate feeding was very occasional and inconsistent, being done mostly for one week in a month only when the farmers had money after they were paid for the milk sales of the previous month. The concentrate feeds used were commercially available and were known as "*Dairy Meal*" (manufactured by *Unga Farmcare Limited, Kenya*). Information provided by the feed milling firm indicated that the average constituents of the concentrate feed were protein (15.5% to 16%), oils (6% to 8%), fibre (9% to 10%) and moisture (12%). The concentrates were in fine ground-grains physical form. In the zero-grazing units that provided regular concentrate feeding, each cow was fed with 6-8 kilograms per day. The zero-grazing units that provided concentrates only occasionally were also inconsistent with the quantities provided per cow. No single zero-grazing unit used any standardized concentrate to forage ratios.

Cows in 47% of the zero-grazing units were fed regularly (once per day) with 50-100 grams of mineral supplements and in 10% of the units they were fed *ad libitum*. In 40% of the units, they were occasionally (at most 1-2 weeks per month) supplemented with 50 grams of minerals per day. In the remaining 3% of the units, the cows were not given any mineral supplements at all. The mineral supplements given were the locally available commercial mineral mixtures for dairy production such as "*Unga High Phosphorus*" (*Unga Farmcare Limited, Kenya*) or "*Supper Maclick*" (*Cooper Kenya limited*). The standard mineral supplements used in these zero-grazing units averagely consisted of sodium chloride (18% to 20%), calcium (16% to 18%), phosphorus (11% to 12%), magnesium (2.5% to 3%), iron (0.5%), copper (0.16%), manganese (0.4%), zinc (0.5%), sulphur (0.4%), cobalt (0.02%), iodine (0.02%) and selenium (0.0015%). The powdered mineral supplements were fed to the cows when mixed with concentrates, but the mineral-lick blocks were provided *ad libitum*.

3.3.3 Slurry management and claw trimming

In 72% of the zero-grazing units, manure (slurry) was removed from the alleyway at least once per day and in 18% it was removed more than once per day, while in 10% of the units it remained for more than one day before being removed (Fig. 19). None of the zero-grazing units had claw trimming programmes and none practiced any claw trimming at all. The farmers were completely ignorant of the need for claw trimming.

3.4 Association between risk factors and claw disorders

The animal-level risk factors found to be significantly associated with chronic laminitis were third or higher parities ($\chi2$ = 11.57, P = 0.009) and lactation stage between 90 to 180 days ($\chi2$

Fig. 19. Three different zero-grazing units A, B, and C in the peri-urban areas of Nairobi, Kenya showing slurry that has been left to accumulate for some days and could predispose the claws of cows to develop lesions.

= 9.75, P = 0.021). But lactation stage between 1 to 90 days was significantly ($\chi2$ = 9.06, P = 0.028) associated with subclinical laminitis. Failure to remove manure (more than one day) from the walk-alleys (r = 0.215, $\chi2$ = 13.85, P<0.001), regular (2-4 times per day) concentrate feeding (r = 0.135, $\chi2$ = 5.45, P = 0.0196) and lack of or inadequate mineral supplementation (r = 0.172, $\chi2$ = 8.9, P = 0.0307) were all found to significantly have unconditional association with subclinical laminitis (Table 2). However stepwise logistic regression analysis (screening interactions of floor types, type of cubicle bedding, manure removal, concentrate feeding, type of forage and mineral supplementation) revealed that the most significant zero-grazing housing-and management-level contributor to the occurrence of subclinical laminitis was regular concentrate feeding (O.R. = 2.08, $\chi2$ = 5.5, 95% C.I: 1.1 to 3.9, P = 0.0212).

Farm variables	$\chi2$	r-value	P-value
Lack of manure removal	13.85	0.21	< 0.001
High concentrate feeding	5.45	0.13	0.0196
Lack of mineral supplementation	8.9	0.17	0.0307
Floor types	5.52	0.14	0.02381
Overstocking (fewer cubicles than cows)	6.2	0.14	0.1844
Type of cubicle bedding	7.2	0.15	0.5155
Presence of a curb	0.13	0.02	0.9361
Fodder type	3.26	0.1	0.5151

Table 2. Association of subclinical laminitis with possible risk factors in a study carried out in 300 dairy cows from 32 smallholder zero-grazing units to evaluate digital characteristics of laminitis and related claw lesions in Nairobi and its environs

Zero-grazing housing- and management-level risk factors that had unconditional association with chronic laminitis were revealed by descriptive statistics. These were overstocking (fewer cubicles than the number of cows) (r = 0.36, $\chi2$ = 38.87, P<0.0001), presence or absence of cubicle bedding (r = 0.312, $\chi2$ = 29.13, P = 0.0003), earthen floor (r = 0.203, $\chi2$ = 12.33, P = 0.0151), presence of a high curb between walk-alley and the cubicles (r = 0.157, $\chi2$ = 7.43, P = 0.0244), leaving manure (more than one day) in the walk-alleys (r = 0. 279, $\chi2$ = 23.33, P<0.0001), regular (2-4 times per day) feeding with concentrates (r = 0.218, $\chi2$ = 14.25, P = 0.0002), and regular (once per day) mineral supplementation (r = 0.321, $\chi2$ = 30.85, P<0.0001) (Table 3).

Farm variables	$\chi2$	r-value	P-value
Overstocking (fewer cubicles than cows)	38.87	0.36	< 0.0001
Absence of cubicle bedding (bare concrete)	29.13	0.31	0.0003
Manure removal (infrequent)	23.33	0.28	< 0.0001
Earthen floor type	12.33	0.2	0.0151
High concentrate feeding	14.25	0.2	0.0002
Lack of mineral supplementation	30.85	0.3	< 0.0001
Presence of a curb	7.43	0.16	0.0244

Table 3. Association of chronic laminitis with possible risk factors in a study carried out in 300 dairy cows from 32 smallholder zero-grazing units to evaluate digital characteristics of laminitis and related claw lesions in Nairobi and its environs

Furthermore, stepwise logistic regression analysis (testing interactions of all these unconditionally associated factors) pointed out the farm-level factors that most significantly contributed to the occurrence of chronic laminitis. These factors were: overstocking (O.R. = 1.7, $\chi2$ = 24.29, 95% C.I : 1.3 to 2.2, P = 0.0002), earthen floor (O.R. = 1.5, $\chi2$ = 8.98, 95% C.I:1.2 to 2.0, P = 0.0006), and presence of a curb between walk-alleys and the cubicles (O.R. = 1.5, $\chi2$ = 5.06, 95% C.I : 1.1 to 2.2, P = 0.0262).

The data and statistical analyses indicated that regular (2-4 times per day) concentrate feeding seemed to significantly (β-estimate = 2.187, s.e. = 0.814, P = 0.007) enhance occurrence of sole bruising. Conversely, earthen floor (bare soil) (β-estimate = -1.796, s.e. = 0.835, P<0.001) and regular (at least once per day) mineral supplementation (β-estimate = -4.59, s.e. = 1.50, P = 0.002) seemed to significantly protect against occurrence of sole bruising. The study also indicated that removal of manure more than once per day significantly (β-estimate = -3.58, s.e. = 1.01, P<0.001) protected the claws against occurrence of white line separation, while regular (at least once per day) mineral supplementation was found to significantly (β-estimate = -4.07, s.e. = 2.05, P = 0.048) reduce the frequency of its occurrence. However, regular concentrate feeding seemed to significantly (β-estimate = 3.69, s.e. = 1.24, P = 0.003) enhance occurrence of white line separation

Furthermore, good concrete floors (non-slip with no ground defects) seemed significantly (β-estimate = -1.969, s.e. = 0.597, P<0.001) protective against occurrence of heel erosion. There were no statistically significant associations found between the rest of the zero-grazing housing-and management-level factors, and the claw disorders. Among the cows with chronic laminitis 89% were zero-grazed and only 11% were pasture-grazed. However, this difference was found to be statistically significant ($\chi2$ = 11.0, P<0.001)

4. Discussion

The strength of this paper is elaborated by the important variations in housing designs, nutritional regimes and management protocol in these smallholder zero-grazing dairy units. The dairy production protocol in these units is so varied to the extent that they can further be subdivided into subunits. It is of paramount importance to note that the prevalent claw disorders affecting cows in these zero-grazing units are non-infective in nature and associated with laminitis syndrome. It is also worth noting that the lesions with high prevalence are mainly non-clinical with the highest proportion being subclinical laminitis. This could be attributed to presence of low-grade but persistent predisposing factors existing in these zero-grazing units that cause insidious claw damage eventually leading to clinical manifestations at an advanced stage that may already have caused irreversible changes in the claws particularly with effects of laminitis (Belge & Bakir, 2005; Nocek, 1997). The significance of high prevalence of subclinical laminitis in these zero-grazing production systems cannot be underrated since it serves as a major contributor to development of other more severe clinical claw lesions with far reaching effects such as sole bruising, heel erosion, white line separation, double soles and eventually chronic laminitis if early remedial measures are not instituted. Such outcomes have been reported previously (Nocek, 1997). A retrospective study carried out on the cases of cattle referred from the same area to the University of Nairobi Large Animal Hospital in the earlier years indicated that previously there was high prevalence of infective claw lesions and negligibly low prevalence of laminitis and associated lesions (Nguhiu-Mwangi, et al., 2008). The pattern of occurrence of claw lesions in the same area has remarkably shifted to high prevalence of non-infective laminitis related lesions since the adoption of confinement of cattle in zero-grazing housing units, whose floors are concreted hence improving on hygiene (Nguhiu-Mwangi, et al., 2009). With the likelihood of improved cleanliness and hygiene of the modern type of cattle houses, the prevalence of infective claw lesions is bound to continually decline.

The importance of the claw lesions prevalent in the cows in these zero-grazing units is that they mainly affect the horn of the claw. When the horn is affected, the claws become vulnerable to many types of insults because it interferes with the protection of the interior parts of the claws. The effects on the horn are due to the fact that in these lesions there underlies the laminitis syndrome as the cause or sequel of the lesions. During this laminitis syndrome, the damaged corium produces poor quality horn that is softer than normal as previously reported (Baggott & Russell, 1981; Nocek, 1997; Weaver, 1979). This explains why the whitish soft crumbling horn was seen in some claws with chronic laminitis in this study. The softer low integrity horn of the sole and the heel predisposes these sites to bruising and erosion, hence the high prevalence of sole bruising and heel erosion. Since the horn quality is poorer in chronic laminitis, sole bruising and heel erosion become more severe during that phase of laminitis. The softened horn easily undergoes bacterial erosion and necrosis (Nocek, 1997) particularly when combined with other physical factors that cause horn softening such as accumulation of slurry on the floor of the cow house (Baggott & Russell, 1981). Sole bruising and heel erosion were found to be deeper and affecting more weight-bearing zones of the claw when the corium was more damaged. When these two lesions are spread more widely on the weight-bearing surface of the claw, it also means they involve the horn more deeply. This makes the prognosis poor because trimming-off the eroded horn leaves very thin layer of normal horn (Nguhiu-Mwangi, 2007) through which external insults easily traverse and damage the underlying corium further. However, when sole bruising and heel erosion are diagnosed early, the necrotic eroded horn can effectively be trimmed-off and the claws dipped in or sprayed with antimicrobial chemicals with either 5% formalin solution or 5-7% copper sulphate solution.

Alterations in the shape of the claws in chronic laminitis has been reported previously (Greenough, 1987; Nocek, 1997) and is attributed to disruption in growth of keratinized horn (Nocek, 1997). This shape alterations associated with chronic laminitis are severe, irreversible and render reshaping of the claw difficult; hence they give chronic laminitis poor prognosis (Baggott & Russell, 1981; Nguhiu-Mwangi et al., 2008). This explains why attempts made to manage deformity conditions by trimming claws of the affected cows during this study, proved to be of little benefit. Therefore, the finding of irregularly shaped (unusual flattening, widening, concaving and rippling of the dorsal wall) claws in chronic laminitis is an indication of poor prognosis (Nguhiu-Mwangi et al., 2008) and attempts to manage them by trimming are therefore only palliative and short-lived. Ignorance and negligence of claw trimming practices in the zero-grazing units that were included in this study have been major cause and an aggravating factor in the progression of claw disorders to chronic stages. This contributes to long-term claw damage and worsens the prognoses (Nguhiu-Mwangi, 2007). It has been shown that trimming of hooves leads to lower locomotion scores and reduced clinical lameness (Manson & Leaver, 1988). When claws are excessively overgrown, weight-bearing is shifted towards the proximal parts near the heel bulb. These parts have softer and weaker horn that inadequately protects the underlying corium (Collins-née Hedges et al., 2004). Excessive claw overgrowth then leads to undue mechanical pressure on the solar corium due to excessive weight-bearing on the pedal bone and inadequate support of the corium by the horny capsule at the sole especially following thinning of the horn of the sole in laminitis (Baggott & Russell, 1981; Toussaint-Raven, 1973). Trimming tends to return the claw to near normal weight-bearing even at chronic stages of regular overgrowth (Weaver, 1993). However, when overgrowth is combined with chronic laminitis, the pedal bone undergoes irreversible alteration in its configuration within the claw capsule (Bargai et al., 1989; Greenough et al., 1990) and this result in poor prognosis.

It is prudent to conclude that regular examination is the most important way of ensuring good claw health. The most practical way to regularly observe the claws is during the milking times. Examination of the claws cannot be conclusive without trimming at least the horn of the sole. This is necessitated by the fact that lesions underneath the horn of the claws are not discernible externally until exposed by trimming (Nguhiu-Mwangi, 2007; Nocek, 1997). Such routine examination will help to recognize and manage claw disorders in the early stages before they progress to unmanageable chronic stages in which the claws are irreversibly damaged. Furthermore, the corium is the tissue that determines the health of the claw. It should therefore be protected from unnecessary insults by ensuring that it is adequately covered by sufficient claw capsule (hoof) thickness whose hardness is maintained by avoiding wet underfoot environment and that distribution of weight-bearing is correctly maintained according to the relative hardness of various parts of the claw (Berry, 1999).

Although regular concentrate feeding in these farms for each cow meant 2 to 4 times per day, nevertheless the feeding frequencies and amounts in each farm were inconsistent and this was similar for mineral supplementation. The results of this study indicated that most of the claw lesions encountered were related to laminitis in nature (Belge & Bakir, 2005; Nocek, 1997) and this could highly suggest the presence of predisposing risk factors in the smallholder zero-grazing units that were studied. Lactation period between 1 to 90 days is the time that subclinical laminitis is most prevalent. This could be attributed to the continuing effects of the transitional period of feeding with high energy concentrates that cause subacute ruminal acidosis and release of vasoactive agents, subsequently triggering laminitic processes (Donovan et al., 2004; Vermunt, 2004). The laminitic processes are further aggravated by the stress of heavy milk production during the first 90 to 120 days post-calving (Rowlands et al., 1985). Gradual progression of subclinical laminitis lesions culminates into chronic laminitis (Nocek, 1997), which could explain the reason for association between the later stages of lactation and chronic laminitis. The association between 3rd or higher parities with chronic laminitis agrees with previous findings by Sogstad et al. (2005) that most lameness is generally associated with third or higher parities. The habit of leaving manure in the walk-alleys for more than one day enhanced development of both subclinical and chronic laminitis. This could be attributed to persistent wetness that causes excessive softening of the claw-horn making it easily susceptible to horn-lesions. With softer solear horn, pressure insult from the ground is easily transmitted to the underlying solear corium which then becomes traumatized, resulting in laminitis. These findings agree with previous reports (Borderas et al., 2004; Vermunt, 2004).

The association of high concentrates feeding with subclinical and chronic laminitis as well as with sole bruising and white line separation is attributed to SARA and production of vasoactive substances. This initially triggers development of subclinical laminitis as has been reported previously (Cook et al., 2004; Vermunt, 2004). Subclinical laminitis then predisposes the cow to other claw lesions (Nocek, 1997). Failure or inadequate supplementation of the cows with mineral mixtures was associated with enhanced development of subclinical and chronic laminitis as well as white line separation. This finding conforms to the reports stating that trace-minerals reduce claw lesion scores (Tomlinson et al., 2004) and biotin supplementation reduces macro-cracks in the claws (Higuchi & Nagahata, 2001). The positive association between overstocking and chronic laminitis may be attributed to the long hours of standing because of fewer cubicles. Standing for long periods causes prolonged pressure injury to the solear corium, thus being an exacerbating factor to the development of claw lesions as reported previously (Leonard et al., 1994). Further to this, the presence of a high curb between the walk-alley and the cubicles was found to be one of the main factors enhancing occurrence of chronic laminitis. This factor may possibly cause additional stress to the claws as the cow enters or disembarks from the cubicle, as has been suggested by others (Leonard et al., 1994; Philipot et al., 1994). Eventually it contributes to compromise of the entire phenomenon of "cow comfort" and some cows prefer standing on the free alleys to occupying the cubicles. Standing for long periods in the free walk-alleys causes further stress to the corium.

The results of this study also found that housing cows on earthen (bare soil) floors was an important risk factor for development of chronic laminitis. This is possibly due to the fact that such floors have soft (yielding) surfaces which tend to promote overgrowth of the claws for lack of constant hoof-wear (Rhebun & Pearson, 1982). Claw overgrowth and change in its angle result in redistribution of weight-bearing that subsequently overloads specific zones of the claw particularly during locomotion (Van der Tol et al., 2003). Consequently, these effects predispose the affected claws to development of laminitis and other claw lesions (Neveux et al., 2006; Sagues, 2002). However, earthen floors were found to be protective against occurrence of sole bruising. This could be attributed to the floor being soft, yielding and non-abrasive which closely relates to a previous report by Somers et al. (2003) stating that cows raised on yielding straw-yard surfaces were found to have very low prevalence of claw disorders. Chronic laminitis is the only claw disorder that showed a significant difference between cows in zero-grazed and in pasture-grazed farms with higher prevalence in zero-grazed cows. The hard concrete surface that was more common in the zero-grazed farms, probably combined with regular concentrate feeding, could have contributed to the difference in the prevalence rates of chronic laminitis between zero-grazed and pasture grazed farms (Bergsten, 1994; Somers et al., 2003; Vokey et al., 2001).

Absence of cubicle bedding and presence of bare concrete in the cubicles was found to be associated with occurrence of chronic laminitis. Some of the cubicles were narrow in such a way that cows remained in a standing posture even within the cubicles. This act of standing in the cubicles further prolongs the cumulative period of stress exerted on the claws while on concrete. The consequence of this stress is the increased incidence of claw lesions, particularly laminitis (Bergsten & Herlin, 1996; Somers et al., 2003). Good concrete (non-slip, without defects or loose stone pebbles) was protective against heel erosion probably due to the absence of traumatic edges. The non-slip nature of the floor would also prevent abrasive trauma on both the heel and sole.

5. Conclusion

The smallholder zero-grazing dairy units in the peri-urban areas of Nairobi, Kenya have uniquely wide variations in their production systems with respect to housing designs, nutritional regime and management protocol. From the statistical associations in this study, it can be concluded that the cow-level factors that strongly contribute to the development of claw lesions are 3rd or higher parities and being in the lactation period between 1 to 90 days. The farm level factors that strongly enhance claw lesion development are frequent high concentrate feeding, lack of regular mineral supplementation, both concrete and earthen floors, overstocking, the presence of a curb between walk-alley and cubicles, and leaving manure in the walk-alley for a long time. These associations are supported strongly by statistics that indicate the direction and strength of the relationship. It can also be concluded that non-infective claw disorders in dairy cows particularly the disorders related to laminitis are insidious in nature, which gradually but progressively damage the integrity of the claw. The subclinical occurrence of these disorders makes them subtle and careful early diagnosis so that remedial measures can be instituted early before these irreversible damages have occurred is essential. The fact that a cow does not show signs of lameness does not necessarily imply her claws are sound, but only calls for further careful scrutiny. Claw trimming is one of the major ways of discerning these underlying claw disorders at the subclinical phase.

6. Future research

Due to the wide variations in the smallholder production units in the study area, there is a need to carry out a controlled experimental study to further elucidate these risk factors. A comparative study should also be carried out in Kenya to establish if similar risk factors exist in the large scale better managed farms. Each of the risk factors incriminated in the current study needs individual experimental evaluation.

7. Acknowledgements

We are grateful to the Deans Committee, University of Nairobi, Kenya for providing a proportion of financial support to carry out the study.

8. References

Baggott, D.G. & Russell, A.M. (1981): Lameness in cattle. *British Veterinary Journal* vol.137, pp 113-132

Bargai, U, Pharr, J.W.& Morgan, J.D. (1989) radiographic changes in chronic laminitis *Proceedings of the 8th International Symposium on Disorders of Ruminant Digit and International Conference on Bovine Lameness*. Saskatoon, Saskachewan, and Banff, Alberta, Canada. pp 207

Belge, A. & Bakir, B. (2005). Subclinical laminitis in dairy cattle: 205 selected cases. *Turkish Journal of Veterinary and Animal Science*, vol. 29, pp 9-15

Bergsten, C. (1994). Haemorrhages of the sole horn of dairy cows as a retrospective indicator of laminitis: an epidemiological study. *Acta Veterinaria Scandinavica*. Vol.35, pp 55-66

Bergsten, C. & Herlin, A.H. (1996). Sole haemorrhages and heel horn erosion in dairy cows: the influence of housing on their prevalence and severity. *Acta Veterinaria Scandinavica* vol37, pp395-408

Berry, S.L. (1999). Hoof health. *Proceedings of the Western Dairy Management Conference*.. Las Vegas, Nevada, April 8-10, 1999, pp 13-17

Borderas, T.F., Pawluczukk, B., de Passille, A.M. & Rushen, J. (2004). Claw hardness of dairy cows: relationship to water content and claw lesions. *Journal of Dairy Science*, vol. 87,pp 2085-2093.

Clarkson, M.J., Downham, D.Y., Faull, W.B., Murray. R.D., Russell, W.B., Sutherst, J. & Ward, W.R. (1996). Incidence and prevalence of lameness in dairy cattle. *Veterinary Record*, vol. 138, pp 563-567

Collins-née Hedges, V.J., Green, L.E., Blowey, R.W., Packington, A.J. & Bonser, R.H.C. (2004). Testing white line strength in the dairy cows. *Journal of Dairy Science*, vol.87, pp 2874-2880

Cook, N.B., Nordlund, K.V. & Oetzel, G.R. (2004). Environmental influences on claw horn lesions associated with laminitis and subacute ruminal acidosis in dairy cows. *Journal of Dairy Science*, Vol. 87, pp E36-E46

Donovan, G.A., Risco, C.A., DeChant Temple, G.M., Tran, T.Q. & van Horn, H.H. (2004). Influence of transition diets on occurrence of subclinical laminitis in Holstein dairy cows. *Journal of Dairy Science*, vol. 87, pp 73-84

Enting, H. D., Kooij, A. A., Dijkhiuzen, R. B. M., Huirne, R. B. M. & Noordhuizen-Stassen, E. N. (1997). Economic losses due to clinical lameness in dairy cattle. *Livestock Production Science*, vol 49, pp 259-267

Garbarino, E.J., Hernandez, J.A., Shearer, J.K., Risco, C.A. & Thatcher, W.W. (2004). Effects of lameness on ovarian activity in postpartum Holstein cows. *Journal of Dairy Science* vol. 87, pp 4123-4131

Gitau, T. (1994). Bovine lameness in small-scale dairy farms in Kikuyu division, Kiambu district, Kenya. *Msc. Thesis*, University of Nairobi, Kenya

Greenough, P. R. (1987). An illustrated compendium of bovine lameness. Part 1. *Modern Veterinary Practice*, vol. 68, pp 6-9

Greenough, P.R., Vermunt, J.J., McKinnon, J.J., Fathy, F.A., Berg, P.A. & Cohen, R.D.H. (1990). Laminitis-like changes in the claws of feedlot cattle. *Canadian Veterinary Journal*, vol. 31, pp 202-208

Greenough, P.R. & Vermunt, J.J. (1991). Evaluation of subclinical laminitis in a dairy herd and observations on associated nutritional and management factors. *Veterinary Record*, vol. 128, pp 11-17

Hernandez, J.A., Garbarino, E.J., Shearer, J.K., Risco, C.A. & Thatcher, W.W. (2005). Comparison of the calving-to-conception interval in dairy cows with different degrees of lameness during the prebreeding postpartum period. *Journal of. American Veterinary Medical Association*, vol. 227, pp1284-1291

Higuchi, H. & Nagahata, H. (2001). Relationship between serum biotin concentration and moisture content of the sole horn in cows with clinical laminitis or sound claws. *Veterinary Record*, vol.148, pp 209-210

Kossaibati, M. A. & Esslemont, R. J. (1997). The cost of production diseases in dairy herds in England. *Veterinary Journal*, vol.154, pp 41-51

Leonard, F.C., O'Connell, J.M. & O'Farrell, K.J. (1994). Effect of different housing conditions on behaviour and foot lesions in Friesian heifers. *Veterinary Record*, vol.134, pp 490-494

Manske, T., Hultgren, J. & Bergsten, C. (2002). The effect of claw trimming on the hoof health of Swedish dairy cattle. *Preventive Veterinary Medicine,* vol. 54, pp 113-129

Manson, F.J., & Leaver, J.D. (1988). The Influence of dietary protein intake and of hoof trimming on lameness in dairy cattle. *Animal Production,*vol. 47, pp 191-199

Mbithi, P.M.F., Mbiuki, S.M., Nguhiu-Mwangi, J. & Kihurani, D.O. (1991). Non-fracture lameness in cattle: A retrospective study. *Bulletin of Animal Health and Production in Africa,* vol. 39, pp 307-309

Melendez, P., Bartolome, J., Archbald, L.F. & Donovan, A. (2003). The association between lameness, ovarian cysts and fertility in lactating dairy cows. *Theriogenology,* vol.59, pp 927-937

Mutugi, J.J. (2004). Various livestock production systems. *Workshop Proceedings on Cattle Production in Kenya-Strategies for Research Planning and Implementation,* KARI Headquarters, Kabete, Kenya, December 2003. pp 3-35

Neveux, S., Weary, D.M., Rushen, J.;, von Keyserlingk, M.A.G. & de Passillé, A.M. (2006). Hoof discomfort changes how dairy cattle distribute their body weight. *Journal of Dairy Science,* vol. 89, pp 2503-2509

Nguhiu-Mwangi, J. (2007). Characteristics of laminitis and associated claw lesions in dairy cows in Nairobi and its environs. *PhD thesis,* University of Nairobi, Kenya.

Nguhiu-Mwangi, J., Mbithi, P.M.F., Wabacha, J.K., Mbuthia, P.G. (2007). Prevalence of sole haemorrhages and its correlation with subclinical and chronic laminitis in dairy cow. *Bulletin of Animal Health and Production in Africa,* vol. 55, pp 232-242

Nguhiu-Mwangi, J., Mbithi, P.M.F., Wabacha, J.K., Mbuthia, P.G. (2008). Factors associated with the occurrence of claw disorders in dairy cows under smallholder production systems in urban and peri-urban areas of Nairobi, Kenya. *Veterinarski Arhiv,* vol. 78, No. 4, pp 343-355, ISSN 0372-5480

Nguhiu-Mwangi, J., Mbithi, P.M.F., Wabacha, J.K., Mbuthia, P.G. (2009). Prevalence of laminitis and the patterns of claw lesions in dairy cows in Nairobi and the peri-urban districts. *Bulletin of Animal Health and Production in Africa,* Vol. 57, pp 199-208.

Nocek, J.E. (1997): Bovine acidosis: implications on laminitis. *Journal of Dairy Science vol* 80: 1005-1028.

Offer, J.E., McNulty, D. & Logue, D.N. (2000). Observations of lameness, hoof conformation and development of lesions in dairy cattle over four lactations. *Veterinary Record,* vol. 147, pp 105-109

Philipot, J.M., Pluvinage, P., Cimarosti, I., Sulpice, P. & Bulgnard, F. (1994). Risk factors of dairy cow lameness associated with housing conditions. *Veterinary Research,* vol. 25, pp 244-248

Rhebun, W.C. & Pearson, E.G. (1982). Clinical management of bovine foot problems. *Journal of American Veterinary Medical Association,* vol. 181, pp 572-579

Rowlands, G.J., Russell, A.M. & Williams, L.A. (1985). Effects of stage of lactation, month, age, and heart girth on lameness in dairy cattle. *Veterinary Record,* vol.117, pp 576-580

Sagues, A.G. (2002). The biomechanics of weight bearing and its significance with lameness. *Proceedings of the 12th International Symposium on Lameness in Ruminants,* Orlando, Florida, U.S.A., 9th-13th January, 2002, pp 117-121

Smilie, R.H., Hoblet, K.H., Eastridge, M.L., Weiss, W.P., Schnitkey, G.L. & Moeschberger, M.L. (1991). Subclinical laminitis in dairy cows: use of severity of hoof lesions to rank and evaluate herds. *Veterinary Record,* vol.144, pp 17-21

Sogstad, A.M., Fjeldaas, T. & Østeras, O. (2005). Lameness and claw lesions of the Norwegian red dairy cattle housed in free-stalls in relation to environment, parity and stage of lactation. *Acta Veterinaria Scandinavica,* vol. 46, pp 203-217

Sogstad, A.M., Østeras, O. & Fjeldaas, T. (2006). Bovine claw and limb disorders related to reproductive performance and production diseases. *Journal of Dairy Science,* vol.89, pp 2519-2528

Somers, J.G.C.J., Frankena, K., Noordhuizen-Stassen, E.N. & Metz, J.H.M. (2003). Prevalence of claw disorders in Dutch dairy cows exposed to several floor systems. *Journal of Dairy Science,* vol.86, pp 2082-2093

Somers, J.G.C.J., Schouten, W.G.P., Frankena, K., Noordhuizen-Stassen, E.N. & Metz, J.H.M. (2005). Development of claw traits and claw lesions in dairy cows kept on different floor systems. *Journal of Dairy Science,* vol. 88, pp 110-120

Sprecher, D.J.; Hostetler, D.E. and Kaneene, J. B. (1997). A lameness scoring system that uses posture and gait to predict dairy cattle reproductive performance. *Theriogenology,* vol. 47, pp 1179-1187

Toussaint-Raven, E. (1973): Lameness in cattle and foot care. *Netherlands Journal of Veterinary Science,* vol.5, pp 105-111

Van der Tol P.P.J., Metz, J.M.H., Noordhuizen-Stassen, E.N., Back, W.; Braam, C.R. & Weijs, W.A. (2003). The vertical ground reaction force and the distribution on the claws of dairy cows while walking on a flat substrate. *Journal of Dairy Science,* vol. 86, pp 2875-2883

Vermunt, J. (2004). Herd lameness: a review, major causal factors, and guidelines for prevention and control. *Proceedings of the 13th International symposium and 5th Conference on Lameness in Ruminants,* Maribor, Slovenija, 11th -15th February, 2004, pp 1-15.

Vokey, F.J., Guard, C.L., Erb, H.N.& Galton, D.M. (2001). Effects of alley and stall surfaces on indices of claw and leg health in dairy cattle housed in a free-stall barn. *Journal of Dairy Science,* vol. 84, pp 2686-2699

Wanyoike, M.M.& Wahome, R.G. (2004). Small-Scale farming systems. *Workshop Proceedings, on Cattle Production in Kenya-Strategies for Research Planning and Implementation,* KARI Head quarters, Kabete, Kenya,. December. 2003, pp 87-133

Weaver, A.D. (1993). Aseptic laminitis of cattle, Interdigital and Digital dermatitis, Interdigital Phlegmon (interdigital necrobacillosis). *Current Veterinary Therapy 3. Food Animal Practice,* W.B. Saunders, Philadelphia. pp 867-870.

Weaver, A.D. (2000). Lameness. *The health of Dairy Cattle,* Blackwell Science, Oxford, U.K. pp 149-202.

22

Pharmacokinetic - Pharmacodynamic Considerations for Bovine Mastitis Treatment

Nora Mestorino and Jorge O. Errecalde
*Department of Pharmacology and Toxicology,
Faculty of Veterinary Science,
Universidad Nacional de La Plata,
Buenos Aires,
Argentina*

1. Introduction

Bovine mastitis is a disease that affects dairy herd production, characterized by considerable economical loss due to diminished milk secretion, potential productive cow damage, increase in production costs and milk contamination. Intramammary infection (IMMI) is the most common reason for the use of antimicrobials in dairy cows. Antimicrobials (ATMs) have been used to treat mastitis for more than fifty years, but consensus about the most efficient, safe, and economical treatment is still lacking.

Staphylococcus aureus is considered one of the main bacteria causing bovine mastitis, which is widely distributed in different countries. *S. aureus* is Gram-positive cocci, catalase-positive and facultative anaerobe. The *Staphylococcus* genus comprises more than thirty species which are able to colonize many environments and are part of the cutaneous or mucousal flora of various animals and humans. The intracellular survival of *S. aureus* is believed to contribute to the recurrence of some infections such as mastitis. Some publications reported the ability of this pathogen to colonize multiple cell types. However the precise fate of intracellular *S. aureus* is still poorly understood.

The general lack of therapeutic success against subclinical mastitis caused by *S. aureus* has prompted a reevaluation of treatment strategies. Despite the availability of several antibiotics with good in vitro activity, cure rates are poor, suggesting that inadequate concentrations of active antibiotic are coming into contact with the infecting bacteria for sufficient time and / or adequate concentrations to be effective.

Over the last few years, much concern has been raised regarding the optimization of antibiotic use, owing to the worrying increase of bacterial resistance. In this context, progress in the field of anti-infective pharmacology has led to the emergence of a new discipline, referred to as pharmacokinetics/pharmacodynamics (PK/PD) of antibiotics, the discipline that strives to understand the relationships between drug concentrations and effects, both desirable (bacterial killing) and undesirable (bacterial resistance). Over the past 15 years, three key PK/PD parameters have been elaborated, which determine how antibiotic concentrations reached in body fluids over time (as predicted from the PK profile of the drug) compare with potentially effective antibiotic concentrations (as deduced from the minimal inhibitory concentration (MIC) or minimal bactericidal concentration (MBC) of antibiotics in vitro). The first parameter is the time during which concentrations of the ATM are above the MIC (t > MIC), it links bactericidal effects to time and is critically dependent on the half-life of the drug, dosage and frequency of administration over a given period of time. The second parameter is the peak plasma concentration divided by the MIC (C_{max}/MIC), it relates bactericidal effects to concentration, and is primarily dependent on the dose. The third parameter is the area under the concentration-time curve divided by the MIC (AUC/MIC), and it combines both types of effects, since it corresponds to the total amount of drug to which bacteria are exposed over the time period, and is directly related to the total dose given during that period and inversely proportional to the drug clearance (Van Bambeke et al., 2006).

These parameters are critical in predicting antibiotic activity and, therefore, in establishing dosages on a rational basis. The application of these parameters, however, have so far been limited to extracellular infections in well-vascularized tissues, because they are all based on serum antibiotic levels.

Antimicrobials exhibit three major patterns of ATM activity. The first pattern in characterized by concentration dependent killing and moderate to prolonged persistent effects. Higher concentrations would kill organisms more rapidly and more extensively than lower levels. The prolonged persistent effects would allow for the administration of large doses with long inter-dose periods. Microorganism regrowth is not immediate at the time in which the drug concentrations fall below the MIC. This called post antibiotic effect is variable between different drug types but always present for ATMs exhibiting this kind of killing. This pattern is observed with aminoglycosides, fluoroquinolones, daptomycin, ketolides, and amphotericin B. The goal of a dosing regimen for these drugs would be to maximize concentrations. The peak level and the AUC should be the pharmacokinetic parameters that would determine in vivo efficacy (Andes et al., 2001; Craig, 2001).

The second pattern is characterized by time-dependent killing and minimal to moderate persistent effects. High drug levels would not kill organisms better than lower concentrations. Furthermore, organism regrowth would start very soon after serum levels fall below the MIC. This pattern is observed with β-lactams, macrolides, clindamycin, and oxazolidinones. The goal of a dose regime for these drugs would be to optimize the duration of exposure. The duration of time that serum levels exceed some minimal value such as the MIC should be the major parameter determining the in vivo efficacy of these drugs (Andes et al., 2001; Craig, 2001).

The third pattern is also characterized by time-dependent killing, but the duration of the persistent effects is much prolonged. This can prevent any regrowth during the dosing interval. This pattern is observed with azithromycin, tetracyclines, quinupristin-dalfopristin, glycopeptides, and fluconazole. The goal of a dose regime is to optimize the amount of drug administered to ensure that killing occurs for part of the time and there is no regrowth during the dosing interval. The AUC should be the primary pharmacokinetic parameter that would determine in vivo efficacy (Andes et al., 2001; Craig, 2001).

Antimicrobial PK/PD relationship reflects a correlation between the drug concentrations in the blood, the concentration of the biologically active drug at the site of the infection, and the microbial or clinical outcome (Levison, 2004). The PK component describes the processing of the drug by the host (absorption, distribution, metabolism, and elimination). The PD component describes the effect of the drug on the bacterial pathogen. By identifying an association between PK and PD for specific host-drug-microbe combinations, the PK/PD approach provides a valuable guide for estimating the doses and dosage regimens that can optimize the bacteriological or clinical outcome.

The extent to which a drug has access into milk when given systemically, or is absorbed and distributes throughout the udder when given intramammarily, depends on its properties: lipid solubility, degree of ionization, and extent of binding to serum and udder proteins. High lipid solubility, poor degree of ionization and less plasma protein binding contributes to a better transfer into milk. With IMM preparations, the type of vehicle is also important.

Weak organic bases like macrolides and sulfonamidestend to accumulate in milk in the ionized form after parenteral administration, and attain concentrations higher than those in blood. On the other hand, concentrations of weak acids like penicillins and cephalosporins in milk are significantly lower than those in blood. In the concentration-dependent group of ATMs (e.g. aminoglycosides and fluoroquinolones) concentration of several times the MIC for the target organisms at the infection site increases the efficacy. In the time dependent group (e.g. penicillins, cephalosporins and classical macrolides) the efficacy depends on the time during which the concentration of the drug exceeds the MIC, but high concentrations do not increase efficacy.

While there are a number of factors that contribute to the PK/PD indexes (e.g. in vitro MIC of the drug, its post-antibiotic effects -PAE-, and sub-MIC effects), there are also several factors that the traditional PK/PD approach does not describe. For example, the in vitro MIC, which is the basis for the PD component of these indices, does not provide information on time to kill, time to maximum kill, log change within a fixed time, or the maximum reduction in viable bacterial counts. There are also many others in vivo factors that influence ATM effectiveness (e.g. anti-inflammatory effects, presence of bacterial biofilms, the drug´s ability to interfere with bacterial colonization on epithelial surface, and influence of the drug on toxin production and release). Furthermore, plasma drug concentrations do not necessarily reflect a compound´s ability to diffuse into the site of infection and into the bacterial cell.

An ideal drug for mastitis therapy should have a low MIC for mastitis pathogens. As treatment should be efficient and targeted towards specific infections, Gram-negative and Gram-positive infections in fact would require different ATMs. Antimastitic drugs should preferably have bactericidal action, as phagocites act normally immediately after milking, but as time elapses they incorporates fat globules becoming "engorged" thus diminishing its phagocytic capacity. As a consequence milk phagocytes are less effective than plasma ones. Milk should not interfere with ATM activity.

The objective of this study was to make a bibliographic compilation on the PK of different ATM agents in milk, based in our experience and some other relevant publications, and to establish the relationship between the PK and the PD interaction with the target bacteria. We consider that non prudent use of ATMs in dairy farms can be fought through the implementation of rational therapeutical procedures based on the following items:

1. Knowledge on the microbiological profile of mastitis. Clinical and laboratory diagnosis (isolation, typification and antibiogram) is one of the basic pillars for the rational use of ATM agents.
2. Knowledge on the PK and PD of antibiotics in milk producing animals, both healthy and mastitic. Understanding the relation PK/PD in these animals will result in an increase of ATM efficiency and a decrease in the selection of resistant strains, two important aspects leading to the success of the therapy.
3. More efficient and safer prescription of antimastitic drugs. The development of new products should involve more than just the discovery of a new substance; it should deal with the full utilization of the effects of the agents in the organism. The way a drug is transported to the target location should be considered; once there, it should be available in the appropriate concentration and time to be effective.
4. Establishment of the correct withdrawal periods for the antibiotics tested, following internationally agreed protocols.

2. General pharmacokinetic considerations

An important question regarding the treatment of mastitis is whether the ATM should accumulate in milk or in the udder tissue (Erskine et al., 2003). The target site may depend on the causative agent: streptococci are known to remain in the milk compartment, but *S. aureus* penetrates udder tissue and causes deep infection (Table 1). The most common route of administration of ATMs in mastitis is the IMM route. Efficacy of IMM treatment varies according to the pathogen, with the best therapeutic response being shown for mastitis caused by streptococci, coagulase-negative staphylococci, and Corynebacterium spp.

	Milk/ducts	Udder tissue	Cow
Streptococcus agalactiae	+++	---	---
Other streptococci	+++	+	---
Staphylococcus aureus	+	+++	---
Coagulase-negative staphylococci	+++	---	---
Arcanobacterium pyogenes (summer mastitis)	---	++	+++
Coliforms	+	--	+++

Table 1. Where to target ATM therapy in clinical mastitis due to different pathogens (Erskine et al., 2003)

2.1 Parenteral antimicrobial treatments

Systemic administration for treatment of mastitis was first used in the 70's (Ziv, 1980a). In acute cases, the IMM administration often fails due to poor and uneven distribution of the drug, either by the growth of breast parenchyma or blockage caused by the products of inflammation. In these circumstances, parenteral therapy is preferred (Mestorino, 1993a).

From the clinical point of view, the success of parenteral therapy depends mainly on the ATM passage from blood to milk (Ziv, 1980a). The ATM concentrations in highly vascularized tissues are equivalent to those determined in blood plasma. By contrast, in places where the irrigation is poor or those which are separated from the central compartment by biological membranes, drug levels are not equivalent (Baggot, 1986). The time during which concentrations in the mammary gland are effective depends largely on the drug characteristics, the dose, the bioavailability of the molecule, the ability to penetrate the mammary gland and the microorganism susceptibility (Ziv, 1980b; Mestorino, 1993a).

The ability to penetrate the mammary gland or milk bioavailability (Fmilk) is determined by the ratio AUC0-∞ milk / AUC0-∞ plasma, as shown in the following equation:

$$F_{milk} = \frac{AUC_{0-\infty(milk)}}{AUC_{0-\infty(serum)}} \tag{1}$$

This equation determines the relationship between the amount of ATM that is absorbed to the central compartment and the amount of ATM that passes through the mammary gland for reaches the milk compartment (Mestorino, 1993a).

Those antibiotics that have a high volume of distribution penetrate better into the mammary gland. However, differences in the degree of penetration blood: milk occur even among compounds that are chemically and structurally related. These differences can be explained by the principle of passive diffusion (Ziv, 1980b). ATMs cross biological membranes by passive diffusion or specialized transport.

Since the surface of the lipid portion of the membrane is extremely high, passive diffusion through membranes can be considered synonymous of diffusion through membrane lipids (Errecalde, 2004). The transfer in this case is directly proportional to the concentration gradient and the lipid-to-water partition coefficient of the ATM (Ziv, 1980b).

Weak organic acids and bases are found in milk and plasma as ionized or nonionized forms. The nonionized fraction is generally more soluble than the ionized one and diffuses better through the biological membrane (Ziv, 1980b; Errecalde, 2004; Mestorino, 1993a). The proportion of the drug in the nonionized form depends on the pKa of the molecule and the pH of the medium in which it is dissolved. When the molecules pass through the membrane by simple diffusion, are distributed according to their degree of ionization, the charge of their ionized form and the extent of protein binding. This is because the molecules bound to proteins or tissues are not able to cross membranes (Ziv, 1980b).

The theoretical relationship between the drug concentrations on both sides of a biological membrane can be calculated according to the Jacobs equation, that for organic acids such as penicillin G (pKa = 2.8) is the following:

$$\text{Ratio milk:plasma} = \frac{1 + 10pH_{milk-}pK_a}{1 + 10pH_{plasma-}pK_a} \tag{2}$$

And in the case of organic bases, such as spiramycin (pKa = 8.2) is:

$$\text{Ratio milk:plasma} = \frac{1 + 10pK_a - pH_{milk}}{1 + 10pK_a - pH_{plasma}} \tag{3}$$

The serum pH is 7.4 and the milk has a pH between 6.6 - 6.8. The organic bases administered by the parenteral route tends to accumulate in milk and remain there in its ionized form (ion trapping), thus achieving milk concentrations which exceed those in plasma. Instead, the concentrations of weak acids in milk are lower than those found in plasma (Erskine, 2002b).

That is to say that the majority of the drugs can be ionized or nonionized according to its pKa and the pH of the surrounding environment. Nonionized compounds have a higher lipid-water partition coefficient than the ionized ones and thus it is easier for them to diffuse through lipidic membranes. The amphoteric molecules, such as danofloxacin (pKa = 6.2 to 9.4) does not depend on the relationship pK/pH and therefore its distribution is essentially determined by the degree of lipid solubility of the molecule and consequently by its lipid-water partition coefficient.

Table 2 shows different ATMs with its theoretical and experimental milk: plasma ratios. Observing this table we can conclude that the diffusion of organic acids into milk is highly predictable, but the diffusion of organic bases can be predicted only when these are largely nonionized in plasma and have a moderate degree of lipid solubility.

Antimicrobial agent	Chemical nature	PK_a	Lipid solubility	Milk to serum concentration ratio Theoretical - Experimental
Sulfanilamide	Acid	10.4	Moderate	1.00 – 0.97
Sulfathiazole	Acid	7.1	Moderate	0.37 – 0.35
Sulfadiazine	Acid	6.5	Moderate	0.28 – 0.21
Penicillin G	Acid	2.8	Moderate	0.16 – 0.20
Cloxacillin	Acid	2.8	High	0.16 – 0.22
Ampicillin	Acid	2.8; 7.2	High	0.26 – 0.26
Amoxicillin	Acid	2.8; 7.2	High	0.26 – 0.26
Cephacetrile	Acid	2.4	Moderate	0.12 – 0.15
Cephapirin	Acid	2.6	Moderate	0.14 – 0.18
Rifamycin SV	Acid	2.8; 6.7	Moderate	0.25 – 0.25
Rifampicin	Acid	7.9	High	0.85 – 1.10
Novobiocin	Acid	4.3	High	0.30 – 0.,33
Penethamate	Base	8.5	High	5.7 - 6.1
Streptomycin, neomycin	Base	8.9	Low	7.5 - 0.5
Polymyxin B, colistin	Base	10.0	Very low	8.0 – 0.3
Erythromycin	Base	8.8	High	6.5 – 8.5
Tylosin	Base	7.1	High	5.0 – 4.5
Spiramycin	Base	8.2	High	4.8 – 4.6
Lincomycin, clindamycin	Base	7.6	High	4.2 – 4.4
Tetracyclines	Amphoteric	-	Moderate	0.4; 0.8 – 0.6; 1.4

Table 2. Partition of ATMs in plasma and milk in lactating animals. From Ziv G. (1980b)

In the case of a weak base, when administered parenterally, e.g intramuscularly, the rapid penetration from blood to milk is characterized by early appearance of measurable concentrations in this fluid, a ratio $C_{max\ (plasma)} / C_{max\ (milk)}$ less than or equal to 1, a t-lag between the $C_{max\ (plasma)}$ and $C_{max\ (milk)}$ short and a ratio $AUC_{\ (plasma)} / AUC_{\ (milk)}$ less than or equal to 1 (Ziv, 1980b).

During the mastitis process, the chemical composition of milk presents changes consequence of the inflammatory process. There are ions, proteins and inflammatory cells passage from blood into the gland lumen, because of a great increase in the vascular permeability. The physical properties of milk are also affected and there is an increase of pH, conductivity and viscosity, while the density and redox potential decrease (Korhonen & Kaartinen, 1995).

In mastitic milk it is expected that the passage of organic bases is diminished as the pH increases in detriment to the ion trapping in milk. However, the lipid solubility of this kind of molecules suggests that the relationship milk:plasma is always greater than 1 (Ziv, 1980b).

In the case of organic acids, the penetration is favored by the increase of milk pH. However, these compounds will reach milk: plasma ratios above 1 only when the milk pH exceeds 7.4. This indicates that lipophilic weak bases parenterally administered may have a certain advantage in distribution into milk in comparison with acids (Ziv, 1980b).

2.2 Intramammary antimicrobial treatments (IMM)

The IMM infusion is the more used administration route in the ATM treatment of mastitis. However, many IMM products have been released to the market without the necessary scientific support about its PK behavior and studies about its clinical and bacteriological efficacy. The benefits of IMM administration are the high concentrations reached in milk and less loss due to drug absorption and transfer processes through biological membranes. While the disadvantages of this route may be the uneven distribution of various compounds within the udder, the risk of mammary contamination by bacteria inoculation through the teat canal and the possible irritation of tissue breast by the formulation (Gruet et al., 2001). Even in vitro studies have shown that ATMs administered by IMM route can negatively affect the phagocytic process in the mammary environment (Nickerson et al., 1985; 1986).

After administration of an IMM infusion, the contact between the ATM agent and the pathogen within the mammary gland is subject to a series of successive events (Ziv, 1980c; Mestorino, 1993a):

1. Pharmaceutical Phase: begins after drug administration including the following steps:
- Disintegration of the formulation
- Drug dissolution
- Liberation of the drug in milk
2. Pharmacokinetic Phase: Assumes the presence of the drug in milk (drug availability) and includes the following events:
- Absorption (milk: plasma)
- Distribution (local and systemic)
- Metabolism (systemic)
- Excretion (local and systemic)
3. Pharmacodynamic Phase: The effect of the drug against bacteria in the infection site.

The kind and proportions of the pharmaceutical excipients contained in the formulation are which determines the rate of drug release. They define the pharmaceutic phase which governs the initial shape of the ATM milk concentration versus time curve.

IMM formulations used for mastitis treatment during lactation contain pharmaceutical excipients that favors quick release of the active components, because the objective is to eradicate the infection of glandular tissue and minimize the withdrawal time (Ziv, 1980c).

On the other hand, when the IMM treatment is during the dry period, the process of release of the active component must be slow and necessarily has great influence on the initial shape of the concentration versus time curve. The excipients used, although very variable, may include vegetable or mineral oils. The mineral oil, such as paraffin or glycerine oils, has the advantage of delaying the release of the active compound and prolonge the permanence of the ATM inside the gland (Ziv, 1980c; Mattie et al., 1997).

In some cases the pharmaceutical excipients contain other agents that increase the retention, such as aluminum monostearate in an oily base, hydroxystearin (Ziv, 1980c). Mercer et al. (1974) studied the absorption in blood and the persistence in milk of penicillin G, after IMM infusion of 400000 IU in mastitic quarters in a rapid (aqueous formulation) and a slow release base (3.6% aluminum monostearate in peanut oil). Penicillin G administered in the slow release formulation could still be detected in the milk 120 h after infusion compared to only 56 h for the aqueous base formulation. These figures must be considered in relation to the MIC of penicillin G for *S. aureus* in milk which is around 0.1 $IU.mL^{-1}$; this means that active levels were maintained above the MIC for 72 h for the oil-based formulation compared to 32 h for the aqueous-based formulation. After the administration of IMM syringes, part of the formulation will be lost with the milk in the first milking post-treatment. This loss is negligible in the case of treatment at the start of the dry period, because the cows are removed from the milking routine. However, the disruptions of the milking routine trigger physiological changes in the mammary gland that may affect the pharmaceutical and PK phases.

The start of the PK phase presupposes the existence of available drug in the milk secretion and depends on the disintegration of the formulation and the dissolution of the drug. These processes determine the drug availability.

The amount of drug recovered in milk is influenced by several factors, among which we can mention the kind of excipient used in the formulation, the milk: plasma passage rate, the size of the udder and the volume of milk contained in the gland (Ziv, 1980c).

The passage of the antibiotic from milk to serum involves the movement of drug across a biological membrane. The factors influencing this passage are the same that affect their transfer from the blood into the milk: the molecule pKa, the lipid solubility of the nonionized fraction, the percentage of udder tissue binding and the binding to milk proteins (Mestorino, 1993a). Hydro soluble drugs pass through the membrane mainly through the protein channels. Lipid soluble drugs cross the membrane through the lipoproteic region .

The ability to cross the milk: plasma barrier is expressed as lipid solubility coefficient for nonionized fraction percentage (Mattie et al., 1997) and the real rate of passage can be quantified by the $C_{max(plasma)}/C_{max\,(milk)}$ or by the $AUC_{(plasma)}$ / $AUC_{(milk)}$ ratios (Ziv, 1980a).

The milk: plasma absorption phenomenon results in the presence of the ATM agent in blood plasma. Once in the systemic circulation, the drug will be object of the PKs phenomena taking place after absorption (distribution, metabolism and excretion) (Ziv, 1980b).

The ATM distribution inside the gland occurs by passive diffusion of the molecules through the lipophilic and hydrophilic components of the glandular secretion. This phenomenon is affected by drug binding to breast tissue and / or components of milk (Mestorino, 1993a). Once the drug establishes contact with the glandular cells, its penetration into tissue continues depending on lipophilicity. As will be detailed later, cell penetration is a critical point in certain infections. The excretion process of the ATM from the mammary gland is governed by the kind of excipient used, the quantity of milk produced, the molecule characteristics, the health of the mammary gland, the number of daily milkings made (Ziv, 1980c), the dose and total volume of formulation among other factors.

After the IMM treatment at cow drying, the ATMs elimination from the mammary gland can be expressed by a monoexponential or biexponential curve. The elimination rate is affected by the dose, the nature of the vehicle, the drug characteristics, and the extent of binding of the antibacterial agent to the gland content and to the mammary tissue (Ziv, 1980c; Mestorino, 1993a; Mattie et al., 1997).

3. Antimicrobial pharmacodynamic concepts

Successful ATM chemotherapy depends on a correct diagnosis, selection of the appropriate ATM agent, and its administration with an adequate dosing scheme.

When considering the choice of ATM agent and dosage regimen, we need to consider the PKs of the chosen drug in the target animal species and the PD indices that drive its clinical effectiveness. For example, penicillins, like all β-lactam ATMs (penicillins, cephalosporins, carbapenems and monobactams), exhibit time-dependent killing. This means that maximum clinical effectiveness is achieved by ensuring that the free serum concentration of the selected β-lactam exceeds the MIC of the pathogen for the appropriate percentage of the dosing interval. If the pathogen is a Gram-positive organism, the targeted duration is usually ≥40% of the dosing interval. Instead, concentrations of most β-lactams should exceed the MIC of the pathogen by ≥80% of the dosing interval when the infectious agent is a Gram-negative organism. In other words, for drugs exhibiting time-dependent killing, increasing the concentration of the drug in excess of the MIC of the pathogen does not increase the killing rate. Rather, the extent of killing is dictated by the duration of time that bacteria are exposed to the drug.

On the other hand, other bactericidal ATM agents, as fluoroquinolones and aminoglycosides, exhibit concentration-dependent killing. In this situation, the rate of killing increases as the drug concentration increases above the MIC of the bacterial pathogen. Thus, ATM agents may be classified as those that exhibit time-dependent killing with null or brief post-antibiotic effect (e.g. β-lactams), time-dependent killing with prolonged post-antibiotic effect (e.g. glycopeptides), concentration dependent killing (e.g. fluorquinolones and aminoglycosides), and those that are generally considered to be bacteriostatic (e.g. tetracyclines, macrolides, lincosamides and phenicols).

Mouton et al. (2005) published an attempt to standardize the interpretation n of these various PK/PD parameters (Fig. 1). Some of the basic definitions are as follow:

- AUC (area under the concentration versus time curve). Should be expressed in terms of unbound drug. If multiple dosing regimens are applied, AUC should be measured over a 24 hour dosing interval at steady state. It should be noted that for compounds exhibiting linear PK, the AUC over a single dosing interval at steady state (AUC0-τ) is equal to AUC extrapolated to infinity (AUC0-∞) following single administration.
- AUC/MIC (AUC divided the minimum inhibitory concentration). Although sometimes given the dimension of time (generally 18 to 24 hours), this ratio can be more conveniently expressed as a dimensionless value.
- T>MIC (period of time during which the drug concentrations exceed the MIC). The cumulative percentage of a 24-hour period that the free drug concentration exceeds the MIC at steady-state pharmacokinetic conditions.
- In vitro PAE (post-antibiotic effect). The period of suppression of bacterial growth after short exposure of an organism to an ATM compound (unit = time).
- In vivo PAE. The difference in time for the number of bacteria in a tissue of treated versus control animals to increase 1 log10 over values when drug concentration in serum or at the infection site fall below the MIC (unit = time). The in vivo PAE includes any effect associated with sub-MIC concentrations. Sub MIC effect. Any effect of an ATM on a microorganism at concentrations below the MIC (unit = time)
- Post-antibiotic sub-MIC effect. The effect of sub-MIC drug concentrations on bacterial growth following serial exposure to drug concentrations exceeding the MIC (unit = time)

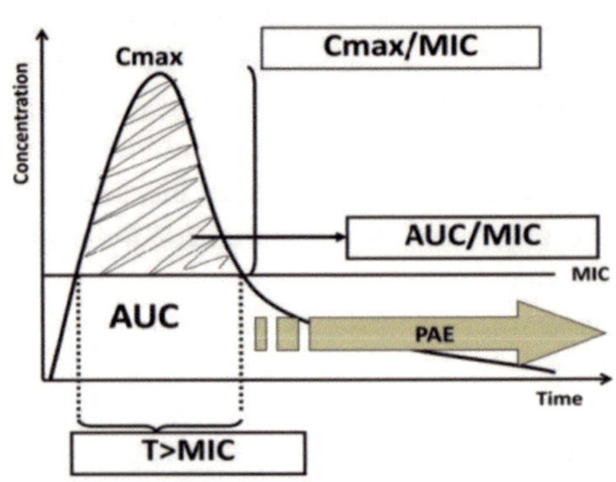

Fig. 1. Illustration of the main PK/PD parameters that correlate with efficacy against extracellular infections.

The units associated with the AUC/MIC ratio are hours, by dividing this value by the dosing interval (e.g. 24 hour), we obtain the average plasma concentration over the steady-state 24-hour dosing interval relative to MIC, which may be far more informative than the traditional method for expressing this value.

The PD parameter providing the most appropriate surrogate for drug effectiveness is dependent on several factors. This includes: mechanism of action of the different drugs, whether its effects are time or concentration dependent, and the PAE duration (Table 3).

Drug	Mechanism	Activity	Bacterial effect	Duration PAE	PD Parameter
Macrolides Erythromycin,	Binds to 50S ribosomes (inhibits proteins synthesis)	Static	Time-dependent	Brief*	T>MIC
Azalides Azithromycin	"			Prolonged	AUC_{24}/MIC
Lincosamides	"			Brief	AUC_{24}/MIC
Ketolides (telithromycin)	Binds to 50S ribosomes and some 30S ribosomal unit activity	Static and cidal (S.pneumoniae, S.pyogenes)	Time-dependent	Prolonged	AUC_{24}/MIC
B-lactams(e.g. penicillins, cephalosporins, carbapenems, monobactams)	Inhibits cell wall synthesis	Cidal	Time-dependent	Gram- bacteria: null or brief, Gram+: may be prolonged	T>MIC
Glycopeptides (e.g.vancomycin)	Inhibits cell wall synthesis	Cidal(slower than β-lactams)	Time-dependent	Prolonged	AUC_{24}/MIC
fluoroquinolone (e.g. enrofloxacin, ciprofloxacin, danofloxacin)	Inhibit DNA gyrase, prents transcription and replication	Cidal	Concentration -dependent	Prolonged	AUC_{24}/MIC C_{max}/MIC
Aminoglycosides (e.g.gentamycin, streptomycin)	Binds to 30S ribosome (inhibits protein synthesis) and disrupts biofilms	Cidal	Concentration -dependent	Prolonged	AUC_{24}/MIC C_{max}/MIC
Tetracyclines	Inhibits protein synthesis (30S)	Static	Time-dependent	Prolonged	AUC_{24}/MIC
Phenicols (e.g.chloramphenicol, florfenicol, thiamphenicol)	Inhibits protein synthesis by inhibition peptydyl-transferase (50S)	Static	Time-dependent	Prolonged	AUC_{24}/MIC
Trimethoprim	Inhibits folic acid synthesis by inhibiting dihydrofolate reductase	Static alone Cidal with sulfonamides	Time-dependent	Brief	T>MIC
Sulfonamides	PABA analogue interferes with folic acid synthesis	Static	Time-dependent	Brief	T>MIC
Oxazolidinones (linezolid)	Inhibits initiation of protein synthesis (50S)	Static (Staph. and Enterococcus) Cidal (most Strep.)	Time-dependent	Brief	T>MIC

Table 3. Relationships among drug, drug effects, and the PD surrogate most closely aligned to its clinical response (Martinez et al, 2006). *Brief: less than one hour. Prolonged: up to six hours.

Whether a drug exhibits concentration-dependent or time-dependent killing is largely a function of the shape of its concentration-effect curve, the steeper the curve, the less will be the impact of increasing drug concentrations on the ATM response. Conversely, the more shallows the curve, the greater the relationships between the rates of bacterial kill versus the ATM drug concentration. This relationship can be described using a sigmoidal Emax model, also known as the Hill model, which can be described as follows (Toutain, 2002):

$$E(t) = E_0 \frac{E\max \times C^h(t)}{EC_{50}{}^h + C^h(t)} \quad (4)$$

Where:

E(t) is the effect observed for a given concentration at time t $(C\,(t))$;
Emax is the maximal effect attributable to the drug;
EC_{50} is the plasma concentration producing 50% of *Emax* ;
h is the Hill coefficient, which adjust the degree of sigmoidicity in the curve; and
E_0 describes the rate of spontaneous cure.
When h= 1, the Hill model reduces to the *Emax* model, which corresponds to a hyperbolic function.

While there are certain characteristics common to all ATMs within a given drug class, there can be important differences in the PK/PD ratios needed to achieve a desired effect. Within the fluoroquinolones (FQ), it has been demonstrated that the rate of kill and the duration of the in vitro PAE (Finberg et al., 2004; Firsov et al., 1998b) can be markedly different across compounds and microbial species. In some cases, the PK/PD relationship necessary to achieve a 2-log kill can also vary as a function of the microbial strain (Andes and Craig, 2002). Similarly, the AUC/MIC ratio of 100-125 frequently quoted as a target for FQ ATM activity may be an appropriate predictor of success for many Gram-negative infections, but lower AUC/MIC ratios (e.g. 35 to 40) may be appropriate for infections due to Gram positive organisms (Wright et al., 2000). With regard to the β-lactams, while there tends to be a substantial in vivo PAE for *S. aureus*, a substantially shorter PAE is associated with Gram negative organisms (Craig, 1993).

Within any given bacterial population, the possibilities of bacterial subpopulations that are less susceptible to the ATM agent exist. As demonstrated by Drusano (2004), unless these less susceptible pathogens are killed, succeeding microbial generations will re-populate the infection site with pathogens whose MIC values are higher than those found within the initial infection.

Accordingly, ensuring adequate exposure following an initial dose of a FQ is as important as insuring that high drug concentrations occur after repeated administration. Drug concentrations need to be adequate to either destroy the existing bacterial population at the site of the infection or to reduce its size to the point where the host defense mechanism can successfully control and eliminate the remaining pathogens.

For drugs exhibiting concentration-dependent killing, C_{max}/MIC ratios may be particularly important when the pathogen has a high MIC value or is rapidly proliferating (Craig and Dalhoff, 1998). Rapidly proliferating bacteria have a greater likelihood of undergoing a mutational event that could lead to the genesis of a less susceptible population. In infectious disease processes where there is a high bacterial burden (inoculum effect), the risk of a mutational event is increased due simply to the laws of probability (Craig and Dalhoff, 1998). In these cases, to ensure maximum killing, the targeted C_{max}/MIC ratios are approximately 10 to 12 (Drusano et al., 1993). Such ratios ensure increased killing of susceptible organisms and an increased killing or inhibition of organisms with higher MICs. The goal in these situations is to reduce bacterial numbers to a level where the host can effectively handle those pathogens not killed by the ATM agent. While a high C_{max}/MIC ratio (e.g. 10) is correlated with a high rate of bacterial kill for compounds exhibiting concentration-dependent killing, there are conditions under which AUC/MIC may be as or more predictive of a sustained ATM activity. AUC/MIC could be considered a major PK/PD parameter when the infection is caused by relatively slow growing bacteria, when there is little or no PAE that will contribute to inhibition of bacterial re-growth or when the MIC for the pathogen is relatively low.

3.1 Post-Antibiotic Effects (PAE)

High drug concentrations relative to the MIC may contribute to an increase in the duration of the in vitro and in vivo PAE. For those bacteria/drug combinations that exhibit a PAE, in vivo PAEs have been shown to be longer than in vitro PAEs for most organisms. Thus, optimizing the C_{max}/MIC ratio will delay the re-growth of the pathogen, sometimes by several hours. This type of dosing schemes results in fewer organisms remaining that can evolve into a resistant subpopulation and can be managed by the host defenses.

For many compounds, the duration of the in vivo and in vitro PAE is substantially greater for Gram-positive than for Gram-negative pathogens. Because the duration of the in vitro and in vivo PAE of β-lactams tends to be negligible for Gram-negative species, it is recommended that concentrations of drug remain above the MIC of the pathogen for >80% of the dosing interval to combat this type of organisms. While a T>MIC of about 40% is sufficient for staphylococcal species.

This difference in the duration of the in vitro and in vivo PAE may also be one of the reasons why the in vivo AUC/MIC for FQ tends to be less for Gram-positive than for Gram-negative pathogens. For Gram-negative organisms, the estimated AUC/MIC ratios needed to ensure effective treatment and prevent the selection of resistant strains is estimated to be approximately 100 to 125 (Forrest et al., 1993). In contrast, the AUC/MIC ratio for Gram-positive bacteria is considerably lower, approximately 30 to 50 for a number of drug-pathogen combinations (Wright et al., 2000). Studies involving the third and fourth generation FQ suggest that for Gram-positive organisms AUC/MIC values are substantially lower when C_{max}/MIC values are ≥10 (Nightingale et al., 2000).

Blood concentrations and MIC data alone cannot predict drug effectiveness. For example, using human and bovine estimated breakpoints for cephapirin and oxytetracycline, Constable and Morin (2002) showed that the MIC values predicted that the causative pathogens would be susceptible to both agents. However, these compounds were not effective in the treatment of acute bovine mastitis. In the same line, compounds effective in the treatment of acute bovine mastitis may be ineffective in the treatment of chronic bovine mastitis (Owens et al., 1997).

Drug potency is often considered in terms of MIC, which is a measure of a drug´s static effect on microbial growth. The MIC may not be the same as a compound´s minimum bactericidal concentration (MBC). Since both the MIC and MBC values are in vitro estimates, they do not reflect the killing rate of a drug, the effect of serum on ATM activity, PAE, or post-ATM sub-MIC (Craig and Dalhoff, 1998). Firsov et al. (1999) demonstrated that even with comparable AUC/MIC ratios, different FQ can have marked differences in in-vitro time-kill profiles. Such differences cannot only influence the selection of an AUC/MIC target value, but also the potential for selecting for resistant bacterial strains.

Traditional in vitro susceptibility data reflect the impact of therapeutic agents on bacteria that are in the active growth phase. These tests do not describe the differential activity across the different life phases of bacteria (Cerca et al., 2005). As an example, in the case of bacteria from biofilms, entering a non-growth phase, many compounds begin to lose their ATM effects. This very important facts cannot be predicted when MIC alone is used as the PD component of the PK/PD relationship.

There are some examples of traditional susceptibility tests failing to adequately reflect in vivo drug activity. For instance, β-lactams are inactivated by purulent material due to the accumulation of bacterial β-lactamases; gentamicin can be inactivated by reversibly binding to DNA released from lysed neutrophils; netilmicin and amikacin are inactivated by disrupted leukocytes (Labro, 2000). Owens et al. (1997) noted that the bacteriologic cure rate for newly (less than two weeks duration) acquired *S. aureus* IMMI, was 70% when treated with penicillin and novobiocin combination. However, the cure rate dropped to less than 35% for chronic infections (those lasting longer than four weeks). Accordingly, they observed that while the successful treatment of acute infections could be predicted on the basis of in vitro susceptibility test results, this was not the case for chronic *S. aureus* infections. It is not know if this outcome is due to the driving of *S. aureus* infections from extracellular to intracellular sites, to biofilm formation, or to some other bacterial pathogen-host interaction (Brouillete et al., 2004).

Staphylococcal bovine mastitis exemplifies a persistent infection that is difficult to treat and where frequent relapses commonly lead to a chronic disease (Sandholm et al., 1990). Antibiotic treatment of this disease is problematic (as mentioned in the preceding paragraphs) partly because in vitro antibacterial susceptibility is a poor predictor of efficacy in chronic cows (Owens et al., 1997). Failure to eliminate the infection from dairy herds is likely due to a combination of an adaptive response of the pathogen to survive in the mammary gland despite the presence of antibiotics (Sandholm et al., 1990), and the inability of host defenses to clear the pathogen. Infectious bacteria in chronic mastitis may survive intracellularly and remain quiescent and protected from the action of antibacterials and host defenses. Although reinfections that follow attempts to treat acute mastitis can be due to newly acquired strains, it is equally possible that the reinfection results from persistence of the original infective organism (Sandholm et al., 1990). It is possible that *S. aureus* having the small-colony variants (SCV) phenotype contribute to chronic mastitis. SCVs have been isolated from cases of bovine mastitis (Ziv & Sompolinsky, 1976; Sompolinsky et al., 1969).

Both lactational and dry cow therapy are part of *S. aureus* control programs. Reported cure rates for *S. aureus* mastitis vary considerably in a range from 4 to 92% (Owens et al., 1988, 1997). The probability of cure depends on cow, pathogen, and treatment factors. Cure rates decrease with increasing age of the cow, increasing somatic cell count, increasing duration of infection, increasing bacterial colony counts in milk before treatment, and increasing number of quarters infected. *S. aureus* mastitis in hind quarters has a low cure rate compared with front quarters. ATM treatment of IMMI with penicillin-resistant *S. aureus* strains results in a lower cure rate for treatment with either β-lactam or non-β-lactam antibiotics. The most important treatment factor affecting cure is treatment duration. Increased duration of treatment is associated with increased chance of cure. Economically, extended treatment is not always justified, even when indirect effects of treatment such as prevention of contagious transmission are taken into consideration.

The β-lactams (penicillins and cephalosporins) have become the first line of ATM agents used for treatment of bovine mastitis in Argentina. Within this class, penicillin, amoxicillin, cloxacillin and ampicillin are the most used agents. In the Nordic countries penicillin is used as the first-line antibiotic treatment of bovine mastitis, because of its potency, low resistance rate and narrow spectrum. This is an important tool to limit the development of antibiotic resistance as much as possible. In a one study performed for us 34% of *S. aureus* were classified as penicillin resistant, while 100% were classified as sensible to the combination amoxicillin/clavulanic acid (Lucas, 2009a). This prevalence of resistance to penicillin was similar than those obtained by others authors in different regions of Argentina (Gentilini et al., 2000; Russi et al., 2008). And this was the same comparing the proportion of resistance (47 %) as obtained by Gianneechini et al. (2002) in Uruguay. The comparison between these results obtained over the years demonstrated that the situation in general has not changed during the last 25 years in relation to penicillin resistance. However, it results higher than in Norway, 4.2% from clinical cases and 18% from sub-clinical cases (Hofshager et al., 1999), and Sweden, 6% (Franklin, 1998). In the Table 4 we present in vitro susceptibility results of strains of *S. aureus* obtained by different studies performed in different countries.

4. Specific antimicrobial agents

4.1 Beta-lactam agents

4.1.1 Amoxicillin

The use of β-lactam antibiotics in the treatment of mastitis remains one of the first elections. Within this group, any penicillin (PEN) such as amoxicillin (AMX) provides a number of advantages. AMX is a β-lactam antibiotic of wide spectrum with a chemical structure and antibacterial activity similar to ampicillin (AMP) (Hunter et al., 1973).

AMX is rapidly converted to AMP in the body. AMX pharmacokinetics is essentially similar to other β-lactam antibiotics. The half-life is short (about 1 hour) and the volume of distribution low. The first AMX pharmacokinetic study was conducted in ruminants after IV and IM administration in cattle and after IM administration in goats by Ziv and Nouws (1979).

Since then AMX disposition was studied after its administration by different routes in production animals (Archimbault et al, 1981; Baggot, 1988, Nouws et al, 1986, Wilson et al, 1988). According to Erskine et al. (2002) about 60% of *S. aureus* isolates are susceptible to AMP. The susceptibility to AMP among mastitis causing streptococci is 100% and for *E.coli* 85%. Therefore, significant improvement by AMX/clavulanic acid treatment compared to AMX or AMP alone can be expected only for *S. aureus*.

	MIC (μg.mL^{-1})				
Antimicrobial agent	**MIC$_{50}$**	**MIC$_{90}$**	**Range**	**Resistance Breakpoint**	**References**
Penicillin	0.5 0.06 0.25 0.06	>8 0.5 1 4	0.12 - > 8 0.06-0.5 0.062-2 0.03-8	≥0.25	Gianneechini et al., 2002 Rubin et al., 2011 San Martín et al., 2002 Russi et al., 2008
Ampicillin	0.5 0.125	8 1	≤0.12->16 0.06-4	≥0.5	Gianneechini et al., 2002 San Martín et al., 2002
Amoxicillin	0.125	1	0.06-4	≥0.5	San Martín et al., 2002
AMX/CLA	2	2	2 – 16	≥8	Rubin et al., 2011
Oxacillin Cloxacillin	≤0.5 0.25 0.25	≤0.5 2 1	≤0.5-1 0.12-0.5 0.125-16	≥4	Gianneechini et al., 2002; Russi et al., 2008 Lucas, 2009 San Martín et al., 2002
Cephalotin	≤4	≤4	4-64	≥32	Gianneechini et al., 2002; Rubin et al., 2011
Cephoperaz.	0.5	2	0.25-16	≥8	San Martín et al., 2002
Gentamicin	≤1 2 1 0.25	≤1 2 4 0.5	≤1-4 2-16 0.25-32 0.03-2	≥16	Gianneechini et al., 2002; Rubin et al., 2011 San Martín et al., 2002 Russi et al., 2008
Neomycin	≤1	≤1	≤1-64	≥64	Gianneechini et al., 2002
Oxitetracy-cline	2 2 2	>64 2 4	≤0.25->64 2-32 0.5-64	≥16	Gianneechini et al., 2002 Rubin et al., 2011 San Martín et al., 2002
Erythromycin	≤0.5 0.25 0.125	≤0.5 0.25 0.25	≤0.5->4 0.25-16 0.125-0.5	≥8	Gianneechini et al., 2002 Rubin et al., 2011 Russi et al., 2008
Azythromycin	0.5	1	0.25-2		Lucas, 2009
Spiramycin	4	4	0.5-4		Lucas, 2009
Clindamycin	≤1 0.25	≤1 0.25	≤1 0.25-16	≥4	Gianneechini (2002) Rubin et al., 2011
Chloramph.	4	8	4-32	≥32	Rubin et al., 2011
Florfenicol	0.5	2	0.25-128	≥32	San Martín et al., 2002
Enrofloxacin	≤0.25 ≤0.25 0.125	≤0.25 ≤0.25 0.5	≤0.25-0.5 0.25-16 0.06-8	≥4	Gianneechini (2002) Rubin et al., 2011 San Martín et al., 2002
Danofloxacin	0.25	1	0.25-2		Lucas, 2009
Trimethoprim/ Sulfamethox	0.25 19	0.5 19	<0.006-8 19-76	≥4 76	Gianneechini (2002) Rubin et al., 2011

Table 4. In vitro susceptibility of strains of *S. aureus* obtained from clinical and sub-clinical bovine mastitis cases by different authors.

We performed some studies with the objective of evaluating the PK behavior of AMX trihydrate in serum and milk after its IM administration at therapeutic dose in healthy lactating cows (Mestorino et al., 1997) (Fig. 2). AMX was absorbed relatively fast after its administration (T $\frac{1}{2}_{ab}$ 0.92 ± 0.10 h). Because AMX trihydrate was used, and that this is not a water soluble salt, the formulation to be administered was a suspension. This formulation allows a fraction of the AMX dose to be rapidly liberated and absorbed whereas another fraction, less soluble, is precipitated at the site of administration, and absorbed more slowly, leading to the slow elimination profile described, which allows the estimation of dosing regimens with long intervals. It should be emphasized that the profile of long-acting medication is not due to slow elimination, but a slow absorption. The drug is eliminated at normal speed, but no molecules are available at a higher rate of absorption; therefore it is the absorption process that commands the elimination half-life.

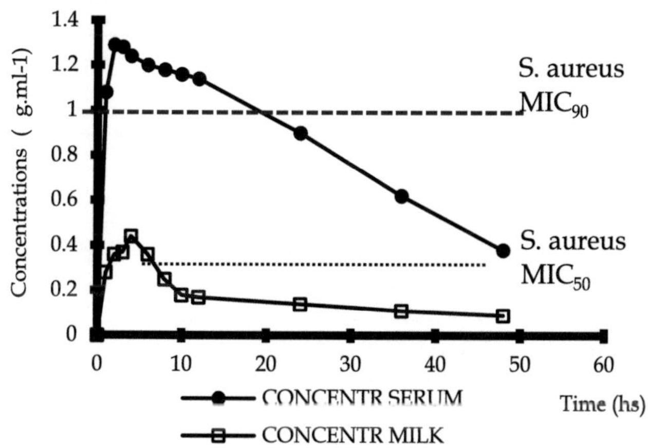

Fig. 2. AMX concentrations in serum and milk after IM administration of 15 mg.kg^{-1} to four healthy dairy cows.

The average time of penetration into milk (T $\frac{1}{2}_{P}$) of 1.24 ± 0.09 h, was considered fast enough for an antimastitic agent. The ratio AUC_{milk}: AUC_{serum} was 0.18, which is a low level of milk availability (Table 5). Beta-lactam antibiotics, due to their acidic character are more dissociated in plasma than in milk, so milk penetration is severely restricted. Serum and milk concentrations were determined until 48 h post-administration, which, considering the mentioned MIC_{90} of *S. aureus*, does not represent a good therapeutic tool because C_{max} in milk was only 0.44 ± 0.13 μg.mL^{-1}. It has to be remarked that the mentioned parameter was obtained in healthy animals. In mastitic animals, on the other hand, it appears logical to expect a greater penetration of AMX in the gland, because raising the pH, would permit greater dissociation of AMX in the cistern, and a process of ion sequestration with higher drug concentrations in milk. In view of the influence of the factors mentioned above, PK studies should always be performed not only on healthy animals but on sick animals too.

The most significant factor affecting the cure rates for clinical *S. aureus* mastitis is the ability of the isolate to produce *ß*-lactamase. This has also been shown by other authors (Sol et al. 2000), and could indicate either that penicillin resistant strains are more virulent than penicillin-susceptible strains, or that the antibiotics used to treat mastitis caused by penicillin-resistant strains are less efficient, due to PK or PD factors, as we explained above. In the study by Sol et al. (2000), clinical *S. aureus* mastitis caused by *ß*-lactamase positive or negative isolates was treated intramammarily with 5 different ATM treatments. All the isolates were found to be in vitro susceptible to the drugs used.

Parameter	Serum	Milk
K_{abs} (h^{-1})	0.75 ± 0.09	0.56 ± 0.08
T ½ $_{abs}$ (h)	0.92 ± 0.10	1.24 ± 0.09
Cmax (µg.ml^{-1})	1.29 ± 0.14	0.44 ± 0.13
Tmax (h)	2.00 ± 0.00	4.00 ± 0.00
ß (h^{-1})	0.025± 0.07	0.03 ± 0.001
T½ ß (h)	27.61± 0.09	22.59± 0.91
AUC (µg.h/ml)	57.72± 3.87	10.34± 1.28
R $Cmax_S/Cmax_M$	2.93	
R AUC_M/AUC_S	0.18	
F_{Milk} (%)	18	

Table 5. PK parameters (mean ± SD, N = 4) obtained in serum and milk after IM administration of AMX at dose of 15mg.kg^{-1} in dairy cows

No difference in the bacteriological cure rates between the different ATM treatments in each group was found, but major differences in bacteriological cure rates between mastitis due to *ß*-lactamase positive and negative strains was significant. *S. aureus* is known to possess many virulence factors, like capsule and slime formation, which make it more resistant to ATMs (Sandholm et al. 1990; Taponen et al., 2003).

Information about the PKs of systemic amoxicillin-clavulanic acid (AMX-CLA) suspension in dairy cows is almost totally lacking. However AMX-CLA is widespreadly used in veterinary medicine. Injectable formulations (IM) have been used in cattle, pigs and sheep, and oral formulations for the treatment of pre-ruminant animals. Like other β-lactams, AMX alone or combined with CLA is commercially available for IMM use in bovine mastitis.

There are few antecedents of combining an IMM treatment with a systemic one, and, in general these studies compare clinical efficacy without PK determinations (Ziv 1980a; 1980c). As this is an interesting strategy to treat mastitis, we decided to compare the pharmacokinetic behaviour of AMX-CLA after their IMM infusion vs the combination therapy IMM-IM in lactating Holstein cows with subclinical mastitis caused by *S. aureus* (Lucas et al., 2009a). The experimental animals were allocated by production level (high and low production level) and the quarters were grouped by health state. In the IMM alone assay, each quarter of all cows received 3 IMM infusions of AMX-CLA (200-50 mg) with a 12 h interval. Individual quarter milk samples were collected until 96 h after 3rd administration (Fig 3). The purpose was to evaluate the effects of the health status of the quarters (SS): mastitic quarters vs. healthy ones and the level of milk production (LP): quarters of high-producing cows vs quarters of low-producing cows. LP had a significant effect on T½*ß* and

MRT, which were higher in quarters of low producing-cows (*P<0.05*). The MIC_{90} for AMX-CLA 4:1 was 8 $\mu g.mL^{-1}$ and the $T>MIC_{90}$ was higher in quarters of low-producing cows. The LP modified the AMX-CLA PK profile and it could be a determinant of efficacy.

It resulted remarkable that levels of AMX were higher and remained for a longer period of time in the mammary quarters of low production cows than in the quarters of high production cows. These differences were not such in the case of CLA, as calculated parameters were similar in both groups. We determined also the $T>MIC_{90}$ in the milk of each quarter (MIC90= 6.4:1.6 $\mu g.ml^{-1}$) (See Table 6). The health status of the mammary quarters had no statistically significant effect on any of the PK parameters. The level of milk production, however, had a significant effect on the AMX $T½_{\lambda}$ (P= 0.0000), $AUC_{0-\infty}$ (P = 0.0057), CL/F (P=0.0057) and MRT (P= 0.0019).

IMM treatment given to supplement systemic administration of ATMs increases drug concentration in the milk compartment, and higher concentrations throughout the mammary gland will follow (Ziv 1980a; 1980c). In theory, combination treatment thus should improve cure rates for deep infections such as *S. aureus* mastitis (Sandholm et al. 1990). On the other hand, *ß*-lactam ATMs are time-dependent drugs, and very high concentrations at the infection site do not increase efficacy (Craig 2001).

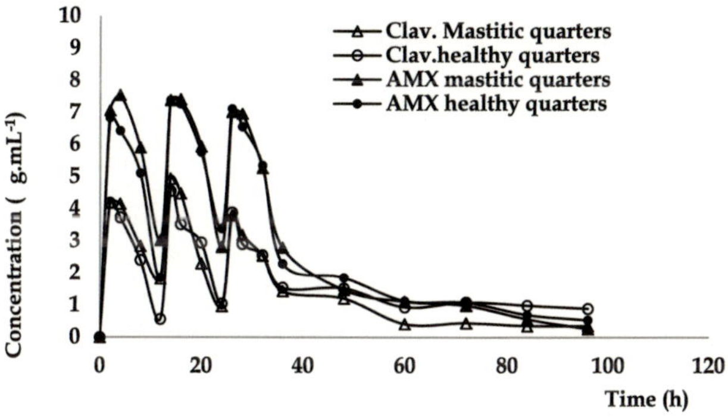

Fig. 3. Mean concentrations of AMX-CLA in milk of healthy and mastitic quarters after administration of 3 IMM syringes (200 mg-50 mg) every 12 h

	$X_{mastitic} \pm SD$	$X_{healthy} \pm SD$	$X_{M\ high\ prod.} \pm SD$	$X_{M\ low\ prod.} \pm SD$
$T>MIC_{90}$ (%)	52.78 ± 30.71	48.60 ± 31.65	46.30 ± 35.80	53.69 ± 26.72

Table 6. $T>MIC_{90}$ (%): percentage of the period between 0 and 12 h post-administration during which the concentration was ≥ the MIC_{90}; Amx:Cla 4:1, the MIC_{90} was considered as the ratio 4:1 (MIC_{90}= 6.4:1.6 $\mu g.mL^{-1}$)

We performed another study where the effect of the combined IMM treatment with systemic administration was evaluated (Lucas et al., 2009a). The experimental animals received 3 IMM infusions of AMX-CLA (200-50 mg) in each quarter in combination with 3.5 $mg.kg^{-1}$ of 15% AMX-CLA by IM route every 12 h. Individual quarter milk samples were collected until 96 h after 3rd administration. The incidence of the health status of the quarters (SS): mastitic quarters vs. healthy ones, and the level of milk production (LP): quarters of high-producing cows vs quarters of low-producing cows in the PKs of the drugs was evaluated (Fig. 4).

We determined once again the T>MIC$_{90}$ in the milk of each quarter (MIC$_{90}$= 6.4:1.6 μg.ml^{-1}) (See Table 7). It was observed that concentrations achieved in cows receiving combination therapy (IM + IMM) were higher than those in cows treated intramammarly alone, which was a logical finding.

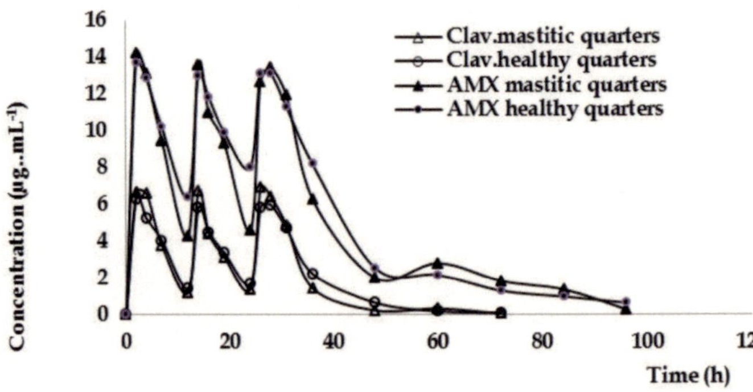

Fig. 4. Mean concentrations of AMX-CLA in milk of healthy and mastitic quarters after administration of 3 IMM syringes (200 mg-50 mg each) in combination with three 3.5 mg.kg^{-1} of 15% AMX-CLA IM administrations every 12 h

Significant effects of quarter health status on the PK parameters were found. The C_{max} resulted higher in milk from sick quarters (P= 0.0384). The level of production had significant effect over milk AMX C_{max} and T_{max}, these were higher in the mammary quarters of low production cows. The level of production also affected the PK profile of CLA, the C_{max}, T_{max} and $AUC_{0-\infty}$ were higher in animals of low production. The differences observed between CL/F_{mam} suggest that the ATMs removal rate was higher in the quarters of high production cows. Significant effect of the mammary quarter health status and production level on the T>MIC$_{90}$ was found. It should be emphasized that the T>MIC determined after the combined treatment (IM + IMM) exceeded those calculated after IMM infusion alone.

	$X_{mastitic}$ ± SD	$X_{healthy}$ ± SD	$X_{M\ high\ prod.}$ ± SD	$X_{M\ low\ prod.}$ ± SD
T>MIC$_{90}$ (%)	62.64 ± 8.87	78.25 ± 23.16	65.31 ± 20.33	79.48 ± 18.67

Table 7. T>MIC$_{90}$ (%): percentage of the period between 0 and 12 h post-administration during which the concentration was ≥ the MIC$_{90}$; Amx:Cla 4:1, the MIC$_{90}$ was considered as the ratio 4:1 (MIC$_{90}$= 6.4:1.6 μg.mL^{-1})

Owens et al. (1988) found higher cure rates in *S. aureus* mastitis with combined treatment as compared with IMM treatment only. Recently, the therapeutical effects of parenteral, IMM and combination treatments with AMX-CLA have been compared by *Perner et al.* 2002 too. They found the combination treatment to be superior than parenteral or IMM treatment only. In this paper, the bacteriological cure rate for all causing agents and mastitis types (acute, subclinical and chronic), was 75.3%. We found also low bacteriological cure rates (62.5%) after AMX-CLA IMM infusion alone, but after the combined treatment the cure rate was complete (100%) (Lucas et al., 2009a, b).

4.1.2 Penethamate

Penethamate hydriodide (PNTM) is a diethylaminoethyl ester of penicillin which, unlike salts of penicillin, is unionised and so exists in a neutral state. It is only weakly water soluble forming a suspension in an aqueous environment. After its intramuscular administration, is rapidly absorbed from the site of injection and on entering the blood, partially dissociates by hydrolysis into penicillin G and diethylaminoethanol. At the blood pH (7.4), equilibrium is established where 90% of the active drug is present in its hydrolyzed form (penicillin G) with the remainder persisting as PNTM. As PNTM leaves the circulation due to its neutral and lipophilic properties and its high affinity to milk, this equilibrium is maintained by re-association of penicillin G and diethylaminoethanol until excretion is complete. PNTM easily passes the milk-blood barrier due to the pH gradient between milk (pH 6.6-6.8) and plasma (pH 7.2-7.4) and its weakly basic properties (pKa = 8.4). This is further facilitated by its highly lipophylic characteristics which facilitates its passage across the lipo-proteic blood-milk barrier. PNTM starts to dissociate as it passes over the barrier and this process continues during diffusion of the drug through the udder, releasing increasing quantities of penicillin G (PEN_G). PEN_G is rapidly ionised in the udder (pKa = 2.8) so limiting its return to the circulation. It therefore becomes "trapped" in the udder in increasing concentrations.

The same pH gradient between blood and milk presides in the case of mild to moderate udder inflammation such as in sub-clinical mastitis, the pH gradient between blood and milk is the same than in healthy animals thus generating similar PK behaviors to those which take place in the healthy udder. In acute mastitis, although the pH of milk is nearer that of blood due to a breakdown of the blood-milk barrier, higher concentrations of PNTM are still found in mastitic milk than in blood due to its lipophilic properties. Not only does undissociated PNTM rapidly and easily penetrate the udder whether inflammed or not, but its liposoluble nature gives it advantage, compared with other beta-lactam antibiotics such as amoxicillin and aminoglycosides to diffuse through the parenchyma of the udder, pass into the milk and penetrate the lactogenic cells. This diffusion through the udder is supported by the mechanism of "ion trapping" mentioned above and so explains the different penetration of PNTM compared with PEN_G (Friton et al., 2003).

It must be remembered, however, that *S. aureus* survives in acidic media, including phagolysosomes. Controversial in vitro/in vivo data exist on its susceptibility to antibiotics in such environments. We performed some studies to evaluate the effect of the pH variation on the antibacterial activity of penicillin against strains of *S. aureus* isolated from mastitic quarters (Moncada Cárdenas et al., 2009). MIC of *S. aureus* field strains and *S. aureus* ATCC 25923 were tested at pH 7.4, 6.5 and 5.0, in order to simulate the conditions of acidity of subcellular structures which are commonly associated with *S. aureus* intracellular persistence. The PEN MIC_{90} at pH 7.4 was consistent with those reported by CLSI 2007 (0.5 µg/mL) but at pH 5.0 (phagolysosomes) the activity of PEN_G increased markedly and almost linearly (~10 fold decrease in MIC -0.06 µg/mL) (Figure 5).

4.1.3 Cloxacillin

Cloxacillin (CLX) is used in the treatment or prevention of staphylococcal bovine mastitis. Its ATM activity against *S.aureus* is higher than that of PEN_G. There are strains of *S. aureus* resistant to isoxazolilpenicillins (oxacillin or methicillin resistant *S. aureus* -MRSA). These strains are a menace to public health.

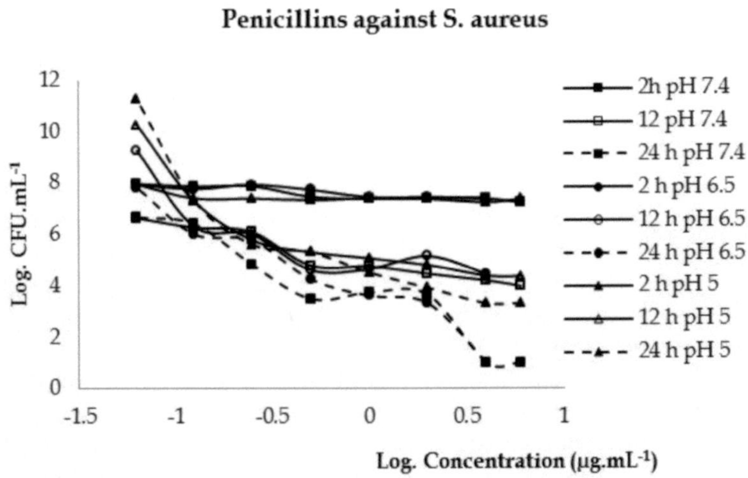

Fig. 5. Inhibition of *S. aureus* exposed to different concentrations (expressed as Log_{10}) of PEN (0.25 MIC; 0.5MIC; 1MIC; 2MIC; 4MIC and 8 x MIC) in function of the time and the medium pH (5; 6.5 and 7.4).

Benzathine CLX is used to treat bovine mastitis by IMM route at drying off and may be associated to other compounds, eg antiinflammatory agents (prednisolone) or other ATMs (eg. AMX, AMP, streptomycin, cephalosporins, etc.). Intramammary products with combinations of two or even three ATMs were introduced due to suggested synergistic action and broad spectrum. The evidence of their efficacy against clinical mastitis is many times lacking and synergistic action has never been proven *in vivo* (Taponen *et al.* 2003; Ødegaard & Sviland, 2001). For mastitis treatment during lactation the formulations contain CLX as sodium salt and, in general, the drug is not combined with other compounds. After IMM administration, sodium CLX binds scarcely to the mammary tissue (25%), with a moderate passage milk:plasma.

We evaluated the pharmacokinetic behavior of sodium CLX in milk after its IMM administration in healthy lactating cows of low production. The experimental cows received syringes containing 250 mg sodium CLX IMM, three times at 12 hour intervals in each quarter. Milk samples were obtained at different post administration times (Fig. 6).

The PK analysis was performed after the last administration. Milk elimination half-life was rather prolonged ($T½\beta = 4.37 \pm 0.458$ h), with levels in milk at 60 h post-last administration ≥ 0.35 µg.mL^{-1}. In low production animals the elimination of the ATM is even slower, with longer milk half life (Table 8). Various benzathine CLX containing IMM suspensions are registered for use in cattle as antibiotics for IMM use at drying off. These formulations are recommended for routine use in cows at drying off to treat existing IMM infections and to provide prolonged protection against new infections during the next lactation.

The excretion through the mammary gland depends on the characteristics of the pharmaceutical excipient, production level (high vs. low), molecule characteristic and udder health status (Mestorino, 1993a). The elimination rate is affected also by the binding level to dry udder secretion, which is very high (80%) (Ødegaard & Sviland, 2001).

In a recent study, we obtained an MIC_{90} of 2 μg.mL^{-1} against *S.aureus* (isolated from cattle with subclinical mastitis) (Lucas et al., 2009a).

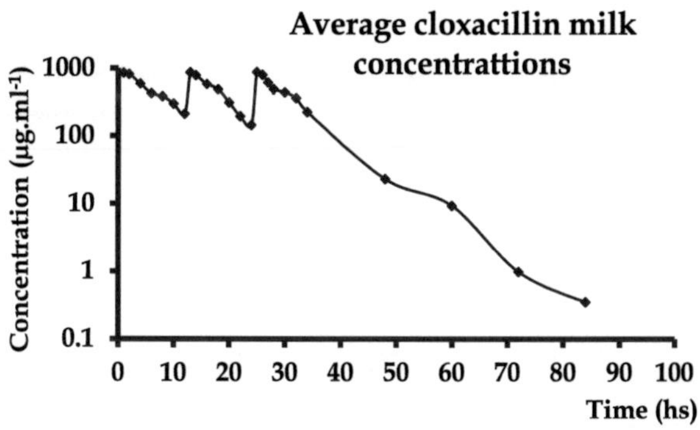

Fig. 6. Mean CLX concentrations in milk of dairy cows after IMM infusion of 250 mg CLX three times at 12 hour intervals in each mammary quarter.

PARAMETER	MEAN	SD
B (μg.mL^{-1})	1288.557	464.800
ß (h^{-1})	0.160	0.016
T 1/2ß (hs)	4.370	0.458
$AUC_{0\text{-all}}$ (μg.h/ml)	7372.005	2062.654
MTR (hs)	6.982	1.469

Table 8. Mean pharmacokinetic parameters obtained in milk after IMM infusion of 250 mg CLX in each quarter three times at 12 hour intervals.

An experiment was conducted in cows with staphylococcal subclinical mastitis at drying off for the purpose of studying the pharmacokinetics behaviour of CLX benzathine in the dry udder secretion after a single IMM infusion (Lucas et al., 2009b) (Fig. 7).

In Table 9 the CLX PK parameters in udder dry secretion are presented. We determined the T>MIC_{90} in the mammary dry secretion of each quarter. Normally, T>MIC_{90} is expressed as the percentage of time of the inter-dose interval during which the concentration remains above the MIC. Since we used a single dose of benzathine CLX in a slow-release formulation, there was no inter-dose interval. Therefore, with the objective of evaluating the PK behaviour based on the MIC, we estimated T>MIC_{90} as hours post-treatment during which the ATM concentration was above the MIC.

The milk:plasma passage rate was very low and the CLX concentrations in plasma could not be measured. For mastitis treatment by IMM route it is desirable the permanence of the ATM in the mammary compartment or glandular tissue for a long time.

Therefore, these findings are acceptable considering that the formulation used is indicated for the IMM treatment of mastitis at drying off. The level of binding of CLX to dry udder secretion is high (86 ± 6.5 %), acting as a reservoir and increasing the persistence of the ATM inside the udder. The experimental animals maintained concentrations above the limit of quantitation by microbiological methods (0.03μg.mL^{-1}) during the first 24-31 days. Oliver et al. (1990) found that after IMM administration of CLX at drying off, the secretion samples had detectable levels until 28-35 days post-treatment in some animals and until 42-49 in others. This finding shows a marked variability in CLX concentrations at drying off. The determinants of this variability are varied and complex: udder size, seasonal effects, body condition of cows, physiological changes in body condition at the start of the dry period and other factors affecting gland and body.

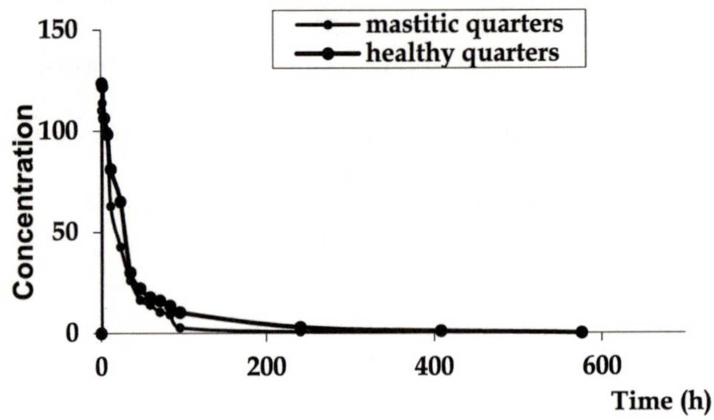

Fig. 7. Average CLX milk concentrations in dry udder secretion (mastitic vs. healthy) after the administration of one IMM syringe in each quarter at drying off (600 mg)

Parameter	Unit	Mean $_{mastitic}$ ± SD	Mean $_{healthy}$ ± SD
T½λ	h	48.13 ± 51.27	39.53 ± 27.29
ABC$_{0\text{-}24h}$	μg.h.mL-1	2130.24 ± 797.31	2144.09 ± 850.06
ABC$_{0\text{-}\infty}$	μg.h.mL-1	5003.07 ± 3292.88	5358.52 ± 3398.00
TMR	h	62.98 ± 47.49	57.42 ± 35.07
T>CIM$_{90}$		257.14 ± 188.45	231.54 ± 153.82

Table 9. CLX PKs in dry udder secretion in mastitic quarters vs. healthy quarters after its administration at dose of 600 mg in each quarter at drying off. T>CIM$_{90}$: post-administration hours during which CLX concentrations remained above the MIC$_{90}$ (2 μg.mL^{-1})

4.1.4 Cephalosporins

Several IMM products for the IMM treatment of dairy cattle contain cephalosporins, most commonly as a single product (e.g., cefquinome; cephapirin), but also in combination (cephalexin and kanamycin). The cephalosporins are semisynthetic antibiotics derived from cephalosporin C (produced by *Cephalosporium acremonium*). There are currently 4 generations of cephalosporins, which vary from the narrow-spectrum first generation through the expanded-spectrum fourth generation (Hornish and Kotarski, 2002), with all generations being used in veterinary medicinal products. When using a first-generation cephalosporin in an IMM preparation, there is a strong rationale for combining it with an additional antibiotic to extend the spectrum of the product. There are cephalosporins that, when administered systemically penetrates in significantly low levels to the mammary gland of the producing cow. Is the case of ceftiofur. Ceftiofur is an interesting tool because it can be used in cows that are producing milk to treat infections that are located out of the mammary gland (metritis, foot infections), without discarding milk.

4.2 Phenicols

The toxicity of chloramphenicol (CAP) has been the cause of its use being banned or strictly regulated worldwide. This has accentuated the need for an effective broad-spectrum antibiotic to be used in food producing animals. Florfenicol (FLF) is a synthetic fluorinated CAP analogue that is exclusively used in food animals for treatment of infectious diseases. PK studies showed that high concentrations, both in serum and milk, are obtained following IMM administration in cows, which led to propose its use for the treatment of bovine mastitis. The FLF MIC_{90} among *S. aureus* obtained by San Martin was 2 $\mu g.mL^{-1}$ (2002). We obtained a thianphenicol (TAP) MIC_{90} higher than FLF (>16 $\mu g.mL^{-1}$) (Lucas, 2009b).

TAP is a structural analogue of CAP with marked toxicological differences (Mestorino et al., 1993b). TAP has a greater in vitro activity against some bacteria that are resistant to CAP. Although TAP has similarities in its antibacterial spectrum with that of CAP, there are marked pharmacological differences between the two drugs. TAP is more stable in solution, is not appreciably protein bound in the body (16%) and does not undergo significant biotransformation. The drug diffuses into intracellular spaces, the central nervous system and the aqueous humour (Mestorino et al., 1993a; b). We performed a complete study about the TAP PK behavior in lactating healthy Holstein cows after its administration by intravenous, intramuscular, subcutaneous routes (Fig. 8) and after IMM infusion in a unique quarter and in the complete gland (in the four quarters) (Mestorino et al., 1993a; 1993b; 1995). Serum PK after IV administration was described following a bicompartmental model. The antibiotic showed a rapid distribution, with a half life of 8.84 ± 4.34 min. Half life of elimination was 1.95 ± 0.55 h, indicating that TAP is rapidly eliminated in the bovine. Volume of distribution was high, 1404.18 ± 428.19 $ml.kg^{-1}$. Milk concentrations were high, when the drug was administered systemically, with a mean time of gland penetration ($T½_P$) rather fast of 36.58 ± 9.72 min and a half life of elimination from the gland of 3.62 ± 0.97 h, indicating that accumulation of TAP takes place in the mammary gland. A milk permanency time of 4.13 ± 1.13 days was determined (Table 10).

A monocompartmental model was used to describe the PKs of TAP after its IM and SC administration. After IM administration, TAP was rapidly absorbed, with a half life of 6.53 ± 6.25 min and a half life of elimination of 2.9 ± 0.50 h. Maximum concentration was 29.38 ± 6.90 $\mu g.mL^{-1}$ and was found at 0.32 ± 0.09 h. Area under the curve was 68.45 ± 16.27 μg.h/mL and bioavailability 86.33 ± 19.08%. Milk therapeutic levels were maintained between 0.5 and 8 h with a C_{max} of 17.05 ± 2.73 $\mu g.mL^{-1}$ reached at 3.42 ± 0.19 h. Mammary gland half life of penetration was 51.11 ± 6.10 min and half life of elimination 5.31 ± 4.27 h. Ratio C_{maxS}:C_{maxM} was 1.81 and the ratio AUC_M:AUC_S 1.19. Theoretical milk permanency time was 4.79 ± 1.40 days.

After SC administration a maximum serum concentration of 19.83 ± 3.57 $\mu g.mL^{-1}$ was obtained at 0.72 ± 0.12 h with a half life of absorption of 10.09 ± 5.37 min. Half life of elimination was 2.60 ± 0.60 h. AUC of 56.17 ± 7.33 µg.h/mL was calculated. Bioavailability was of 82.33 ± 13.36 %. Milk concentrations after SC administration were maintained above the MIC between 1 and 8 h. The C_{max} was of 13.19 ± 3.47 µg.mL^{-1} at 3.42 ± 0.19 h. The half life of penetration was 41.97 ± 4.55 min and the half life of elimination 7.57 ± 4.47 h. ratio C_{maxS}:C_{maxM} and AUC_M:AUC_S were 1.57 and 1.03 respectively.

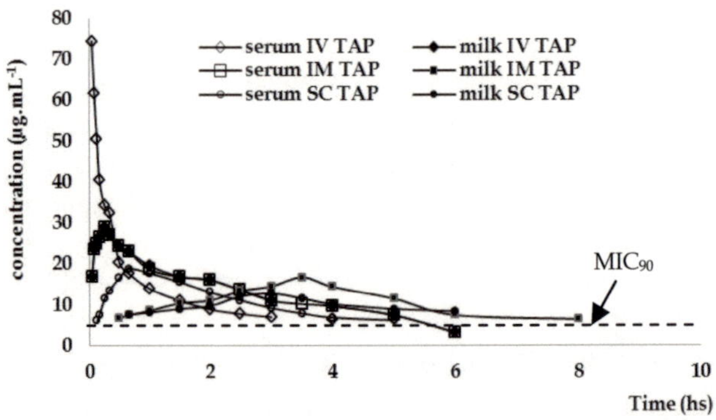

Fig. 8. Averaged serum and milk TAP concentrations for six lactating cows following intravenous, intramuscular and subcutaneous TAP administration at dose of 25 mg.kg^{-1}

Parameter	Serum IV	Serum IM	Serum SC	Milk IV	Milk IM	Milk SC
β (h^{-1}) – K_e (h^{-1})$_M$	0.40±0.16	0.24±0.05	0.28±0.06	0.21±0.067	0.18±0.07	0.15±0.10
$T½_β$ (h)- $T½_E$(h)$_M$	1.95±0.55	2.88±0.50	2.60±0.60	3.62±0.97	5.31±4.27	7.57±4.47
K_{ab}(h^{-1})- K_P(h^{-1})$_M$		14.26±8.72	5.06±2.00	1.23±0.37	0.83±0.10	1.00±0.10
$T½_{ab}$-$T½_{PM}$ (min)		6.53 ±6.25	10.09±5.37	36.58±9.72	51.11±6.10	41.97±4.55
AUC (µg.mL/h)	50.59 ±7.56	68.45±16.27	56.17±7.33	85.77±21.58	75.31±9.85	52.85±17.50
C_{max} (µg.mL^{-1})		29.38±6.90	19.83±3.57	23.09±3.12	17.05±2.73	13.19±3.47
T_{max} (h)		0.32±0.09	0.72±0.12	2.50±0.29	3.42 ±0.19	3.42±0.19
Lag time (h)		0.05±0.02	0.15±0.02	1.08±0.51	0.78 ±0.23	1.42±0.34
F (%)		86.33±19.08	82.33±13.36			
C_{max-S}/C_{max-M}					1.81±0.64	1.57±0.39
AUC_M/AUC_S				1.72±0.47	1.19±0.41	1.03±0.26
Withdrawal (d)				4.13±1.13	4.79±1.40	3.23±0.86

Table 10. TAP PKs in serum and milk after its administration at dose of 25 mg.kg-1 by IV, IM and SC routes in lactating Holstein cows. Withdrawal is the withdrawal time needed to obtain milk with TAP levels of less than 20 ng.mL^{-1}.

As a conclusion, TAP is a highly bioavailable antibiotic, independently of the administration route, it arrives with ease to the mammary gland, reaching high levels there. That is why it can be considered an excellent option to use in mastitis therapeutics, especially against Gram negative pathogens. It has a bacteriostatic action against a broad range of microorganisms, although it may be bactericidal for some species under some conditions, and in concentrations 3 to 5 times higher than the bacteriostatic ones. Among the bacteria inhibited *in vitro* by relatively low concentrations of TAP are *Clostridium*, *Corynebacterium diphtheriae, Diplococcus pneumoniae, Staphylococcus albus, Streptococcus pyogenes, Streptococcus viridans, Bacteroides, Fusobacterium, Bordetella, Brucella, Haemophilus, Neisseria, Pasteurella, Shigella* and some vibrio strains. Some Bacilli, *Erysipelothrix, Staphylococcus aureus* and *Streptococcus faecalis* are sensitive to moderate concentrations of TAP (FAO: JECFA 47, 1997).

When TAP was administered by IMM route in an individual quarter high concentrations in the nonmedicated quarters and in serum were determined. Although the experimental animals were milked every 12 h, TAP levels above the minimum inhibitory concentration for the majority of pathogens were maintained during 36 h in the treated quarter and between 8 and 12 h in the nontreated quarters and during less than 4 h in the case of blood (Mestorino et al., 1995). TAP passes easily and completely from the treated quarter to the homolateral quarter by simple diffusion, to the heterolateral quarters, on the other hand, it arrives through circulation. Time elapsed between the administration and the first measured concentration (Lag Time) was shorter in the case of the homolateral quarter (rear-left -RL-: 0.25 h), intermedial in the case of serum (0.54 h) and rear-right -RR- (0.58 h) and longer in the case of the front-right quarter -FR- (1.92 h). In the homolateral quarter, the main C_{max} and the highest bioavailability were recorded. Maximum TAP concentrations of 75.30, 15.06, 19.03 and 15.79 $\mu g.mL^{-1}$ in the RL, FR, RR and serum respectively, and the ratios of AUC_{treat}:$AUC_{nontreat}$ of 10.61, 1.81, 2.78 and 1.35 in the RL, FR, RR and serum respectively, suggests that the largest passage takes place from the administered quarter to the homolateral one, and that the arrival of TAP to the contralateral quarters is via the blood due to that the longitudinal medial intermammary septum is a highly impermeable (Fig. 9).

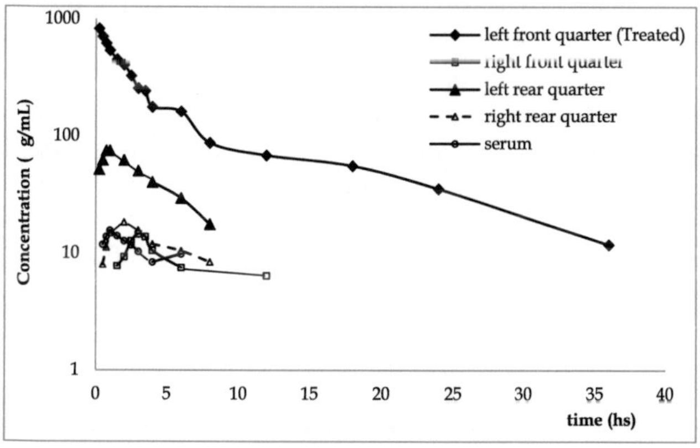

Fig. 9. Averaged milk and serum TAP concentrations for six lactating cows following IMM infusion in the front-left mammary quarter at dose of 25 $mg.kg^{-1}$.

4.3 Macrolides

The prototype of the macrolide class of antibiotics was introduced in 1952 under the name of erythromycin. Macrolides are frequently used in Argentina for bovine mastitis treatment, since high concentrations in milk are obtained following parenteral administration. Erythromicin, oleandomycin, tylosin and the modern tilmycosin, azithromycin and tulathromycin have been extensively used in bovines. The last three are drugs with substantial improvements respect the older ones.

4.3.1 Spiramycin

Spiramycin (SPM) is a macrolide antibiotic that is active against most of the microorganisms isolated from the milk of mastitic cows. SPM is a very soluble weak base (pKa 8.2) which crosses lipid membranes with ease.

Our team investigated the disposition of SPM in plasma and milk after intramuscular administration. Lactating Holstein cows with subclinical mastitis caused by *S. aureus* were given two injections of SPM adipate at a dose of 16,000 IU/kg (3.75mg.kg^{-1}) with 24 hours of intervals. The experimental animals were allocated by production level (high vs. low production) and the quarters were grouped by health state (Fig. 10 A and B).

In susceptibility studies, usually erythromycin is used as a representative of the macrolide group of compounds. However, when evaluating SPM susceptibility, it is not recommended to use erythromycin disks, because it has been shown that proteins involved in macrolide-ribosome interactions are different. The binding site of erythromycin is the L22 protein while for SPM the protein is the L27 (Lucas et al., 2007; 2009b). The MIC_{90} calculated for *S. aureus* field isolates was 4µg.mL^{-1}(17 IU.mL^{-1}) (Lucas et al., 2009b). Other authors have reported MICs of 3.25 µg.mL^{-1} (15 IU.mL-1) (Renard et al., 1996) and 8 µg.mL^{-1} (Friis et al., 1988). After IM SPM administration, milk concentrations were higher than those in plasma. The maximum concentration in milk (C_{maxM}) was 3.91µg.ml^{-1} at 9 hours post-first administration and 3.46 µg.ml^{-1} 12 hours post-second administration. Whereas in plasma the measured concentrations were far below those in milk (0.12µg.ml^{-1} at 0.5 h and 0.097µg.ml^{-1} at 8 h post first and second administration respectively). An average milk-to-plasma ratio of 46.35 ± 11.09 was calculated by comparison of the areas under the concentration vs time curves. Our results were coincident with other authors (Renard et al., 1996). The level of production factor exert a significant effect on C_{max}, $T_{½λ}$, AUC_{0-24}, MRT and Fmilk (Table 11). SPM in milk of high production animals reached higher concentration with AUC_{0-24h} also higher in this group. However, $T_{½λ}$ was longer in the milk of low production cows. The average milk-to-plasma ratio was higher in high producing cows. Since the milk production level has a significant effect on efficacy predictors such as $T>MIC_{90}$ (41.60) and AUC_{0-24}/MIC_{90} (15.87) milk SPM concentrations obtained were insufficient to achieve optimal PK/PD relationships.

4.3.2 Azithromycin

Azithromycin (AZT) is a semisynthetic compound in which the lactone ring has been expanded to a 15-member structure and is considered the prototype of the new macrolide structures identified as azalides (Lucas et al., 2007; 2009c). AZT formulations are not available for use in production animals but could possess advantages for the treatment of certain bovine infections such as those produced in the mammary gland. We performed a complete study about its pharmacokinetic behaviour in blood and milk following administration by different routes, both in healthy cows and cows with mastitis. The results observed after its administration by any of the analyzed routes was typical of the macrolides, with low plasma concentrations and very high concentrations in milk and soft tissues; a great volume of distribution and a prolonged terminal half-life both in blood and milk. The fact that AZT tends to accumulate in inflammatory cells, has a kinetic incidence in the results, especially after IMM administration.

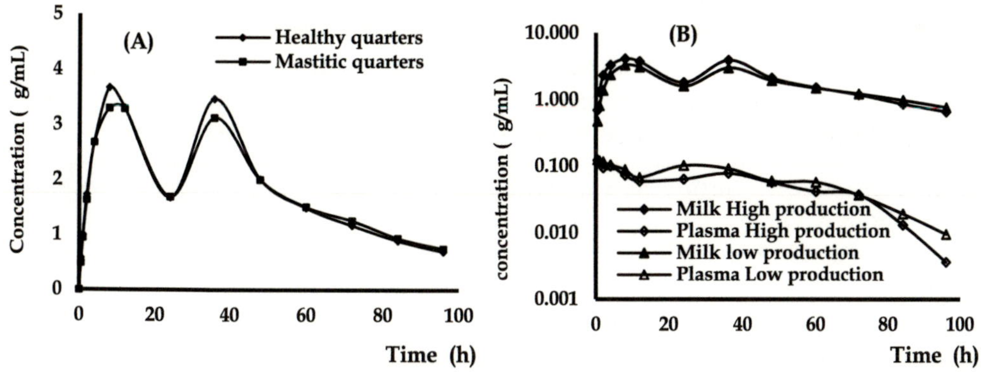

Fig. 10. **(A)** Mean concentrations of SPM in mastitic quarters and healthy quarters after two doses of 3.75 mg.kg^{-1} IM with a 24-h interval. **(B)** Mean concentrations of SPM in plasma in high-producing and low-producing cows; and quarters of high-producing cows and low-producing cows after two doses of 3.75 mg.kg^{-1} IM with a 24-h interval

Parameter		Milk High-Pr	Milk low-Pr	Plasma High.Pr	Plasma Low-Pr
$C_{máx1}$	μg.mL^{-1}	4.26 ± 0.66	3.57 ± 0.61	0.12 ± 0.01	0.12 ± 0.01
$T_{máx1}$	h	8.67 ± 2.31	9.33 ± 1.97	0.50 ± 0.00	0.50 ± 0.00
$C_{máx2}$	μg.mL-1	3.92 ± 0.71	3.01 ± 0.71	0.08 ± 0.02	0.11 ± 0.04
$T_{máx2}$	h	36.00 ± 0.00	36.00 ± 0. 00	32.00 ± 6.93	32.00 ± 6.93
$T½\lambda_1$	h	12.26 ± 2.11	13.64 ± 4.20	27.83 ± 18.,38	25.19 ± 11.03
$T½\lambda_2$	h	21.17 ± 8.65	36.60 ± 7.61	7.70 ± 2.06	14.07 ± 5.56
$ABC_{0\text{-}24h}$	μg.h/mL	70. 54 ± 10.62	56.44 ± 8.41	1.77 ± 0.30	2.14 ± 0.42
$ABC_{0\text{-}\infty}$	μg.h/mL	169.76 ± 30.66	171.80 ± 25.61	3.10 ± 1. 14	4.04 ± 0.83
TMR	h	43.54 ± 10.93	55.96 ± 9.83	22.66 ± 0.99	28.78 ± 5.83
F_M		57.73 ± 13.61	43.91 ± 10.30	-	-

Table 11. Milk and plasma PK parameters obtained after two doses of SPM IM administration (3.75 mg.kg^{-1} every 24h) in high and low producing lactating cows.

Due to the described PK behaviour, AZT proved to be an interesting tool with potential for infections in soft tissues, although this should be backed by efficacy trials. Classically, macrolides have been considered drugs of choice to treat IMM infections not only because of their antibacterial efficacy but because of their favourable PK profile. On the basis of these antecedents we decided to investigate the distribution of AZT in plasma and milk after IV, IM and IMM administration in healthy and mastitic lactating Holstein cows and cows at drying-off (Turic et al., 2003a; 2003b; 2006; Errecalde et al., 2003;Lucas et al., 2009b; 2009c).

There are no AZT susceptibility studies on S. aureus isolated from bovine mastitis cases. However, there are data for other macrolides. Usually when running antibiograms, an erythromycin disk is used to evaluate susceptibility to the macrolide family. Variable percentages of macrolide resistant S. aureus ranging from 1.9% to 26.3% have been reported in several studies (Ziv., 1980a; Owens et al., 1997; Andrade et al., 2000). At present there are highly specific methods for susceptibility testing of veterinary pathogens (CLSI, 2008). The AZT MIC_{50} calculated for the 51 S. aureus isolations was 0.5 μg.mL^{-1} and the MIC_{90} was 1 μg.mL^{-1} (Lucas et al, 2009c). Although it is not advisable to compare MICs of different ATM agents, it is worth stating that erythromycin MIC_{90} for bovine isolated *S. aureus* was reported as 0.5 μg.mL^{-1} in several publications (Gianneechini et al., 2002; Russi et al., 2008).

The most prominent pharmacokinetic characteristic of AZT is the presence of high tissue concentrations which are maintained a long time after serum concentrations decline to very low levels (Fig 11 A y B). This characteristic was demonstrated after its IV and IM administration (Tables 12 and 13) (Turic et al., 2003a; 2003b), and was coincident with some authors in goats (Cárceles et al., 2005) and humans (Foulds et al., 1991). AZT $T_{½λ}$ was long, an expected finding, according to the characteristics of this ATM.

Milk AZT levels resulted much higher than those found in serum (Fig. 11 A y B), after each IV and IM administration which is a logical finding according to the lipophilicity and wide distribution of the drug.

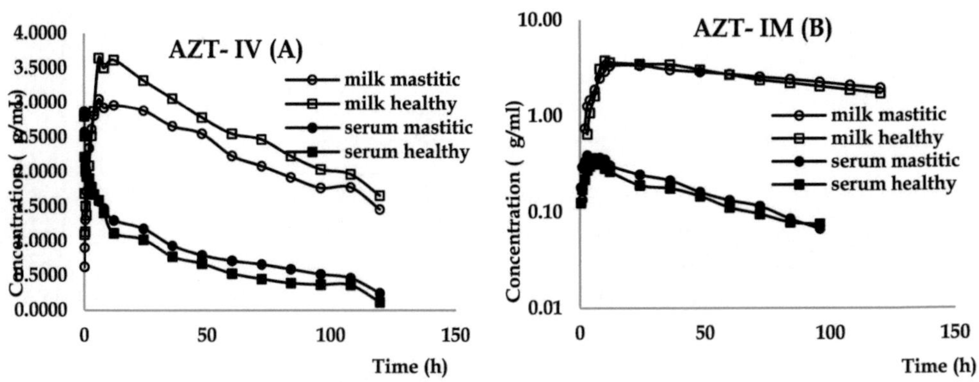

Fig. 11. Mean concentration of AZT in serum and milk of healthy and mastitic lactating cows after IV (A) and IM (B) administration (10 mg.kg^{-1})

AZT exhibited major penetration into milk and it was cleared rather slowly. PK parameters indicated a high retention of the drug in peripheral compartments. The $T_{½λ}$ in milk after each route of administration was always at least four times longer that in plasma. AZT $T_{½λ}$ clearly suggested that milk concentrations decrease more slowly than plasma ones.

Later we performed other assay with AZT, but in this case it was administered intramuscularly in two doses of 10 mg.kg^{-1} body weight with a 48 h interval (Lucas et al., 2009c). The experimental animals were allocated by production level (high and low production levels) and the quarters were grouped by health state (Fig. 12 A and B). The $T_{½λ}$ in milk after first administration was at least four times longer than that in plasma. AZT $T_{½λ}$ suggested that milk concentrations exhibited a tendency to decrease more slowly than plasma ones. The same pattern was observed after a single 10 mg.kg^{-1} IM dose of AZT to lactating Holstein cows (Turic et al., 2003a).

Parameter	Serum healthy	Serum mastitic	Milk healthy	Milk mastitic
λz (h^{-1})	0.019±0.006	0.013±0.001	0.007±0.003	0.007±0.001
$t_{½}\lambda z$ (h)	38.6±9.11	53.07±3.36	113.3±39.5	111.2±21.1
V_{area} (ml.kg^{-1})	6325.6±1467.9	6486.3±318.6		
$AUC_{0-\infty}$ (h.µg/ml)	89.0±13.4	84.8±2.65	599.4±141.3	511.2±113.3
MRT (h)	52.1±8.94	69.5±7.61	163,6±58,6	159,6±34,7
T_{max} (h)			9.0±3.29	12.0±9.86
C_{max} (µg/ml)			3.82±0.49	3.13±0.66
AUC_M/AUC_S			389.4±69.4	232.3±52.7

Table 12. Serum and milk PK parameters obtained after one IV administration of AZT at a dose of 10 mg.kg^{-1} in healthy and mastitic lactating Holstein cows

Parameter	Serum healthy	Serum mastitic	Milk healthy	Milk mastitic
λz (h ^{1})	0.018±0.004	0.015±0.003	0.008±0.003	0.005±0.001
$t_{½}\lambda z$ (h)	39.11±7.64	47.31±9.75	105.19±40.18	135.96±39.49
$AUC_{0-\infty}$ (h.µg/ml)	36.26±8.59	26.09±1.71	561.89±60.55	713.36±249.99
MRT (h)	53.43±6.80	63.22±10.92	158.38±54.21	201.03±56.06
T_{max} (h)	3.00±1,549	2.17±0.41	17.33±10.56	13.67±5.13
C_{max} (µg.ml-1)	0.92±0.11	0.66±0.03	4.35±2.17	3.48±0.42
F (%)	41.38±13.98	18.860±2.51		
AUC_M/AUC_S			1041.57±516.7	1422.57±350.
$C_{max\,M}/C_{max\,S}$			4.85±2.67	5.26±0.73

Table 13. Serum and milk PK parameters obtained after one IM administration of AZT at a dose of 10 mg.kg^{-1} in healthy and mastitic lactating Holstein cows

When comparing PK parameters by grouping quarters according to health status, it was observed that AZT was eliminated more slowly from and $AUC_{0-\infty}$ was substantially higher in mastitic quarters. Although this was an unexpected finding (the pKa partition hypothesis suggests the opposite), it is coincident with previously reported data (Turic et al., 2003a). Milk pH in the experimental animals ranged between 6.5 and 7.5 with the majority of values around 7.0. Average pH from all mastitic quarters was 7.13 ± 0.23 and from healthy quarters was 6.90 ± 0.21 (see Table 14). This is a normal finding for animals carrying subclinical mastitis. AZT is a weak base with a pKa value of 8.74, as a consequence, by application of the Henderson–Hasselbach equation, there would be approximately double AZT molecule dissociation in mastitic milk and more than three times in milk of healthy animals in comparison with plasma (see Table 14). Alkaline drugs (like AZT) are trapped in acidic compartments. This theoretical considerations could not, however, be confirmed by the experimental findings reported here. Azithromycin (IM) gave rise to very low plasma AUCs, which could be explained by its very high liposolubility and penetration into tissues. Although higher AUC was expected in milk of healthy animals (more acidic), which is a common finding with the classic macrolide antibacterials, we found exactly the opposite. In our experiment, the highest concentrations were determined in the milk of mastitic animals, with an AUC_{milk} / AUC_{plasma} ratio extremely high. Our explanation for this unexpected finding is the amount of somatic cells (SCC) present in mastitic milk in comparison with the normal milk. In the former case, the number of SCC was several times above those in normal milk. Mastitic milk normally exhibits very high cell counts as consequence of the inflammatory reaction. As it is known, AZT is able to reach high concentrations at infected sites, as a result of increased delivery from phagocytes (Lucas et al, 2009c). On this basis, we consider that the inflammatory reaction (and the high amount of cells) in the infected quarters is the main reason for the differences found between mastitic and healthy quarters.

	Plasma	Mastitic milk	Healthy milk
pH	7.40	7.13	6.90
Dissociated/non dissociated molecules	21.88/1	40.74/1	69.18/1
Theoretical ratio of dissociated molecules	1	2.86	3.16
Experimental Milk/plasma AUC		743.60	482.72

Table 14. Average plasma pH, milk pH (mastitic vs. healthy), dissociation as a function of AZT pKa (8.74), theoretical ratio milk/plasma and experimental AUC ratio milk/plasma

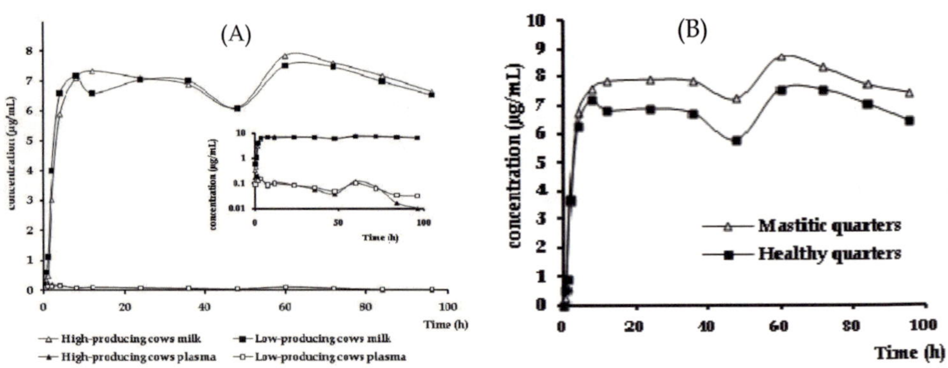

Fig. 12 **(A)** Mean plasma and milk concentrations of AZT in high-producing and low-producing cows after two 10 mg.kg-1 IM doses with a 48-h interval. **(B)** Mean concentrations of AZT in mastitic quarters and healthy quarters after two 10 mg.kg-1 IM doses with a 48-h interval

A significant AZT fraction could be trapped in the milk-cell compartment without participating of the plasma:milk equilibrium, largely dependent on the pKa - pH relationship. The $AUC_{0-\infty}$ ($P < 0.05$) and the MRT were higher in whole milk from mastitic quarters, which may indicate that the drug is present in higher amounts and persist during longer time in mastitic quarters than in healthy ones. At the same time, the F_{milk} of AZT was higher in the mastitic quarters indicating a different PK profile of AZT depending on the quarter status. The previous data, reported after a single 10 mg.kg-1 IM dose of AZT to lactating Holstein cows, support our observations (Turic et al., 2003a). The Cl_{mam}/F showed that AZT elimination was faster in healthy quarters than in mastitic quarters. Separating the groups according to median production allowed us to identify observations that may be extrapolated to situations of major differences in productive levels (Fig. 12 A). The $AUC_{0-\infty}$ was higher in quarters of low-producing cows than in quarters of high-producing cows. It is possible that a lower milk production causes a slower antibiotic elimination, with a lower Cl_{mam}/F value and more prolonged $T_{½\lambda}$. On the basis of these results, we could suggest that low-producing cows have a high tendency to exceed LMR in milk while high-producing cows can eliminate (and dilute) the drug fast enough so as to diminish "contact time" and clinical efficacy possibilities.

AZT is a time dependent bacterial killing antibiotic with prolonged persistence (PAE, PASME). Thus, for this group, both t>MIC and AUC_{24h}/MIC ratio play an important role in planning the dosage regimens. According to microbiological results, the MIC for *S. aureus* of bovine udder origin was 0.5µg.mL^{-1}. The AUC_{24}/MIC was greater than 100 in both healthy as mastitis cows (Table 15). The times above MIC were longer than 95 h for all cases.

Six healthy and six mastitic lactating cows received one IMM AZT syringe containing 125 mg in each mammary quarter for three consecutive milkings; and six healthy and six mastitic Holstein cows at drying-off received one IMM AZT syringe containing 500 mg in each mammary quarter (Turic et al., 2003a; Errecalde et al., 2003).

For the study in lactating cows, serum profiles evolved with three peaks and a final elimination phase after the last infusion. Serum penetration from milk was fast and the three peaks were in the same order of magnitude both in healthy and mastitic animals (Fig. 13 A). AUC serum concentration was however higher in healthy animals than mastitic animals. This resulted an unexpected finding, since AZT is a basic drug. No significant differences were found either in $T_{½\lambda}$ or MRT between healthy and mastitic animals.

Parameters	Mastitic ± SD	Healthy ± SD	H-prod ± SD	L-prod ± SD
AUC_{0-24h}/MIC_{90}	156.53 ± 39.68	152.46 ± 33.19	152.53 ± 38.93	155.11 ± 31.51
$T>MIC_{90}$	96.09 ± 0.74	96.48 ± 1.20	95.83 ± 0.00	96.88 ± 1.80

*MIC_{90} of AZT against the 51 S. aureus isolated

Table 15. PK/PD parameters of AZT in mastitic and healthy quarters; and in quarters of high-producing cows and quarters of low-producing cows after two 10 mg. Kg^{-1} IM doses with a 48-h interval

4.3.2.1 AZT IMM administration

Milk profiles, on the other hand, showed interesting differences. Peak concentrations resulted much higher in mastitic animals (Fig. 13 A). $T_{½\lambda}$ and MRT resulted in the same order of magnitude in both groups and the differences lacked statistical significance. Areas under the curves, however, showed major differences. In the case of the healthy animals $AUC_{0-\infty}$ was in the order of 503.03 and 1615.65 µg.h.mL^{-1} in mastitic animals, the difference being statistically significant (Table 16). This difference in AUC was coincident with the differences found in serum, although these were not statistically significant.

Once again we found higher AZT concentrations in milk of mastitic animals than in milk of healthy animals after IMM administration, similarly as the situation occurred after IM administration. As AZT is so penetrating in tissues and cells, and mastitic milk is so rich in somatic and inflammatory cells, the drug becomes included into the more acidic cellular compartment, thus hindering its participation in the milk-serum diffusion process (despite its pKa), and this retains high amounts of AZT in milk by cell trapping. All this evidence supports the fact that AZT is a penetrating azalide with concentrations several times higher in tissues and milk than in plasma. If we observe the ratios $AUC_{(milk)}/AUC_{(serum)}$ and we compare them with those obtained when the drug was administered intramuscularly we could conclude that the high availabilities in milk compared to those of serum are independent of the route of administration.

But, in the study with cows at drying-off, we found some differences: AZT milk concentrations in mastitic and healthy animals were similar (Fig. 13 B and Table 17). One explanation for this is that physiology of the mammary gland during the dry period differs markedly from that during lactation. Very few cells (less than 2% are epithelial cells) and total leukocyte concentration increase rapidly in early involution and the milk fat and casein may decrease the leukocytes phagocytic function.

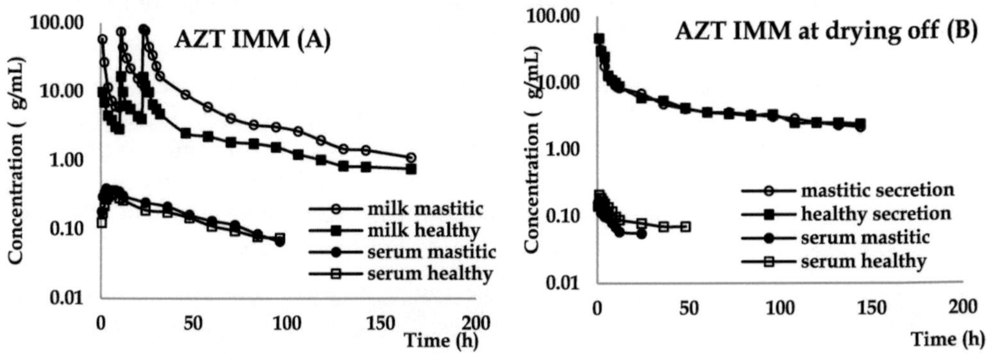

Fig. 13. Mean serum and milk AZT concentrations after one IMM AZT syringe containing 125 mg in each mammary quarter for three consecutive milkings in healthy and mastitic lactating cows (A). Mean serum and milk AZT concentrations after IMM syringe containing 500 mg in each mammary quarter of healthy and mastitic cows at drying off (B).

Finally, the excellent milk availability observed allows us to consider AZT as a potential antimastitic drug, although we have to fit the dosage through PK-PD modeling and corroborate with efficacy studies in the future. As was mentioned in previous paragraphs, the site where the pathogen is located represents one of the real challenges of ATM chemotherapy of the mammary gland. The pathogen can be in milk, or tissues. In the last case it can be in the interstitium or in cells. And in this case it can be in the cytoplasm or in fagolysosomes. As can be easily understood the deeper the location of the microorganism the more difficult will be to reach it by the ATM. Furthermore, if the ATM reaches the site of the microorganism, it has to exert its antibacterial effect and to do this it needs some special conditions, being the pH a critical one. And the pH becomes more acidic the deeper in tissues and cells it is measured. As illustrative figures we can mention the pH of plasma of 7.4, of interstitium 7.0, of cytoplasm 6.5 and of fagolysosome of 5.0. We evaluated the effect of the pH variation on the antibacterial activity of AZT against strains of *S. aureus* isolated of mastitic quarters. *S. aureus* strains isolated and *S. aureus* ATCC 25923 were tested at pH 7.4,

6.5 and 5.0, in order to simulate the conditions of acidity of plasma, tissue and subcellular structures which are commonly associated with *S. aureus* intracellular persistence. The results at pH 7.4, were consistent with those reported by CLSI 2008 (MIC: 1μg.mL^{-1}). However, MIC was approximately 16 times higher at pH 5.0 than at pH 7.4 (Fig. 14).

Parameter	Serum healthy	Serum mastitic	Milk healthy	Milk mastitic
λz (h^{-1})	0.078±0.04	0.050±0.02	0.01±0.00	0.02±0.00
$t_{½}$λz (h)	11.76±7.31	16.20±7.20	59.67±12.48	46.85±20.87
$AUC_{0-\infty}$ (h.μg/ml)	3.51±2.25	2.74±1.11	503.03±49.47	1615.65±501.9
MRT (h)	17.37±10.56	24.69±11.20	76.37±18.32	52.41±28.39
T_{max} (h)	22.75±0.27	22.67±0.26		
C_{max} (μg/ml)	0.23±0.05	0.13±0.02		
F (%)	2.03±1.50	1.06±0.43		
AUC_M/AUC_S			143.31	589.65

Table 16. Serum and milk PK parameters obtained after one IMM AZT syringe containing 125 mg in each mammary quarter for three consecutive milkings in healthy and mastitic lactating Holstein cows

Parameter	Serum healthy	Serum mastitic	Milk healthy	Milk mastitic
λz (h^{1})	0.02±0.01	0.08±0.03	0.01±0.00	0.01±0.00
$t_{½}$λz (h)	34.73±10.96	10.18±4.77	107.60±27.72	99.09±28.46
AUC_{0-tlas} (h. μg/ml)	3.53±1.28	1.35±0.81	760.00±196.3	747.16±223.94
$AUC_{0-\infty}$ (h.μg/ml)	6.73±2.31	2.12±1.14	1136.11±262.	1056.71±171.4
MRT (h)	15.09±3.15	6.87±1.82	131.18±35.31	122.51±46.32
T_{max} (h)	1.25±0.61	2.08±2.18		
C_{max} (μg/ml)	0.22±0.08	0.21±0.19		
F (%)	4.36±1.52	1.14±0.67		
AUC_M/AUC_S				

Table 17. Serum and milk pharmacokinetic parameters obtained after one IMM AZT syringe containing 500 mg in each mammary quarter of healthy and mastitic Holstein cows at drying off

4.3.3 Tylosin

Tylosin (TYL), other antibiotic of the macrolide group, is commonly used in food animal practice. Because it is an organic base (pKa = 7.1), moderately bound by serum proteins (40%), with a high degree of lipid solubility (Lucas et al., 2007), TYL would be expected to be widely distributed in body fluids and tissues. The MIC of TYL for *S. aureus* was <1 μg.ml^{-1} for most isolates studied. We determined the elimination milk profile of TYL after IM administration at multiple dose schemes.

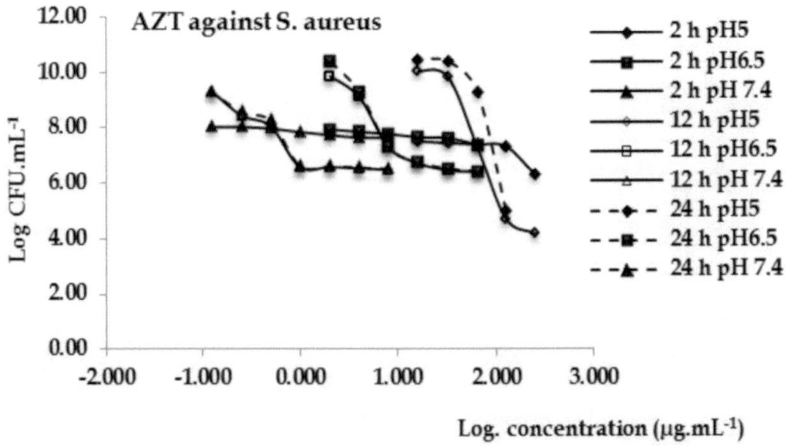

Fig. 14. Inhibition of *S. aureus* exposed to different concentrations (expressed as Log_{10}) of AZT (0.25 MIC; 0.5MIC; 1MIC; 2MIC; 4MIC and 8 x MIC) in function of the time and the medium pH (5; 6.5 and 7.4)

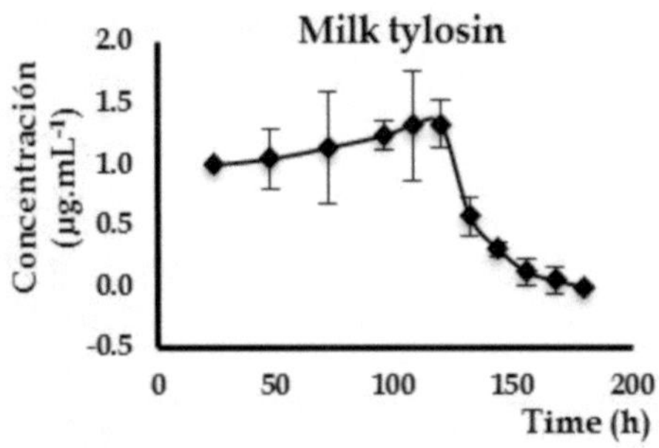

Parameter		Mean	SD
λ	h^{-1}	0.095	0.028
T½λ	h	7.873	2.232
$AUC_{0\text{-}t}$	µg.h/mL	141.360	23.290
$AUC_{0\text{-}\infty}$	µg.h/mL	142.458	23.512
MRT	h	78.395	3.380

Fig. 15. Milk concentrations of TYL after its multiple (five doses every 24 hours) IM administration (10 $mg.kg^{-1}$)

The objective was to calculate the withdrawal time in milk necessary for TYL to reach acceptable limits for human consumption. Healthy Holstein lactating cows received 5 intramuscular doses of TYL at 10$mg.kg^{-1}$ every 24 hours (Figure 15). Milk samples were obtained from the four mammary quarters before the start of treatment, during and post-treatment every 12 hours until to 216 hours. The withdrawal time was calculated using the harmonized Time to Safe Concentration (TTSC) recommended by the European Union. The time required for milk to carry TYL concentration below the maximum residue limits (MRL 50 ppb) was 120 h post last TYL dose.

4.4 Tetracyclines

Oxytetracycline (OTC) is the most representative member of this class of ATMs. OTC is widely used in veterinary medicine worldwide. Its use is widespread due to its broad spectrum (not only against bacteria but also against some chlamydia, rickettsia and protozoa), and certain PK properties as its wide distribution throughout the body and prolonged therapeutic effects of some long-acting formulations. Our team has performed a review of clinical aspects and major PK characteristics of the preparations commercially available (Mestorino & Errecalde, 1997). The PK of OTC has been studied by several authors (Schifferli et al, 1982, Errecalde, 1992; Errecalde et al, 1997). The new long-acting formulations have allowed the maintenance of high concentrations for periods much longer than the classical preparations, which represented a significant change in the practicity and in the usage habits, but also in maintaining residual levels for long periods in different tissues. Moreover, these formulations result in a significantly longer persistence of milk concentrations. The popularity of these preparations is extended to virtually all types of exploitation. Although the use of these long acting formulations is justified especially in extensive cattle exploitations, in other kind of management (i.e. more intensive), where the animals can be daily treated and controlled, is not justified. The use of tetracyclines with long elimination half-lives in dairy cows has impacted on repeated violations of maximum permitted levels of these antibiotics in milk. Problems that may emerge from the misuse of antibiotics are especially related to the presence of sub-therapeutic concentrations and the resultant emergence and dissemination of bacterial resistant strains in both animals and humans. Indisputably, the possibility of transference of portions of DNA carrying genes encoding resistance between bacteria is one of the issues of major concern to people with different responsibilities in the area. Direct toxicity must not be discarded.

Tetracyclines achieve milk concentrations that are approximately in the range of those of blood. They are second-choice parenteral antibiotics for serious infections of the udder caused by Gram positive organism and possibly by coliforms, although susceptibility among the later is controversial. In previous studies we evaluated the serum and milk PK behaviour of OTC after administration of therapeutic doses of three commercially available preparations at 5, 10 and 20 % by the IM route. In Figure 16 A and B, the plasma and milk OTC concentrations after IM administration of three different formulations is represented.

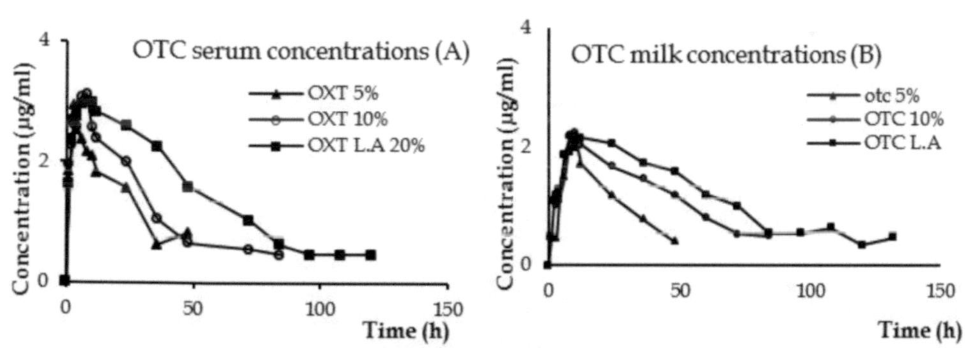

Fig. 16. Serum (A) and milk (B) OTC concentrations after IM administration of three different formulations to milk cows in production.

The 20% solution presented the higher area under the curve milk concentration versus time and the longest elimination half-life (Table 18). The withdrawal periods from milk for the 5%, 10% and 20% solutions resulted 3.5, 5 and 8 days respectively.

Addition of milk to Mueller-Hinton susceptibility test medium permitted measurement of milk effect on agar disc diffusion zone diameters obtained from *S. aureus* field isolates and stock strains. Milk reduced the activity of different ATM agents, as novobiocin, streptomycin, gentamycin, tetracycline, and vancomycin (Owens & Watts, 1987). As was previously said, MIC is used in conjunction with PK data to determine the more appropriate surrogate PK/PD parameter. Problems arise in vivo, however, due to the physiological status of the host, site of infection, and properties of the ATM. Often, the concentration of active ATM at the infection site is quite different from that in serum or milk. Antimicrobial concentrations are lower in ischemic areas, scar tissue, and abscess contents. Also, pH, protein binding, and also normal metabolism and renal excretion mechanism reduce concentration of available ATM (Thornsberry and Sherris, 1985). Standard susceptibility tests measure ATM effectiveness in conditions extremely different from those in the udder.

Parameters	OXT 5%		OXT 10%		OXT 20%	
	Serum	Milk	Serum	Milk	Serum	Milk
	Mean±SD	Mean ±SD	Mean ± SD	Mean ±SD	Mean ±SD	Mean ±SD
Cmax (μg.ml^{-1})	2.94±0.74	2.46±0.49	3.37±0.54	2.35±0.27	3.39±0.43	2.30±0.33
Tmax (h)	3±0	9.50±1.91	7.50±1.91	9.00±1.15	6.25±3.30	17.50±7.55
ß (h^{-1})	0.035 ±0.02	0.049±0.018	0.037±0.018	0.024±0.006	0.022±0.007	0.025±0.017
T½ ß (h)	23.06±8.53	16.16±7.31	22.94±13.04	31.16±10.16	34.07±11.00	36.74±17.61
MRT (h)	33.68±12.01	29.97±11.96	33.35±16.07	51.88±16.19	52.21±11.85	64.94±19.57
AUC (μg.ml/h)	92.72 ±26.77	62.67±13.60	114.64 ±32.81	125.21 ±21.18	196.05 ±32.50	166.43 ±28.10
AUC_{0-24} (μg.ml/h)		32.51±7.59		40.49±6.03		43.14 ± 6.88
C_{maxS}/C_{maxM}		1.19		1.43		1.47
AUC_M/AUC_S		0.68		1.09		0.85

Table 18. Serum and milk PK parameters obtained after administration of three OTC formulations in Holstein healthy lactating cows

In bovine mastitis, the unique environment of the mammary gland presents a problem for determining usefulness of chemotherapeutic agents against a given organism. Milk proteins, lipids, pH, and ionic characteristics can reduce concentration of active ATM (Ziv, 1980a). A simple in vitro method for determining the effect of milk on ATM activity could be of use when examining ATMs for possible use in mastitis therapy (Owens & Watts, 1987).

The MIC_{90} of tetracycline (used as group representative) among *S. aureus* determined by different authors was between 0.5 - 1 μg.mL^{-1} (Pol and Ruegg, 2007). If we consider than the MIC of tetracycline in milk could increase 4 to 32 times that in MH, the *S. aureus* mastitis treatment with OTC has not real possibilities of success, because the OTC concentrations reached in milk are insufficient for achieving the appropriate PK/PD predictors of therapeutic efficacy against *S. aureus* (AUC_{24}/MIC>100). An irreversible binding between tetracycline and large molecules of milk, which might be due to a hydrophobic interaction, was demonstrated by a dialysis test, suggesting the observed impairing effect was due to the action of milk on the tetracycline being tested. Further investigation revealed that much of the reduction of tetracycline activity in milk was attributable to the milk protein casein, while other heat-sensitive components in milk also play some roles (Kuang, et al., 2009).

4.5 Fluoroquinolones

Enrofloxacin (ENR) is a quinolone widely used for treatment of various infectious diseases in cattle caused both by Gram-positive and Gram-negative bacteria, but is not specifically recommended for bovine mastitis treatment; although high concentrations are reached and maintained in milk following parenteral administration. The MIC_{90} value found by Russi et al., (2008) was similar than the reported for isolates from Uruguay and (Gianneechini et al., 2002; San Martín et al., 2002).

4.5.1 Danofloxacin

Danofloxacin (DAF) is a fluoroquinolone (FQ) ATM drug developed for use in veterinary medicine. DAF shows a broad spectrum of activity against most Gram-negative, Gram-positive bacteria and mycoplasma, but has poor activity against anaerobes (Shojaee Aliabadi & Lees, 2003). FQ share some characteristics such as a broad spectrum of bactericidal activity, a large volume of distribution, low plasma protein binding and relatively low minimal inhibitory concentrations (MICs) against target microorganisms (Otero et al., 2001a; 2001b; Mestorino et al., 2009). Danofloxacin 18% was demonstrated to be effective in the treatment of bacterial pneumonia caused by *P. multocida, M. hemolytica* and *H. somnus* or bacterial enteritis (Mestorino et al., 2009) given as a single injection at a dose rate of 6 $mg.kg^{-1}$ of body weight, or two doses 48 hours apart, as needed. This formulation has the advantage of being safe and effective with a single dose or at maximum two doses, as handling animals many times for treatment is not practical. The concept of the high dosage in a single injection is that, after injection, the drug is available in high concentrations sufficient to kill all the sensitive bacteria during a relatively short period of time (Mestorino et al., 2009). This reduces the selection pressure for resistance. These characteristics suggest that it could be useful for the treatment of bovine mastitis caused by *S. aureus*. In order to investigate this possibility, the PK profile of DAF 18% was studied in plasma, milk and various tissues in dairy cows, when administered as a single subcutaneous injection at a dose of $6mg.kg^{-1}$ (Fig. 17 A, B, C and D). DAF was rapidly absorbed and reached peak plasma concentrations of 0.53 $\mu g.ml^{-1}$ within 2 hours after injection. Distribution of DAF from plasma to all sampled tissues and milk was extensive (Table 19). Peak milk concentrations of $1.37\mu g.ml^{-1}$ were achieved 8 hours post-injection. Maximum concentrations (9.73 $\mu g.g^{-1}$) in udder tissue (average of 4 quarters) and uterus (2.53 $\mu g.g^{-1}$) were achieved 6 and 4 hours after injection, respectively. Maximum concentrations in intestinal tract tissue and lymph nodes ranged from 3.6 $\mu g.g^{-1}$ (duodenum) to 10.22 $\mu g.g^{-1}$ (lymph nodes), and were reached between 2 h and 12 hours after injection. Plasma area under the curve (AUC_{pl}) was 9.69 μg.h/ml. Plasma elimination half life ($T_{½ß}$) was 12.53 h. DAF has a very high volume of distribution (Vd = 5.79 $l.kg^{-1}$). AUC values for the various tissues and milk greatly exceeded AUC_{pl}. Elimination half life from milk and tissues varied between 4.57 hours to 21.91 hours and the milk withdrawal time was 73.48 h (Table 19). The reported results support the potential use of DAF in the treatment of mastitis and other infections in milk cows with three days of withdrawal.

Later we performed other study in cows with subclinical mastitis, who received 10 $mg.kg^{-1}$ subcutaneous (SC) 18% DAF(Fig. 18). The MIC_{50} calculated for *S. aureus* isolated from bovines with subclinical mastitic was 0.25 $\mu g.mL^{-1}$ and the MIC_{90} was 1 $\mu g.mL^{-1}$. Gram-negative bacteria have a MIC range of 0.05-2.5 $\mu g.mL^{-1}$, while for Gram-positive bacteria it was 0.25-5 $\mu g.mL^{-1}$ (Shem-Tov et al, 1998). Although, there are no references for DAF MIC against *S. aureus*, it has been proved that enrofloxacin MIC is ≥1 $\mu g.mL^{-1}$ (Cester et al, 1992). The MIC_{90} was determined in order to calculate PK/PD parameters. Milk maximum concentration (C_{max}) was 2.99 ± 0.88 $\mu g.mL^{-1}$ (6.13 h post-administration) whereas plasma C_{max} was 1.45 ± 0.26 $\mu g.mL^{-1}$ (1.17 h post-administration). DAF was eliminated with mean half-lives of 7.56 ± 2.53, and 4.43 ± 1,36 h from milk and plasma respectively. The mean area under the concentration versus time curve from 0 to 24 hours (AUC_{0-24h}) was 34.84 ± 10.97 µg.h/mL in milk and 8.71 ± 1.14 µg.h/mL in plasma.

Parameter	Plasma X	SD	Milk X	SD
ß (h^{-1})	0.06	0.01	0.15	0.02
T½ ß (h)	12.53	1.47	4.57	0.46
K_{abs} (h^{-1})	1.46	0.91	0.32	0.06
$T½_{abs}$ (h)	0.64	0.39	2.27	0.48
MRT (h)	18.38	2.52	8.34	1.31
Cmax (µg.mL^{-1})	0.53	0.13	1.37	0.73
$Cmax_M/Cmax_{PL}$	---	---	2.68	1.35
$T½ß_M/T½ß_{PL}$	---	---	0.37	0.08
Tmax (h)	2.17	0.98	8.67	2.07
AUC (µg.mL/h)	9.69	1.41	15.46	5.42
WT (h)			73.48	

Table 19. DAF PK parameters (mean ± SD) obtained in plasma and after its subcutaneous administration (6 mg.kg^{-1})

When comparing PK parameters by grouping quarters by level of production, it was observed that DAF was eliminated more slowly from quarters of low-producing cows (Fig 18 B). Concomitantly, the MRT in milk from quarters of low-producing cows was higher than the MRT determined in milk from quarters of high-producing cows.

FQ are considered to have a concentration-dependent effect, although a time dependent bactericidal effect against some Gram-positive bacteria has also been described (Cester et al. 1996). DAF exhibited the same behavior against *S.aureus*, because it was strongly bactericidal at higher concentrations and when increases contact time. Importantly, the potency of DAF, evaluated by the MIC, is not affected by pH changes (Fig.19). The WT resulted longer in milk of low production than in milk of high production cows. Separating the groups according to the health status (Fig 18 A), it was observed that there were no statistically significant differences between the PK behaviour in both groups. The level of production had a significant effect ($P < 0.05$) on the milk elimination half-life ($T½\lambda$) and on the mean residence time (MRT) of DAF in milk. The infection presence in the mammary quarters had a significant effect ($P < 0,05$) on $T_{máx}$ and MRT in milk. The incidence of infection and level of production on the PK of danofloxacin in milk cows have to be considered when designing dosage regimens for this drug.

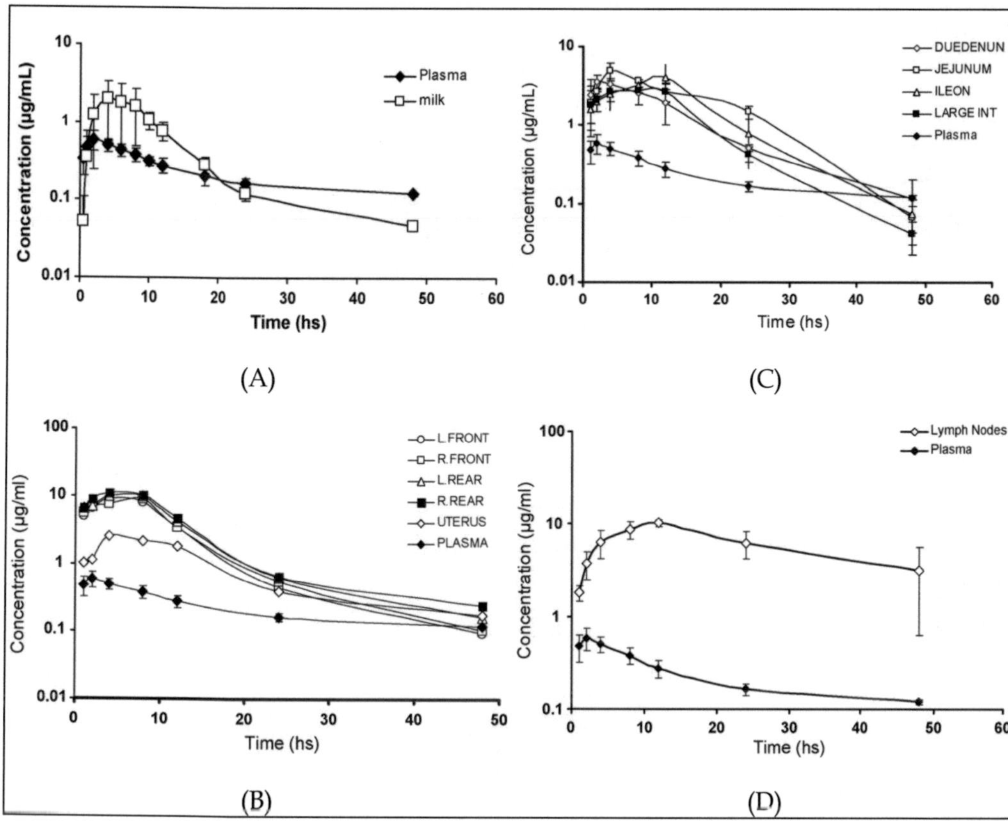

Fig. 17. Mean DAF plasma and milk concentrations (A). Mean DAF mammary, uterus and plasma concentrations (B), mean DAF duodenum, jejunum, ileum, large intestine vs. plasma concentrations (C) and mean DAF mesenteric lymph nodes vs. plasma concentrations (D), each after its SC administration in lactating cows

5. Perspectives for new therapeutic formulations

As described, *S.aureus* represents a major problem in bovine mastitis because of the poor cure rate despite the in vitro susceptibility of the acting bacteria. One possible reason for this is its intracellular location within the udder phagocytes, a place difficult to access for the majority of ATMss, wich combines with reduced activity at the acidic pH of lysosomes. Other possible obstacle for antibacterial efficacy is the non-diffusion of acidic antibiotics through the lysosomal membrane due to their ionic presentation at extracellular (7.4) or cytoplasmic (6.5) pH and the very poor retention in cells of antibiotic which penetrate freely such lincomycins. Consequently, there is a clear need for more specialized dosage forms to be developed for use in the treatment of *S. aureus* bovine mastitis. Indeed, in view of the magnitude of the economic losses, every effort to develop new dosage forms should be undertaken. Such new formulations should be designed to counteract the causes of failure of antibiotic therapy for *S.aureus* and should have the following features (Gruet et al., 2001):

- Ability to penetrate phagocytes and to be retained in cells for adequate time;
- No substrate or low metabolism in the cells;

- Effective at low pH against S. aureus;
- Administration via local infusion through the teat canal

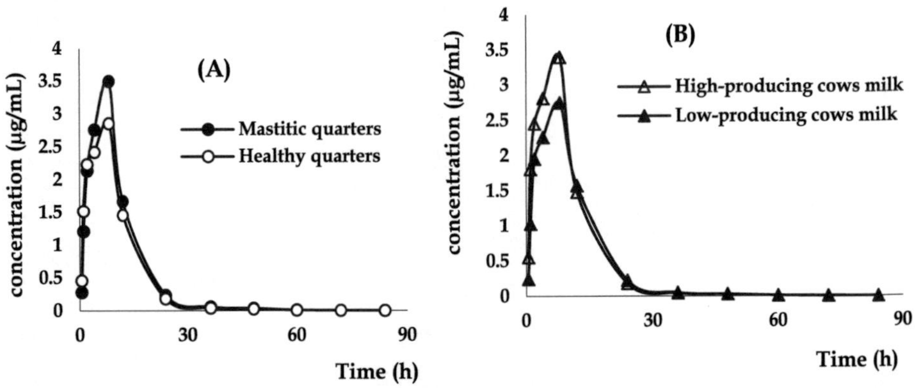

Fig. 18. **(A)** Mean concentrations of DAF in mastitic quarters and healthy quarters after 10 mg.kg-1 IM dose **(B)** Mean plasma and milk concentrations of DAFin high-producing and low-producing cows after 10 mg.kg-1 IM dose.

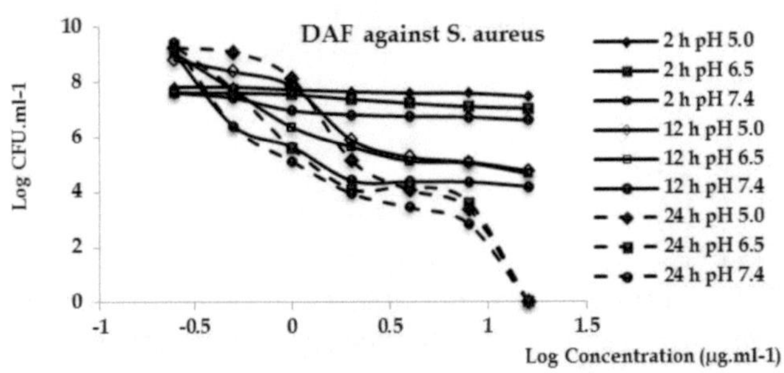

Fig. 19. Inhibition of *S. aureus* exposed to different concentrations (expressed as Log_{10}) of DAF (0.25 MIC; 0.5MIC; 1MIC; 2MIC; 4MIC and 8 x MIC) in function of the time and the medium pH (5; 6.5 and 7.4)

Delivery systems such as injectable microparticles or colloidal suspensions could fulfill some of the expectations for a new dosage form for IMM infusion. Microparticles are small, spherical particles with a diameter larger than 5µm, and can be made out of natural (gelatin or alginate) or synthetic material. Only the smallest could potentially be taken up by the phagocytes and therefore offer therapeutic value against intracellular bacteria. Colloidal suspensions include liposomes and nanoparticles. Both types are spherical particles with an average diameter of between 0.05 µm and <5µm. Liposomes are phospholipidic particles with an aqueous core. Due to their amphiphilic structure, they can incorporate either lipophilic or hydrophilic compounds. Nanoparticles are polymeric particles.

Thus liposomes, microparticles and nanoparticles may be considered potential delivery systems in the treatment of bovine mastitis by *S. aureus* since they may be taken up by the phagocytes liberating the active once inside. From the available literature, microparticles appear to be the most appropriate candidates for clinical trials in cows. Bodmeier et al. (1997) have reported the preparation of ceftiofur microparticles for the purpose of administration to cows at drying off. The microparticles exhibited a good encapsulation ratio and were shown in vitro to release the drug slowly over a long period of time. This dosage form could be considered an example of possible paths to follow looking for suitable new therapeutic systems.

6. Conclusion

Milk producing cows are a very special kind of animal. An average good level of milk production is in the order of 5% of body weight daily, but everyday more "top" producers are rounding 10%. These are values that years ago were considered impossible to reach. The huge milk production increase was due to a very efficient process of selection. But these processes of increase of milk production pushed these animals to the border between physiology and pathology. In these conditions slight modifications in the diet, environment and management can result in defensive disequilibrium and disease. It has to be noted that the defensive aspects of the mammary gland did not receive the same attention than production itself, and, as a consequence, the higher the level of production, the higher the risks of mammary pathology. The mammary gland is a rather immunologically weak organ, the phagocitic activity in the mammary gland is poor. During the first few hours after milking, neutrophils coming into the gland are active, however, in a short time, they become to "engorge" with lipid molecules and their activity slows. At the end of the intermilking interval, phagocytes are almost inactive. Dilution of defensive elements in great producers is another factor. A special consideration has to be made when *S. aureus* is present. This bacterium has the ability of penetrate cells and to reproduce very slowly there. The majority of ATMs have serious difficulties to penetrate cells, and, if they reach the site where bacteria are, they fail acting against slowly growing organisms or low pH. The combination of these factors constitutes a complex problem. The knowledge of PK and PD of the different ATMs in consideration of the physiological characteristics of the mammary gland can contribute to a more rational use of ATMs. In the present paper we presented some data on the behaviour of some ATMs in milk producing cows. The consideration of the available data in a frame of prudent use surely will increase the number of therapeutically positive results with less possibilities of emergence and dissemination of resistant bacterial strains.

Antibiotic therapy is an essential component in programs of mastitis control, but does not replace preventive hygienic measures. A rational approach to ATM treatment involves knowing the different variables that influence the outcome of therapy. Accordingly, it is of fundamental importance to determine what type of etiological agents are predominant in each farm, which are their susceptibilities and what drugs can be used to combat them, having a special consideration on the penetration and tissue distribution in the mammary compartment, penetration into cells, and ATM activity once in the target place.

The increase in the knowledge of the microbiology and pathophysiology of mastitis by *S. aureus*, together with the new tools to model PK and PD of ATMs will contribute to the design of new administration systems and therapeutic plans with improvement of clinical and microbiological efficacy without the emergence and dissemination of resistant microorganisms.

7. References

Andes D. Stamsted T. & Conklin R. (2001) In-vivo pharmacodynamics of amphotericin B against Candida albicans. *Antimicrobial agents and chemotherapy* 2001 March; 45:922-926 ISSN 1098-6596

Andes K. & Craig W.A. (2002) Pharmacodynamics of the new fluoroquinolone gatifloxacin in murine thigh and lung infection models. *Antimicrobial Agents and Chemotherapy*, Vol. 46, No. 6 (June 2002), p. 1665-1670, ISSN: 1098-6596

Andrade, M.A., de Carvalho Dias Filho, F., de Mesquita, A.J. & Tironi Rocha, P. (2000) Sensibilidade in vitro de Staphylococcus aureus isolados de amostras de leite de vacas com mastite subclínica. Ciencia Animal Brasileira, June 2000, 1, 53–57. ISSN: 1089-6891

Archimbault, P. & Boutier C. (1981). Amoxicilline chez le veau: Biodisponibilité comparée de préparations injectables. Revue Med Vet. 132, 51-56. ISSN 0035-1555, OCLC: 6274197

Baggot J.D. (1986) *Principios de farmacología clínica veterinaria.* 1 ed. España: Acribia S.A.; 1986. ISBN.: 978-84-200-0566-9

Baggot JD, Love DN, Stewart J & Raus J. (1988). Bioavailability and disposition kinetics of amoxicillin in neonatal foals. *Equine veterinary journal* (Mar 1988) 20, 125-127. ISSN: 0425-1644.

Bodmeier R., Huagang C., Davidson R.G.W., & Hardee G.E. (1997) microencapsulation of antimicrobial ceftiofur drug, *Pharmaceutical development and technology*. 4, 323-334. ISSN 1097-9867

Brouillete E., Grondin G., Lefebvre C., Talbot B.G. & Malouin F. (2004) Mouse mastitis model of infection for antimicrobial compound efficacy studies against intracellular and extracellular forms of *Staphylococcus aureus*. *Veterinary microbiology* 101(4) (2004 Aug): 253-262. ISSN: 0378-1135.

Cárceles, C.M., Font, A., Espuny, A., Fernández-Varón, E., Serrano, J.M. & Escudero, E. (2005) Pharmacokinetics of azithromycin after intravenous and intramuscular administration in goats. *Journal of veterinary pharmacology and therapeutics*, 28, 51–55. ISSN: 0140-7783

Cerca N., Martins S., Cerca F., Jefferson K., Pier G., Oliveira R., & Azeredo J. (2005) Comparative assessment of antibiotic susceptibility of coagulase-negative staphylococci in biofilms versus planktonic culture as assessed by bacterial enumeration or rapid XTT colorimetry. *Journal of Antimicrobial Chemotherapy* 56 (2) (May 2005): 331-336. ISSN 1460-2091

Cester, C., Schneider, M. and Toutain, P. 1996. Comparative kinetics of two orally administered fluoroquinolones in dog: Enrofloxacin versus Marbofloxacin. Revue de Medicine Veterinaire; 147:703-716. ISSN: 0035-1555

CLSI (2008) Clinical Laboratory Standards Institute. Performance Standards for Antimicrobial Disk and Dilution Susceptibility Tests for Bacteria Isolated From Animals; Approved Standard, 3rd edn. CLSI document M31-A3. CLSI, Wayne, USA.

Constable P.D. & Morin D.E. (2002). Use of antimicrobial susceptibility testing of bacterial pathogens isolated from the milk of dairy cows with clinical mastitis to predict response to treatment with cephapirin and oxytetracycline. *J.Am Vet med Assoc* 221 (1) (July 2002):103-108. ISSN 0003-1488

Craig W.A. & Dalhoff A. (1998). Pharmacodynamics of fluoroquinolones in experimental animals, p. 208-232. In J. Kuhlman, A. Dalhoff, and H. J. Zeiller (ed.), Handbook of experimental pharmacology, vol. 127. Quinolone antibacterials. Springer-Verlag, Berlin, Germany. ISBN: 3540625127

Craig W.A. (1993). Post-antibiotic effects in experimental infection models: relationship to in vitro phenomena and to treatment of infections in man. *Journal of Antimicrobial Chemotherapy* 31, Suppl D, 149-158. ISSN 1460-2091

Craig W.A. (2001) *Antimicrobial Pharmacodynamics in theory and Clinical Practice.* Edited by Ch.H. Nightingale, T. Murakawa and P.G. Ambrose. Marcel Dekker, Inc. Cap. 1: Pharmacodynamics of Antimicrobials: General Concepts and Applications, p1-28. ISBN 0-203-90837-6

Drusano G.L. (2004). Antimicrobial pharmacodynamics: critical interactions of "bug and drug". *Nat Rev Microbiol* (April 2004) 2(4):289-300. ISSN 1740-1534

Drusano G.L., Johnson D.E., Rosen M. & Standiford H.C. (1993) Pharmacodynamics of a fluoroquinolone antimicrobial agent in a neutropenic rat model of Pseudomonas sepsis. *Antimicrobial Agents and Chemotherapy.* 37, 1993 March: 483-490. ISSN: 1098-6596

Errecalde J.O. (1992). Tissue concentrations of a long-acting oxytetracycline formulation after intramuscular administration in cattle. *Revue scientifique et technique (International Office of Epizootics)* 1992 Sep;11 (3):909-15. ISSN:0253-1933

Errecalde JO, Mestorino ON, Mariño E (1997). Effects of the calculation methodology in the evaluation of oxytetracycline pharmacokinetic parameters after its intravenous administration to calves. *Veterinary Research Communications* 21(9): 273-281. ISSN 0165-7380

Errecalde J.O., E Turic, J Pesoa, & N Mestorino. (2003). Azithromycin Serum and Milk Concentrations After Its Intramammary Administration In Healthy And Mastitic Lactating Cows. *Journal of Veterinary Pharmacology and Therapeutics.* Volume 26, supplement 1, 135. ISSN: 0140-7783

Errecalde J.O. (2004) Uso de antimicrobianos en animales de consumo. Incidencia del desarrollo de resistencias en salud pública. Organización de las naciones unidas para la agricultura y la alimentación (FAO - Roma), Study 162, 57 pp. ISBN 92-5-305150-7.

Erskine R.J., S. Wagner; F.J DeGraves (2003) Mastitis therapy and pharmacology. *The Veterinary clinics of North America. Food animal practice.* Mar 2003; 19(1):109-38, vi. ISSN 1558-4240

Erskine R.J., Walker R.D., Bolin C.A., Bartlett P.C., and D.G. White. 2002 a. Trends in antibacterial susceptibility of mastitis pathogens during a seven-year period. J Dairy Sci. May; 85(5):p.1111-8

Erskine RJ. (2002b) Empleo de Fármacos antimicrobianos en la mastitis bovina. Terapéutica Antimicrobiana en Medicina Veterinaria. En: Prescott JF, Baggot JD, Walter RD, editores. Terapéutica antimicrobiana. 3 ed. Buenos Aires, Argentina: Intermédica; 2002b. p. 613-642. ISBN: 9505552564

FAO. Food and Nutrition Papers. 1997. Residues of some veterinary drugs in animals and foods.Abamectin, chlortetracycline and tetracycline, clenbuterol, cypermethrin, moxidectin neomycin oxytetracycline, spiramycin, thiamphenicol, tilmicosin, xylazine. (JECFA 47). ISSN 0254-4725

Finberg R.W., Moellering R.C., Tally F.P., Craig W.A., Pankey G.A., Dellinger E. P., West M.A., Joshi M., Linden P.K., Rolston K.V., Rotschafer J.C. & Rybak M.J. (2004). The importance of bactericidal drugs: future directions in infectious disease. *Clinical Infectious Diseases* 39 (9), November 2004: 1314-1320. ISSN 1537-6591.

Firsov A.A. Shevchenko A.A., Vostrov S.N. & Zinner S.H. (1998). Inter- and intraquinolone predictors of antimicrobial effect in an in vitro dynamic model: new insight into a widely used concept. *Antimicrobial Agents and Chemotherapy*, Vol. 42, No. 3, March 1998, p. 659-665, ISSN: 1098-6596

Firsov A.A. Vostrov S.N., Kononenko O.V., Zinner S.H. & Portnoy Y.A. (1999) Prediction of the effects of inoculums size on the antimicrobial action of travafloxacin and ciprofloxacin against Staphylococcus aureus and Escherichia coli in an in vitro dynamic model. *Antimicrobial Agents and Chemotherapy*, Vol. 43, No. 3, March 1999: 498-502, ISSN: 1098-6596

Forrest A, Nix DE, Ballow CH, Goss TF, Birmingham MC, Schentag JJ. 1993. Pharmacodynamics of intravenous ciprofloxacin in seriously ill patients. Antimicrobial Agents and Chemotherapy 1993 May; 37(5): 1073-1081. ISSN: 1098-6596

Foulds G., Chan, K.H.; Johnson, J.T.; Shepard, R.M.; Johnson, R.B. (1991). Concentrations of azithromycin in human tonsillar tissue. European Journal Clinical Microbiology Infections Disease. 10 (10):853-856. ISSN:1198-743X

Franklin A. (1998). Antibiotic policy and ocurrence of resistance in Sweden. Proceedings of 25th International Dairy Congress, Aarhus, Denmark, 229-234.

Friis C, Erhardsen E, Bisgaard Madsen E, Nielsen P, Raun K (1988) Spiramycin: oral bioavailability and its tissue and bronchial secretions in calves. Proc IVth Congress of Eurolean Association for Veterinary Pharmacology and Toxicology. Budapest, 63

Friton, G. HattumVan J.J.C. & Hörstermann D. (2003) Pharmacokinetics in plasma and milk of benzylpenicillin following repeated intramuscular administration of mamyzin (penethamate hydriodide) in lactating cows - Blackwell Publishing, *Journal of Veterinary Pharmacology and Therapeutics*. Proceedings of 9th Intenational Congress of the EAVPT, Lisbon, Portugal 26 (Suppl 1, august 2003) 100 - 101. ISSN: 0140-7783ftp://ftp.fao.org/docrep/fao/007/y5468s/y5468s00.pdf

Gentilini E., Denamiel G., Llorente P., Godaly S., Rebuelto M. & De Gregorio O. (2000). Antimicrobial susceptibility of Staphylococcus aureus isolated from bovine mastitis in Argentina. *J. Dairy Sci.* (June 2000) 83:1224-1227. ISSN 1525-3198

Gianneechini, R.E., Concha, C. & Franklin, A. (2002) Antimicrobial susceptibility of udder pathogens isolated froma dairy herds in the West littoral region of Uruguay. *Acta vet. Scand.* (March 2002) 43 (1): 31-41. ISSN 1751-0147

Gruet P, Maincent P, Berthelot X. & Kaltsatos V. (2001). Bovine mastitis and intramammary drug delivery: review and perspectives. *Advanced drug delivery reviews,* (Sep 2001); 50: 245-259. ISSN 1872-8294.

Hofshager M., Kruse H., Lassen J., Stavnes T-L., Essun K., Holstad G., Mork T., Schau J. & Grave K. (1999). Resistance in bacteria from infections in animals. In: *Usage of antimicrobial agents in animlas and occurrence of antimicrobial resistance in bacteria from animals, feed, and food in Norway*. NORM-VET, Ed. Hilde Kruse, ISSN-1502-4695, Oslo, Norway. 1999, pp 16-18.

Hornish RE & Kotarski SF. (2002) Cephalosporins in veterinary medicine—Ceftiofur use in food animals. *Current topics in medicinal chemistry,* 2002;2: 717–731. [PubMed: 12052187] ISSN 1873-4294

http://sedici.unlp.edu.ar/search/request.php?id_document=ARG-UNLP-TPG-0000001053

Hunter P.A., G.N. Rolison & D.A. Witting. (1973). Comparative activity of amoxycillin and ampicillin in an experimental bacterial infection in mice. *Antimicrobial agents and chemotherapy*, Sep. 1973, 4(3), 285-293. ISSN 1098-6596

Korhonen H, Kaartinen L. Changes in the composition of milk induced by mastitis. En: Sandholm M, Honkanen-Buzalski T, Kaartinen L, Pyörälä S, editors. *The bovine udder and mastitis*. 1 ed. Helsinki: Facultad de Medicina Veterinaria Universidad de Helsinki; 1995. p. 76-82. ISBN 9518340471

Kuang, Y., Jia, H., Miyanaga, K., & Tanji, Y. (2009). Effect of milk on antibacterial activity of tetracycline against Escherichia coli and Staphylococcus aureus isolated from bovine mastitis. *Applied Microbiology and Biotechnology*, *84*(1), 135-142. ISSN 1738-8872

Labro M.T. (2000). Interference of antibacterial agents with phagocyte functions: immunomodulation or "immunofairy tales". *Clinical Microbiology Reviews*, Vol. 13 (4) (October 2000): 615-650, ISSN: 1098-6618

Levison M.E. (2004). Pharmacodynamics of antimicrobial drugs. Infectious disease clinics of North America (Sep 2004) 18:451-465. ISSN 1557-9824.

Lucas M.F. (2009b) Alternativas terapéuticas para el manejo racional de la mastitis subclínica por *Staphylococcus aureus*. June 2009. Thesis Dra. Nora Mestorino and Dr Jorge Errecalde.
http://www.fcv.unlp.edu.ar/index.php?option=com_content&view=article&id=665&Itemid=1299

Lucas M.F; Moncada Cárdenas, A; Marchetti, L.; Lambertini, A.; Mestorino, N. & Errecalde, J. (2009a) Pharmacokinetics and T>MIC of amoxicillin-clavulanic acid in milk of lactating dairy cows with S. aureus subclinical mastitis. Proceeding BII-50. BIOCELL 2009, 33(1): A1-A52. ISSN 0327 - 9545.

Lucas, M., Mestorino N., Errecalde J.O. (2009c) Pharmacokinetics and pk/pd integration of azithromycin in lactating dairy cows with subclinical mastitis caused by Staphylococcus aureus. Journal of Veterinary Pharmacology and Therapeutics, 33, 132-140. ISSN: 0140-7783. http://dx.doi.org/10.1111/j.1365-2885.2009.01128.x

Lucas, Mariana F. Nora Mestorino & Jorge O. Errecalde (2007) *Macrólidos: novedades de un clásico grupo de antimicrobianos*. Analecta Veterinaria 27 (1):36-45. ISSN 0365-5148

Martinez M., Toutain P.L.& Walker R.D. (2006) The Pharmacokinetic-Pharmacodynamic (PK/PD) Relationship of Antimicrobial Agents. *Antimicrobial Therapy in Veterinary Medicine*, 4th ed. Giguère S, Prescott JF, Baggot JD, Walker RD, Dowling PM. Blackwell Publishing, Ames, Iowa, 2006. 626 pp. Chapter 5, pag 91. ISBN 0-8138-0656-9.

Mattie H, Zhang L, van Strijen E, Sekh BR, Douwes-Idema AEA. (1997). Pharmacokinetic and pharmacodynamic models of the antistaphylococcal effects of meropenem and cloxacillin in vitro and in experimental infection. *Antimicrobial Agents and Chemotherapy* (October 1997); 41(10): 2083-2088. ISSN 1098-6596

Mercer H.D., Geleta J.N. & Carter G.G. (1974). Absorption and excretion of penicillin G from the mastitis bovine udder. *Journal of the American Veterinary Medical Association*, (Mar. 1974) 164. 613-617. ISSN 1943-569X

Mestorino, N. (1993a). Thesis: Estudio Farmacocinético del Tianfenicol en Vacas Lecheras Holando Argentino. Universidad Nacional de La Plata.
http://sedici.unlp.edu.ar/search/request.php?id_document=ARG-UNLP-TPG-0000001053

Mestorino N., Alt M., Errecalde J. (1993b). The pharmacokinetics of thiamphenicol in lactating cows. *Veterinary Research Communications*, 17: 295-303. ISSN: 0165-7380

Mestorino N., Speroni J. & Errecalde JO. (1995) Thiamphenicol concentrations in serum and milk after its intramammary administration to Holando Argentino cows. Proceedings II of The 3rd International Mastitis Seminar, ISBN 965-222-636, Tel Aviv, Israel, June 1995.

Mestorino O.N, Turic E., Ive P., Errecalde J. (1997a) Concentraciones séricas y lácteas de oxitetraciclina luego de la administración de tres formulaciones por la vía intramuscular. VI Congreso Panamericano de la Leche. Buenos Aires. Argentina.

Mestorino O.N, Turic E., Ive P., Errecalde J. (1997b). Concentraciones séricas y lácteas de amoxicilina luego de la administración de una formulación por la vía intramuscular. VI Congreso Panamericano de la Leche. Buenos Aires. Argentina.

Mestorino, N.; Marchetti, M. L.; Turic, E.; Pesoa, J.; Errecalde J. (2009). Concentrations of danofloxacin 18% solution in plasma, milk and tissues after subcutaneous injection in dairy cows. Analytica Chimica Acta. Vol 637, 33-39. ISSN 0003-2670

Moncada Cárdenas A, Marchetti L, Daniele M, Errecalde J, Mestorino N. Effect of pH on the antibacterial activity of azithromycin and penicillin against *Staphylococcus aureus. BIOCELL,* 33(1), 2009

Mouton J.W., Dudley M.N., Cars O., Derendorf H. & Drusano G.L. (2005). Standardization of pharmacokinetic/pharmacodynamic (PK/PD) terminology for anti-infective drugs: an update. *Journal of antimicrobial Chemotherapy*, 55, (March 2005), 601-607. ISSN 0305-7453

Nickerson SC, Paape MJ, Harmon RJ. & Ziv G. (1986). Mammary leukocyte response to drug therapy. *Journal of dairy science*; (June 1986) 69 (6):1733-1742. ISSN 1525-3198

Nickerson SC, Paape MJ. & Dublin AM. 1985. Effect of antibiotics and vehicles on bovine mammary polymorphonuclear leukocyte morphologic features, viability, and phagocytic activity in vitro. *American journal of veterinary research*; 46 (November 1985):2259-2265. ISSN 1943-5681

Nightingale C.H., Grant EM & Quintiliani R (2000). Pharmacodynamics and pharmacokinetics of levofloxacin. *Chemothera*py 46 Suppl 1:6-14. ISSN1421-9794

Nouws, J.F.M.; Guelen P, Mevius D. & Driessens F. (1986). Age difference in pharmacokinetics of an amoxycillin trihydrate-15% formulation administered intramuscularly to ruminants. The Vet. Quarterly. 8, 339-342. ISSN 0165-2176, OCLC: 5393794

Ødegaard S, Sviland S. 2001 Comparison of intramammary antibiotic preparations for the treatment of clinical bovine mastitis caused by bacteria sensitive to penicillin. Proc. 2nd Int. Symp. Mast Milk Quality, 2001:502-503. Vancouver, Canadá

Oliver SP, Lewis TM, Lewis MJ, Dowlen HH, Maki JL. (1990). Persistence of antibiotics in bovine mammary secretions following intramammary infusion at cessation of milking. *Preventive Veterinary Medicine* 1990, 9:301-311. ISSN 1873-1716

Otero, J.L.; Mestorino, O.N.; Errecalde, J.O. (2001a). Enrofloxacina: Una Fluorquinolona de uso exclusivo en veterinaria. Parte I: Química, Mecanismo de acción, Actividad Antimicrobiana y Resistencia Bacteriana. *Analecta Veterinaria* 21 (1): 31-41. ISSN: 0365-5148

Otero, J.L.; Mestorino, O.N.; Errecalde, J.O. (2001b). Enrofloxacina: Una Fluorquinolona de uso exclusivo en veterinaria. Parte II: Farmacocinética y Toxicidad. *Analecta Veterinaria* 21 (1): 42-49. ISSN: 0365-5148

Owens W.E. & Watts J.L. (1987) Effects of Milk on Activity of Antimicrobics Against *Staphylococcus aureus* Isolated from Bovine Udders. *J Dairy Sci.* 70: 1946—1951. ISSN 1525-3198

Owens W.E., Ray C.H., Watts J.L and Yancey R.J. (1997). Comparison of success of antibiotic therapy during lactation and results of antimicrobial susceptibility test for bovine mastitis. *Journal of Dairy Science* 80 (February 1997); 313-317. ISSN: 1525-3198

Owens WE, Watts JL, Boddie RL. & Nickerson SC. (1988) Antibiotic treatment of mastitis: Comparison of intramammary and intramammary plus intramuscular therapies. *Journal of Dairy Science*. November 1988, *71*, 3143-3147. ISSN 1525-3198

Perner J, Winter P, Baumgartner W: Retrospektive Studie zum Einsatz von Synulox® in der Mastitistherapie (Retrospective study using Synulox® in mastitis therapy). *Tierärztliche Praxis*. 2002, *30*, 286-294. ISSN:0303-6286

Pol M, Ruegg PL. (2007). Relationship between antimicrobial drug usage and antimicrobial susceptibility of gram-positive mastitis pathogens. *Journal of Dairy Science.;* 90(1):262-73. ISSN 1525-3198

Renard L, Sanders P, Laurentie M, Delmas JM (1996) Pharmacokinetic-pharmacodynamic model for spiramycin in staphylococcal mastitis. *Journal of veterinary pharmacology and therapeutics* (April 1996), 19 (2): 95-103. ISSN:0140-7783

Rubin J.E., Ball K.R. & Chirino-Trejo, M. (2011). Antimicrobial susceptibility of Staphylococcus aureus and Staphylococcus pseudintermedius isolated froma various animals. *The Canadian Veterinary Journal,* (February 2011) 52:153-157. ISSN:0008-5286

Russi N.B.; Bantar C. & L.F. Calvinho (2008). Antimicrobial susceptibility of Staphylococcus aureus causing bovine mastitis in Argentine dairy herds. *Revista Argentina de Microbiología,* (June 2008) 40 (2):116-119. ISSN: 0325-7541

San Martín B., Kruze J., Morales MA., Leon B.; Esppinoza S.; Iraguen D., Puga J. & Borie C. (2002) Bacterial resistance of mastitis pathogens isolated from dairy cows in the Vth region, Metropolitan Region and Xth Region, Chile. *Archivos de Medicina Veterinaria* 34(2): 1-12. ISSN0301-732X <http://www.scielo.cl/scielo.php? doi: 10.4067/S0301-732X2002000200008.

Sandholm M., Kaartinen L. & Pyörälä S. (1990). Bovine mastitis - why does antibiotic therapy not always work? An overview. *Journal of Veterinary Pharmacology and Therapeutic.* 13 (3) (September 1990), 248-260. ISSN 0140-7783

Schifferli, D., Galeazzi, R.L., Nicolet, J. and Wanner, M. (1982) Pharmacokinetics of oxytetracylcine and therapeutic omplications in veal calves. *Journal of veterinary pharmacology and therapeutics,* December 1982, 5 :274-257. ISSN: 0140-7783

Shem-Tov, M., Rav-Hon, O., Ziv, G., Lavi, E., Glickman, A. and Saran, A., 1998. Pharmacokinetics and penetration of danofloxacin from the blood into the milk of cows. *Journal of veterinary pharmacology and therapeutics,* 21:209-213. ISSN: 0140-7783

Shojaee Aliabadi F. & P. Lees (2003) Pharmacokinetic–pharmacodynamic integration of danofloxacin in the calf. Research in Veterinary Science, Volume 74, Issue 3, June 2003, Pages 247-259. ISSN 1532-2661

Sol J, Sampimon OC, Barkema HW & Schukken YH (2000) Factors associated with cure after therapy of clinical mastitis caused by *Staphylococcus aureus. Journal of dairy science.* (February 2000), 83, 278-284. ISSN 1525-3198

Sompolinsky D., Gluskin I. & Ziv G. (1969). Pantothenate-requiring dwarf colony variants of Staphylococcus aureus as the etiological agent in bovine mastitis. *J. Hyg. (Camb.)* Sep; 67(3):511-6. ISSN: 0022-1724

Taponen S., Jantunen A., Pyoral E. & S. Pyörälä. (2003). Efficacy of targeted 5-day combined parenteral and intramammary treatment of clinical mastitis caused by penicillin-susceptible or penicillin-resistant Staphylococcus aureus. *Acta veterinaria Scandinavica*; 44(1-2):p.53-62. ISSN 1751-0147

Thornsberry, C., and J. C. Sherris. 1985. Laboratory tests in chemotherapy, general considerations. Page 959 *in Manual of clinical microbiology*. 4th ed. E. H. Lennette, ed. Am. Soc. Microbiol., Washington, DC.

Toutain P.L. (2002). Pharmacokinetic/pharmacodynamic integration in drug development and dosage-regimen optimization for veterinary medicine. *AAPS Pharma Sci* 4(4) December 2002:160-188. ISSN: 1522-1059

Turic E, N Mestorino, J Pesoa, M Colantonio, J.O. Errecalde (2003a). Azithromycin Serum and Mammary Secretion Concentrations after its Intramammary Administration in Healthy and Mastitic Cows at Drying off. *Journal of Veterinary Pharmacology and Therapeutics*. Volume 26, supplement 1, 131. ISSN:0140-7783

Turic E, N Mestorino, J Pesoa, M Colantonio, J.O. Errecalde (2003b). Azythromicin Serum and Milk Concentrations After Intramuscular Administration In Healthy And Mastitic Milk Cows. Journal of *Veterinary Pharmacology and Therapeutics*. Volume 26, supplement 1, 130. ISSN:0140-7783

Turic, E.; Mestorino, N.; Errecalde, J.O. (2006) Farmacocinética intramuscular de azitromicina en vacas lecheras, impacto en salud pública. *Sociedad Argentina de Farmacología y Terapéutica, Asociación Médica Argentina*. N°37 vol 4: 17-20. www.ama-med.org.ar

Van Bambeke F., M. Barcia-Macay, S.Lemaire & P.M. Tulkens. (2006). Cellular pharmacodynamics and pharmacokinetics of antibiotics: Current views and perspectives. *Current Opinion in Drug Discovery & Development*, (Mar 2006) 9(2):218-230. ISSN 2040-3437

Wilson, D.; Spensley, M.S.; Baggot, J.D. & Hietala, Sh.K. (1988). Pharmacokinetics and estimated bioavailability of amoxycillin in mares after intravenous, intramuscular, and oral administration. Am.J.Vet.Res. (October 1988) 49 (10) 1688-1694. ISSN: 0002-9645.

Wright D.W. Brown G.H., Peterson M.L. & Rotschafer J.C. (2000). Application of flouoroquinolone pharmacodynamics. *Journal of Antimicrobial Chemotherapy* 46 (July 20, 2000):669-683, ISSN 1460-2091

Zelenitsky S.A. Ariano R.E., Iacovides H., Sun S. & Harding G.K.M. (2003) AUC_{0-t}/MIC is a continuous index of fluoroquinolone exposure and predictive antibacterial response for Streptococcus pneumonia in an in vitro infection model. *Journal of Antimicrobial Chemotherapy* 51 (March 2003):905-911. ISSN 1460-2091

Ziv G. & Sompolinsky D. (1976) Studies on the pathogenicity of thiamineless dwarf-colony variants of Staphylococcus aureus isolated from the bovine udder. *Res. Vet. Sci.*20 (3) (May 1976), 288-294. ISSN: 0034-5288

Ziv G. (1980a) Drug selection and use in mastitis: systemic vs. local therapy. *Journal of the American Veterinary Medical Association* (May 1980); 176:1109-1115. ISSN: 0003-1488

Ziv G. (1980b) Practical pharmacokinetic aspects of mastitis therapy: Parenteral treatment. *Veterinary medicine, Small Animal Clinician* (Feb.1980); 75:277-290. ISSN: 0042-4889

Ziv G. (1980c). Practical pharmacokinetic aspects of mastitis therapy: Intramammary treatment. *Veterinary medicine, small animal clinician* (April 1980); 75:657-670. ISSN: 0042-4889

Ziv, G. & Nouws, J.F.M. (1979). Serum and milk concentrations of ampicillin and amoxycillin in ruminants. *Refuah Veterinarith* 36: 104-110. ISSN 0034-3153 0034-3153

23

Protozoan Diseases in Farm Ruminants

Sima Sahinduran
University of Mehmet Akif Ersoy,
Faculty of Veterinary Medicine,
Department of Internal Medicine,
Turkey

1. Introduction

Protozoa are ubiquitous throughout aqueous environments and the soil, and play an important role in their ecology. Protozoa occupy a range of trophic levels.They also play a vital role in controlling bacteria population and biomass. Farm animals are usually infected with several species of parasites and they are also confined to pasture or pens. Parasite eggs, larvae, and cysts are intense in soil. One of the most important aspect of animal protozoology is transmission to humans. These are called zoonosis. Some of zoonotic diseases are common and important to public health. This chapter is about of the etiology, epidemiology, pathogenesis, clinical findings, diagnosis, treatment and control of the pathogenic protozoa in farm ruminants.

2. Babesiosis

2.1 Etiology

Babesiosis is an infectious tick- borne disease of livestock that characterised by fever, anemia, haemoglobinuria and weakness. The disease also is Known by such names as bovine babesiosis, piroplasmosis, Texas fever, redwater, tick fever, and tristeza (Zaugg, 2009). The disease also is a hemoparasitic disease caused by protozoa of the genus Babesia (Phylum: Apicomplexa), which infects mainly ruminants (Melendez, 2000). Infection of a vertebrate host is initiated by inoculation of sporozoite form of parasites into the blood stream during the taking of a blood meal (Radostits et al. 2008). The list of babesia species is shown in table 1.

2.2 Epidemiology

2.2.1 Bovine babesiosis

Bovine babesiosis disease is caused by at least six Babesia species (Table 1). Bovine babesiosis associated with B. bigemina and B. bovis is the most important disease of tropical and subtropical regions between 40°N and 32°S. Both species are transmitted transovarially by Boophilus ticks, but only tick larvae transmit B. bovis, whereas nymphs and adults transmit B. bigemina. (Radostits et al. 2008). In Europe, babesiosis is caused by Babesia divergens, an intraerythrocytic parasite that can persist for >13 months in the organs of infected cattle. The distribution of B. divergens reflects its triphasic telotropic tick vector, Ixodes ricinus, which is widespread across Western Europe and North Africa (Jammes, 2009). B. major occur in Europe, North Africa and South America. B. major is transmitted by the three host tick Haemaphysalis punctata (Taylor et al., 2007).

Organism	Livestock affected	Geographic distribution
B. bigemina	Cattle	Americas, Europe, Africa, Australia, Middle East
B. bovis (B. berbera, B. argentina)	Cattle	Americas, Europe, Russia, Africa, Asia Australia
B. divergens	Cattle	Europe
B. major	Cattle	Europe, Russia, North Africa, Middle East
B. jakimovi	Cattle	Asia
B. ovata	Cattle	Japan
B. motasi	Sheep and goats	Southern Europe, Russia, southest Asia, Middle East, Africa
B. ovis	Sheep and goats	Southern Europe, Russia, Asia, Middle East

Table 1. Babesia species (Babesiosis)

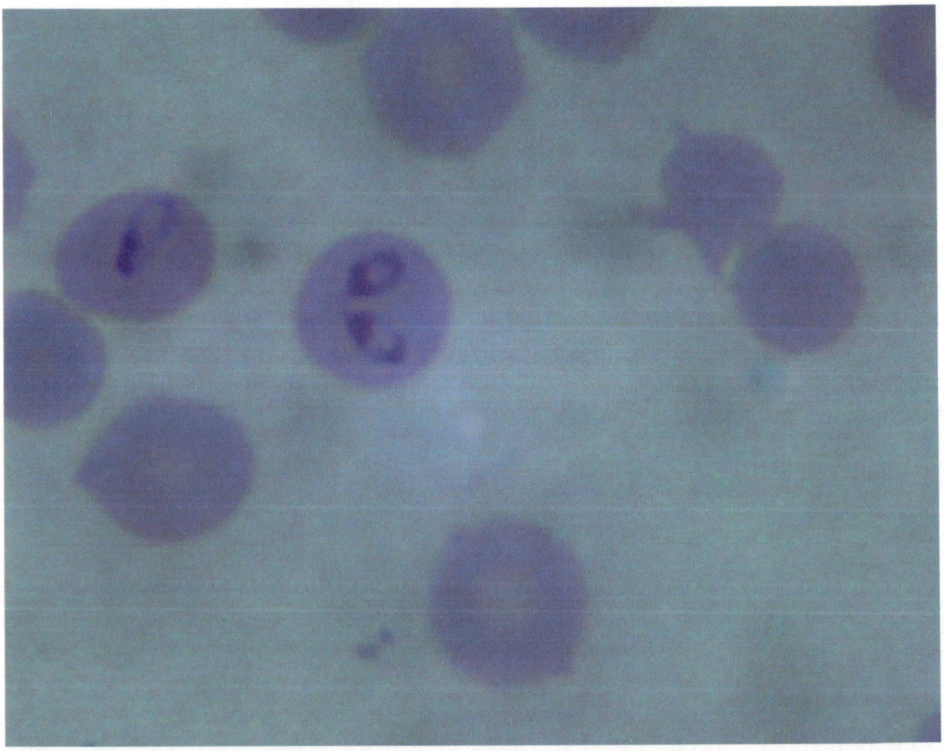

Fig. 1. Babesia bigemina in blood smear from a cow, Giemsa staining, immersion objective

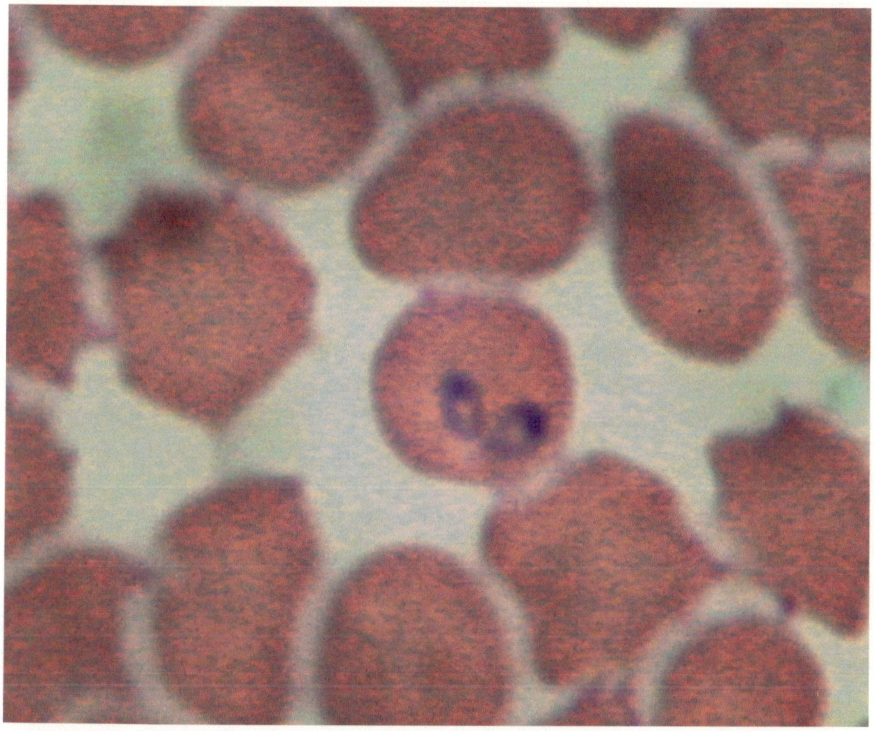

Fig. 2. B. bovis in blood smear from a cow, Giemsa staining, immersion objectiv

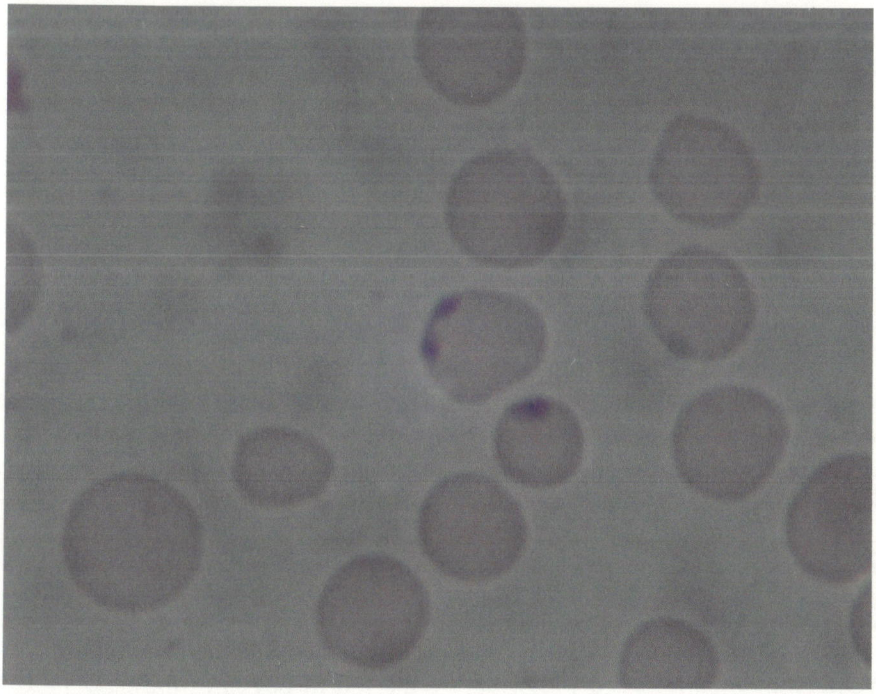

Fig. 3. B. divergens in blood smear from a cow, Giemsa staining, immersion objective

2.2.2 Sheep and goats

In sheep and goats, babesiosis is associated with B. ovis and B. motası. Babesia ovis occur in Southern Europe, former Soviet State, Middle East and Asia. Rhipicephalus bursa has been shown to be a vector for B. ovis. Babesia motasi occur in Southern Europe, the Middle East, the former Soviet State, southeast Asia and Africa. The parasite transmitted by ticks of the genus Haemaphysalis (H. Punctata, H. Otophila), Dermacentor (D. silvarum) and Rhipicephalus (R. Bursa)(Taylor et al., 2007).

2.3 Pathogenesis

Babesia spp. are a various group of tickborne, obligate, intra-erythrocytic Apicomplexan parasites infecting a wide variety of animals. Ticks are most often infected transovarially. The female tick becomes infected by the ingestion of parasites during engorgement. After it drops off the host, the babesial agents reproduce within the tick's tissues. Some of the reproducing organisms are incorporated within developing tick embryos, and the disease agents are transmitted to new hosts by the feding of ensuing tick larvae, nymphs, or adults (Zaugg, 2009).

B. bovis is the most pathogenic of the bovine Babesia. B. bigemina infections are not as virulent as those of B. bovis, however the parasites may infect 40% of the red cells (Taylor et al., 2007).

Babesia affecting small ruminants are generally less pathogenic than their bovine likes (Cebra, C., and Cebra, M., 2002a).

2.4 Clinical findings

2.4.1 Cattle

Incubation period is 2-3 weeks. B. bigemina and B. bovis produce acute syndromes which are clinically indistinguishable, and are characterized by high fever (41°C), anorexia, depression, weakness, cessation of rumination, and a fall in milk yield. Hemoglobinuria can be seen, the color of urine is dark-red to brown. Respiratory and heart rates are increased, and the red conjunctivate and mucous membranes change to the extreme pallor of severe anemia. Abortion ocur in pregnant animals (Radostits et al., 2000). Subacute syndrome also occurs in young animals, but fever is mild and hemoglobinuria is absent (Radostits et al., 2008).

In cerebral babesiosis, hyperexcitability, convulsions, opisthotonos, coma, and death, may be observed in cattle infected with either B. bigemina or B. bovis, but especially with the B. bovis. Central nervous system signs are caused by brain anoxia resulting from severe anemia (Zaugg, 2009).

2.4.2 Sheep

In sheep develop fever and parasitemia within 2 to 4 days; the clinical signs of the disease include anorexia, listlessness, anemia, moderate jaundice and hemoglobinuria. In intact animals, hyperthermia returned to normal on the fourth day after the peak pyrexia, and parasitemia is eliminated within the course of the disease (Rahbari et al., 2008).

2.5 Diagnosis

Blood smears and clinical findingd are useful in acute cases of piroplasmosis, but are not sufficient in subclinical cases. The complement fixation test is used serological test for bovine babesiosis. The most commonly used tests are ELISA, PCR and a DNA probe, which can detect specific parasitemias at very low levels of infection (Radostits, 2008). Recently, the 'reverse line blot (RLB) is a versatile technique for simultaneous detection and identification of small ruminant piroplasm species, based on the recognition of specific gene regions by oligonucleotide probes (Nagore et al., 2004; Qingli et al., 2009; Inci et al., 2010;).

2.6 Treatment and control

After the hemoglobinuria or cerebral signs, prognosis is not well. In acute cases that PVC values are above 12%, treatment will be successful. Supportive therapy such as blood transfusions (4 L of whole blood per 250 kg of body weight), fluids, hematinics, and prophylactic antibiotics are important (Zaugg, 2009). Babesiosis can be treated using diminazene aceturate (3-5 mg/ kg), phenemidine diisethionate (8-13 mg/ kg), imidocarb dipropionate (1-3 mg/kg), and amicarbalide diisethionate (5-10 mg /kg) (Cebra, C., and Cebra, M., 2002a; Taylor, 2007; Radostits, 2008; Zaugg, 2009).

The control of the disease depends on effective quarantine to prevent the introduction of the vector tick. The control of ticks by dipping or spraying animals at risk with recommended acaricides. In routine surgery, Care should be taken to prevent accidental transfer of blood from one animal to another (e.g., castration, dehorning). In addition, in cattle, the selection and breeding of cattle which acquire a high degree of resistance to ticks is practiced. Widespread use of tick vaccines may also have a significant influence on the incidence of infection in cattle (Taylor et al., 2007; Radostits et al., 2008; Zaugg, 2009).

3. Theileriosis

3.1 Etiology

Theileriosis is caused by Theileria spp. in cattle, goats, sheep and wild and captive ungulates (Radostits et al., 2000). Theileriosis is a hemoparasitic disease caused by protozoa of the genus Theileria (Apicomplexa). Theileria species affect domestic and wild ruminants, especially in Africa, Europe, Australia, and Asia (Allison and Meinkoth, 2010). The parasites are transmitted by tick. These parasites, undergo repeated merogony in the lymphocytes ultimately releasing small merozoites, which invade the red cells to become piroplasms. Theileriosis, have a variety of tick vectors which cause infections ranged from clinically inapparent to rapidly fatal (Taylor et al., 2007).

The list of theileria species is shown in table 2.

3.2 Epidemiology

3.2.1 Bovine theileriosis

Bovine theileriosis is caused by at least six Theileria species (Table 2). In bovine, East coast fever caused by T. parva. The tick vector is Rhipicephalus appendiculatus. East coast fever generally ocur in cattle but also buffalo (OIE., 2004), and occurs in East Africa. ECF is prevalent throughout the wetter areas, but is absent from the wet highlands in the horn of

Africa (Norval et al., 1991). In Africa, T. parva is a highly fatal disease in cattle (Lawrence et al., 2004; Carlson, 2009a) and this disease is a great threat on the development of the livestock industry, with losses of about 10,000 cattle per year in Zambia (Nambota et al., 1994).

Organism	Livestock affected	Geographic distribution
T. parva	Cattle, African buffalo	East and central Africa
T. annulata	Cattle, domestic buffalo	Mediterranean countries (Portugal and Spain, the Balkans), Middle East, Indian subcontinent and China
T. orientalis complex	Cattle	Southern Europe, Middle East, Asia, Australia
T. velifera	Cattle, zebu	Africa
T. taurotragi	Cattle, antelope, particularly the eland	Africa
T. mutans	Cattle, deer	Africa, Caribbean Islands
T. hirci	Sheep and goats	North and East Africa, Middle East, India
T. ovis	Sheep and goats	Europe, Africa, India

Table 2. Theileia species (Theileriosis)

In bovine, Tropical Theileriosis (Mediterranean coast fever), occurs in Mediterranean countries, Middle East, Indian and China, and is caused by T. annulata (Radostits et al., 2008). The tick vectors are Hyalomma dentritum in north Africa and in the Mediterranean countries, H. dentritum and H. dromedaii in central Asia, and H. marginatum in India (Taylor et al., 2007; Radostits et al., 2008). In endemic areas indigenous cattle are relatively resistant while modified cattle, especially European breeds, are highly susceptible (Taylor et al., 2007).

Theileria orientalis complex is a milder disease than East coast fever and Tropical Theileriosis, and called benign theileriosis in cattle (Radostits et al., 2008). The tick vectors are Ambylomma variegatum, A. cohaerens, A. haebraeum, Haemaphysalis bispinosa that these vectors are the probable vectors in Australia (Taylor et al., 2007).

Theileria velifera has been described in cattle and zebu and is very mild theileriosis in Africa (Taylor et al., 2007; Radostits et al., 2008). The tick vectors are Ambylomma variegatum, A. lepidu and A. haebraeum (Taylor et al., 2007).

Theileia taurotragi has been described in cattle, antelope and eland. Geographical distribution of this disease is in Africa. The tick vectors are Rhipicephalus appendiculatus and Rhipicephalus pulchellus (Taylor et al., 2007).

Thieleria mutans has been seen in cattle and deer (Carlson, 2009a).

Theileria hicri is malignant theileriosis of small ruminants (Lawrence, 2004;), and is enzootic in North Africa, Middle East, China and India. The tick vectors are Rhipicephalus bursa and Hyalomma anatolicum (Taylor et al., 2007).

Theileria ovis is benign theileriosis in Africa (Radostits et al., 2008). But in China Theileria ovis is a serious disease (Miling et al., 2009). The incidence and mortality rates of Theileria ovis in adult sheep and goats were have found 17.12% and 65.78%; 8.06% and 73.33% respectively in China (Shuzhen et al., 2002). In Turkey, prevalence of infection of T. ovis in sheep and goats were have been 64.19% and 12.36%, respectively (Sayın et al., 2009). The tick vectors of disease are Rhipicephalus bursa in the Mediterranean and Rhipicephalus evertsi in Africa(Taylor et al., 2007).

3.3 Pathogenesis

Pathogenesis of different form of theileriosis is based on the production of schizonts in lymphocytes and piroplasms in erythrocyte. T. parva, T. annulata and T. hicri produce highly schizonts and piroplasma and are very pathogenic; T. mutans, T. buffeli, and T. ovis unusually produce schizonts but may cause varying degree of anemia when piroplasms are highly in red blood cells (Radostits et al., 2008).

3.4 Clinical findings

Theileia spp. are classified in to 2 groups. In first group (T. parva and T. annulata), proliferate is seen in lymphocytes but in the second group (T. orientalis) it is seen in erythrocytes that causes hemolytic anemia (Magona et al., 2010).

In Theileria parva incubation period is approximately1-3 weeks (Radostits et al., 2008). One or two days later, the first clinical sign is generalised swelling of the superficial lymph nodes, eyes, ears and submandibular regions. After few days there is anorexia, decreased milk production, loses condition, ceases rumination, rapid hearth beat, petechial haemorrhages under the tongue and on the vulva (Taylor et al., 2007). In cerebral theileriosis there are localized nervous signs and convulsions, tremor, salivation and head pressing (Radostits et al., 2008).

In Theileria annulata, pyrexia, anorexia, enlargement of superficial lymph nodes, nasal and ocular discharges and salivation is the most common signs. Constipation is recorded in some cases. Respiratory distress, coughing and pulmonary oedema are seen (Osman and Gaabary, 2007).

In Theileria orientalis clinical signs are associated with anemia, jaundice and lymphadenopathy (Taylor et al., 2007).

Clinical signs in Theileria velifera is not reported but in Theileria taurotragi mild fever and anemia are reported (Taylor et al., 2007).

In Theileria hicri which is very pathogenic in sheep and goats, in the acute form there is fever, inappetence, ceases rumination, weakness, enlargement of superficial lymph nodes, diarrhoea, jaundice, haemorrhage in submucous may ocur. In chronic form there is fever, inappetence, emaciation, anemia and jaundice (Taylor et al., 2007). Theileria hicri is not reported in Turkey (Sayın et al., 2009).

Theileria ovis that is benign theileriosis in sheep and goats, clinically is mild (Taylor et al., 2007).

3.5 Diagnosis

Investigation of giemsa-stained blood smears and lymph node biopsy will reveal piroplasms in erythrocytes and schizonts in lymphocytes. For diagnosis, indirect fluorescent antibody test (IFAT) and indirect enzyme-linked immunosorbent assay (ELISA) are the most commonly used techniques. The ELISA tests is more sensitive than IFAT (Taylor et al., 2007; Radostits et al., 2008).

3.6 Treatment and control

Buparvaquone is the most effective drug and the recommended dose in cattle, sheep and goat is 2.5mg / kg BW (Taylor et al., 2007; Radostits et al., 2008).

In control of the disease use of genetically resistant breed, a judicious and selective application of acaricides at 3- week intervals and the use of vaccines are recommended (Radostits et al., 2008).

4. Sarcocystosis

4.1 Etiology

Sarcocystosis caused by Sarcocystis species in cattle, sheep and goats. The names of Sarcocystis species are according to their intermediate and final hosts (Taylor et al., 2007; Radostits et al., 2008). The list of sarcocystis species is shown in table 3.

Species	**Synonyms**	**Intermediate host**	**Definitive host**	**Geographic distribution**
S. bovicanis	S. cruzi	Cattle	Dog, wolf	Worlwide
S. bovifelis	S. hirsuta	Cattle	Cat	Worlwide
S. bovihominis	S. hominis	Cattle	Human	Worlwide
S. ovicanis	S. tenella	Sheep	Dog, fox	Worlwide
S. ovifelis	S. gigantica	Sheep	Cat	Worlwide
S. capracanis	-	Goats	Dog, wolf	Worlwide
S. hircicanis	-	Goats	Dog	Worlwide
S. hircifelis	S. moulei	Goats	cat	Worlwide

Table 3. Sarcocystis species (Sarcocystosis)

4.2 Epidemiology

Sarcocystosis occur in all of countries (Taylor et al., 2007; Radostits et al., 2008).

4.3 Pathogenesis

Sarcocystis spp, are protozoon parasites with a requisite two-host life cycle. Sexual reproduction phase occur in the intestine of a carnivore (dogs, cats) later, cysts in the muscles of cattle, sheeo or goats. Sporocysts are shedded by carnivore's feces and then ingested by cattle, sheep, or goats. Then sporocysts hatch in the ruminant small bowel and invade the vascular endothelium during three phases of asexual reproduction. After third phase, merozoites enter the ruminant's muscle tissue and encyst as sarcocysts (Smith and George, 2009; Reilly et al., 2002).

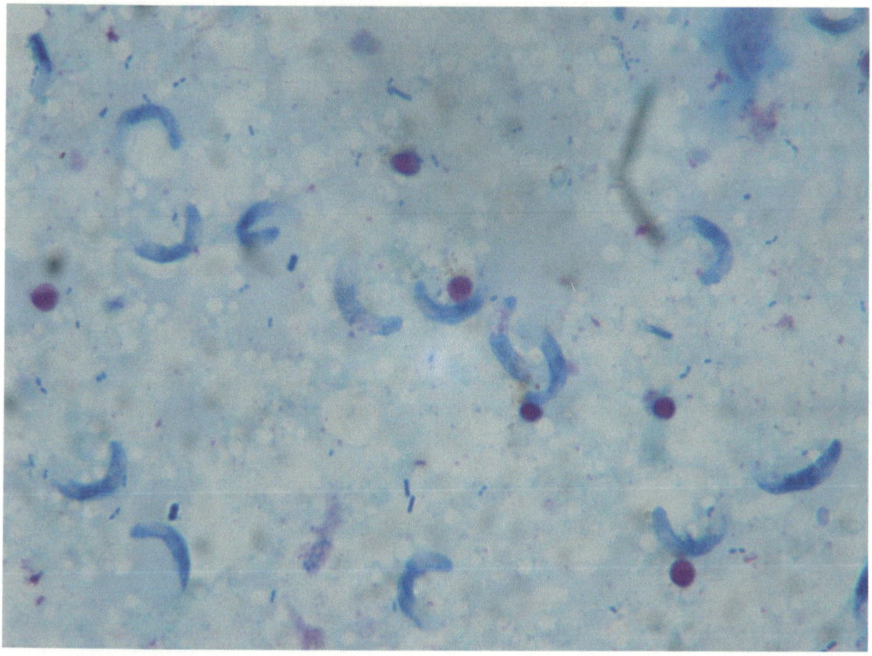

Fig. 4. Sarcocystis spp. bradizoit smear, impression smear, Giemsa staining, immersion objective

4.4 Clinical findings

In cattle is usually asymptomatic, but in heavy infections and nonimmune cattle, clinical signs include fever, anorexia, anemia, weight loss, lameness, abortion and diarrhea (Taylor et al., 2007; Radostits et al., 2008). Neurologic signs are occasionally seen in cattle that include ataxia, tremors, muscular weakness, hypersalivation, blindness, opisthotonos and nystagmus (Radostits et al., 2008; Smith and George, 2009). In infections that caused by Sarcocystis bovicanis in cattle there is usually loss of hair at the end of the tail (Taylor et al., 2007).

In sheep and goats clinical signs may be asymptomatic. In heavy infections there is fever, loss of weight, anemia and weakness. Abortion may occur (Taylor et al., 2007).

In chronic infections clinical signs include poor weight gain, edema of the limbs, anemia and abortion (Smith and George, 2009). Neurological disorder that is described in sheep caused by S. Tenella and S. Arieticanis (Radostits et al., 2008). Neurological symptoms, such as depression, in coordination, hind leg paralysis and coma can be seen in Encephalitic Sarcocystosis (Ozmen et al., 2009).

4.5 Diagnosis

Generally Sarcocystis infections are diagnosed at meat inspection with grossly visible sarcocysts in the animala's muscle. When infection is very heavy in intermediate hosts, clinical signs and histological evidence of schizonts in the blood vessels of organs and the presence of cysts in the muscles at necropsy will be used for diagnosis (Urquhart et al., 1987).

Indirect hemagglutination test (IHA) and ELISA test are used for serological diagnosis. In acute form of disease titers of antibodies are not high but 1week to 3 months later will be at diagnostic levels. For certain diagnosis immunohistochemistry, electron microscopy and PCR techniques are available (Radostits et al., 2008).

4.6 Treatment and control

There is no effective treatment for sarcocystosis described in the literature (Urquhart et al., 1987; Radostits et al., 2008; Smith and George, 2009). Treatment of infected calves and sheep with salinomycin (4 mg /kg and 1-2 mg /kg BW; respectively) has been recommended. Amprolium 100 mg /kg, for 30 days reduces the severity of infection (Radostits et al., 2008; Smith and George, 2009).

After beginning nervous system symptoms in sheep, Encephalitic Sarcocystosis recovery has not been observed (Ozmen et al., 2009).

Control of disease is based on protection the food supply of ruminants. Feed bunk should be kept clean, also farm dogs and cats that have access to the feed or pastures should not be fed uncooked meat (Urquhart et al., 1987; Smith and George, 2009).

5. Neosporosis

5.1 Etiology

Neosporosis caused by Neospora caninum in cattle, sheep and goats. Neospora caninum is a protozoan parasite of the phylum Apicomplexa in the family Sarcocystidae (Radostits et al., 2008; Matthewa, 2009).

The list of sarcocystis specie is shown in table 4.

Specie	Intermediate host	Definitive host	Geographic distribution
N. caninum	Cattle, sheep, goats	Dog	Worlwide

Table 4. Neosporosis

5.2 Epidemiology

The protozoa Neospora caninum is an important parasite that cause abortion in cattle, sheep and goats (Gumber et al., 2002). The majority of N. caninun- positive cattle prenatally infected via their dams. Transplacental transmission is considered the major route of transmission of N. caninum in cattle (Schares and Conraths, 2007).

In nonfatal infection in the fetus, the fetus is born with neurologic disorder (Smith and George, 2009).

N. caninum has a worldwide distribution, the prevalence of infection in cattle and sheep approaches 100% with a lower ((Taylor et al., 2007; Radostits et al., 2008).

5.3 Pathogenesis

Definitive host: Dogs are the final host and sexual phase occur in them but, they are also intermediate host in prenatal infections (Taylor et al., 2007; Matthews, 2009).

Intremediate host: Cattle are the major intermediate hosts and asexual reproduction phase ocur in them (Matthews, 2009). Infection can be transmitted from dam to calf in utero and lactogenically. Infection of cattle can also occur from the ingestion of food or water contaminated with dog feces containing Neospora caninum oocysts (Taylor et al., 2007; Radostits et al., 2008). Neospora caninum is a major cause of abortion in cattle, however, sporadic abortions can occur in beef cows that have been infected congenitally (Dubey et al., 2006; Taylor et al., 2007; Radostits et al., 2008; Matthewa, 2009).

Asexual phase has 2 stages:

1. Tachyzoites: Tachyzoites penetrate host cell like central nervous system, muscles, macrophages and other cells, where they divide rapidly. Tachyzoites can also be transmitted either with contaminated food and water or transplacentally to the fetus in pregnant animals. Tissue cyst containing bradyzoites that these are found only in the nervous system. After the asexual phase, sexual phase ocur in definite host. It results in production of oocysts, which is shed in the dog feces.
2. Tissue cysts (Taylor et al., 2007).

Infection in sheep and goats is infrequently (Radostits et al., 2008).

5.4 Clinical findings

In cattle, Neosporosis causes stillbirth, fetal resorption, mummification, abortion and decreases in their milk production(Gumber et al., 2002; Radostits et al., 2008). Abortions in cows are seen between 5-7 month gestations (Gumber et al., 2002). Fetus may born alive but congenitally diseased (Gumber et al., 2002; Radostits et al., 2008).

Neurological symptoms are different because of the widespread distribution of the parasite in the central nervous system. Calves are born with neurological symptoms, which these symptoms initially are mild but after birth become progress. In calves with neurologic dysfunction clinical signs are included of unable to stand, unable to suckle, domed skull and torticollis (Smith and George, 2009).

5.5 Diagnosis

The diagnosis of neosporosis is based on the examination of maternal and fetal sera ideally combined with the examination of fetal tissues.

- Immunofluorescent antibody test (IFAT) and indirect enzyme-linked immunosorbent assay (ELISA) are used for diagnosis.
- Histopathology of fetus: In histopathological examination of brain characteristic non-suppurative encephalitis is suggestive of Neospora infection and also the lesions in the heart are characteristic for diagnosis (Gumber et al., 2002; Taylor et al., 2007; Matthews, 2009).

5.6 Treatment and control

At present, there is no effective treatment for bovine neosporosis.

Control of abortion in infected cattle depends on saving food and water sources and the grazing environment from feces of any animal. Aborted fetuses and placentas should be removed or incinerated. The feces of dogs should be prevented from contaminating animal foodstuffs.

Congenitally infected cows are at high risk for abortion thus seropositive animals should be culling from a herd (Taylor et al., 2007; Radostits et al., 2008).

6. Coccidiosis

6.1 Etiology

Coccidiosis is a protozoan parasitic (genus Eimeria) disease that cause diarrhea in calves, lambs and kids (Radostits et al., 2008).

There are 12 species of Eimeria in cattle but all of them may not be pathogenic (Table 5).

The species that identified in sheep and goats is shown in table 6 and table 7.

Specie	Affinity site	Hosts
E. zuernii	Small and large intestine	Cattle, zebu, water buffalo
E. bovis	Small and large intestine	Cattle, zebu, water buffalo
E. auburnensis	Small intestin	Domestic cattle, zebu, buffalo
E. alabamensis	Small and large intestine	Domestic cattle, zebu
E. bukidnonensis	unknown	Domestic cattle, zebu, buffalo
E. cylindrica	unknown	Domestic cattle, zebu, buffalo
E. canadensis	unknown	Domestic cattle, zebu
E. ellipsoidalis	Small intestine	Domestic cattle, zebu
E. pellita	unknown	Bovine
E. subspherica	unknown	Domestic cattle, zebu, water buffalo
E. wyomingensis	unknown	Cattle, zebu, water buffalo
E. brasiliensis	unknown	Cattle, zebu, water buffalo

Table 5. Eimeris species identified in cattle.

Specie	Affinity site	Hosts
E. crandallis	Small and large intestine	Rocky Mountain big-horn sheep, domestic sheep, moufflon
E. ahsata	Small intestine	Domestic sheep, Rocky Mountain big-horn sheep, moufflon
E. faurei	Small and large intestine	Sheep, Rocky Mountain big-horn sheep
E. intricata	Small and large intestine	Domestic sheep, Rocky Mountain big-horn sheep
E. ovinoidalis	Small and large intestine	Domestic sheep, big-horn sheep, moufflon
E. pallida	unknown	Domestic sheep
E. parva	Small and large intestine	Domestic sheep, Rocky Mountain big-horn sheep, Barbary sheep
E. granulosa	unknown	Domestic sheep, Rocky Mountain big-horn sheep
E. marsica	unknown	Domestic sheep
E. bakuensis	Small intestine	Domestic sheep
E. weybridgensis	Small intestine	Domestic sheep

Table 6. Eimeris species identified in sheep.

Specie	Affinity site	Hosts
E. alijevi	Small and large intestine	Domestic goat
E. caprina	Small and large intestine	Domestic goat
E. ninakohlyakimovae	Small and large intestine	Domestic goat, Ibex, persian gazelle
E. arloingi	Small intestine	Domestic goat, Ibex, angora
E. christenseni	Small intestine	Domestic goat
E. aspheronica	unknown	Domestic goat
E. caprovina	unknown	Domestic goat
E. hirci	unknown	Domestic goat
E. jolchijevi	unknown	Domestic goat

Table 7. Eimeris species identified in goats.

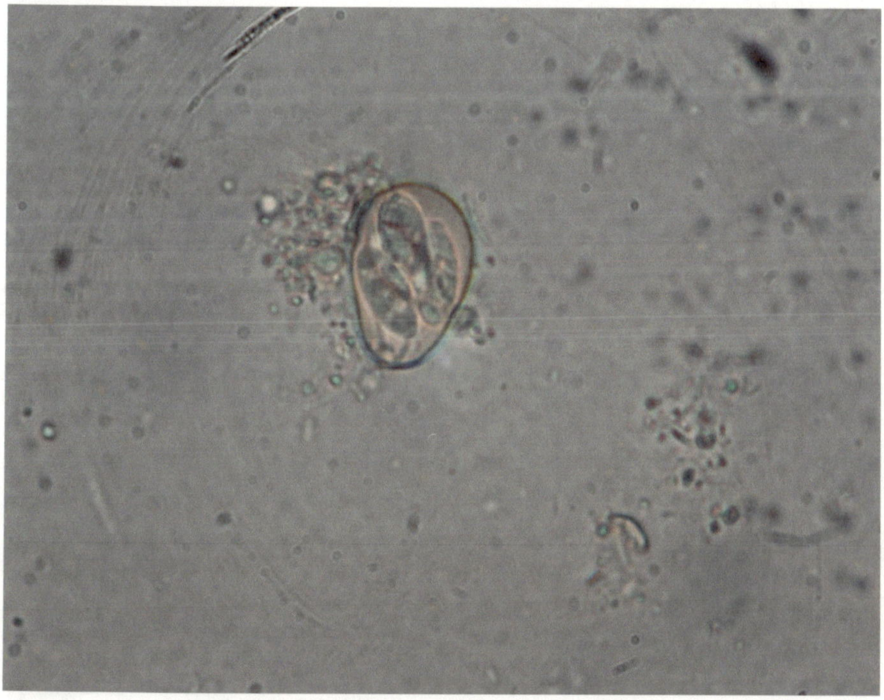

Fig. 5. Eimeria spp. sporulated oocyst, calve, 40X

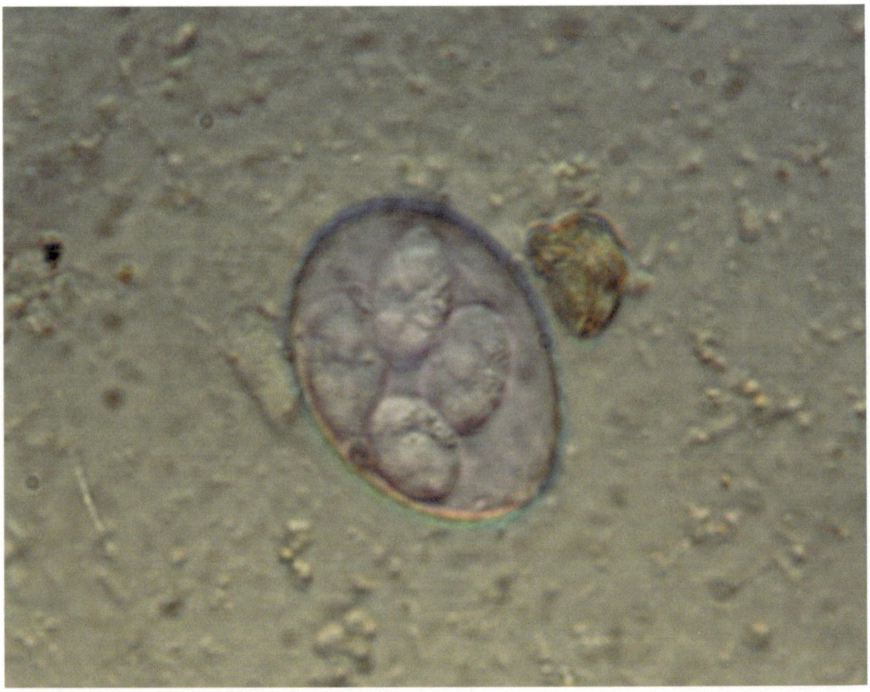

Fig. 6. Eimeria spp. sporulated oocyst, lamb, 40X

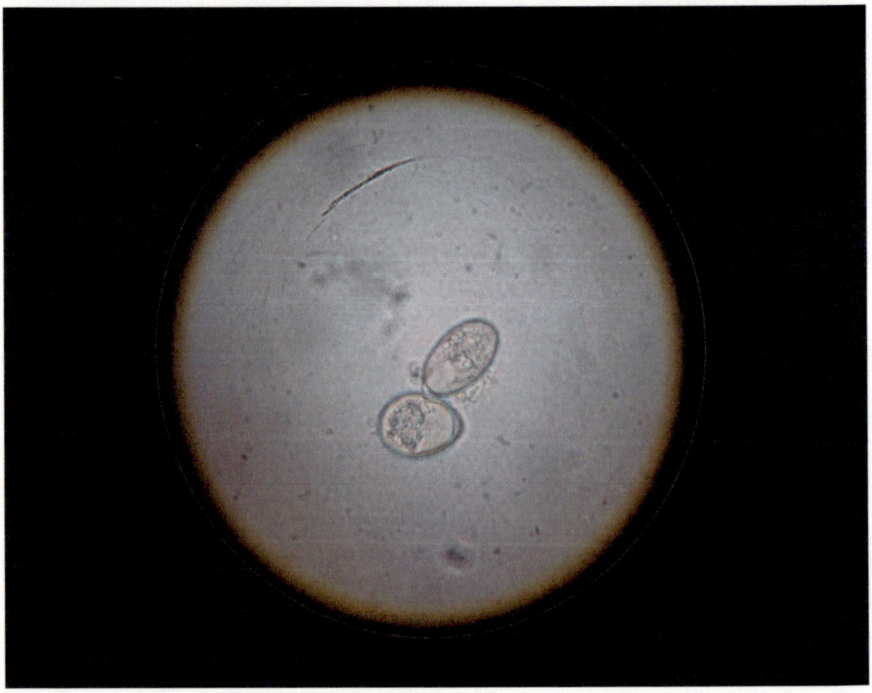

Fig. 7. Eimeria spp. sporulated oocyst, kid, 40X

6.2 Epidemiology

6.2.1 Bovine coccidiosis

Coccidiosis occurs universally in young animals normally in calves between 3 weeks and 6 months of age and caused important economic losses but also has been reported in cattle aged 1 year or more (Radostits et al., 2008; Ballweber, 2009). The disease occur in animals that kept in crowded and contaminated pens. The prevalence in calves is 46%, in yearling is 43% and in adult cows is 16% (Radostits et al., 2008). Eimeria zuernii is the most pathogenic species that causing haemorrhagic diarrhoea (Taylor et al., 2007).

In Turkey the most prevalent species are E. bovis (17.6-58.5%), E. zuernii (12.7-47.3%). E. auburnensis (4.5-64.7%) and E. canadensis (4.1-64.7%) (Ozer and Saki, 2001). The prevalence of infection in calves 60-90% have been reported in Tuekey (Arslan and Sari, 2010).

6.2.2 Sheep and goats coccidiosis

Coccidiosis in sheep and goats is caused by protozoa of the genus Eimeria. Coccidiosis is one of the most common, important and higher infection rates disease in lambs and kids that kept in small areas contaminated with oocysts (Radostits et al., 2008).

In Turkey, the identified species in sheep are E. ahsata, E. bakuensis, E. crandallis, E. granulosa, E. intricata, E. marsica, E. ovinoidalis, E. pallida, E. parva, E. punctata and E. weybridgensis and in goats are E. arloingi, E. christenseni, E. alijevi, E. hicri, E. caprina, E. caprovina, E. jolchijevi, E. apsheronica, E. crandallis, E. fauri, E. granulosa, E. ninakohlyakimovae, E. pallida, E. parva and E. punctata (Arslan and Sari, 2010).

6.3 Pathogenesis

The pathogenesis of the disease is dependent on the destroy of the crypt cells of the intestinal mucosa because in the ruminant small intestine is very long and providing a high number of host cells and is a potential for parasite replication with minimal damage. Some E. species that invade the large intestin, cause pathological changes, specially when large numbers of oocysts are ingested in a short period of time (Taylor et al., 2007).

In healthy nonimmune animals when number oocysts ingested is low, the animals show no clinical signs of disease but in many oocysts condition, rupture and exfoliation of intestinal cells triggers intestin function and causes loss of blood, fluid, albumin and electrolytes into the intestin. Disaster of mucosal capillaries of intestin can cause to hypoproteinemia and anemia. Secondary bacterial infection may cause severe enteritis(Ballweber, 2009)

6.4 Clinical findings

In most clinical cases the temperature is normal or subnormal. Disaster of epithelial cells of intestine cause sometimes bloody diarrhea. Dehydration may occur but anemia occasionally may be seen. In severe infections, diarrhea which may be mucoid or bloody, tenesmus, abdominal pain, anorexia, dehydration and weight loss are the common clinical signs in coccidiosis (Radostits et al., 2008; Ballweber, 2009).

In the calves that infected with high number of oocysts may developed dysentery. Clinical coccidiosis occurs rarely in adult cattle (Radostits et al., 2008).

Coccidiosis in lambs is widely similar to that in calves, but dysentery do not usually ocur. In infected kids clinical signs are associated with watery diarrhea, dehydration, anorexia, and weight loss (Ballweber, 2009).

6.5 Diagnosis

Diagnosis is based on history, clinical and necropsy findings, and microscopic examination of feces. Acute coccidiosis can be diagnosed by direct examination of feces but in chronic coccidiosis that very low oocysts number are seen in feces, direct examination of feces may not be adequate (Navarre and Pugh, 2002).

6.6 Treatment and control

Coccidiostats drugs are used for treatment. All animals in the flock should be treated. Trimetoprim+Sulfadoxine (16-24 mg/kg) can be used for three days by intramuscularly in lambs and kids (Ozmen et al., 2004). Amprolium (10 mg/kg BW) for 5 days or 65 mg/kg BW one dose can be effective in calves (Radostits et al., 2008). Diclazuril and toltrazuril are also coccidiocidal drugs (Matthews, 2009). Ghanem et al., (2008), suggested that toltrazuril is more effective than sulphadimine and amprolium in E. Bbovis and E. zuernii oocysts shedding in buffalo calves.

Because of the low serum vitamin C levels in lambs with coccidiosis, administration of vitamin C is suggested to be useful combined with the classical treatment (Sahinduran et al., 2006).

Hygiene in the house, minimisation of predisposing factors, avoiding the overcrowding of animals in the pens, control of the pastures for parasite are important factors for the control of coccidiosis.

7. Cryptosporidiosis

7.1 Etiology

Cryptosporidium are protozoan parasite with at least 13 species occuring in mammals, birds, and fish, but only 2 of these species are important in livestock animals (Taylor et al., 2007; Radostits et al., 2008). Cryptosporidium spp. are intracelluler protozoan parasites causing gastrointestinal disease and diarrhea, and they are important infection in young ruminants.

The list of cryptosporium specie is shown in table 8.

Specie	Affinity site	Hosts
Cryptosporidium parvum	Small intestine	Cattle, sheep, goats, deer, man
Cryptosporidium andersoni	Abomasum	Cattle

Table 8. Cryptosporidia species

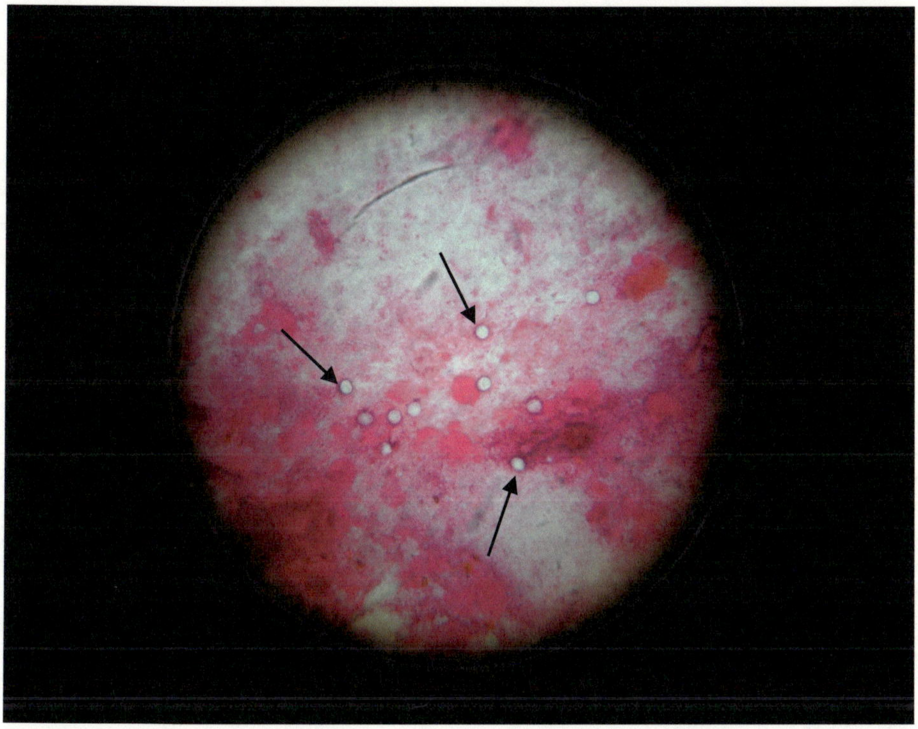

Fig. 8. Cryptosporidium spp. oocyst, calve, carbol- fuchsin staining, 40X

7.2 Epidemiology

Cryptosporidiosis occur primarily in neonatal calves, but also in lambs and kids (Radostits et al., 2008). One major species, C. Parvum, infects both farm animals and humans (Jones et al., 1997). Cryptosporidiosis is a fairly prevalent disease in many countries, and the disease is one of the most economically important diseases especially in calves and kids in Turkey (Sevinc, 2004). Calves at 1-15 days are at the highest risk (El- Khodery and Osman, 2008). The higher percentage of oocysts excreted is observed in 7-day-old calves (Feitosa et al., 2008). Animals of all ages can be infected, but diarrhea occur only in young animals (Gunn et al., 2009). In young calves there is a significant relationship between season and infection (El- Khodery and Osman, 2008). Infection peaks occur in spring and autumn (Taylor et al., 2007). Castro-Hermida et al.,(2007), suggested that asymptomatic adult sheep and goats are the source of environmental contamination by C. Parvum.

7.3 Pathogenesis

Cryptosporidium parvum infection, cause a malabsorptive diarrhea (Navarre and Pugh, 2002). Cryptosporium life cycle consist of six developmental stages. After ingestion of the oocyt there is excystation, merogony, gametogony, fertilization, oocyst wall formation, and sporogony.In contrast to other protozoal agents, cryptosporidia do not require fecal excretion for sporulation to infective stages, and they sporulate in the intestine (Navarre and Pugh, 2002; Radostits et al., 2008).

7.4 Clinical findings

Infections that caused by Cryptosporidium andersoni, are usually asymptomatic, although depressed weight gain in calves and milk yields in milking cows have been reported (Taylor et al., 2007).

In Cryptosporidium parvum infections, clinical signs are characterised by anorexia, depression, weight loss, diarrhea, dehydration, high morbidity and possibly death because of dehydration. Cryptosporidium parvum can cause diarrhea in calves 5to15, and in lambs and kids 5to10 days of age (Navarre and Pugh, 2002; Radostits et al., 2008). Relapses of diarrhea are quite common, and Cryptosporidium parvum usually occurs as a component of mixed infections (Navarre and Pugh, 2002).

7.5 Diagnosis

Oocysts can be detected using Ziehl-Nielsen stained fecal smears Cryptosporidia can be diagnosed by fecal flotation. A number of molecular and immunological assays are available for diagnosis for example immunofluorescence (IF) or enzyme-linked immunosorbent assays (ELISA) (Taylor et al., 2007).

Cryptosporidiosis is a zoonotic disease, and human can become infected from handling infected animals or feces.

7.6 Treatment and control

There is no effective drug for treatment of cryptosporidiosis in ruminants (Navarre and Pugh, 2002).

Halofuginone has efficacy in calves with diarrhea due to Cryptosporidium parvum. Halofuginone is reported to reduce oocyst shedding and the intensity of diarrhea. Oral dose of 0.1mg / kg BW daily for 7 days can reduce oocyte excretion (Klein, 2008).

Decoquinate, 2.5 mg/ kg, can be effective in prevention of cryptosporidiosis in kids and goats (Navarre and Pugh, 2002; Matthews, 2009).

Cryptosporidiosis in young animals should be treated with fluid therapy and correction of acid-base disturbance.

The control of disease is based on the minimize transmission between the source of the organism and young animals. During the course of the diarrhea, diarrheic young animals should be isolated from healthy animals (Radostits et al., 2008). The best control of cryptosporidiosis comes from calves, lambs, and kids getting adequate immunity through colostrum after birth.

8. Giardiasis

8.1 Etiology

Giardiasis is the infection caused by the flagellate intestinal protozoan, Giardia intestinalis, which is also known as G. duodenalis, or G. lamblia. Infection occur in a wide range of domestic and wild animals and also humans. G. intestinalis is a potential pathogen in livestock, causing diarrhea, weight loss, poor condition and lethargy (Hunter and Thompson, 2005). The parasite is a noninvasive and colonizes the mucosal surface of the small intestine.

The Giardia specie is shown in table 9.

Specie	Affinity site	Hosts	Geographic distribution
Giardia intestinalis	Small intestine	Cattle, sheep, goats, man	Worlwide

Table 9. Giardiasis

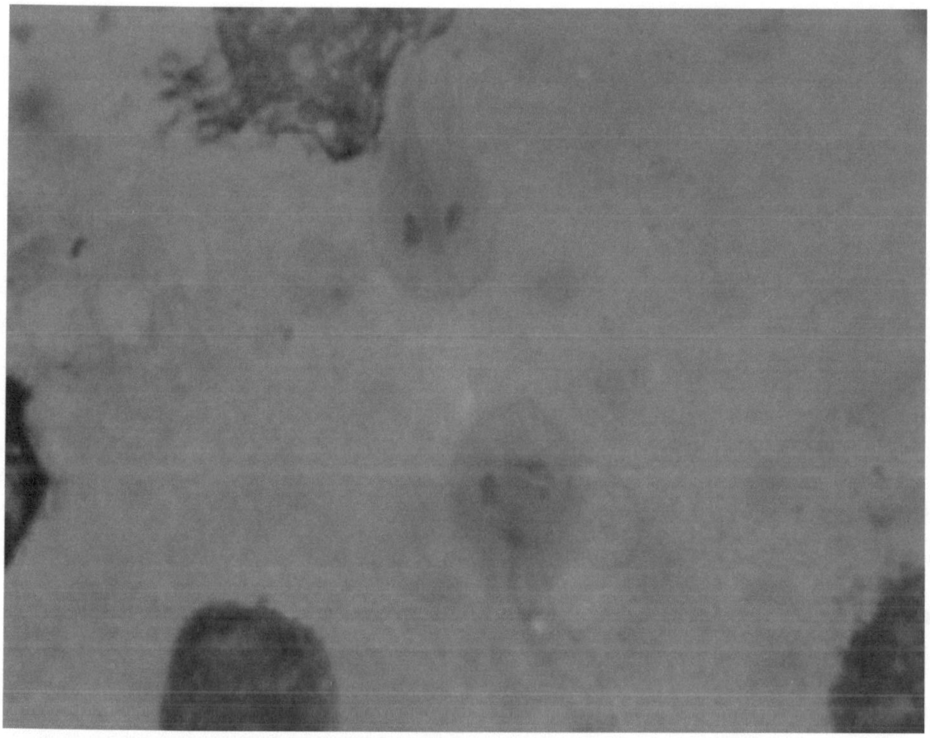

Fig. 9. Giardia spp. trophozoit, lamb, Giemsa staining, immersion objective

8.2 Epidemiology

Giardia intestinalis is an important parasite in domestic ruminants also, in human with high levels of infection. Asymptomatic adult animals may act as healthy transporters and may be a source of environmental contamination and infection for young animals. Infection is commonly exhibited by the animals within 2 weaks of birth (Xiao, 1994). Diarrhea may be transient, but infected animals can continue to shed cysts for long times.

Olson et al., (2004) reported a high levels of infection of Giardia in young livestock, specially calves. Giardia occur in both beef and dairy cattle, and studies have demonstrated prevalence rates of 100% (Xiao and Herd, 1994).

8.3 Pathogenesis

The pathogenesis of giardiasis is similar to cryptosporidiosis. The parasite cause diffuse injure to the small intestinal mucosal surface. Giardia infection cause a reduction in the height of the villus and microvillus, and loss of the absorptive surface area throughout the small intestine. If disease occur in animals, it is believed to result from nutrient malabsorption and due to diarrhea.

8.4 Clinical findings

Clinical signs of giardiasis include acute or chronic diarrhea, dehydration, weight loss, abdominal pain and reduction in weight gain. The diarrhea may be continuous or intermittent.

8.5 Diagnosis

Diagnostic methods for introduce of giardia include direct microscopy, fluorescent antibody staining, ELISA, and PCR(Gunn et al., 2009).

8.6 Treatment and control

Infected sheep and goats can be treated with fenbendazole (5-10 mg /kg) for 3 days (Navarre and Pugh, 2002). Fenbendazole (10 mg / kg BW daily for 3 days) and albendazole (20 mg / kg BW daily for 3 days), have been shown also to be effective in calves (Thompson, 2004; Radostits et al., 2008). Geurden et al., (2006), demonstrated the efficacy and safety of a 5 day treatment with paromomycin against an experimental Giardia infection in calves. Ozmen et al., (2006), suggested that five day treatment with Trimethoprim + Sulfodoxine and multivitamin complexes can be effective in the prevention and treatment of Giardia intestinalis in lambs and kids.

Re-infection frequently occurs if the sources of contamination are not eliminated and the frequency of transmission is high (Thompson, 2004).

9. Besnoitiosis

9.1 Etiology

Besnoitia are protozoan parasite with classified 7 species, which only 2 of these species are important in livestock animals. Besnoitiosis is a emergent parasitic disease that caused by Besnoitia besnoiti in cattle, and Besnoitia caprae in goats. These parasites are in the family Sarcocystidae (Radostits et al., 2008), and they are an obligate intracellular apicomplexan protozoan parasites (Majzoub et al., 2010).

The Besnoitia species are shown in table 10.

Specie	**Affinity site**	**Intermediate hosts**	**Definitive host**
Besnoitia besnoiti	Skin, conjunctiva	Cattle	Cat, wild cats (Lion, leopard)
Besnoitia caprae	Skin, conjunctiva	Goats	Cat, wild cats (Lion, leopard)

Table 10. Besnoitia species

9.2 Epidemiology

Disease occur as outbreaks in tropical and subtropical countries and sporadically in other countries (Radostits et al., 2008). Bovine besnoitiosis is common in Africa, Asia and in the South of Euroup (Jacquiet et al., 2010), Caprine besnoitiosis in Kenya, Uganda, Iran, and Kazakhastan. In livestock, cattle and goats are intermediate host, and cats are their definitive hosts (Radostits et al., 2008). It is thought that agent is transmitted mainly by hematophagous insects (Jacquiet et al., 2010).

9.3 Pathogenesis

Following infection in cattle or in goats, the tachyzoites proliferate in macrophages, fibroblasts, and endothelial cells and causing vasculitis. Subsequently they develop to form bradyzoite cysts in fibroblasts in the dermis, subcutaneous tissues, fascia, nasal and laryngeal mucosa (Taylor et al., 2007; Radostits et al., 2008). After developing cysts in the skin, painful subcutaneous swelling, thickening of the skin, alopecia, and necrosis can be occur (Taylor et al., 2007).

9.4 Clinical findings

9.4.1 Bovine besnoitiosis

Clinical signs occur in two stages: acute and chronic. In acute stage clinical signs include fever, increase in pulse and respiratory rates, painful swelling on the ventral aspects of the body, edema of the skin, sometimes diarrhea, lacrimation and nasal discharge. In chronic stage, there is a severe dermatitis over most of the body (Radostits et al., 2008).

9.4.2 Caprin besnoitiosis

Thickening of the skin over the lower limbs, around the eyes, nose, face and scrotum, alopecia in all of these areas, and white granular cysts in the sclero-conjunctiva can be seen (Oryan and Azizi, 2008).

9.5 Diagnosis

Besnoitiosis can be diagnosed by demonstration of B. bradyzoites in skin biopsy smears or scleral conjunctival scrapings (Radostits et al., 2008).

9.6 Treatment and control

There is no known treatment for this disease (Taylor et al., 2007).

Control is effected by the elimination of infected animals and carriers, and early diagnosis (Antonio Castillo et al., 2009).

10. Toxoplasmosis

10.1 Etiology

Toxoplasma gondii is a systemic coccidian, a universal parasite that causes rarely abortion in cattle but is a major abortifacient in sheep and goats (Radostits et al., 2008; Christensen et al., 2009). Felids are the definitive hosts, and there are wide range of intermediate hosts in this disease (Radostits et al., 2008).

Toxoplasma specie is shown in table 11.

Specie	Affinity site	Intermediate hosts	Definitive host
Toxoplasma gondii	Muscle, liver, lung, reproductive system, central nervous system	Cattl, sheep, goats, man, and all warm blooded animals	cats

Table 11. Toxoplasmosis

10.2 Epidemiology

Toxoplasmosis is one of the most common parasitic infections of man and other warm-blooded animals in most parts of the world (Hill et al., 2002). The disease is a major cause of ovine abortion in many countries (Christensen et al., 2009). Seroprevalence studies show high rates of infection in farm animals. Major importance of disease in farm animals is its zoonotic potential. Source of infection for sheep, and cattle is the oocyst passed in the feces of the cats. Cats shed oocysts by their feces. Cats infected by ingesting tissues of intermediate hosts (Radostits et al., 2008).

10.3 Pathogenesis

The whole developmental process of Toxoplasma gondi requires two hosts, the definitive host, cat and the intermediate host (all warm blooded animals).

There are three stages of Toxoplasma gondii that are infectious for all hosts: tachyzoites, bradyzoites, and oocysts.

- Tachyzoites: This form of the parasite present during the acute stage of infection in the intermediate host.
- Bradyzoites: present in the tissue cysts. tissue cysts can develop in lungs, liver, and kidneys, and they are more prevalent in muscular and neural tissues, including the brain, eye, skeletal, and cardiac muscle.
- Oocysts: Present only in cat feces and containing sporozoites. Cats shed oocysts after ingesting tachyzoites, bradzyoites, or oocysts. Oocysts are the infective form of importance in livestock animals (Dubey et al., 1998; Radostits et al., 2008).

10.4 Clinical findings

There are a few reports of clinical toxoplasmosis in cattle associated with fever, dyspnea, nervous signs and abortion (Taylor et al., 2007).Toxoplasma is not significant in bovine abortion (Radostits et al., 2008). In small ruminants, particularly in sheep clinical signs include fever, abortion, prenatal mortality in lambs, generalized tremor, and dyspnea (Taylor et al., 2007; Radostits et al., 2008).

10.5 Diagnosis

Diagnosis is based on the clinical signs, presence of the characteristic small white necrotic foci in cotyledons. ELISA test is also used to detect antibodies in fetal and maternal serum. Indirect fluorescent antibody test can be used for fluid collrcted from the fetal thorax or abdomen. Serological examination of dam can be used for diagnosis (Matthews, 2009).

10.6 Treatment and control

Combination of sulfamethazine (SMZ) and pyrimethamine (PMA) can be effective in pregnant ewes. The drugs are effective against the proliferating parasites in the acute stage of the disease (Radostits et al., 2008). The combined dose of (SMZ) and (PMA) is administered over 3 days for 3 periods, commencing on days 110, 115 and 130 of gestation. SMZ is injected s.c. at an initial dose of 5 ml/ 10kg body wt on the first day of a given treatment priod with subsequent doses of 2.5 ml/10kg on each of the next 2 days. PMA is administered at 2 mg/kg on the first day and at 1mg/kg on the subsequent 2 days of each treatment period (Buxton et al., 1993).

Because the cats are the only definitive host of toxoplasmosis, it should be kept away from animals farm, and also feeding and drinking water should be prevented from contamination by cat feces. Aborted fetus and excreta, must be strictly disinfected.

It is important to reduce the risk for human disease associated with consumption of infected meat (Radostits et al., 2008).

11. Tryponosomosis

11.1 Etiology

Member of the genus Trypanosoma are lived in the bloodstream and tissues of cattle, sheep, and goats in all of the world. These are flagellated protozoal agents that produce a variety of very serious disease of human and animals, but many of them are nonpathogenic (Carlson, 2009b). Trypanosomosis is one of the most important diseases of animals and human. Most African species are transmitted by the tsetse flies (Taylor et al., 2007).

There are 7 species of Trypanosoma species in cattle, sheep, and goat but all of them may not be pathogenic (Table 12).

Specie	Affinity site	Hosts
T. brucei brucei	Blood, myocardium, central nervous system, reproductive tract	Cattle, sheep, goats,camel
T. brucei evansi	Blood	Cattle, zebu, goats, camel, deer
T. congolense	Blood	Cattle, sheep, goats, camel
T. vivax	Blood	Cattle, sheep, goats, camel, antelope
T. theileri	Blood	Cattle
T. simiae	Blood	Sheep, goats, camel
T. melophagium	Blood	Sheep, goats

Table 12. Trypanosoma species identified in cattle, sheep, and goats.

11.2 Epidemiology

11.2.1 African trypanosomosis (Nagana)

African trypanosomoses caused by T. brucei brucei, T. congolense, T. vivax, and T. simiae, these species affect all domestic mammals and transmitted by tsetse flies.

The most pathogens species in cattle are T. congolense and T. vivax. T. congolense and T. vivax caused severe disease in cattle, sheep, and goat but T. brucei brucei causes a subclinical infection in cattle, and severe disease in sheep, and goat (Radostits et al., 2008). Trypanosoma evansi causes surra disease of cattle in India.

In Ethiopia the prevalence of infection in bovine which is caused by by T. species are reported as below:

T. vivax 90.5%; T. congolence 4%; and mixed infection of T. vivax and T. congolence 5.5% (Mihret and Mamo, 2007).

T. theileri is nonpathpgeic and reported in Nourth Americabut occasionally occur parasitemia that causes fever, depression, and decreased milk production (Carlson, 2009b).

11.3 Pathogenesis

In African Trypanosomosis, after entering through the skin, parasite reach to the bloodstream via the lymphatic system. Infection characterized with parasitemia. Some trypanasoma spp. invade extravascular spaces such as the ocular aqueous humor and cerebral spinal fluid. Some trypanasoma spp. may produce hemolysin that causes anemia in the host. Then, phagocytic activity increased because of the massive erythrocyte failure. (Cebra, C., and Cebra, M., 2002b).

11.4 Clinical findings

Clinical findings are based on the speed of onset of anemia and the grade of organ impairment. Trypanasomosis can be acute, subacute, or chronic. In acute form abortion, drop in milk, depression, anorexia can be seen. Hyperemic mucous membranes and lacrimation also can be occur. In subacute form clinical signs include weight loss, enlargement of lymph nodes and dry hair coat. In chronic form dull, dry hair coat, inelastic skin, lethargy, pale mucous membranes and execise intolerance may be seen (Cebra, C., and Cebra, M., 2002b) .

T. theileri rarely causes clinical signs, but sometimes parasitemia may develope (Carlson, 2009b).

11.5 Diagnosis

Diagnosis can be based on the clinical findings, presence of vectors, appearance of trypanasomes on a fresh blood smear, or a Giemsa-stained blood smear. Indirect fluorescent antibody test (IFA) and the enzyme-linked immunospecific assay (ELISA) test are used for diagnosis (Cebra, C., and Cebra, M., 2002b) .

11.6 Treatment and control

The most common drugs that is used for treatment of trypanosomosis are shown below.

- Diminazene aceturate 3.5-7 mg/ kg BW.
- Homidium bromide and chloride 1mg / kg BW -Pyrithidium bromide 2mg / kg BW.
- Isometamidium 0.25-1mg / kg BW (Radostits et al., 2008).

Vector control can help to control or prevent trypanosomosis. Insecticides can be used for prevent bites by tsetse flies and other flies (Cebra, C., and Cebra, M., 2002b) .

12. References

Allison, R.W., and Meinkoth, J.H., 2010. Anemia Caused by Rickettsia, Mycoplasma, and Protoza. In: Weiss, D.J., and Wardrop, K.J. (Eds): Schalm's Veterinary Hematology.6th Ed. Wiley- Blackwell.

Antonio Castillo, J., Miguel Marcen, J., Ortega Mora, L. M., and Alvarez Garcia, G., 2009. Bovine besnoitiosis, presented as an emerging disease in Europe. Albeitar (127) Zaragoza: ASIS Veterinaria, 24-25.

Arslan, M.O., and Sari, B., 2010. Eimeridae (Memelilerde Coccidiosis). In: Dumanli, N., and Karaer, Z. (Eds): Veteriner Protozooloji. Medisan Ltd, Turkiye.

Ballweber, L.R., 2009. Coccidiosis in food animals. In: Smith, B.P. (Eds): Large Animal Internal Medicine. Mosby, Elsevier, St. Louis, pp: 1645-1647.

Buxton, D., Thomson, K.M., and Maley, S., 1993. Treatment of ovine toxoplasmosis with a combination of sulphamezathine and pyrimethamine. Veterinary Record. 132 (16), 409-411.

Carlson, G.P., 2009a. Theileriasis. In: Smith, B.P. (Eds): Large Animal Internal Medicine. Mosby, Elsevier, St. Louis, p 1160.

Carlson, G.P., 2009b. Trypanosomiasis. In: Smith, B.P. (Eds): Large Animal Internal Medicine. Mosby, Elsevier, St. Louis, p 1160.

Castro-Hermida, J.A., Gonzalez-Warleta, M., and Mezo, M., 2007. Natural infection by Cryptosporidium parvum and Giardia duodenalis in sheep and goats in Galicia (NW Spain). Small Ruminant Research. 72, 96-100.

Cebra, C., and Cebra, M., 2002a. Diseases of the Hematologic, Immunologic, and Lymphatic Systems (Multisystem Diseases). In: Pugh, D.G. (Eds): Sheep and Goat Medicine. Saunders, An Imprint of Elsevier. Philadelphia, Pennsylvania.

Cebra, C., and Cebra, M., 2002b. Diseases of the Cardiovascular System. In: Pugh, D.G. (Eds): Sheep and Goat Medicine. Saunders, An Imprint of Elsevier. Philadelphia, Pennsylvania.

Christensen, B.W., Drost, M., Troedsson, M.H.T., 2009. Female reproductive disorders. In: Smith, B.P. (Eds): Large Animal Internal Medicine. Mosby, Elsevier, St. Louis, pp: 1419-1469.

Dubey, J.P., Buxton, D., and Wouda, W., 2006. Pathogenesis of bovine neosporosis. Journal of Comparative Pathology. 134 (4), 267-289.

Dubey, J.P., Lindsay, D.S., and Speer, C.A., 1998. Structure of Toxoplasma gondii tachyzoites, bradyzoites and sporozoites, and biology and development of tissue cysts. Clinical Microbiology Reviews 11, 267–299.

El- Khodery, S.A., and Osman, S.A., 2008. Cryptosporidiosis in buffalo calves (Bubalus bubalis): prevalence and potential risk factors. Tropical Animals Health and Production. 40(6), 419-426.

Feitosa, F.L.F., Shimamura, G.M., Roberto, T., Mendes, L.C.N., Peiro, J.R., Feres, F.C., Bovino, F., Perri, S.H.V., and Meireles, M.V., 2008. Importance of Cryptosporidium spp. As a cause of diarrhea in calves. Pesquisa Veterinaria Brasileira. 28(10), 452-456.

Geurden, T., Claerebout, E., Dursin, L., Deflandre, A., Bernay, F., Kaltsatos, V., and Vercruysse, J., 2006. The efficacy of an oral treatment with paromomycin against an experimental infection with Giardia in calves. Veterinary Parasitology. 135, 241-247.

Ghanem, M.M., Radwaan, M.E., Moustafa, A.M.M., and Ebeid, M.H., 2008. Comparative therapeutic effect of toltrazuril, sulphadimine and amprolium on Eimeria bovis and Eimeria zuernii given at different times following infection in buffalo calves (Bubalus bubalis). Preventive Veterinary Medicine. 84, 161-170.

Gumber, S., Aradhana, Sharma, D.R., and Singh, J., 2002. Neosporosis- an emerging cause of bovine abortion- a review. Veterinary Practitioner. 3(2), 94-100.

Gunn, A.A., Naylor, J.A., and House, J.K., 2009. Diarrhea. In: Smith, B.P. (Eds): Large Animal Internal Medicine. Mosby, Elsevier, St. Louis, pp: 340- 363.

Hill, D.E., Sreekumar, C., Dubey, J.P., 2002. Toxoplasma gondii. In: Friend, M., Brand, C.J. (Eds.), Field Manual of Wildlife Zoonosis. USGS National Wildlife Health Center, Madison, WI, pp. 357–382.

Hunter, P.R., and Thompson, R.C.A., 2005. The zoonotic transmission of Giardia and Cryptosporidium. International Journal of Parasitology. 35, 1181-1190.

Inci, A., Ica, A., Yildirim, A., and Duzlu, O., 2010. Identification of Babesia and Theileria species in small ruminants in Central Anatolia (Turkey) via reverse line blotting. Turkish Journal of Veterinary Science. 34(2): 205-210.

Jacquiet, P., Lienard, E., and Franc, M., 2010. Bovine besnoitiosis: epidemiological and clinical aspects. In: Franc, M., Kramer, L., and Jacobs, D. (Eds): Veterinary Parasitology. 174 (1/2) Oxford: Elsevier Ltd, 30-36.

Jammes, C., 2009. Bovine anaplasmosis caused by Anaplasma marginale and bovine babesiosis. The situation in metropolitan France and on the island of Reunion. Bulletin de la Societe Veterinaire Pratique France. 93 (2), 34-41.

Jones, T.C., Hunt, R.D., and King, N.W., 1997. Diseases due to protozoa. In: Veterinary Pathology, Ed. Williams & Wilkins Company, Pennsylvania, 575-581.

Klein, P., 2008. Preventive and therapeutic efficacy of halofuginone-lactate against Cryptosporidium parvum in spontaneously infected calves: acentralised, randomized, double-blind, placebo-controlled study. Veterinary Journal 177 (3), 429-431.

Lawrence, J.A., 2004. Theileriosis of sheep and goats. In: Coetzer, J.A.W., and Tustin, R.C. (Eds): Infectious diseases of livestock, Volum One 2th Edn, Oxford: Oxford University Press, pp: 498-499.

Lawrence, J.A.W., Perry, B.D., and Williamson, S.M., 2004. Zimbabwe theileriosis. In. Coetzer, J.A.W., and Tustin, R.C. (Eds): Infectious diseases of livestock, Volum One 2th Edn, Oxford: Oxford University Press, pp: 472-474.

Lingli, N., Jianxum, L., Guiquan, G., Miling, M., Zhijie, L., Aihong, L., Zhisheng, D., Jinliang, G., Qiaoyun, R., Youquan, L., Junlong, L., and Hong, Y., 2009. Detection and differentiation of ovine Theileria and Babesia by reverse line blotting in China. Parasitology Research. 104(6), 1417-1423.

Magona, J.W., Walubengo, J., Olaho-Mukani, W., Jonsson, N.N., Welburn, S.C., and Eisler, M.C., 2010. Clinical features associated with seroconversion to Anaplasma marginale, Babesia bigemina and Theileria parva infections in African cattle under natural tick challenge. Veterinary Parasitology. 171(3-4): 207- 215.

Majzoub, M., Breuer, W., Gollnick, N.S., Rostaher, A., Schares, G., and Hermanns, W., 2010. An outbreak of bovine besnoitiosis in Germany; pathomorphological, ultrastructural and molecular-biological investigations. Wiener Tierarztliche Monatsschrift. 97 (1/2) Wien: BWK Public Relations- Brigitte Weber- Kraus, 9-15.

Matthewa, J., 2009. Abortion. 3th Edn. In: Diseases of the goat.Blackwell Publishing Ltd. Pp: 23- 41.

Melendez, R.D., 2000. Babesiosis: An emerging zoonosis in temperate and tropical zones. a review. Revista Cientifica- Facultad De Ciencias Veterinarias. 10 (1), 13-18.

Mihret, A., Mamo, G., 2007. Bovine trypanosomosis in three districts of East Gojjam zone bordering the Blue Nile River in Ethiopia. Journal of Infection in Developing Countried 1 (3) Sassari: Open Learning on Enteric Pathogens (OLOEP), 321-325.

Miling, M., Jianxum, L., Guiquan, G., Zhijie, L., Youquan, L., Jinliang, G., Aihong, L., Junlong, L., Qiaoyun, R., and Hong, Y., 2009. A serological survey of ovine Theileriosis by ELISA. Chinese Journal of Veterinary Sciences. 29 (12), 1575- 1577.

Nagore, D., Garcia-Sanmartin, J., Garcia-Perez, A.L., Juste, R.A., and Hurtado, A., 2004. Identification, genetic diversity and prevalence of Theileria and Babesia species in a sheep population from Northern Spain. International Journal of Parasitology. 34:1059-1067.

Nambota, A., Samui, K., Sugimoto, C., Kakuta, T., and Onuma, M., 1994. Theileriosis in Zambia: etiology, epidemiology and control measures. Japanese Journal of Veterinary Research. 42 (1), 1-18.

Navarre, C.B., and Pugh, D.G., 2002. Diseases of the Gastrointestinal System. In: Pugh, D.G. (Eds): Sheep and Goat Medicine. Saunders, An Imprint of Elsevier. Philadelphia, Pennsylvania.

Norval, R.A.I., Perry, B.D., Gebreab, F., and Lessard, P., 1991. East coast fever: a problem of the future fort he horn of Africa? Preventive Veterinary Medicine. 10,3, 163-172.

OIE., 2004. Manual of diagnostic tests and vaccines for terrestrial animals, 5th Edn, chapter 2.3.11.

Olson, M.E., Ryan O'Handley, R., Ralston, B., and Thompson, R.C.A., 2004. Emerging issues of Cryptosporidium and Giardia infections in cattle. Trends in Parasitology. 20, 185–191.

Oryan, A., and Azizi, S., 2008. Ultrastructure and pathology of Besnoitia caprae in the naturally infected goats of Kerman, East of Iran. Parasitology Research. 102 (6), 1171-1176.

Osman, S.A., Al-Gaabary, H., 2007. Clinical, haematological and therapeutic study on tropical theileriosis in water buffaloes (Bubalus bubalis) in Egypt. Veterinary Parasitology. 146, 337-340.

Ozer, E., and Saki, C.E., 2001. Turkiye'de Coccidiosis. In: Dincer, S. (Eds): Coccidiosis. Turkiye Parazitoloji Dernegi, Yayin no: 17. Izmir, pp:151-162.

Ozmen, O., Sahinduran, S., and Yukari, B.A., 2004. Comparative evaluation on clinicopathological findings and treatment in acute and chronic Coccidiosis of lambs and kids. Indian Veterinary Journal. 81, 1206-1209.

Ozmen, O., Sahinduran, S., Haligur, M., Yukari, B.A., and Dorrestein, G.M., 2009. Encephalitic Sarcocystosis and its Prophylactic Treatment in Sheep. Turkish Journal of Veterinary & Animal Sciences. 33(2): 151-155.

Ozmen, O., Yukari, B.A., Haligur, M., and Sahinduran, S., 2006. Observations and immunohistochemical detection of Coronavirus, Cryptosporidium parvum and Giardia intestinalis in neonatal diarrhea in lambs and kids. Schweizer Archiv Fur Tierheilkunde. 148, 357- 364.

Radostits, O.M., Gay, C.C., Blood, D.C., and Hinchcliff, K.W., 2000. Veterinary Medicine. A Textbook of the Diseases of Cattle, Sheep,Pigs, Goats and Horses, 9th Ed. W.B. Saunders, Philadelphia.

Radostits, O.M., Gay, C.C., Hinchcliff, K.W., and Constable, P.D., 2008. Diseases associated with protozoa. 10th Edn. In: Veterinary Medicine: A Textbook of Diseases of cattle, horses, sheep, pigs, and goats. Saunders Elsevier; pp: 1483-1540.

Rahbari, S., Nabian, S., Khaki, Z., Alidadi, N., and Ashrafihelan, J., 2008. Clinical, haematologic and pathologic aspects of experimental ovine babesiosis in Iran. Iranian Journal of Veterinary Research. 9(1), 59-64.

Reilly, L.K., Baird, A.N., and Pugh, D.G., 2002. Diseases of the Musculoskeletal System. In: Pugh, D.G. (Eds): Sheep and Goat Medicine. Saunders, An Imprint of Elsevier. Philadelphia, Pennsylvania.

Sahinduran, S., Sezer, K., Buyukoglu, T., Yukari, B.A., and Albay, M.K., 2006. Plasma ascorbic acid levels in lambs with coccidiosis. Turkish Journal of Veterinary & Animal Sciences. 30, 219-221.

Sayın, F., Nalbantoglu, S., Yukari, B.A., Cakmak, A., and Karaer, Z., 2009. Epidemiological studies on sheep and goat Theileria infection. Ankara Universitesi Veteriner Fakultesi Dergisi. 56 (2), 127-129.

Schares, G., and Conraths, F.J., 2007. Neospora caninum as a cause of bovine abortion- news regarding pathogenesis, epidemiology and diagnosis. Praktische Tierarzt. 88 (9) Hannover: Schlütersche Verlagsgesellschaft mbH & Co. KG, 730-740.

Sevinc, F., 2004. Cryptosporidiosis in ruminants. Veteriner Bilimleri Dergisi, Konya,Turkey. 20 (4) , 79-84.

Shuzhen, G., Zhengphu, Y., Guoxue, W., Wenbiao, W., Denglu, M., and Hongde, D., 2002. Epidemiology of ovine theileriosis in Ganan region, Gansu Province, China. Parasitology Research. 88 (Suppl. 1). Berlin: Springer- Verlag, 536-537.

Smith, M.O., George, L.W., 2009. Diseases of the nervous system. In: Smith, B.P. (Eds): Large Animal Internal Medicine. Mosby, Elsevier, St. Louis.

Taylor, M.A., Coop, R.L., and Wall, R.L., 2007. Veterinary Parasitology. Third Edn. Blackwell Publishing.

Thompson, R.C.A., 2004. The zoonotic significance and molecular epidemiology of Giardia and giardiasis. Veterinary Parasitology. 126, 15- 35.

Urquhart, G.M., Armour, J., Duncan, J.L., Dunn, A.M., and Jennings, F.W., 1987. Veterinary Parasitology. Longman Group UK Limited, Essex CM20 2JE, England.

Xiao, L., 1994. Giardia infection in farm animals. Parasitology Today. 10, 436-438.

Xiao, L., and Herd, R.P., 1994. Infection pattern of Cryptosporidium and Giardia in calves. Veterinary Parasitology. 55, 257-262.

Zaugg, J.L., 2009. Babesiosis. In: Smith, B.P. (Eds): Large Animal Internal Medicine. Mosby, Elsevier, St. Louis, p 1157.

Part 4

Research in Veterinary Medicine

24

Vitronectin and Its Receptor (Integrin $\alpha_v\beta_3$) During Bovine Fertilization *In Vitro*

Mirjan Thys[1], Hans Nauwynck[2], Leen Vandaele[1], Jo Bijttebier[1], Dominiek Maes[1], Herman Favoreel[2] and Ann Van Soom[1]

[1]Dept. of Reproduction, Obstetrics and Herd Health,
[2]Dept. of Virology, Immunology and Parasitology,
Faculty of Veterinary Medicine,Ghent University,
Belgium

1. Introduction

The complex series of molecular interactions between male and female gametes required for successful fertilization has captured the interest of several research groups for many years (Benoff 1997). Even though substantial insights into these interactions have been elucidated in particular in human and mouse (Fusi *et al.* 1992; 1996*a*; 1996*b*; Almeida *et al.* 1995; Evans *et al.* 1995; Bronson & Fusi 1996), a clear identification of the underlying mechanisms and molecules implicated in bovine fertilization is still required. The involvement of numerous carbohydrates and glycoproteins in adhesion and binding events during several reproductive processes has been described in ruminants, ranging from roles in sperm-oviduct adhesion (Revah *et al.* 2000; Talevi & Gualtieri 2001; Sostaric *et al.* 2005; Gwathmey *et al.* 2006; Ignotz *et al.* 2007), sperm-oocyte interactions (Gougoulidis *et al.* 1999; Amari *et al.* 2001; Tanghe *et al.* 2004*a*; 2004*b*) to embryo implantation (Spencer *et al.* 2004).

A convenient way to study receptor-ligand interactions is to incubate sperm and/or oocytes with possible ligands in order to inhibit fertilization. Using this approach, Tanghe *et al.* (2004*b*) demonstrated that vitronectin - among other glycoproteins and carbohydrates - when present during bovine *in vitro* fertilization (IVF) inhibited sperm penetration.

Vitronectin (Vn) is a multifunctional 75 kDa glycoprotein - rather exclusively secreted by the liver into the plasma in a monomeric form - and abundantly stored in an essentially multimeric form in diverse extracellular matrices (Stockmann *et al.*1993; Gechtman *et al.* 1997; Francois *et al.* 1999). Like other adhesive proteins (e.g. fibronectin), Vn possesses a heparin binding site and interacts via its Arg-Gly-Asp (RGD) amino acid sequence with integrin receptors (mainly the $\alpha_v\beta_3$ integrin) at the cell surface (Bronson *et al.* 2000). Interaction of this glycoprotein with a wide range of macromolecules has been described allowing it to participate in several physiological processes, among which complement-mediated cell lysis, cell surface proteolysis, cell adhesion, coagulation and fibrinolysis (Gibson *et al.* 1999; Bronson *et al.* 2000).

RGD sequences, present in Vn as well as in other extracellular matrix proteins (Fusi *et al.* 1992), are believed to take part in various integrin-mediated recognition systems involved in cell-to-cell and cell-to-matrix adhesion (Ruoslahti & Pierschbacher 1986). Since - in human - integrins have been detected on both male and female gametes and spermatozoa express Vn on their surface following capacitation (Bronson & Fusi 1996), Vn may be involved in sperm-egg interaction. The present study was conducted to determine whether the inhibitory effect of exogenously supplemented Vn on bovine IVF appeared during a) the sperm penetration of the cumulus oophorus, b) the sperm-zona binding, c) the sperm-oolemma binding or d) the sperm-oocyte fusion. Subsequently, the expression of endogenous Vn and integrin subunit α_v (subunit of the Vn receptor) on bovine oocytes and sperm cells was evaluated using indirect immunofluorescence, and the effect of exogenous Vn on sperm membrane integrity and sperm motility was assessed.

2. Materials and methods

2.1 Oocyte and semen preparation

Oocytes were derived from bovine ovaries randomly collected at a local abattoir and prepared following the protocol of Tanghe *et al.* (2004*a*). Immature cumulus-oocyte complexes (COC) were aspirated from follicles with a diameter ranging from 2 to 8 mm. Only COCs displaying a multilayered compact cumulus and a homogeneous ooplasm were selected. Frozen-thawed bull semen from the same ejaculate was used for all inhibition experiments. Straws were thawed in a water bath (37°C) for 60 s. Subsequently, the semen was centrifuged on a discontinuous Percoll gradient (90% and 45%; Pharmacia, Uppsala, Sweden) as described by Thys *et al.* (2009*a*).

Media and chemicals were analogous to those used by Tanghe *et al.* (2004*a*). Vitronectin from bovine plasma (V9881) used in all experiments was purchased from Sigma-Aldrich (Bornem, Belgium).

2.2 Removal of the zona pellucida

The cumulus oophorus of the matured COCs was removed mechanically by vortexing (8 min). Subsequently, the cumulus-denuded oocytes were incubated in 0.1% protease (P5147, Sigma-Aldrich, Bornem, Belgium) in phosphate-buffered saline (PBS) for 5 to 15 min at 37°C to dissolve their zona pellucida (ZP). Afterwards the oocytes were washed and transferred to the incubator to allow recovery of the oolemma for at least 30 min (Tanghe *et al.* 2004*a*).

2.3 Fixation and staining of oocytes

After fertilization, the presumed zygotes were fixed overnight (2% paraformaldehyde – 2% glutaraldehyde in PBS) and stained with 10 μg mL^{-1} Hoechst 33342 (Molecular Probes, Leiden, The Netherlands) for 10 min. Zygotes were mounted in a droplet of glycerol with (25 mg mL^{-1}) 1,4-diazabicyclo (2.2.2) octane (DABCO; Acros, Ghent, Belgium) and evaluated using a Leica DMR fluorescence microscope (Leica Microsystems, Groot-Bijgaarden, Belgium). The presence of two pronuclei was indicative for a successful fertilization of the oocyte. Penetration percentage was defined as the sum of the fertilization and polyspermy (more than two pronuclei) percentage per experimental group.

2.4 Dose-response effect of Vn on sperm penetration after bovine IVF

To reconfirm whether the Vn concentration (500 nM) applied in the study of Tanghe *et al.* (2004*b*) was the most appropriate, a preliminary experiment was conducted to assess the dose-response effect of Vn on sperm penetration after bovine IVF. *In vitro* matured COCs were randomly assigned to 6 different fertilization media (Tanghe *et al.* 2004*a*): fertilization medium supplemented with 0 nM, 10 nM, 100 nM, 250 nM, 500 nM and 1 µM Vn. The oocytes were co-incubated with sperm at a final concentration of 10^6 spermatozoa mL^{-1} for 20 h (39°C; 5% CO_2). Prior to overnight fixation (2% paraformaldehyde - 2% glutaraldehyde in PBS) and staining with 10 µg mL^{-1} Hoechst 33342 for 10 min, the presumed zygotes were vortexed for 3 min to remove excess spermatozoa and/or cumulus cells. Zygotes were mounted in glycerol with 25 mg mL^{-1} DABCO and evaluated for sperm penetration using a Leica DMR fluorescence microscope.

2.5 Effect of Vn on sperm penetration of the cumulus oophorus

In vitro matured COCs were randomly assigned to 4 groups (3 replicates). Half of the oocytes (2 groups) were denuded by vortexing (cumulus denuded or CD) and the other half (2 groups) were kept cumulus-enclosed (CE). Both CD and CE oocytes were fertilized under control conditions (in standard fertilization medium) and in the presence of 500 nM Vn (Tanghe *et al.* 2004*b*). The oocytes were co-incubated with sperm at a concentration of 10^6 sp mL^{-1} for 20 h (39°C; 5% CO_2). Each group contained 76 to 100 oocytes. Prior to fixation the presumed zygotes were vortexed to remove excess spermatozoa. After staining, the oocytes were evaluated for fertilization and polyspermy.

2.6 Effect of Vn on sperm-zona binding

In vitro matured COCs were denuded by vortexing and randomly assigned to two groups (4 replicates). The first group was fertilized under control conditions (in standard fertilization medium), the second group in the presence of 500 nM Vn. The oocytes were co-incubated with sperm at a concentration of 10^5 sp mL^{-1}. This sperm concentration was previously determined in order to allow unambiguous counting of the number of spermatozoa bound to the ZP. Each group consisted of 77 to 100 oocytes. After 20 h of co-incubation, the oocytes were washed 3 times to remove loosely attached spermatozoa, and subsequently fixed and stained with Hoechst 33342. Per presumed zygote the number of spermatozoa bound to the ZP was determined.

2.7 Effect of Vn on sperm-oolemma binding

The sperm pellet - obtained after Percoll® centrifugation - was diluted with fertilization medium to a concentration of 5 x 10^5 sp mL^{-1} and incubated for 30 min (39°C; 5% CO_2) to allow capacitation of the sperm cells. Subsequently, the acrosome reaction was induced by incubation of the sperm suspension for 15 min (39°C; 5% CO_2) in 100 µg mL^{-1} lysophosphatidyl choline (LPC; L5004, Sigma-Aldrich, Bornem, Belgium).

In vitro matured ZP-free oocytes (3 replicates) were randomly assigned to two different media: standard fertilization medium or fertilization medium supplemented with 500 nM Vn. The female gametes were co-incubated with sperm at a final concentration of 2.5 x 10^5 sp mL^{-1} in 50 µl droplets of medium (10 oocytes per droplet) covered with paraffin oil (Tanghe *et al.* 2004*a*). The number of oocytes per experimental group ranged from 28 to 44. One hour after insemination the oocytes were washed 3 times to remove loosely attached spermatozoa, fixed and stained. Of each presumed zygote the number of spermatozoa bound to the oolemma was evaluated.

2.8 Effect of Vn on sperm-oocyte fusion

The experimental setup was identical to the one described in the previous experiment, except that the ZP-free oocytes were fixed 20 h after insemination. All presumed zygotes were evaluated for sperm-oolemma fusion (defined as the presence of two or more pronuclei).

2.9 Localization of endogenous Vn on female and male bovine gametes

With respect to the female bovine gamete, immature COCs, *in vitro* matured COCs, CD oocytes as well as ZP-free oocytes were sampled following the protocol of Tanghe *et al.* (2004*a*). Cumulus cells were removed mechanically by vortexing (8 min), and the ZP was dissolved by incubation of the CD oocytes in 0.1% protease in PBS for 5 to 15 min (at 37°C). Since Vn is an extracellular matrix protein, expression of Vn was also analyzed in cumulus monolayers that were grown *in vitro* for a week (as described by Vandaele et al. 2007). All female gamete samples were fixed with 4% paraformaldehyde (P6118, Sigma-Aldrich, Bornem, Belgium) in PBS for 1 h (4°C) and permeabilized with 0.5% Triton X-100 (Sigma-Aldrich, Bornem, Belgium) in PBS for 30 min at room temperature (RT). Subsequently, they were incubated with 10% goat serum (16210-064, Invitrogen, Merelbeke, Belgium) in polyvinyl pyrrolidone (PVP, 0.1% in PBS) solution for 30 min (37°C), with mouse monoclonal antibody A18 to Vn (Abcam, Cambridge, UK) (1/300) for 1 to 2 h (37°C) and with goat-anti-mouse FITC antibody (Molecular Probes, Leiden, The Netherlands) (1/100) for 1 h (37°C). To stain the nuclei, all oocyte types were treated with 2% Propidium Iodide (Molecular Probes, Leiden, The Netherlands) in PBS for 30 min. Between each treatment the samples were washed in PVP. They were mounted in a droplet of glycerol with 25 mg mL^{-1} DABCO and evaluated for the presence of Vn using a Leica DM/RBE laser scanning confocal fluorescence microscope (Leica Microsystems, Groot-Bijgaarden, Belgium).

To evaluate the expression of Vn on the male gamete, frozen-thawed semen was thawed in water of 37°C for 60 s and centrifuged on a discontinuous Percoll® gradient. Next, the sample was split into 3 fractions. Fraction 1 was diluted to a concentration of 10x10^6 sp mL^{-1} (with medium consisting of a HEPES-buffered Tyrode balanced salt solution supplemented with 25 mM $NaHCO_3$, 10 mM sodium lactate, 0.2 mM sodium pyruvate, 10 µg mL^{-1} gentamycin sulphate) prior to indirect immunofluorescence, and represented non-treated (NT) sperm. Fraction 2 and 3 were diluted to a concentration of 5x10^6 sp mL^{-1} (with medium containing Tyrode balanced salt solution supplemented with 25 mM $NaHCO_3$, 10 mM sodium lactate, 0.2 mM sodium pyruvate, 10 µg mL^{-1} gentamycin sulphate, 6 mg mL^{-1} fatty acid-free BSA, and 20 µg mL^{-1} heparin) and subsequently incubated for 30 min (39°C; 5% CO_2) to induce capacitation. Then, fraction 2 (representing capacitated - CAP - sperm) was processed in the same way as fraction 1. Fraction 3 was supplemented with 100 µg mL^{-1} LPC and incubated for 15 min (39°C; 5% CO_2) in order to induce the acrosome reaction (acrosome reacted - AR - sperm). All three sperm fractions were fixed with 1% paraformaldehyde (in PBS) for 30 min (at 4°C) and permeabilized with 0.5% Triton X-100 in

PBS for 30 min (at RT). Subsequently, they were incubated with 10% goat serum in PVP for 30 min (37°C), with mouse monoclonal antibody A18 to Vn (1/300) for 1 to 2 h (37°C) and with goat-anti-mouse FITC antibody (1/100) for 1 h (37°C). To stain the nuclei, all sperm fractions were treated with 10 µg mL^{-1} Hoechst 33342 for 10 min (RT). Between each treatment the sperm fractions were centrifuged (10 min, 200g) and re-suspended in PVP. They were mounted in glycerol with 25 mg mL^{-1} DABCO and evaluated for the presence of Vn using fluorescence microscopy (Olympus IX81 inverted fluorescence microscope and a Hamamatsu Orca B/W camera using Olympus Cell*R software, Aartselaar, Belgium) and flow cytometry (FacsCanto II, BD, Belgium). Additionally, frozen-thawed semen originating from the same ejaculate was stained to evaluate Vn-expression (as described above) without previous fixation and permeabilization. The latter samples were processed on ice.

The mouse monoclonal antibody A18 is claimed to be highly specific for Vn, since there is no evidence for cross reactivity with other connective tissue proteins (fibronectin, elastin, collagen, laminin). Nevertheless, two negative controls were additionally included: 1) a sample processed without primary antibody prior to the incubation with the secondary FITC-labeled goat-anti-mouse antibody, and 2) a sample incubated with an isotype-matched mouse IgG1 antibody prior to the FITC-labeled secondary antibody treatment.

2.10 Localization of α_v (subunit of the Vn integrin receptor) on female and male bovine gametes

With respect to the female bovine gamete, *in vitro* matured CD oocytes were sampled (as described above) and fixed with 2% paraformaldehyde in PBS for 30 min (4°C) prior to indirect immunofluorescence.

To assess the presence of α_v on the male gamete, frozen-thawed semen originating from the same ejaculate was centrifuged on a discontinuous Percoll gradient, and the sperm pellet was diluted to a concentration of $10x10^6$ sp mL^{-1}. Subsequently, the sample was split into 3 fractions. Each fraction was processed as described before, resulting in a non-treated (NT), capacitated (CAP) and acrosome reacted (AR) sperm fraction. All sperm samples were fixed with ice-cold methanol during 15 min.

This time, the primary antibody used was rabbit polyclonal antibody to integrin subunit α_v (AB1930; Chemicon – Millipore, Belgium) (1/100), which was fluorescently labeled with goat anti-rabbit FITC antibody (Molecular Probes, Leiden, The Netherlands) (1/100). The primary antibody is guaranteed to have no-cross reactivity with α_1, α_2, α_3, α_4 or α_6 integrin subunits. To evaluate the specificity of the rabbit polyclonal antibody to integrin subunit α_v, a sample processed without primary antibody prior to the incubation with the FITC-labeled secondary antibody was included as negative control.

2.11 Effect of sperm incubation with Vn on membrane integrity and sperm motility

Frozen-thawed bull semen originating from the same ejaculate (3 replicates) was centrifuged on a discontinuous Percoll gradient and diluted to a concentration of $60x10^6$ spermatozoa mL^{-1} (with medium containing Tyrode balanced salt solution supplemented with 25 mM $NaHCO_3$, 10 mM sodium lactate, 0.2 mM sodium pyruvate, 10 µg mL^{-1} gentamycin sulphate, 6 mg mL^{-1} fatty acid-free BSA, and 20 µg mL^{-1} heparin). Subsequently, the sperm suspension was split into three fractions, which were diluted (1:1) respectively with the modified

Tyrode balanced salt solution (control), modified Tyrode balanced salt solution supplemented with 200 nM Vn (100 nM Vn) and modified Tyrode balanced salt solution supplemented with 1 μM Vn (500 nM Vn). Three aliquots from each sperm fraction were incubated (39°C; 5% CO_2), and at three different time points of incubation (1 h, 3 h and 6 h, respectively) one aliquot per fraction was evaluated for membrane integrity and total versus progressive sperm motility.

Membrane integrity was evaluated using a fluorescent SYBR14-Propidium Iodide (PI) staining technique (L7011; Molecular Probes, Leiden, The Netherlands). A stock solution of 1 mmol L^{-1} SYBR14 reagent was diluted (1:50) in HEPES-TALP, stored frozen at -20°C and thawed just before use. From each sperm aliquot, 100 μl was used and 1 μL SYBR14 was added. After 5 min of incubation (at 37°C), 1 μL PI was added prior to another 5 min incubation (at 37°C). Per aliquot 200 spermatozoa were examined using a Leica DMR fluorescence microscope. Three populations of sperm cells were identified: living (membrane intact; stained green), dead (membrane damaged; stained red), and moribund (double stained; green-orange) spermatozoa. The moribund sperm cells were considered to be part of the dead sperm population.

Total and progressive motility were determined by means of computer-assisted sperm analysis (Hamilton-Thorne CEROS 12.3) (Tanghe *et al.* 2004*a*).

2.12 Statistical analyses

Differences in fertilization and penetration percentages, and differences in number of Vn-positive cells were analyzed by means of binary logistic regression (including the effect of replicate). To evaluate the differences in mean number of spermatozoa bound to the ZP, the non parametric Kruskal Wallis test was applied, since the concerning variable was not normally distributed. Differences in mean number of sperm cells binding the oolemma were analyzed using ANOVA. Differences in membrane integrity and (total and progressive) sperm motility were evaluated using repeated measures analysis of variance. Hypothesis testing was performed using a significance level of 5% (2-sided test) and results were cited as mean ± S.E.M. (SPSS 15.0).

3. Results

3.1 Dose-response effect of Vn on sperm penetration after bovine IVF

Compared to the control (0 nM Vn), sperm penetration significantly improved when supplementing 100 nM Vn during IVF ($P<0.05$; Fig.1). Sperm penetration was significantly inhibited in a concentration-dependent manner starting from 500 nM Vn. This suggests that at higher concentrations, the inhibiting effect of Vn dominates the beneficial effect observed at lower concentrations. In order to determine the mechanism underlying the inhibitory effect of Vn on sperm penetration, subsequent experiments were performed using 500 nM Vn.

3.2 Effect of Vn on sperm penetration of the cumulus oophorus

When 500 nM Vn was supplemented to the fertilization medium, sperm penetration percentages and fertilization percentages in both CE and CD oocytes decreased significantly compared to the respective control group (Table 1). The difference in reduction of sperm penetration was not statistically significant when comparing the CE and the CD groups (P=0.106). Nevertheless, considering the small sample size (n=6), the mean difference of 30.2% in inhibition of penetration between cumulus-enclosed and cumulus-denuded groups suggests a relevant effect of cumulus denudation.

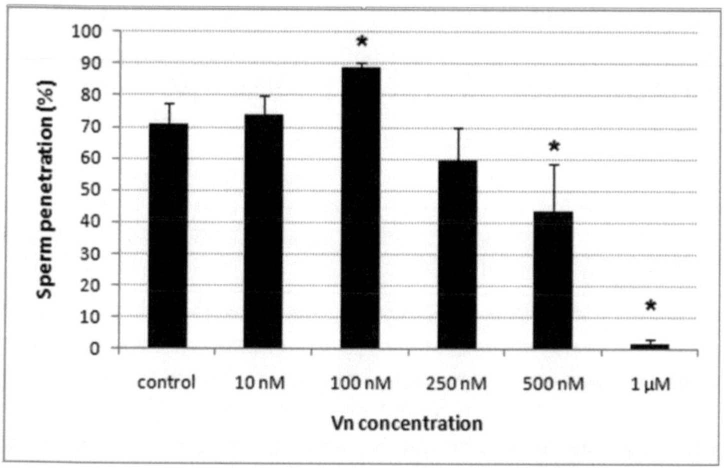

Fig. 1. Dose-response effect of vitronectin (Vn) on sperm penetration after bovine IVF. Data represent mean ± SEM. *Values significantly different from control with 0 nM Vn (P < 0.05).

Oocytes	Vn (nM)	No.	Fertilization (%)	Polyspermy (%)	Penetration (%)	Inhibition of penetration (%)
CD	0	249	$15.0^{a} \pm 2.06$	$0.4^{a} \pm 0.43$	$15.4^{a} \pm 1.86$	-
	500	258	$1.6^{b} \pm 1.60$	$0.4^{a} \pm 0.40$	$2.0^{b} \pm 1.44$	87.0
CE	0	296	$62.9^{a} \pm 7.23$	$7.7^{a} \pm 1.04$	$70.6^{a} \pm 7.71$	-
	500	273	$26.8^{b} \pm 12.16$	$3.7^{b} \pm 0.87$	$30.5^{b} \pm 12.42$	56.8

Table 1. Fertilization, polyspermy and penetration percentages of cumulus-denuded (CD) and cumulus-enclosed (CE) oocytes inseminated in standard fertilization medium and in fertilization medium supplemented with 500 nM of vitronectin (Vn). [a,b] Values with a different superscript in the same column within the CD and the CE groups differ significantly (P<0.05).

3.3 Effect of Vn on sperm-zona binding

The number of spermatozoa bound to the ZP in the Vn supplemented group was slightly but significantly higher than that in the control group (50.1 ± 2.1 versus 42.9 ± 1.9 spermatozoa per oocyte; P<0.05).

3.4 Effect of Vn on sperm-oolemma binding

Vitronectin supplementation did not significantly influence the sperm-oolemma binding. However, a slight numerical decrease in sperm adherence (from 27.4 ± 1.9 to 23.0 ± 2.8 spermatozoa per oocyte) was observed in the presence of Vn (P>0.05).

3.5 Effect of Vn on sperm-oocyte fusion

A significant decrease in fertilization percentage (from 25.4% to 14.2%) and sperm penetration percentage (from 28.0% to 16.0%) was found when 500 nM Vn was supplemented during IVF ($P<0.05$). Compared to the ZP-free control group, the sperm penetration was inhibited with 42.9%.

3.6 Localization of endogenous Vn on female and male bovine gametes

After fixation and permeabilization, the percentage of Vn positive sperm cells was very high and consistent in all three sperm fractions (≥ 99.4% positive cells for NT, CAP and AR sperm). However, the intensity of fluorescence was 3 times higher in the AR group compared to the NT and CAP group (Table 2) and the predominant fluorescent pattern observed in the AR fraction (Fig.2*e* – fluorescence at the acrosomal region and midpiece) also differed from the one mainly observed in the NT and CAP groups (Fig.2*a* and 2*c* – fluorescence at the postacrosomal region and midpiece). When using fluorescence microscopy for the evaluation of unfixed spermatozoa, fluorescence was only observed in the AR fraction, displaying a green signal at the apical sperm head region (Fig.2*f*). No fluorescence could be visualized in the NT and CAP sperm fractions (Fig.2*b* and *d*). Subsequently, flow cytometry (a far more sensitive technique) was applied, resulting in detection of fluorescent spermatozoa in all three sperm fractions: 5.8% in NT sperm, 14.4% in CAP sperm and 49.5% in AR sperm respectively. The intensity of fluorescence was much lower compared to the equivalent fixed and permeabilized sperm fractions (Table 2).

Treatment	Sperm fraction	No.	Positive cells (%)	Mean	Median
+ fixation + permeabilization	NT	10 000	99.4	1586	1452
	CAP	10 000	99.7	1458	1226
	AR	10 000	99.8	4635	4728
- fixation - permeabilization	NT	10 000	5.8	134	116
	CAP	10 000	14.4	172	139
	AR	10 000	49.5	284	214

Table 2. Percentage of vitronectin positive cells, mean and median of relative fluorescence intensity per sperm fraction after flow cytometric evaluation of frozen-thawed bovine sperm cells labeled with mouse monoclonal to vitronectin and goat-anti-mouse FITC. NT: non-treated spermatozoa; CAP: capacitated spermatozoa; AR: acrosome-reacted spermatozoa.

With respect to the cumulus monolayer, the cytoplasm of approximately 100% of the cumulus cells stained positively for Vn, and fluorescent mesh forming structures were observed in the extracellular matrix (Fig.3*a*). In immature COCs a relatively small number of cumulus cells expressed Vn in their cytoplasm. After *in vitro* maturation, the number of Vn positive cumulus cells was considerably increased in the COCs (Fig.3*b*). Cumulus-denuded oocytes appeared to express Vn on the surface of the ZP, including fluorescent spurs penetrating the ZP (Fig.4). Protease treated (ZP-free) oocytes did not show membrane expression.

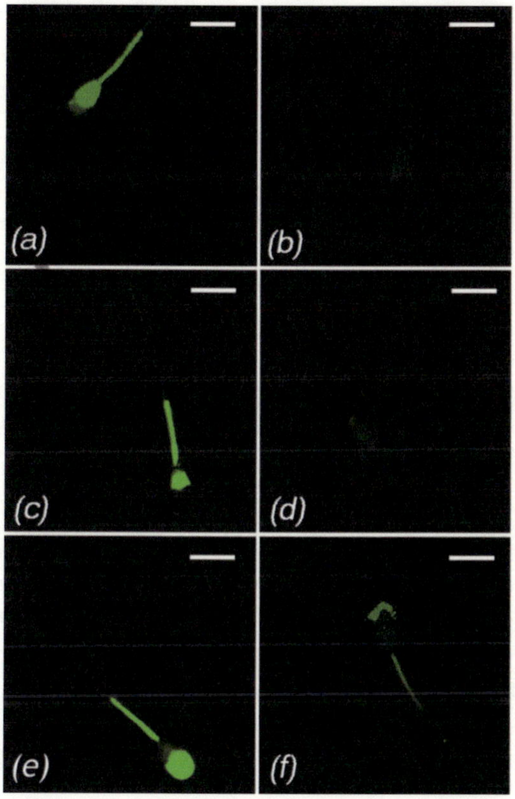

Fig. 2. Fluorescent images of frozen-thawed bovine semen labeled with mouse monoclonal to vitronectin and goat-anti-mouse FITC after fixation and permeabilization (*a*, *c* and *e*) and without prior fixation and permeabilization (*b*, *d* and *f*) (Original Magnification x600; Bar = 10 μm) (*a-b*) non-treated sperm; (*c-d*) capacitated sperm; (*e-f*) acrosome reacted sperm.

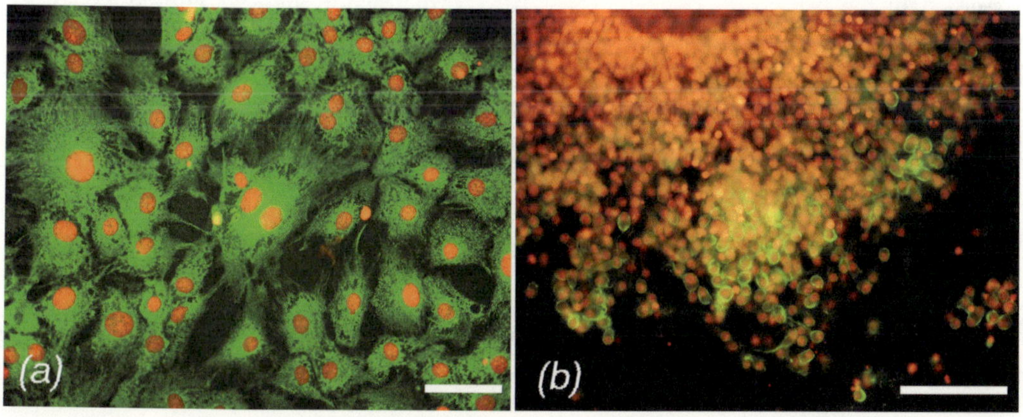

Fig. 3. Indirect immunofluorescent staining with mouse monoclonal to vitronectin and goat-anti-mouse FITC combined with Propidium Iodide staining after fixation and permeabilization (*a*) cumulus cell monolayer (Original Magnification x600; Bar = 25 μm), (*b*) *in vitro* matured COC (Original Magnification x400; Bar = 50 μm).

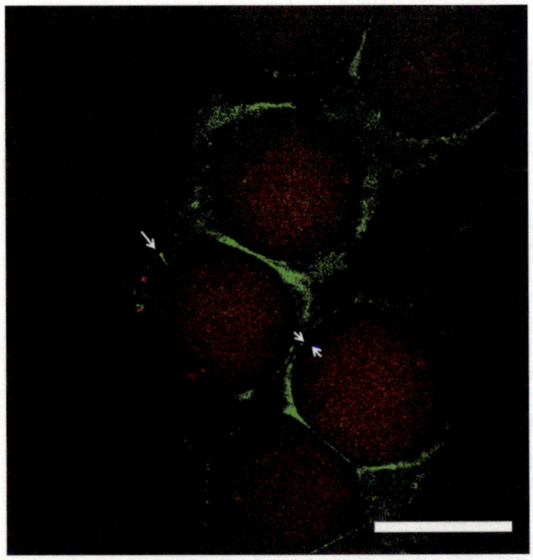

Fig. 4. Confocal fluorescent image of cumulus-denuded ZP intact bovine oocytes labeled with mouse monoclonal to vitronectin and goat-anti-mouse FITC combined with Propidium Iodide staining after fixation and permeabilization (Original Magnification x200; Bar = 100 μm). ⇨ fluorescent spur penetrating the ZP; ⇨ ⇦ ZP

3.7 Localization of α_v (subunit of the Vn integrin receptor) on female and male bovine gametes

After fixation and permeabilization, all sampled sperm cells were positive for integrin subunit α_v, irrespective of their functional state (NT, CAP or AR sperm). All spermatozoa displayed the same fluorescent pattern (Fig.5*a*, 5*c* and 5*e*). When staining NT, CAP and AR sperm without prior fixation and permeabilization, no integrin expression was visually observed in the NT group (Fig.5*b*). The CAP sperm fraction showed faint fluorescence (Fig.5*d*), whereas a bright signal was detected in the AR sperm cells (Fig.5*f*). Flow cytometric evaluation of unfixed sperm cells confirmed these subjective observations (Table 3). The number of integrin subunit α_v positive cells increased after heparin treatment, whereas the relative fluorescence intensity was substantially increased after artificial induction of the acrosome reaction.

Sperm fraction	No.	Positive cells (%)	Mean	Median
NT	10 000	55.7	5105	4964
CAP	10 000	69.4	6373	6273
AR	10 000	69.9	8018	7808

Table 3. Percentage of integrin subunit α_v positive cells, mean and median of relative fluorescence intensity per sperm fraction after flow cytometric evaluation of unfixed frozen-thawed bovine sperm cells labeled with rabbit polyclonal to integrin subunit α_v and goat-anti-rabbit FITC. NT: non-treated spermatozoa; CAP: capacitated spermatozoa; AR: acrosome-reacted spermatozoa.

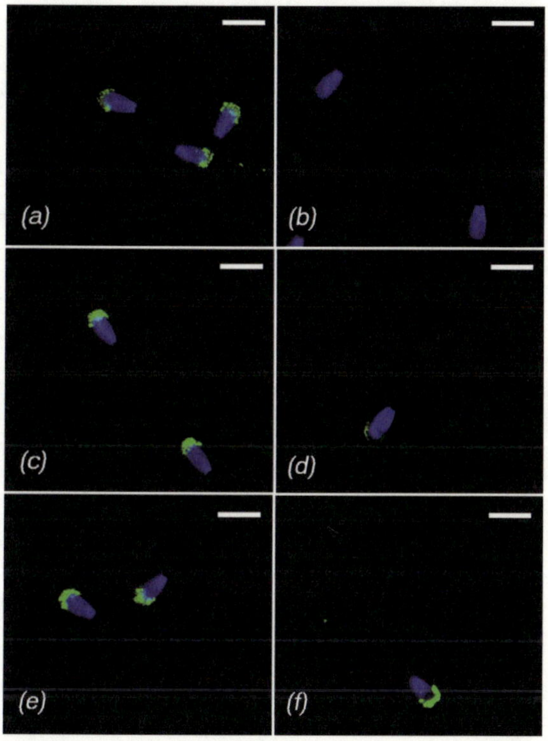

Fig. 5. Fluorescent images of frozen-thawed bovine semen labeled with rabbit polyclonal antibody to integrin subunit α_v and goat-anti-rabbit FITC combined with Hoechst 33342 staining after fixation (*a-c-e*) and without prior fixation (*b-d-f*) (Original Magnification x600; Bar = 10 µm) (*a-b*) non-treated sperm; (*c-d*) capacitated sperm; (*e-f*) acrosome reacted sperm.

However, incubation of the same *fixed* sperm fractions with heat-inactivated rabbit serum - instead of the primary rabbit polyclonal antibody - appeared to induce a similar fluorescent pattern in NT and CAP sperm cells (data not shown).

Integrin subunit α_v was detected in *in vitro* matured CD bovine oocytes (Fig.6). The fluorescent pattern varied between the sampled cells: some oocytes (12.5%) expressed integrin subunit α_v only at their oolemma (Fig.6*a*), whereas the greater part (87.5%) appeared to express the receptor molecule also at (the exterior side of) the ZP, including fluorescent spurs penetrating the ZP (Fig.6*b*). ZP-free bovine oocytes displayed green fluorescent spots at their surface.

3.8 Effect of sperm incubation with Vn on membrane integrity and sperm motility

Sperm membrane integrity was negatively affected in the presence of 500 nM Vn ($P < 0.05$), but was not altered by sperm incubation with 100 nM Vn (Fig.7). Total and progressive motility differed significantly ($P < 0.05$) between all sampled groups (control, 100 nM Vn and 500 nM Vn; Fig.8). However, especially the substantial twofold decrease in progressive motility in presence of 500 nM Vn should be considered as an important factor contributing to the inhibitory effect of Vn supplementation on sperm penetration during IVF.

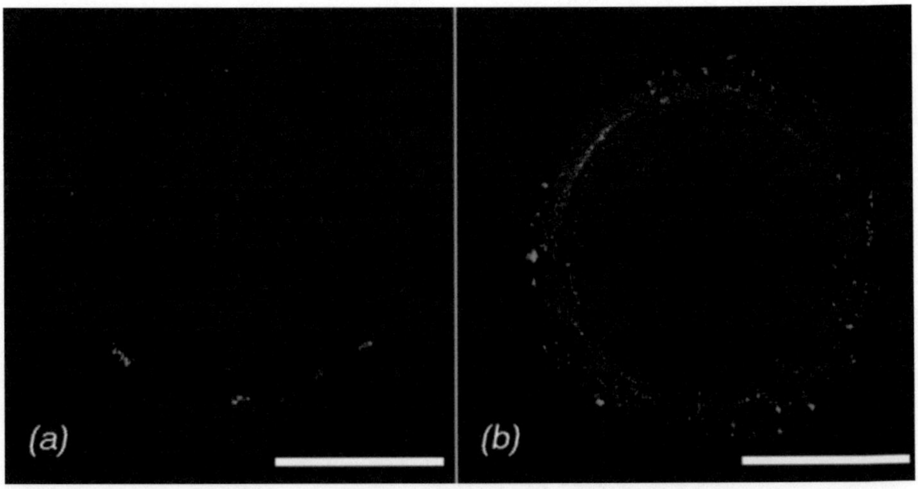

Fig. 6. Confocal fluorescent images of cumulus-denuded ZP intact bovine oocytes labeled with rabbit polyclonal antibody to integrin subunit α_v and goat-anti-rabbit FITC after fixation (*a*) fluorescent signal confined to the oolemma, (*b*) fluorescent signal at the level of the oolemma, the ZP and exterior side of the ZP (Original Magnification x400; bar = 50 μm).

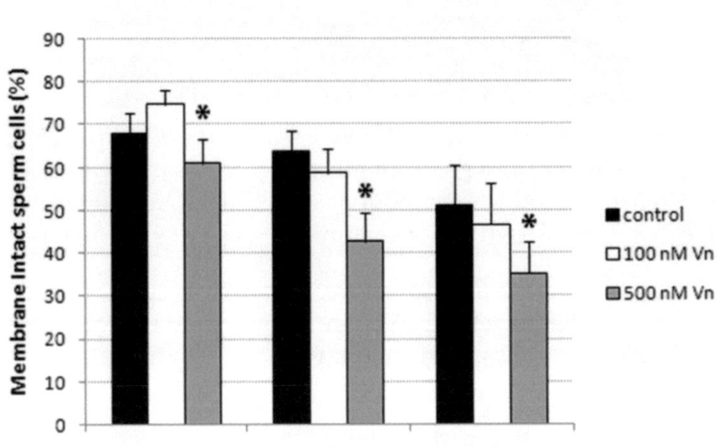

Fig. 7. Effect of 100 nM and 500 nM Vn on membrane integrity of bovine frozen-thawed spermatozoa during incubation (evaluated by means of SYBR14-PI staining). Data represent mean ± SD. *Values significantly different from control with 0 nM Vn ($P < 0.05$).

4. Discussion

In cattle, cumulus cells are shed from the oocyte in the oviduct within a few hours (Lorton & First 1979) to 10 h after ovulation (Hyttel *et al.* 1988). However, it is not entirely clear whether the cumulus cells or the matrix are necessary for bovine fertilization *in vivo*: sperm cells probably need to penetrate the cumulus matrix first, before they can pass through the ZP and subsequently fuse with the oolemma (Van Soom *et al.* 2002). The molecular basis of each of these processes has not been resolved yet, but it is now accepted that sperm-egg

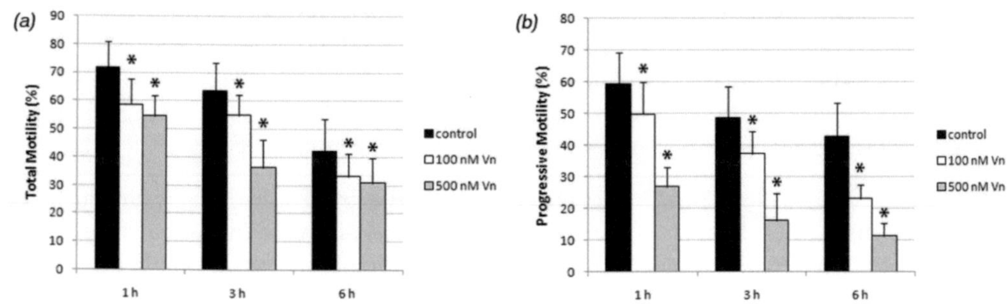

Fig. 8. Effect of 100 nM and 500 nM vitronectin (Vn) on total motility (*a*) and progressive motility (*b*) of bovine frozen-thawed spermatozoa during incubation (evaluated by means of CASA). Data represent mean ± SD. *Values significantly different from control with 0 nM Vn ($P < 0.05$).

binding requires a multitude of receptor-ligand interactions (Lyng & Shur 2007). In the present study, we have confirmed that high concentrations of the extracellular matrix glycoprotein Vn had a negative effect on bovine sperm-oocyte interaction (Tanghe *et al.* 2004*b*). Furthermore, the main inhibitory effect of 500 nM exogenous Vn was observed at the level of sperm-oolemma fusion, implicating that Vn might be one of the ligands involved in sperm-egg recognition in cattle.

Vitronectin strongly reduced sperm penetration in CD oocytes, but did only moderately so in CE oocytes. These results suggest that the cumulus oophorus is able to capture a substantial part of the supplemented Vn allowing still a fair sperm penetration rate. Vitronectin (serving as a cell-to-substrate adhesion molecule) is likewise known to interact with glycosaminoglycans and proteoglycans and is recognized by certain members of the integrin family (http://www.uniprot.org/uniprot/P04004). Integrin expression has been observed in bovine cumulus cells before (Sutovsky *et al.* 1995).

A slightly augmented number of spermatozoa binding the ZP was noted in the presence of 500 nM Vn. This was an intriguing finding, since despite the increased sperm binding in the presence of Vn, oocyte penetration was substantially reduced. In mice, it has been demonstrated that sperm binding to the ZP is not sufficient to induce acrosomal exocytosis (Baibakov *et al.* 2007). The actual sperm passage through the pores of the ZP is believed to mechanically trigger the acrosome reaction: contact between motile sperm and the small ZP-pores would generate sufficient shear force to bring forth a mechanosensory signal and acrosomal exocytosis. Binding to the ZP is suggested to slow down the forward progression of motile sperm and the forceful thrusting of the tail in order to transduce a mechanosensory signal mobilizing acrosomal Ca^{2+} stores and - consequently - to induce the acrosome reaction (Baibakov *et al.* 2007). The substantial twofold decrease in progressive sperm motility noted in presence of 500 nM Vn may therefore well be responsible for a defective sperm-ZP penetration. Furthermore, considerable head-to-head agglutination was observed when incubating bovine spermatozoa in the presence of Vn, especially at the high concentration of 500 nM (data not shown). Probably, the sperm is able to bind the ZP (assisted by the exogenous Vn connecting the integrin α_v at the sperm cell surface to the integrin α_v at the exterior side of the ZP), but is not capable of successful penetration of the

ZP. When supplementing low concentrations of Vn (100 nM), sperm penetration of COCs was enhanced, possibly through the increased sperm-ZP binding. Compared to 500 nM Vn, such low doses did not affect sperm membrane integrity and did not have the same impact on progressive sperm motility. Possibly, the forward progression of these ZP bound sperm cells was still sufficient for proper penetration.

A reversible dual binding function connecting both the male and female gamete - as suggested in human by Fusi *et al.* (1996*b*) - is even more plausible in the bovine species, since ruminant Vn apparently displays two integrin binding RGD sequences, in contrast to only one RGD site in human Vn (Suzuki *et al.* 1985; Mahawar & Joshi 2008). Furthermore, the presumed reversible dual binding function could additionally be exerted through spontaneous multimerization of several Vn molecules as described by Stockmann *et al.* (1993). The C-terminal half of the molecule comprises two hemopexin-like domains, able to interact with the acidic residues of the connecting segment of the same molecule, consequently allowing intramolecular and intermolecular linking to form Vn polymers (Royce & Steinmann 2002).

Vitronectin supplementation to the fertilization medium did not lead to a statistically significant inhibition of the sperm-oolemma binding in our experiment. Fusi *et al.* (1996*b*) previously suggested that Vn was well suited to play a significant role in human sperm-egg adhesion. These authors found a promotion of oolemmal adherence of spermatozoa, following addition of Vn to the medium over a certain concentration range. Supplementation of Vn enhanced oolemmal adherence of spermatozoa over a concentration range of 2.2 nM to 100 nM (Fusi *et al.* 1996*b*). Higher Vn concentrations - like in the present study - reduced the number of spermatozoa adhering to the egg (Fusi *et al.* 1996*b*), possibly due to Vn-mediated sperm aggregation within the culture dish. During the sperm incubation experiment, substantial head-to-head agglutination was observed after 4 h incubation in the presence of 100 nM and the agglutination was even more distinct when supplementing 500 nM Vn. Nevertheless, we could merely detect a slight - statistically insignificant - decrease in sperm-oolemma binding during bovine IVF when 500 nM Vn was supplemented to the fertilization medium. However, the physiological relevance of the number of sperm cells bound to a ZP-free oocyte is debatable (Talbot *et al.* 2003). The underlying assumption is that spermatozoa are bound to the oolemma by a mechanism that can result in sperm-egg fusion. This *in vitro* assay may - though - include a heterogeneous population of bound sperm cells, including non-physiologically bound sperm, besides the specific population of physiologically relevant sperm that are tethered or docked before fusion. Some of the sperm cells may well be bound via interactions that will not result into fusion. Acrosome-intact sperm are - for instance - known to be able to bind to ZP-free oocytes, but not to fuse. If the non-specifically bound sperm fraction outnumbers the specifically bound fraction, variations in the number of physiologically relevant bound sperm will not be measured (Talbot *et al.* 2003).

In the present experiment, Vn supplementation inhibited 42.9% of sperm-oolemma fusion. Accordingly, Fusi *et al.* (1996*b*) observed a decrease in the number of penetrating sperm cells in human oocytes in the presence of increasing Vn concentrations starting from 100 nM.

Furthermore, a dual (concentration-dependent) effect of exogenous Vn supplementation on bovine IVF was observed. Low Vn concentrations (10 nM - 100 nM) appeared to enhance sperm penetration, whereas a negative influence was noted in the presence of high concentrations (500 nM - 1 μM). This inhibitory effect may well be - at least partially - due to compromised membrane integrity. Furthermore, the observed substantial twofold decrease in progressive motility in the presence of 500 nM Vn (compared to the control) should most likely be considered as an important factor contributing to the inhibitory effect of Vn supplementation on sperm penetration during IVF. Combining these results with the explicit head-to-head agglutination noted when incubating sperm with 500 nM of Vn, high concentrations of Vn should obviously be regarded as detrimental for successful fertilization.

As in human (Fusi *et al.* 1994), Vn also appears to be an intrinsic protein of bovine sperm cells (Fig.9*a-b*). After fixation and permeabilization, practically all sampled spermatozoa did show fluorescence when applying indirect immunofluorescence and flow cytometry. The observed shift in fluorescent pattern after LPC treatment suggests that the Vn sequestered inside the sperm head is exposed at the sperm cell surface after acrosomal reaction (Fig.9*c*). With respect to the biological relevance of the moderate increase of Vn expression in unfixed sperm cells, we have to note that LPC treatment does not induce acrosomal reaction in all spermatozoa of the samples. Only penetration of the sperm cells through the ZP leads to nearly 100 % acrosome reaction. LPC treatment typically induces an increase in the percentage of acrosome reacted sperm ranging from about 26% to 38% (compared to a negative control group) in frozen-thawed bovine semen (O'Flaherty *et al.* 2005). With respect to the expression of the α_v subunit of the Vn receptor, all fixed and permeabilized sperm cells seemed to display fluorescence at their apical surface, irrespective of their functional state (NT, CAP or AR). However, incubation of the same fixed sperm fractions with heat-inactivated rabbit serum (instead of the primary rabbit polyclonal antibody to integrin subunit α_v) appeared to induce a similar fluorescent pattern in NT and CAP sperm cells (data not shown). Some degree of non-specific binding of the secondary FITC-labeled antibody should therefore be acknowledged. Nevertheless, the finding that AR sperm cells did not show non-specific fluorescence (when using heat-inactivated rabbit serum) indicates that there is specific binding of the primary anti-integrin subunit α_v antibody as well. In addition, the equivalent unfixed, not permeabilized, sperm fractions displayed a different pattern. Only the AR fraction was visually fluorescent, which also supports the specificity of the primary antibody to integrin subunit α_v. The present findings, including a clear quantitative increase in fluorescence in unfixed AR spermatozoa measured by means of flow cytometry, are in accordance with the work of Fusi *et al.* (1996*a*) stating that α_v expression was maximal following ionophore-exposure, used as inducer of the acrosome reaction. They assumed that α_v may be located on the inner acrosomal membrane, which becomes accessible to anti-integrin subunit antibody both during the acromose reaction and following permeabilization of capacitated spermatozoa.

A fascinating observation was the abundant presence of endogenous Vn at the exterior side of the ZP including fluorescent spurs penetrating the ZP. Probably, the glycoprotein is bound to the $\alpha_v\beta_3$ integrin, of which the α_v subunit was also detected at the level of the ZP in a similar pattern. These fluorescent signals are assumed to be remnants of molecules present at the cell surface of corona radiata cells which have been shown to have cellular projections traversing the ZP and terminating upon the oolemma (Tanghe *et al.* 2002). Integrin expression has been observed in bovine cumulus cells before (Sutovsky *et al.* 1995), and

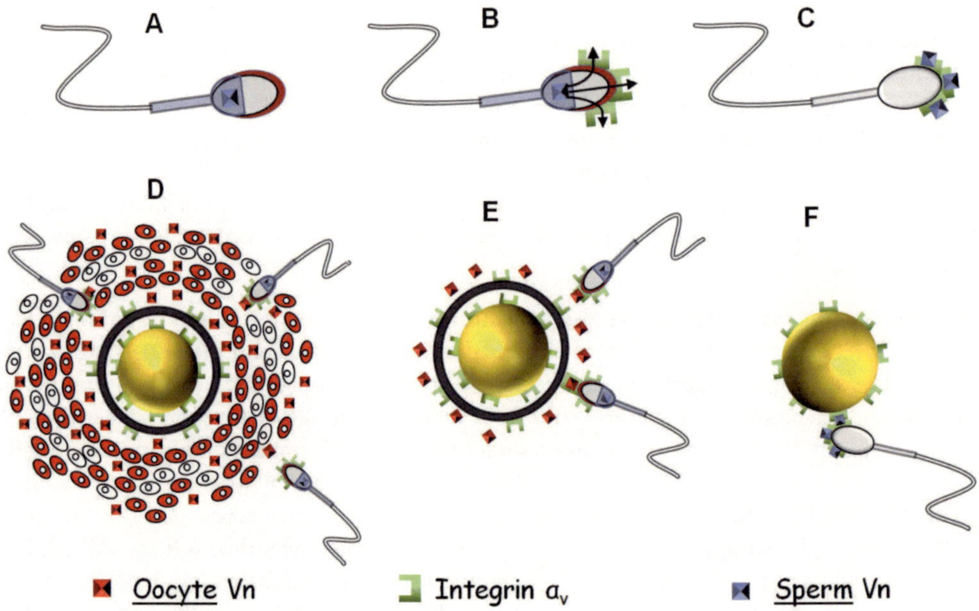

Fig. 9. Hypothetical model concerning the function of endogenous vitronectin (Vn) during bovine IVF. (A) non-treated sperm cell, (B) capacitated sperm cell, (C) acrosome-reacted sperm cell, (D) capacitated spermatozoa traversing the cumulus oophorus, (E) spermatozoa binding the zona pellucida through endogenous oocyte Vn, (F) sperm-oocyte interaction through endogenous sperm Vn liberated after acrosome reaction.

these transmembrane receptors are assumed to be easily ripped out of the cell with bits of attached membrane, when internal anchorage with the cytoskeleton is disturbed (Alberts *et al.* 1994). This disconnection may have occurred through the mechanical force exerted on the cell by vortexing. Since integrin subunit α_v was already observed to some extent in CAP spermatozoa (69.4% of that sperm fraction), it could be speculated that capacitated sperm cells reversibly bind to the endogenous Vn at the level of the ZP resulting in penetration of this acellular egg vestment. The finding that low concentrations of exogenously supplemented Vn improve sperm penetration during bovine IVF, might be attributed to an additional reversible dual adhesion function exerted by this molecule. A similar function could be exerted by the endogenous Vn located at the level of the extracellular cumulus matrix, in this case assisting the sperm cell in traversing the cumulus oophorus (Fig.9*d*). In order to elucidate the beneficial effect of low Vn concentrations on sperm penetration, further studies are required.

Since integrin subunit α_v was present at the oolemma of ZP-free bovine oocytes and spermatozoa express Vn at their surface after acrosomal reaction, this receptor-ligand after might play a role in sperm-oocyte interaction. To confirm this hypothesis additional experiments investigating whether supplementation of low Vn concentrations to an IVF system with ZP-free oocytes and acrosome reacted sperm effectively inhibit sperm penetration are necessary. Previously, detection of fibronectin (another extracellular matrix glycoprotein) underneath the ZP together with observed expression of the α_5 subunit of its corresponding receptor on oolemma and acrosome reacted sperm cell surface already indicated a reversible dual binding interaction between the fibronectin ligand and corresponding receptors on both (acrosome reacted) sperm cell and oolemma, initiating bovine sperm-egg binding (Thys *et al.* 2009*b*).

Bearing in mind the present findings, the following hypothesis can be put forward concerning the interaction of vitronectin during bovine fertilization. Detection of endogenous Vn and integrin subunit α_v at the exterior side of the ZP and integrin subunit α_v on the sperm cell surface, combined with an increased sperm-ZP binding in presence of exogenous Vn, suggests at least some intervention of this glycoprotein in initial sperm-ZP interaction (Fig.9*e*). Since the α_v subunit of the Vn receptor was identified on the oolemma, and spermatozoa appeared to express both integrin subunit α_v and Vn after acrosomal reaction, this receptor-ligand mechanism may play a role in sperm-oocyte interaction (Fig.9*f*). The inhibitory effect of exogenously supplemented Vn on sperm penetration of ZP-free oocytes could then be explained by the competition between the exogenous Vn and the Vn liberated from the sperm cell following the acrosome reaction. Further research is required to distinguish the dual effect of low versus high concentrations of exogenous Vn during bovine IVF.

5. References

Alberts, B.; Bray, D; Lewis, J.; Raff, M.; Roberts, K. & Watson, J. D. (1994). Cell Junctions, Cell Adhesion, and the Extracellular Matrix. In: *Molecular Biology of the Cell*, Third Edition, Alberts, B.; Bray, D.; Lewis, J.; Raff, M.; K. Roberts, K. & Watson, J. D., 949-1009, Garland Publishing, ISBN 0-8153-1620-8, New York, USA

Almeida, E. A.; Huovila, A. P.; Sutherland, A. E; Stephens, L. E.; Calarco, P. G.; Shaw, L. M.; Mercurio, A. M.; Sonnenberg, A.; Primakoff, P.; Myles, D. G. & White, J. M. (1995). Mouse egg integrin alpha 6 beta 1 functions as a sperm receptor. *Cell,* Vol.81, No.7, (June 1995), pp. 1095-1104, ISSN 0092-8674

Amari, S.; Yonezawa, N.; Mitsui, S.; Katsumata, T.; Hamano, S.; Kuwayama, M.; Hashimoto, Y.; Suzuki, A.; Takeda, Y. & Nakano, M. (2001). Essential role of the nonreducing terminal alpha-mannosyl residues of the N-linked carbohydrate chain of bovine zona pellucida glycoproteins in sperm-egg binding. *Molecular Reproduction and Development,* Vol.59, No.2, (June 2001), pp. 221-226, ISSN 1098-2795

Baibakov, B.; Gauthier, L.; Talbot, P.; Rankin, T. L. & Dean, J. (2007). Sperm binding to the zona pellucida is not sufficient to induce acrosome exocytosis. *Development,* Vol.134, No.5, (March 2007), pp. 933-943, ISSN 1477-9129

Benoff, S. (1997). Carbohydrates and fertilization: an overview. *Molecular Human Reproduction,* Vol.3, No.7, (July 1997), pp. 599-637, ISSN 1460-2407

Bronson, R. A. & Fusi, F. M. (1996). Integrins and human reproduction. *Molecular Human Reproduction,* Vol.2, No.3, (March 1996), pp. 153-168, ISSN 1460-2407

Bronson, R.; Peresleni, T.; Golightly, M. & Preissner, K. (2000). Vitronectin is sequestered within human spermatozoa and liberated following the acrosome reaction. *Molecular Human Reproduction,* Vol.6, No.11, (November 2000), pp. 977-982, ISSN 1460-2407

Evans, J. P.; Schultz, R. M. & Kopf, G. S. (1995). Mouse sperm-egg plasma membrane interactions: analysis of roles of egg integrins and the mouse sperm homologue of

PH-30 (fertilin) beta. *Journal of Cell Science,* Vol.108, No.10, (October 1995), pp. 3267-3278, ISSN 1477-9137

François, P. P.; Preissner, K. T.; Herrmann, M.; Haugland, R. P.; Vaudaux, P.; Lew, D. P. & Krause, K. H. (1999). Vitronectin interaction with glycosaminoglycans. Kinetics, structural determinants, and role in binding to endothelial cells. *The Journal of Biological Chemistry,* Vol.274, No.53, (December 1999), pp. 37611-37619, ISSN 1083-351X

Fusi, F. M.; Lorenzetti, I.; Vignali, M. & Bronson, R. A. (1992). Sperm surface proteins after capacitation. Expression of vitronectin on the spermatozoan head and laminin on the sperm tail. *Journal of Andrology,* Vol.13, No.6, (November-December 1992), pp. 488-497, ISSN 1939-4640

Fusi, F. M.; Lorenzetti, I.; Mangili, F.; Herr, J. C.; Freemerman, A. J.; Gailit, J. & Bronson, R. A. (1994). Vitronectin is an intrinsic protein of human spermatozoa released during the acrosome reaction. *Molecular Reproduction and Development,* Vol.39, No.3, (November 1994), pp. 337-343, ISSN 1098-2795

Fusi, F. M.; Tamburini, C.; Mangili, F.; Montesano, M.; Ferrari, A. & Bronson, R. A. (1996*a*). The expression of alpha v, alpha 5, beta 1, and beta 3 integrin chains on ejaculated human spermatozoa varies with their functional state. *Molecular Human Reproduction,* Vol.2, No.3, (March 1996), pp. 169-175, ISSN 1460-2407

Fusi, F. M.; Bernocchi, N.; Ferrari, A. & Bronson, R. A. (1996*b*). Is vitronectin the velcro that binds the gametes together? *Molecular Human Reproduction,* Vol.2, No.11, (November 1996), pp. 859-866, ISSN 1460-2407

Gechtman, Z.; Belleli, A.; Lechpammer, S. & Shaltiel, S. (1997). The cluster of basic amino acids in vitronectin contributes to its binding of plasminogen activator inhibitor-1: evidence from thrombin-, elastase- and plasmin-cleaved vitronectins and anti-peptide antibodies. *Biochemical Journal,* Vol.325, No.2, (July 1997), pp. 339-349, ISSN 0264-6021

Gibson, A. D.; Lamerdin, J. A.; Zhuang, P.; Baburaj, K.; Serpersu, E. H. & Peterson, C. B. (1999). Orientation of heparin-binding sites in native vitronectin. Analyses of ligand binding to the primary glycosaminoglycan-binding site indicate that putative secondary sites are not functional. *The Journal of Biological Chemistry,* Vol.274, No.10, (March 1999), pp. 6432-6442, ISSN 1083-351X

Gougoulidis, T.; Trounson, A. & Dowsing, A. (1999). Inhibition of bovine sperm-oocyte fusion by the carbohydrate GalNAc. *Molecular Reproduction and Development,* Vol.54, No.2, (October 1999), pp. 179-185, ISSN 1098-2795

Gwathmey, T. M.; Ignotz, G. G.; Mueller, J. L.; Manjunath, P. & Suarez, S. S. (2006). Bovine seminal plasma proteins PDC-109, BSP-A3, and BSP-30-kDa share functional roles in storing sperm in the oviduct. *Biology of Reproduction,* Vol.75, No.4, (October 2006), pp. 501-507, ISSN 1529-7268

Hyttel, P.; Greve, T. & Callesen, H. (1988). Ultrastructure of in-vivo fertilization in cattle. *Journal of Reproduction and Fertility,* Vol.82, No.1, (January 1988), pp. 1-13, ISSN 1741-7899

Ignotz, G. G.; Cho, M. Y. & Suarez, S. S. (2007). Annexins are candidate oviductal receptors for bovine sperm surface proteins and thus may serve to hold bovine sperm in the oviductal reservoir. *Biology of Reproduction,* Vol.77, No.6, (December 2007), pp. 906-913, ISSN 1529-7268

Liu, Y. X. *(2007). Involvement of plasminogen activator and plasminogen activator inhibitor type 1 in spermatogenesis, sperm capacitation, and fertilization.* Seminars in Thrombosis and Hemostasis, Vol. 33, No.1, (February 2007), pp. 29-40, ISSN *1098-9064*

Lorton, S. F. & First, N. L. (1979). Hyaluronidase does not disperse the cumulus oophorus surrounding bovine ova. *Biology of Reproduction,* Vol.21, No.2, (September 1979), pp. 301–308, ISSN 1529-7268

Lyng, R. & Shur, B. D. (2007). Sperm-egg binding requires a multiplicity of receptor-ligand interactions: new insights into the nature of gamete receptors derived from reproductive tract secretions. *Society of Reproduction and Fertility,* Supplement 65, pp. 335-351

Mahawar, M. & Joshi, P. (2008). Goat vitronectin: characterization and binding to Staphylococcus aureus. *Comparative Biochemistry and Physiology, Part B, Biochemistry & Molecular Biology,* Vol.149, No.3, (March 2008), pp. 410-418, ISSN 1096-4959

O'Flaherty, C.; Breininger, E.; Beorlegui, N. & Beconi, M. T. (2005). Acrosome reaction in bovine spermatozoa: role of reactive oxygen species and lactate dehydrogenase C4. *Biochimica et Biophysica Acta,* Vol.1726, No.1, (October 2005), pp. 96-101, ISSN 0006-3002

Revah, I.; Gadella, B. M.; Flesch, F. M.; Colenbrander, B. & Suarez, S. S. (2000). Physiological state of bull sperm affects fucose- and mannose-binding properties. *Biology of Reproduction,* Vol.62, No.4, (April 2000), pp. 1010-1015, ISSN 1529-7268

Royce, P. M. & Steinmann, B. (2002). *Connective tissue and its heritable disorders. Molecular, Genetic, and Medical Aspects,* Second Edition, 305, Wiley, ISBN 978-0-471-25185-9, New York, USA

Ruoslahti, E. & Pierschbacher, M. D. (1986). Arg-Gly-Asp: a versatile cell recognition signal. *Cell,* Vol.44, No.4, (February 1986), pp. 517-518, ISSN 0092-8674

Sostaric, E.; van de Lest, C. H.; Colenbrander, B. & Gadella, B. M. (2005). Dynamics of carbohydrate affinities at the cell surface of capacitating bovine sperm cells. *Biology of Reproduction,* Vol.72, No.2, (February 2005), pp. 346-357, ISSN 1529-7268

Spencer, T. E.; Johnson, G. A.; Bazer, F. W. & Burghardt, R. C. (2004). Implantation mechanisms: insights from the sheep. *Reproduction,* Vol.128, No.6, (December 2004), pp. 657-668, ISSN 1741-7899

Stockmann, A.; Hess, S.; Declerck, P.; Timpl, R. & Preissner, K. T. (1993). Multimeric vitronectin. Identification and characterization of conformation-dependent self-association of the adhesive protein. *The Journal of Biological Chemistry,* Vol.268, No.30, (October 1993), pp. 22874-22882, ISSN 1083-351X

Sutovsky, P.; Flechon, J. E. & Pavlok, A. (1995). F-actin is involved in control of bovine cumulus expansion. *Molecular Reproduction and Development,* Vol.41, No.4, (Augustus 1995), pp. 521-529, ISSN 1098-2795

Suzuki, S.; Oldberg, A.; Hayman, E. G.; Pierschbacher, M. D. & Ruoslahti, E. (1985). Complete amino acid sequence of human vitronectin deduced from cDNA. Similarity of cell attachment sites in vitronectin and fibronectin. *The EMBO Journal,* Vol.4, No.10, (October 1985), pp. 2519-2524, ISSN 0261-4189

Talbot, P.; Shur, B. D. & Myles, D. G. (2003). Cell adhesion and fertilization: steps in oocyte transport, sperm-zona pellucida interactions, and sperm-egg fusion. *Biology of Reproduction,* Vol.68, No.1, (January 2003), pp. 1-9, ISSN 1529-7268

Talevi, R. & Gualtieri, R. (2001). Sulfated glycoconjugates are powerful modulators of bovine sperm adhesion and release from the oviductal epithelium *in vitro*. *Biology of Reproduction*, Vol.64, No.2, (February 2001), pp. 491-498, ISSN 1529-7268

Tanghe, S.; Van Soom, A.; Nauwynck, H.; Coryn, M. & de Kruif, A. (2002). Minireview: Functions of the cumulus oophorus during oocyte maturation, ovulation, and fertilization. *Molecular Reproduction and Development*, Vol.61, No.3, (March 2002), pp. 414-424, ISSN 1098-2795

Tanghe, S.; Van Soom, A.; Duchateau, L. & de Kruif, A. (2004*a*). Inhibition of bovine sperm-oocyte fusion by the p-aminophenyl derivative of D-mannose. *Molecular Reproduction and Development*, Vol.67, No.2, (February 2004), pp. 224-232, ISSN 1098-2795

Tanghe, S.; Van Soom, A.; Duchateau, L.; Nauwynck, H. & de Kruif, A. (2004*b*). Carbohydrates and glycoproteins involved in bovine fertilization *in vitro*. *Molecular Reproduction and Development*, Vol.68, No.4, (Augustus 2004), pp. 492-499, ISSN 1098-2795

Thys, M.; Vandaele, L.; Morrell, J. M.; Mestach, J.; Van Soom, A.; Hoogewijs, M. & Rodriguez-Martinez, H. (2009*a*). *In vitro* Fertilizing Capacity of Frozen-thawed Bull Spermatozoa Selected by Single-layer (Glycidoxypropyltrimethoxysilane) Silane-coated Silica Colloidal Centrifugation. *Reproduction in Domestic Animals*, Vol.44, No.3, (June 2009), pp. 390-394, ISSN 1439-0531

Thys, M.; Nauwynck, H.; Maes, D.; Hoogewijs, M.; Vercauteren, D.; Rijsselaere, T.; Favoreel, H. & Van Soom, A. (2009*b*). Expression and putative function of fibronectin and its receptor (integrin $\alpha_5\beta_1$) in male and female gametes during bovine fertilization *in vitro*. *Reproduction*, Vol.138, No.3, (September 2009), pp. 471-482, ISSN 1741-7899

Vandaele, L.; Mateusen, B.; Maes, D. G.; de Kruif, A. & Van Soom, A. (2007). Temporal detection of caspase-3 and -7 in bovine *in vitro* produced embryos of different developmental capacity. *Reproduction*, Vol.133, No.4, (April 2007), pp. 709-718, ISSN 1741-7899

Van Soom, A.; Tanghe, S.; De Pauw, I.; Maes, D. & de Kruif, A. (2002). Function of the cumulus oophorus before and during mammalian fertilization. *Reproduction in Domestic Animals*, Vol.37, No.3, (June 2002), pp. 141-151, ISSN 1439-0531

Analysis of 3'UTR of *Prnp* Gene in Mammals: Possible Role of Target Sequences of miRNA for TSE Sensitivity in Bovidae and Cervidae

Daniel Petit, Jean-Michel Petit and François Gallet
UGMA, UMR 1061 INRA/University of Limoges
France

1. Introduction

Transmissible spongiform encephalopathies (TSEs) or prion diseases are neurodegenerative diseases with an inexorably fatal outcome. They affect both human and other mammals, and are especially frequently found in Bovidae and Cervidae families (Aguzzi & Sigurdson, 2004). In the first family, sheep and goats can spontaneously develop the prion disease or scrapie (Grenn et al., 2008), whereas cattle only develop a disease, known as Bovine Spongiform Encephalopathy (BSE), after a contact with the infectious prion protein (Ducrot et al., 2008). In the second family, the chronic wasting disease (CWD) has been described in mule deer and elk. Besides these two Ruminant families, TSE was also retrieved in Carnivora, as Felidae (feline spongiform encephalopathy) and Mustelidae (transmissible mink encephalopathy) (Miller et al., 2008; Sigurdson & Miller, 2003). All these diseases are characterized by the accumulation of PrPSc, the abnormally folded isoform of the cellular prion protein (PrPC) in the brain (Prusiner 1998; Norrby 2011).

The cellular prion protein is a glycosylphosphatidylinositol anchored glycoprotein of 256 amino acids in sheep (Bovidae, subfamily Caprinae) and Cervidae. The N-ter of the protein is mainly unstructured while the C-terminal domain is globular. The C-ter domain is highly structured and is stabilized by an intramolecular disulfide bound. It contains three α-helices and a short β-sheet. TSE is a conformational disease, where change from PrPC to PrPSc involves an increase in β-sheet content from 3% to 40%, and a decrease in α-helical structure from 40% to 30% that modify irreversibly the protein folding (Cohen et al., 1998; Smirnovas et al., 2009).

In a previous work, we showed that the ARQ/ARQ genotype is rather sensitive to the TSE in sheep and goat but protective in pig and rabbit, known as resistant species [Martin et al., 2009]. Otherwise, several studies point out that the expression level of the prion gene could modulate the onset of TSE, particularly in cattle (Sander et al., 2004-2006; Haase et al., 2007). The involvement of the 5' promoter region has been investigated in sheep (Saunders et al., 2009) and cattle (Xue et al., 2008) and it appears that polymorphism in this region could induce different responses to scrapie (Marcos-Carcavilla et al., 2008) and BSE (Brunelle et al., 2008). The analysis of a 23 pb insertion/deletion polymorphism in German and Swiss breed cattle revealed that the deletion is associated to a higher expression level, more frequently found in cattle affected with classical BSE. Moreover, in vitro studies showed that the possibility to infect neurosphere cultures with scrapie prion is linked to an over-expression of PrPC (Giri et al., 2006). Although the determinism of prion disease is multifactorial (prion strain and prion protein sequence, see Doherr, 2003), it seems that a high expression of PrPC gene is necessary (Krejciova et al., 2011).

In numerous organisms, post-transcriptional gene regulation involves small (around 18-25 nucleotides long) non-coding RNA molecules, the microRNAs (miRNAs). They recognise specific target sequences in the 3'UTR of some transcripts, mediating their silencing (Fabian et al., 2010) by inducing their degradation or by inhibiting their translation. According to the accuracy of the complementarity between both sequences, each miRNA can regulate up to hundreds of genes. Using microarrays, Saba et al. (2008) evidenced 15 miRNAs, potentially controlling the transcript amount of more than 100 genes. These micro-RNAs are de-regulated in the brain of scrapie affected mouse.

In their comparative analysis of *Prnp* organization in human, mouse and sheep, Lee et al. (1998) showed the high conservation of the 3'-UTR regions and suggested their role in mRNA stability. Moreover, they evidenced insertions of transposable elements in the sheep gene. In their more recent work, Premzl and Gamulin (2007) could easily align the part close to the poly-adenylation signal in a wide range of mammals. Taken together, these data suggest that alignments of mammalian 3'UTRs contain reliable information. The regulatory sequences borne by the 3'-UTR are involved in mRNA processing, transport, stability, and translation. As the 3'-UTRs harbor recognition sites for diverse RNA-binding proteins that regulate gene expression as well as active microRNA target sites, our strategy was to compare the 3'UTR of the *Prnp* gene, from the sequence terminus codon to the poly-adenylation signal in different Mammal lineages. We expect to find in the most frequently families affected by TSE, i.e., Bovidae and Cervidae, oddities in their potential miRNA targets. Three types of results are predicted to be obtained. In the first type, based on the miRNA-related gene silencing, we predict that Bovidae and Cervidae families lack some targets widely present in other Mammals. In the second type, we predict that target sequences are present in DNA from most mammalian lineages, including the both critical families, but a loss of corresponding miRNA may occur. The last type, although more unlikely, would be to detect targets only harboured by Bovidae and Cervidae, which would lead to the activation of gene expression, as shown by Vasudevan et al. (2007) in a model of cellular stress. In brief, we search for potential targets that could affect repression mechanisms of *Prnp* gene expression, and thus enhance the sensitivity of Bovidae and Cervidae to the disease.

2. Material and methods

A Blast search of 3'UTR sequences in traces (Whole Genome Shotgun) of NCBI genome database was undertaken. We restricted the species sampling to 2 to 3 species for main division within Eutherians, and the outgroup was constituted by 2 Marsupials (*Macropus eugenii* and *Didelphis domestica*). The alignment of sequences was performed with Dialign software (Morgenstern, 2004), and then manually edited. The table 1 gathers the 27 analyzed sequences and their accession numbers. The repeat elements were identified with RepeatMasker site (http://www.repeatmasker.org/cgi-bin/WEBRepeatMasker). The target sequences of mi-RNAs were detected with mirBase (www.mirbase.org/), following the papers of Griffiths-Jones (2004), Griffiths-Jones et al. (2006 and 2008), and Kozomara & Griffiths-Jones (2011).

Species names	Access numbers in NCBI
Bos taurus	NM_181015
Canis familiaris	NC_006606.2
Cervus elaphus	EU032284
Equus caballus	AAWR02026367.1
Felis catus	ACBE01091825.1
Homo sapiens	NC_000020.10, U29185
Lama pacos	ABRR01260217
Loxodonta africana	AAGU03041236.1
Macropus eugenii	ABQO010016931.1
Monodelphis domestica	AAFR03004045.1
Mus musculus	NM_011170, U29186
Mustella putorius	AEYP01079589.1
Myodes glareolus	EF455012
Myotis lucifugus	AAPE02032418.1
Ochotona princeps	AAYZ01152834.1
Odocoileus hemionus	AY330343
Oryctolagus cuniculus	NW_003159242.1
Ovis aries	NM_00100948.1, U67922
Procavia capensis	ABRQ01131540.1
Pteropus vampyrus	ABRP01225173.1
Rattus norvegicus	NC_005102.2
Saimiri sciureus	ti | 1176383476
Sorex araneus	AALT01429291.1
Spermophilus tridecemlineatus	AAQQ01175405.1
Sus scrofa	NM_001008687
Tarsius syrichta	ABRT010290102.1
Tursiops truncatus	ABRN01336586.1

Table 1. List of mammalian studied species

In order to visualise the most conservative sites of the 3'UTR, we used the method developed in Petit et al. (2006) and Martin et al. (2009). Briefly, the alignment of sequences, cleaned from insertions, was treated by Parsimony program of Phylip package vers. 3.69 (Felsenstein, 2004), to export the number of changes site by site. For each site, this number was divided by the number of sequences, allowing drawing the profile of site change rates.

3. Results

3.1 Position of insertions and deletions

The length of the 3'UTR region is highly variable as the murine sequence comprises about 1250 pb whereas the bovine one is more than 3500 pb long and possesses two potential poly-adenylation signals that are separated by about 1300 bp. Such a result is also observed in sheep and mule deer. The difference is mainly explained by a series of transposable elements (SINE/RTE-BovB of about 380 pb length, SINE/BovA about 160 bp length and DNA/TcMar-Mariner with at least 1220 bp). All are inserted in the 3′ end of the bovine sequence, and also shared by sheep, red deer, and mule deer (Fig. 1). A striking common organization is indicative of events that occurred in the ancestors of Bovidae and Cervidae. In spite of extensive searches, these elements were not retrieved in the other studied mammals. Upstream the transposable elements, we observe a good overall conservation of the sequences, allowing to clearly evidence insertion and deletion events in some species.

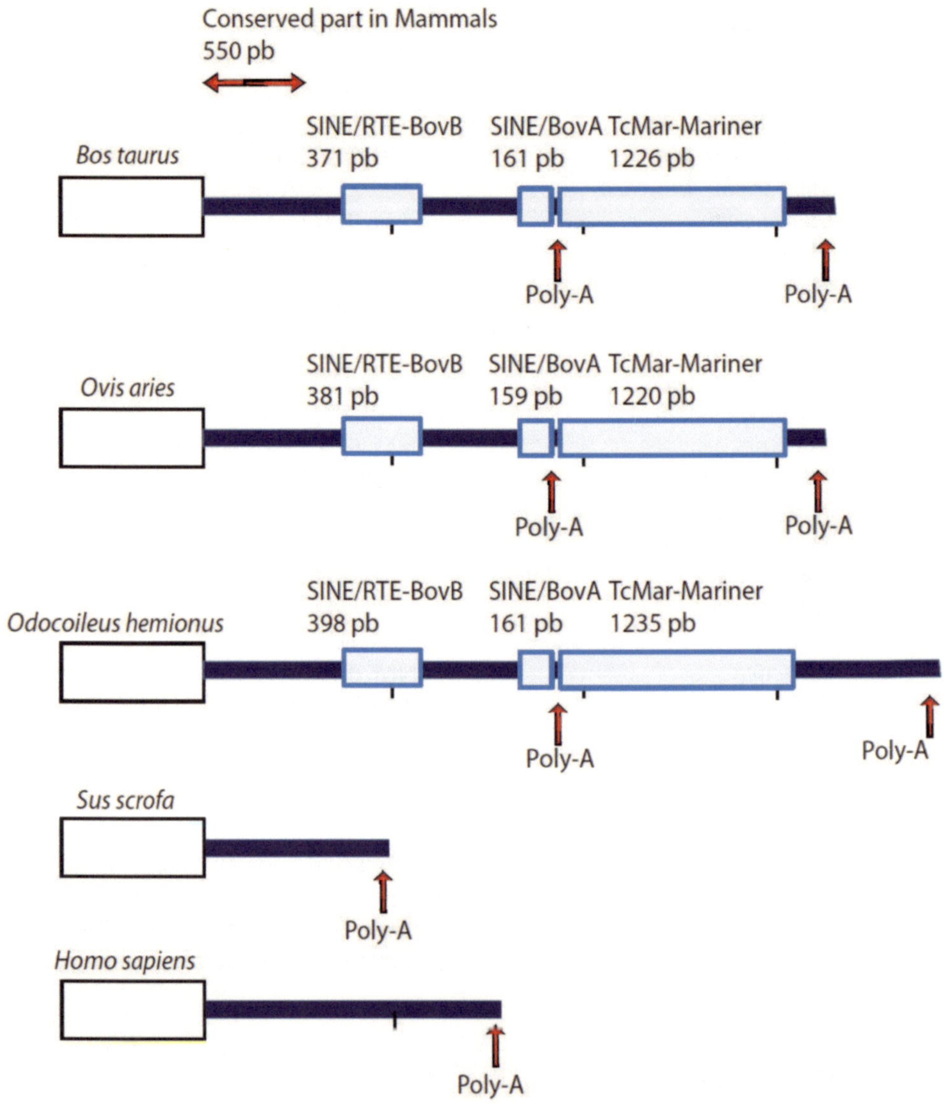

Fig. 1. Organization of the *Prnp* 3′UTR in several Mammals. The coding sequence of the gene is boxed in thin black. The inserted elements are boxed in thick blue. There are two poly-adenylation signals (Poly-A) in Bovidae-Cervidae, and only one in human and porcine. Short vertical lines are drawn every 1000 pb.

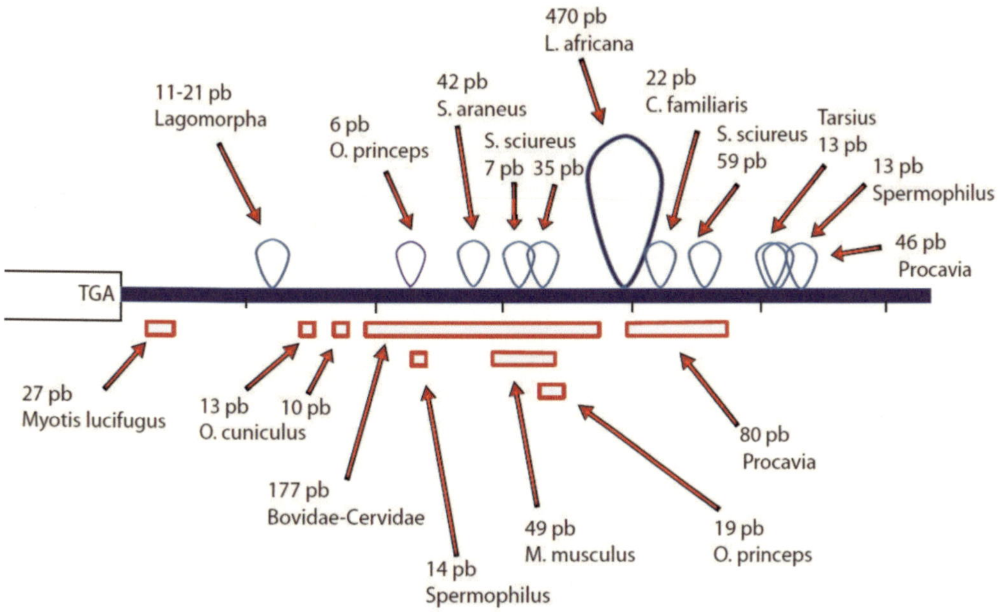

Fig. 2. Position of insertions (loops above the thick blue line) and deletions (wide red rectangles below) in the 550 pb conserved part of Eutherians. The hypothetical ancestral sequence for Eutherians was taken as reference. Short vertical lines are drawn every 100 pb.

The most interesting features are a 177 pb long deletion specific to Cervidae-Bovidae, and a 470 pb long insertion in African Elephant. No indel particular to human could be recorded.

3.2 Profile of site change rates along the 3'UTR

The most conserved sites are between positions 70 and 115 downstream the stop codon. In the part corresponding to the deletion shared by Bovidae and Cervidae, we observed little variations in Eutherians. From the 550th position on, due to a lowest identity between species, the sequences cannot be significantly aligned (fig. 3).

3.3 Position of potentially target sites of miRNAs

We focused on 3 regions: (i) between the stop codon and the last nucleotide before the large deletion affecting Bovidae-Cervidae (proximal region), (ii) inside this large deletion, and (iii) in the transposable elements shared by Bovidae-Cervidae.

The table 2 gathers the target sites present in more than one species, in the region proximal to codon stop. It should be noted that several potential targets are overlapping (fig. 4-5). In two cases, the distribution of target is identical in the studied mammals and they were considered as identical (Mir-569-155 on one hand and Mir-3918-509 on the other). In contrast, the presence of other overlapping targets varies among mammals (Mir-3102, 376, 3918/509, 519 and 3432). Several targets are shared by only Bovidae-Cervidae species (Mir-569/155, 376, 15b), or Bovidae-Cervidae and other mammals (Mir-3102 with *M. musculus* and *S. scrofa*, Mir-3918/509 with *H. sapiens*, *M. putorius*, and *O. princeps*, Mir-519 with *C familiaris, M. putorius* or *S. scrofa*). Mir-3432 and Mir-140 have targets shared by *C. familiaris* and several species, including *H. sapiens, M. musculus* or *S. scrofa*. In summary, the widest distribution of a given target concerns Mir-3918/509, but not any target is common to all mammals excluding the Bovidae-Cervidae. If the region encompassing targets to Mir-3918/509-519 is taken as a whole, it should be noted that most mammals (except murine) share this region. Otherwise, no target was found to be shared by *F. catus* and non-Felidae mammals, or by *E. caballus* and non-Equidae mammals. In contrast, there are several targets shared by only Bovidae-Cervidae.

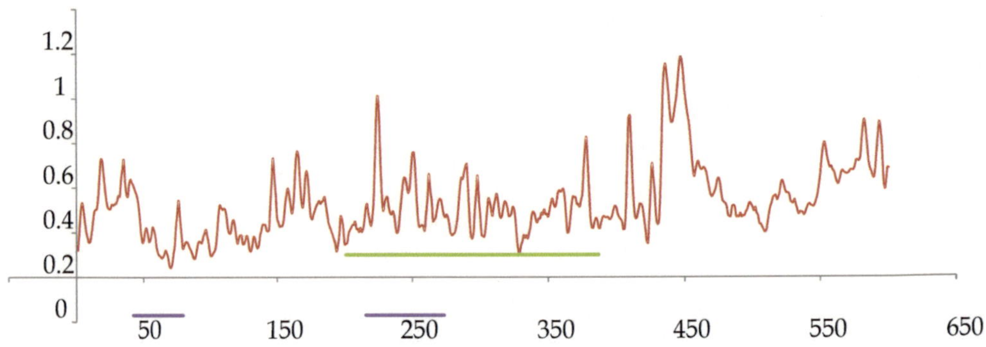

Fig. 3. Profile of site change rates within a 600 pb long segment downstream the stop codon. The numbers in abscissa correspond to the hypothetical ancestral sequence for Eutherians. The violet thick lines correspond to the position of putative targets for Mir-509 group and Mir-369 group. The green thick line corresponds to the position of the large deletion in Bovidae-Cervidae.

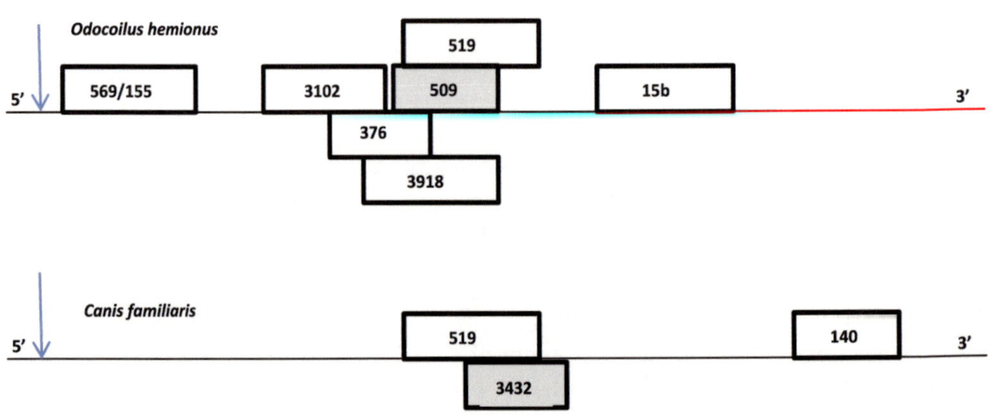

Fig. 4. Position of the putative targets for Mirs in the proximal region. The arrows indicate stop codon position. The grey boxes correspond to targets present in human.

```
C. familiar.   TGAGGAGGGCCTTCCCATTCCCTGC---ACCA   //   GGTT--GGGGGAGGGGGTGTCTACCTACAGCCCTTT----AG---
Felis catus    TGAGGGCGACCTTGCCGTCTTCCCC---ATCT   //   GGTT--GGGGGAGGGGGTGTCTACCTACAGCCCTGT----AGTGG
M. putorius    TGAGGATGGCCTTCCCATT---CTCTCCATCG   //   GGTT--GGGGGAGGGGGTGTCTACCTACAGCCCTGT----AGTGA
T. truncatus   TGAGGAAGGCCTTCCTGTT---TTC---ATCA   //   GGTT--GGGGGAGGGGGTATCTACCTGCAGCCCTTT----AGTGG
E. caballus    TGAGGATGGCCTTCCTGTT---T-C---ACCA   //   GGGT--TGGGGAAGGGCTGTCTACCTGCGGCCCTTT---CAGTGG
P. vampyrus    TGAGGGTGGCCTTCTGTTT---T-C---ACCA   //   GGTT--GGGGAAGGGGGTATCTACCTACAGCCCTTT----AGTGG
Bos taurus     TAGGGGCAACCTTCCTGTTTTCATT---ATCT   //   GGTT--GGGGGAGGGAGTATCTACCTGCAGCCCCGT----AGTGG
Ovis aries     TAGGGGCAACCTTCCTGTTTTCATT---ATCT   //   GGTT--GGGGGAGGGAGTGTCTACCTGCAGCCCTGT----AGTGG
Lama pacos     TGATGGCAGCCTTCCTGTTTTCATC---ATCT   //   CAGT--TGGGGAGGGGGTATCTACCTGCAGCTCTTT----AGTGG
O. hemionus    TAGGGGCAACCTTCCTGTTTTCATT---ATCT   //   GGTT--GGGGGAGGGAGTATCTACCTGCAGCCCTGT----AGTGG
C. elaphus     TAGGGGCAACCTTCCTGTTTTCATT---ATCT   //   GGTT--GGGGGAGGGAGTATCTACCTGCAGCCCTGT----AGTGG
Sus scrofa     TGAGGGTGGCCTTTCTGTCGGCATC---ATCT   //   GGTT--GGGGGAGGGAATATCTACCTGCAGCCCTTT----AGTGG
Mus musculus   TGAGGGAGGCCTTCCTGCTTGTTCC---TTCG   //   GGCT--GGGGGAGGGGTTATCCACCTGTAGCTCTTT---CAATGG
Rattus rattus  TGAGGAGGCCTTCCTGCTT---G-T---TCCT   //   GGCT--GGGGGAGGGGTTACCCACCTGTAGCTCTTTcaaTTGTTG
S. trideceml.  TGAGGAAGGACTTCCTATTTCCA-A---GTCG   //   AGGG--TTGGGAGGGGGTATCCACCTGCAGCCCTTT---TAGAGG
O. cuniculus   TGAGGGAGGCCTCCCCGCCTGCGCC---GTCT   //   TCGC--CGGGGGAGGGGTGTCCGCCTGCAGCCCTTT---TCGTGG
Homo sapiens   TGAGGAAGGTCTTCCTGTTTTCACC---ATCT   //   GCTT--GAGGGAGGCGGTATCCACCTGCAGCCCTTT---TAGCGG
S. sciureus    TGAGGAAGGTCTTCCTGTTTTCACC---ATCT   //   GGTT--GAGGGAGGGGGTATCTACATGCAGCCCTTT---TAGTGG
T. syrichta    TGAGGAAGGTCTTCCTGTTTTCACC---ATCT   //   AGGTtgGGGAAGGGGTATATCCACCTGCAGCCTTTT---TAGTGG
Sorex araneus  TGAGGACGGCCCTCCCGCCCGCGCC---GACT   //   AGGT--TGGGGAGGGGGGGTCTACCTGCGGCCCTTG---TAGTGG
L. africana    TGAGGCAGTCTTCCCACTGTCAT-C---ATCT   //   CAGG--TGGGGAGGGGGTATCTACCTGCAGCCCTTT---TAGTGA
P. capensis    TGAGGCAATCTTCCAATTGTCAC-C---ATTT   //   AGGT--GGGGAGGGGGTTATCTACCTGCAGCCCTTT---TAGTGG
                   <---Mir-569/155--->             <---Mir-3102--->   <---Mir-519--->
                                                          <---Mir-3918--->
```

Fig. 5. Alignment of targets for Mirs in the proximal region of stop codon in mammals. Grey background: stop codon, green background: conservation in ≥ 75 % sequences; yellow background: conservation in ≥ 50 % sequences.

Mir name	*Bos taurus*	*Canis familiaris*	*Cervus elaphus*	*Homo sapiens*	*Mus musculus*	*Mustela putorius*	*Ochotona princeps*	*Odocoilus hemionus*	*Ovis aries*	*Sus scrofa*
569/155	X		X					X	X	
3102	X		X		X			X	X	X
376	X		X					X		
3918/509	X		X	X		X	X	X	X	
519		X	X			X		X	X	X
3432		X		X						X
15b			X					X	X	
140		X			X					

Table 2. Micro RNAs targets common to several species between the stop codon and the large deletion.

Within the sequence corresponding to the large deletion region, several targets (Mir-4763, 1587, 744, 15b, and 369) are overlapping (fig. 6-7). In two couples of targets (Mir 569/155 and Mir-3918/509), the distributions in mammals are identical and each couple was considered as single. The remaining overlapping targets have their own distribution within mammals, and they were considered separately (table 3). It appears that the number of putative targets common to at least 2 species varies among the different mammals. There is none in *M. musculus*, but four in *E. caballus* and even six in *F. catus*. There is no target common to all orders of mammals, but two are shared by at least four groups: Mir-4763 is common to *E. caballus, F. catus, L. pacos* and *S. scrofa*, and Mir-15B/369 to *E. caballus, H. sapiens, F. catus,* and *S. scrofa*.

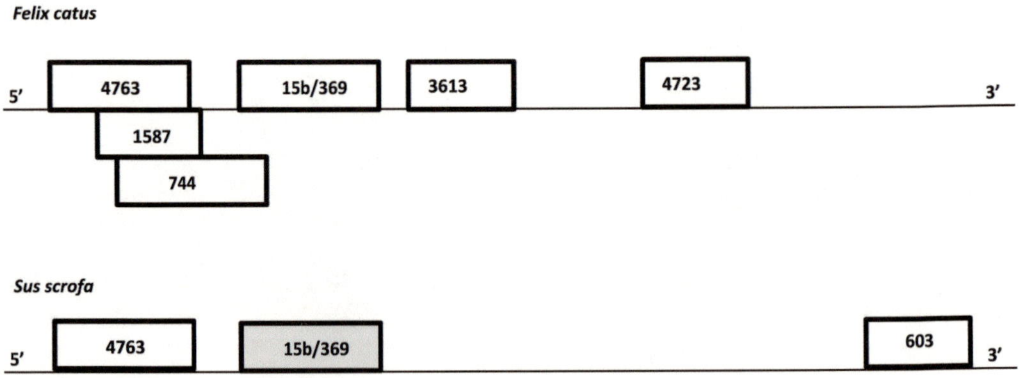

Fig. 6. Position of putative targets within the large deletion common to Bovidae-Cervidae. The grey box corresponds to the target present in human.

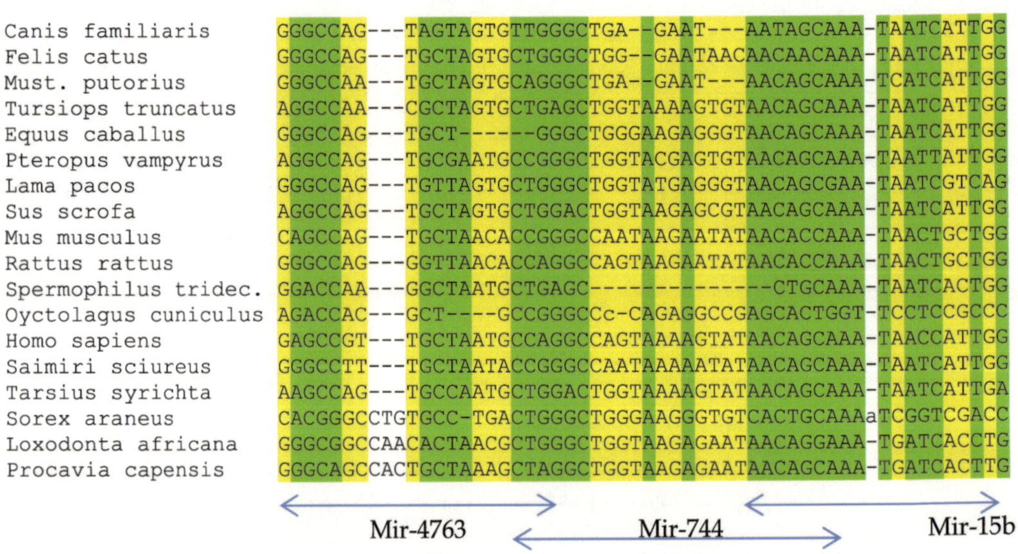

Fig. 7. Alignment of overlapping target sequences in the region of the large deletion common to Bovidae-Cervidae. Green background: conservation in ≥ 75 % sequences; yellow background: conservation in ≥ 50 % sequences.

Bovidae and Cervidae are the only species in which transposable elements are present in the transcript of *Prnp*. The poly-adenylation site common with other mammals is situated downstream from Mariner element. However, there is an extra potential poly-adenylation site, downstream to the SINE/BovA. Few targets for Mirs are present in these regions and common to *B. taurus, O. aries* and *O. hemionus* (Table 4).

Mir name	*Canis familiaris*	*Equus caballus*	*Homo sapiens*	*Felix catus*	*Lama pacos*	*Mustela putorius*	*Myodes glareolus*	*Sus scrofa*
4763		X		X	X			X
1587		X		X				
744				X		X	X	
15B/369		X	X	X				X
3613	X	X		X				
603					X			X
4723	X			X				

Table 3. Micro RNAs targets common to several species in the large deletion region

Transposable elements	miRNAs
SINE/RTE-BovB	miR-4311
SINE/BovA	miR-137
	miR-596
	miR-2277-3p
TcMar-Mariner	miR-2284d
	miR-701
	miR-3143
	miR-2294

Table 4. Putative targets for Mirs, common to *B. taurus*, *O. aries* and *O. hemionus*.

4. Discussion

The most interesting feature in the comparative analysis of 3'UTR region of mammalian *Prnp* genes is the occurrence of 2 events common to Bovidae and Cervidae, i.e., a 177 pb-long deletion and 3 insertions of transposable elements. These insertions were previously described in sheep by Lee et al. (1998), but not in Cervidae. To simplify the discussion, the large deletion common to Bovidae-Cervidae will be abbreviated as LDCBC in the following. The mammalian distribution of insertions and LDCBC fit with the phylogeny within Artiodactyla, since Bovidae and Cervidae were shown to be sister families by Kuznetsova et al. (2005) from 12S and 16S rRNA analyses. In contrast, the more basally genera *Lama* (Camelidae), *Sus* (Suidae) and *Tursiops* (Delphinidae) within Cetartiodactyla do not possess the LDCBC and the 3 insertions. Unfortunately, none studied species harboured either the LDCBC or the 3 insertions. The sequencing of *Giraffa camelopardalis* could bring an interesting issue, given its intermediary position between Bovidae-Cervidae group and the Camelidae. Consequently, it is not possible to assign a specific role of the LDCBC and or of the 3 insertions to the stability of *Prnp* RNA. Their potential influence will be assessed independently.

As explained in introduction, several types of potential target sites of micro-RNAs are to be explored according to their position relatively to the LDCBC: (i) between the stop codon and the LDCBC (all Mammals), (ii) in the sequence corresponding to the LDCBC (all Mammals except Bovidae-Cervidae), and (iii) in the inserted elements particular to Bovidae-Cervidae. Generally, miRNAs inhibit protein synthesis either by the repression of the translation and/or by the activation of the deadenylation leading to the degradation of the targeting mRNA (Eulalio et al., 2008; Chekulaeva & Filipowicz, 2009). Multiples sites, either for the same or different miRNAs, are usually required for a more effective repression, and they tend to act cooperatively when they are close to each other (Doench et al., 2004; Grimson et al., 2007). As most microRNA play a role in repression, a special attention will be brought to targets present in most mammals but absent in Bovidae-Cervidae. However, targets restricted to Bovidae-Cervidae could have significance through an activating role in stability of transcript (Vasudevan et al., 2007).

In the sequence comprised between stop codon and LDCBC, the most spread target within mammals corresponds the group Mir-3918/509-519. According to fig. 3, this target is situated in one of the most conservative regions in 3'UTR of PRNP gene. Its possible involvement could be interpreted as the result of an under-expression of the corresponding Mirs in Bovidae-Cervidae, according to the second type mentioned in introduction section. It would be interesting to test experimentally this idea.

In the sequence corresponding to the large deletion common to Bovidae-Cervidae, two targets are interesting to consider, corresponding to Mir4763 and Mir-15B/369. According to literature, Mir15b belong to a small family of Mirs comprising also Mir15a, Mir16a, Mir16b, and Mir195 in Mammals. Interestingly, Mir15b as well as Mir15a have proved to be involved as repressing cancers in human. It has been shown an association between deletion of Mir15a gene, together with Mir16 in the same cluster in HSA chromosome 13q14, and pituitary tumors (Bottoni et al., 2005). The same authors showed that the expression level of these 2 Mirs is inversely correlated to the pituitary adenoma growth. Another study of Calin et al., 2002 showed that deletions in the same cluster of Mir 15b-Mir16 are associated to chronic lymphocytic leukemia. Similarly, a high expression level of Mir15b is linked to a low proliferation of hepatocellular carcinoma (Chung et al., 2010). As for Mir369, Williams et al. (2007) showed that members of family Mir-154 are especially expressed in mouse and human foetal, but not adult, lung.

The presence of transposable elements in the *Prnp* transcript of Bovidae and Cervidae is an interesting feature. It is unlikely that they contribute to the stability of mRNA, as no publications support this view. However, these transposable elements contain putative targets for Mirs that confer new possibility of regulation, as shown in mammals (Smalheiser & Torvik, 2005). These putative Mirs could play a role in the stability of *Prnp* gene. We could bring the same hypothesis to the targets of Mir-569/155, 376, 15b, present in the region proximal to stop codon.

The sensitivity to TSE has certainly multiple origins according to mammals. For example, the 3'UTR sequence of *M. putorius*, known as sensitive (Miller et al., 2008), presents no indel in the first 600 pb. Interestingly, this species as well as *Myodes glareolus*, another sensitive species (Di Bari et al., 2008), lack both targets for Mir 4763 and 15B/369, in contrast to the species known as resistant..

The sensitivity of a given Mammal to TSE depends on many factors, as shown by the network analyses conducted during the development of the disease. But even if many genes have been proved to be involved, the expression level of *Prnp* gene plays a crucial role and many approaches still need to be explored such as miRNAs.

5. Conclusion

The 3'-UTR sequence of *Prnp* gene is well conserved with Eutherians, and especially in the first 600 pb downstream to stop codon, suggesting a role in regulation of stability of the mRNA. In the 3'-UTR part of *SPRN*, a gene close to *Prnp*, Premzl & Gamulin (2007) showed a target for Mir-34a shared by several orders of mammals. It is unlikely that such a unique target is present in the case of mammalian *Prnp*. Our analyses rather suggest an interaction between several targets which could modulate the stability of this transcript. The combination of acting Mirs could vary among mammals and the high sensitivity of Bovidae and Cervidae could be explained by the lack of several key targets.

6. Acknowledgments

We thank the Région Limousin for financial support.

7. References

Aguzzi, A & Sigurdson, C (2004). Antiprion immunotherapy: to suppress or to stimulate? Nat. Rev. Immunol., 4(9), 725-736.

Bottoni, A, Piccin, D, Tagliati, F, Luchin, A, Zatelli, MC, & degli Uberti, EC. (2005). miR-15a and miR-16-1 down-regulation in pituitary adenomas. J. Cell Physiol., 204(1), 280-285.

Brunelle, BW, Greenlee, JJ, Seabury, CM, Brown, CE 2nd, & Nicholson, EM. (2008). Frequencies of polymorphisms associated with BSE resistance differ significantly between Bos taurus, Bos indicus, and composite cattle. BMC Vet. Research, 4, 36-43.

Calin, GA, Dumitru, CD, Shimizu, M, Bichi, R, Zupo, S, Noch, E, Aldler, H, Rattan, S, Keating, M, Rai, K, Rassenti, L, Kipps, T, Negrini, M, Bullrich, F, & Croce, CM. (2002). Frequent deletions and down-regulation of micro- RNA genes miR15 and miR16 at 13q14 in chronic lymphocytic leukemia. Proc Natl Acad Sci U.S.A., 99(24), 15524-15529.

Chekulaeva, M, & Filipowicz, W. (2009) Mechanisms of miRNA-mediated post-transcriptional regulation in animal cells. Curr. Opin. Cell Biol., 21, 452–60.

Chung, GE, Yoon, JH, Myung, SJ, Lee, JH, Lee, SH, Lee, SM, Kim, SJ, Hwang, SY, Lee, HS, & Kim, CY. (2010). High expression of microRNA-15b predicts a low risk of tumor recurrence following curative resection of hepatocellular carcinoma. Oncol. Rep., 23(1), 113-119.

Cohen, FE, & Prusiner, SB. (1998). Pathologic conformations of prion proteins. Annu. Rev. Biochem., 67, 793-819.

Di Bari, MA, Chianini, F, Vaccari, G, Esposito, E, Conte, M, Eaton, SL, Hamilton, S, Finlayson, J, Steele, PJ, Dagleish, MP, Reid, HW, Bruce, M, Jeffrey, M, Agrimi, U, & Nonno, R. (2008). The bank vole (*Myodes glareolus*) as a sensitive bioassay for sheep scrapie. J Gen Virol., 89(Pt 12), 2975-2985.

Doench, JG, & Sharp, PA. (2004). Specificity of microRNA target selection in translational repression. Genes Dev. 18:504–11.

Doherr, MG. (2003). Bovine spongiform encephalopathy (BSE) infectious, contagious, zoonotic or production disease? Acta Vet Scand Suppl., 98, 33-42.

Ducrot, C, Arnold, M, de Koeijer, A, Heim, D, & Calavas, D. (2008). Review on the epidemiology and dynamics of BSE epidemics. Vet. Research, 39(4), 15.

Eulalio, A, Huntzinger, E, Izaurralde, E. (2008) Getting to the root of miRNA-mediated gene silencing. Cell, 132, 9–14.

Grimson A, Farh KK, Johnston WK, Garrett-Engele P, Lim LP, Bartel DP. (2007). MicroRNA targeting specificity in mammals: determinants beyond seed pairing. Mol. Cell., 27, 91–105.

Fabian, RF, Sonenberg, N, & Filipowitz, W, (2010). Regulation of mRNA translation and stability by microRNAs. Annu. Rev. Biochem., 79, 351-379.

Felsenstein, J. (2004). PHYLIP (Phylogeny Inference Package) version 3.6. Distributed by the author. Department of Genome Sciences, University of Washington, Seattle.

Giri, R.K, Young, R., Pitstick, R., DeArmond, S.J., Prusiner, S.B., & Carlson, G.A. (2006). Prion infection of mouse neurospheres Proc Natl Acad Sci U.S.A, 103(10), 3875–3880.

Green KM, Browning SR, Seward TS, Jewell JE, Ross DL, Green MA, Williams ES, Hoover EA, & Telling GC. (2008). The elk PRNP codon 132 polymorphism controls cervid and scrapie prion propagation. J. Gen. Virol., 89(2), 598-608.

Griffiths-Jones, S. (2004). The microRNA Registry. Nucleic Acids Research, 32, D109-D111.

Griffiths-Jones, S, Grocock, RJ, van Dongen, S, Bateman, A, & Enright, AJ. (2006). miRBase: microRNA sequences, targets and gene nomenclature. Nucleic Acids Research, 34, D140-D144.

Griffiths-Jones, S, Saini, HK, van Dongen, S, & Enright, AJ. (2008). miRBase: tools for microRNA genomics. Nucleic Acids Research 36, D154-D158

Groschup, MH, Lacroux, C, Buschmann, A, Lühken, G, Mathey, J, Eiden, M, Lugan, S, Hoffmann, C, Espinosa, JC, Baron, T, Torres, JM, Erhardt, G, & Andreoletti, O. (2007). Classic scrapie in sheep with the ARR/ARR prion genotype in Germany and France. Emerg Infect Dis., 13(8), 1201-1207.

Haase, B, Doherr, MG, Seuberlich, T, et al. (2007). PRNP promoter polymorphisms are associated with BSE susceptibility in Swiss and German cattle. BMC Genetics, 8, 15.

Kozomara, A, & Griffiths-Jones, S. (2011). miRBase: integrating microRNA annotation and deep-sequencing data. Nucleic Acids Research, 39, D152-D157

Krejciova, Z, Pells, S, Cancellotti, E, Freile, P, Bishop, M, Samuel, K, Barclay, GR, Ironside, JW, Manson, JC, Turner, ML, De Sousa, P, & Head, MW. (2011). Human embryonic stem cells rapidly take up and then clear exogenous human and animal prions in vitro. J. Patholology, 223(5), 635-645.

Kuznetsova, M. V., Kholodova, M. V, & Danilkin, A. A. (2005). Molecular Phylogeny of Deer (Cervidae: Artiodactyla). Russian Journal of Genetic; 41(7), 742–749.

Lee, I.Y., Westaway, D., Smit, A.F.A., Wang, K., Seto, J., Chen, L., Acharya, C., Ankener, M., Baskin, D., Cooper, C., Yao, H., Prusiner, S.B. & Hood, L.E. (1998). Complete genomic sequence and analysis of the prion protein gene region from three mammalian species. Genome Res. 8 (10), 1022-1037.

Marcos-Carcavilla, A, Calvo, JH, González, C, Moazami-Goudarzi, K, Laurent, P, Bertaud, M, Hayes, H, Beattie, AE, Serrano, C, Lyahyai, J, Martín-Burriel, I, & Serrano, M. (2008). Structural and functional analysis of the HSP90AA1 gene: distribution of polymorphisms among sheep with different responses to scrapie. Cell Stress Chaperones,13(1), 19-29.

Martin, R., Gallet, P.-F., Rocha, D., & Petit, D. (2009). Polymorphism of the Prion Protein in Mammals: A Phylogenetic Approach. Recent Patents on DNA & Gene Sequences, 3(1), 1-9.

Miller, MW, Swanson, HM, Wolfe, LL, Quartarone, FG, Huwer, SL, Southwick, CH, & Lukacs, PM. (2008). Lions and prions and deer demise. PLoS One, 3(12), e4019.

Morgenstern, B. (2004). DIALIGN: Multiple DNA and Protein Sequence Alignment at BiBiServ. Nucleic Acids Research, 32, W33-W36.

Norrby, E. (2011). Prions and protein-folding diseases. J Intern Med., 270(1), 1-14.

Petit, D., Maftah, A., Julien, R., & Petit, J.M. (2006). En bloc duplications, mutation rates and densities of amino acid changes clarify the evolution of vertebrate α1,3/4 fucosyltransferases. Journ. Molecular Evolution, 63(3), 353–364.

Premzl, M., & Gamulin, V. (2007). Comparative genomic analysis of prion genes. BMC Genomics.; Jan 2;8:1.

Prusiner, SB. (1998). Prions. Proc Natl Acad Sci USA, 95, 13363–13383.

Saba, R, Goodman, CD, Huzarewich, RLCH, Robertson, C, & Booth, SA. (2008). A miRNA Signature of Prion Induced Neurodegeneration. PLoS ONE, 3(11), e3652.

Sander, P, Hamann, H, Pfeiffer, I, et al. (2004). Analysis of sequence variability of the bovine prion protein gene (PrnP) in German cattle breeds. Neurogenetics, 5, 19–25.9

Sander, P, Hamann, H, ller, CDG, Kashkevich, K, Schiebel, K, & Leeb, T. (2006). Bovine Prion protein gene (PrnP) promoter polymorphisms modulate PrnP expression and may be responsible for differences in bovine spongiform encephalopathy susceptibility. J Biol Chem., 280 (45), 37408–37414.

Saunders, GC, Cawthraw, S, Mountjoy, SJ, Tout AC, Sayers AR, Hope J, & Windl O. (2009). Ovine PRNP untranslated region and promoter haplotype diversity. J Gen Virol., 90(Pt 5),1289-1293.

Sigurdson, CJ, & Miller, MW. (2003). Other animal prion diseases. Br. Med Bull., 66, 199-212.

Smalheiser, NR, & Torvik, VI. (2005). Mammalian microRNAs derived from genomic repeats. Trends Genet. (6),322-6.

Smirnovas, V, Kim, JI, Lu, X, Atarashi, R, Caughey, B, & Surewicz, WK. (2009). Distinct structures of scrapie prion protein (PrPSc)-seeded versus spontaneous recombinant prion protein fibrils revealed by hydrogen/deuterium exchange. J. Biol. Chem., 284(36), 24233-24241.

Vasudevan, S, Tong, Y, & Steitz, JA. (2007). Switching from repression to activation: MicroRNAs can up-regulate translation. Science, 318(5858), 1931–1934.

Williams, AE, Moschos, SA, Perry, MM, Barnes, PJ, & Lindsay, MA. (2007). Maternally imprinted microRNAs are differentially expressed during mouse and human lung development. Dev. Dyn., 236(2), 572-580.

Xue, G, Sakudo, A, Kim, CK, & Onodera, T. (2008). Coordinate regulation of bovine prion protein gene promoter activity by two Sp1 binding site polymorphisms. Xue G, Sakudo A, Kim CK, Onodera T. Biochem Biophys Res. Commun., 372(4), 530-535.

Causes of Death of Rhesus Monkeys Undergoing Liver Transplantation

Jiang-hua Ran
*The Department of Hepato-Biliary-Pancreatic Surgery
The First Hospital of Kunming
and The Ganmei Hospital Affiliated of
Kunming Medical College Kunming,
China*

1. Introduction

As non-human primates have similar pathophysiological reactions to humans, experimental data evaluating acute rejection reactions following liver transplantation in rhesus monkeys are clinically significant. However, the success rate and long-term survival rate are low, and post-transplant death is one of the major factors influencing survival in rhesus monkeys undergoing experimental liver transplantation.Non-human primates provide the ideal model for clinical studies of liver transplantation. A large number of factors can influence the establishment of a stable and reproducible non-human primate model. Post-transplant death is a major problem in experimental liver transplantation in rhesus monkeys. This study investigates causes of rhesus monkey death following liver transplantation.

2. Materials and methods

2.1 Materials

Healthy rhesus monkeys provided by the Laboratory Animal Center of Kunming Medical University were used as donor and recipient. Recipients were male rhesus monkeys weighing 7.2–11.5 kg, and donors were of either gender weighing 5.3–8.1 kg. The animals were housed in the Laboratory Animal Center of Kunming Medical University, and were allowed free access to food and water. Food access for donors was restricted, and recipients were starved of food for 12 hours and of water for 6 hours preoperatively. Recipients were given cefazolin sodium (0.1 g/kg) before transplantation. Experimental procedures were performed in accordance with the Guidance Suggestions for the Care and Use of Laboratory Animals formulated by the Ministry of Science and Technology of the People's Republic of China.

2.2 Surgery for donor and recipient animals

We operated on 9 rhesus monkeys using our original surgical model, and then modified our model for the remaining 16 monkeys. In the original model, the hepatic vein, portal vein and hepatic artery were directly anastomosed, and a supporting tube was placed in the biliary tract. The modified model is described below.

2.3 Donor surgery and liver perfusion

Donor animals were anesthetized by intravenous injection of 3% pentobarbital sodium dissolved in normal saline (0.5 mL/kg). Under sterile conditions, a large, crucial incision was made in the abdominal wall and the liver was harvested. One cannula was placed in the portal vein and another was placed in the inferior vena cava below the liver. The splenic and renal veins were ligated. The liver was perfused with HTK solution at 4 °C, and bleeding tissues were ligated. A 2 mm diameter supporting tube was placed in the bile duct.

2.4 Recipient surgery

Recipient animals were anesthetized by intravenous injection of 3% pentobarbital sodium dissolved in normal saline (0.5 mL/kg), followed by subcutaneous injection of atropine (0.03–0.04 mg/kg). Recipient surgery was undertaken while the donor liver was undergoing Histidine- Tryptophan- Ketoglutarate (HTK) perfusion. Briefly, a large, cross-shaped incision was made in the abdominal wall. The perihepatic ligaments and inferior vena cava above and below the liver were separated from adjacent structures. The hepatic artery, portal vein and biliary tract were separated at the porta hepatis. Tissues surrounding the inferior vena cava below the liver between the right renal and right adrenal veins were separated over 0.5–1.0 cm. The right suprarenal and lumbar veins were ligated adjacent to the inferior vena cava using a 4-0 suture. Blood was collected from the liver according to the autotransfusion method described for rat liver transplantation. The inferior vena cava below the liver and the portal vein were clamped, and 60–100 mL of sterile balanced salt solution was slowly injected into the portal vein until the liver became khaki in color. The inferior vena cava above the liver was then immediately clamped. The inferior vena cava above the liver was cut adjacent to the liver, and was trimmed into a bellmouth shape at the bifurcation of the portal vein. The inferior vena cava was cut below the liver with some liver tissue included. The donor liver was transplanted using standard orthotopic liver transplantation techniques (double-cuff and one support tube). The inferior vena cava above the liver was anastomosed using 5-0 prolene, the cuff of the portal vein was anastomosed and the portal vein was declamped. When blood was observed flowing from the inferior vena cava below the liver, the cuff was anastomosed. The inferior vena cava above and below the liver were declamped to terminate the anhepatic phase. The liver and gastrointestinal tract were perfused with 0.9% sodium chloride injection at 40–50 °C for rewarming until the color of the liver was restored. The common hepatic artery was anastomosed and a supporting tube was placed in the common bile duct. The abdominal cavity was washed with warm saline. If no hemorrhage or bile leakage was detected, the abdominal wall was closed.

2.5 Postoperative observation and treatment

Animal activities, facial expressions, food and water intake and reactions to stimulation were observed. Animals who died were immediately dissected to obtain samples and to analyze the cause of death. Each monkey was caged separately at 22–25 °C and was allowed access to water after 24 hours and food after 48 hours. Intramuscular cefazolin sodium (0.1 g/kg) was administered twice a day for 2 days. Colloid and sugar water (500–1 000 mL per day) was administered postoperatively to maintain electrolyte and acid-base balance.

3. Conclusion

We successfully performed liver transplantation in 25 pairs of rhesus monkeys. In the early postoperative period (within 6 hours after portal vein opening), seven animals (25%) died; five (20%) due to abdominal hemorrhage, one (4%) due to primary nonfunction and 1 (4%) due to pneumothorax-induced respiratory failure. In the short-term postoperative period (12–72 hours after portal vein opening), seven animals (28%) died; one due to hyperacute rejection within 12 hours, one due to hyperacute rejection and arterial thrombosis within 12 hours, one due to pulmonary infection and one due to accidental death at 72 hours. In the long-term postoperative period (> 72 hours after portal vein opening), eleven animals (44%) died; six due to acute rejection, three due to arterial thrombosis and two due to pulmonary infection. Abdominal hemorrhage occurred mainly in the early and short-term postoperative periods, and acute rejection occurred mainly in the long-term postoperative period.

4. Discussion

Liver transplantation in the rhesus monkey frequently uses the classical model because the inferior vena cava is embedded in the parenchyma of the posterior segment of the liver, making the anatomy unsuitable for piggyback liver transplantation. The rhesus monkey is fragile and often dies early following experimental transplantation. In this study, seven animals (66.7%) died within 6 hours after portal vein opening, seven died at 12–72 hours and eleven died at > 72 hours. The double-cuff method has been used extensively in established liver transplantation models. This model significantly shortens the duration of the anhepatic phase, decreases the incidence of portal vein bleeding and stenosis, and decreases the incidence of early death.

4.1 Abdominal hemorrhage in rhesus monkeys following liver transplantation

In this study, the main cause of death was abdominal hemorrhage in the early postoperative period (within 6 hours after portal vein opening). In small animals such as rats, abdominal hemorrhage is also the major cause of death after reduced-size liver transplantation. This bleeding is often from the inferior vena cava anastomosis above the liver but has also been observed from the ligation points, liver capsule, right adrenal vein, lumbar veins, portal vein and inferior vena cava below the liver. In this study, abdominal hemorrhage was most often from the anastomoses of the portal vein and the inferior vena cava below the liver, and was also observed from the anastomosis of the inferior vena cava above the liver, liver bed, liver capsule, right adrenal vein and lumbar veins, similar to the bleeding points observed in rats. Abdominal hemorrhage sometimes involved multiple sites in one animal. In this study, the animals did not tolerate bleeding well and showed signs of decreased peripheral circulation after a blood loss of 100 mL. Animals with abdominal hemorrhage commonly died within 6 hours after portal vein opening. Rhesus monkeys may also have a preoperative hypercoagulable state and a postoperative hypocoagulable state, greatly influencing the stability of this model. Hemostasis and fluid balance are therefore very important for successful liver transplantation in the rhesus monkey. As a variety of factors contribute to the development of abdominal hemorrhage, surgeons should be familiar with the surgical procedures used including microsurgical techniques. Our original model prior to modification used direct anastomosis, which is a complex and time-consuming process, increasing the duration of the anhepatic phase and resulting in circulatory and other systemic problems following portal vein opening. Liver transplantation may also lead to coagulation disorders, resulting in wound hematoma. In this study, seven animals (28%) died of abdominal hemorrhage following transplantation; five (71.4%) in the early postoperative period and two (28.6%) in the short-term postoperative period.

4.2 Rejection

Hyperacute rejection occurs within 24 hours after anastomosis of the major vessels. It is a complement-mediated response caused by the recipient having pre-existing IgM antibodies to donor antigens (such as ABO, platelet and HLA antigens). Hyperacute rejection of unmatched blood type is mainly mediated by IgG antibodies. In this study, hyperacute rejection caused two deaths. Acute rejection is the most common rejection reaction after transplantation. It is clinically manifested by fever, general malaise, pain, transplant swelling and functional impairment. CD4 Th1 cell-mediated delayed hypersensitivity is the main cause of transplant injury. Transplant HLA stimulates T lymphocyte differentiation and proliferation in the recipient, producing large numbers of sensitized lymphocytes which can damage or kill target cells by releasing lymphokines. In this study, acute rejection caused six deaths in the long-term postoperative period. CD4+T and CD8+T cells may change significantly in acute rejection, followed by worsening of liver function and sometimes death. The pathological characteristics of acute rejection are: (1) inflammatory cell infiltration of the portal area, including activated lymphocytes, neutrophils and eosinophilic granulocytes, (2) endothelial cell inflammation of the portal vein or central vein and (3) bile duct inflammation and injury. Animals with at least two of these characteristics in association with liver dysfunction can be diagnosed with acute rejection. Inflammatory cell infiltration involving > 50% of the bile duct or involving the portal area or central vein are evidence of acute rejection.

4.3 Hepatic artery thrombosis

Studies have reported that hepatic artery anastomosis in a rat model of orthotopic liver transplantation does not influence survival rate or survival time. However, hepatic artery anastomosis is critical in clinical liver transplantation, because hepatic artery thrombosis can cause transplant loss in a short period of time, requiring emergency surgery or even repeat transplantation. Dissection of rhesus monkeys shows that the outer diameter of the proper hepatic artery is 3–4 mm and the inner diameter is 1–2 mm, supplying a large amount of blood to the liver. Liver parenchyma and bile duct necrosis may therefore occur if the hepatic artery is not reconstructed, affecting post-transplant acute rejection and decreasing the survival rate. The cut end of the common hepatic artery can be trimmed into a bellmouth shape to allow full anastomosis. This simplifies the arterial anastomosis for clinical liver transplantation. In this study, the hepatic artery was reconstructed using microsurgical techniques. There were no deaths due to hepatic artery thrombosis or stenosis in the early postoperative period, but hepatic artery thrombosis caused one death in the short-term postoperative period and three deaths in the long-term postoperative period. Hepatic artery thrombosis is mainly caused by the following factors: (1) the artery is very thin and the intima is fragile and easily injured during surgery, which significantly increases the incidence of thrombosis, (2) the rhesus monkey has a relatively low body mass, (3) a hypercoagulable state is common in the rhesus monkey and (4) rejection causes damage to the tunica intima, resulting in degeneration or necrosis and subsequent arterial thrombosis.

4.4 Pulmonary infection

Pulmonary infection is one of the causes of early death following reduced-size liver transplantation in rats, and may be due to an infection focus prior to surgery or to aspiration during surgery. Kamada et al proposed that an anhepatic phase of 26 minutes was safe. Shortening of the anhepatic phase to restore organ perfusion and maintain hemodynamic function is one method to reduce the rate of pulmonary infection following reduced-size liver transplantation, as pulmonary infection is associated with prolonged blood vessels clamping. Prolonged clamping of the portal vein causes prolonged intestinal tract congestion, increasing the likelihood of enteric bacteria entering the circulation and of inflammation-induced lung injury. Preoperative intramuscular atropine to reduce respiratory secretions, small tidal volume anesthesia to reduce aspiration and comfortable living environment and surgical conditions can help to prevent pulmonary infection in rats. In this study, pulmonary infection caused one death in the short-term postoperative period and two deaths in the long-term postoperative period.

4.5 Other causes of death following liver transplantation

One animal in this study died due to primary nonfunction and one due to pneumothorax-induced respiratory failure. Both these animals underwent transplantation using our original surgical model. The primary nonfunction may have been due to the significant fatty degeneration (> 50%) of the donor liver and the differences in weight between donor and recipient causing microhepatia. This animal underwent an anhepatic phase of approximately 1 hour with significant blood loss after portal vein opening, and deteriorated postoperatively with high bilirubin levels, hypoventilation, mydriasis, respiratory arrest and cardiac arrest. Autopsy showed no bleeding, a large amount of ascites in the abdominal cavity and gaseous distension of the gastrointestinal tract. In one animal, the diaphragm was damaged during surgery, resulting in pneumothorax. This animal died due to respiratory failure despite attempted treatment. Autopsy showed a normal liver, no abdominal bleeding or ascites, a bulging diaphragm, gas in the abdominal cavity and collapse of both lungs.

Compared with the rat model, establishment of an orthotopic liver transplantation model in large animal such as the monkey is more difficult. There are some important issues to consider to improve animal survival rate following liver transplantation. The quality of the donor liver is a key factor, and donation of an unhealthy liver is not appropriate. The weight of recipient and donor livers should be similar for the donor liver to function well. Intraoperative blood loss and injury to tissues and organs should be minimized. The modified cuff technique can minimize the duration of the anhepatic phase and of anesthesia, reducing circulatory and other systemic problems. Care should also be taken in perioperative management. This study analyzes the causes of death of rhesus monkeys at different stages following liver transplantation, which can help to modify models of liver transplantation to improve survival rate and to increase the quality of future experimental studies.

5. References

Ariyakhagorn V, Schmitz V, Olschewski P,et al. Improvement of microsurgical techniques in orthotopic rat liver transplantation[J].J Surg Res, 2009, 153(2):332-9

Boillot O, Patrzalek D, Cardoso J, et al. Delay of xenogeneic hyperacute rejection by BN 52021 in a model of rabbit to pig liver transplantation[J].Transplant Proc, 1996, 28(2):767

Cherqui D, Emond JC,et al.Orthotopic liver transplantation and living donors. An experimental study in the dog[J].Chirurgie, 1990, 116(8-9):711-20

D'Hooghe TM, Kyama CM, et al.Nonhuman primate models for translational research in endometriosis [J].Reprod Sci, 2009, 16(2):152-61

Guo H, Wu YJ, Zheng SS, et al.Application of modified two-cuff technique and multiglycosides tripterygium wilfordii in hamster-to-rat liver xenotransplant model[J].World J Gastroenterol, 2003, 9(7):1550-1553

Goto M, Takei Y, Kawano S, et al. Tumor necrosis factor and endotoxin in the pathogenesis of liver and pulmonary injuries after orthotopic liver transplantation in rat[J]. Hepatology, 1992, 16:487 Hammer C, Thein E. Physiological aspects of xenotransplantation, 2001[J].Xeno-transplantation, 2002, 9(5):303-305

Kamada N, Calne RY. A surgical experience with five hundred thirty liver transplantation in the rat[J]. Surgery,1983,93:64

Liu J, Li J, Zhang SN, et al. Modified model of reduced-size liver transplantation in rats. Zhongguo Zuzhi Gongcheng Yanjiu yu Linchuang Kangfu, 2010, 14(18):3252-3257

Liu J, Li J, Zhang SN, et al. Cause of early death after reduced-size liver transplantation in rats [J]. Zhongguo Quanke Yixue Zaizhi, 2010, 14(18):3252-3257

Liu J, Li J, Zhang SN, et al. A causal analysis of intra-abdominal hemorrhage after reduced-size liver transplantation in rats [J]. Nanfang Yike Daxue Xuebao, 2010, 30(5):1075-1078

Liu LX, Peng WX, Zhang YN, et al. Analysis of cause of death in recipient rats after modified orthotopic liver transplantations and the countermeasure [J]. Zhongguo Puwai Jichu yu Linchuang, 2005, 12(2): 120-122

Li T, Sun X, Peng ZH. Preventive measures of postoperative complications in rat liver transplantation [J]. Gandanyi Waike Zazhi, 2007, 19(4):215-219

López-Benítez R, Schlieter M, et al.Successful arterial thrombolysis and percutaneous transluminal angioplasty for early hepatic artery thrombosis after split liver transplantation in a four-month-old baby.Pediatr Transplant, 2008 Aug, 12(5):606-10

Ma Y, Wang GD, Guo ZY,et al.Surgical techniques of arterialized orthotopic liver transplantation in rats.Chin Med J (Engl), 2007 Nov 5, 120(21):1914-7

Mansfield KG, Kemnitz JW. Introduction:challenges in microbial quality control for nonhuman primate[J].ILARJ,2008,49(2):133-136

Matevossian E ; Doll D ; Huser N ; et al.Liver transplantation in the rat: single-center experience with technique, long-term survival, and functional and histologic findings[J]. Transplant Proc, 2009, 41(6):2631-2636

Mever K, Brown MF, Zobari G, et al. ICAM-1 upregulation in distant tissue after hepatic ischemia/reperfusion: a clue to the mechanism of multiple organ failure[J]. J Pediatr Surg, 1998, 33:350

Morelli AE, Coates PT, Shufesky WJ, et al. Growth factor-induced mobilization of dendritic cells in kidney and liver of rhesus macaques :implications for transplantation[J].Transplantation, 2007, 83(5):656-662

Peng Y, Gong JP, Yan LN;et al. Improved two-cuff technique for orthotopic liver transplantation in rat[J].Hepatobiliary Pancreat Dis Int, 2004, 3(1):33-37

Powelson J, Cosimi AB, Austen W Jr,et al. Porcine-to-primate orthotopic liver transplantation[J]. Transplant Proc, 1994,26(3):1353-4

Ran JH, Zhang SN, Liu J, et al. Improvements of Orthotopic Liver Transplantation Model in Rhesus Monkey [J]. Zhongguo Puwai Jichu yu Linchuang Zazhi, 2010,17(7): 694-698

Roberts JA, Andrews K.Nonhuman primate quarantine: its evolution and practice[J].ILAR J, 2008, 49(2):145-56

Steinhoff G, Jonker M, Gubernatis G,et al.The course of untreated acute rejection and effect of repeated anti-CD3 monoclonal antibody treatment in rhesus monkey liver transplantation[J].Transplantation, 1990, 49(4):669-74

Schleimer K, Stippel DL, Tawadros S, et al.Improved technique of heterotopic auxiliary rat liver transplantation with portal vein arterialization.Langenbeck-s Arch Surg, 2006 Apr, 391(2):102-7

Taniai N, Onda M, Tajiri T, et al.Graft survival following three occurrences of hepatic arterial thrombosis after living-related liver transplantation. A case report. Hepatogastroenterology, 2002 Sep-Oct, 49(47):1420-2

The Ministry of Science and Technology of the People's Republic of China. Guidance Suggestions for the Care and Use of Laboratory Animals. 2006-09-30.

Wang Y, Yu RS, Zhou XJ, et al. Complications in settng up the model of orthotopic liver transplantation in rat [J]. Gandan Waike Zazhi, 2008, 16(5):355-357

Wang HX, Zhang PJ, Wang YQ, et al. Technique established for animal model of partial living donor liver transplantation in rat [J]. Zhongguo Xiandai Yixue Zazhi, 2006, 16(8):1160-1163

Williams JW, Peters TGS Vera SR, et al. Biopsy-directed immunosuppression follwing hepatic transplantation in man[J].Transplantation,1985,39(6):589-596

Yan S, Zhang Q, Cai M, et al.A novel model of portal vein transplantation in mice using two-cuff technique[J].Microsurgery, 2007, 27(6):569-574

Yokota K, Uchida H, Satoh K, et al. Orthotopic liver transplantation in dogs -using a cuff technique for vascular anastomoses, with special reference to long term survival dogs treated with mizorbine [J]. Jpn J Transplant, 1987, 22(1): 8-11

Zhang SN, Li L, Ran JH, et al. Hepatic Artery Reconstruction in Orthotopic Liver Transplantation and Postoperative Complications [J]. Zhongguo Puwai Jichu yu Linchuang Zazhi, 2007, 4(5): 564-568

Zhang SN, Li L, Ran JH, et al. Strategies of hepatic artery in living donor liver transplantation [J]. Guoji Waike Zazhi, 2008, 35(2): 92-95

Zhong R, Tucker J, Grant D,et al.Long-term survival and functional tolerance of baboon to monkey kid -ney and liver transplantation: a preliminary report[J].Transplant Proc, 1996, 28(2):762

Zhong R, Tucker J, Zhang Z, et al. The long-term survival of baboon-to-monkey kidney and liver xeno-grafts[J].Xenotransplantaion, 2003, 10(5):398-409

Zahorsky-Reeves JL, Gregory CR, Cramer DV, et al.Similarities in the immunog-lobulin response and VH gene usage in rhesus monkeys and humans exposed to porcine hepatocytes[J].BMC Immunol,2006,7:3

Zhou XP, Du M, Li JY, et al. Steady livability after orthotopic liver transplantation in rats [J]. Shanghai Jiaotong Daxuebao, 2006, 26(6):603-606

27

Molecular Characterization of Hypothalamo–Pituitary-Thyroid Genes in Pig (Sus Scrofa)

Ningying Xu and Xiaoling Jiang
Zhejiang University,
China

1. Introduction

Almost all the body functions of vertebrate animals including swine are regulated by the nervous system and endocrine system. Especially the hormones released from the endocrine system have effective biological amplifying effects. Minor changes in the hormone level could cause huge alternations in physiology.

Hypothalamus, extensively connected with other brain regions, is the vital bridge between nervous system and endocrine system. It exerts its regulating function on endocrine system mainly via the pituitary, which is the central endocrine organ in vertebrate animals. Except the growth hormone, most of the hormones secreted by pituitary have their specific target organs. For example, thyrotropin target thyroid. The hypothalamus, pituitary and target organs are always described together as "axis", e.g. hypothalamo-pituitary-gonadal axis, hypothalamo-pituitary-adrenal gland axis, and hypothalamo-pituitary-thyroid (HPT) axis.

2. The axis of hypothalamo-pituitary-thyroid

Hypothalamus synthesizes and secrets thyrotropin-releasing-hormone (TRH) to send regulating information to pituitary (see Figure 1). TRH binds to the thyrotropin-releasing-hormone receptors (TRHR) on the thyrotroph cells of pituitary to activate the intracellular signal pathways and induce the secretion and synthesis of thyrotropin (TSH). Circulating TSH in blood then binds to the thyrotropin receptors (TSHR) on the follicular cells of thyroid to activate the synthesis and secretion of thyroid hormone (TH).

TH secreted by thyroid is important to growth, development, and protein, fat, and carbohydrate metabolisms (Porterfield and White, 2007). It acts on almost all the organs and tissues. Each individual has a unique thyroid function set-point, and this set-point was suggested to be genetically determined (Hansen et al., 2004). Genetic variations of the hormones of HPT axis and their respective receptors could be the excellent candidates as the causing of related phenotype variations.

2.1 Thyrotropin releasing hormone gene (*TRH*)

Thyrotropin releasing hormone, produced in the paraventricular nucleus of the hypothalamus, is fully conserved in all species from human to bony fish that have been investigated so far (Harder, 2001). It is the tripeptide pyro-Glu-His-Pro-NH_2 derived from the preprohormone gene *TRH*, which also produces other non-TRH active peptides e.g. $ppTRH_{160-169}(pST_{10})$ and $ppTRH_{178-199}$ (pFE_{22}).

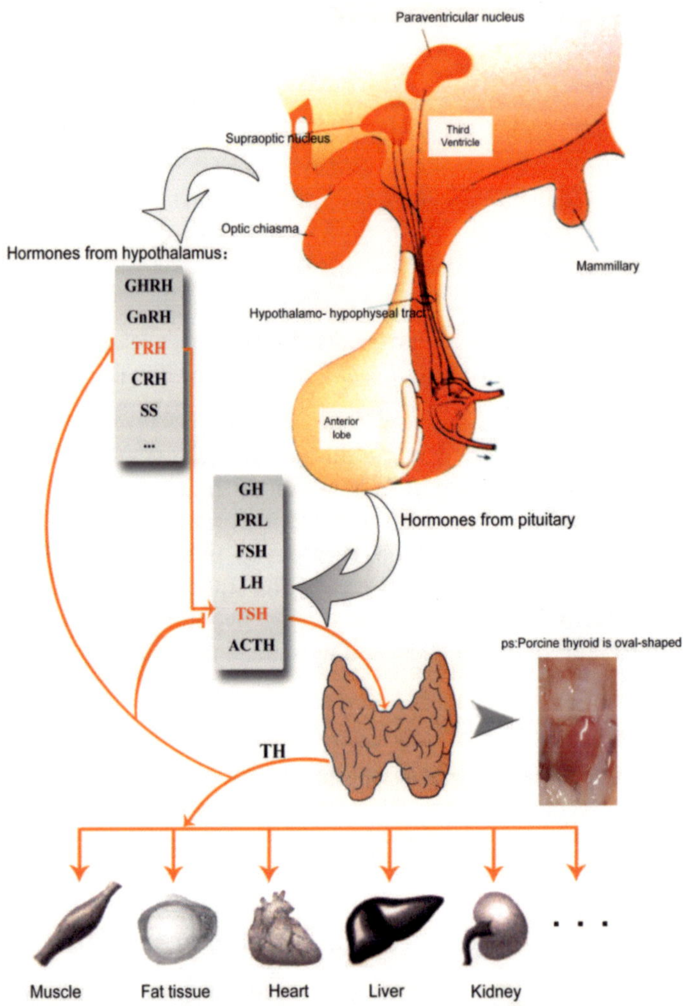

Fig. 1. Schematic map of hypothalamo-pituitary-thyroid axis regulatory network, according to Porterfield and White (2007).

No experiments on the porcine *TRH* gene have been done previously. However genomic information of this gene has became available as a result of the porcine whole genome sequencing, and comprehensive sequence analysis on it with bioinformatics method has been done (Wallis, 2010). Porcine *TRH* gene was found to be 3, 136 bp with 3 exons and two introns. A conserved signal peptide of 24 amino acids was predicted to be present, and there existed six copies of TRH sequences in the preprohormone peptide.

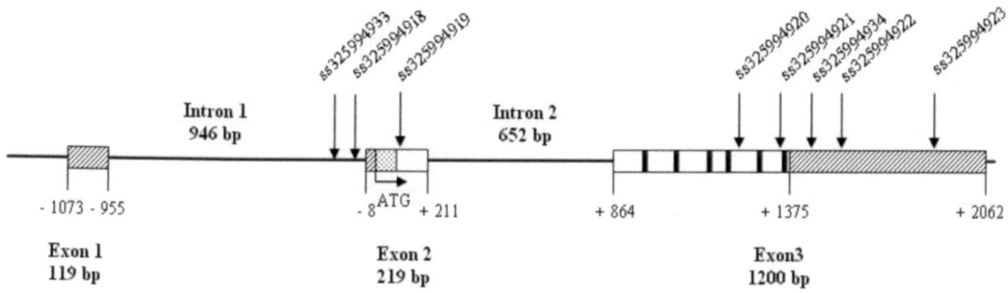

Fig. 2. The schematic map of the porcine *TRH* gene. Boxes represented the exons (according to Wallis), of which the regions filled with diagonal lines were un-translated regions (UTR), the region filled with dots represented the signal peptide, and the regions full filled with black color represented the six copies of TRH precursor peptides.

2.2 Thyrotropin Releasing Hormone Receptor gene (*TRHR*)

TRH initiates its effects by interacting with its receptor TRHR in the anterior pituitary. TRHR is a seven transmembrane spanning receptor and belongs to the G protein-coupled receptor superfamily. Actually, to date in total of three TRH receptors have been reported, each encoded by their specific genes. The first subtype of TRHR found in 1990 have been described in many species, e.g. mouse, rat, human, cow, chicken, frog and fish (Sun et al., 2003). However, the second subtype of TRHR identified in 1998 has only been reported in rodents, frog and fish. Further information of human genome sequence does not support the existence of the second TRHR subtype (Pfleger et al., 2004). The third TRHR subtype has only been reported in frog and fish (Mekuhi et al., 2010) so far. Figure 3 shows the evolution tree of the TRHR proteins.

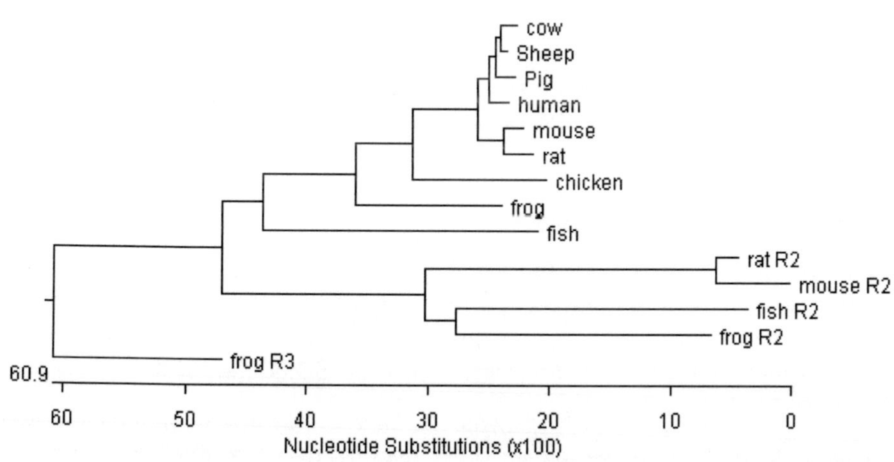

Fig. 3. Polygenetic tree of TRHR proteins, aligned with method Clustal W. The branch of fish used the TRHR sequence of *Catostomus commerson*.

Porcine *TRHR* gene (the first subtype) has been cloned and characterized recently (Jiang et al., 2011). It contains an open reading frame encoding 398 amino acids and shares 96.2%

amino acid identity to human TRHR. An intron disrupts the open reading frame in the sequence encoding the putative third intracellular loop which is between the fifth and sixth transmembrane domain. Besides, alternative spliced transcript variants and multiple transcription start sites have been observed in porcine *TRHR* gene (Jiang, 2011). The biological functions of these variants remain investigations.

2.3 Thyrotropin alpha subunit gene (*CGA*)

Thyrotropin is a member of the glycoprotein hormone family which consisting thyrotrophin, follicle-stimulating hormone and luteinizing hormone in anterior pituitary, and chorionic gonadotrophin in the placenta. Each of these hormones is a heterodimer composed of a common α subunit and a hormone specific β subunit. The common α subunit is encoded by the *CGA* gene, while the TSH specific β subunit is encoded by the *TSHB* gene.

Porcine *CGA* gene is 14 kb in length and consists of four exons and three introns (Kato et al., 1991). The coding region starts from the exon 2, and the premature peptide contains a signal peptide of 24 amino acids. Two intragenic microsatellites were found in intron 1 and intron 2. The microsatellite located in the intron 1 has been named as PGHAS or ALPHA and widely used in genetic studies as genetic marker.

2.4 Thyrotropin beta subunit gene (*TSHB*)

The porcine *TSHB* gene has at least two exons and one intron, and is 939 bp in length. Its product consists of a signal peptide of 24 amino acid residues, a mature beta subunit protein of 112 residues and an additional extension of six amino acid residues at the carboxyl terminus (Hirai et al., 1989). The TSH beta subunit noncovalently links to the alpha subunit by wrapping with its "seat belt" structure around the alpha subunit's long loop αL2.

TSH β subunit is responsible for the hormone specificity, and its translation rate directly determines the secretion and synthesis rates of the hormone TSH. Additionally, it has been reported that mature Chinese meishan pigs had 3 fold greater expressions of *TSHB* gene and greater plasma TSH concentrations than mature Western white composite pigs (Li et al., 1996).

2.5 Thyrotropin receptor gene (*TSHR*)

The receptor of the thyrotorpin,TSHR, belongs to the glycoprotein hormone receptor (GPHR) family which is a subset of the G-protein coupled receptor (GPCR) superfamily. It contains a seven-transmembrane domain and a large ectodomain composed of many leucine-rich repeats forming TSH-binding surface (Farid & Szkudlinski, 2004).

The mRNA sequence of the porcine *TSHR* gene has been cloned in 2003 (Igarashi & Nagata, 2003) and characterized to containing an open reading frame coding 764 amino acids. However, as the genomic sequence of porcine *TSHR* gene is still unavailable so far, the genomic structure of the gene is not clear. Of human, the *TSHR* gene was found to be 190 kb in length, and contains ten exons and nine introns, of which all the first nine exons encodes the large extodomain while the last exon is in charge of the seven transmembrane domains and the intracellular tail.

3. The gene mapping and expression analysis of these genes

Quantitative trait loci (QTLs) are regions of the chromosome that are found to be associated with particular phenotypic traits by statistical analysis. Thus far, the pig QTL database (Pig QTLdb) has collected 6,344 QTLs from 281 publications in the past more than ten years. Any genes in these regions might be the positional candidate genes underlying the respective traits. Mapping the genes of HPT axis would help us to evaluate the potential genetic effects of the variations of these genes.

3.1 The electrical mapping

In April 2009, the International Swine Genome Sequencing Consortium has completed and released a 4× sequence depth draft (Sscrofa9) by a minimal tile path BAC by BAC approach. This assembly can be conveniently accessed by the web-based query on the Ensemble (http://www.ensembl.org/Sus_scrofa/Info/Index). Though a more recent assembly Sscrofa 10, a mixed BAC and WGS-based assembly of the porcine genome, has been released in April 2011, assemble errors in it remain to be resolved. Furthermore, the Pig QTLdb offered GBrowse map view of QTLs (http://www.animalgenome.org/cgi-bin/gbrowse/pig/) based on the Sscrofa 9. Thus, with the chromosome position of the candidate gene obtained by querying the Ensemble, one can easily get the information of the QTLs which were mapped onto the same genomic region (Figure 4).

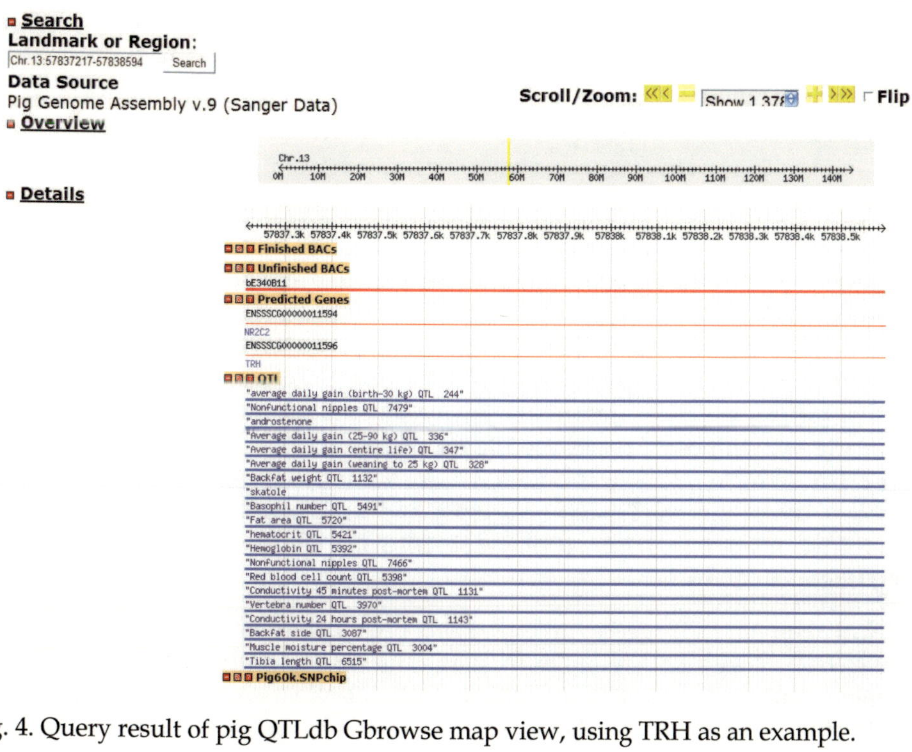

Fig. 4. Query result of pig QTLdb Gbrowse map view, using TRH as an example.

Porcine *TRH* gene is located between 57,837,217 and 57,838,594 on chromosome 13. Based on the pig QTLdb, there are 20 QTLs including average daily gain (ADG), backfat weight, meat quality traits and blood parameters have been reported in this region. Porcine *CGA* gene is mapped between 58,190,063 and 58,192,283 on chromosome 1, and 21 QTLs have been mapped to the corresponding region, most of which are fat related traits. The position of *TSHB* gene on porcine chromosome 4 is from 109,815,329 to 109,816,276. 49 QTLs including ADG, backfat thickness (BFT) and meat quality related traits have been mapped to this position. Unfortunately, *TRHR* and *TSHR* genes had not yet been annotated in Sscrofa 9 in the year 2009.

3.2 The RH mapping

Radiation hybrid (RH) mapping is a kind of somatic cell genetics tool for mapping candidate genes into a framework of microsatellites/sequence-tagged sites. First, RH panels are constructed by using a lethal dose of X-rays to fragment the chromosomes of the donor cell and the chromosome fragments with known genetic markers are retained randomly in the host cells. The frequency of breakage between pairs of markers in a panel of RH clones can be used to calculate the relative physical distances of the genetic markers or genes. Several porcine RH panels have been produced, among which the whole-genome high-resolution IMpRH panel constructed by Yerle et al. (1998) is the most frequently used one.

The porcine *TRHR* gene was localized to the microsatellite *SW969* on chromosome 4 by RH mapping using the IMpRH panel (Jiang et al., 2011). 44 Quantitative trait loci affecting BFT, daily gain, and carcass and meat quality traits have been mapped to the same region. *TSHR* gene has been mapped between microsatellites *SW1083* and *SW581* on porcine chromosome 7 (Sato et al., 2006), and 50 QTLs including BFT, ADG, carcass length and muscle related traits have been mapped to the same position.

3.3 Expression analysis

TRH is present not only in the hypothalamus but also in many other brain loci as a neuromodulator / neurotransmitter. Besides, expression of *TRH* gene is found in gastrointestinal tract, pancreas, reproductive tissues, heart, spleen, adrenals and thymus. Its receptor, *TRHR* is also found in gastrointestinal tract, pancreas, testis and adrenals. Paracrine effects of TRH on gastrointestinal tract and pancreas have been suggested. Furthermore, *TSHR* is widely expressed in many organs and tissues such as pituitary, heart, skeletal muscle and adipose. Species-specific expression of HPT genes in certain tissues might exist. For example, *TRHR* was detected in medaka spleen (Mekuchi et al., 2010) but undetectable in bovine spleen (Takata et al., 1998), and *TSHB* has been reported in the liver and reproductive organs of red drum (Cohn et al., 2010) but undetectable in the same organs of ducks (Hsieh et al., 2007).

Tissue distribution of HPT genes in pigs have been investigated by using real-time quantitative RT-PCR with RNA samples from 15 different organs/tissues including brain, hypot4halamus, pituitary, thyroid, lung, kidney, gastrointestinal tract, muscle and fat tissue (Jiang 2011). Highest expression level of *TRH* was found in the brain, and considerable level of *TRH* also expressed in testis, spleen, fat tissue, small intestine and pancreas in addition to hypothalamus. Porcine *TRHR* mRNA was detected in almost all the investigated organs/tissues except the spleen, with high expression levels in the brain, hypothalamus, pituitary, testis and fat tissue. The two subunit genes of hormone TSH showed different tissue distribution patterns. While *CGA* was detectable in almost all the tissues, *TSHB* was mainly existed in the pituitary and undetectable or in extremely low level in the other tissues except in the fat tissue and stomach. Expression of porcine *TSHR* existed in all the fifteen organs/tissues, with highest expression level in the thyroid, then the testis and spleen.

3.4 The polymorphisms of these genes

Polymorphism in human *TRH* gene has been reported to be associated with blood pressure variations and hypertension (Kokubo et al., 2006). By re-sequencing the whole coding region of porcine *TRH* gene, eight sequence variations were identified (Jiang 2011). Among these polymorphisms, only ss325994920 (Figure 2) is a missense mutation which would bring a Val to Met amino acid mutation into pFE22 and pSE14 peptides.

Mutations in the *TRHR* could result in central hypothyroidism which causes growth retardation, pudginess and sluggishness (Collu et a., 1997). Further *TRHR* polymorphism has been identified responsible for human lean body mass variations in a genome-wide association study (Liu et al., 2009). Re-sequencing the coding sequences and flanking regions of porcine *TRHR* gene identified seven polymorphic loci (Figure 5a). Two of them locate in the intron 1 which has been proven to encompass important regulatory elements.

Mice with *CGA* gene knockout were viable, but exhibited severe growth insufficiency and infertility (Kendall et al., 1995). While no inactivating mutations of *CGA* gene has been detected in humans and mice, cases of nonsense mutations of *TSHB* gene have been continuously reported in humans which cause growth retardation and fat metabolism disorders (McDermott et al., 2002; Baquedano et al., 2010). 14 new polymorphic loci of porcine *CGA* gene (Figure 5b) and 5 polymorphisms of *TSHB* gene (Figure 5c) were identified and confirmed by re-sequencing (Jiang et al., 2011b). Four of the *CGA* polymorphisms locate in the promoter region. The single nucleotide polymorphism ss181129018 of *TSHB* gene locates on the first coding exon and brings amino acid change to the signal peptide of TSH β subunit.

TSHR is sensitive to point mutations and most of the reported human mutations existed in the last exon 10 (Davis et al., 2006). Re-sequencing the last exon of porcine *TSHR* gene detected three polymorphisms in the exon (Figure 5d), but all were synonymous mutations.

4. The association study with economic traits in the crossbred of Jinhua and Pietrain

A crossbred was established by mating the Chinese Jinhua pigs (Central China swine type) with European Pietrain pigs, and obvious segregation of growth, carcass and meat quality characters were observed in this population. Association study of the five HPT key genes' polymorphisms with the economic important traits in the crossbred of Jinhua and Pietrain were carried out by analyzing fourteen polymorphic loci with thirty traits of 463 individuals (Jiang 2011). Both PCR-RFLP and tetra-primer ARMS PCR procedure were utilized for genotyping, and haplotypes were considered as genetic effects.

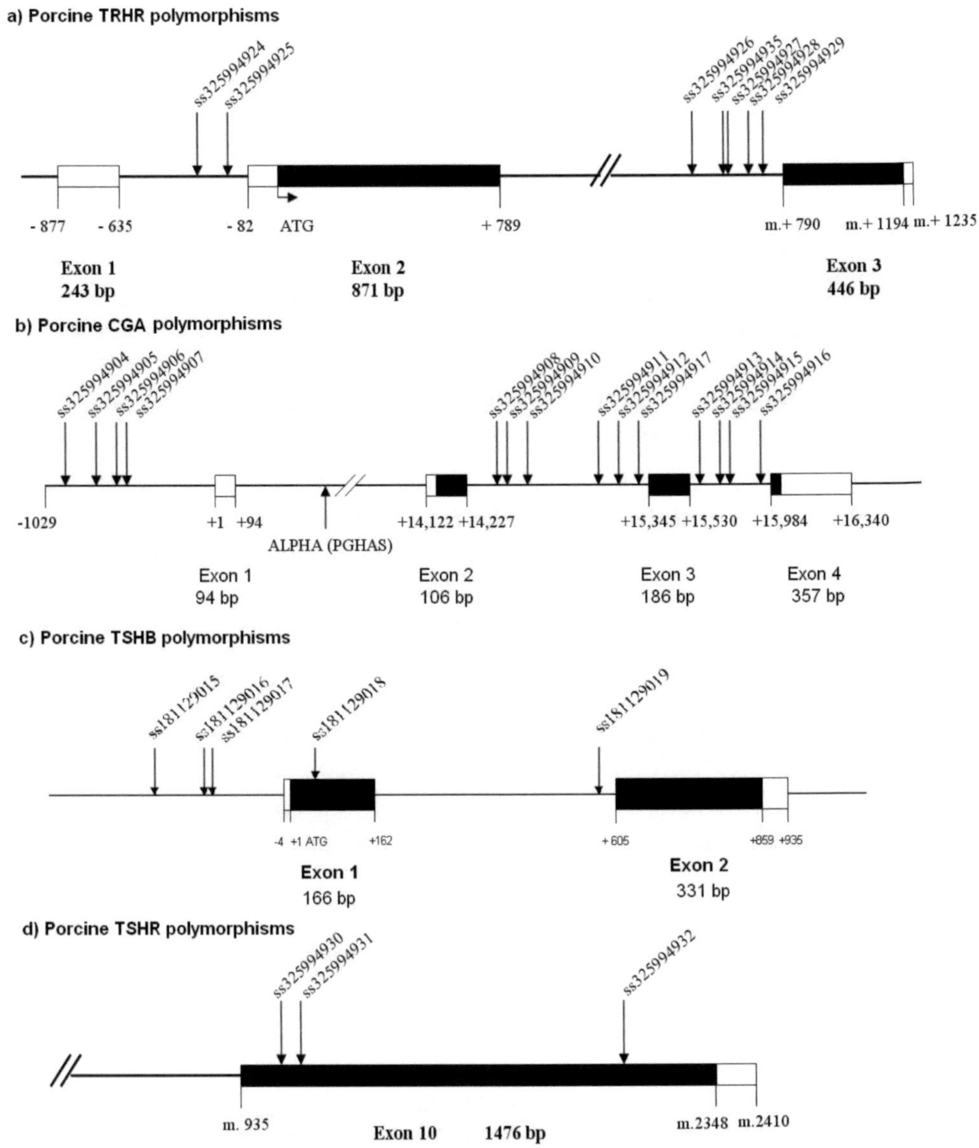

Fig. 5. Genetic Polymorphisms of porcine HPT axis

4.1 Thyrotropin releasing hormone gene (*TRH*)

Polymophisms of porcine *TRH* gene were divided into two linkage disequilibrium (LD) blocks. One of the LD block contained locus ss325994933 which was significantly ($P < 0.05$) associated with daily weight gain. Homozygous individuals with the insertion allele showed higher daily weight gain than the ones with the deletion allele. TRH could be considered functional and positional candidate gene of the ADG QTL reported by de Koning et al. (de Koning et al., 2001). However, further investigation on the allele frequencies in different purebreds showed that the insertion allele mainly existed in the

Chinese breeds, while the Western pig breeds those normally grow faster than the Chinese breeds were almost fixed with the deletion allele. Another LD block consisted of the other *TRH* polymorphisms investigated. No significant associations of this LD block with the economical traits were observed.

4.2 Thyrotropin releasing hormone receptor gene (*TRHR*)

All the four *TRHR* polymorphisms studied existed in one LD block. Significant associations of porcine *TRHR* gene with head weight, carcass length, and electric conductivity, pH value, water holding capacity, intramuscular water and fat content of loin muscle were observed. QTLs of carcass length (Wimmers et al., 2002; Murani et al., 2006) and intramuscular fat content (de Koning et al., 1999) have been reported previously. Individuals with the haplotypes originating from the Pietrain had smaller head weight, longer carcass length and lower conductivity of loin muscle. However, no significant difference among the haplotypes was observed after bonferroni corrections.

4.3 Thyrotropin alpha subunit gene (*CGA*)

As the existence of microsatellites in the porcine *CGA* gene, QTL mapping results around the *CGA* gene from different studies were instable. In the crossbred of Jinhua and Pietrain, haplotypes of *CGA* was only associated with intramuscular water content. But if further dissecting by dividing the gene with the microsatellites, the promoter region of this gene showed significant associations with growth and backfat thickness traits.

4.4 Thyrotropin beta subunit gene (*TSHB*)

Haplotypes of porcine *TSHB* gene were significantly associated daily weight gain between 90 to 120 days, ham muscle weight, and extremely significantly (P<0.01) associated with carcass weight, carcass length and average backfat thickness. Corresponding QTLs have also been reported in this chromosome region (Grindflek et al., 2001; Knott et al., 2002; Malek et al., 2001; Nagamine et al., 2003; van Wijk et al., 2007; Walling et al., 2000). The homozygous individuals with haplotype originating from Jinhua pig had significantly lower daily weight gain and thicker average BFT than the homozygous individuals with the haplotype originating from Pietrain, which were in accordance with the observations in the two purebreds. Further verification in a Duroc population confirmed the association of *TSHB* gene with the average BFT trait and showed that the allele A of ss181129015 which was popular in Western breeds was associated with lower average BFT. Porcine *TSHB* gene, especially the locus ss181129015 is worthwhile for further study and to be developed as genetic marker for the BFT trait.

4.5 Thyrotropin receptor gene (*TSHR*)

Individuals with allele A of *TSHR* ss325994932 showed higher daily gain between 120 to 180 days old and lower muscular pH value than the ones with allele G in the crossbred of Jinhua and Pietrain. Several QTLs on ADG around the *TSHR* gene have been reported previously (Edwards et al., 2008; Nagamine et al., 2003). Though individuals with allele A showed better performance in growth and meat quality in the crossbred, investigation on the allele frequencies in different purebreds showed that the allele A appeared in fairly low frequencies in all the purebreds studied. Verification in Landrace population confirmed the association of *TSHR* with ADG, but allele A was associated with slower growth rate in this breed. Possible explanations were interactions of alleles of *TSHR* with other genes' might exist, or recombination might have occurred between the causative polymorphism and ss325994932 in some breeds.

5. Acknowledgment

This work was supported by National Natural Science Foundation of China (NO. 30972078) and Main Research Projects of Zhejiang Province (NO.2006C004-2).

6. References

Baquedano MS, Ciaccio M, Dujovne N, Herzovich V, Longueira Y, Warman DM, et al. Two novel mutations of the TSH-beta subunit gene underlying congenital central hypothyroidism undetectable in neonatal TSH screening. J Clin Endocrinol Metab 2010; 95: E98-103.

Cohn WB, Jones RA, Valverde RA, Leiner KA, MacKenzie DS. Molecular cloning and regulation of mRNA expression of the thyrotropin beta and glycoprotein hormone alpha subunits in red drum, Sciaenops ocellatus. Fish Physiol Biochem 2010; 36: 1277-90.

Collu R, Tang J, Castagne J, Lagace G, Masson N, Huot C, et al. A novel mechanism for isolated central hypothyroidism: inactivating mutations in the thyrotropin-releasing hormone receptor gene. J Clin Endocrinol Metab 1997; 82: 1561-5.

Davies TF, Ando T, Lin RY, Tomer Y, Latif R. Thyrotropin receptor-associated diseases: from adenomata to Graves disease. J Clin Invest 2005; 115: 1972-83.

de Koning DJ, Janss LL, Rattink AP, van Oers PA, de Vries BJ, Groenen MA, et al. Detection of quantitative trait loci for backfat thickness and intramuscular fat content in pigs (Sus scrofa). Genetics 1999; 152: 1679-90.

de Koning DJ, Rattink AP, Harlizius B, Groenen M, Brascamp EW, van Arendonk J. Detection and characterization of quantitative trait loci for growth and reproduction traits in pigs. Livestock Production Science 2001; 72: 185-198.

Edwards DB, Ernst CW, Tempelman RJ, Rosa GJ, Raney NE, Hoge MD, et al. Quantitative trait loci mapping in an F2 Duroc x Pietrain resource population: I. Growth traits. J Anim Sci 2008; 86: 241-53.

Farid NR, Szkudlinski MW. Minireview: structural and functional evolution of the thyrotropin receptor. Endocrinology 2004; 145: 4048-57.

Grindflek E, Szyda J, Liu ZT, Lien S. Detection of quantitative trait loci for meat quality in a commercial slaughter pig cross. Mammalian Genome 2001; 12: 299-304.

Hirai T, Takikawa H, Kato Y. Molecular cloning of cDNAs for precursors of porcine pituitary glycoprotein hormone common alpha-subunit and of thyroid stimulating hormone beta-subunit. Mol Cell Endocrinol 1989; 63: 209-17.

Hansen PS, Brix TH, Sørensen TIA, Kyvik KO and Hegedüs L. Major genetic influence on the regulation of the Pituitary-thyroid axis: a study of healty Danish twins. J Clin Endocrinol Metab 2004; 89: 1181-1187.

Harder S, Dammann O, Buck F, Zwiers H, Lederis K, Richter D, et al. Cloning of two thyrotropin-releasing hormone receptor subtypes from a lower vertebrate

(Catostomus commersoni): functional expression, gene structure, and evolution. Gen Comp Endocrinol 2001; 124: 236-45.

Hsieh YL, Chowdhury I, Chien JT, Chatterjee A, Yu JY. Molecular cloning and sequence analysis of the cDNA encoding thyroid-stimulating hormone beta-subunit of common duck and mule duck pituitaries: in vitro regulation of steady-state TSHbeta mRNA level. Comp Biochem Physiol B Biochem Mol Biol 2007; 146: 307-17.

Igarashi M, Nagata A. Molecular cloning, sequencing and functional expression of porcine thyrotropin (TSH) receptor cDNA1). Clin Chem Lab Med 2003; 41: 796-803.

Jiang XL. Study on five key genes of porcine hypothalamo-pituitary-thyroid axis. Dissertation of Zhejiang University, 2011.

Jiang XL, Cai ZW, Zhao XF, Zhang LF, Chen Z, Wang Y, Guo XL, Xu NY. Mapping, cDNA cloning and tissue expression of the porcine thyrotropin-releasing hormone receptor gene. Animal Biotechnology, 2011, 22 (1): 30-36.

Kato Y, Ezashi T, Hirai T, Kato T. The gene for the common alpha subunit of porcine pituitary glycoprotein hormone. J Mol Endocrinol 1991; 7: 27-34.

Kendall SK, Samuelson LC, Saunders TL, Wood RI, Camper SA. Targeted disruption of the pituitary glycoprotein hormone alpha-subunit produces hypogonadal and hypothyroid mice. Genes Dev 1995; 9: 2007-19.

Knott SA, Nystrom PE, Andersson-Eklund L, Stern S, Marklund L, Andersson L, et al. Approaches to interval mapping of QTL in a multigeneration pedigree: the example of porcine chromosome 4. Animal Genetics 2002; 33: 26-32.

Kokubo Y, Tomoike H, Tanaka C, Banno M, Okuda T, Inamoto N, et al. Association of sixty-one non-synonymous polymorphisms in forty-one hypertension candidate genes with blood pressure variation and hypertension. Hypertens Res 2006; 29: 611-9.

Li MD, Matteri RL, Macdonald GJ, Wise TH, Ford JJ. Overexpression of beta-subunit of thyroid-stimulating hormone in Meishan swine identified by differential display. J Anim Sci 1996; 74: 2104-11.

Liu XG, Tan LJ, Lei SF, Liu YJ, Shen H, Wang L, et al. Genome-wide association and replication studies identified TRHR as an important gene for lean body mass. Am J Hum Genet 2009; 84: 418-23.

Malek M, Dekkers JC, Lee HK, Baas TJ, Rothschild MF. A molecular genome scan analysis to identify chromosomal regions influencing economic traits in the pig. I. Growth and body composition. Mamm Genome 2001; 12: 630-6.

McDermott MT, Haugen BR, Black JN, Wood WM, Gordon DF, Ridgway EC. Congenital isolated central hypothyroidism caused by a "Hot Spot" mutation in the thyrotropin-β gene. Thyroid 2002; 12:1141-1146.

Mekuchi M, Saito Y, Aoki Y, Masuda T, Iigo M, Yanagisawa T. Molecular cloning, gene structure, molecular evolution and expression analyses of thyrotropin-releasing hormone receptors from medaka (Oryzias latipes). Gen Comp Endocrinol 2010.

Murani E, Muraniova M, Ponsuksili S, Schellander K, Wimmers K. Molecular characterization and evidencing of the porcine CRH gene as a functional-positional candidate for growth and body composition. Biochem Biophys Res Commun 2006; 342: 394-405.

Nagamine Y, Haley CS, Sewalem A, Visscher PM. Quantitative trait loci variation for growth and obesity between and within lines of pigs (Sus scrofa). Genetics 2003; 164: 629-635.

Pfleger KD, Kroeger KM, Eidne KA. Receptors for hypothalamic releasing hormones TRH and GnRH: oligomerization and interactions with intracellular proteins. Semin Cell Dev Biol 2004; 15: 269-80.

Porterfield SP, White BA. Endocrine Physiology 3rd. Philadelphia, PA: Mosby Inc., 2007.

Sato S, Hasebe H, Sato S, Asahi Y, Hayashi T, Kobayashi E, et al. High-resolution physical mapping and construction of a porcine contig spanning the intramuscular fat content QTL. Anim Genet 2006; 37: 113-20.

Sun Y, Lu X, Gershengorn MC. Thyrotropin-releasing hormone receptors -- similarities and differences. J Mol Endocrinol 2003; 30: 87-97.

Takata M, Shimada Y, Ikeda A, Sekikawa K. Molecular cloning of bovine Thyrotropin-Releasing Hormone Receptor gene. Journal of Veterinary Medical Science 1998; 60: 123-127.

van Wijk HJ, Buschbell H, Dibbits B, Liefers SC, Harlizius B, Heuven HC, et al. Variance component analysis of quantitative trait loci for pork carcass composition and meat quality on SSC4 and SSC11. J Anim Sci 2007; 85: 22-30.

Walling GA, Visscher PM, Andersson L, Rothschild MF, Wang L, Moser G, et al. Combined analyses of data from quantitative trait loci mapping studies. Chromosome 4 effects on porcine growth and fatness. Genetics 2000; 155: 1369-78.

Wallis M. Molecular evolution of the thyrotrophin-releasing hormone precursor in vertebrates: insights from comparative genomics. J Neuroendocrinol 2010; 22: 608-19.

Wimmers K, Murani E, Ponsuksili S, Yerle M, Schellander K. Detection of quantitative trait loci for carcass traits in the pig by using AFLP. Mamm Genome 2002; 13: 206-10.

Yerle M, Pinton P, Robic A, Alfonso A, Palvadeau Y, Delcros C, et al. Construction of a whole-genome radiation hybrid panel for high-resolution gene mapping in pigs. Cytogenet Cell Genet 1998; 82: 182-8.

28

Detection of Poxvirus Using Transmission Electron Microscopy Techniques During Outbreak in Bay-Winged Cowbird (*Gnorimopsar Chopi*)

M.H.B. Catroxo[1], A.M.C.R.P.F. Martins[1],
S. Petrella[2] and L. Milanelo[3]
[1] *Biological Institute of São Paulo, São Paulo,SP,*
[2] *Adolfo Lutz, Institute São Paulo, SP,*
[3]*Tietê Ecological Park, São Paulo, SP,*
Brazil

1. Introduction

Gnorimopsar chopi is a bird with blackish plumage, smooth and silky look, found throughout the Brazilian territory, except in the Amazon State. It is one of the singing birds most wanted by illegal trade and is listed as endargered to extinction by the Brazilian Institute of Environment and Renewable Natural Resources (IBAMA). Avian pox or variola avium is a infectious disease of worldwide distribution, that infect poultry, pet and wild birds of many species which result in economic losses to the poultry industry and commercial aviaries (Trypathy et al., 2000; Weli & Tryland, 2011).

Poxvirus infection has been considered one of important extinction factors for the endangered avian species in Hawaii and in North America (Friend & Franson, 1999; Smits et al., 2005; Kim & Tripathy, 2006).

The etiologic agent is a member of the genus *Avipoxvirus, Chordopoxvirinae* subfamily, *Poxviridae* family (Van Riper & Forrester, 2007). The genome is linear, double-stranded DNA molecule rariying in size from 260 to 309 kpb and encodes 260 putative genes (Moss, 2007). Virus particles measure 270 x 450 nm and are composed of an electron dense, centrally located core and two lateral bodies that are visible in ultrathin sections. In negative stained preparations the membrane displays an outer coat composed of a random arranjement of tubules (Carter & Cheville, 1981).

The disease is characterized by cutaneous or wet pox, diphteric or wet pox and septicemic forms. The cutaneous form is the most common in Passeriformes. In this form occurs the development of nodular proliferative skin lesions on the unfeathered parts of the body, legs, feet, face at the base of the beak and eyelids. In the diphtheric form, fibronecrotic lesions occur in the membranes of the upper respiratory tract and esophagus. Anorexia, letargy, ruffled plumage, respiratoty distress, sonnolence, cyanosis and death characterize septicemic poxvirus infections (Ritchie et al., 1994).

The virus may be mechanically transmited by insect vectors such as mosquitoes, mites or ticks, by direct contact with another infected bird, by contact with contamined food, water, semen or surfaces (Metz et al., 1985; Ritchie & Carter, 1995).

The curse of the disease is influencied by strain, route of infection, and the species of bird (Ensley et al., 1978).

The incubation period of Avipoxvirus is usually less than a week but may be up to 30 days. The morbidity rate during an outbreak may reach 100% (Ritchie et al., 1994).

Non-specific stress factors are associated with viral reactivation (Ritchie et al., 1994).

Transmission electron microscopy is a classic tool for the diagnosis of poxviruses, where the viral particles with characteristic morphology are present in large numbers in swabs, biopsies or dry crusts (OIE, 2008). This method has typically been used by national reference or research laboratories to identify avianpoxvirus (Weli et al., 2004).

The aim of this study was to detect the presence of avianpoxvírus particles in samples of skin lesions of *Gnorimopsar chopi* using negative staining (rapid preparation) immunocytochemistry (immunolabeling with colloidal gold particles) and resin embedding techniques.

2. Material and methods

2.1 Description of the outbreak

In the period of the 2005 to 2007, during illegal commercialization of Brazilian birds, 60 bay-winged cowbird (*Gnorimopsar chopi*) were apprehended and being forwarded to the CRAS (Wild Animals Recovery Center, Ecological Park of Tietê). After one or two months, 45 birds presented cutaneous lesions in the feet. Some of them also had lesions on the beak and wings. The animals also been presented clinical signs of anorexia, emaciation, locomotion difficulties, diarrhea, dehydration and death. The birds were sent to the Electron Microscopy Laboratory of the Biology Institute of São Paulo to investigate viral agents. Scabs and fragments of skin lesions collected of these birds were processed for transmission electron microscopy utilizing negative staining (rapid preparation), resin embedding and immunocytochemistry techniques.

2.2 Negative staining technique (rapid preparation)

In the negative staining the scabs and fragments of skin lesions were suspended in phosphate buffer 0.1 M, pH 7.0. Drops of the obtained suspension were placed in contact with metallic copper grids with carbon stabilized supporting film of 0.5% collodium in amyl acetate. Next, the grids were drained with filter paper and negatively stained at 2% ammonium molybdate, pH 5.0 (Brenner & Horne, 1959; Hayat & Miller, 1990; Madeley, 1997).

2.3 Resin embedding technique

Thin slices of scabs and fragments of skin lesions were fixed in 2.5% glutaraldehyde in 0.1M, pH7.0 phosphate buffer and pos-fixed in 1% osmium tetroxide in the same buffer. After dehydration in cetonic series, the fragments were embedded in Spurr resin (González-

Santander, 1969; Luft, 1961). Ultrathin sections were cut on the LKB ultratome and mounted on copper grids. The sections were stained by combination of uranyl acetate-lead citrate (Watson, 1958; Reinolds, 1963).

2.4 Immunocytochemistry technique

At the immunolabeling technique with colloidal gold particles for negative staining, the copper grids were placed in contact with viral suspension and, after removing excess with filter paper, the same were put on specific primary antibody drops. After successive washings in PBS drops, the grids were incubated in protein A drops in association with 10 nm gold particles (secondary antibody). Grids were then contrasted at 2% ammonium molybdate, pH 5.0 (Knutton, 1995). Observations were made in a Philips EM 208 electron microscope, at 80 kV.

3. Results

Among the 45 analyzed animals, all of them (100%) presented in feet, and occasionally in beaks and wings, small yellowish-brown proliferative nodules or scabs of different sizes, characteristic of cutaneous form (fig. 1, arrow).

Fig. 1. Nodular and crusted lesions on the feet of *Gnorimopsar chopi* (arrow).

3.1 Negative staining technique (rapid preparation)

On the transmission electron microscopy by the negative staining technique, two types of poxvirus particles were visualized in all the analyzed samples. the M form, with regular spaced thread-like ridges comprising the exposed surface (fig. 2, big arrow), measuring 280 x 230 nm and the C form or stain-penetrated particle showing the dumbbell-shaped core (fig. 3, arrow) measuring 360 x 330 nm.

3.2 Resin embedding technique

Using the resin embedding technique (positive staining) were visualized in the ultrathin sections, three types of intracytoplasmic inclusion bodies. The type A or Bollinger body (fig. 4), outlined by membrane, containing in its interior a great number of mature particles (fig. 4,

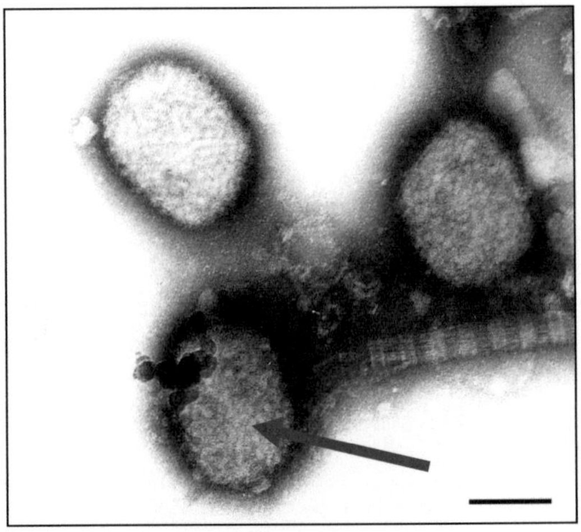

Fig. 2. Negatively stained avian poxvirus particles showing the M form, with regular spaced thread-like ridges comprising the exposed surface (arrow). Bar: 180 nm.

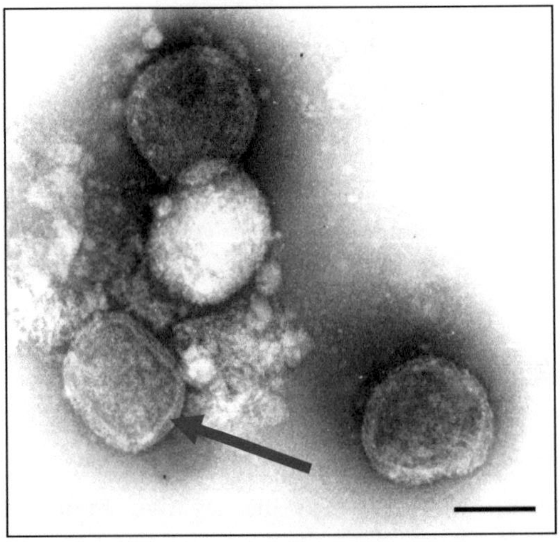

Fig. 3. Negatively stained avian poxvirus particles showing the C form or stain-penetred particle reveling the dumbell-shaped core (arrow). Bar: 200 nm.

arrow), measuring 200 x 300 nm, revealing the inner dumbbell-shaped core (fig. 5, big arrow), two lateral bodies (fig. 5 minor arrow) and an external envelope (fig. 5, blue arrow). In the type B electron dense inclusions bodies, viral particles budding of dense amorphous material were observed (figs. 6 and 7, arrow). Fibrillar inclusions constituted by groups of fibrils (fig. 8, arrow) were disposed in groups varying from 2 up to 7 and sometimes showed cross striations.

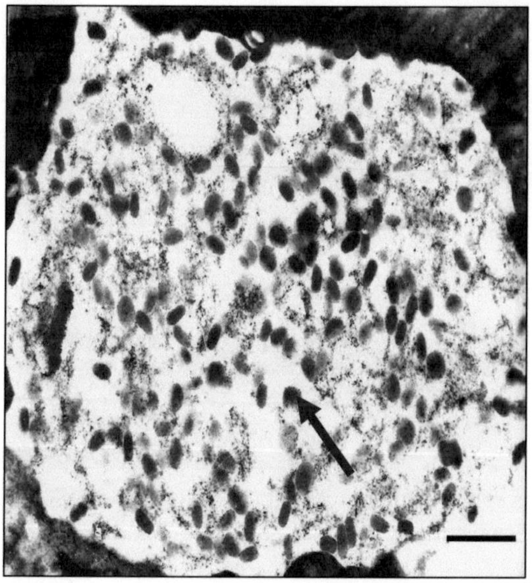

Fig. 4. Ultrathin sections of the scabs fragments showing type A or Bollinger intracytoplasmic inclusion bodies, containing mature particles (arrow). Bar: 1040 nm.

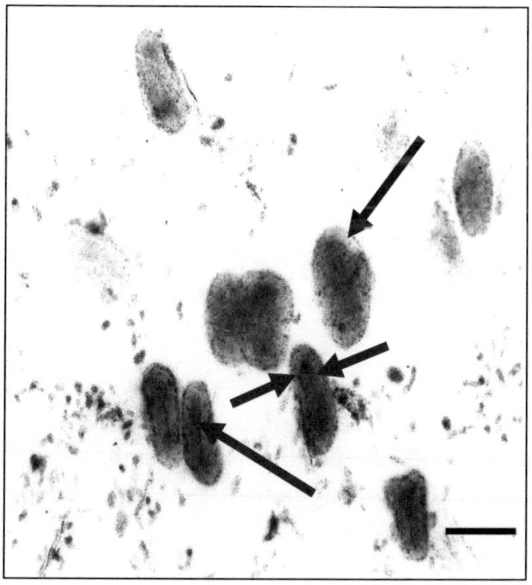

Fig. 5. Ultrathin sections of the skin lesions fragments, showing mature particles with inner dumbell shaped core (big arrow), lateral bodies (minor arrow) and external envelope (blue arrow). Bar: 280 nm.

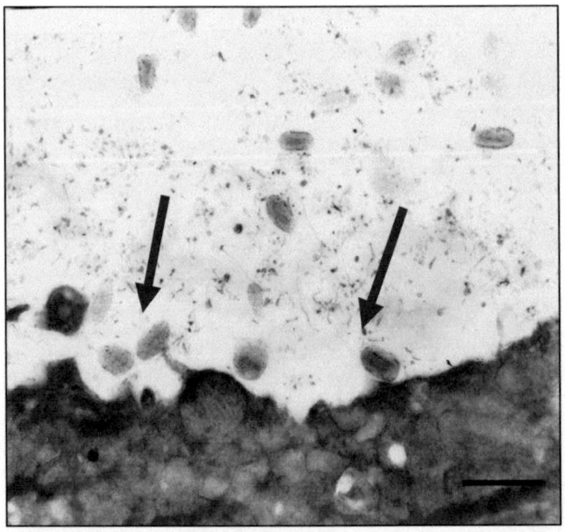

Fig. 6. Ultrathin sections of the scabs fragments reveling type B electron dense inclusions bodies with viral particles budding of dense amorphous material (arrow). Fig. 6 - Bar: 600 nm.

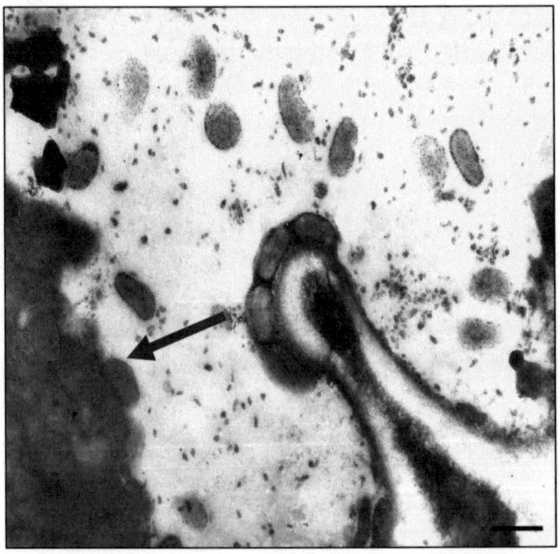

Fig. 7. Bar: 430 nm.

3.3 Immunocytochemistry technique

In the immunocytochemistry technique, the antigen-antibody interaction was strongly enhanced by the dense gold particles over the viruses (Fig. 9, arrow).

4. Discussion

In this paper the presence of nodular proliferative and crusted lesions, was observed in feet de 45 (100%) de 60 bay-winged cowbird (*Gnorimopsar chopi*) examined. Some of them also had lesions on the beak and wings.

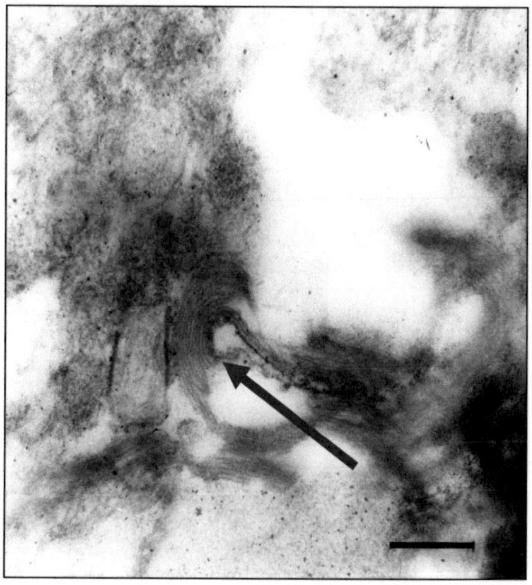

Fig. 8. Ultrathin sections of the skin fragments revealing fibrillar inclusions constituted by groups of fibrils or lamelae (arrow). Bar: 320 nm.

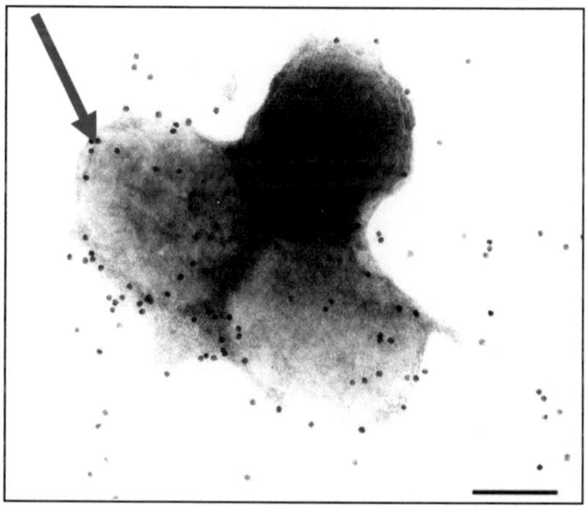

Fig. 9. In the immunocitoquemistry technique, the antigen-antibody interaction was strongly enhanced by the dense colloidal gold particles (arrow) over the avipoxvirus particles. Bar: 130 nm.

Lesions similar have been described in other birds (Goodpasture & Anderson, 1961; Trapp, 1980; Van Riper et al., 2002; Catroxo et al., 2009).

Lesions can interfere wich perching when occur on the feet or legs and frequently lead to development of secondary infections (Van Riper III & Forrester, 2007).

Clinical signs of lack appetite, weight lost, difficulties of locomotion, diarrhea, dehidratation and death were observed in all birds of this study. This has been documented in similar occurrence of avianpox in other avian species (Mc Donald et al., 1981; Docherty et al., 1991; Ostrowski et al.., 1995; Gerlach et al., 1998; Kreuder et al., 1999; Catroxo et al., 2009).

By means of the negative staining technique, poxvirus particles were identified in suspension of skin fragments, as been seen in other cases (Tantawi et al., 1981; Terragino et al., 1981; Docherty et al.,1991; Allwright et al., 1994; Gulbahar et al., 2005; Kulich et al., 2008; Catroxo et al., 2009; Shivaprasad et al., 2009).

In the present ultrastructural study, intracytoplasmic inclusions bodies of different types were found. The analysis of ultrathin sections of skin lesions showed the presence of type A or Bollinguer intracytoplasmic inclusion bodies, containing oval, mature or complete viral particles, measuring on the average, 330 x 260 nm, with an inner dumbell-shaped core. These morphological characteristics described by us were similar to those found in other birds species (Tripathy & Reed, 2003; Smits et al., 2005; Beyut & Haligur, 2007; Kulich et al., 2008; Saito et al., 2009; Shivaprasad et al., 2009; Catroxo et al., 2009).

The observation of immature intracytoplasmic particles, budding of a dense amorphous material or type B inclusion body, agrees with findings of other authors (Purcell et al., 1972; Thiele et al., 1979, Sadosiv et al., 1985; Catroxo et al., 2009).

We also observed fibrilar intracytoplasmic inclusions, constituted by groups of fibrils or lamellae, disposed in groups varying from 2 to 7, occasionally showing cross striations. These ultrastructural aspects were also reported by other authors in avipoxvirus studies (Catroxo et al., 2009).

The avian poxvirus were intensely labeled by collodial gold, when we applied the immunocitoquemistry method, also used to detect these viruses in skin lesions in other passerines (Catroxo et al., 2009).

All birds in our study were sudden death. During avipox outbreak , mortality can reach 80 to 100% in canaries and other finches in contrast with lower mortality observed in chicken and turkey (Tripathy & Reed, 2003).

The increased population of mosquitoes during the summer at the place of the outbreak probably contributed to the sudden onset of the outbreak. Other sucking insects such flea and mite are considered to be important vectors for the spread of avian poxvirus (Ritchie et al., 1994; Gustafson et al., 1997).

Prevention and control measures, such vaccination, quarantine for new birds, control of arthropods, prevention and reduction of stress, cleaning and disinfecting of cages should be instituted to farms and zoos (Ritchie et al., 1994).

It is very important to confirm the presence of the ethiological viral agent in the samples. Infection as other agents, such pappiloma virus, mites, flies and mycotoxins may produce like lesions in the skin (Trypathy & Reed, 2003).

Our results obtained through the techniques mentioned in this study to detect avipoxvirus particles in these birds, allowed rapid preventive measures to control the disease, preventing spread to commercial breeder farms and aviaries with economic losses thus protecting and preserving these endangered species.

To our knowledge, this is the first report of avian poxvirus infection in bay-winged cowbird (*Gnorimopsar chopi*).

5. References

Allwright, D.M.; Burger, W.P.; Geyer, A. & Wessles, J. Avian pox in ostriches. J. S. Afri. Vet. Assoc., 65 (1): 23-5, 1994.

Brenner, S. & Horne, R. W. A negative staining method for high resolution electron microscopy of viruses. Biochem. Biophys. Acta., 34:103, 1959.

Bohls, R.L.; Linares, J.A.; Gross, S.L.; Ferro, P.J.; Silvy, N.J.; Collisson, E.W. Phylogenetic analyses indicate little variation among reticuloendotheliosis viruses infecting avian species, including the endangered Attwater's prairie chicken. *Virus Res., 119:*187-94, 2006.

Carter, J.K.V. & Cheville, N.F. Isolation of surface tubules of fowlpox virus. *Avian Dis., 25:*454-62, 1981.

Catroxo, M. H. B.; Pongiluppi T.; Melo, N. A.; Milanelo, L.; Petrella, S.; Martins, A. M. C. P. F.; Rebouças, M. M. Identification of poxvirus under transmission electron microscopy during outbreak period in wild birds, in São Paulo, Brazil. *Int. J. Morphol., 27(2):*577-585, 2009.

Docherty, D.E.; Long, R. I.; Flickinger, E.L. & Locker, L.N. Isolation of poxvírus from debilitating cutaneous lesions on four immature grackles (*Quiscalus sp*). *Avian Dis., 35 (1):* 244-7, 1991.

Ensley, P.K.; Anderson, M.P.; Costello, M.L.; Powell, H.C. & Cooper, R. Expornitic of avian pox in a zoo. *J. Am. Vet. Med. Assoc., 173:*1111-4, 1978.

Friend, M.; Franson, J.C. & Ciganovich, E.A. (Eds). *Field Manual of Wildlife Diseases: General Field Procedures and Diseases of Birds.* Washington, Ed.Washington D.C., United States Geological Survey, 1999. pp.163-170.

Gonzales-Santander, R. Técnicas de microscopia eletrônica en biología. Madrid, Ed. Aguilar, 1969. 666p.

Goodpasture, E.W. & Anderson, K. Isolation of a wild avian pox virus inducing both cytoplasmatic and nuclear inclusions. Am. J. Pathol., 40:437-53, 1961.

Gülbahar, M.Y.; Çabalar, M. & Boynukara, B. Avipoxvirus infection in quails. Turk. J. Vet. Anim. Sci., 29: 449-4, 2005.

Gustafson, C.R.; Bickford, A.A.; Cooper, G.L. & Charlton, B.R. Sticktight fleas associated with fowl pox in a backya chicken flock in California. Avian. Dis., 41: 1006-9, 1997.

Hayat, M. A. & Miller, S. E. Negative Staining. Mc. Graw-Hill Publ. Company., 1990, 235p.

Kim, T. & Tripathy, D.N. Evatuation of pathogenicity of avian poxvirus isolates from endangered Hawaiian wild birds in chickens. Avian Dis., 50(2):288-91, 2006.

Knutton, S. Electron microscopical methods in adhesion. Meth. Enzymol., 253:145-58, 1995.

Kulich, P.; Roubalová, E.; Dubská, L.; Sychra, O.; Smid,. B.; Literák, I. Avipoxvirus in blakcaps (Sylvia atricapilla). Avian Pathol., 37(1):101-7, 2008.

Luft, J. A. Improvements in an epoxy resin embedding methods. J. Biophys. Biochem. Cytol., 9:409-14, 1961.

Madeley, C. R. Electron microscopy and virus diagnosis. J. Clin. Pathol., 50:454-6, 1997.

Metz, A.L.; Hatcher, L.; Newman, A.J. & Halvorson, D.A. Venereal pox in breeder turkeys in Minnesota. Avian Dis., 29 (3): 850-3, 1985.

Moss, B. Poxviridae: The viruses and their replication. In: Virology, Philadelphia, USA, 5 th. Ed. Lippincot Williams & Wilkins, 2007. pp.2905-75.

OIE. Manual of Diagnostic Tests & Vaccines for Terrestrial Animal. In: Fowlpox. França, Paris, 6 th ed., 2008. pp. 531-7.

Ostrowski, S.; Dorrestein, G.M.; Ancrenas, M. & Saint-Jalme, M. Debilitanting cutaneous poxvirus lesion on two captive houbara bustards (Chlamydotis undulata). Avian Dis., 39(4): 904-11, 1995.

PurcelL, D.A.; Clarke, J.K.; McFerran, B.J. & Hughes, D.A. The morphogenesis of pigeonpox virus. J. Gen. Virol., 15: 79-83, 1972.

Reinolds, E.S. The use of lead citrate at high pH as an electron-opaque stain in electron microscopy. J. Cell. Biol., 17:208-12, 1963.

Ritchie, B.W. & Carter, K. Avian viruses: Function and control. Lake Worth, Florida, Ed. Publishing Incorporated, 1995. pp.285-311.

Ritchie, B.W.; Harrison, G.J. & Harrison, L.R. Avian Medicine: Principles and application. Florida, Ed. Wingers Publishing Inc., 1994. pp.865-74.

Sadosiv, E.C.; Chang, P.W. & Gulka, G. Morphogenesis of canary poxvirus and its entrance into inclusion bodies. Am. J. Vet. Res., 46 (2): 529-35, 1985.

Shivaprasad, H.L.; Kim, T.; Tripathy, D.; Woolcook, P.R.; Uzal, F.; Unusual pathology of canary poxvirus infection associated with high mortality in young and adult breeder canaries (Serinus canaria). Avian Pathol., 38(4):311-6, 2009.

Smits, J.E.; Tella, J.L.; Carrete, M.; Serrano, D. & López, G. Na epizootic of avian pox in endemic short-toed larks (Calandrella rufescens) and berthelot's pipits (Anthus berthelotti) in the Canary Island, Spain. Vet. Pathol., 42: 59-65, 2005.

Terregino, C.; Catelli, E.; Delogu, M.; Capua, I. & Tonelli, A. Poxvirus infection in a blue bonnet owls from Florida. J. Wildl. Dis., 145 (9): 264, 1999.

Thiele, J.; Kiel, H. & Adolphs, H.D. Avien pox virus. An ultrastructural study on a cherrug falcon. Brief report. Arch. Virol., 62 (1): 77-82, 1979.

Trapp, J.L. Avian pox in the gray-crowned rosy finck in Alaska. North Am.Birds Bander., 5 (2): 146-7, 1980.

Tripathy, D.N.; Schnitzlein, W.M.; Morris, P.J.; Janssen, D.L.; Zuba, J.K.; Massey, G.; Atkinson, C.T. Characterization of poxviruses from forest birds in Hawaii. *J. Wildl. Dis., 36*:225-30, 2000.

Tripathy, D.N., and W.M. Reed. 2003. Pox. In *Diseases of Poultry* , 11th Ed., Y.M. Saif, H.J. Barnes, J.R. Glisson, A.M. Fadly, and L.R. McDougald (eds.). Iowa State University Press, Ames, Iowa, U.S.A., pp. 253–269.

Van Riper III, C.V.; Riper, S.G.V. & Hansen, W.R. Epizootiology and effect of avian pox on havaiian forest birds. *The Auk., 119(4):* 929-42, 2002.

Van Riper III, C. & D. J. Forrester. *Avian pox.* In: Infectious Diseases of Wild Birds. Ames, IA, Ed. Blackwell Publishing, 2007. pp.131–176.

Watson, M. L. *Staining of tissue sections for electron microscopy with heavy metals. J. Biophyis. Biochem. Cytol.*, 4:475-8, 1958.

Weli, S.C.; Okeke, M.I.; Tryland, M.; Nilssen, O. & Traavik, T. Characterization of avipoxviruses from wild birds in Norway. Can. J. Vet. Res., 68: 140-5, 2004.

Weli, S, & Tryland, M. Avipoxviruses: infection, biology and their use as vaccine vectors. *Virol. J.*, 8(49):1-15, 2011.

29

Doppler Ultrasonography for Evaluating Vascular Responses to Ergopeptine Alkaloids in Livestock

G.E. Aiken and J.R. Strickland
USDA-Agricultural Research Service,
Forage-Animal Production Research Unit,
Lexington, KY,
USA

1. Introduction

Ergot alkaloids are produced by non-spore producing fungal endophytes that infect certain species of grasses, most notably tall fescue [*Lolium arundinaceum* (Schreb.) Darbysh.] and perennial ryegrass (*Lolium perenne* L.), and the spore producing *Claviceps* spp. that infect seed heads of certain grasses and particularly the cereal grains [rye (*Secale cereal* L.), barley (*Hordeum vulgare* L.), wheat (*Triticum aestivum* L.), , and oats (*Avena sativa* L.)] (Strickland et al., 2011). Ergot alkaloids induce a toxicosis in grazing livestock, with symptoms in cattle that include rough hair coats during the warm season, severe heat stress in warm and humid temperatures, reduced dry matter intake, agalactia, and poor reproductive and weight gain performance (Porter & Thompson, 1992; Paterson et al., 1995). Sheep grazing endophyte-infected fescue also can have elevated core body temperatures in warm and humid environments (Zanzalari et al., 1989; Hanna et al., 1990), and reduced dry matter intakes (Chestnut et al., 1992; Aldrich et al., 1993). The most pronounced effect on horses is observed with broodmares, which can exhibit prolonged gestation and agalactia (Cross et al., 1995). Symptomatology of the malady are reflective of alterations in hormone profiles (Porter & Thomson, 1992; Browning et al., 1997, 1998) and reductions in blood flow to peripheral tissues caused by interactions of ergopeptine ergot alkaloids with biogenic amine receptors in the vasculature (Oliver et al., 1998) to induce persistent vasoconstriction and restrict regulation of core body temperature by the sympathetic nervous system (Oliver, 1997). Consequently, lack of thermoregulation by livestock exposed to ergopeptines are extremely vulnerable to heat and cold stresses (Hemken et al., 1981; Al-Haidary et al., 2001).

Ergot alkaloids contain a common tetracycline ergoline ring structure and there are 3 different classes: 1) clavine alkaloids, 2) lysergic acid and its derivatives, and 3) ergopeptines (Lyons et al., 1986; Bush and Fannin, 2009; Strickland et al. 2011). Ergopeptines exert the greatest influence on the vasculature, with ergovaline being the ergopeptine of highest concentration in tall fescue (Lyons et al., 1986) and with a demonstrated high potency as a vasoconstrictor (Klotz et al., 2006).

Reduced blood flows in cattle and sheep consuming endophyte-infected tall fescue was first reported by Rhodes et al. (1991) who used radioisotope labeled microspheres to estimate flow rates between those consuming endophyte-infected and endphyte-free tall fescue. Vascular dysfunction is also evidenced by in vitro findings of ergovaline induced constriction of bovine lateral saphenous veins (Klotz et al., 2006, 2007), uterine and umbilical arteries (Dyer, 1993), and rat tail and guinea pig iliac arteries (Schoning, et al., 2001). *Caviceps* spp. do not produce ergovaline, but can produce high quantities of ergotamine. Blaney et al. (2009) indentified ergotamine as the dominant ergopeptine produced by *Claviceps purpurea* sclerotia, which is the ergot that infects Australian rye. Blaney et al. (2011) reported severe hyperthermia in steers fed feedyard rations containing sorghum infected with Claviceps Africana, as compared to steers consuming non-infected sorghum.

Color Doppler ultrasonography has been used with humans as a noninvasive technique for real time diagnoses of aberrant blood flow (Someda et al, 1995; Whelan and Barry, 1992), stenosis (Hatle et al., 1980; Olin et al., 1995; Schmidt et al., 1997), and occlusions (Moneta et al., 1992; Müller et al., 1995). It also has been used to determine vasoconstrictive and blood flow responses in cattle (Aiken et al., 2007; 2009b) and sheep (Aiken et al., 2011) to ergot alkaloids, and was performed with steers to quantify artery lumen area and blood flow responses to heat and cold challenges (Kirch et al., 2008). Color doppler ultrasonography has potential use as a diagnostic or research tool in identifying and quantifying aberrant constriction caused by ergot alkaloids or other toxicants. This chapter will discuss: 1) blood flow aspects of thermoregulation in livestock, 2) effects of ergot alkaloids on blood flow and thermoregulation and, 3) procedures and sources of error in using Doppler ultrasonography as a research tool in evaluating vasoconstrictive responses in livestock exposed to ergot alkaloids.

2. Vascular control of blood flow for regulating core body temperature

Blood flow is necessary for effective thermoregulation in all mammals. Core body temperature is maintained and stabilized through constriction or dilation of vasculature to control blood flow in peripheral tissues for the regulation of heat dissipation from capillary beds in the skin (Chotani et al., 2000; Johnson & Kellogg, 2010). Excessive declines in core body temperature during cold ambient temperatures are controlled by vasoconstriction of blood flow that reduces heat transfer in the skin, whereas increases in core body temperature during warm ambient temperatures are mitigated through vasodilation to increase heat transfer (Chotani et al., 2000). During heat stress, there is decreased activity by sympathetic vasoconstrictor nerves and increased activity by a sympathetic cutaneous vasodialator nervous system (Johnson and Proppe, 1996). Expansion of the luminal area of blood vessels delivering blood flow to the skin during heat stress also provides the volume of water needed to drive sweating for evaporative cooling of the skin (Gagge and Gonzalez, 1996).

Blood flow rate is a function of cross-sectional luminal area of the vessel and flow velocity, with vessel luminal area having the greatest influence on flow rate. Based on Poiseuille's Law, volume flow rate (mL/min.) is directly proportionate to the fourth power of the radius of the cross-section of the vessel's lumen. Carter (2000) noted that a 10% decrease in radius of a tube under steady state conditions will reduce flow rate by approximately 35%. Constriction and dialation of arteries and veins are mediated in response to hot or cold environments by endogenous biogenic amines: primarily serotonin, norepinephrine, and epinephrine (Johnson and Proppe, 1996; Strickland et al., 2011).

3. Ergopeptnes and their effects on vascular circulation

Ergopeptines are ergot alkaoids that are produced by an endophyte (*Neotyphodium coenphialum*) that infects tall fescue plants (Bacon, 1995), and by the *N. lolii* endophyte that infects perennial ryegrass (Easton & Tapper, 2005). Although the endophyte that infects perennial ryegrass produces small amounts of ergot alkaloids that can induce vasoconstriction (Aiken et al., 2011), it is its production of lolitrem B that causes ryegrass staggers cattle and sheep which is of greater concern (Fletcher and Harvey, 1981). Livestock exhibiting staggers are incapacitated due to tremors, but the malady also affects animal performance (Siegel et al., 1985). All plant parts of tall fescue and perennial ryegrass contain ergot alkaloids, but alkaloid concentrations differ among plant parts. Rottinghaus et al. (1991) determined the ranking of plant parts from highest to lowest ergot alkaloid concentrations is seed, stem, leaf sheath, and leaf blade. There is a distinction between the mutual relationships between the wild-type, toxic endophytes that infect naturalized populations of tall fescue and perennial ryegrass and non-ergot alkaloid producing novel endophytes that are artificially infected into commercially released cultivars of each grass (Bouton et al., 2002). Claviceps spp. also produce ergot alkaloids, but fungal colonization and alkaloid concentrations are restricted to the seed or grain (Bandyopadhyay et al., 1998).

Ergovaline has been proposed as the likely causal agent in the fescue toxicosis syndrome (Lyons et al, 1986). In vitro electromyograph studies have reported ergovaline to cause contractile responses of bovine uterine and umbilical arteries (Dyer, 1993), rat tail and guinea pig iliac arteries (Schoning, et al., 2001), and lateral saphenous weins of cattle (Klotz et al., 2007). Klotz et al. (2007) reported similar contractile responses between ergovaline and ergotamine, with contractile responses being initiated at 1 x 10^{-8} M concentrations for both ergopeptines. An earlier experiment by Klotz et al. (2006) determined a weak in vitro contractile response of the lateral saphenous vein to lysergic acid (a structurally simpler ergot alkaloid (reviewed by Strickland et al., 2011) that did not mediate contraction until concentrations reached supraphysiological levels (1 x 10^{-4} M). Dyer (1993) and Schoning et al. (2001) both showed that ergovaline elicited its contractile effects through activation of $5HT_{2A}$ serotonergic receptors. However, in contrast to Dyer (1993) who showed that the α_1-adrenergic receptor was not important in the contractile effects of ergovaline on the bovine uterine and umbilical arteries; Schoning et al. (2001) clearly demonstrated that the α_1-adrenergic receptors were important to vascular regulation by ergovaline in their blood vessel models. Similar findings have been noted for other ergot alkaloids produced by both Neotyphodium and Claviceps spp. (reviewed by Strickland et al., 2009a, b; Strickland et al., 2011).

Although in vitro models are useful tools for investigating and identifying the modes by which the ergot alkaloids may effect vascular dysfunction, the data from these models must be interpreted with care until fully validated by in vivo models. Partial validation is provided by the results of earlier in vivo studies. Lewis and Gelfand (1935) demonstrated that ergotamine treatment of chickens resulted in cessation of blood flow to the comb and subsequent gangrene. They postulated that the gangrene developed as a result of endothelia damage. Shappell (2003) demonstrated that ergovaline was, in fact, directly cytotoxic to a number of cell types, including endothelial cells. Additional morphological changes and blood flow dysfunction of peripheral tissues in cattle have been reported (Walls and Jacobson, 1970; Julien et al., 1974; Williams et al., 1975), including: blood vessel congestion, perivascular hemorrhage, blood vessel distension, and thickened blood vessel walls with small lumens. Strickland et al. (1996), provided evidence, using bovine vascular smooth muscle cells in vitro, that the thickened blood vessel walls were likely the result of hyperplasia and not hypertrophy.

In addition to the noted effects in the peripheral vascular beds, Rhodes et al. (1991) reported ergot alkaloid induced in vivo constrictive responses in several vascular beds of steers. Using radiolabeled microspheres, steers fed diets with high ergot alkaloid concentrations (2.6 mg egovaline/kg DM) had reduced blood flow to rib skin, cerebellum, duodenum, and colon, as compared to those fed low ergot alkaloid diets (< 0.05 mg ergovaline/kg DM). They also reported similar findings in sheep. The aforementioned in vivo studies have provided partial validation of the in vitro findings to date. Further, the in vitro data coupled with the in vivo findings support a strong role of the ergot alkaloids in vascular dysfunction. However, none of the aforementioned studies have been capable of studying, in real time, the effects of the ergot alkaloids on vascular function in the intact and unaltered animal. Color Doppler Ultrasonography offers the opportunity to do exactly that as is evidenced by recent publications (Aiken et al., 2007, 2009b, and 2011) concerning blood flow changes in cattle and sheep exposed to ergot alkaloids.

4. Color Doppler ultrasonography in measuring vasoconstriction

Color Doppler ultrasonography was introduced in the 1980s, but it was not until the 1990s that B-mode (two dimensional, gray scale) and color Doppler scans were combined as duplex images (Fig. 1) and used as a diagnostic or research tools in evaluating hemodynamics and increasing our knowledge of the characteristics and physiology of normal and abnormal blood flows. Color Doppler ultrasonography was used to evaluate hemodynamics and cross-sectional luminal areas of the cerebral artery in humans during

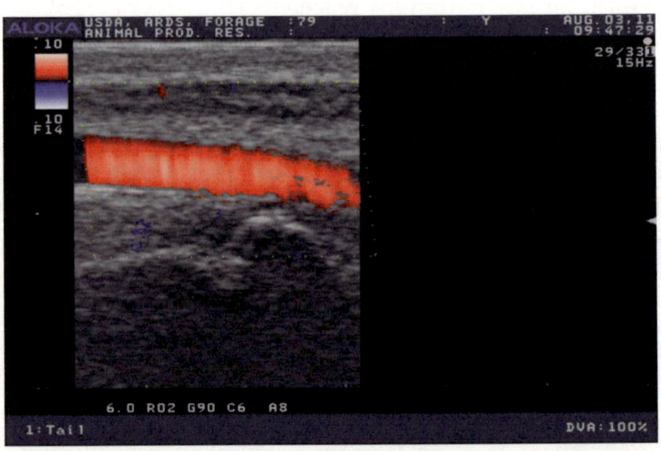

Fig. 1. A duplex image with B-mode image of a longitudinal section of the caudal artery of a heifer combined with color Doppler to exhibit blood flow.

hypoxia and hypercapnia (Poulin and Robbins, 1996). Wendelhag et al. (1991) utilized B-mode ultrasonography to measure carotid luminal and plaque areas in humans. Doppler spectra have provided estimates of blood flow characteristics in arteries of the dog (Lee et al., 2004) and horses (Raisis et al., 2000; Hoffman et al., 2001). Recently, color doppler ultrasonography was performed in evaluating vasoconstriction and blood flow responses to ergot alkaloid exposure in cattle (Aiken et al. 2007; 2009b) and sheep (Aiken et al., 2011). Therefore, the technology has been utilized to acquire baseline measures of normal flood flow characteristics and assessing vascular constriction as a factor of abnormal blood flow.

It is beyond the scope of this chapter to provide a detailed description of the mechanics and physics associated with color Doppler ultrasonography, but some basic fundamentals are necessary for understanding the procedures and potential errors when using the technology. Basically, the transducer transmits a frequency of ultrasonic sound waves (1 to 30 MHz) through tissues and, depending on the densities of the tissues the ultrasound beam contacts, a portion of the waves pass through a particular tissue and the remainder is reflected. Piezoelectric crystals in the transducer transmit and receive wave frequencies and the ultrasound unit compiles and converts the wave energy to electrical energy. There is a progressive loss of acoustic energy as ultrasonic beams passes through tissues, which is referred to as attenuation. Attenuation is the decrease in the intensity of returning sound waves as ultrasonic beams are transmitted into deeper tissues. Amount of attenuation, measured in dB/cm, is dependent on depth of ultrasonic beam, density of the tissue the beam passes through, and the frequency of the transmitted beam (Sites et al., 2007). Gain settings (i.e., separate settings for B-mode, color Doppler, and Doppler spectra) affect sensitivity. Amplitude and velocity information of the transmitted versus received frequencies are used to compute and delineate anatomical structures viewed in B-mode images, blood flow identification, and velocity within the sample volume cursor placed in the real-time color Doppler images.

There are two types of transducers used in color Doppler ultrasonography, continuous- and pulse-wave Doppler transducers. Continuous-wave transducers use 2 crystal units to continuously transmit and receive ultrasound waves. They can accurately measure a wide range of flow velocities within a vessel, but lacks an ability to control location of measurements within the vessel lumen. Most Doppler ultrasound units use pulsed-wave transducers that alternate groups of crystals in transmitting and receiving ultrasound wave in pulses to provide Doppler shift data from a specified area within the vessel lumen. This specified area, referred to as the sample volume, is set by the operator. Ultrasonic beans also can be angled with curvilinear and linear phased array transducers. For pulse Doppler scanning, the same mechanisms apply as with B-mode scans, but the sample volume cursor is placed in a B-mode image of a longitudinally scanned vessel and within the color generated by blood flow (red for flow towards the transducer and blue for flow away from the transducer). Doppler spectra derived from blood flow within the sample volume cursor can be automatically or manually traced for a given cardiac cycle to measure systolic velocity, end diastolic velocity, mean velocities, and resistance and pulsatility indices using algorithms in the computer of most Doppler ultrasound units. Most Doppler ultrasound units also will provide measures of heart rate, time, and acceleration. The longitudinal color Doppler image and Doppler spectra are typically combined (Fig. 2), with the color Doppler scan being utilized for placement of the sample volume cursor for the observed Doppler spectra.

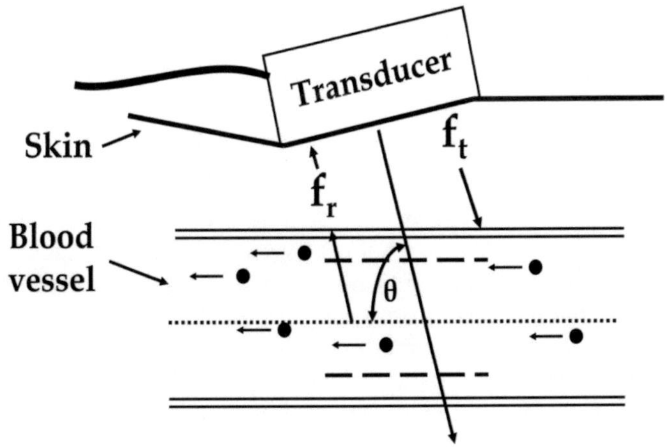

Fig. 2. Transmission (ft) and receiving (fr) ultrasound beams for the detection of Doppler signals from blood flow. The Doppler angle, θ, is the angle between the directions of blood flow and the beam. Space within the dashed lines represents the sample volume (Adapted from Zagzebski, 2000).

Blood flow velocity is determined by calculating Doppler Frequency (f_D):

$$f_D = 2f_t v\cos\theta / c$$

where ft is the transmitted frequency, fr is the received frequency, v is the blood flow velocity, θ is the Doppler angle between the axis of the ultrasound beam and the direction of flow , and c is the speed of sound. The Doppler effect allows the calculation of velocity of a moving object as the frequency of sound waves generated by the object increases as it moves from a far to a near distance. Calculation of blood flow velocity is not possible if the ultrasound beam is parallel with blood flow (i.e., θ = 90°). Angling of the ultrasonic bean must be done through angling of the transducer or using the beam steering with phased array transducers.

Pulse-wave transducers are designed to receive frequencies of scattered sound waves bounced from moving particles (e.g., hemoglobin) within the sample volume cursor of the targeted vessel (Fig. 3). Cross-sectional color Doppler images can be used to trace luminal areas of vessels and combined with blood flow velocities derived from Doppler spectra from a longitudinal image of the vessel to calculate blood flow rate (mean velocity X cross-sectional luminal area). Luminal areas determined from cross-sectional scans of vessels can be used to evaluate contractile responses of smooth muscle within artery and vein walls to environmental conditions, pharmaceuticals, or toxicants. Therefore, duplex imaging of B-mode and color Doppler ultrasound images has provided a tool for conducting either diagnoses or scientific research in real time that is noninvasive, and can be repeatable and objective by following good scanning technique and interpretation procedures.

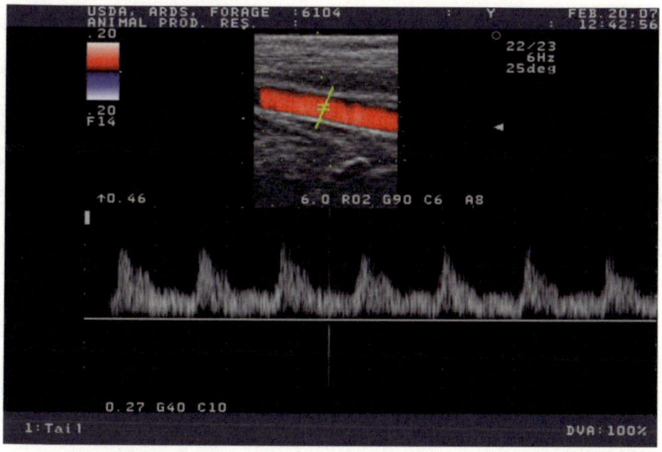

Fig. 3. Duplex image of a longitudinal scan of the median caudal artery of a heifer showing the sample volume cursor (upper image) and the Doppler spectra (lower image).

4.1 Doppler imaging the vasculature of livestock

Determinations of constriction and blood flows in small arteries or veins are possible with Doppler ultrasonography if comparisons are made between with and without toxicant treatment groups of animals, or between baseline measures when animals are without toxicant exposure versus measures after exposure to toxicants. Walls of small veins possess smooth muscle and endothelium at the vessel intima that contain biogenic amide receptors; therefore, veins will elicit a vasoconstrictrive response to ergot alkaloids and facilitate a comparison between animals with and without exposure to ergot alkaloids. Unfortunately, elasticity of veins negates obtaining measures of blood flow resistance, which is valuable information on down-flow resistance. Small arteries possess thicker walls and layers of smooth muscle and are positioned upstream from capillary beds, which facilitate measurement of blood flow resistance, a relevant blood flow characteristic. Aiken et al. (2011) did not observe constriction in the carotid arteries of lambs over a 9-d period of grazing perennial ryegrass infected with a novel endophyte that produced ergovaline, but the carotid showed linear increases in pulsatility, a measure of blood flow resistance (Petersen et al., 1997). It was concluded that constriction of cranial arteries were the source of resistance detected in the carotid arteries.

Emphasis on blood flow to peripheral tissues facilitates use of high frequency (≥ 5 MHz) transducers that will provide high resolution imaging of vessels at low depths from the transducer. Hemodynamics can be studied for vessels less than 2 to 3 cm from the skin surface and supplying blood to the tissues of the extremities. Cross-sectional images of the medial caudal artery (Fig. 4) of cattle has been imaged at the base of the tail to evaluate constrictive responses of the artery to ergot alkaloids (Aiken et al., 2007; 2009b; Kirch et al., 2008). For sheep, flow through the caudal artery is affected from the docking of tails and of questionable reliability, but cross-sectional images of the posterior auricular artery (Fig. 5) that supplies blood to the ear can reliably be used for assessing vasoconstriction (Aiken et al., 2011). Cross-sections of the medial palmer artery of the forelimb also have been imaged (Fig. 6; unpublished data collect by K. McDowell; Dept. of Veterinary Sciences; University of Kentucky).

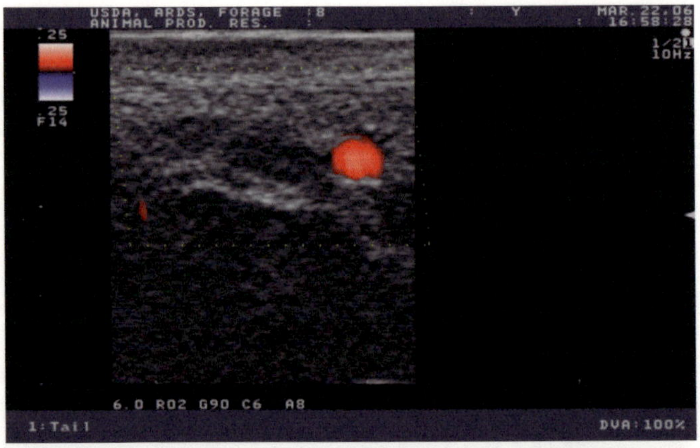

Fig. 4. Cross-sectional image of the medial caudal artery in a heifer calf.

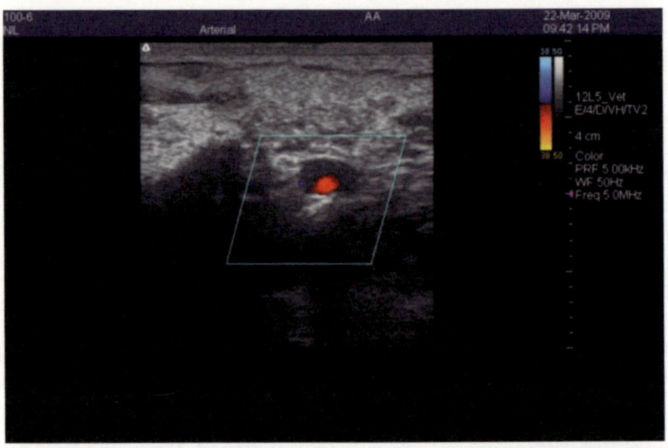

Fig. 5. Cross-sectional image of the right auricular artery of a ewe lamb.

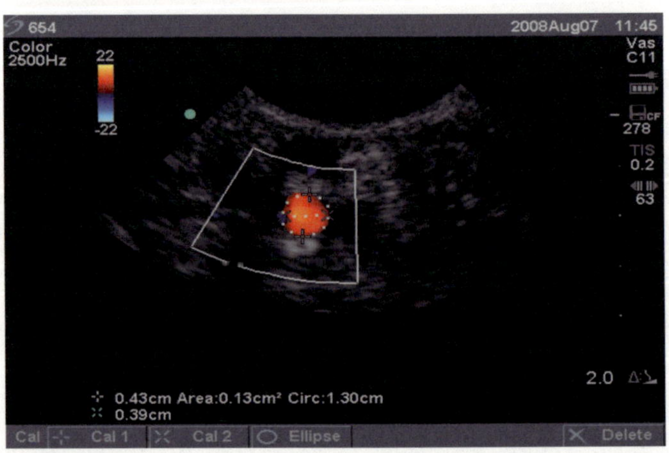

Fig. 6. Cross-sectional image of the medial palmer artery of a mare.

Other vessels can be scanned, but decisions of which vessels to study in livestock animals should be partly based on ease and safety for a particular image site. Imaging of larger livestock that must be handled in chutes (i.e., pastured animals with minimal exposure to humans) or stanchions should be restricted to exposed upper regions. Ultrasounding vessels in legs should be restricted to haltered animals at ease with human contact because of the vulnerability to operator injury and equipment damage. Major arteries (e.g., aorta, carotid, mysenteric, etc.) and those supplying blood to major organs (e.g., hepatic, pulmonary, cerebral, etc.) can be studied to ascertain toxicant effects on organ function. These evaluations at greater tissue depths must be done with low pulse frequencies (2 to 5 Mhz) to obtain the needed resolution and with high end, high resolution ultrasound units. It is further recommended these studies be done with haltered, calm animals in controlled environments. Further, obese animals should be avoided because thick layers of subcutaneous adipose will cause interference in achieving high resolution with B-mode Doppler images taken at low frequencies (Pozniak, 2000).

Lumen areas of vessels can be traced from B-mode scans with or without color Doppler imaging (Fig. 7). Aiken et al. (2009a) concluded that artery luminal areas can be measured with similar precision by tracing the intima of connective tissue in the artery wall or the outer boundary of the color Doppler flow signal. Color Doppler imaging for measuring luminal area will require images depicting the maximum flow signal which will coincide with peak systole. These images can be identified using the cine memory of the ultrasound unit. There are two methods of imaging color blood flow. Color Doppler uses the Doppler frequency shift to detect flow within the vessel and shifts the color shade from dark to light as frequency and velocity increases. Power Doppler imaging uses the power or intensity of the Doppler signal to show differences in blood flow velocities within a vessel. Power Doppler is generally more sensitive than color Doppler in detecting blood flow. Without color Doppler, luminal areas should be at certain points of a cardiac cycle (e.g., peak systole or end diastole) identified using Doppler spectra. The ultrasound beam should be perpendicular to blood flow when measuring cross-sectional luminal areas, as deviations from perpendicular will enlarge luminal areas. A 10° departure from perpendicular will increase luminal area by approximately 5%. Lumen area diameter also can be measured from longitudinal images of vessels, but the ultrasonic beam must be centered over the vessel. Departures from this center point will be biased towards smaller luminal areas.

As previously discussed, Doppler spectra are derived from blood flow within the sample volume cursor or gate, which can be adjusted to various sizes. A small sample volume can be used and positioned in the middle of the vessel, which has the highest velocity with luminar blood flow; however, inconsistent positioning with this approach can inflate between image variations. A wider sample volume can provides a measure of mean flow velocity within the vessel and result in less between-image variability. The sample volume should be consistent width between animals in a study to reduce this source of variation.

Aliasing of color displays and Doppler spectra can occur when using pulse Doppler. Aliasing is caused by a wraparound of the display, with high velocity flow moving towards

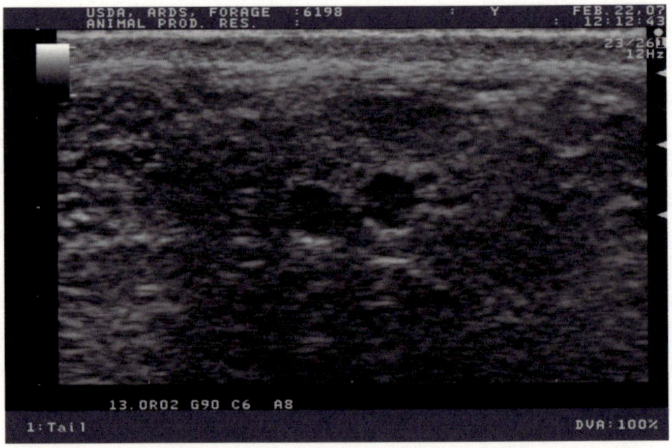

(a)

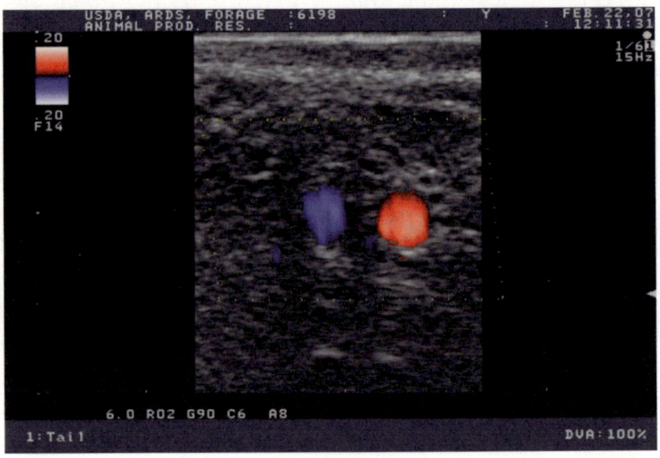

(b)

Fig. 7. Cross-sectional B-mode images of the medial caudal artery (vessel on right) and vein (vessel on left) in a heifer calf, without (a) or with (b) color Doppler.

the probe (i.e., expected to be red) appearing that is moving away from the probe (Fig. 8; i.e., displayed to be blue). This aliasing can be eliminated by increasing the pulse repetition frequency, which increases the color flow velocity scale. Aliasing also occurs with Doppler spectra, with the spectra of high velocity blood flow being converted to reverse flow spectra (Fig. 9). As with the color display, the pulse repetition frequency can be increased to increase the flow velocity scale. The operator also can reduce the spectra baseline down until the spectra represents flow in one direction.

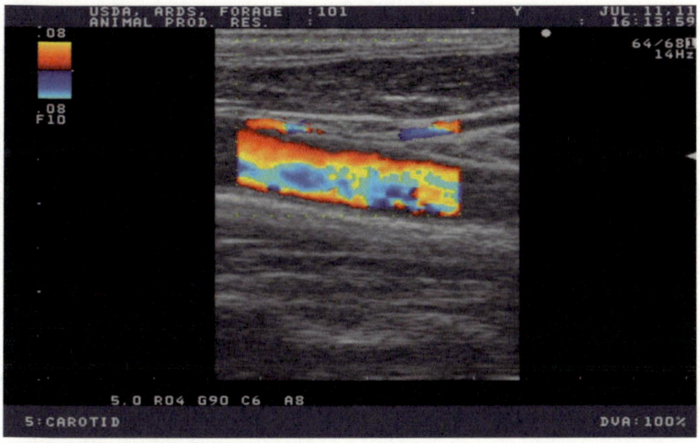

Fig. 8. Aliasing of the flow display with a color Doppler image

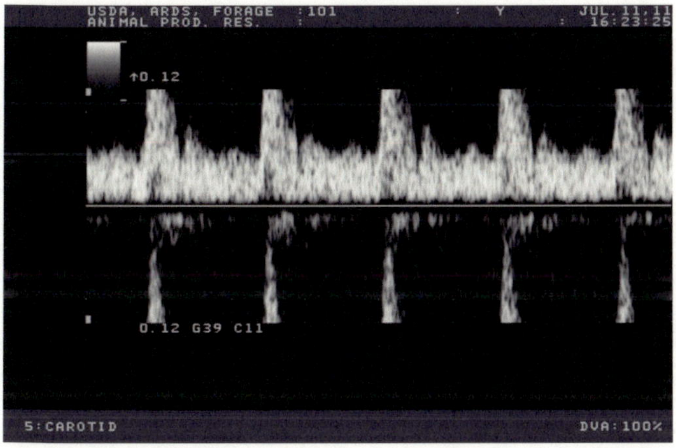

Fig. 9. Aliasing of the Doppler spectra.

4.2 Using serial images to determine sensitivities and response to toxicants

Direct comparisons in luminal vessel areas and blood flow characteristics can be made with color Doppler ultrasonography between groups of animals that are with or without exposure to toxicants. Of greater importance, however, can be in measuring vascular sensitivity to toxicants. The research tool has successfully measured vascular sensitivity to ergot alkaloids in cattle (Aiken et al. 2007; 2009b) and sheep (Aiken et al., 2011), and to determine vascular recovery of sheep after being switched from toxic endophyte-infected pastures of perennial ryegrass to endophyte-free ryegrass (Aiken et al., 2011). Following an initial feeding of a diet containing 0.8 mg ergovaline/kg dry matter to heifers, Aiken et al. (2007) observed a tendency for a vasoconstrictive response by the caudal artery in 4 hours. A similar experiment that fed treatment diet concentrations of 0, 0.2, and 0.8 mg ergovaline/kg dry matter to heifers determined there were vasoconstrictive responses by the caudal artery in 27 hours for the 0.8 mg ergovaline/kg dry matter diet concentration, and in 51 hours for the 0.4 mg ergovaline/kg DM dry matter.

Serial imaging also can be used to determine the extent and persistence of the vascular constriction response to toxicants. Following a reduction in luminal area and blood flow rate in heifers consuming diets consuming 0.4 or 0.8 mg ergovaline/kg DM, Aiken et al. (2007) reported adjustments in luminal area and blood flow velocity to original baseline measures, taken when heifers were on a 0 mg ergovaline/kg DM diet concentration adjustment diet, as air temperatures increased (Fig. 10). However, animals on control diets have shown greater variation in blood flow than those on treatment diets containing ergot alkaloids. Aiken et al. (2007; 2009b) concluded there was greater variation in caudal artery luminal areas and blood flow rates among control cattle compared to those on treatment diets (refer to standard error bars in Fig. 10), making it necessary to evaluate deviations from baseline measures for determine alkaloids effects. Low variation in caudal artery luminal areas among cattle on the treatment diet was suggested to be associated with persistent vasoconstriction causing an inability to adjust their vasculature to changes in ambient temperature.

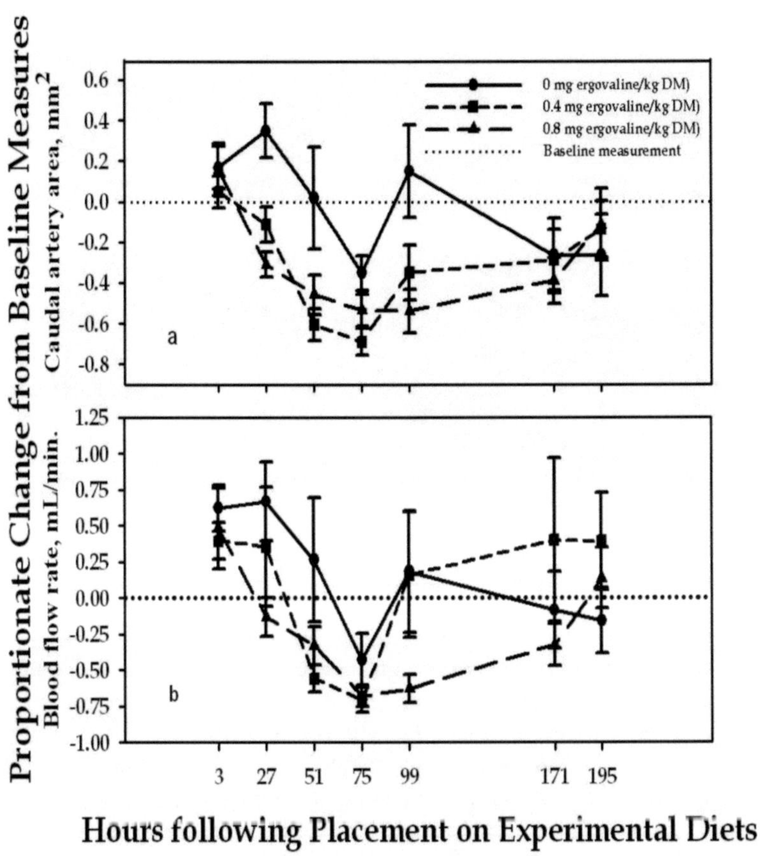

Fig. 10. Trends in luminal area of the caudal artery (a) and blood flow rate through the caudal artery (b) in heifer calves flowing placement of cattle on experimental diets (0, 0.4, or 0.8 mg ergovaline/kg DM) (from Aiken et al., 2009b).

Aiken et al. (2011) used color Doppler ultrasound to determine constriction of the auricular and carotid arteries in ewe lambs after they were switched from endophyte-free perennial ryegrass to ryegrass infected with a novel endophyte that produces high concentrations of ergovaline. A linear decline in luminal area of the auricular artery was observed over a 9-day after grazing was initiated. Lambs also were switched from AR6 novel endophyte ryegrass to endophyte-free pasture. Over the 9 days, luminal area of the auricular artery increased linearly and pulsatillity indices for the carotid decreased linearly. There was relaxation of the vasculature from alkaloid clearance, but a longer duration of data collection was needed to determine the time needed for complete clearance of alkaloids from the vasculature.

4.3 Special considerations in animal handling

Accuracy and precision of measurements of blood flow characteristics in livestock are vastly improved if the measurements are collected in controlled environments with halter broke animals that are at ease with human contact. Furthermore, controlled diets fed in pens with know concentrations of toxicants will be necessary if research objectives are focused on determining sensitivities and strength of response to the toxicant. Blood flow measurements for pastured animals with minimal contact with humans will be affected by animal nervousness (e.g. increases in heart rate). However, measurements from ultrasound images are possible in uncontrolled environments and without halter-broke animals if control animals (e.g., animals grazing toxicant-free pastures) are measured and special animal handling procedures are followed.

Pastured cattle are typically handled in squeeze chutes with narrow alleys (Fig. 11) leading to the chute. Cattle are gregarious and individuals become nervous when singly separated from the group. When imaging cattle in squeeze chutes, the time individuals spend in the alley prior to being imaged should be minimized. Rather than crowding the alley, cattle should be placed in the alley in groups of three, with subsequent groups of three being placed in the alley such that an animal is not alone in the alley. Once in the chute, the sides of the chute should be reduced to not squeeze the animal to a point of discomfort, but reduce the area containing the animal. Potential nervousness of cattle worked in chutes makes it difficult to image cranial regions; therefore, imaging is better done if restricted to caudal regions (e.g., caudal artery or vein). The area should also be as quiet of sound, as possible, to minimize animal startling and raising of heart rates. Further, any sample collections (e.g., blood) or measurements (e.g., rectal or skin temperatures) should be done after imaging.

Imaging of horses can typically be done in stanchions with those that are haltered and at ease with humans (Fig. 12). Sheep and goats tend to also be gregarious, but do not present the same degree of nervousness as cattle. Small ruminants can be worked on cradles (Fig. 13) and provide opportunities to reliably image the cranial vessels (Aiken et al., 2011).

Imaging animals in an uncontrolled environment requires that images be collected over a period of time during the day with minimal changes in air temperature and humidity (Kirch et al., 2008). Imaging quality also is a major consideration in providing reliable interpretations and measures. Time must be taken to correctly place tranducers over the vessels, with sufficient gain and focus settings to optimize B-mode and color Doppler images, and Doppler spectra, Therefore, time of day and time necessary to collect quality images must be considered. For each animal, five images for each cross-sectional and longitudinal section should be collected at each scanning session. A minimum of 5 animals per treatment should be scanned to account for the variation in blood flow characteristics between animals.

Fig. 11. Squeeze chute and alley used for to retrain cattle.

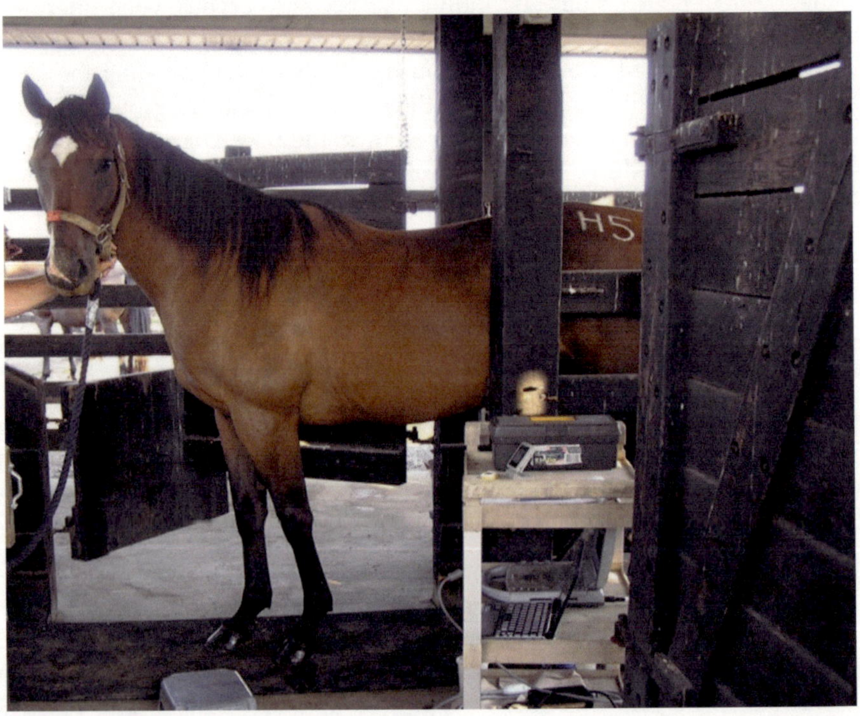

Fig. 12. A haltered horse in a wooden stanchion.

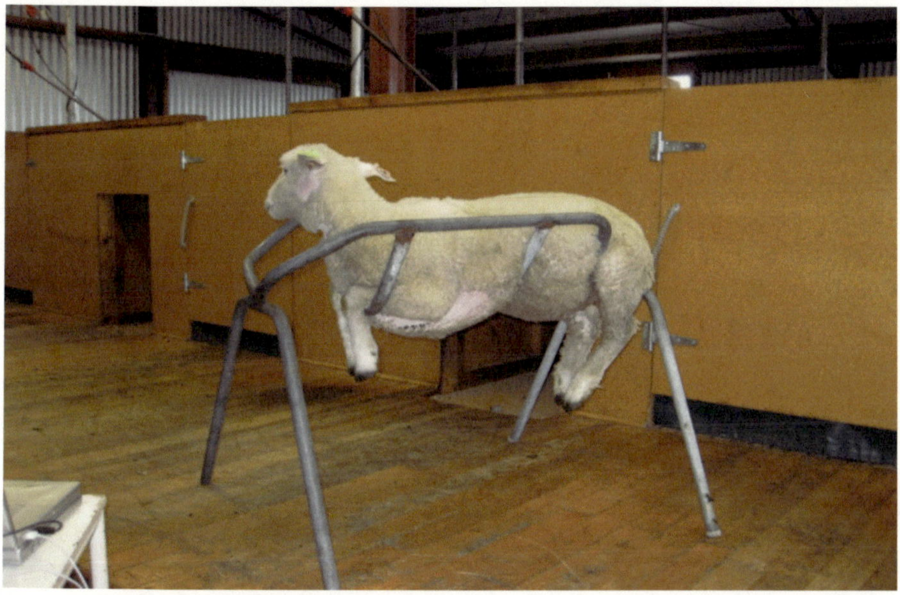

Fig. 13. Cradle used to restrain sheep.

5. Summary

Color Doppler ultrasonography is a non-invasive procedure that is routinely used to assess normal and abnormal blood flow in human and veterinary medicines. It also has application as a research tool in determining vascular responses to toxicants that can effect the vasculature. These studies are ideally performed in controlled environments with animals that are at ease with human contact and do not require restraint that can cause discomfort. However, reliable measures can be collected with pastured animals with minimal human contact if control animals with no exposure to toxicants are measured for comparisons with treatment animals, and special handling procedures are employed to reduce animal nervousness and tension effects on vasculature blood flow.

6. References

Aiken, G.E., Sutherland, B.L., & Fletcher, L.R. (2011). Haemodynamics of lambs grazing perennial ryegrass (*Lolium perenne* L.) either infected with AR6 novel, wild-type endophyte, or endophyte-free. *New Zealand Veterinary Journal*, Vol. 59, (June 2011), 179-184, ISSN 0048-0169

Aiken, G.E., Kirch, B.H., Strickland, J.R., Bush, L.P., Looper, M.L., & Schrick, F.N. (2007) Hemodynamic responses of the caudal artery to toxic tall fescue in beef heifers. *Journal of Animal Science*, Vol. 85, (September 2007), 2337-2345, ISSN 0021-8812

Aiken, G.E., Klotz, J.L., Kirch, B.H., Strickland, J.R., & Boykin, D.L. (2009a). Technical note: Comparison between two tracing methods with ultrasonography to determine lumen area of the caudal artery in beef cattle. *Journal of Animal Science*, Vol. 87, (January 2009), 371-374, ISSN 0021-8812

Aiken, G.E., Strickland, J.R., Looper, M.L., Bush, L.P., & Schrick, F.N. (2009b) Hemodynamic are altered in the caudal artery of beef heifers fed different ergot alkaloid concentrations. *Journal of Animal Science*, Vol. 87 (February 2009), 2142-2150, ISSN 0021-8812.

Aldrich, C.G., Rhodes, M.T., Miner, J.L., Kerley, M.S., & Paterson, J.A., (1993). The effects of endophyte-infected tall fescue consumption and use of a dopamine antagonist on intake, digestibility, body temperature, and blood constituents in sheep. *Journal of Animal Science*, Vol. 71, (January 1993), 158-163, ISSN 0021-8812

Al-Haidary, A., Spiers, D.E., Rottinghaus, G.E., Garner, G.B., & Ellersieck, M.R. (2001). Thermoregulatory ability of beef heifers following intake of endophyte-infected tall fescue during controlled heat challenge. *Journal of Animal Science*, Vol. 79, (July 2001), 1780-1788, ISSN 0021-8812

Bandyopadhyay, R., Frederickson, D.E., McLaren, N.W., Odvody, G.N., & Ryley, M.J. (1998). Ergot: A new disease threat to sorghum in the Americas and Australia. *Plant Disease*, Vol. 82, (April 1998), 356-367, ISSN 0191-2917

Bacon, C.W. (1995) Toxic endophyte-infected tall fescue and range grasses: Historic perspectives. *Journal of Animal Science*, Vol. 73, (March 1998), 861-870, ISSN 0021-8812

Blaney, B.J., McLennan, S,R, Kidd, J.F., Connell, J.A., McKenzie, & Downing, J., Jr. (2011). Effect of sorghum ergot (*Claviceps Africana*) on the performance of steers (Bos Taurus) in a feedyard. *Animal Production Science*, Vol. 51, (January 2011), 156-166, ISSN 1836-0939

Blaney, B.J., Molloy, J.B., & Brock, I.J. (2009). Alkaloids in Austraiian rye ergot (*Claviceps purpurea*) sclerotia: Implications for food and stockfeed regulations. *Animal Production Science*, Vol. 49, (October 2009), 975-982, ISSN 1836-0939

Bouton, J.H., Latch, C.G., Hill, N.S., Hoveland, C.S., & Wood, D.T. (2002). Reinfection of tall fescue cultivars with non-ergot alkaloid-producing endophytes. *Agronomy Journal*, Vol. 94, (May 2002), 567-574, ISSN 0002-1962

Browning, R., Jr., Leite-Browning, M.L., Smith, H.M., & Wakefield, T., Jr. (1998). Effect of ergotamine and ergonovine on plasma concentrations of thyroid hormones and cortisol in cattle. *Journal of Animal Science* 76, (June 1998) 1644-1650, ISSN 0021-8812

Browning, R., Jr., Thompson, F.N., Sartin, J.L., & Leite-Browning, M.L. (1997) Plasma concentrations of prolactin, growth hormone, and luteinizing hormone in steers administered ergotamine or ergonovine. *Journal of Animal Science*, Vol. 75, (March 1998), 796-802, ISSN 0021-8812

Bush, L.P., & Fannin, F.F. (2009). Alkaloids, In: *Tall Fescue for the Twenty-First Century*. Fribourg, H.A., Hannaway, D.B., & West, C.P. (Eds.) 229-249, ASA, CSSA, SSSA. ISBN 978-0-89118-172-9, Madison, WI

Carter, S.A. (2000). Hemodynamic considerations in peripheral vascular and cerbrovascular disease, In: *Introduction to Vascular Ultrasonography*, Zwiebel, W.J. (Ed.), 3-16. W.B. Saunders Co., ISBN 0-7216-6949-2, Philadelphia, PA

Chestnut, A>B. Bernard, J.K., Harstin, J.B., & Reddick, B.B. (1992). Performance of growing lambs fed *Acremonium coenophilum* infested tall fescue (*Festuca arundinacea* Schreb.) hay. *Small Ruminant Research*, Vol. 7 (February 1992), 9-19, ISSN 0921-4488

Chotani, M.A., Flavahan, S, Mitra, S., Danunt, D., & Flavahan, N.A. (2000). Silent α2c-adrenergic receptors enable cold-induced vasoconstriction in cutaneous arteries.

American Journal of Heart and Circulatory Physiology, Vol. 278, (April 2000), 1075-1083, ISSN 0363-6135

Cross, D. L., Redmond, L. M., & Strickland, J. R. (1995). Equine fescue toxicosis: Signs and solutions. Journal of Animal Science, Vol. 73, (March 1995), 899-908, ISSN 0021-8812

Dyer, D. C. (1993). Evidence that ergovaline acts on serotonin receptors. *Life Sciences*, Vol. 53 (July 19913), PL223-PL228, ISSN 0024-3205

Fletcher, L.R., & Harvey, I.C. (1981). An association of a *Lolium* endophyte with ryegrass staggers. *New Zealand Veterinary Journal*, Vol. 29, (October 1981), 185-186, ISSN 0048-0169

Gagge, A.P., & Gonzalez R.R. (1996). Mechanisms of heat exchange: Biophysics and physiology, In: *Handbook of Physiology, Section 4: Environmental Physiology*, Fregley M.J. & Blatteis, C.M. (Eds.), 45-83, Oxford University Press, ISBN 0-19-509174-4, New York, NY

Hannah, S.M., Paterson, J.A., Williams, J.E., Kerley, M.S., & Miner, J.L. (1990). Effects of increasing levels of endophyte-infected tall fescue seed on diet digestibility and ruminal kinetics in sheep. *Journal of Animal Science*, Vol. 68, (June 1990), 1693-1701, ISSN 0021-8812

Hatle, L., Angelsen, B.A., & Tromsdale, A. (1980). Non-invasive assessment of aortic stenosis by Doppler ultrasonography. *British Heart Journal*, Vol. 43, (March 1980), 284-292, ISSN 0007-0769

Hemken, R.W., Boling, J.A., Bull, L.S., Hatton, R.H., Buckner, R.C., & Bush, L.P. (1981) Interaction of environmental temperature and anti-quality factors on the severity of summer fescue toxicosis. *Journal of Animal Science*, Vol. 73, (April 1981), 1954-1961, ISSN 0021-8812

Hoffman, K.L., Wood, A.K.W.,Griffiths, K.A., Evans, D.L., Gill, R.W., & Kirby, A.C. (2001) Doppler sonographic measurements of arterial blood flow and their repeatability in the equine foot furing weight bearing and non-weight bearing. *Research in Veterinary Science*, Vol. 70, (June 2001), 199-203, ISSN 0034-5288

Johnson, J.M. & Kellogg, D.L., Jr. (2010). Local thermal control of the human cutaneous circulation. *Journal of Applied Physiology*, Vol. 109, (October 2010), 1229-1238, ISSN 8750-7587.

Johnson, J.M. & Proppe, D.W. (1996). Cardovascular adjustments to heat stress, In. *Handbook of Physiology, Section 4: Environmental Physiology*, Fregley M.J. & Blatteis, C.M. (Ed.), 215-243, Oxford University Press, ISBN 0-19-509174-4, New York, NY

Julien, W. E., Martz, F. A., Williams, M., and Garner, G. B. (1974). Feed intake in Hereford calves infused intraperitoneally with toxic fescue extract. *Journal of Dairy Science*, Vol. 57, (August 1974) 1385-1387, ISSN 1811-9743

Kirch, B.H., Aiken, G.E., & Spiers, D.E. (2008). Temperature influences on vascular dynamics in cattle measured by Doppler ultrasonography. *Journal of Thermal Biology*, Vol. 33, (October 2008), 375-379, ISSN 0306-4565

Klotz, J.L., Bush, L.P., Smith, D.L., Shaffer, W.D., Smith, L.L., Vevoda, A.C., Craig, A.M. Arrington, B.C., & Strickland, J.R. (2006). Assessment of vasoconstrictive potential of D-lysergic acid in an isolated bovine lateral saphenous vein bioassay. *Journal of Animal Science*, Vol. 84, (September 2006), 3167-3175, ISSN 0021-8812

Klotz, J.L., Bush, L.P., Smith, D.L., Shaffer, W.D., Smith, L.L., Arrington, B.C., & Strickland, J.R. (2007). Ergovaline-induced vasoconstriction in an isolated bovine lateral saphenous vein bioassay. *Journal of Animal Science*, Vol. 85, (September 2007), 2330-2336, ISSN 0021-8812

Lee, K., Mincheol, M., Junghee, Y., Juhyun, J. 2004. Spectral waveform analysis of major arteries in conscious dogs by Doppler ultrasonography. *Veterinary Radiology and Ultrasonography*, Vol. 45, (March 2004), 166-171, ISSN 1058-8083

Lewis, T., & Gelfand, B. (1935). The manner in which necrosis arises in the fowl's comb under ergot poisoning. *Clinical Science.*, Vol. 2, 43-55, ISSN 0143-5221

Lyons, P.C., Plattner, R.D., Bacon, C.W. (1986). Occurrence of peptide and clavinet ergot alkaloids in tall fescue grass. *Science*, Vol. 232, (April 1986), 487-489, ISSN 0036-8075

Moneta, G.L., Yeager, R.A., Dalman, R., Antonovic, R., Hall, L.D., & Porter, J.M. (1992). Duplex ultrasound criteria for diagnosis of splanchnic artery stenosis or occlusion. *Journal of Vascular Surgery*, Vol. 16, (November 1992), 511-518, ISSN 0741-5214

Müller, M., Hermes, M., Brückmannm, H., & Schimrigk, K. (1995). Transcranial Doppler ultrasound in the evaluation of collateral blood flow in patients with internal carotid artery occlusion: Correlation with cerebral angiography. *American Journal of Neuroradiology*, Vol. 16, (January 1995), 195-202, ISSN 0195-6108

Olin, J.W., Piedmonte, M.R., Young, J.R., DeAnna, S., Grubb, M., & Childs, M.B. (1995). The utility of duplex ultrasound scanning of the renal arteries for diagnosing significant renal artery stenosis. *Annals of Internal Medicine*, Vol. 122, (June 1995), 833-838, ISSN 0003-4819

Oliver, J.W. (1997). Physiological manifestations of endophyte toxicosis in ruminant and laboratory species. In: *Neotyphodium/Grass Interactions*, Bacon, C.W., & Hill, N.E. (Ed.), 311-346, Plenum Press, ISBN 978-0-306-45688-6, New York, NY

Oliver, J.W., Strickland, J.R., Waller, J.C., Fribourg, H.A., Linnabary, R.D., & Abney, L.K. (1998). Endophytic fungal toxin effect on adrenergic receptors in lateral saphenous veins (cranial branch) of cattle grazing tall fescue. *Journal of Animal Science*, Vol. 76, 2853-2856, (November 1998), ISSN 0021-8812

Paterson, J., Forcherio, C., Larson, B. Samford, & Kerley, M. (1995). The effects of fescue toxicosis on beef cattle productivity. *Journal of Animal Science*, Vol. 73, (March 1992), 889-898, ISSN 0021-8812

Petersen, L.J. Petersen, J.R., Talleruphuus, U., Ladefoged, S.D., Mehlsen, J., & Jensen, H.E. (1997). The pulsatility index and the resistive index in renal arteries. Associations with long-term progression in chronic renal failure. *Nephrology Dialysis Transplantation*, Vol. 12 (July 1997), 1376-1380, ISSN 0931-0509

Porter, J.K., & Thompson, F.N, Jr. (1992). Effects of fescue toxicosis on reproduction in livestock. *Journal of Animal Science*, Vol. 70, (May 1992), 1594-1603, ISSN 0021-8812

Poulin, M.J., & Robbins, P.A. (1996). Indexes of flow and cross-sectional area of the middle cerebral artery using doppler ultrasound during hypoxia and hypercapnia in humans. *Stroke, Vol.* 27, (December 1996), 2244-2250, ISSN 0039-2499

Pozniak, M.A. (2000). Doppler ultrasound of the liver. In: *Clinical Doppler Ultrasound*, Allan, P.L. (Ed.), 141-184, Churchill Livingstone, ISBN 84-8174-566-9, London, UK

Raisis, A.L., Young, L.E., Meire, H.B., Taylor, P.M., Walsh, K., & Lekeux P. (2000). Variability of Doppler ultrasound measurements of hindlimb blood flow in conscious horses. *Equine Veterinary Journal*, Vol. 32 (March 2000), 125-132, ISSN 2042-3306

Rhodes, M.T., Paterson, J.A., Kerley, M.S., Garner, H.E., & Laughlin, M.H. (1991). Reduced blood flow to peripheral and core body tissues in sheep and cattle induced by endophyte-infected tall fescue. *Journal of Animal Science*, Vol. 69, (May 1991) 2033-2043, ISSN 0021-8812

Rottinghaus, G.E., Garner, G.B., Cornell, C.N., & Ellis J.L. (1991). HPLC method for quantitating ergovaline in endophyte-infected tall fescue: Seasonal variation of ergovaline levels in stems with leaf sheaths, leaf blades, and seed heads. *Journal of Agricultural Food and Chemistry*, Vol. *39*, (January 1991), 112-115, ISSN 0021-8561

Schmidt, W.A., Kraft, H.E., Vorpahl, K., Völker, L., & Gromnica-Ihle, J. (1997). Color duplex ultrasonography in the diagnosis of temporal arteritis. *The New England Journal of Medicine*, Vol. 337 (November 1997), 1336-1340, ISSN 0028-4793

Schoning, C., Flieger, M., & Pertz, H. H. (2001). Complex interaction of ergovaline with 5-HT2A, 5-HT1B/1D, and alpha 1 receptors in isolated arteries of rat and guinea pig. *Journal of Animal Science*, Vol. 79, (August 2001) 2202-2209, ISSN 0021-8812

Shappell, N. W. (2003). Ergovaline toxicity on CACO-2 cells as assessed by MTT, alamarblue and DNA assays. *In Vitro Cellular Developmental Biology Animal*, Vol. 39, (July 2003) 329-335, ISSN0143-5221

Siegel, M.R., Latch, G.C.M., & Johnson, M.C. (1985). *Acremonium* fungal endophytes of tall fescue and perennial ryegrass: Significance and control. *Plant Disease*, Vol. 69, (February 1985), 179-183, ISSN 0191-2917

Sites, B.D., Brull, R., Chan, V.W.S. Spence, B.C., Gallagher, J., Beach, M.L., Sites, V.R., & Hartman, G.S. (2007). Artifacts and pitfalls errors associated with ultrasound-guided regional anesthesia. Part I: Understanding the basic principles of ultrasound physics and machine operations. *Regional Anesthesia and Pain Medicine*, Vol. 32, (September 2007), 419-433, ISSN 1098-7339

Someda, H., Moriyasu, F., Masazumi, F, Hamata, N., Nabeshima,M., Nishikawa, K., Okuma, M., Tanaka, K., & Ozawa, K. (1995). Vascular complications in living related liver transplantation detected with intraoperative and postoperative Doppler US. *Journal of Hepatology*, Vol. 22, (June 1995), ISSN 0168-8278

Strickland, J. R., Aiken, G. E., & Klotz, J. L. (2009a). Ergot alkaloid-induced blood vessel dysfunction contributes to fescue toxicosis. In*: Forage and Grazinglands*. Date of access: 8/2/2011. Retrieved from:
http://www.plantmanagementnetwork.org/sub/fg/review/2009/ergot/

Strickland, J. R., Aiken, G. E., Spiers, D.E., Fletcher, L. R., & Oliver, J. W. (2009b). Physiological Basiss of Fescue Toxicosis. In: *Tall Fescue for the Twenty-First Century*. Fribourg, H.A., Hannaway, D.B., & West, C.P. (Eds.) 203-227, ASA, CSSA, SSSA. ISBN 978-0-89118-172-9, Madison, WI

Strickland, J. R., Bailey, E. M., Abney, L. K., & Oliver, J. W. (1996). Assessment of the mitogenic potential of the alkaloids produced by endophyte Acremonium coenophialum-infected tall fescue (*Festuca arundinacea*) on bovine vascular smooth muscle in vitro. *Journal of Animal Science*, Vol. 74 (July 1996) 1664-1671, ISSN 0021-8812

Strickland, J.R., Looper, M.L., Matthews, J.C., Rosenkrans, C.F., Jr. Flythe, M.D., & Brown, K.R. (2011). St. Anthony's Fire in livestock: Causes, mechanisms, and potential solutions. *Journal of Animal Science*, Vol. 89, (May 2011), 1603-1626, ISSN 0021-8812

Walls, J. R., & Jacobson, D. R. (1970). Skin temperature and blood flow in the tail of dairy heifers administered extracts of toxic tall fescue. *Journal of Animal Science*, Vol. 30, (March 1970), 420-423, ISSN 0021-8812

Wemdelhag, I, Gustavsson, T.,Suurküla, Berglund, G., & Wikstrand, J. (1991). Ultrasound measurement of wall thickness in the carotid artery: Fundamental principles and description of a computerized analyzing system. *Clinical Physiology*, Vol. 11, (November 1991), 566-577, ISSN 1475-0961

Whelan, J.F., & Barry, M.H. (1992). Color flow doppler ultrasonography with peripheral arteriography for the investigation of peripheral vascular disease. *Journal of Clinical Ultrasound*, Vol. 20, (July 1992), 369-374, ISSN 0091-2751

Williams, M., Shaffer, S. R., Garner, G. B., Yates, S. G., Tookey, H. L., Kintller, L. D., Nelson, S. L., & McGinity, J. T. (1975). Induction of fescue foot syndrome in cattle by fractionated extracts of toxic fescue hay. *American Journal of Veterinary Research*, Vol. 36, (September 1975) 1353-1357, ISSN 0002-9645

Zagzebski, J.A. (2000). Physics and instrumentation in Doppler and B-mode ultrasonography, In: *Introduction to Vascular Ultrasonography*, Zwiebel, W.J. (Ed.), 17-43, W.B. Saunders Co., ISBN 0-7216-6949-2, Philadelphia, PA

Zanzalari, K.P., Heitmann, R.N., McLaren, J.B., Fribourg, H.A. (1989). Effects of endophyte-infected fescue seed and cimetidine on respiration rates, rectal temperatures and hepactic mixed function oxidase activity as measured by hepatic antipyrine metabolism in sheep. *Journal of Animal Science*, Vol. 67, (December 1989), 3370-3378, ISSN 0021-8812

30

Pathogenicity of Avian Paramyxovirus Serotype-3 in Chickens and Turkeys

Sachin Kumar and Siba K. Samal
Virginia-Maryland Regional College of Veterinary Medicine,
University of Maryland, College Park,
Maryland,
USA

1. Introduction

Members of the family *Paramyxoviridae* are pleomorphic enveloped viruses that contain a single-stranded, non-segmented RNA genome of 13-19 kilobases (Lamb and Parks, 2007). These viruses have been isolated from many species of avian, terrestrial and aquatic animals around the world. Some members of the family *Paramyxoviridae* are well known human (measles, mumps, respiratory syncytial virus and various human parainfluenza viruses) and animal (rinderpest virus, canine distemper virus and Newcastle disease virus [NDV]) pathogens, while the pathogenic potential of many others is still unknown. The family is divided into two subfamilies, *Paramyxovirinae* and *Pneumovirinae*. Subfamily *Paramyxovirinae* is further divided into five genera: *Rubulavirus, Respirovirus, Henipavirus, Morbillivirus* and *Avulavirus*. Subfamily *Pneumovirinae* is divided into two genera: *Pneumovirus* and *Metapneumovirus* (Mayo, 2002).

All paramyxoviruses that have been isolated to date from avian species are placed in the genus *Avulavirus* except for avian metapneumoviruses which are placed in the genus *Metapneumovirus* due to antigenic differences and differences in genetic organization. The avian paramyxoviruses (APMVs) that comprise genus *Avulavirus* have been divided into nine different serotypes (APMV-1 through -9) based on haemagglutination inhibition (HI) and neuraminidase inhibition (NI) assays (Alexander and Collins, 1984). APMV-1 comprises all strains of NDV and is the most characterized serotype due to the severity of disease caused by virulent NDV strains in chickens. (Alexander, 1980a). The complete genome sequences and the molecular determinants of virulence have been determined for representative NDV strains (Krishnamurthy and Samal, 1998; de Leeuw and Peeters, 1999; Huang et al., 2004; Panda et al., 2004; Rout and Samal, 2008). As a first step in characterizing the other APMV serotypes, complete genome sequences of one or more representatives of APMV serotypes 2 to 9 were recently determined, expanding our knowledge about these viruses (Chang et al., 2001; Kumar et al., 2008; Nayak et al., 2008; Subbiah et al., 2008; Paldurai et al., 2009; Samuel et al., 2009; Xiao et al., 2009).

NDV causes respiratory, neurological or enteric disease in birds. Among poultry, chickens are the most susceptible while geese and ducks are the least susceptible (Wakamatsu et al.,

2006). NDV strains are classified into low virulent (lentogenic), moderately virulent (mesogenic), and highly virulent (velogenic) pathotypes based on severity of disease produced in chickens. In contrast very little is known about pathogenicity of APMV-2 to -9 in chickens and turkeys. APMV-2 has been shown to cause mild disease and drop in egg production in chickens and turkeys (Warke et al., 2008). APMV-6 and -7 have been associated with respiratory disease in turkeys (Shortridge et al., 1980; Saif et al., 1997). APMV-4, -8 and -9 have been isolated from different species of birds but the clinical signs of the disease in those birds were not apparent (Alexander et al., 1983; Gough and Alexander, 1984; Stallknecht et al., 1991; Maldonado et al., 1995; Capua et al., 2004). Recently, experimental infection of 1-day-old chicks with APMV-2, -4 and -6 showed viral infection in gastrointestinal tract, respiratory tract and pancreas (Warke et al., 2008).

APMV-3 has been isolated from wild and domesticated birds in different parts of the world (Tumova et al., 1979b; Alexander, 1980b; Macpherson et al., 1983; Alexander, 2003). Recently, APMV-3 was isolated from ostrich, indicating a wide host range for the virus (Kaleta et al.). The virus has been isolated from diseased turkeys associated with coughing, nasal discharge and swelling of the infra-orbital sinus (Redmann et al., 1991). APMV-3 has been associated with encephalitis and high mortality in caged birds (Tumova et al., 1979a). The virus causes acute pancreatitis and central nervous system (CNS) symptoms in *Psittacine* and *Passerine* birds (Beck et al., 2003). APMV-3 also infects chickens at an early age, with evidence of stunting growth that may be more marked in broiler chicken breeds (Alexander and Collins, 1982a). In terms of pathogenicity to domestic poultry birds, APMV-3 probably is second in importance to NDV. The exact economic impact of APMV-3 infection in poultry industry is not known. This is partly because the pathogenicity of APMV-3 in poultry is not well studied. There is a high degree of amino acid sequence variation between APMV-3 and APMV-1 (NDV), but by the HI test there is low level of cross reaction between APMV-3 and APMV-1 serum samples, which often leads to misdiagnosis of APMV-3 as APMV-1.

APMV-3 strain Wisconsin was first isolated from a turkey in 1968 in Wisconsin (Tumova et al., 1979b). APMV-3 strain Netherlands was isolated from a parakeet in 1975 in the Netherlands and is the prototype for the entire serotype (Alexander and Chettle, 1978). Initially, these two viruses were considered as two different strains of APMV-3 based on cross HI test using monoclonal antibodies (Anderson et al., 1987; Anderson and Russell, 1988). Recently, reciprocal cross HI and cross neutralization assays using post infection serum from chicken indicated that these two strains are antigenically distinct, although the difference was modest (Kumar et al., 2010b). However, complete genome sequence analysis revealed substantial genome-wide nucleotide and amino acid sequence differences that are consistent with the two strains representing distinct antigenic subgroups (Kumar et al., 2008; Kumar et al., 2010b). We have performed experimental infection of APMV-3 strains Wisconsin and Netherlands in 9-day-old embryonated chicken eggs, 1-day-old chicks and turkeys, and 2-week-old chickens and turkeys in order to investigate their tropism and pathogenicity. Birds were infected by the oculonasal route and the viral tropism and replication efficiency were evaluated by quantitative virology and immunohistochemistry of a wide range of possible target organs. In addition, a separate group of 1-day old chicks were infected intracerebrally to evaluate the potential of these viruses to replicate in neuronal tissue.

2. Methods used in the study

2.1 Viruses and cells

APMV-3 strains parakeet/Netherlands/449/75 and turkey/Wisconsin/68 (obtained from National Veterinary Service Laboratory, Ames, Iowa, USA) was propagated in 9-day-old specific pathogen free (SPF) embryonated chicken eggs via allantoic route of inoculation. The allantoic fluids from infected embryonated eggs were collected 96 h post-inoculation and virus titer was determined by hemagglutination (HA) assay with 0.5% chicken RBC. The virus titer in infected tissue samples was determined by the tissue culture infective dose ($TCID_{50}$) method and by plaque assay in chicken embryo fibroblast (DF-1) cells (ATCC, Manassas, VA, USA) (Reed and Muench, 1938). For the plaque assay, DF-1 cell monolayers in 12 well plates were infected with different dilutions of tissue homogenates. The tissue homogenates were allowed to adsorb for 1 h at 37 °C, washed with phosphate-buffered saline (PBS), and then overlaid with 1 mL DMEM containing 0.8% (wt/vol) methylcellulose. The virus plaques in the DF-1 cell monolayer were visualized 96 h PI after staining with 1% crystal violet.

2.2 Pathogenicity index tests

The pathogenicity of the APMV-3 strains Netherlands and Wisconsin was determined by two standard pathogenicity tests. These included the mean death time (MDT) in 9-day-old embryonated SPF chicken eggs and the intracerebral pathogenicity index (ICPI) in 1-day-old SPF chick (Alexander, 1998).

MDT value was determined following the standard procedure (Alexander, 1998). Briefly, a series of 10-fold (10^{-6} to 10^{-12}) dilutions of fresh infective allantoic fluid in sterile PBS were made and 0.1 mL of each dilution was inoculated into the allantoic cavities of five 9-day-old embryonated SPF chicken eggs, which were then incubated at 37 °C. Each egg was examined three times daily for 7 days, and the times of embryo deaths were recorded. The minimum lethal dose is the highest virus dilution that caused death of all the embryos. MDT is the mean time in hours for the minimum lethal dose to kill all inoculated embryos. The MDT has been used to characterize the NDV pathotypes as follows: velogenic (less than 60 h), mesogenic (60 to 90 h), and lentogenic (more than 90 h) (Alexander, 1980b).

For determining ICPI value, 0.05 mL of 1:10 dilution of fresh infective allantoic fluid (2^8 HA units) of each virus was inoculated into groups of 10 one-day-old SPF chicks via the intracerebral route. The birds were observed for clinical symptoms and mortality every 8 h for a period of 8 days. At each observation, the birds were scored as follows: 0, healthy; 1, sick; and 2, dead. The ICPI is the mean score for all of the birds in the group over the 8-day period. Highly virulent NDV (velogenic) viruses give values approaching 2 and avirulent NDV (lentogenic) viruses give values close to 0 (Alexander, 1998).

2.3 Pathogenesis assessment in chickens and turkeys

1-day-old and 2-week-old SPF chickens (Charles River, North Franklin, Connecticut, USA) and 1-day-old and 2-week-old turkeys (Murry-McMurry hatchery, Webster City, Iowa, USA) were housed in positive pressure isolators in our BSL-2 facility. Birds confirmed to be negative for APMV-3 specific antibody by HI assay were further selected for pathogenesis experiments and housed in negative pressure isolators. The 1-day-old and 2-week-old chickens and turkeys, in groups of 12 for each species and age, were infected with 0.1 mL (10^3 PFU) per bird of APMV-3 strain Netherlands or Wisconsin through the occulonasal route. Infections with the different strains were performed at separate times to avoid cross infections. An additional six birds of each species/age group remained as uninfected controls and sacrificed 14 day post infection (DPI) after collection of serum. Birds were provided with food and water *ad libitum* and monitored daily for any visible signs and symptoms twice daily. Three birds from each infected group were euthanized on 3, 5, 7 and 14 DPI by rapid asphyxiation in a CO_2 chamber. The birds were swabbed orally and cloacally just before euthanasia. The following tissue samples were collected both for IHC and for virus isolation: brain, trachea, lung, spleen, kidney and pancreas. In addition serum samples were collected on day 14 when the three remaining birds in each group and the control birds were euthanized. Seroconversion was evaluated by HI assay (Alexander, 1980b).

3. Pathogenicity of avian paramyxovirus serotype-3

The mean death time (MDT) for APMV-3 strain Netherland was 112 hr while that of APMV-3 strain Wisconsin was > 168 hr. The intracerebral pathogenicity index (ICPI) value for APMV-3 strain Netherlands was 0.39, while the ICPI for APMV-3 strain Wisconsin was zero, consistent with a lentogenic virus. Although, the ICPI and MDT values of strain Netherlands were higher than those of strain Wisconsin; their values indicate that both the APMV-3 strains are lentogenic viruses. These results indicated that APMV-3 strain Netherlands and Wisconsin are probably not highly pathogenic to chickens, similar to lentogenic NDV strains (Kumar et al., 2010a).

In order to evaluate the ability of APMV-3 to replicate in neuronal tissue, $1X10^3$ PFU of each APMV-3 strain was inoculated intracerebrally into 1-day-old chicks. Our results showed that neither of the two APMV-3 strains produced any clinical signs nor did they kill any chicks during 5 DPI. Both of the APMV-3 strains were isolated from the brain homogenates of all of the chicks on days 1 to 5 day post inoculation, grew with similar kinetics, and reached similar maximum titers (Kumar et al., 2010a). Interestingly, while both the APMV-3 strains replicated more slowly than moderately virulent NDV, they achieved a similar final titer compared to NDV, but without causing noticeable neurological disease and with no mortality (Kumar et al., 2010a).

Infection of either APMV-3 strain in 1-day-old chickens and turkeys resulted in mild clinical signs that included altered gait, respiratory distress, dullness, ruffled-feathers, loss of appetite and weight loss. The other visible signs included diarrhea that was evident in 1-day-old chickens and turkeys at 4 DPI. The neurological signs were more evident in 1-day-old chicks and turkeys infected with APMV-3 strain Netherlands as compared to strain Wisconsin. In contrast, infection of 2-week-old chickens and turkeys with the APMV-3 strains did not result in any apparent clinical signs of disease. APMV-3 infection did not kill any of the 1-day-old or 2-week-old chickens and turkeys. The most remarkable finding upon postmortem examination of birds was enlargement of the pancreas with focal necrosis in the 2 week-old chickens and turkeys (Fig. 1). The foci of necrosis were distributed along the entire length of pancreas and the extent of necrosis was similar in both chickens and turkeys. Gross examination of other organs including the brain, trachea, liver, kidney, spleen and lung showed normal tissue morphology with no noticeable gross lesions.

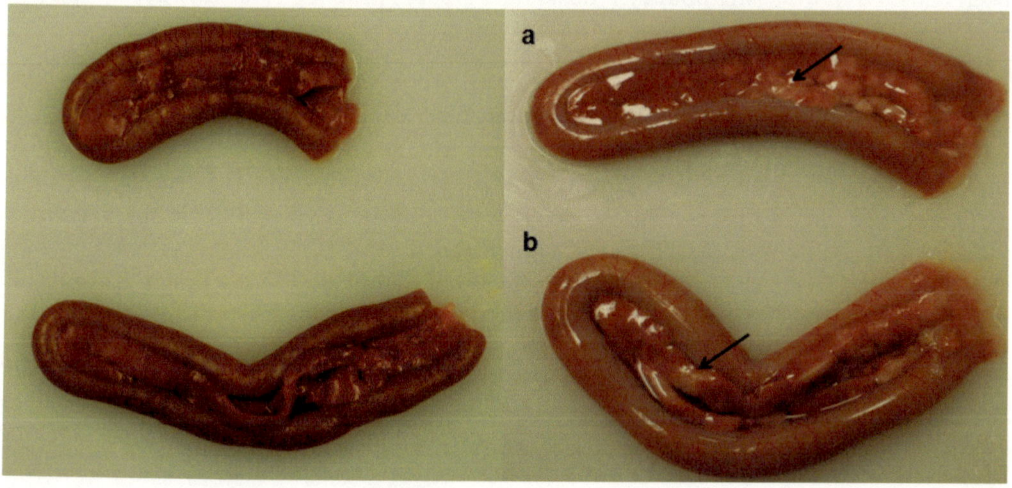

Fig. 1. Gross morphology of pancreas collected from 2-week-old turkeys infected with APMV-3 strain Netherlands (a) or Wisconsin (b): comparison with pancreas from uninfected controls (left panels) reveals necrotic foci along the length of the infected organ.

The brain, trachea, lung, spleen, kidney and pancreas showed the presence of viral antigen upon inoculation. Although APMV-3 strain Wisconsin replicated less efficiently in 2-week-old chickens, the titer of virus in lungs was somewhat higher than that of strain Netherlands, suggesting that it might be more respirotropic, whereas strain Netherlands appeared to be more neurotropic (Kumar et al., 2010a).

In general, infected turkeys had virus replication in fewer organs and for shorter duration than chickens of same age groups. Strain Netherlands was isolated from all of the tested organs except for the pancreas and kidneys in turkeys. Strain Wisconsin was isolated from all of the tested organs in turkeys. Results also suggested that turkeys are less susceptible to infection with either APMV-3 strain compared to chickens (Kumar et al., 2010a).

All the chickens and turkeys infected with either strain seroconverted 14 days post infection confirming virus replication. Extensive immunofluorescence was observed in the brain, trachea, lung and pancreas, suggesting extensive virus replication in these organs (Kumar et al., 2010a). The study also suggested that the epithelial lining in the lungs and trachea are more privileged site for virus replication.

4. Outcome of the study

APMVs are frequently isolated from wild and domestic birds around the world and have been placed into nine serotypes based on antigenic relatedness. Among these nine serotypes, APMV-1 (NDV) is the best studied because virulent NDV strains cause severe disease in chickens. APMV-2 to -9 are present in both free living and domestic birds, but their disease potential is neither known in wild birds nor in domestic birds. It is probable that wild birds can transmit APMV-2 to -9 to domestic birds. APMV-2, -3, -4, -6, -7, -8, -9 have been isolated from domestic poultry birds and antibodies to these viruses have also been detected in domestic poultry (Alexander, 1980b; Alexander and D.J., 2003; Warke et al., 2008).

APMV-3 strains Netherlands and Wisconsin were first characterized by standard pathogenicity tests (MDT and ICPI). The results of the MDT test showed that APMV-3 strain Netherland was slightly pathogenic (112 hr) compared to strain Wisconsin (>168 hr). Similar findings were also observed in ICPI test: APMV-3 strain Netherland showed an ICPI value of 0.39, while the ICPI value of strain Wisconsin was zero. Although the results of pathogenicity index tests (MDT and ICPI) showed that APMV-3 strain Netherlands was slightly pathogenic to chickens, the pathogenicity of both the APMV-3 strains is similar to that of lentogenic NDV strains (Alexander, 1998).

The growth kinetics of APMV-3 strains in chicken brain showed that both the strains are equally competent to grow in brain tissue and the highest virus titer was observed 120 hr post inoculation. These results suggest that the low ICPI values of APMV-3 strains were due to nondestructive replication of the virus in brain tissue (Kumar et al., 2010a).

It has been shown previously by experimental infection that APMV-3 strain Netherlands is more virulent than strain Wisconsin (Alexander, 1980b). In another study, it was shown that the extent of disease and death in the birds after infection with APMV-3 strain Netherlands depends on dose as well as route of infection (Alexander and Collins, 1982b). The clinical signs of illness were more evident in 1-day-old birds compared to 2-weeks-old birds. In chickens, both the strains of APMV-3 were also isolated from tissues of respiratory, digestive and nervous systems. Similar findings were also observed in turkeys infected with both the APMV-3 strains. However, the titers of APMV-3 strain Netherlands in different organs were higher than those of strain Wisconsin in all the infected groups of birds. These results confirmed previous findings that chickens and turkeys are susceptible to APMV-3 infection. Furthermore, younger birds are more susceptible than older birds (Alexander and Collins, 1982b; Russell et al., 1989).

An interesting finding observed was the presence of large amounts of viral antigens at the epithelial cell linings, suggesting that these cells are highly permissive to APMV-3 replication. The study also suggested that APMV-3 has a tropism towards both epithelial and sub-epithelial cells (Kumar et al., 2010a). In addition, the detection of viral antigens, and in most cases infectious virus, in multiple internal organs of the birds indicates that both the viruses are capable of replicating in multiple organs rather than being restricted only to respiratory and alimentary tracts. Presumably, the virus reached the various internal organs through the blood stream. Nonetheless, this extensive amount of virus replication was not accompanied by severe disease in birds. These results show that APMV-3 strains are capable of infecting young and adult chickens and turkeys using an oculonasal route of infection. Serologic assays demonstrated a humoral response in all the birds inoculated with either APMV-3 strain, a further indication of successful replication. However, our findings suggest that chickens are comparatively more susceptible than turkeys to APMV-3 infection.

5. Conclusions

APMV-3 is capable of infecting young and adult SPF chickens and turkeys. The disease caused by APMV-3 is more severe in younger birds than in older birds, where it is mostly inapparent. Our results also showed that APMV-3 strains vary in pathogenicity. Strain Netherlands is more virulent than strain Wisconsin. This study has demonstrated that APMV-3 has an affinity for both epithelial as well as sub-epithelial cells of respiratory and alimentary tracts. In terms of pathogenicity for poultry, APMV-3 is probably second in importance only to NDV. However, in commercial chickens and turkeys the disease picture could be quite different depending on management practices, environmental conditions and other concomitant infections. Further studies are needed to understand the disease potential of this virus to commercial poultry.

6. Acknowledgments

We thank Daniel Rockemann and all our laboratory members for their excellent technical assistance and help.

7. References

Alexander and D.J., Avian paramyxoviruses 2–9. Iowa State University Press, Ames (2003).

Alexander, D., Paramyxoviridae, 11th ed. Iowa State University Press, Iowa (2003).

Alexander, D.J. Avain paramyxoviruses *Vet Bull* 50 (1980a), pp. 737-752.

Alexander, D.J. Avian paramyxoviruses. *Veterinary Bulletin* 50 (1980b), pp. 737-752.

Alexander, D.J.: Alexander DJ: Newcastle disease and other avian paramyxoviruses In: Swayne DE, G.J., Jackwood MW, Pearson JE, Reed WM, (Swayne DE, G.J., Jackwood MW, Pearson JE, Reed WM,)Swayne DE, G.J., Jackwood MW, Pearson JE, Reed WM,s), *In: A Laboratory Manual for the Isolation and Identification of Avian Pathogens.* . American Association of Avian Pathologists, .University of Pennsylvania, Kennett Square, PA, (1998), pp. 156-163.

Alexander, D.J. and Chettle, N.J. Relationship of parakeet/Netherland/449/75 virus to other avian paramyxovirus. *Res Vet Sci* 25 (1978), pp. 105-106.

Alexander, D.J. and Collins, M.S. Pathogenecity of PMV-3/Parakeet/Netherland/449/75 for chickens. *Avian Pathol* 11 (1982a), pp. 179-185.

Alexander, D.J. and Collins, M.S. Pathogenicity of PMV-3/parakeet/Netherlands/449/75 for chickens. *Avian Pathol* 11 (1982b), pp. 179-85.

Alexander, D.J. and Collins, M.S. Characterization of avain paramyxoviruses of serotype PMV-3 isolated from commercial turkey in Great Britain. . *Avian Pathol* 13 (1984), pp. 215-221.

Alexander, D.J., Hinshaw, V.S., Collins, M.S. and Yamane, N. Characterization of viruses which represent further distinct serotypes (PMV-8 and PMV-9) of avian paramyxoviruses. *Arch Virol* 78 (1983), pp. 29-36.

Anderson, C., Kearsley, R., Alexander, D.J. and Russell, P.H. Antigenic variation in avian paramyxovirus type 3 isolates detected by mouse monoclonal antibodies. *Avian Pathol* 16 (1987), pp. 691-8.

Anderson, C.L. and Russell, P.H. Monoclonal antibodies against avian paramyxovirus-3: antigenic mapping and functional analysis of the haemagglutinin-neuraminidase protein and the characterization of nonspecific monoclonal antibodies. *Arch Virol* 101 (1988), pp. 49-63.

Beck, I., Gerlach, H., Burkhardt, E. and Kaleta, E.F. Investigation of several selected adjuvants regarding their efficacy and side effects for the production of a vaccine for parakeets to prevent a disease caused by a paramyxovirus type 3. *Vaccine* 21 (2003), pp. 1006-22.

Capua, I., De Nardi, R., Beato, M.S., Terregino, C., Scremin, M. and Guberti, V. Isolation of an avian paramyxovirus type 9 from migratory waterfowl in Italy. *Vet Rec* 155 (2004), p. 156.

Chang, P.C., Hsieh, M.L., Shien, J.H., Graham, D.A., Lee, M.S. and Shieh, H.K. Complete nucleotide sequence of avian paramyxovirus type 6 isolated from ducks. *J Gen Virol* 82 (2001), pp. 2157-68.

de Leeuw, O. and Peeters, B. Complete nucleotide sequence of Newcastle disease virus: evidence for the existence of a new genus within the subfamily Paramyxovirinae. *J Gen Virol* 80 (Pt 1) (1999), pp. 131-6.

Gough, R.E. and Alexander, D.J. Avian paramyxovirus type 4 isolated from a ringed teal (Calonetta leucophrys). *Vet Rec* 115 (1984), p. 653.

Huang, Z., Panda, A., Elankumaran, S., Govindarajan, D., Rockemann, D.D. and Samal, S.K. The hemagglutinin-neuraminidase protein of Newcastle disease virus determines tropism and virulence. *J Virol* 78 (2004), pp. 4176-84.

Kaleta, E.F., Werner, O. and Hemberger, Y. Isolation and characterization of avian paramyxovirus type 3b from farmed Namibian ostriches (Struthio camelus f. dom.). *Berl Munch Tierarztl Wochenschr* 123, pp. 103-10.

Krishnamurthy, S. and Samal, S.K. Nucleotide sequences of the trailer, nucleocapsid protein gene and intergenic regions of Newcastle disease virus strain Beaudette C and completion of the entire genome sequence. *J Gen Virol* 79 (Pt 10) (1998), pp. 2419-24.

Kumar, S., Militino Dias, F., Nayak, B., Collins, P.L. and Samal, S.K. Experimental avian paramyxovirus serotype-3 infection in chickens and turkeys. *Vet Res* 41 (2010a), p. 72.

Kumar, S., Nayak, B., Collins, P.L. and Samal, S.K. Complete genome sequence of avian paramyxovirus type 3 reveals an unusually long trailer region. *Virus Res* 137 (2008), pp. 189-97.

Kumar, S., Nayak, B., Samuel, A.S., Xiao, S., Collins, P.L. and Samal, S.K. Complete genome sequence of avian paramyxovirus-3 strain Wisconsin: evidence for the existence of subgroups within the serotype. *Virus Res* 149 (2010b), pp. 78-85.

Lamb, R. and Parks, G., In Paramyxoviridae: The Viruses and Their Replication. Lippincott Williams & Wilkins, Philadelphia (2007).

Macpherson, I., Watt, R.G. and Alexander, D.J. Isolation of avian paramyxovirus other than Newcastle disease virus from commercial poultry in Great Britain. *Vet Rec* 112 (1983), pp. 479-80.

Maldonado, A., Arenas, A., Tarradas, M.C., Luque, I., Astorga, R., Perea, J.A. and Miranda, A. Serological survey for avian paramyxoviruses from wildfowl in aquatic habitats in Andalusia. *J Wildl Dis* 31 (1995), pp. 66-9.

Mayo, M.A. A summary of taxonomic changes recently approved by ICTV. *Arch Virol* 147 (2002), pp. 1655-63.

Nayak, B., Kumar, S., Collins, P.L. and Samal, S.K. Molecular characterization and complete genome sequence of avian paramyxovirus type 4 prototype strain duck/Hong Kong/D3/75. *Virol J* 5 (2008), p. 124.

Paldurai, A., Subbiah, M., Kumar, S., Collins, P.L. and Samal, S.K. Complete genome sequences of avian paramyxovirus type 8 strains goose/Delaware/1053/76 and pintail/Wakuya/20/78. *Virus Res* 142 (2009), pp. 144-53.

Panda, A., Huang, Z., Elankumaran, S., Rockemann, D.D. and Samal, S.K. Role of fusion protein cleavage site in the virulence of Newcastle disease virus. *Microb Pathog* 36 (2004), pp. 1-10.

Redmann, T., Zeydanli, M.M., Herbst, W. and Kaleta, E.F. [Isolation of a paramyxovirus-3 from turkeys with respiratory tract disease in Germany]. *Dtsch Tierarztl Wochenschr* 98 (1991), pp. 138-41.

Reed, L.J. and Muench, H. A simple method of estimation of 50% end points. *Am J Hyg* 27 (1938), pp. 493-497.

Rout, S.N. and Samal, S.K. The large polymerase protein is associated with the virulence of Newcastle disease virus. *J Virol* 82 (2008), pp. 7828-36.

Russell, P.H., Awang, I.P., Parsons, G. and Alexander, D.J. Pathogenicity of PMV-3/Turkey/England/Midland poultry holdings/1981 for chickens and turkeys. *Avian Pathol* 18 (1989), pp. 125-34.

Saif, Y.M., Mohan, R., Ward, L., Senne, D.A., Panigrahy, B. and Dearth, R.N. Natural and experimental infection of turkeys with avian paramyxovirus-7. *Avian Dis* 41 (1997), pp. 326-9.

Samuel, A.S., Kumar, S., Madhuri, S., Collins, P.L. and Samal, S.K. Complete sequence of the genome of avian paramyxovirus type 9 and comparison with other paramyxoviruses. *Virus Res* 142 (2009), pp. 10-8.

Shortridge, K.F., Alexander, D.J. and Collins, M.S. Isolation and properties of viruses from poultry in Hong Kong which represent a new (sixth) distinct group of avian paramyxoviruses. *J Gen Virol* 49 (1980), pp. 255-62.

Stallknecht, D.E., Senne, D.A., Zwank, P.J., Shane, S.M. and Kearney, M.T. Avian paramyxoviruses from migrating and resident ducks in coastal Louisiana. *J Wildl Dis* 27 (1991), pp. 123-8.

Subbiah, M., Xiao, S., Collins, P.L. and Samal, S.K. Complete sequence of the genome of avian paramyxovirus type 2 (strain Yucaipa) and comparison with other paramyxoviruses. *Virus Res* 137 (2008), pp. 40-8.

Tumova, B., Robinson, J.H. and B.C.Easterday A hitherto unreported paramyxovirus of turkey. *Res in Vet Sci* 27 (1979a), pp. 135-140.

Tumova, B., Robinson, J.H. and Easterday, B.C. A hitherto unreported paramyxovirus of turkeys. *Res Vet Sci* 27 (1979b), pp. 135-40.

Wakamatsu, N., King, D.J., Kapczynski, D.R., Seal, B.S. and Brown, C.C. Experimental pathogenesis for chickens, turkeys, and pigeons of exotic Newcastle disease virus from an outbreak in California during 2002-2003. *Vet Pathol* 43 (2006), pp. 925-33.

Warke, A., Stallknecht, D., Williams, S.M., Pritchard, N. and Mundt, E. Comparative study on the pathogenicity and immunogenicity of wild bird isolates of avian paramyxovirus 2, 4, and 6 in chickens. *Avian Pathol* 37 (2008), pp. 429-34.

Xiao, S., Paldurai, A., Nayak, B., Subbiah, M., Collins, P.L. and Samal, S.K. Complete genome sequence of avian paramyxovirus type 7 (strain Tennessee) and comparison with other paramyxoviruses. *Virus Res* 145 (2009), pp. 80-91.

31

In Vitro Antimicrobial Activity of Crude Extracts of *Erythrina abyssinica* and *Capsicum annum* in Poultry Diseases Control in the South Western Agro-Ecological Zone of Uganda

Charles Lagu[1] and Kayanja I.B. Frederick[2]
[1]*Department of Biology, Faculty of Science, Mbarara University of Science and Technology,*
[2]*Department of Anatomy, Faculty of Medicine, Mbarara University of Science and Technology, Mbarara*

1. Introduction

Many small holder farmers in the south western agro-ecological zone (SWAEZ) of Uganda have for a very long time been using medicinal plants especially *Erythrina abyssinica* and *Capsicum annum* for the management of worms, Newcastle disease and other microbial infections respectively in local poultry (Nsubuga-Mutaka *et al.*, 2005 and Lagu and Kayanja (2010). Commonly the root barks of *Erythrina abyssinica* are picked crushed and mixed with water and administered to the birds. The ripped fruits of *Capsicum annum* are picked crushed and mixed with solutions of ash and water (ITDG and IIRR) (1996); Lagu and Kayanja (2010). These medicinal combinations are mainly given to birds to treat them against worms and other microbial infections and worms (ITDG and IIRR, 1996). The farmers have been using these medicinal plant extracts for a very long time Katunguka-Rwakishaya *et al.*, 2004; Olila *et al.*, 2007; Ejobi *et al.*, (2007). It is however, not clear if the *Erythrina abyssinica and Capsicum annum* have activity against the common microbes that affect the poultry. This study aims to investigate the anti-microbial activities and minimum inhibitory concentrations (MICs) of *Erythrina abyssinica* and *Capsicum annum* used by the farmers in the control and treatment of common poultry infections in the south western agro-ecological zone of Uganda.

2. Materials and methods

2.1 Identification and description of medicinal plants (for Erythrina abyssinica and Capsicum annum)

2.1.1 Erythrina abyssinica (Leguminosae)

Erythrina abyssinica (Leguminosae)is a species of leguminous tree as seen in Figure 1. It is distributed in Congo Republic (ex Belgian), the Sudanese Republic, Ethiopia, Eritrea,

Uganda, Kenya, Tanzania and Zimbabwe. In northern and western Ethiopia, it is found at elevations between 1600 and 2100 m. *Erythrina abyssinica* (Luganda name: Muyirigiti or Jjirikiti; Runyakole name: Ekiko), is a deciduous savannah legume. It grows in open woodland and grassland. It has characteristic red overflowing flowers. It can be propagated through seedlings, cuttings and truncheons. In the south western rangelands of Uganda, it is sometimes planted along fences of paddocks to support barbed wires. It has various traditional medicinal applications in livestock. It is also used in traditional human medicine Figure 2.

Fig. 1. Erythrina abyssinica tree

Fig. 2. Dry root barks of *Erythrinna abyssinica*

2.1.2 Capsicum annum (*solanaceae*)

Capsicum annum (solanaceae) was identified in the study areas by a botanist from Mbarara University of Science and Technology. *Capsicum annum* belongs to the kingdom plantae plants, subkingdom of tracheobiota (vascular plants), super-division of spermatophyta (seed plants), division of magnoliophyta (flowering plants), class magnoliopsida (dicotyledons), subclass, asteridae, order, solanels, family of solanaceae (potato family), genus capsicum L. (pepper), species *Capsicum annum* L. (cavenne pepper) and variety *Capsicum annum* L. var annum (Cayenne pepper).

Capsicum annum is a perennial shrub growing up to 2 m (6') in height, with woody a trunk? Its leaves have various shapes usually elliptical up to 10 cm (4") long as seen in Figure 1. The flowers are white to yellowish in groups of 2 or 3 followed by small, upright, fiery, green fruits that ripen to red. The active ingredient in the plant is *capsaicin* that is used for management of various medical conditions. The varieties of this "fruit" vary greatly in size, color and pungency. The plant extract that provides therapeutic action is the seed oil.

Fig. 3. *Capsicum annum* plant

2.2 Sample collection and post harvest handling (for Erythrina abyssinica and Capsicum annum)

Information gathered included vernacular names and parts used in the preparation of herbal remedies. The plants were identified by botanist from Mbarara University of Science and Technology. Voucher specimens were deposited in the University. Fresh samples of the plant materials (root barks, stem barks and leaves) of *Erythrina abyssinica* were collected from 500m away from Rubare town in Ntungamo district on 19th May, 2010. Four (4) km away from Rubindi sub county in Mbarara district on 22nd May, 2010. Seven (7) km away from Bugongi sub county in Bushenyi district on 23rd May, 2010 and 4 km away from Lwanda Sub County GPS location S00040.379'; E031028.7779, Rakai district on 16th June, 2010. The seeds and leaves of *Capsicum annum* were collected during early morning to late afternoon and placed in a plastic bag and stored in the vehicle

During field collection of the samples, the sites where the plants were found were geographically and ecologically described. Ejobi *et al.*, 2007 found out that these plants were abundantly found in all areas in the SWAEZ. The quantity of each plant biomass collected depended on approximately with the amount of plant biomass needed to yield enough concentrates for in vitro studies. The plant materials were kept in a plant press in the department of Biology, Mbarara University of Science and Technology. The plants were dried at room temperature at Mbarara Zonal Agricultural Research and Development Institute (Mba ZARDI). The partially dried plants samples were collected and packed in a plastic bag and then transported to Natural Chemothepeutics Research laboratories (NCRL) Wandegeya, Kampala for extraction and concentration of plant extracts.

The concentrated samples were taken to the Microbiology and Parasitology Laboratory of the School of Veterinary Medicine, Makerere University for the activity studies. Voucher specimens of the plants collected were as follows; Root barks Ntu001Ea, Bus 002Ea, Mbra 003 Ea, Rakai 004 Ea; the stem barks included Ntu005Ea, Bus 006Ea, Mbra 007Ea, Rakai 008 Ea. The leaves sample include Ntu009Ea, Bus 010Ea, Mbra 011Ea, Rakai 012 Ea.

The Voucher specimens for *Capsicum annum* viz; Leaves Ntu001Ca, Bus002Ca, Mbra003Ca, Rakai 004Ca. Fruits Ntu005Ca, Bus006Ca, Mbra007Ca, Rakai008Ca of the plants studied was kept in the departmental laboratory.

Drying and milling

The leaves, root bark and stem bark of *Erythrina abbysinica* and the leaves and fruits of *Capsicum annum* from Mbarara, Bushenyi, Ntungamo and Rakai districts were dried at 50-60°C in a vacuum oven for 24 hours. The dry plant material samples were milled using an electric grinder in fine particles.

2.3 Obtaining process of crude extracts of the plant parts used, including the determination of 'extract yield" and "Standardization of dosages"

The leaves, root bark and stem bark of *Erythrina abbysinica* leaves and fruits of *Capsicum annum*

2.3.1 Extraction, filtration and concentration

The milled samples (100-500g) were soaked in 70% ethanol (5L) for 48hours with frequent shaking. Thereafter the extracts were filtered first with cotton wool followed by Whatman filter paper® and stored in at room temperature. The filtrate was then concentrated using vacuum rotary evaporator. The concentrates were then dried in vacuum oven at 60°C to dryness.

2.3.2 Determination of extract yield

The percentage yield of the extract was determined gravimetrically using the dry weight of extract (x) and soaked samples material (y) as follows

$$\text{Percentage yield} = x/y * 100 \quad (1)$$

2.3.3 Standardization of dosages

The information gained on the percent yields of crude extracts was used for standardizing dosage rates of fine powder preparations of the plant materials. For example, the amount of crude extract contained in a known weight of fine powder of plant materials was calculated from the formulae given above.

2.4 Determination of antimicrobial activity of the leaves, root bark and stem bark of Erythrina abbysinica, the leaves and fruits of Capsicum annum

The concentrated extracts of the roots, stem and leaves of *Erythrina abyssinica* and *Capsicum annum,* from the National Chemotherapeutics Laboratory, Wandegeya were transported to the department of microbiology laboratory at the School of Veterinary Medicine, Makerere University.

2.4.1 Materials required antimicrobial activity

Bacterial culture; Mueller Hinton agar medium; Nutrient broth; Plant extracts; Bunsen flame; Micro pipettor (10µl-200µl), adjustable; Spreader/wire loop; Agar borer and Incubator.

2.4.2 Method antimicrobial activity

Approximately 2-5 freshly grown bacterial colonies of the test organism were emulsified into Nutrient broth and incubated for 10-15 minutes at room temperature. With a spreader of wire loop, Mueller Hinton agar plate was evenly inoculated and plate allowed to stand for 5 minutes at room temperature (i.e. 25°C).

Using a sterile agar borer (Sterilized using a bunsen flame), wells were dug into the inoculated agar at reasonable distance apart (approximately 5 cm).

The plant extract(s) were transferred into the created agar wells till when full. Extract were not allowed to float on the agar surface. The plate lid was replaced and did not turn the petri-dish upside down. The setup was incubated at 37°C overnight. The presence for bacterial inhibition zones around each well looked for. Appearance of clear zones around a well was indicative of the anti bacterial activity of an extracts (Bizimenyera *et al.,* 2005).

2.4.3 Determination of the Minimum Inhibitory Concentration (MIC) of extract

2.4.3.1 Materials for Minimum Inhibitory Concentration (MIC) of extract

Broth culture of test organism (s); known concentration of plant extracts (e.g. 0.5g/ml, 1g/ml etc); set of test tubes; micro-pipettor (adjustable 100µl-1000µl); Nutrient broth; Sterile physiological saline; incubator.

2.4.3.2 Method for Minimum Inhibitory Concentration (MIC) of extract

A set of test tubes were dispensed 0.5 ml (500µl) of physiological saline. An equal volume (i.e. 0.5ml) of test plant extract were added to the saline in first tube and mixed the two thoroughly well.

This was repeated throughout all the test tube and the last aliquot 0.5 ml of solution from the last tube discarded so as to have uniform volume. This constituted a two fold serial dilution whereby each step moved to the right reduced the concentration of the extract by a factor of 2. About 100µl of a 24 hour culture of the test organism were added to each of the test tubes (containing the already serially diluted extract). This was mixed thoroughly well, plugged with cotton wool and incubated the preparation at 37°C for 16-24 hours.

At microbiology laboratory, the samples were each weighed and dissolved in Dimethylesulfoxide (DMSO) at a final concentration of 0.5g/ml. Mueller Hinton Agar (MHA) plates for antibiotic sensitivity testing were prepared and inoculated with pure colonies of *E. coli, Staphylococcus spp., Streptococcus spp., Salmonella spp., and Pseudomonas spp.,* which were known to be the common causes of poultry diseases.

The wells were bored in the inoculated plates and the samples from the extracts were impregnated into the wells and incubated overnight. A control plate inoculated with E. coli and impregnated with Dimethylesulfoxide (DMSO) and a known antibiotic, Ciprofloxacin was set up as negative and positive controls respectively.

After 24hrs of incubation, the plates were examined for antibacterial activity on the different sample extracts, and the results were as follows:

2.4.4 Observation/ interpretation of results

The inoculated test tubes were examined for inhibition of growth, where there is antibacterial activity; there was inhibition of growth, thus no turbidity in the test tube. The minimum inhibitory concentration (MIC) of a compound (extract) in this case the least concentration to have inhibited bacterial growth.

2.5 Data analysis

Data collected were entered in Excel windows 2007 (Microsoft Corporation). Frequencies, means and graphs were derived to explain phenomenon on significance and relationships to activities and minimum inhibitory concentrations of crude extracts of root, stem barks and leaves of *Erythyrina abyssinica and leaves and* leaves and fruits of *Capsicum annum.*

2.6 Results

2.6.1 Crude extracts of root, stem barks and leaves of *Erythrina abyssinica*

It is clear from Figure 4 and Table 1 that root barks and stem barks of *Erythrina abbyssinica* have better yields than the leaves.

In-vitro microbiological studies on *Erythrinna abbyssinica* indicate that root and stem bark extracts have activities against *Staphyloccus aureus* in Ntungamo, Mbarara, Bushenyi and Rakai. *Pseudomonas auroginosa* in Mbarara, Bushenyi except leaves from Bushenyi. No activities were noted on *E. coli* and *Salmonella species* as detailed in table 2 and Figure 5.

It has been demonstrated that root and stem barks have activity against *Staphylococcus aureus* in all districts except stem bark for Rakai. The root barks, stem barks and leaves of Mbarara,

Bushenyi respectively were effective against *Pseudomonas aeruginosa demonstrated in Table 3.* The study found that the leaves extracts did not have antimicrobial activities.

Plant Sample	Location	Plant part	Weight of sample (g)	Dry weight of concentrate (g)	% yield
E. abbysinica	Ntungamo	Root barks	450.0	39.6	8.8
	Mbarara	Root barks	295.6	37.8	12.8
	Bushenyi	Root barks	689.5	53.0	7.7
	Rakai	Root barks	500.0	24.3	4.9
	Ntungamo	Stem barks	537.0	11.6	2.2
	Mbarara	Stem barks	435.8	10.7	2.5
	Bushenyi	Stem barks	483.7	13.4	2.8
	Rakai	Stem barks	500.0	44.6	8.9
	Ntungamo	Leaves	500.1	3.2	0.6
	Mbarara	Leaves	451.7	3.2	0.7
	Bushenyi	Leaves	761.6	29.4	3.9
	Rakai	Leaves	500.0	31.1	6.2

Table 1. Percentage yield extract of leaves, root bark and stem bark of *Erythrina abbysinica* from Mbarara, Bushenyi, Ntungamo and Rakai districts

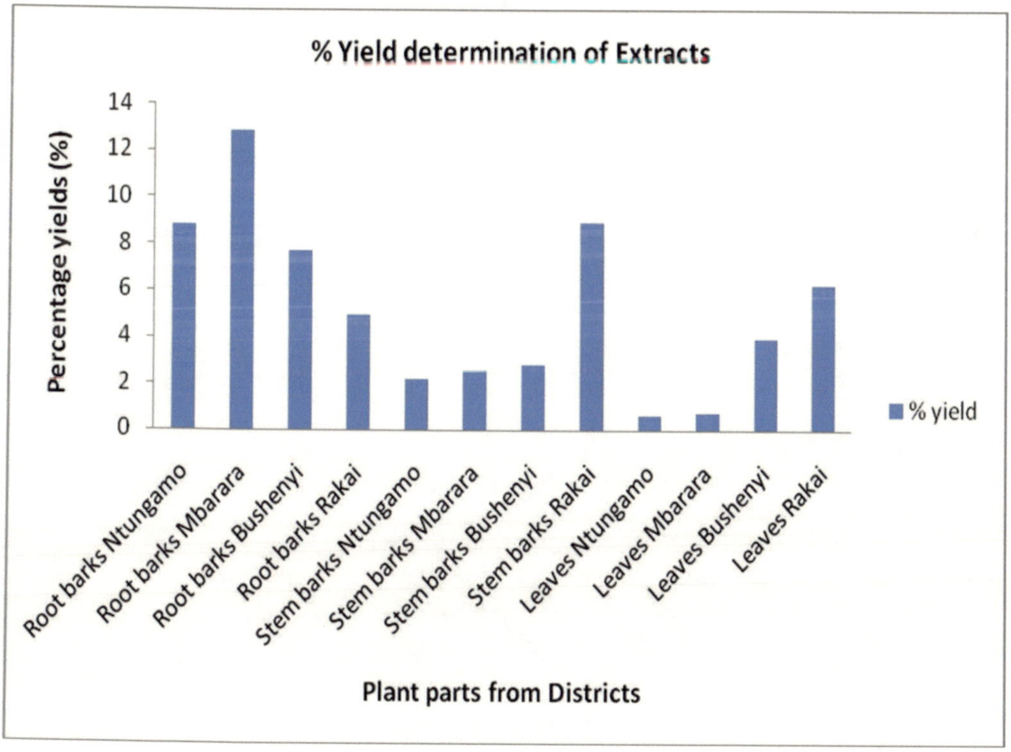

Fig. 4. Percentage yield determination of plant extracts (*Erythrinna abbyssinica*)

Antimicrobial assay of plant extracts against selected bacteria as indicated below

		Inhibition zone (mm)				
No.	**Identity/ Name of Extracts**	*E. coli*	*Salmonella*	*Staphlococcus aureus*	*Pseudomonas aeruginosa*	**District**
1	Root bark Ea Ntungamo	0	0	13	0	Ntungamo
2	Root bark Ea Bushenyi	0	0	14	0	Bushenyi
3	Root bark Ea Mbarara	0	0	12	8	Mbarara
4	Root bark Ea Rakai	0	0	15	0	Rakai
5	Stem bark Ea Ntungamo	0	0	11	0	Ntungamo
6	Stem bark Ea Bushenyi	0	0	12	7	Bushenyi
7	Stem bark Ea Mbarara	0	0	14	0	Mbarara
8	Stem bark Ea Rakai	0	0	0	0	Rakai
9	Leaves Ea Ntungamo	0	0	0	0	Ntungamo
10	Leaves Ea Bushenyi	0	0	0	8	Bushenyi
11	Leaves Ea Mbarara	0	0	0	0	Mbarara
12	Leaves Ea Rakai	0	0	0	0	Rakai

Table 2. Antimicrobial assay of *Erythrina abyssinica* plant extracts against selected bacteria

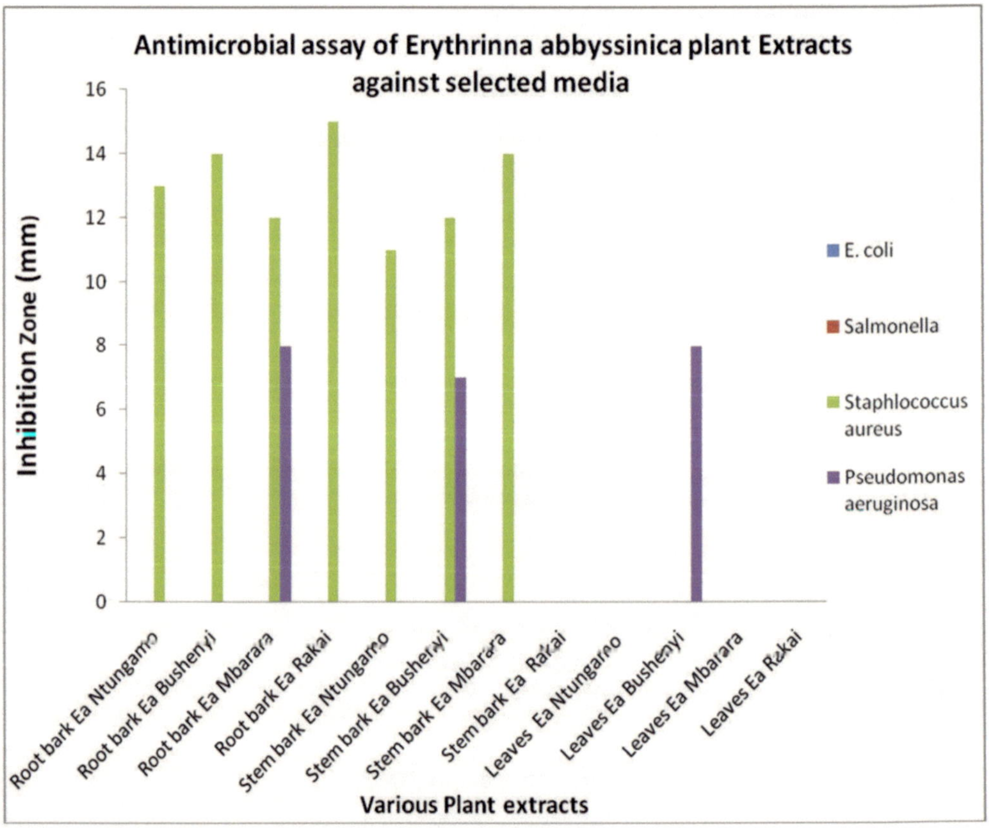

Fig. 5. Antimicrobial assay of plant extracts from four districts against selected media

		Minimum Inhibitory concentration of different micro-organisms (g/ml)				
No.	**Identity/ Name of Extracts**	**Original sample concentration (g/ml)**	*E. coli*	*Salmonella*	*Staphylo-coccus aureus*	*Pseudo-monas aeruginosa*
1	Root bark Ea Ntungamo	0.3	0	0	0.0047	0
2	Root bark Ea Bushenyi	1	0	0	0.0313	0
3	Root bark Ea Mbarara	1	0	0	0.0313	1
4	Root bark Ea Rakai	1	0	0	0.0156	0
5	Stem bark Ea Ntungamo	0.25	0	0	0.0039	0
6	Stem bark Ea Bushenyi	0.41	0	0	0.0256	0.41
7	Stem bark Ea Mbarara	0.224	0	0	0.0035	0
8	Stem bark Ea Rakai	1	0	0	0	0
9	Leaves Ea Ntungamo	0.39	0	0	0	0
10	Leaves Ea Bushenyi	1	0	0	0	1
11	Leaves Ea Mbarara	0.412	0	0	0	0
12	Leaves Ea Rakai	1	0	0	0	0

Table 3. Minimum Inhibitory concentration of different micro-organisms

2.6.2 Crude extracts of leaves and fruits of *Capsicum annum*

The yields of *Capsicum annum* leaves and fruits are detailed in Figure 4. It is clear from the tabular representation that leaves had better yield than fruits.

Studies on *Capsicum annum* indicate that fruits extracts of Mbarara have activities on *Salmonella* species and *Psudomonas auroginosa* for the case of fruit extracts from Rakai. No microbial activities were noted against *Staphylococcus aureus* and *E.coli*. It was demonstrated that leaves extracts have no microbial activities as illustrated in Table 5 and Figure 7.

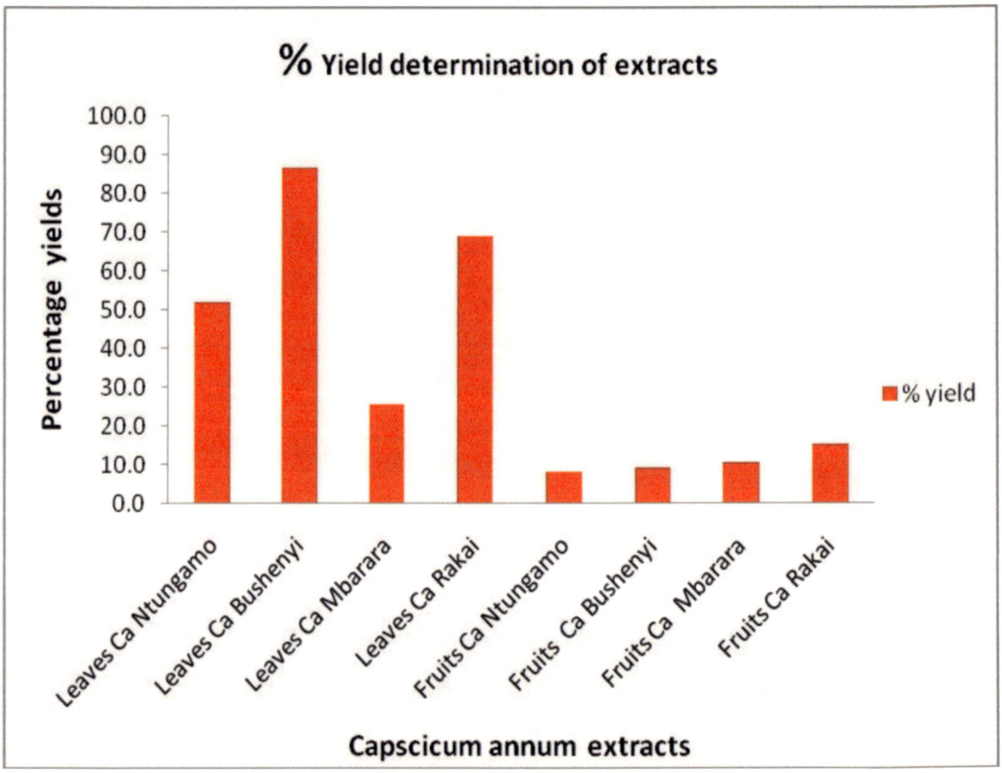

Fig. 6. Percentage yield determination for the plant *Capsicum annum*

Results indicated that leaves of *Capsicum annum* have better yields than fruits of *Capsicum annum* in all the four districts Mbarara, Ntangamo, Rakai and Bushenyi.

Antimicrobial assay of plant extracts against selected bacteria as indicated below

		Inhibition Zone (mm)				
No.	**Identity/ Name of Extracts**	*E. coli*	*Salmonella*	*Staphylococcus aureus*	*Pseudomonas aeruginosa*	**District**
1	Leaves Ca Ntungamo	0	0	0	0	Ntungamo
2	Leaves Ca Bushenyi	0	0	0	0	Bushenyi
3	Leaves Ca Mbarara	0	0	0	0	Mbarara
4	Leaves Ca Rakai	0	0	0	0	Rakai
5	Fruits Ca Ntungamo	0	0	0	0	Ntungamo
6	Fruits Ca Bushenyi	0	0	0	0	Bushenyi
7	Fruits Ca Mbarara	0	8	0	0	Mbarara
8	Fruits Ca Rakai	0	0	0	7	Rakai

Table 5. Antimicrobial assay of *Capsicum annum plant* extracts against selected bacteria

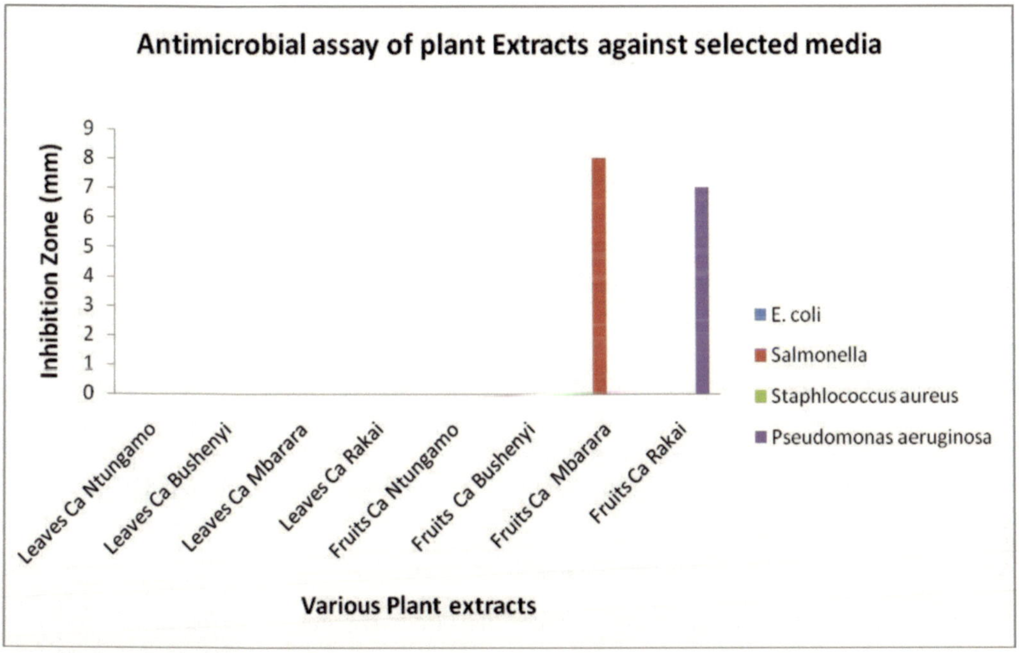

Fig. 7. Antimicrobial assay of plant extracts against selected bacteria.

It was demonstrated that root and stem barks had activity against *Staphylococcus aureus* and *Pseudomonas aeruginosa* demonstrated in Table 3.

Sample ID	Original sample concentration (g/ml)	MICs of Extracts to different microorganisms (g/ml)			
		Staphylococcus aureus	*Pseudomonas aeruginosa*	*Salmonella*	*E.coli*
Leaves Ca Ntungamo	1	0	0	0	0
Leaves Ca Bushenyi	1	0	0	0	0
Leaves Ca Mbarara	1	0	0	0	0
Leaves Ca Rakai	1	0	0	0	0
Fruits Ca Ntungamo	0.328	0	0	0	0
Fruits Ca Bushenyi	1	0	0	1	0
Fruits Ca Mbarara	1	0	0	0	0
Fruits Ca Rakai	1	0	1	0	0

Table 6. Minimum inhibitory concentration of different micro-organism

Fruits of Capsicum annum exhibited minimum inhibitory concentration against *Salmonella species* and *Pseudomonas aeruginosa* detailed in table 6.

2.7 Discussion

2.7.1 Crude extracts of root, stem barks and leaves of *Erythrina abyssinica*

Yields of the plants extracts

The results show that the yields of root barks perform better than the stem barks and leaf extracts detailed in Table 1 and Figure 3. There is an average 8.6%, 5.5% and 2.9% yield for the root barks, stem barks and leaves respectively in the districts of Ntungamo, Mbarara, Bushenyi and Rakai. The information gained on the percent yields of crude extracts were used for standardizing dosage rates of fine powder preparations of the plant materials. For example, the amount of crude extract contained in a known weight of fine powder of plant materials can then be calculated. This agrees with findings by (Ejobi F and Olila D 2004; Olila *et al.*, 2007).

Antibacterial activities of the plant extracts

The antimicrobial activity of crude extracts of root barks and stem barks of *Erythrina abyssinica* had activity against *Staphylococus aureus in* Ntungamo, Bushenyi, Mbarara and Rakai districts except stem bark from Rakai. There was activity against *Pseudomonas aeuroginosa* in Mbarara and Bushenyi respectively as detailed in table 2 and figure 4. Generally, leaves extracts showed no microbial activity except for leaf extracts from Bushenyi district. No antibacterial activities were noted against *Salmonella species* and *E.coli.*

The attributing factor for anti bacterial activity of the crude fruit extracts could be due to presence of bioactive constituents' in the extracts. This agrees with findings by Masola *et al.*, 2009 and Ogundare *et al.*, (2006).

Moriyasu et al., (1998) collected stem bark E.abyssinica from Meru district of Kenya. They carried structural elucidation of new flavanones isolated from E.abyssinica and found presence of prenylated flavanones (abyssinin I(**1**), II (**2**), III(**3**), along with abyssinone V, sigmoidin A, B, C and F, and sigmoidin B 4′-(methyl ether). These compounds exhibited antimicrobial activities.

Further still, the presence of bioactive compounds viz; abyssinoflavanone IV, V and VI posses some antimicrobial activities as reported by (Ichimaru *et al.*, 1996).

The stem woods of *Erythrina latissima* another species of Erythrina have two isoflavones and a flavanone with isolates Isoflavone (erylatissin A, B, C) in addition to 10 known flavonoids. These compounds exhibited antimicrobial activity against *Escherichia coli, Staphylococcus aureus, Bacillus subtilis* and *Candida mycoderma* (Chacha *et al.*, 2004). This study found that *Erythrina abyssinica* was effective against *Staphyloccus aureus* and *Pseudomonas aeuroginosa.*

The root and stem extracts of *Erythrina abyssinica* however did not show any microbial activity against Salmonella species and *Escherichia coli.*

A study by Masola *et al.*,(2009) in Mpwapwa, in the semi arid central zone of Tanzania found out that bioactive constituents of terpenoids, tannins, phlobotannins, saponins and cardiac glycosides were found to be present in the stem barks of the plant *Adansonia digitata* (Bombacaceae) (African baobab). It would therefore mean similar bioactive compounds were present in the plant *Erythrina abysinnica.*

Ikigai et al., (1993) indicated that purified tannins, saponins and terpenoids have anti microbial activity. These bioactive compounds were reported to be effective against gram positive and gram negative bacteria. They exert bactericidal and bacteriostatic effects. This explained why there were activities on bacteria by the *Erythrina abyssinica* plant extracts.

Understanding the spectrum of antibacterial activity indicated that bioactive substances had broad spectrum or narrow spectrum of activity. The degree of susceptibility (diameter of inhibition zone) showed that gram positive bacteria were more susceptible compared to gram negative bacteria. There was urgent need to undertake phytochemical analysis of *Erythrina abyssinica* extracts to determine the chemical bioactive substances present in the extracts.

Minimum inhibitory concentration (MICs)

The root and stem barks of *Erythrina abyssinica* had activity against *Staphylococcus aureus* in all districts except stem bark for Rakai. The root barks, stem barks and leaves of Mbarara, Bushenyi respectively were effective against *Pseudomonas aeruginosa* demonstrated in Table 3. This demonstrated that the leaves extracts do not have antimicrobial activities.

Traditionally, the results from in vitro antimicrobial tests are written as qualitative or quantitative. Qualitative results are reported as susceptible, intermediate or resistant, whereas quantitative results are reported as minimum inhibitory concentration (MIC) in µg/ml or g/ml or mg/l (Walker, 2006).

Walker,(2006), indicated that in-vitro antimicrobial susceptibility tests were predictive of in vivo therapeutic efficacy. However the ability of an in-vitro test to predict the clinical effectiveness of an antimicrobial agent is dependent on that test being performed correctly.

In vitro tests involve the continuous exposure of a relatively small concentration of bacteria to a constant level of antimicrobial agent under standardized testing conditions (Walker, 2006).

The selection of an appropriate dose can be driven by the result of quantitative susceptibility tests (Ambrose, 2005). Studies in human medicine have demonstrated the clinical value of in vitro susceptibility tests.

The interpretation of susceptibility testing depended on the relationship between in vitro susceptibility and factors involved in relation to tissue drug concentration (which depended on factors such as dose and pharmacokinetic and pharmaco-dynamic properties of the drug or drug class.

2.7.2 Crude extracts of leaves and fruits of *Capsicum annum*

2.7.2.1 Yields of the plants extracts

This study found out that the yields of the leaves extracts of *Capsicum annum* were better than the fruits as seen in Table 1 and Figure 2. There is an average 58% yield for the leaves compared to 10-11% yields for the fruits in all the districts of Ntungamo, Mbarara, Bushenyi and Rakai. The information gained on the percent yields of crude extracts are used for standardizing dosage rates of fine powder preparations of the plant materials. For example, the amount of crude extract contained in a known weight of fine powder of plant materials can then be calculated. This agrees with findings by (Ejobi F and Olila D 2004; Olila et al., 2007).

2.7.2.2 Antibacterial activities of the plant extracts

The antimicrobial activity of crude extracts of fruits of *Capsicum annum* was positive for *Salmonella species* and *Pseudomonas aeuroginosa* in Mbarara and Rakai respectively as detailed in table 2. There were no microbial activity by leaves of *Capsicum annum* in all the districts of Ntungamo, Mbarara, Bushenyi and Rakai. The fruit extracts did not have any activity on *Staphylococcus aureus* and *E.coli*.

The attributing factor for anti bacterial activity of the crude fruit extracts could be due to presence of bioactive constituents' in the extracts. This agrees with findings by Masola *et al.*, 2009 and (Ogundare et al., 2006).

A study by Masola *et al.*,2009 in Mpwapwa, in the semi arid central zone of Tanzania found out that bioactive constituents of terpenoids, tannins, phlobotannins, saponins and cardiac glycosides were found to be present in the stem barks of the plant *Adansonia digitata* (Bombacaceae) (African baobab).

Further still, studies by Ikigai et al., 1993 indicate that purified tannins, saponins and terpenoids have anti microbial activity. These bioactive compounds were reported to be effective against gram positive and gram negative bacteria. They exert bactericidal and bacteriostatic effects.

Understanding the spectrum of antibacterial activity will indicate if given bioactive substance had broad spectrum or narrow spectrum of activity. The degree of susceptibility (diameter of inhibition zone) will show that gram positive bacteria are more susceptible compared to gram negative bacteria. The need to undertake phytochemical analysis of fruit extracts of *Capsicum annum* to know the chemical bioactive substances present in these extracts.

2.7.2.3 Minimum inhibitory concentration (MICs)

The fruit extracts of *Capsicum annum* from Bushenyi and Rakai exhibited minimum inhibitory concentration against *Salmonella species* and *Pseudomonas aeruginosa* respectively as detailed in table 4. There was no minimum inhibitory concentration noted for the case of leaves of *Capsicum annum.*

Traditionally, qualitative and quantitative results from in vitro antimicrobial tests are reported. Qualitative results are reported as susceptible, intermediate or resistant, whereas quantitative results are reported as minimum inhibitory concentration (MIC) in µg/ml or g/ml or mg/l (Walker, 2006).

Walker,(2006), noted that in-vitro antimicrobial susceptibility tests are predictive of in vivo therapeutic efficacy. However the ability of an in-vitro test to predict the clinical effectiveness of an antimicrobial agent is dependent on that test being performed correctly.

Plant Sample	Location	Plant part	Weight of sample (g)	Dry weight of concentrate (g)	% yield
C. annum	Mbarara	Leaves	146.8	37.1	25.3
	Bushenyi	Leaves	62.4	53.9	86.4
	Ntungamo	Leaves	162.5	84.2	51.8
	Rakai	Leaves	253.0	174.0	68.8
	Mbarara	Fruits	85.3	8.8	10.3
	Bushenyi	Fruits	127.1	11.6	9.1
	Ntungamo	Fruits	61.3	4.9	8.0
	Rakai	Fruits	248.9	37.0	14.9

Table 4. Percentage yield extracts of the leaves and fruits of *Capsicum annum* from Mbarara, Bushenyi, Ntungamo and Rakai districts

In vitro tests cannot always predict the efficacy of an antibacterial agent in vivo (Walker, 2006). In vitro tests involve the continuous exposure of a relatively small concentration of bacteria to a constant level of antimicrobial agent under standardized testing conditions. The selection of an appropriate dose can be driven by the result of quantitative susceptibility tests (Ambrose, 2005). Despite these considerable differences, studies in human medicine have demonstrated the clinical value of in vitro susceptibility tests (Ambrose, 2005).

By generating full range MICs, a laboratory can give clinicians information that may allow them to individualize the therapeutic regimen, especially with regard to dose and dosing frequency. For example, if the MIC is low, the dose or frequency of dosing may be decreased. On the other hand, if the MIC is higher, but the organism is still considered susceptible and the drug has a wide pharmacotoxicity margin, a higher dose of the drug may be used.

Interpretation of susceptibility testing depends on knowing the relationship between in vitro susceptibility and factors involved in relation to tissue drug concentration (which depend on factors such as dose and pharmacokinetic and pharmaco-dynamic properties of the drug or drug class.

3. Conclusions

The root and stem barks of *E. abbyssinica* have better yields than their leaves. The root and stem barks exhibit antimicrobial activities against *Staphylococus aureus* and *Pseudomonas aeruginosa.* The study had demonstrated that the leaves extracts generally did not have antimicrobial activities. The MICs of the root and stem bark extracts ranged from (3.5-31.3) mg/ml for *Staphylococcus aureus* and (410-1000) mg/ml for *Pseudomonas aeruginosa.* The study agreed that farmers were right in using root barks of *E. abyssinica* to treat various ailments in poultry diseases including other livestock diseases. There is need for further research to do phyto-chemical analysis to analyze the bioactive constituents of the extracts, undertake acute toxicity tests of the extracts.

It was clear that *Capsicum annum* have antibacterial activities against *Salmonella species* and *Pseudomonas aeruginosa.* The results reveal that the fruit extracts of *Capsicum annum* have activity against *Salmonella species, Pseudomonas aeruginosa* in Mbarara and Rakai districts respectively. No activity was noted against E. coli and *Staphylococcus aureus*. The leaves of *Capsicum annum* did not show any antibacterial activities. The MICs of the fruit extracts was 1000mg/ml. The study agreed that farmers may be right in using fruits of *Capsicum annum* to treat various ailments in poultry diseases. There is need for further research to do phytochemical analysis to analyze the bioactive constituents of the extracts, undertake acute and chronic toxicity tests for *Capsicum annum* extracts.

4. Acknowledgements

We acknowledge the support Belgian Technical Cooperation (BTC) for funding this research. We thank NARO for their assistance in transportation by availing means of transport during plant collection process. Francis Umujal, Moses Agwaya and Henry Tumusime staff of Natural Chemotherapeutics Research Laboratories (NCRL) Wandegeya thanked for facilitating the process of extraction and concentration of plant extracts. We applaud Mr. Nathan L. Musisi a staff of the Department of Microbiology and Parasitology, Makerere University, Ms. Betty Ayoo Laura and Dr. Nsubuga Mutaka of Mbarara Zonal Agriculture Research and Development Institute (Mba ZARDI) for assisting to get the work done. We thank NARO for their assistance in transportation by availing means of transport during plant collection process.

5. References

Ambrose P (2005). Antimicrobial susceptibility breakpoints.PKPD and susceptibility break points. *Treat Respiratory Medicine* (Suppl 4) 1:5

Bizimenyera S.E.,Swan G.E., Chikoto H&Eloff J N (2005). Rationale for using *Peltophorum africanum* (Fabaceae) extracts in Veterinary Medicine. *Journal of South African Veterinary Association*, 76:54-58.

Chacha M., Bojase-Moleta G and Majinda R.T.R (2004). Antimicrobial and radical scavenging flavonoids from the stem wood of *Erythrina latissima. Phytochemistry* Vol 66, Issue 1, 2005, pp 99-104.

Ejobi F and Olila D (2004). On-farm validation of selected ethno-veterinary medical practices in the Teso farming system. A technical report of NARO/DFID COARD. National Agriculture Research Organization. from http://docsdrive.com/pdfs/medwelljournals/rjpharm/2007/56-60.pdf

Ichimaru M., Moriyasu M., Nishiyama Y., Kato A (1996). Structural Elucidation of New Flavanones Isolated from Erythrina abbyssinica. *Journal of Natural Products*, 1996, Vol.59, No.12

Ikigai H, Nakae T, Hara Y, Shimamura T (1993). Bacterial catechins damage the lipid bilayer. Biochin. Biophys. Act, 1147: 132-136.

ITDG and IIRR (1996) .Ethno veterinary medicine in Kenya: A field manual of traditional animal of health care practices. Intermediate Technology Development Group and International Institute of Rural Reconstruction, Nairobi, Kenya.

Katunguka-Rwakishaya, Nalule S K and Sabiiti E N (2004). Indigenous Knowledge in Ethno Veterinary Medicine in South Western Uganda. DHP Publications series No.9, March 2004, Dry land Husbandry Project (DHP).

Lagu C and Kayanja F I B ,2010: Medicinal plant extracts widely used in the control of Newcastle disease (NCD) and helminthosis among village chickens of South Western Uganda. *Livestock Research for Rural Development. Volume 22, Article #200.*Retrieved May 18, 2011, from http://www.lrrd.org/lrrd22/11/lagu22200.htm

Masola. S.N., Mosha.R.D., Wambura.P.N (2009). Assessment of antibacterial activity of crude extracts of stem and root barks from *Adansonia digitata* (Bombacaceae) (African baobab). *African Journal of Biotechnology* Vol.8 (19), pp 5076-5083. ISSN 1684-5315© 2009 Academic Journals.

Moriyasu M., Ichimaru M., Nishiyama Y., Kato A and Mathenge G.S (1998). Minor flavanones from *Erythrina abyssinica. Journal of Natural Products* 1998, Vol 61, No.2

Nsubuga-Mutaka R, Ssenteza J and Niwagaba B (205). Livestock Indigenous Technical Knowledge, Mbarara zonal Agriculture Research and Development Institute book Part 1, Volume 1. Unpublished report.

Nsubuga-Mutaka R, Ssenteza J and Niwagaba B (205). Livestock Indigenous Technical Knowledge, Mbarara zonal Agriculture Research and Development Institute book Part 1, Volume 1. Unpublished report.

Ogundare A. O., Adetuyi F.C, Akinyosoye F. A (2006). Antimicrobial activity of *Vernonia tenoreana. Afr.J. Biotech.*5 (18):1663-1668.

Olila D, Bukenya-Ziraba R and Kamoga D (2007). Bio-prospective studies on medicinal plants used to manage poultry diseases in the Mount Elgon region of Uganda. Research Journal of Pharmacology 1: 56-60.retrieved from http://docsdrive.com/pdfs/medwelljournals/rjpharm/2007/56-60.pdf

Walker .R.D (2005). Antimicrobial susceptibility testing methods and interpretation of results, In: *Antimicrobial therapy in Veterinary medicine,* Giguere S, Prescott J.F, Baggot J.D, Walker R.D &Dowling P.M (Eds).pp 22-38